Acoustic, Electromagnetic and Elastic Wave Scattering — Focus on the T-Matrix Approach

Acoustic, Electromagnetic and Elastic Wave Scattering — Focus on the T-Matrix Approach

International Symposium held at The Ohio State University, Columbus, Ohio, USA, June 25-27, 1979

Edited by
V.K. Varadan
V.V. Varadan

Sponsored by
U.S. Office of Naval Research
and The Ohio State University

Pergamon Press
New York ☐ Oxford ☐ Toronto ☐ Sydney ☐ Frankfurt ☐ Paris

Pergamon Press Offices:

U.S.A.	Pergamon Press Inc., Maxwell House, Fairview Park, Elmsford, New York 10523, U.S.A.
U.K.	Pergamon Press Ltd., Headington Hill Hall, Oxford OX3 0BW, England
CANADA	Pergamon of Canada, Ltd., Suite 104, 150 Consumers Road, Willowdale, Ontario M2J 1P9, Canada
AUSTRALIA	Pergamon Press (Aust.) Pty. Ltd., P.O. Box 544, Potts Point, NSW 2011, Australia
FRANCE	Pergamon Press SARL, 24 rue des Ecoles, 75240 Paris, Cedex 05, France
FEDERAL REPUBLIC OF GERMANY	Pergamon Press GmbH, Hammerweg 6, Postfach 1305, 6242 Kronberg/Taunus, Federal Republic of Germany

This work relates to the Department of Navy Contract No. N00014-79-C-0111 issued by the Office of Naval Research. The United States Government has a royalty-free license throughout the world on all copyrightable material contained herein.

Library of Congress Cataloging in Publication Data

Main entry under title:

Acoustic, Electromagnetic and Elastic Wave
Scattering—Focus on the T-Matrix Approach

Includes index.
1. Scattering (Physics)—Congresses. 2. T-matrix—Congresses. I. Varadan, V.K., 1943- II. Varadan, V.V., 1948- III. United States. Office of Naval Research. IV. Ohio. State University, Columbus.
QC20.7.S3R4 1980 530.1′24 79-24134
ISBN 0-08-025096-3

Printed in the United States of America

Dedicated to our little girl

Haima

Contents

Preface

The scattering and propagation of waves is a classical area which has engaged the interest of several generations of mathematicians, physicists and engineers. Extensive research in the areas of acoustics, electromagnetics and elastodynamics has resulted in many mathematical and computational techniques. Recent advances in digital computer technology have made it possible to solve complex scattering and wave propagation problems which are not amenable to analytical solution. Due to the underlying unity in the mathematical description of the three wave fields and the spurt of recent developments, the Organizing Committee felt the need for bringing together, for the first time, experts from all three areas for meaningful interaction and exchange of ideas. The Symposium, that was held at The Ohio State University, June 25 - 27, 1979, was very successful in establishing close rapport among the many participants. The group had an excellent opportunity to discuss critically the merits and limitations of several new theoretical and computation techniques that have been developed in the recent past and report on new methods and results. The main theme was to present unified theoretical and computational methods for the scattering of acoustic, electromagnetic and elastic waves.

A major development in the last ten years has been the T-Matrix method. This method which incorporates certain elegant analytical properties has also proven to be an efficient computational technique. Even though the method has received scant attention in the past, recent advances by small research groups throughout the world have demonstrated its potential applicability in many areas of single and multiple scattering of all three wave fields. The Symposium was focused on the T-Matrix so that its full potential and/or limitations could be properly understood in the context of other well known methods such as geometrical theory of diffraction, moment method, finite element method, singularity expansion method, etc.

The papers presented at the Symposium are collected in this volume which may serve as a basic reference for unified approaches to scattering problems in all three wave fields. At the suggestion of Dr. M.A. Chaszeyka, Dr. N.L. Basdekas and Dr. L. Flax, and the interest expressed by many of the participants, an introductory chapter on the differential and integral representations for acoustic, electromagnetic and elastic fields, has been included. The book is organized into ten sections : the first few sections consist of papers dealing exclusively with the T-Matrix and its many applications, the subsequent sections consist of papers in other methods, and the last section consists of papers on special topics.

The Symposium brought together 35 invited speakers and about 125 participants from all over the world. The lectures were very stimulating and led to lively discussions. Recommendations for future research areas in the T-Matrix method as summarized by a panel of experts are included in the book.

The Organizing Committee responsible for the planning of the Symposium consisted of Vijay K. Varadan and Vasundara V. Varadan (Engineering Mechanics Department and Atmospheric Sciences Program), Viswanathan N. Bringi (Electrical Engineering Department and Atmospheric Sciences Program); who are members of the Wave Propagation Group at The Ohio State University, and the Conference Coordinator

Prem R. Kumar.

The Session Chairmen were : Jan D. Achenbach (Northwestern University), R.H.T. Bates (University of Canterbury, New Zealand), W.M. Boerner (University of Illinois, Chicago Circle), L. Flax (Naval Research Laboratory), J.T. Hall (Atmospheric Sciences Laboratory, WSMR), R. Harrington (Syracuse University),J.M. McKisic (U.S. Office of Naval Research), R. Mittra (University of Illinois), T.A. Seliga (The Ohio State University), A. Sankar (T.R.W. Systems), S. Ström, Institute of Theoretical Physics, Sweden), C.H. Walter (The Ohio State University) and P.C. Waterman (Center for Science and Technology, Inc.).

We wish to thank the U.S. Office of Naval Research for its generous financial support. We would, in particular, like to express our sincere appreciation to Dr. N.L. Basdekas of the Structural Mechanics Section of ONR, Wasington, and Dr. M.A. Chaszeyka of ONR, Chicago, for their active interest and encouragement in the Symposium.

Additional financial support from the Graduate School, the College of Engineering through the Dr. Thomas Alvin Boyd Lectureship funds and the Department of Engineering Mechanics, is gratefully acknowledged. Facilities provided by The Ohio State University for the conduct of the Symposium is also acknowledged.

The editors would like to express their gratitude to Ms. Prem Kumar, who was largely responsible for the smooth running of the Symposium, for her help in organizing and editing the tape recording of the discussions and the contributed manuscripts. Dr. V.N. Bringi's assistance in the planning and conduct of the Symposium is appreciated. Finally, the editors wish to thank Dr. Bo A. Peterson, Mr. P.H. Cheng, Mr. J.H. Su, Mr. R. Peindl, Ms. Sreedevi Bringi, Ms. Marty Moffa, and Ms. Mary Trufant for their help during the Symposium. We are also indebted to Professor Sunder H. Advani for his encouragement and help in the planning of the Symposium. Thanks are also due to Pergamon Press, who succeeded in giving the text such an appealing appearance and in getting it printed on time.

The contributions of the speakers, the session chairmen, and the participants in the discussion were greatly appreciated. This book would not have been possible but for the generous co-operation of the invited speakers in sending their manuscripts on schedule.

Columbus, Ohio
August, 1979

Vijay K. Varadan
Vasundara V. Varadan

Acoustic, Electromagnetic and Elastic Wave Scattering — Focus on the T-Matrix Approach

Introduction to Acoustic,
Electromagnetic and
Elastic Field Equations

ACOUSTIC, ELECTROMAGNETIC AND ELASTIC FIELD EQUATIONS PERTINENT TO SCATTERING PROBLEMS

Vasundara V. Varadan and Vijay K. Varadan
Wave Propagation Group
Department of Engineering Mechanics
The Ohio State University, Columbus, Ohio 43210

ABSTRACT

For the convenience of researchers in different areas of classical wave scattering a brief introduction is given to the differential equations, potential decomposition, Green's functions, integral representations, boundary conditions, reciprocity and optical theorems for acoustic, electromagnetic and elastic wave fields. References at the end of the chapter will direct the reader to more detailed, pedagogical treatments of these ideas.

I. INTRODUCTION

Due to a curious set of circumstances surrounding 18th century ideas regarding the propagation of light, the elastic aether was invented. At that time it was generally understood that acoustic waves were longitudinal and that light was polarized transverse to the direction of propagation. But the elastic aether by its very nature had to admit both irrotational (curl free) and solenoidal (divergence free) solutions. Thus basic ideas regarding the nature of the elastic field were established long before the application of elasticity theory to the behavior of structural materials. However, the solution of the elastic aether equation by Poisson sounded the deathknell of the elastic aether theory since Poisson showed that the irrotational and solenoidal parts of the elastic field propagate with distinct wave speeds. Later Cauchy, Stokes and Love laid the foundations of the mathematical theory of elasticity as we know it today.

Although the elastic and electromagnetic field parted company in the beginning of the 19th Century, there is a mathematical unity that is present in the description of these fields that is particularly relevant in unified approaches to the solution of scattering and boundary value problems for acoustic, electromagnetic and elastic fields. The common link in the description of these fields is the decomposition into vector and scalar potentials which describe the solenoidal and irrotational parts of the field respectively. Further it can be shown that for time harmonic fields the scalar and vector potentials satisfy the scalar and vector Helmholtz equations for all three classical fields.

The acoustic field is purely longitudinal and hence the particle velocity can be expressed as the gradient of a scalar potential. The electromagnetic field in the absence of free charges can be expressed as the curl of a vector potential. The elastic field on the other hand requires both types of potentials for a complete description. In addition, although the differential equations for the scalar and vector potentials are uncoupled, the two potentials will get coupled due to boundary conditions at a free surface or a bimaterial interface. Thus acoustic, electromagnetic and elastic wave problems are in increasing order of difficulty. The striking difference between the three fields is that acoustic and electromagnetic wave propagation are characterized by a single speed of propagation for each field whereas two components of the elastic field propagate with two distinct wave speeds.

Researchers in the three different areas can benefit greatly from the existing literature for the solution of certain types of boundary value and scattering problems that have common mathematical features. The barrier in most cases is that researchers in one field are unfamiliar with the description and properties of other fields. Since the thrust of this symposium proceedings is a unified approach to classical wave scattering, this introductory chapter summarizing the description of acoustic, electromagnetic and elastic wave fields is included as part of the proceedings. By necessity, the treatment is quite brief, but it is hoped that it will still benefit all readers.

II. ACOUSTIC FIELD

Field Equations

The propagation of sound in an infinite non-viscous (or perfectly elastic) fluid can be described by the variation of the pressure p as a function of space and time or by the variation of the particle velocity $\dot{\vec{u}}$ or the particle displacement $\vec{u}$. The particle velocity can be expressed as the gradient of a scalar potential ϕ and the pressure is related to the time derivative of ϕ. The formalism can thus be a completely scalar formalism or a vector formalism. The material properties involved in the description of the field in an inviscid fluid are the compressibility λ_f and the mass density ρ_f.

From Euler's equation or the conservation of momentum equation, neglecting terms quadratic in $\dot{\vec{u}}$, p and ρ_f we obtain

$$\left(\nabla^2 - \frac{1}{c_f^2}\frac{\partial^2}{\partial t^2}\right)\phi(\vec{r}, t) = 0 \qquad (2.1)$$

where

$$\dot{\vec{u}} = \nabla\phi \; ; \; p = -\rho_f \frac{\partial\phi}{\partial t} \; ; \; c_f = \sqrt{\frac{\lambda_f}{\rho_f}} \qquad (2.2)$$

In Eqs. (2.1) and (2.2), c_f is the speed of sound in an ideal fluid and linearizing Euler's equation corresponds to assuming $|\dot{\vec{u}}| \ll c_f$.

Confining the rest of this discussion to time harmonic fields of frequency ω, we find that

$$\phi(\vec{r}, t) = \phi(\vec{r})\, e^{-i\omega t} \qquad (2.3)$$

so that Eq. (2.1) reduces to the scalar Helmholtz equation,

$$(\nabla^2 + k_f^2)\,\phi(\vec{r}) = 0 \qquad (2.4)$$

where $k_f = \omega/\bar{c}_f$ is the wavenumber

If there are sources of sound in the infinite fluid then the right hand sides of Eqs. (2.1) and (2.4) will be non-zero. The Green's function $g(\vec{r}, \vec{r}')$ is the response at $\vec{r}$ due to a point source at $\vec{r}'$ and is the solution of

$$(\nabla^2 + k_f^2)\, g(\vec{r}, \vec{r}') = -\delta(\vec{r} - \vec{r}') \qquad (2.5)$$

The solution of Eq. (2.5) is well known and is given by

$$g(\vec{r}, \vec{r}') = e^{ik_f|\vec{r} - \vec{r}'|}/4\Pi|\vec{r} - \vec{r}'| \qquad (2.6)$$

Integral Representations for Scattering Problems

In the solution of scattering problems we are interested in finding $\phi(\vec{r})$ exterior to a surface S that is excited by a source exterior to S that we shall denote by ϕ^0 (see Fig. 1). The additional pressure field created outside S due to the scattering of sound waves from S is denoted by ϕ^s. Thus

$$\phi(\vec{r}) = \phi^0(\vec{r}) + \phi^s(\vec{r})\ ;\ \vec{r} \text{ outside S} \qquad (2.7)$$

The scattered field ϕ^s satisfies the source free Helmholtz equation exterior to S, while ϕ^0 satisfies the source free Helmholtz equations in the region occupied by the scatterer.

$$(\nabla^2 + k_f^2)\,\phi^0(\vec{r}) = 0\ ;\ \vec{r} \text{ inside S} \qquad (2.8)$$

$$(\nabla^2 + k_f^2)\,\phi^s(\vec{r}) = 0\ ;\ \vec{r} \text{ outside S} \qquad (2.9)$$

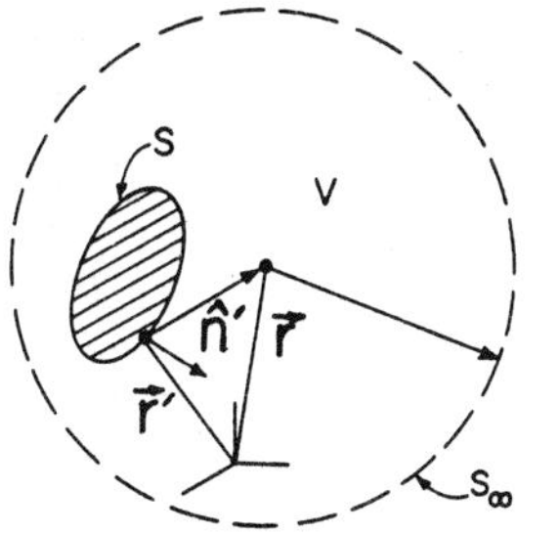

Fig. 1. Scattering geometry

Multiplying Eq. (2.8) by g and subtracting from Eq. (2.5) multiplied by ϕ^0 and integrating over the volume occupied by S and using Green's theorem, we obtain

$$-\int\!\!\int_S \left\{ \phi^{0\prime}\hat{n}'.\nabla' g - g\,\hat{n}.\nabla'\,\phi^{0} \right\} dS = \begin{cases} 0 \;; \; \vec{r} \text{ outside } S \\ \phi^{0}(\vec{r}) \;; \; r \text{ inside } S \end{cases} \tag{2.10}$$

Similarly from Eqs. (2.9) and (2.5) and integrating over the region exterior to S, we obtain

$$\int\!\!\int_S \left\{ \phi^{s\prime}\,\hat{n}'.\nabla'\,g - g\,\hat{n}'.\nabla'\,\phi^{s\prime} \right\} dS + \oint\!\!\!\!\int_{S_\infty} \text{--------} = \begin{cases} \phi^{s}(\vec{r}) ; \; \vec{r} \text{ outside } S \\ 0 \;; \; \vec{r} \text{ inside } S \end{cases} \tag{2.11}$$

(with the integral over S_∞ equal to 0.)

Adding Eqs. (2.10) and (2.11), we obtain the Helmholtz integral representation

$$\int\!\!\int_S \left\{ \phi'\,\hat{n}'.\nabla' g - g\,\hat{n}'.\nabla'\phi \right\} dS = \begin{cases} \phi^{s}(\vec{r}) ; \; \vec{r} \text{ outside } S \\ -\phi^{0}(\vec{r}) , \; \vec{r} \text{ inside } S \end{cases} \tag{2.12}$$

In Eqs. (2.10) - (2.12), the prime on ϕ', $\hat{n}'$ etc. denotes that the argument is $\vec{r}'$ which is a point on S. The Green's function $g = g(\vec{r}, \vec{r}')$ where $\vec{r}$ is the field point and $\vec{r}'$ is the source point which in this case is on the surface. In deriving Eq. (2.11), the surface integral at S_∞ is equal to zero because g satisfies proper radiation conditions. The first of Eq. (2.12) is simply Huygen's principle that states that the scattered field ϕ^s is created by point sources induced on the surface S of the scatterer. The second of Eq. (2.12) has been called the null field equation, the extinction theorem of Ewald-Oseen or the extended boundary condition. It simply states that the incident field ϕ^0 is extinguished at all points to the interior of the scatterer by the negative of the surface integral that creates the scattered field to the exterior of S.

If the surface S encloses a different fluid with material properties ρ_{f1} and λ_{f1} and if ϕ_1 is the velocity potential inside S, then for the region inside S

$$-\int\!\!\int_S \left\{ \phi_1\hat{n}'.\nabla' g_1 - g_1\,\hat{n}'.\nabla'\phi_1 \right\} dS = \begin{cases} 0 \;; \; \vec{r} \text{ outside } S \\ \phi_1(\vec{r}) ; \; \vec{r} \text{ inside } S \end{cases} \tag{2.13}$$

The derivation of Eq. (2.13) is exactly similar to that of Eq.(2.10)using in this case ϕ_1 instead of ϕ^0 and g_1, the Green's function for the fluid inside S instead of g.

Boundary Conditions

(a) Sound soft obstacle : If the region occupied by S is void, then the boundary condition at S is that the pressure field must vanish at all points on S and thus

$$\phi'_+ = 0 \text{ on } S \tag{2.14}$$

Equation (2.14) is simply a Dirichlet boundary condition.

(b) Sound hard Obstacle : If the obstacle enclosed by S is rigid relative to the surrounding fluid, then the particle velocity must be zero everywhere within and on S. Only the normal component of the particle velocity appears in the integral representation and hence

$$\hat{n}'\cdot\nabla'_{+}\phi = 0 \text{ on } S \tag{2.15}$$

Thus Neumann boundary conditions imposed on S corresponds to a sound hard obstacle.

(c) Permeable Obstacles : The region within S may be occupied by a fluid with properties ρ_{f1} and λ_{f1} or it may enclose a solid. In either case, continuity conditions require that the normal component of the particle velocity and the pressure must be continuous across the bimaterial interface at S.

$$\left.\begin{aligned} \rho_f\,\phi'_{+} &= \rho_{f1}\,\phi'_{1-} \\ \hat{n}\cdot\nabla'_{+}\phi &= \hat{n}\cdot\nabla'_{-}\phi_1 \end{aligned}\right\} \text{ on } S \tag{2.16}$$

For a solid fluid interface, if $\vec{u}$ and $\vec{\tau}_1$ are the displacement and stress fields in the solid, then the continuity of pressure and normal component of the particle velocity imply that

$$\left.\begin{aligned} i\omega\rho_f\,\phi'_{+} &= \hat{n}'\cdot\vec{\vec{\tau}}'_{1-}\cdot\hat{n}' \\ \hat{n}'\cdot\nabla'_{+}\phi &= -i\omega\,\hat{n}'\cdot\vec{u}'_{1-} \end{aligned}\right\} \text{ on } S \tag{2.17}$$

In equations (2.14) - (2.17) the $\pm$ signs indicate that the surface S is approached from the outside and inside respectively.

Reciprocity and Optical Theorems

At distances far from the surface of the scatterer, the scattered field potential ϕ^s consists of outgoing spherical waves propagating along the radius vector r

$$\phi^s(\vec{r}) \xrightarrow[|\vec{r}|\to\infty]{} f(\hat{k}_0,\hat{r})\, e^{ik_f r}/r \tag{2.18}$$

In Equation (2.18), $f(\hat{k}_0, \hat{r})$ is the scattered far field amplitude that depends on the geometry and nature of the scatterer, the frequency of the incident wave, the direction of incidence $\hat{k}_0$ and the direction of observation $\hat{r}$.

The reciprocity theorem states that independent of the obstacle

$$f(\hat{k}_0, \hat{r}) = f(-\hat{r}, -\hat{k}_0) \tag{2.19}$$

Equation (2.19) is called the principle of reciprocity and states that the amplitude of the far field is unchanged by interchanging the position and direction of source and receiver. This principle leads directly to the symmetry property of the scattering matrix.

The total scattering cross section σ_{tot} is the total energy scattered by the obstacle into a solid angle of 4Π in 3-D,

$$\sigma_{tot} = \frac{1}{4\Pi}\int |f(\hat{k}_o, \hat{r})|^2 d\hat{r} \tag{2.20}$$

Using the principle of conservation of energy it can be shown that

$$\sigma_{tot} + \sigma_{abs} = \frac{1}{k_f} \text{ Im } f(\hat{k}_o, \hat{k}_o) \tag{2.21}$$

Equation (2.21) is called the optical theorem and σ_{abs} is the energy if any that is dissipated inside the scatterer. If the material constants of the scatterer are real and there is no viscosity, then $\sigma_{abs} = 0$. The optical theorem is sometimes also known as the forward amplitude theorem.

III. ELECTROMAGNETIC FIELD

Field Equations

In a charge free medium, using the electric field $\vec{E}$ and the magnetic field $\vec{H}$ as the field variables, Maxwell's equations for the electromagnetic field are given by

$$\nabla \cdot \vec{H} = 0 \tag{3.1}$$

$$\nabla \cdot \vec{E} = 0 \tag{3.2}$$

$$\nabla \times \vec{H} - \frac{1}{c}\frac{\partial \vec{E}}{\partial t} = 0 \tag{3.3}$$

$$\nabla \times \vec{E} + \frac{1}{c}\frac{\partial \vec{H}}{\partial t} = 0 \tag{3.4}$$

In Eqs. (3.1) - (3.4), c is the speed of light in the medium and is given by $c = 1/\sqrt{\varepsilon\mu}$ where ε is the dielectric constant and μ is the permeability. Equations (3.1) and (3.2) state that $\vec{E}$ and $\vec{H}$ are solenoidal fields i.e. thay can be represented by the curl of a vector potential. Maxwell's equations for a charge free medium can be simplified to

$$\left(\nabla^2 - \frac{1}{c^2}\frac{\partial^2}{\partial t^2}\right)\left\{\begin{matrix}\vec{E}\\ \vec{H}\end{matrix}\right\} = 0 \tag{3.5}$$

The potential decomposition of the $\vec{E}$ and $\vec{H}$ field can be written as

$$\vec{E} = - \nabla \times \vec{A} - \frac{1}{c}\frac{\partial \vec{B}}{\partial t} \tag{3.6}$$

and

$$\vec{H} = \nabla \times \vec{B} - \frac{1}{c}\frac{\partial \vec{A}}{\partial t} \tag{3.7}$$

with the guage condition

$$\nabla . \vec{A} = \nabla . \vec{B} = 0$$

If the ensuing discussion is confined to time harmonic fields of frequency ω, then $\vec{E}$ and $\vec{H}$ satisfy the vector Helmholtz equation. The Green's function for the vector Helmholtz equation is a dyadic which is the solution of

$$(\nabla^2 + k^2) \vec{\vec{G}} (\vec{r}, \vec{r}') = - \vec{\vec{I}} \delta(\vec{r} - \vec{r}') \tag{3.8}$$

and hence by comparison with Eq. (2.5)

$$\vec{\vec{G}} (\vec{r}, \vec{r}') = \vec{\vec{I}} g(\vec{r}, \vec{r}') \tag{3.9}$$

where $\vec{\vec{I}}$ is the idemfactor and g is the Green's function of the scalar Helmholtz equation as given by Eq. (2.6).

Integral Representations for Scattering Problems

Consider the geometry of Fig. 1, where S now encloses a perfectly conducting obstacle or a dielectric with dielectric constant ε_1 and permeability μ_1. A monochromatic wave of frequency ω giving rise to an electric field $\vec{E}^0$ and magnetic field $\vec{H}^0$ is incident on S. The total electric and magnetic fields outside S can be written as

$$\vec{E} (\vec{r}) = \vec{E}^0 (\vec{r}) + \vec{E}^s (\vec{r}) \tag{3.10}$$

$$\vec{H} (\vec{r}) = \vec{H}^0 (\vec{r}) + \vec{H}^s (\vec{r}) \tag{3.11}$$

$\vec{E}^o$ and $\vec{H}^0$ satisfy the source from Helmholtz equation in the region occupied by S and $\vec{E}^s$ and $\vec{H}^s$ satisfy the source free Helmholtz equation exterior to S.

In order to derive an integral representation for the $\vec{E}$ field, we note the following vector identities

$$(\nabla'^2 \vec{H}') . \vec{\vec{G}} - (\nabla'^2 \vec{\vec{G}}) . \vec{H} = \vec{H}' \delta(\vec{r} - \vec{r}'), \tag{3.12}$$

$$\nabla \times \left[\nabla'^2 \vec{H}' . \vec{\vec{G}} - (\nabla'^2 \vec{\vec{G}}) . \vec{H}' \right]$$
$$= \nabla \times \left\{ \nabla' . \left[\vec{\vec{G}} \times (\nabla' \times \vec{H}') - \vec{H}' \times (\nabla' \times \vec{\vec{G}}) \right] \right\}$$

and

$$= \nabla \times (\vec{H}' \delta(\vec{r} - \vec{r}')), \tag{3.13}$$

$$\nabla' \times \vec{\vec{G}} = -\nabla \times \vec{\vec{G}} \tag{3.14}$$

Using the procedure used in deriving Eqs. (2.10) - (2.12), using $\vec{H}^0$ and $\vec{H}^s$ successively in Eq. (3.13) and using the divergence theorem and radiation conditions on g and adding, yields

$$\nabla \times \iint_S \left\{ \left[\hat{n}' \times (\nabla' \times \vec{H}')\right] g - (\hat{n}' \times \vec{H}') \cdot (\nabla' \times \vec{\vec{I}} g) \right\} dS = \begin{cases} \nabla \times \vec{H}^s(\vec{r}) \; ; \; \vec{r} \text{ outside } S \\ -\nabla \times \vec{H}^0(\vec{r}) \; ; \; \vec{r} \text{ inside } S \end{cases} \tag{3.15}$$

Using Eqs. (3.14) and (3.3) in (3.15), manipulating the vector products in the second term of the surface integral, we finally obtain

$$k\nabla \times \int_S (\hat{n}' \times \vec{E}') g \, dS + i\nabla \times \nabla \times \int_S (\hat{n}' \times \vec{H}') g \, dS = \begin{cases} \vec{E}^s(\vec{r}) ; \; \vec{r} \text{ outside } S \\ -\vec{E}^0(\vec{r}) ; \; \vec{r} \text{ inside } S \end{cases} \tag{3.16}$$

Equation (3.16) is the Helmholtz integral representation for the electric field. The first part of Eq. (3.16) is once again a statement of Huygen's principle and the second part is the null field equation.

If the surface S encloses a dielectric and if $\vec{E}_1$ and $\vec{H}_1$ are the electric and magnetic fields interior to S, then an equation similar to Eq. (2.13) can be written for $\vec{E}_I$

$$k\nabla \times \int_S (\hat{n}' \times \vec{E}') g \, dS + i\nabla \times \nabla \times \int_S (\hat{n}' \times \vec{H}') g \, dS = \begin{cases} 0 \; ; \; \vec{r} \text{ outside } S \\ -\vec{E}_1(\vec{r}) ; \; \vec{r} \text{ inside } S \end{cases} \tag{3.17}$$

Boundary Conditions

(a) Perfect Conductor : If the surface S encloses a perfect conductor, then the $\vec{E}$ field everywhere inside and on S must vanish. In particular, the tangential component of $\vec{E}$ on S, i.e.

$$(\hat{n}' \times \vec{E}')_+ = 0 \text{ on } S \tag{3.18}$$

(b) Permeable Dielectric : If the scatterer is a dielectric, then the tangential components of the $\vec{E}$ and $\vec{H}$ fields must be continuous across the interface at S,

$$\left.\begin{aligned} (\hat{n}' \times \vec{E}')_+ &= (\hat{n}' \times \vec{E}_1')_- \\ (\hat{n}' \times \vec{H}')_+ &= (\hat{n}' \times \vec{H}_1')_- \end{aligned}\right\} \text{at } S \tag{3.19}$$

Reciprocity and Optical Theorem

At distances far from S, the scattered field $\vec{E}^S$ consists of outgoing spherical waves and can be written in the form

$$\vec{E}^s(\vec{r}) \xrightarrow[|\vec{r}| \to \infty]{} \vec{F}(\hat{k}_o,\hat{r})e^{ikr}/4\Pi r \tag{3.20}$$

where $\hat{k}_o$ is the direction of incidence and $\hat{r}$ the direction of observation. Let $\hat{u}_o$ be the polarization of the incident wave. If we interchange source and receiver and reverse the direction of the waves, then the reciprocity condition states that

$$\hat{u}.\ \vec{F}(\hat{k}_o,\ \hat{r}) = \hat{u}_o.\ \vec{F}(-\hat{r},\ -\hat{k}_o) \tag{3.21}$$

where $\hat{u}$ is the polarization of the wave incident along $-\hat{r}$, i.e. after interchange of source and receiver. Equation (3.21) is independent of the boundary conditions at S and the geometry of the scatterer.

The optical theorem is again a statement of the conservation of energy. This theorem relates the far field scattered amplitude in the forward direction to the total energy scattered and absorbed by the obstacle, i.e.

$$\sigma_{tot} + \sigma_{abs} = \frac{1}{k}\ \mathrm{Im}\ \hat{u}_o.\ \vec{F}(\hat{k}_o,\ \hat{k}_o) \tag{3.22}$$

where

$$\sigma_{tot} = \frac{1}{4\Pi}\int |\vec{F}(\hat{k}_o,\hat{r})|^2\ d\hat{r} \tag{3.23}$$

The absorption crosssection is zero if the dielectric constant of the scatterer is real.

IV. ELASTIC FIELD

Field Equations

The field variables that describe the elastic field are the displacement vector $\vec{u}$ and the stress tensor $\vec{\tau}$ that are related by generalized Hooke's law

$$\tau_{ij} = C_{ijk\ell}\ \partial_k\ u_\ell \tag{4.1}$$

where C_{ijkl} is the fourth rank stiffness tensor. For a linear isotropic elastic solid, there are only two independent elastic constants λ and μ called the Lame' constants. The stiffness tensor then reduces to

$$C_{ijk\ell} = \lambda\ \delta_{ij}\ \delta_{k\ell} + \mu\ (\delta_{ik}\ \delta_{j\ell} + \delta_{i\ell}\delta_{jk}) \tag{4.2}$$

so that the stress tensor can be written in dyadic form as

$$\overleftrightarrow{\tau} = \lambda \overleftrightarrow{I}\nabla.\vec{u} + \mu (\nabla\vec{u} \quad \vec{u}\nabla) \tag{4.3}$$

The stress tensor is symmetric and has only six independent components. The displacement equation in the solid can be derived from Newton's law for a source free elastic continuum and is given by

$$\nabla.\overleftrightarrow{\tau} - \rho \frac{\partial^2 \vec{u}}{\partial t^2} = 0 \tag{4.4}$$

where ρ is the mass density.

Since there are no additional restrictions on $\vec{u}$, the Helmholtz decomposition of the displacement field should include both irrotational and solenoidal parts. Thus

$$\vec{u} = \nabla\phi + \nabla \times \vec{\psi} \; ; \; \nabla.\vec{\psi} = 0 \tag{4.5}$$

where ϕ and $\vec{\psi}$ are the scalar and vector potentials of the elastic field. Substituting Eq. (4.5) into (4.4) and (4.3) it can be seen that for harmonic fields of frequency ω

$$(\nabla^2 + k_p^2)\phi = 0 \tag{4.6}$$

$$(\nabla^2 + k_s^2)\vec{\psi} = 0 \tag{4.7}$$

where

$$k_p^2 = \omega^2\rho/(\lambda + 2\mu) \; ; \; k_s^2 = \omega^2\rho/\mu \tag{4.8}$$

Even though the potentials satisfy Helmholtz equation, the important feature of Eqs. (4.6) and (4.7) is that the irrotational and solenoidal parts of the elastic field propagate with distinct speeds c_p and c_s respectively where $c_p = \sqrt{\frac{\lambda + 2\mu}{\rho}}$ and $c_s = \sqrt{\mu/\rho}$. The subscript p denotes primary and s denotes secondary null $c_p > c_s$. These waves are variously referred to as P-waves or pressure waves or compressional waves and S-waves or shear waves.

The two components of the shear wave are further classified as SV- and SH-waves for problems where there is a preferred axis. Shear waves polarized parallel to the preferred axis are called horizontally polarized shear waves or SH waves.

Although the differential equations for the scalar and vector potentials are uncoupled, whenever boundaries are present, the two potentials will be coupled by boundary conditions making problems of elastodynamics quite difficult.

The Green's stress dyadic $\overleftrightarrow{\Sigma}$ and the displacement dyadic $\overleftrightarrow{G}$ are the response to a uniform point force applied at $\vec{r}'$ in an infinite medium and satisfy

$$\nabla.\overleftrightarrow{\Sigma} + \rho\omega^2\overleftrightarrow{G} = - I\delta\ (\vec{r} - \vec{r}') \tag{4.9}$$

The dyadic $\overleftrightarrow{\Sigma}$ is of third rank and is related to $\overleftrightarrow{G}$ by Hooke's law

$$\overleftrightarrow{\Sigma} = \lambda\overleftrightarrow{I}\ \nabla.\ \overleftrightarrow{G} + \mu\ (\nabla\overleftrightarrow{G} + \overleftrightarrow{G}\nabla) \tag{4.10}$$

$\overleftrightarrow{\Sigma}$ is defined so that it is symmetric in the first two indices.

The Green's dyadic $\overleftrightarrow{G}$ can be constructed by noting the potential decomposition of $\vec{u}$. The Green's dyadic will also have irrotational and solenoidal parts and is given by

$$\overleftrightarrow{G}\ (\vec{r}, \vec{r}') = \frac{1}{4\Pi\rho\omega^2}\left\{k_s^2\,\overleftrightarrow{I}\ g(k_s|\vec{r} - \vec{r}'|) + \nabla\left[g(k_p|\vec{r} - \vec{r}'|) - g(k_s|\vec{r} - \vec{r}'|)\right]\nabla'\right\} \tag{4.11}$$

where g is the scalar Green's function given by Eq. (2.6).

Integral Representations for Scattering Problems

Consider the geometry shown by Fig. 1 where the region exterior to S is an infinite elastic medium with material properties ρ, λ, μ. An incident wave of frequency ω giving rise to a displacement $\vec{u}^0$ excites the scatterer S. The total field outside S is given by

$$\vec{u}(\vec{r}) = \vec{u}^0\ (\vec{r}) + \vec{u}^s\ (\vec{r}) \tag{4.12}$$

where $\vec{u}^s$ is the scattered displacement field. The sources giving rise to $\vec{u}^0$ are outside S and those giving rise to $\vec{u}^s$ are located on the surface of the scatterer. Thus both $\vec{u}^0$ and $\vec{u}^s$ satisfy Eq. (4.4) inside and outside S respectively. The procedure for deriving the integral representation for $\vec{u}^s$ and $\vec{u}^0$ is exactly similar to that used for the acoustic and electromagnetic fields. From Eqs. (4.4) and (4.9), using the following vector identities

$$(\nabla.\ \overleftrightarrow{\tau}).\ \overleftrightarrow{G} = \nabla.\ (\overleftrightarrow{\tau}\ .\ \overleftrightarrow{G}) - \overleftrightarrow{\tau} : \nabla\ \overleftrightarrow{G}, \tag{4.13}$$

$$\vec{u}\ .\ (\nabla\ .\overleftrightarrow{\Sigma}) = \nabla.(\vec{u}\ .\overleftrightarrow{\Sigma}) - \nabla\vec{u} : \overleftrightarrow{\Sigma}, \tag{4.14}$$

$$\overleftrightarrow{\tau} : \nabla\overleftrightarrow{G} - \nabla\vec{u} : \overleftrightarrow{\Sigma} = 0, \tag{4.15}$$

so that

$$\nabla.\left[\vec{\tau}.\,\overleftrightarrow{G} - \vec{u}.\overleftrightarrow{\Sigma}\right] = \vec{u}\,\delta(\vec{r} - \vec{r}') \tag{4.16}$$

we obtain

$$\iint\left\{\vec{u}'.\,(\hat{n}'.\overleftrightarrow{\Sigma}) - \hat{n}'.\,\overleftrightarrow{\tau}'.\,\overleftrightarrow{G}\right\}dS = \begin{cases}\vec{u}^s\,(\vec{r});\ \vec{r}\text{ outside } S\\ -\vec{u}^o\,(\vec{r});\ \vec{r}\text{ inside } S\end{cases} \tag{4.17}$$

The first part of Eq. (4.17) is a statement of Huygen's principle for the scattered elastic field and the second part is the extinction theorem or the null field equation.

If the surface S encloses a different elastic material of peoperties ρ_1, λ_1 and μ_1, then the integral representation for $\vec{u}_1$, the displacement field in S is given by

$$\iint_S\left\{\vec{u}'.\,(\hat{n}'.\overleftrightarrow{\Sigma}_1) - \hat{n}'.\,\overleftrightarrow{\tau}'.\,\overleftrightarrow{G}_1\right\}dS \quad \begin{cases}0 \quad ;\ \vec{r}\text{ outside } S\\ -\vec{u}\,(\vec{r})\ ;\ \vec{r}\text{ inside } S\end{cases} \tag{4.18}$$

In Eq. (4.18), $\overleftrightarrow{\Sigma}$ and $\overleftrightarrow{G}_1$ are the Green's stress and displacement dyadics for the material inside S.

It is useful to observe that although sound propagation in an ideal fluid can be given an entirely scalar formulation, for many problems involving solid-fluid boundaries, it becomes convenient to use a vector formalism for the acoustic field similar to that used for the elastic field. We set the shear modulus of the ideal fluid to zero so that the stress tensor in the fluid is completely diagonal and simply corresponds to the pressure. Both the displacement field $\vec{u}_f$ and Green's dyadic $\overleftrightarrow{G}_f$ in the fluid have only irrotational parts. Thus the integral representations given in Eqs. (4.17) and (4.18) may be used for the fluid also keeping these factors in mind.

Boundary Conditions

(a) Cavity : If S encloses a void or cavity then the normal component of the stress tensor should vanish everywhere on S and thus

$$\vec{t}' = \hat{n}'.\,\overleftrightarrow{\tau}' = 0 \text{ on } S \tag{4.19}$$

The vector $\vec{t}$ defined in Eq. (4.19) is called the traction vector. This boundary condition on the traction can be considered as a generalized Neumann boundary condition if we think of the Hookian differential operator $C_{ijkl}\,\partial_k$ as the analogue of the gradient operator in scalar problems. However, for the acoustic field the Neumann boundary condition corresponds to a sound hard obstacle whereas in the elastic case it refers to a soft obstacle.

(b) Rigid fixed scatterer : In this case the scatterer is considered to have material properties λ_1, $\mu_1 \to \infty$, so that it is rigid relative to the surrounding material. In addition to this, the scatterer should also have inifinite mass i.e. $\rho_1 \to \infty$ or be fixed. This is difficult to realize in practice except perhaps in the case of infinitely long cylinders. The boundary condition in this case is a Dirichlet boundary condition and requires the displacement field $\vec{u}$ to be zero everywhere on S

$$\vec{u}'_+ = 0 \text{ on S} \tag{4.20}$$

It is again interesting to note that for the acoustic field the Dirichlet boundary condition corresponds to the sound soft obstacle and for the elastic field to a rigid obstacle.

(c) Elastic Solid Inclusion :

If the contact at the bimaterial interface S is to be perfectly welded, then all components of the displacement and traction vectors must be continuous, i.e.

$$\left.\begin{aligned} \vec{u}'_+ &= \vec{u}'_{1-} \\ \vec{t}'_+ &= \vec{t}'_{1-} \end{aligned}\right\} \text{ on S} \tag{4.21}$$

(d) Elastic Fluid Inclusion :

If the surface S encloses a perfect fluid, then only the normal components of the displacement and traction are continuous across the interface at S and since the fluid does not support any shear stresses, the tangential component of the traction when approached from the outside (the solid side) must be zero and hence

$$\left.\begin{aligned} \hat{n}'\cdot\vec{u}'_+ &= \hat{n}'\cdot\vec{u}'_{1-} \\ \hat{n}'\cdot\vec{t}'_+ &= \hat{n}'\cdot\vec{t}'_{1-} \\ (\vec{t}')_{+\text{tangential}} &= 0 \end{aligned}\right\} \text{ on S} \tag{4.22}$$

It must be noted that Eq. (4.22) does not specify anything about the tangential component of the displacement at the fluid solid interface.

Reciprocity and Optical Theorems

One if the distinct features of elastic wave scattering is the phenomenon of mode conversion. When a P wave hits a free surface or a bimaterial interface, then both P-waves and S-waves are generated on scattering and S-waves when scattered give rise to both S-waves and P-waves. For the elastic field in addition to the usual reciprocity relations for the non-mode converted scattered field amplitudes, we need an additional relation for the mode conversion amplitudes.

At distances far from the scatterer the displacement field $\vec{u}$ consists of outgoing spherical P- and S- waves given by

$$\vec{u}^s(\vec{r}) \xrightarrow[|\vec{r}|\to\infty]{} \hat{r}\, f_p\,(\hat{k}_o, \hat{r})\, e^{ik_p r}/r + \vec{f}_s\,(\hat{k}_o, \hat{r})\, e^{ik_s r}/r \tag{4.23}$$

where $\vec{f}_s \cdot \hat{r} = 0$ denoting that $\vec{f}_s$ is polarized transverse to the direction of propagation, $\hat{k}_o$ is the direction of incidence and $\hat{r}$ the direction of observation and hence that of the propagating spherical P- and S- waves.

Using the Betti- Rayleigh reciprocal identity for two elastodynamic states, we can derive reciprocity relations between the scattered field amplitudes when the source and receiver are interchanged and reversed. In the second case the wave is incident along $-\hat{r}$ and observed along $-\hat{k}_o$. Let $\hat{u}_o$ be the polarization of the shear wave incident along $\hat{k}_o$ and $\hat{u}$ that of the shear wave incident along $-\hat{r}$. Then

$$f_{pp}\,(\hat{k}_o, \hat{r}) = f_{pp}(-\hat{r}, -\hat{k}_o) \tag{4.24}$$

$$\hat{u}_o \cdot \vec{f}_{ss}\,(-\hat{r}, -\hat{k}_o) = \hat{u} \cdot \vec{f}_{ss}\,(\hat{k}_o, \hat{r}) \tag{4.25}$$

and

$$f_{sp}(-\hat{r}, -\hat{k}_o) = \frac{c_s^2}{c_p^2}\; \hat{u} \cdot \vec{f}_{ps}(\hat{k}_o, \hat{r}) \tag{4.26}$$

In Eqs. (4.24) - (4.26), the first subscript on f denotes the polarization of the incident wave and second subscript the polarization of the scattered wave. Equations (4.24) and (4.25) are similar to reciprocity relations for the acoustic and electromagnetic field respectively whereas Eq. (4.26) which relates the mode converted amplitudes is unique to elastic wave scattering.

The optical theorem for elastic waves can be derived using the principle of conservation of energy. If σ_{tot} is the total energy in the scattered field and σ_{abs} the energy if any that is dissipated in the obstacle then, for P-wave incidence

$$\sigma_{tot} + \sigma_{abs} = \frac{1}{k_p}\, \text{Im}\,(f_{pp}(\hat{k}_o, \hat{k}_o)) \tag{4.27}$$

and for S-wave incidence

$$\sigma_{tot} + \sigma_{abs} = \frac{1}{k_s}\, \text{Im}\,[\hat{u}_o \cdot \vec{f}_{ss}(\hat{k}_o, \hat{k}_o)] \tag{4.28}$$

For an elastic obstacle σ_{abs} is zero.

V. CONCLUDING REMARKS

In this brief summary, some of the important features of acoustic, electromagnetic and elastic waves relevant to scattering problems have been briefly discussed. Many of the details are missing and some features like volume integral formulations for the three fields have been omitted for the sake of brevity. The list of references given at the end of this section will point the reader to more detailed treatments of the ideas summarized here.

After a first reading, it will be realized that the mathematical unity in the description arises from the potential decomposition of the fields and although the field equations are quite different to begin with, the scalar and vector potentials for all three fields satisfy the wave equation. Another interesting observation is that in the surface integral representation for scattering problems, only those surface quantities that are prescribed by boundary conditions natural to the problem occur in the integrand. For the acoustic field it is the pressure and particle velocity, for the electromagnetic field it is the tangential component of the $\vec{E}$ and $\vec{H}$ fields and for the elastic field it is $\vec{u}$ and $\vec{t}$.

Lastly there is one particular scattering problem that is mathematically identical for all three wave fields. This is the scattering of waves normally incident to the axis (z-axis) of an infinitely long cylinder. For the electromagnetic field the polarization of the incident field in the z-direction and for the elastic field, the incident wave is an SH-wave polarized in the z-direction. For this particular problem there is no depolarization of both vector fields upon scattering and the differential equation for both reduces to the scalar wave equation that describes the acoustic field. Hence this particular example serves as a convenient check for all unified approaches to classical wave scattering problems.

ACKNOWLEDGEMENTS

The authors wish to thank Dr. M.A. Chaszeyka, Dr. N.L. Basdekas and Dr. L. Flax for their suggestion to include this chapter in the book.

REFERENCES

Acoustic Waves

J.J. Bowman, T.B.A. Senior and P.L.E. Uslenghi, Electromagnetic and acoustic scattering by simple shapes, North Holland, (1969).

Electromagnetic Waves

W.R. Smythe, Static and dynamic electricity, McGraw Hill, New York, (1939).

J.A. Stratton, Electromagnetic theory, McGraw Hill, New York, (1941).

S.R. de Groot and L.G. Suttrop, Foundations of Electrodynamics, North Holland, Amsterdam, (1972).

J. Van Bladel, Electromagnetic fields, McGraw Hill, New York (1966).

M. Born and E. Wolf, Principles of Optics, Pergamon Press, New York, (1975).

J.D. Jackson, Classical Electrodynamics, John Wiley, New York (1976).

D.S. Jones, The theory of Electromagnetism, Pergamon Press, New York, (1964).

H. Hönl, A.W. Maue and K. Westpfahl in Handbuch der Physik edited S. Flügge, Springer-Verlag, Berlin (1961).

P.C. Waterman, Symmetry, unitarity, and geometry in electromagnetic scattering, Phys. Rev. D 3, 825 (1971).

Elastic Waves

J.D. Achenbach, Wave Propagation in elastic solids, North Holland, Amsterdam, (1973).

A. Eringen and E. Suhubi, Elastodynamics, Volume II, Academic Press, (1974).

Y.H. Pao and C.C. Mow, The diffraction of elastic waves and dynamic stress concentrations, Crane Russak, New York, (1973).

V.D. Kupradze, Dynamical problems in elasticity, Progress in Solid Mechanics, Volume III, edited by I.N. Sneddon and R. Hill, North Holland, (1963).

Y.H. Pao and V. Varatharajulu (Varadan), Huygens' principle, radiation conditions, and integral formulas for the scattering of elastic waves, J. Acoust. Soc. Am. 59, 1361 (1976)

T.H. Tan, Diffraction theory for time-harmonic elastic waves, doctoral thesis, Delft University of Technology, Delft, The Netherlands (1975).

General References

P.M. Morse and H. Feshbach, Methods of Theoretical Physics, Volumes I and II, McGraw Hill, New York (1953)

I. Stakgold, Boundary value problems of mathematical Physics, Volume II, Macmillan, New York (1968).

B.B. Baker and E.T. Copson, The mathematical theory of the Huygens' principle, Clarendon Press, Oxford, (1939).

R.G. Newton, Scattering theory of waves and particles, McGraw Hill, New York, (1966).

Part 1

T-Matrix Approach
Single Scattering

GENERAL INTRODUCTION TO THE EXTENDED BOUNDARY CONDITION
(Thomas Alvin Boyd Lecture*)

R.H.T. Bates
Electrical Engineering Department, University of Canterbury,
Christchurch, New Zealand

ABSTRACT

To review the essentials of P.C. Waterman's extended boundary condition approach to classical wave diffraction, the scattering of a scalar monochromatic cylindrical wave by a homogeneous cylinder of arbitrary cross-section is discussed in some detail. The conventional volume-source (or polarisation-source) approach is described first, and it is pointed out that improved numerical efficiency is to be expected if the volume sources are replaced by surface sources. Waterman's extended boundary condition, which can be thought of as a surface-source formulation, and which is a specialisation of Love's equivalence principle or the Ewald-Oseen optical extinction theorem, is then introduced. It is explained how Waterman computes the scattered field without explicitly evaluating either the field inside the body or the field on the body's surface - this is the T-matrix method, which implies restrictions on the shape of the body's surface. The null-field methods are then outlined - these employ Waterman's approach to evaluate surface fields, on bodies whose surfaces can have arbitrary shapes, and they can be formulated in terms of any separable coordinate system - improved numerical efficiency is realised when elliptic-cylinder or spheroidal coordinates are used for bodies of appreciable aspect ratio. Finally, it is indicated how the null-field methods lead to recently discovered approximate diffraction formulations of Kirchoff's type. The conventional Kirchoff approximation, which is called planar physical optics, is exact for homogeneous bodies having plane surfaces. The new approximations, which are called circular physical optics, spheroidal physical optics, etc., are exact for homogeneous bodies having the forms of circular cylinders, spheroids, etc.

I. INTRODUCTION

I wish to discuss the theoretical and computational usefulness of Waterman's (1), (2), (3), (4) approach to the solution of diffraction problems. While I must not over-simplify my argument, I am determined to avoid obscuring essentials with complicated technicalities. Although the differences between vector and scalar wave motion, and between fields varying in three and two dimensions, are far from trivial (especially from a computational viewpoint, and even more so when elastic as opposed to electromagnetic wave motion is examined), I hold that all the basic considerations are retained by restricting the discussion to monochromatic, two-dimensional wave motion. I also assert that little that is fundamental is missed by examining diffraction by a homogeneous body. **The recognition of the value of Waterman's approach for elastic waves is due to the authors of Refs. 13 through 15.** Consider a two-dimensional space Ω consisting of the union of a closed curve C and its interior Ω_- and its exterior Ω_+. Choose a point 0 within Ω_- to be the origin for cylindrical polar coordinates (ρ,ϕ) and (r,θ). An arbitrary point P in Ω has the coordinates (ρ,ϕ) and an arbitrary point Q on C has the coordinates (r,θ). Denote the minimum and maximum values of r by r_- and r_+ respectively. It is convenient to partition Ω_- and Ω_+ into pairs of non-intersecting parts:

$$\Omega_- = \Omega_{--} \cup \Omega_{++} \text{ and } \Omega_+ = \Omega_{+-} \cup \Omega_{++} \tag{1}$$

where $\rho<r_-$ for all points in Ω_{--} and $\rho>r_+$ for all points in Ω_{++}.

The space Ω is to be thought of as a plane in a three-dimensional space. The curve C is the intersection with Ω of the surface of a body having the form of a cylinder of arbitrary cross-section whose generators are perpendicular to Ω. The media embedded in the spaces outside and inside the body are homogeneous and of refractive indices unity and ν respectively. A monochromatic line source, parallel to the generators of the cylinder, intersects Ω at P_0, which point lies within Ω_{++}. The angular frequency of the source is ω and its suppressed time dependence is $\exp(i\omega t)$. The source induces a scalar wave motion which is scattered by the body. Within Ω_+, the wavelength and the wave number of the wave motion are λ and k respectively. It is convenient to affix a particular point Q_0 to C and to draw from O through Q_0 the datum from which are measured the polar angles ϕ and θ. Arc length along C is measured from Q_0 to Q. The outward normal to C at Q is $\hat{n}$ and the angle between OQ and $\hat{n}$ is α.

The total wave motion at any point P is represented by the wave function $\psi(P)$. The cylindrical wave emitted by the line source (which is of unit amplitude) is the incident wave function $\psi_0(P)$, which has the form (Ref. 5)

$$\psi_0(P) = (-i/4)\ H_0^{(2)}(kR_0) \tag{2}$$

where $H_0^{(2)}(\cdot)$ denotes the Hankel function of the second kind of zero order and R_0 is the distance from P_0 to P.

When $\psi_0(P)$ impinges on the cylinder a scattered wave function $\tilde{\psi}(P)$ is produced, so that the total wave function is written as

$$\psi(P) = \psi_0(P) + \tilde{\psi}(P) \tag{3}$$

Denoting a two-dimensional delta function at P by δ_P, the wave equations obeyed by ψ are

$$\nabla^2\psi + k^2\psi = -\delta_{P_0} \tag{4}$$

when P is in Ω_+, and

$$\nabla^2\psi + k^2\nu^2\psi = 0 \tag{5}$$

when P is in Ω_-. The forms of equations (4) and (5) ensure that both $\psi(P)$ and $\nabla\psi(P)$ are continuous across C, which are the boundary conditions I have chosen to impose upon the wave motion. Note from Eqs. (2) through (4) that

$$\nabla^2\psi_0 + k^2\psi_0 = -\delta_{P_0} \text{ and } \nabla^2\tilde{\psi} + k^2\psi = 0 \tag{6}$$

when P is in Ω_+.

The scattering problem is posed thus:

PROBLEM: Given the position of P_0, the position and shape of C and the value of ν, evaluate $\tilde{\psi}$ throughout Ω.

A conceptually convenient and theoretically rigorous approach to the evaluation of $\tilde{\psi}$ is the volume-source, or polarization-source approach, in which all fields are taken to propagate everywhere at the fundamental velocity and inhomogeneities of media are replaced by equivalent source densities (Ref. 6). The wave equations (4) through (6) are rewritten as

$$\nabla^2\psi_0 + k^2\psi_0 = -\delta_{P_0} \text{ and } \nabla^2\tilde{\psi} + k^2\tilde{\psi} = -k^2[\mu^2-1]\psi \tag{7}$$

for P throughout Ω, where $\mu=\nu$ when P is in Ω_-, and $\mu=1$ when P is in Ω_+. It follows from equations (3) and (4) and the general solution to the monochromatic reduced wave equation (Ref. 5) that

$$\psi(P) = \psi_0(P) - (i[\nu^2-1]k^2/4)\iint_{\Omega_-}\psi(P')\, H_0^{(2)}(kR')\, d\Omega(P') \tag{8}$$

where R' is the distance to P from an arbitrary point P' in Ω_-. In equation (8) the point P can be considered fixed, whereas the point P' spans all of Ω_-, implying that $d\Omega(P')$ is the element of area enclosing P'.

Equation (8) is a Fredholm integral equation of the second kind for ψ. Once this equation is solved for all P in Ω_-, the integral on the right side can be evaluated for any P in Ω, which means that the scattering problem is solved. Numerical solutions are obtained by expanding ψ within Ω_- in terms of a finite set of basis functions:

$$\psi(P) = \sum_{m=1}^{N} D_m\, \Psi_m(P) \tag{9}$$

where N is a finite positive integer, the D_m are expansion coefficients (to be determined) and the $\Psi_m(P)$ are basis functions. When equation (9) is substituted into equation (8), the latter can be manipulated straightforwardly into the form

$$d_\ell = \sum_{m=1}^{N} D_m\, K_{\ell m} \tag{10}$$

where the integer ℓ ranges from unity to $L \geq N$ and where the d_ℓ and $K_{\ell m}$ are known. The D_m are found by any appropriate numerical inversion of equation (10). Since the number of required storage locations is proportional to N^2 and the number of necessary computational operations is proportional to N^3, it is obviously worthwhile trying to arrange matters so that N (which is necessarily proportional to r_+^2/λ^2) is as small as is consistent with evaluating $\tilde{\psi}$ to whatever accuracy is desired.

If the diffraction problem could be reformulated as an integral over C, rather than as an integral over Ω_- as in equation (8), the required number of basis functions would be expected to be markedly reduced - i.e. the number should be proportional to r_+/λ since C is a one-dimensional curve.

By appealing to Green's theorem, $\tilde{\psi}$ can be expressed in terms of radiations from sources existing only on C. This formulation, which is developed in detail below, is theoretically equivalent to equation (8), and it has the computational advantage mentioned in the final sentence of the previous paragraph. Define $g(P',P|\mu)$ to be the Green's function at the point P' in a homogeneous space of refractive index μ due to a point source of unit amplitude at P. Reference to equation (2) then indicates that

$$g(P',P|\mu) = g(P,P'|\mu) = (-i/4)\,H_o^{(2)}(kR') \tag{11}$$

Note also that

$$\psi_o(P) = g(P,P_o|1) \tag{12}$$

because P_o lies in Ω_+ where $\mu=1$.

Use the symbol $\tilde{\Omega}$ to denote either Ω_- or Ω_+. Denote by $\tilde{C}$ the closed curve(s) bounding $\tilde{\Omega}$. An arbitrary point on $\tilde{C}$ is $\tilde{Q}$. Denote the outward unit normal to $\tilde{C}$ at $\tilde{Q}$ by $\tilde{n}$. It follows from elementary vector calculus and from the Divergence Theorem that

$$\iint_{\tilde{\Omega}} [g(P,P'|\mu)\ \nabla^2\psi(P') - \psi(P')\ \nabla^2 g(P,P'|\mu)]\ d\Omega(P')$$

$$= \iint_{\tilde{\Omega}} \nabla\cdot(g(P,P'|\mu)\ \nabla\psi(P') - \psi(P')\nabla\ g(P,P'|\mu))\ d\Omega(P')$$

$$= \int_{\tilde{C}} [g(P,\tilde{Q}|\mu)\ \nabla\psi(\tilde{Q}) - \psi(\tilde{Q})\nabla g(P,\tilde{Q}|\mu)]\cdot\tilde{n}\ dC(\tilde{Q}) \tag{13}$$

where $dC(\tilde{Q})$ denotes the element of arc length along $\tilde{C}$. Since P_o is in $\tilde{\Omega}$ when the latter is Ω_+, both $\psi(P)$ and $g(P,P'|1)$ are outgoing on the part of $\tilde{C}$ that bounds Ω_+ at infinity. It therefore follows from the Sommerfeld radiation condition (Ref. 5) that the integral over $\tilde{C}$ in equation (13) always reduces to an integral over C. However, at each point $\tilde{Q}$ on C the direction of $\tilde{n}$ reverses when $\tilde{\Omega}$ changes from Ω_- and Ω_+. Note that $\tilde{n}=\hat{n}$ when $\tilde{\Omega}=\Omega_-$. So, on defining

$$h(P|\mu) = \int_C [g(P,Q|\mu)\psi(Q) - \psi(Q)\nabla g(P,Q|\mu)]\cdot\hat{n}\ dC(Q) \tag{14}$$

it is seen that

$$\int_{\tilde{C}} [g(P,\tilde{Q}'|\mu)\nabla\psi(\tilde{Q}) - \psi(\tilde{Q})\nabla g(P,\tilde{Q}|\mu]\ \cdot\ \tilde{n}\ dC(\tilde{Q})$$

$$= -h(P\ \mu),\ \text{when}\ \tilde{\Omega} = \Omega_-$$

$$= h(P|\mu), \text{ when } \tilde{\Omega} = \Omega_+ \tag{15}$$

It is now convenient to consider the following set of equations:

$$\nabla^2\psi_o + k^2\mu^2\psi_o = [(\mu-\nu)/(\nu-1)]\delta_{P_o}$$

$$\nabla^2\tilde{\psi} + k^2\mu^2\tilde{\psi} = 0 \tag{16}$$

$$\nabla^2 g + k^2\mu^2 g = -\delta_P$$

Refer to equation (3). The first two of the equations (16) are seen to be consistent with equation (5) when P is in Ω_- and with the equations (6) when P is in Ω_+.

Note that the right side of the first of the equations (16) reduces to zero when $\mu=\nu$. Consequently, when the equations (16) are substituted into equations (13) and (15), the correct form for $\psi(P)$ is obtained if μ is given the values unity and ν appropriately. It follows from equations (3), (11), (12) and (16) that

$$\iint_{\tilde{\Omega}} [g(P,P'|\mu)\nabla^2\psi(P') - \psi(P')\nabla^2 g(P,P'|\mu)]\, d\Omega(P')$$

$$= \iint_{\tilde{\Omega}} [\Psi(P')\delta_P - (\nu-\mu)g(P,P'|\mu)\delta_{P_o}/(\nu-1)]\, d\Omega(P')$$

$$= \psi(P), \text{ when P in } \Omega_-,\ \mu = \nu,\ \tilde{\Omega} = \Omega_-$$

$$= 0, \text{ when P in } \Omega_+,\ \mu = \nu,\ \tilde{\Omega} = \Omega_-$$

$$= -\psi_o(P), \text{ when P in } \Omega_-,\ \mu = 1,\ \tilde{\Omega} = \Omega_+$$

$$= \tilde{\psi}(P), \text{ when P in } \Omega_+,\ \mu = 1,\ \Omega = \Omega_+ \tag{17}$$

which, when substituted into equations (13) and (15) give

$$h(P|\nu) = -\psi(P) \text{ and } h(P|1) = -\psi_o(P) \tag{18}$$

when P is in Ω_-, and

$$h(P|\nu) = 0 \text{ and } h(P|1) = \tilde{\psi}(P) \tag{19}$$

when P is in Ω_+.

Equations (18) and (19) are particularly illuminating from a physical point of view. The second of the equations (19) states that the scattered field can be thought of as being radiated by the equivalent sources on C. At first sight it may be difficult to see how the two equations (19) are mutually compatible. However, remember that the body becomes indistinguishable from the space surrounding it when $\nu=1$, in which case $\tilde{\psi}=0$ because there is nothing to cause any scattering. The second equation in (18) is a statement of Love's equivalence principle, or the Ewald-Oseen optical extinction theorem, or Waterman's extended boundary condition (Ref. 7). When the body is indistinguishable from its surroundings, the field within the body must have the same form as the incident field, which explains why the two equations (18) are mutually compatible.

It is convenient to introduce Waterman's approach by expanding the Green's function in multipoles (Ref. 5):

$$g(P,Q|\mu) = (-i/4)\sum_{m=-\infty}^{\infty} J_m(k\mu\rho)H_m^{(2)}(k\mu r)\exp(im[\phi-\theta\ ]) \qquad (20)$$

for $\rho\leq r$, where $J_m(\cdot)$ and $H_m^{(2)}(\cdot)$ denote, respectively, the Bessel function of the first kind and the Hankel function of the second kind, both of order m. The expansion of $g(P,Q|\mu)$ has the above form for P anywhere in Ω_{--}, because $\rho<r_-$ there. Similarly, if the symbols J and H are interchanged, the expansion of $g(P,Q|\mu)$ has the above form for P anywhere in Ω_{++}, because $\rho>r_+$ there. It is further convenient to introduce the notation

$$\tilde{Z}_m(\xi,r) = [Z_{m-1}(\xi r)\exp(i\alpha) - Z_{m+1}(\xi r)\exp(-i\alpha)]\ \xi/2 \qquad (21)$$

where the symbol Z stands for either J or $H^{(2)}$. Since P_o is defined to be in Ω_{++}, it follows that $\rho<\rho_o$ when P is in Ω_{-}. So, on substituting equations (2), (14), (20) and (21) into the second of the equations (18), for P in Ω_{--}, on invoking the recurrence relations for Bessel functions (Ref. 5) and on recognising that ϕ spans the range $[0,2\pi]$ within Ω_{--} (so that the functions $\exp(im\phi)$ are orthogonal within Ω_{--} when m is any integer), it follows (for all integers m) that

$$-H_m^{(2)}(k\rho_o)\exp(-im\theta_o) = \int_C [f(Q)H_m^{(2)}(kr) + \psi(Q)\tilde{H}_m^{(2)}(k,r)]\exp(-im\theta)\ dC(Q) \qquad (22)$$

which it is convenient to call a null-field equation (Ref. 7), where (ρ_o,θ_o) are the coordinates of P_o and

$$f(Q) = \partial\Psi(Q)/\partial n \qquad (23)$$

The infinite set of equations (22) only ensures that the second of the equations (18) is explicitly satisfied within Ω_{--}. However, $\psi(P)$ is analytic throughout Ω_{-}, so the uniqueness of analytic continuation ensures that the second equation in (18) is implicitly satisfied throughout Ω_{-+}. Nevertheless, this point has computational significance because, in order to evaluate $\tilde{\psi}$ to a desired accuracy, the precision with which f(Q) and $\psi(Q)$ must be computed is greater, the greater is the ratio of the area of Ω_{-+} to the area of Ω_{--}. The consequences of this have been studied in detail for totally reflecting bodies (Ref. 7). Appropriate to any separable coordinate system there is a multipole expansion of $g(P,Q|\mu)$ having a form similar to the right side of equation (20). In a two-dimensional space, elliptic-cylinder coordinates can be used instead of cylindrical polar coordinates, in which case the Bessel and Hankel functions in equation (20) are replaced by Mathieu functions and the boundary between Ω_{-+} and Ω_{--} is an ellipse instead of a circle. It is clear that in any particular case there is a particular ellipticity which minimises the ratio of the area of Ω_{-+} to the area of Ω_{--}. When coordinates having this ellipticity are used in practice, a marked increase in computational efficiency is realized (Ref. 7). In a three-dimensional space, spheroidal or ellipsoidal coordinates can be used (although it is worth remembering that the only three-dimensional coordinate systems in which vector wave motion is generally separable are Cartesian and spherical polar). D.J.N. Wall and I say that we are using the circular, spheroidal, elliptical, etc., null-field methods when we invoke cylindrical-polar, spheroidal, elliptic-cylinder, etc., coordinates (Ref. 7).

Before numerical solutions to (22) can be obtained, a connection between f(Q) and $\psi(Q)$ must be found. It is clear that an expansion of the form

$$\psi(P) = \sum_{\ell=-\infty}^{\infty} A_\ell\, J_\ell(k\nu\rho)\exp(i\ell\phi) \tag{24}$$

must be valid for any P in Ω_{--}. When C is such that the internal Rayleigh Hypothesis (Ref. 8) is valid then equation (24) applies, by definition, for any P in Ω_{-}. Consequently, it follows from equations (21) and (23) and the aforementioned recurrence relations for Bessel functions that

$$\psi(Q) = \sum_{\ell=-\infty}^{\infty} A_\ell J_\ell(k\nu r)\exp(i\ell\theta) \tag{25}$$

and

$$f(Q) = -\sum_{\ell=-\infty}^{\infty} A_\ell\, \tilde{J}_\ell(k\nu r)\exp(i\ell\theta) \tag{26}$$

Waterman (3) assumes that the right side of (24) is a valid expansion of $\psi(P)$ throughout an interior neighbourhood of C, for a wider class of curves C than the class for which the internal Rayleigh hypothesis is valid - this is an important, as yet seemingly unresolved, aspect of diffraction theory (Refs. 8 and 9). Of course, equations (25) and (26) can be used whenever Waterman's assumption is valid. On interchanging J and $H^{(2)}$ in equation (20) and applying it, within Ω_{++}, to equation (14) and manipulating the second of the equations (19) in the same way as is done above for the second of the equations (18), it is found that $\tilde{\psi}(P)$ can be expressed as

$$\tilde{\psi}(P) = \sum_{m=-\infty}^{\infty} B_m H_m^{(2)}(k\rho)\exp(im\phi) \tag{27}$$

for any P in Ω_{++}, where

$$B_m = \sum_{\ell=-\infty}^{\infty} A_\ell \int_C [J_\ell(k\nu r)\tilde{J}_m(kr) - \tilde{J}_\ell(k\nu r)J_m(kr)]\exp(i[\ell-m]\phi)\ dC(Q) \tag{28}$$

By manipulating equations (22) and (28), Waterman obtains his T-matrix which permits the scattered field to be computed without any explicit evaluation of the fields in or on the body being necessary.(Refs. 1 through 4).

On manipulating the first of the equations (19) for P in Ω_{++}, in the same way as the second of the equations (18) is manipulated for P in Ω_{--}, it is seen that

$$\int_C [f(Q)J_m(k\nu r) + \psi(Q)\tilde{J}_m(k\nu r)]\exp(-im\phi)\; dC(Q) = 0 \qquad (29)$$

for all integers m. Provided that the summations on the right sides of equations (25) and (26) exist, they can be substituted for $\psi(Q)$ and $f(Q)$, respectively, in equation (29). It is then found, as Waterman (3), (4) intimates and Peterson and Ström (10) discuss in some detail, that the equation (29) is satisfied identically: an easy way to establish this is to appeal to the Divergence Theorem and the nature of Bessel's equation. This confirms the correctness of Waterman's T-matrix for those bodies for which the right sides of equations (25) and (26) exist.

Ström and his colleagues (10), (11), (12) have developed the T-matrix formalism so that it can conveniently and efficiently (from a numerical point of view) be applied to complicated geometries incorporating several homogeneous scatterers of arbitrary shape. Boström (12) has further extended the formalism so that it can handle a body whose refractive index is, effectively, continuously variable.

The expansions on the right sides of equations (25) and (26) are not the most convenient if it is desired to evaluate the surface fields explicitly. It is preferable to choose basis functions that vary along C in the general manner to be expected from the mathematical physics of the situation - e.g. the basis functions should exhibit the right kinds of singularity at corners and they should be appropriately smooth where C is smooth, as Wall and I argue for our null-field methods, whose numerical efficiency we have demonstrated for totally reflecting bodies (Ref. 7). The main defect of the basis functions appearing in equations (25) and (26) is that they depend upon the distance of Q from O, rather than (as is desireable) on the distance along C of Q from Q_o.

In the null-field methods, which are applicable to bodies of arbitrary shape, appropriate basis functions are chosen and $f(Q)$ and $\psi(Q)$ are written as truncated summations:

$$f(Q) = \sum_{\ell=-M}^{M} E_\ell^{(1)}\,\Phi_\ell^{(1)}(Q) \quad \text{and} \quad \psi(Q) = \sum_{\ell=-M}^{M} E_\ell^{(2)}\,\Phi_\ell^{(2)}(Q) \qquad (30)$$

where the $E_\ell^{(1)}$ and $E_\ell^{(2)}$ are expansion coefficients (to be determined). Now, the equations (30) are substituted into equations (22) and (29), for (2M+1) values of m, thereby permitting the expansion coefficients to be evaluated. This completes the demonstration that the extended boundary condition can conveniently be invoked to formulate the diffraction problem in terms of an integral over C, instead of as an integral over Ω_-.

The use of equation (29), as proposed in the previous paragraph, to allow the null-field methods to be applied to penetrable bodies seems to have gone unnoticed previously. It should now be possible to handle diffraction by such bodies as straightforwardly as diffraction by totally reflecting bodies (for the latter see Ref. 7). **Morita (Ref. 16) has recently reached similar conclusions.**

A promising aspect of the null-field methods is that they have led to the development of new approximate formulations of the Kirchoff type. Wall and I call these approximations circular physical optics, elliptical physical optics, spheroidal physical optics, etc., when they are based on the circular, elliptical, spheroidal, etc., null-field methods. Conventional physical optics (i.e. the Kirchoff approximation) becomes exact when the body reduces to a homogeneous half space with a plane surface - so Wall and I call it planar physical optics. Our new approximate methods become exact when the surface of the body reduces to an appropriate coordinate surface - e.g. a spheroid of the correct eccentricity for spheroidal physical optics. We have demonstrated the power of these new physical optics methods for totally reflecting bodies (Ref. 7). I conclude this review with a brief discussion of circular physical optics for a totally reflecting body.

Either $f(Q)\equiv 0$ or $\psi(Q)\equiv 0$ when the body is totally reflecting, in which case equation (22) can be rewritten as

$$2\pi\beta_m = \int_C F(Q)\, I_m(k,r)\exp(-im\theta)\; dC(Q) \tag{31}$$

where $2\pi\beta_m = -H_m^{(2)}(k\rho_o)\exp(-im\theta_o)$ and $F(Q)$ is either $f(Q)$ or $\psi(Q)$, and I denotes, respectively, either $H^{(2)}$ or $\tilde{H}^{(2)}$. The nature of the asymptotic expansion of Hankel functions (Ref. 5) is such that the r-dependence of $I_m(k,r)$ becomes increasingly independent of m as $|m|$ decreases below $(kr_- -2)$. This suggests that a useful approximation to $F(Q)$ is

$$F(Q) \simeq (d\theta/dC(Q)) \sum_{\ell=-\infty}^{\infty} \beta_\ell \exp(i\ell\theta)/I_\ell(k,r) \tag{32}$$

provided that C is such that $r=r(\theta)$ is single-valued. Further discussion of equation (32) is deferred until after the following paragraph.

When $r(\theta)$ is not single-valued it is convenient to partition C into the non-intersecting parts C_- and C_+. Draw all straight lines from 0 to all points Q on C. Those lines which contain only one point Q are defined to intersect the part of C_+ called C_{++}. For each line which contains more than one point Q, denote by Q_+ the Q for which r is largest. The part of C spanned by all points Q_+ is defined to be the part of C_+ called C_{+-}. The whole of C_+ is defined to be the union of C_{++} and C_{+-}.

In general, $F(Q)$ is defined to have the form of the right side of equation (32) for all Q on C_+, whereas $F(Q)=0$ for all Q on C_-. It is seen that C_+ and C_- are analogous to the illuminated and shadowed parts of surfaces in conventional physical optics. However, the partitioning of C into C_+ and C_- is determined by the coordinate system and not by the incident field. This may seem physically ridiculous. However, it should be remembered that conventional physical optics is most satisfactory when the body has a large lateral extension and the incident field is a plane wave. This means that the origin 0 of coordinates can be positioned a long way behind the surface of the body, so that straight lines drawn from 0 are, when they intersect the body's surface, effectively parallel to the rays of which the incident plane wave is composed. In such a case it is clear that C_+ and C_- correspond to the conventional definitions of the illuminated and shadowed parts of C. This demonstrates the physical reasonableness of the above choice of where $F(Q)$ is taken to be zero.

Substituting equation (32) into equation (31), and making $F(Q)=0$ for all Q on C_-, gives

$$2\pi\beta_m \simeq \sum_{\ell=-\infty}^{\infty} \beta_\ell \int_0^{2\pi} (I_m(k,r)/I_\ell(k,r)\exp(i[\ell-m]\theta)\; d\theta \tag{33}$$

For sufficiently low values of $|\ell|$ and $|m|$, the integral on the right side of equation (33) is very small, unless $\ell=m$ in which case the integral reduces to 2π. For $|\ell|>(kr_+ +2)$ the integral decreases rapidly with increasing $|\ell|$ provided that $|\ell|> m$. It follows that the terms on the right side of equation (32) that introduce the most significant errors are those for which

$$kr_- -2 < |\ell| < kr_+ + 2 \qquad (34)$$

Wall has devised an improved physical optics, whose accuracy and numerical efficiency appear impressive (Ref. 7), in which an extra term is added to the right side of (32). The explicit θ-dependence of this extra term, which is not identically zero on C_-, contains no factors $\exp(i\ell\theta)$ for which $|\ell|$ lies outside the inequality (34).

In order to extend the new physical optics to partially transparent bodies, it is necessary to devise a scheme whereby equation (29) is combined with equation (22) in such a way as to lead to approximate expressions, similar to the right side of equation (32), for both $f(Q)$ and $\psi(Q)$. I leave this as a challenge to the reader.

REFERENCES

(1) P.C. Waterman, New formulation of acoustic scattering, J. Acoust. Soc. Am. 45, 1417 (1969).

(2) P.C. Waterman, Scattering by dielectric obstacles, Alta Frequenza 38 (Speciale), 348 (1969).

(3) P.C. Waterman, Symmetry, unitarity and geometry in electromagnetic scattering, Phys. Rev. D 3, 825 (1971).

(4) P.C. Waterman, Matrix theory of elastic wave scattering, J. Acoust. Soc. Am. 60, 567 (1976).

(5) D.S. Jones, The theory of electromagnetism, Pergamon Press, Oxford (1964).

(6) R.H.T. Bates and F.L. Ng, Polarization-source formulation of electromagnetism and dielectric-loaded waveguides, Proc. IEE 119, 1568 (1972).

(7) R.H.T. Bates and D.J.N. Wall, Null-field approach to scalar diffraction: I. General method; II. Approximate methods; III. Inverse methods, Phil. Trans. Roy. Soc. London A 287, 45 (1977).

(8) R.H.T. Bates, Analytic constraints on electromagnetic field computations, Trans. IEEE MTT-23, 605 (1975).

(9) S. Ström, private communication (1979).

(10) B. Peterson and S. Ström, Matrix formulation of acoustic scattering from multilayered scatterers, J. Acoust. Soc. Am. 57, 2 (1975).

(11) G. Kristensson and S. Ström, Scattering from buried inhomogeneities - a general three-dimensional formalism, J. Acoust. Soc. Am. 64, 917 (1978).

(12) A. Boström, Multiple-scattering of elastic waves by bounded obstacles, submitted to J. Acoust. Soc. Am. (1979).

(13) Y.H. Pao and V. Varatharajulu, Huygens principle, radiation conditions and integral formulas for the scattering of elastic waves, J. Acoust.Soc. Am. 59, 1361 (1976).

(14) V. Varatharajulu and Y.H. Pao, Scattering matrix for elastic waves. I.Theory. J. Acoust. Soc. Am. 60, 556 (1976).

(15) V.K. Varadan, V.V. Varadan and Y.H. Pao, Multiple scattering of elastic waves by cylinders of arbitrary cross section, I. SH-waves, J. Acoust. Soc. Am. 63 1310 (1978).

(16) N. Morita, Another method of extending the boundary condition for the problem of scattering by dielectric cylinders, Trans. IEEE AP-27, 97 (1979).

* Thomas Alvin Boyd Lecture:

Dr. Thomas Alvin Boyd, B.Ch.E. '18, is an internationally known authority on combustion and fuel chemistry. He was an early co-worker with Dr. Charles F. Kettering in the original General Motors Research Laboratories in Dayton and is credited with being one of the early pioneers in the development of tetraethyl lead. As a life-time associate and close friend of Dr. Charles F. Kettering, he authored "Professional Amateur", a biography of Charles Franklin Kettering, and "Prophet of Progress," selections from the speeches of Charles F. Kettering. Dr. Boyd has thoughtfully made available to the University the income from his book, "Professional Amateur," to be used in sponsoring lectures which will bring together persons with a variety of backgrounds but mutual interests.

ELASTIC WAVE SCATTERING

Vasundara V. Varadan
Wave Propagation Group
Department of Engineering Mechanics
The Ohio State University
Columbus, Ohio 43210

ABSTRACT

The scattering of elastic waves by voids, elastic, visco-elastic and fluid inclusions embedded in an infinite elastic solid is analyzed at wavelength comparable to the size of the scatterer. The limitations of the separation of variables method is first discussed by considering the simple examples of a circular cavity and an elliptical cavity. The method fails in the latter case for elastic wave scattering. The T-matrix or null field method is formulated for elastic waves and applied to various types of scatterers. The general properties of the T-matrix which are independent of the geometry of the scatterer are then discussed. Numerical applications to elliptic cylindrical cavities and inclusions, prolate and oblate spheroidal cavities and inclusions, rough spheroidal cavities and stress free strips and penny shaped cracks are discussed.

I. INTRODUCTION

The scattering of elastic waves has been a subject of interest from the times of Rayleigh and Lamb. The practical applications of this subject are many and wide ranging. To mention a few examples - elastic wave scattering problems arise in geophysical exploration, seismology, the response of structures to seismic waves, acoustic response of elastic objects immersed in water, response of aircraft parts and high speed machinery to dynamic loads, non-destructive testing of flaws in structural materials, applications of ultrasound to biological systems for diagnostic and therapeutic purposes. In short, theories of diffraction, reflection, refraction, scattering and propagation come into play whenever the response of an elastic or viscoelastic structure to applied dynamic loads is to be studied.

Although the first few simple problems of elastic wave scattering from spheres and circular cylinders using the separation of variables approach were solved many years ago, solutions of elastic wave problems have not kept pace with their cousins in electromagnetic and acoustic waves. In spite of the mathematical unity present in the description of the three classical fields, due to the coupling by boundary conditions of the rotational and irrotational parts of the elastic field that propagate at distinct wave speeds, many problems have defied solution. Till a few years ago no results were available in the literature at wave lengths comparable to the size of the scatterer for non-spherical and non-circular obstacles. The only other method that had some degree of success was the integral equation approach for the scattering of elastic waves normally incident on strips and penny shaped cracks.

In the last six to seven years, tremendous developments have taken place in this field thanks largely to a giant effort in the non-destructive testing of flaws in structural materials using ultrasonic waves (1). The moment method was developed by Tan (2) for elastic wave scattering from elliptical cavities and strips. The T-matrix or the null field method was developed by Varadan and Pao (3) and applied to elliptical cavities and inclusions (4), spheroidal cavities and inclusions (5) and statistical distribution of scatterers (6). This method has proved to be a powerful computation technique for long and intermediate wavelengths. Recently, it has been extended to layered scatterers and finite number of scatterers (7) and (8). Simultaneously, Achenbach and co-workers have generalized Keller's geometrical theory of diffraction (GTD) to elastodynamic (9) and obtained extensive results for scattering from strips, penny shaped and lens shaped cracks.

This article focuses on the T-matrix method for elastic wave scattering and summarizes the various types of problems to which this method has been applied, the numerical results that have been obtained and the limitations on the numerical computations. The actual results have already been published elsewhere and are not reproduced here. References direct the interested readers to the appropriate sources.

II. ELASTIC FIELD EQUATIONS

The propagation of elastic waves in an infinite elastic medium is described by the displacement vector $\vec{u}$ and the stress tensor $\vec{\vec{\tau}}$ which are related by Generalized Hooke's law. For an isotropic linear elastic medium

$$\vec{\vec{\tau}} = \lambda \vec{\vec{I}}\, \nabla . \vec{u} + \mu\, (\nabla\vec{u} + \vec{u}\nabla) \tag{2.1}$$

where λ and μ are Lame's constants for the medium. From Eq. (2.1), it is seen that $\vec{\vec{\tau}}$ is a symmetric second rank tensor with six independent components and is the unit second rank tensor. The displacement equation in a source free medium for time harmonic waves of frequency ω is given by

$$\nabla . \vec{\vec{\tau}} + \rho\omega^2\vec{u} = 0 \tag{2.2}$$

where ρ is the mass density. The displacement vector can be resolved into scalar and vector potentials which describe the irrotational and solenoidal parts of the field respectively:

$$\vec{u} = \nabla\phi + \nabla \times \vec{\psi} \;;\; \nabla \vec{\psi} = 0 \tag{2.3}$$

Substituting Eqs. (2.1) and (2.3) into (2.2), we find that ϕ and $\vec{\psi}$ satisfy the scalar and vector Helmholtz equations, respectively

$$(\nabla^2 + k_p^2)\, \phi\, (\vec{r}) = 0 \quad , \tag{2.4}$$

$$(\nabla^2 + k_s^2)\, \vec{\psi}(\vec{r}) = 0 \quad , \tag{2.5}$$

where

$$k_p = \omega/c_p \;;\; k_s = \omega/c_s \tag{2.6}$$

are the compressional (P-) and shear (S-) wave numbers and c_p and c_s are the P- and S- wave speeds given by

$$c_p = \sqrt{(\lambda + 2\mu)/\rho} \;;\; c_s = \sqrt{\mu/\rho} \tag{2.7}$$

The existance of two wave speeds in the elastic medium causes a number of mathematical difficulties in the solution of scattering and boundary value problems.

III. SEPARATION OF VARIABLES APPROACH

The vector and scalar Helmholtz equations are separable in several different coordinate systems - spherical, circular cylindrical, elliptical, spheroidal etc. to mention some. Except for spherical and circular cylindrical coordinates, the angular parts of the eigenfunctions explicitly depend on the wavenumbers. For elastic waves since there are two wave speeds, there are no orthogonality relations for the product of P- wave and S- wave angular functions. This difficulty does not arise in acoustic and electromagnetic wave scattering problems when the field inside the scatterer is zero as in the case of sound soft, sound hard and perfectly conducting obstacles because there is only one wave speed that enters the formalism. Thus one can find fairly extensive results for such problems that are collected in a book edited by Senior, Bowman and Uslenghi (10).

The scattering of elastic waves by circular cylinders have been discussed by White (11) and the sphere problem has been studied by Ying and Truell (12). In this section, the simple example of a circular cylindrical cavity for normally incident P- and S- waves is illustrated using the separation of variables approach and then the problem of a elliptical cavity is considered.

Circular Cavity

Consider an infinitely long cylinder of circular cross section (Fig. 1a). A plane monochromatic wave incident normal to the axis (z-axis) of the cylinder. The wave may be a P- wave, SV- wave (polarized in the x-y plane) or an SH- wave (polarized along the z-axis). Without any loss of generality, the direction of incidence can be taken as the x-axis. The displacement $\vec{u}^{0}(r)$ due to the incident wave may be given by

$$\vec{u}^{0}(\vec{r}) = \begin{cases} \hat{i}\; e^{ik_p x}\, e^{-i\omega t} & \text{P-wave} \\ \hat{j}\; e^{ik_s x}\, e^{-i\omega t} & \text{SV-wave} \\ \hat{k}\; e^{ik_s x}\, e^{-i\omega t} & \text{SH-wave} \end{cases} \tag{3.1}$$

where $\hat{i}$, $\hat{j}$, $\hat{k}$ are unit vectors along the coordinate axes and ω is the frequency of the incident wave. Due to the symmetry of the problem, the incident and hence the scattered displacement field $\vec{u}^{s}$ is independent of the z-coordinate. For oblique incidence this is not so, as discussed by White (11).

The total displacement field outside the cavity is given by

$$\vec{u}(\vec{r}) = \vec{u}^{0}(\vec{r}) + \vec{u}^{s}(\vec{r}) \tag{3.2}$$

To solve the problem using separation of variables, the incident and scattered fields are expanded in vector cylindrical functions given by

$$\vec{u}^{0}(\vec{r}) = \sum_{\tau=1}^{3} \sum_{n=0}^{\infty} \sum_{\sigma=1}^{2} a_{\tau n\sigma}\, \mathrm{Re}\vec{\psi}_{\tau n\sigma}(\vec{r}) \tag{3.3}$$

$$\vec{u}^{s}(\vec{r}) = \sum_{\tau=1}^{3} \sum_{n=0}^{\infty} \sum_{\sigma=1}^{2} f_{\tau n\sigma}\, \vec{\psi}_{\tau n\sigma}(\vec{r}) \tag{3.4}$$

where

$$\left\{\begin{matrix} \mathrm{Re}\ \vec{\psi}_{1n\sigma} \\ \vec{\psi}_{1n\sigma} \end{matrix}\right\} = \sqrt{\varepsilon_n}\ \nabla \left\{\begin{matrix} J_n(k_p r) \\ H_n(k_p r) \end{matrix}\right\} \left\{\begin{matrix} \cos n\theta ; \sigma = 1 \\ \sin n\theta ; \sigma = 2 \end{matrix}\right\} \tag{3.5}$$

$$\left\{\begin{matrix} \mathrm{Re}\ \vec{\psi}_{2n\sigma} \\ \vec{\psi}_{2n\sigma} \end{matrix}\right\} = \sqrt{\varepsilon_n}\ \nabla \times \hat{k} \left\{\begin{matrix} J_n(k_s r) \\ H_n(k_s r) \end{matrix}\right\} \left\{\begin{matrix} \cos n\theta ; \sigma = 1 \\ \sin n\theta ; \sigma = 2 \end{matrix}\right\} \tag{3.6}$$

$$\left\{\begin{matrix} \mathrm{Re}\ \vec{\psi}_{3n\sigma} \\ \vec{\psi}_{3n\sigma} \end{matrix}\right\} = \sqrt{\varepsilon_n}\ \nabla \times \nabla \times \hat{k} \left\{\begin{matrix} J_n(k_s r) \\ H_n(k_s r) \end{matrix}\right\} \left\{\begin{matrix} \cos n\theta ; \sigma = 1 \\ \sin n\theta ; \sigma = 2 \end{matrix}\right\} \tag{3.7}$$

In Eqs. (3.5) - (3.7), J_n and H_n are the cylindrical Bessel and Hankel functions, ε_n is the Neumann factor. The position vector $\vec{r}$ is in the x-y plane and θ is measured counterclockwise from the x-axis. It is important to note that the angular parts of the basis functions $\cos n\theta$ and $\sin n\theta$ are completely independent of the wave number.

For an incident wave of the form given in Eq. (3.1), the expansion coefficients $a_{\tau n\sigma}$ in Eq. (3.3) are given by

$$a_{\tau n\sigma} = \sqrt{\varepsilon_n}\ i^n\ \delta_{\sigma,1} \begin{cases} \delta_{\tau,1} & ; \quad \text{P-wave} \\ \delta_{\tau,2} & ; \quad \text{SV-wave} \\ \delta_{\tau,3} & ; \quad \text{SH-wave} \end{cases} \tag{3.8}$$

The boundary conditions at the surface of a cavity of radius a are

$$\hat{r} \cdot \vec{\vec{\tau}}(\vec{r}) \Big|_{|\vec{r}|=a} = 0 \tag{3.9}$$

Substituting the expansions for $\vec{u}^0$ and $\vec{u}^s$ in the stress tensor, noting that $\vec{\psi}_1$ and $\vec{\psi}_2$ are entirely in the x-y plane whereas $\vec{\psi}_3$ has only a z-component, and using the orthogonality of the angular functions, we obtain the following set of equations relating f and a

$$[T^3_{11}\ \delta_{\sigma,1}\ \delta_{\sigma,1} + T^3_{12}\ \delta_{\sigma,2}\ \delta_{\tau,2}]\ f_{\tau n\sigma} = -[T^1_{11}\ \delta_{\sigma,1}\ \delta_{\tau,1} + T^1_{12}\ \delta_{\sigma,2}\ \delta_{\tau,2}]\ a_{\tau n\sigma} \tag{3.10}$$

$$[T^3_{41}\ \delta_{\sigma,2}\ \delta_{\tau,1} + T^3_{42}\ \delta_{\sigma,1}\ \delta_{\tau,2}]\ f_{\tau n\sigma} = -[T^1_{41}\ \delta_{\sigma,2}\ \delta_{\tau,1} + T^1_{42}\ \delta_{\sigma,1}\ \delta_{\tau,2}]\ a_{\tau n\sigma} \tag{3.11}$$

$$T^3_{53}\ \delta_{\sigma,1}\ \delta_{\tau,3}\ f_{\tau n\sigma} = -\ T^1_{53}\ \delta_{\sigma,1}\ \delta_{\tau,3}\ a_{\tau n\sigma} \tag{3.12}$$

where the T^i_{mn} are stress components (see Eringen and Suhubi (13)) given by

$$T^i_{11} = (n^2 + n - 1/2\ k_s^2 a^2)\ C^i_n - k_p a\ C^i_{n-1} \tag{3.13a}$$

$$T^i_{12} = \mp\ n\ [-(n+1)\ C^i_n + k_s a\ C^i_{n-1}] \tag{3.13b}$$

$$T^i_{41} = \mp\ n\ [-(n+1)\ C^i_n + k_p a\ C^i_{n-1} \tag{3.13c}$$

$$T^i_{42} = -(n^2 + n - 1/2\ k_s^2 a^2)\ C^i_n + k_s a\ C^i_{n-1} \tag{3.13d}$$

$$T^i_{53} = k_s^2\ a\ (-n\ C^i_n + k_s a\ C^i_{n-1}) \tag{3.13e}$$

where $C^1_n = J_n$ and $C^3_n = H_n$. In Eqs. (3.13), if the second subscript of T is 1, the argument of C^i_n is $k_p a$; if the second subscript of T is 2 or 3, the argument of C^i_n is $k_s a$.

From Eqs. (3.10) - (3.12), we observe that an incident P-wave gives rise to both P- and SV-waves on scattering and vice versa. However, SH-waves are completely uncoupled as given by Eq. (3.12). This is called anti-plane strain in elasticity and the case of P- and SV-waves is called the plane strain problem. Equations (3.10) - (3.12) may be easily solved for the scattered field coefficients.

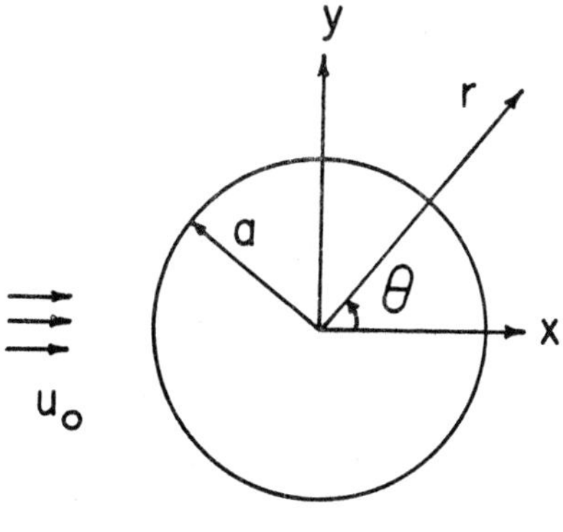

Fig. 1a. Circular scatterer

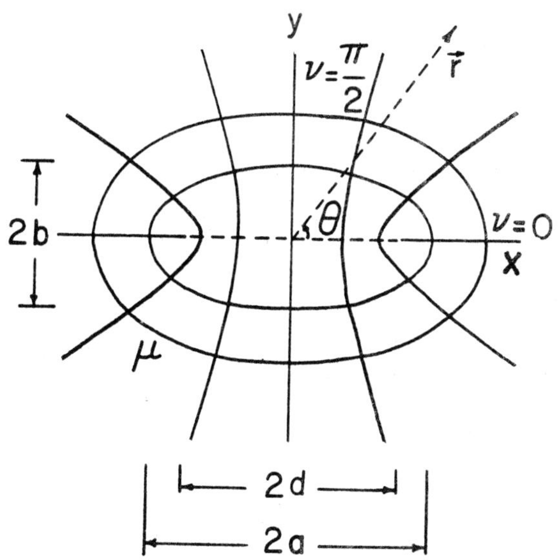

Fig. 1b. Elliptical scatterer

Elliptical Cavity

Consider an infinitely long cylinderical cavity of elliptical cross section (see Fig. 1b) with major axes 2a along the x-direction and minor axis 2b along the y-direction. P-, SV- or SH-waves are incident in the x-y plane at an angle α to the x-axis. As discussed before, the SH-wave problem uncouples and reduces to the scalar wave equation. In elliptic coordinates, μ, ν, the Helmholtz equation is separable and the solutions are given by the radial and angular Mathieu functions (see Morse and Feshbach (14)). The angular Mathieu functions are $Se_m(h, \nu)$, $So_m(h,\nu)$ and the radial functions are $Je_m(h,\mu)$, $Jo_m(h,\mu)$, $Ne_m(h,\mu)$ and $No_m(h,\mu)$ where $h = kd$ where 2d is the interfocal distance and $k = \omega/c$ is the wave number. The Mathieu functions have a completely different series expansion depending on whether the index m is an even or odd integer. The radial functions J correspond to the Bessel functions in polar coordinates and the N functions to the Neumann functions. The angular functions for a given value of h form a complete set of eigenfunctions which are mutually orthogonal.

Let $Re\phi_{n\sigma}(h,\mu,\nu)$ where $n = 0, 1, \ldots,\infty$ and $\sigma = e,o$ denote the product of radial (J) functions and angular functions. The 'Re' denotes that the J functions are regular and well behaved at $\mu = 0$ and let $Ou\phi_{n\sigma}(h,\mu\ \nu)$ denote the product of radial (N) functions and angular functions.

We can expand the incident field $\vec{u}^0$ and the scattered field $\vec{u}^s$ as

$$\vec{u}^0 = \sum_{n,\sigma} [a_{1n\sigma} \nabla Re\phi_{n\sigma} (h_p,\mu,\nu) + a_{2n\sigma} \nabla x\ \hat{k}\ Re\phi_{n\sigma}(h_s,\mu,\nu)] \quad (3.14)$$

$$\vec{u}^s = \sum_{n,\sigma} [f_{1n\sigma} \nabla Ou\phi_{n\sigma} (h_p, \mu,\nu) + f_{2n\sigma} \nabla \times \hat{k}\ Ou\phi_{n\sigma} (h_s,\mu,\nu) \quad (3.15)$$

The boundary condition at $\mu = \mu_o$, which defines the boundary of the ellipse is given by

$$\hat{u} \cdot \vec{\tau}(\vec{u}^0 + \vec{u}^s)|_{\mu = \mu_o} = 0 \quad (3.16)$$

where $\hat{\mu}$ is the unit normal to the elliptic boundary. Since Eq. (3.15) is a vector boundary condition, we get two separate equations similar to Eqs. (3.10) and (3.11) for the circular boundary. However, as before we cannot use the orthogonality of the angular functions to obtain expressions for $f_{\tau n\sigma}$ in terms of $a_{\tau n\sigma}$ since there is no orthogonality for products of the form $Se_m(h_p,\nu)\ Se_m(h_s,\nu)$.

One of the ways of circumventing this difficulty is to expand the angular functions in an infinite series of trigonometric functions cos $n\nu$ and sin $n\nu$ and the coefficients of expansion depend on the wavenumber h_p or h_s as the case may be. Now the orthogonality of the trigonometric functions can be invoked to obtain expressions for $f_{\tau n\sigma}$. However, instead of obtaining a simple expression for $f_{\tau n\sigma}$ in terms of $a_{\tau n\sigma}$ for the circular cavity, the scattered field coefficients are given in the form of an infinite series which are poorly convergent and not easy to evaluate numerically. This difficulty will also be encountered when one solves the SH-wave or acoustic wave problem when the scatterer is permeable. In this case instead of dealing with two distinct wavenumbers in the exterior as for the elastic cavity is concerned, one wavenumber for the exterior and a different one for the interior enter into the boundary conditions, again causing the same problems. This will explain the reason for the dearth of results in the literature for acoustic wave scattering by permeable scatterers and electromagnetic wave scattering by dielectric scatterers using the separation of variables approach.

IV. OTHER APPROACHES FOR NON-SPHERICAL AND NON-CIRCULAR SCATTERERS

Very few problems have been solved other than the sphere and the circular cylinder. Integral equation methods have been used for the rigid strip (15) and the stress free strip (16) as well as for the penny shaped crack (17). Harumi (18) has also obtained the first few scattered field coefficients for the rigid strip and the stress free strip using Mathieu functions. The Wiener-Hopf method, restricted to two dimensional problems (19) has also been used for the strip. The method of matched asymptotic expansions has been successfully used by Datta (20) to obtain low frequency expansions of the scattered field for spheroidal inclusions and cavities. Krumhansl, Gubernatis and Domany (21) have used the Born approximation starting with a volume integral formula for the scattered field. Recent extensions include the quasi-static and the extended quasi-static approximations. Domany (22) has recently tried the distorted wave Born approximation. These methods have been used extensively in quantum mechanical scattering problems. The work of Tan (2) and Achenbach et. al (9) has already been referred to in the introduction.

V. T-MATRIX FORMULATION IN SPHERICAL COORDINATES

This approach was first introduced by Waterman for acoustic (23) and electromagnetic (24) waves for single scatterers. Later it was applied to elastic wave scattering by Varadan and Pao (3) and Waterman (25). The basic difference between the two formulations is the integral representation for the elastic field. In (3), the formulation begins with integral representations (26) that are analogous to Helmholtz type integral formulas for the acoustic and electromagnetic field that involve only the field quantities that are naturally prescribed by boundary conditions pertinent to the problem occur in the integrand. Waterman (25) has used the integral representation given by Morse and Feshbach (27) which involves the divergence and curl of the displacement field $\vec{u}$, whereas the stress tensor, Eq. (2.1), involves the divergence and gradient of $\vec{u}$.

The Green's function for the infinite elastic medium is the response to a uniformly applied point force at $\vec{r}'$. The response at $\vec{r}$ is denoted by the Green's displacement dyadic $\vec{G}$ and the stress dyadic $\vec{\Sigma}$ which are related to each other by Hookes law as described in the introductory chapter of these proceedings.

Consider a scatterer embedded in an infinite medium whose surface is described by surface S with a continuously turning unit normal $\hat{n}$ satisfying the restrictions of the divergence theorem (see Fig. 2). A wave of frequency ω giving rise to a displacement field $\vec{u}^0$ is incident on S. The total field outside S is given by

$$\vec{u} = \vec{u}^0 + \vec{u}^s \quad . \tag{5.1}$$

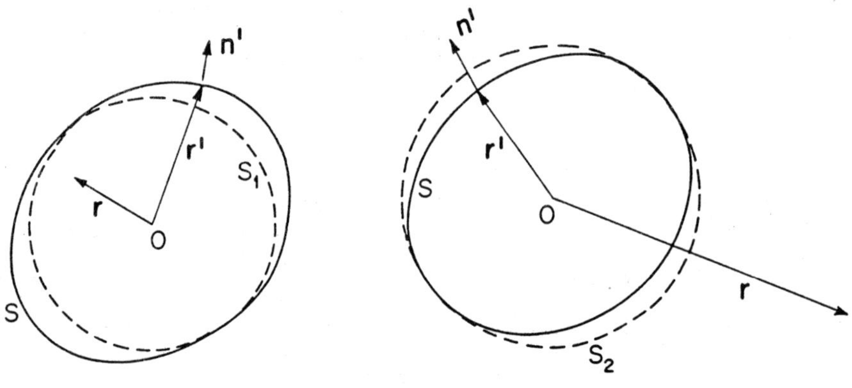

Fig. 2. Scattering geometry (a) for points $\vec{r}$ interior to S (b) for points $\vec{r}$ exterior to S

The surface integral representation for the elastic field is given by

$$\int_S \left\{ \vec{u}'_+ \cdot \hat{n}' \cdot \vec{\Sigma}\,(\vec{r},\vec{r}') - \vec{t}'_+ \cdot \vec{G}\,(\vec{r},\vec{r}') \right\} dS' = \begin{cases} \vec{u}^S\,(\vec{r}) & ;\ \vec{r} \text{ outside S} \\ -\vec{u}^0\,(\vec{r}) & ;\ \vec{r} \text{ inside S} \end{cases} \tag{5.2}$$

where $\vec{u}_+$ and $\vec{t}_+$ are the displacement and traction vectors on the surface S, the + sign indicating that the surface is approached from the outside. The primes on the variables indicate that they are functions of $\vec{r}'$ which is a point on S.

The basic philosophy of the appproach is to expand every field variable in Eq. (5.2) in terms of a convenient set of orthogonal basis functions, not necessarily orthogonal on the actual surface S of the scatterer. As discussed in Section III, since the elastic field involves two distinct wavenumbers, only the vector cylindrical and vector spherical functions are suitable.

Vector Spherical Functions

These are solutions of the vector Helmholtz equation in spherical coordinates r, θ, ϕ and form a complete orthogonal set. They can be constructed directly from the solutions of the scalar Helmholtz equation which are given by

$$\begin{Bmatrix} \mathrm{Re}\ \phi_{nm\sigma}\,(k\vec{r}) \\ \phi_{nm\sigma}\,(k\vec{r}) \end{Bmatrix} = \xi_{nm} \begin{Bmatrix} j_n\,(kr) \\ h_n\,(kr) \end{Bmatrix} Y_{nm\sigma}\,(\theta,\phi) \tag{5.3}$$

where j_n are the spherical Bessel functions which are finite or regular at the origin and the h_n are spherical Hankel functions which behave as outgoing spherical waves for large values of the argument. The function Y is the spherical harmonic given by

$$Y_{nm\sigma}(\theta,\phi) = P_n^m(\cos\theta)\begin{Bmatrix}\cos m\phi & \sigma = 1\\ \sin m\phi\ ; & \sigma = 2\end{Bmatrix} \tag{5.4}$$

where the P_n^m are the associated Legendre polynomials. The normalization constant ξ_{nm} is given by

$$\xi_{nm} = \sqrt{\frac{4\pi\varepsilon_n(n-m)!}{(2n+1)(n+m)!}} \tag{5.5}$$

with $\varepsilon_0 = 1$, $\varepsilon_n = 2$ for $n > 0$ and $n = 0, 1, \ldots, \infty$; $m = 0, 1, \ldots, n$.

There are three vector spherical functions, the first one describing the irrotational field and the second and third describing the two components of the solenoidal field:

$$\vec{\psi}_{1nm\sigma}(k_p\vec{r}) = \sqrt{k_p/k_s}\ \nabla\phi_{nm\sigma}(k_p\vec{r}) \tag{5.6}$$

$$\vec{\psi}_{2nm\sigma}(k_s\vec{r}) = k_s / \sqrt{n(n+1)}\ \nabla \times [\vec{r}\ \phi_{nm\sigma}(k_s\vec{r})] \tag{5.7}$$

$$\vec{\psi}_{3nm\sigma}(k_s\vec{r}) = \frac{1}{k_s}\nabla \times \vec{\psi}_{2nm\sigma} \tag{5.8}$$

For convenience, we use the abbreviation

$$\vec{\psi}_{\tau nm\sigma} = \vec{\psi}_{\tau n}\ ;\ \tau = 1, 2, 3 \tag{5.9}$$

The vector spherical functions corresponde to the $\vec{L}$, $\vec{M}$ and $\vec{N}$ functions defined by Morse and Feshbach (27). It should be noted that for $\vec{\psi}_{2n}$ and $\vec{\psi}_{3n}$, the index n starts from 1 whereas for $\vec{\psi}_{1n}$ it starts from 0. The functions can also be expressed in terms of the vector spherical harmonics. The vector function $\vec{\psi}_2$ is orthogonal to $\vec{\psi}_1$ and $\vec{\psi}_3$, $\vec{\psi}_1$ and $\vec{\psi}_3$ are orthogonal with respect to themselves when their orders differ, but $\vec{\psi}_1$ is not orthogonal to $\vec{\psi}_3$ if their orders are the same, see for example (3). In elastic wave problems where all three functions are involved, quantities involving $\vec{\psi}_1$ and $\vec{\psi}_3$ have the same properties because the angular part of these two functions is the same whereas quantities involving $\vec{\psi}_2$ form a separate group.

Expansion of Fields in Vector Spherical Functions

The incident displacement field $\vec{u}^0$ is involved in the interior Helmholtz formula and hence expanded in terms of Bessel functions that are finite at the origin

$$\vec{u}^0(\vec{r}) = \sum_{\tau=1}^{3}\sum_{n=0}^{\infty}\sum_{m=0}^{n}\sum_{\sigma=1}^{2} a_{\tau nm\sigma}\ \mathrm{Re}\vec{\psi}_{\tau nm\sigma}(\vec{r})$$

$$= \sum_{\tau\, n} a_{\tau n}\ \mathrm{Re}\vec{\psi}_{\tau n}(\vec{r}) \tag{5.10}$$

The scattered displacement field $\vec{u}^s$ is expected to be outgoing at infinity and hence

$$\vec{u}^s(\vec{r}) = \sum_{\tau n} f_{\tau n} \vec{\psi}_{\tau n}(\vec{r}) \quad . \tag{5.11}$$

From the construction of the Green's dyadic G (Chapter I of these proceedings)

$$\vec{G}(\vec{r}, \vec{r}') = (ik_s/\rho\omega^2) \sum_{\tau n} \vec{\psi}_{\tau n}(r_>) \mathrm{Re}\, \vec{\psi}_{\tau n}\ (r_<) \tag{5.12}$$

where $r_>$ and $r_<$ are greater and lesser of $\vec{r}$ and $\vec{r}'$, respectively.

Substituting Eqs. (5.10) - (5.12) into the integral representation given in Eq. (5.2) and considering points $\vec{r}$ outside the sphere circumscribing the surface S of the scattering and using the orthogonality of the angular parts of the basis function on the sphere $|\vec{r}|$ = constant, we obtain for $|\vec{r}| > |\vec{r}'|$

$$f_{\tau n} = (ik_s/\rho\omega^2) \int_S \left\{ \vec{u}'_+ \cdot \vec{t}' (\mathrm{Re}\, \vec{\psi}_{\tau n}) - \vec{t}'_+ \cdot \mathrm{Re}\, \vec{\psi}'_{\tau n} \right\} dS' \ . \tag{5.13}$$

Similarly, considering points $\vec{r}$ inside the sphere inscribing S, for $|\vec{r}| < |\vec{r}'|$

$$a_{\tau n} = -(ik_s/\rho\omega^2) \int_S \left\{ \vec{u}_+ \cdot \vec{t}'(\vec{\psi}_{\tau n}) - \vec{t}_+ \cdot \vec{\psi}'_{\tau n}\ dS' \right\}. \tag{5.14}$$

In Eqs. (5.13) and (5.14), we use the notation

$$\vec{t}(\vec{A}) = \hat{n} \cdot [\lambda\ \vec{I}\ \nabla . \vec{A} + \mu\ (\nabla\vec{A} + \vec{A}\nabla)] \tag{5.15}$$

and the primes again indicate that the variables are functions of $\vec{r}'$.

From this step onwards, since the dependence on $\vec{r}$ has disappeared from the expressions for $f_{\tau n}$ and $a_{\tau n}$, we will omit the prime on $\vec{r}'$ and it will be understood that $\vec{r}$, the variable of integration is a point on the surface.

Recently, Wall (28) has suggested the possibility of using basis functions more natural to the boundary of the scatterer for the expansion of $\vec{u}^0$, $\vec{u}^s$ and $\vec{G}$. But as we have seen for the elastic wave problem only cylindrical and spherical functions can be used.

In the integrand of Eqs. (5.13) and (5.14), we still have to specify the surface displacement and the surface traction. This is prescribed by boundary conditions at S which depend on the type of scatterer. So far we have not restricted the scatterer properties. Our aim is to obtain a matrix which is called the T-matrix to relate the incident and scattered field coefficients.

Cavity

If S encloses a void, then the boundary condition at S is that all components of the traction must vanish at S, i.e.,

$$\vec{t}_+(\vec{r}) = 0 \quad , \vec{r} \text{ on } S \quad . \tag{5.16}$$

The second term in the integrand of Eqs. (5.13) and (5.14) are zero and the only field variable to be specified is the unknown displacement $\vec{u}_+$ on the surface S which can be expanded as follows:

$$\vec{u}_+(\vec{r}) = \sum_{\tau n} \alpha_{\tau n} \text{ Re } \vec{\psi}_{\tau n} (\vec{r}) \quad , \vec{r} \text{ on S} . \tag{5.17}$$

Substituting the above expansion in the integral expressions for $f_{\tau n}$ and $a_{\tau n}$, we obtain

$$f_{\tau n} = i \sum_{\tau' n'} Q^c_{\tau n, \tau' n'} (\text{Re},\text{Re})\alpha_{\tau' n'} \tag{5.18}$$

$$a_{\tau n} = -i \sum_{\tau' n'} Q^c_{\tau n, \tau' n'} (\text{Ou},\text{Re}) \alpha_{\tau' n'} \tag{5.19}$$

whence in vector matrix notation

$$f = -T^c a \tag{5.20}$$

$$T^c = Q^c(\text{Re},\text{Re}) [Q^c(\text{Ou},\text{Re})]^{-1} \tag{5.21}$$

where

$$Q^c_{\tau n, \tau' n'} \begin{bmatrix} \text{Ou} \\ \text{Re}, & \text{Re} \end{bmatrix} = (k_s/\rho\omega^2) \int_S \vec{t} \begin{bmatrix} \text{Ou} \\ \text{Re} \end{bmatrix} \vec{\psi}_{\tau n} \cdot \text{Re } \vec{\psi}_{\tau' n'} \, dS. \tag{5.22}$$

In Eqs. (5.21) and (5.22), we have used the notation Ou $\vec{\psi}$ to denote spherical functions containing Hankel functions and Re $\vec{\psi}$ to denote functions containing Bessel functions.

Rigid Fixed Scatterer

The boundary condition in this case is that the displacement should vanish everywhere inside and on S. Thus,

$$\vec{u}_+(\vec{r}) = 0 \quad , \vec{r} \text{ on S} . \tag{5.23}$$

The unknown quantity in the integrand of Eqs. (5.13) and (5.14) is the surface traction $\vec{t}_+$ which may be expanded in regular functions as follows

$$\vec{t}_+(\vec{r}) = \sum_{\tau n} \alpha_{\tau n} \, \vec{t} \, [\vec{\psi}_{\tau n}(\vec{r})]. \tag{5.24}$$

Substitution of Eq. (5.24) in Eqs. (5.13) and (5.14) results in

$$f = -T^R a \tag{5.25}$$

$$T^R = -Q^R(\text{Re},\text{Re}) [Q^R(\text{Ou},\text{Re})]^{-1} \tag{5.26}$$

where the Q-matrix of the rigid inclusion is given by

$$Q^R_{\tau n, \tau' n'} \begin{bmatrix} \text{Ou} \\ \text{Re}, \text{Re} \end{bmatrix} = (k_s/\rho\omega^2) \int_S \begin{bmatrix} \text{Ou} \\ \text{Re} \end{bmatrix} \vec{\psi}_{\tau n} \cdot \vec{t}(\text{Re } \vec{\psi}_{\tau n}) \, dS \tag{5.27}$$

Elastic Inclusion

If the contact at the bimaterial interface is perfectly welded, then we require that all components of the displacement and traction must be continuous across the surface S. Let $\vec{u}^1$ be the displacement field inside the elastic inclusion of material properties ρ^1, λ^1 and μ^1. Wavefunctions pertaining to the interior are distinguished by the superscript 1. The displacement field in the interior of S is represented by

$$\vec{u}^1(\vec{r}) = \sum_{\tau n} \alpha_{\tau n} \operatorname{Re} \vec{\psi}^1_{\tau n}(\vec{r}), \ \vec{r} \text{ inside S.} \tag{5.28}$$

The proof of the convergence of expansions of the above type have been discussed by several authors, notably Millar (29), Bates and Wall (30), Waterman (23) and Pao (31).

The expansion for the displacement can be differentiated to yield the stress at all points within S and on S approached from the inside. The boundary conditions at S are then

$$\begin{aligned} \vec{u}_+(\vec{r}) &= \vec{u}^1_-(\vec{r}) \\ \vec{t}_+(\vec{r}) &= \vec{t}^1_-(\vec{r}) \ . \end{aligned} \tag{5.29}$$

Substituting the values of $\vec{u}^1_-$ and $\vec{t}^1_-$ from Eq. (5.28) into (5.13) and (5.14), we obtain

$$f = -T^I a \tag{5.30}$$

$$T^I = Q^I(\text{Re},\text{Re}) \ [Q^I(\text{Ou},\text{Re})]^{-1} \tag{5.31}$$

where the Q-matrix of the inclusion is given by

$$Q^I_{\tau n,\tau' n'} \begin{bmatrix} \text{Ou} \\ \text{Re} \end{bmatrix} \text{Re} = (k_s/\rho\omega^2) \int_S \left\{ \vec{t} \left[\begin{matrix} \text{Ou} \\ \text{Re} \end{matrix} \vec{\psi}_{\tau n} \right] \cdot \operatorname{Re} \vec{\psi}^1_{\tau' n'} - \begin{matrix} \text{Ou} \\ \text{Re} \end{matrix} \vec{\psi}_{\tau n} \cdot \vec{t}(\operatorname{Re} \vec{\psi}^1_{\tau' n'}) \right\} dS \tag{5.32}$$

Fluid Inclusion

If the surface S encloses an invicid or perfect fluid of density ρ_f and compressibility λ_f, then the problem is very much more complicated. This is because the exterior solid supports both compressional and shear waves, whereas the perfect fluid supports only compressional waves. The boundary conditions at the surface S require the normal component of the particle velocity and hence the displacement to be continuous and also the normal component of the traction or pressure to be continuous. In addition since the fluid cannot support shear stresses, the tangential component of the traction on the solid side of the surface is zero. However, this leaves the tangential component of the displacement on the solid side unspecified. Hence, even if an expansion of the type given in Eq. (5.28) for the displacement $\vec{u}^1$ in the fluid is assumed in terms of irrotational functions alone and an expansion in terms of Re $\vec{\psi}_{\tau n}$, $\tau = 1,2,3$ for the tangential displacement on S is assumed, Eqs. (5.13) and (5.14) are not sufficient to yield the desired relation between the incident and scattered field coefficients.

We denote the fluid wave function by

$$\vec{\psi}^{1}_{1n} \equiv \vec{\psi}_{fn} \tag{5.33}$$

and hence

$$\vec{u}^{1}(\vec{r}) = \sum_{n} \alpha_{fn} \, \mathrm{Re}\, \vec{\psi}_{fn}(\vec{r}), \quad \vec{r} \text{ inside S.} \tag{5.34}$$

The tangential component of the displacement on S_+ is given by

$$\vec{u}_{+}(\vec{r})\big|_{\text{tangential}} = \sum_{\tau\, n} \beta_{\tau n} \, \mathrm{Re}\, \vec{\psi}_{\tau n}(\vec{r}), \quad \vec{r} \text{ on S.} \tag{5.35}$$

The boundary conditions at S require

$$\left.\begin{aligned} \hat{n}\cdot\vec{u}_{+}(\vec{r}) &= \hat{n}\cdot\vec{u}^{1}_{-}(\vec{r}) \\ \hat{n}\cdot\vec{t}_{+}(\vec{r}) &= \hat{n}\cdot\vec{t}^{1}_{-}(\vec{r}) \\ \vec{t}_{+}(\vec{r})\big|_{\text{tangential}} &= 0 \end{aligned}\right\} \quad \vec{r} \text{ on S.} \tag{5.36}$$

Substituting for $\hat{n}\cdot\vec{u}$ and $\hat{n}\cdot\vec{t}$ from Eq. (5.34) and using Eq. (5.35) and the boundary conditions in (5.36), we obtain

$$f_{\tau n} = i\left[\sum_{n'} Q^{FS}_{\tau n,n'}(\mathrm{Re},\mathrm{Re})\,\alpha_{fn'} + \sum_{\tau'}\sum_{n'} R_{\tau n,\tau' n'}(\mathrm{Re},\mathrm{Re})\,\beta_{\tau' n'}\right] \tag{5.37}$$

$$a_{\tau n} = -i\left[\sum_{n'} Q^{FS}_{\tau n,n'}(\mathrm{Ou},\mathrm{Re})\,\alpha_{fn'} + \sum_{\tau'}\sum_{n'} R_{\tau n,\tau' n'}(\mathrm{Ou},\mathrm{Re})\,\beta_{\tau' n'}\right] \tag{5.38}$$

where

$$Q^{FS}_{\tau n,n'}\left[{}^{\mathrm{Ou}}_{\mathrm{Re}}\,\mathrm{Re}\right] = (k_s/\rho\omega^2)\int_S \Big\{ \vec{t}\left[{}^{\mathrm{Ou}}_{\mathrm{Re}}\,\vec{\psi}_{\tau n}\right]\cdot\hat{n}\,[\hat{n}\cdot\mathrm{Re}\,\vec{\psi}_{fn'}] - \left[{}^{\mathrm{Ou}}_{\mathrm{Re}}\,\vec{\psi}_{\tau n}\right]\cdot\hat{n}\,\lambda_f\,\nabla\cdot\vec{\psi}_{fn'} \Big\}\, dS \tag{5.39}$$

$$R_{\tau n,\tau' n'}\left[{}^{\mathrm{Ou}}_{\mathrm{Re}}\,\mathrm{Re}\right] = (k_s/\rho\omega^2)\int_S \Big\{ \vec{t}\left[{}^{\mathrm{Ou}}_{\mathrm{Re}}\,\vec{\psi}_{\tau n}\right]\cdot\mathrm{Re}\,\vec{\psi}_{\tau' n'}\big|_{\text{tangential}} \Big\}\, ds \tag{5.40}$$

To obtain additional equations connecting α and β, consider the integral representation for the field in the fluid,

$$\int_S \left\{ \lambda_f(\hat{n}' \cdot \vec{u}'_-)\ (\nabla' \cdot \vec{G}_f) - (\hat{n}' \cdot \vec{t}'_-)\ (\hat{n}' \cdot \vec{G}_f) \right\} dS'$$

$$= \begin{cases} 0 & , \vec{r} \text{ outside } S \\ -\vec{u}^1(\vec{r}) & , \vec{r} \text{ inside } S \end{cases} \tag{5.41}$$

where $\vec{G}_f$ is the Green's function in the fluid given by

$$\vec{G}_f(\vec{r},\vec{r}') = (k_f/\rho\omega^2)\sum_n \vec{\psi}_{fn}(r_>)\ \mathrm{Re}\ \vec{\psi}_{fn}(r_<). \tag{5.42}$$

Considering points $\vec{r}$ outside the sphere circumscribing S, i.e., for $|\vec{r}| > |\vec{r}'|$, we obtain using Eqs. (5.34), (5.35) and (5.36) in (5.41):

$$\sum_{\tau'}\sum_{n'} P_{n,\tau' n'}\ (\mathrm{Re},\mathrm{Re})\ \beta_{\tau' n'} - \sum_{n'} M_{n,n'}\ (\mathrm{Re},\mathrm{Re})\ \alpha_{fn'} = 0 \tag{5.43}$$

where

$$P_{n,\tau' n'}\ (\mathrm{Re},\mathrm{Re}) = \int_S (\nabla \cdot \mathrm{Re}\ \vec{\psi}_{fn})\ (\hat{n} \cdot \mathrm{Re}\ \vec{\psi}_{\tau' n'})\ dS \tag{5.44}$$

$$M_{n,n'}\ (\mathrm{Re},\mathrm{Re}) = \int_S (\nabla \cdot \mathrm{Re}\ \vec{\psi}_{fn})\ (\hat{n} \cdot \mathrm{Re}\ \vec{\psi}_{fn})\ dS \tag{5.45}$$

From Eq. (5.43), we can write

$$\alpha = M^{-1}\ P\ \beta. \tag{5.46}$$

Substituting the above equation into (5.37) and (5.38), we finally obtain

$$f = -T^{FS}\ a \tag{5.47}$$

where T^{FS} is the T-matrix of a fluid inclusion in an elastic solid given by

$$T^{FS} = [Q^{FS}(\mathrm{Re},\mathrm{Re})\ M^{-1}P + R\ (\mathrm{Re},\mathrm{Re})]\ [Q^{FS}(\mathrm{Ou},\mathrm{Re})\ M^{-1}P + R\ (\mathrm{Ou},\mathrm{Re})]^{-1}. \tag{5.48}$$

The formalism given here for the fluid inclusion in the solid is analogous to that proposed by Boström (32) for the solid inclusion in a fluid medium.

VI. GENERAL PROPERTIES OF THE T-MATRIX

Symmetry and Unitarity

Independent of the geometry and nature of the scatterer, some very general properties of the T-matrix of a scatterer embedded in an infinite medium can be proved. From the Betti-Rayleigh reciprocal identity for two elastodynamic states, the T-matrix can be proved to be symmetric, i.e.,

$$T = T^t \tag{6.1}$$

where the superscript t denotes matrix transposition. The details of the proof have been given in (3).

If no energy is dissipated in the scatterer or the embedding medium, then from the conservation of energy, we can show that

$$TT^* = -\mathrm{Re}\ T \qquad (6.2)$$

where the asterisk denotes complex conjugate and Re stands for real part. In quantum mechanics, it is more common to refer to the S-matrix which is related to T by S = 1 - 2T. Then, from (6.1) and (6.2) it follows that

$$S = S^t \quad ; \quad SS^\dagger = 1 \qquad (6.3)$$

where the dagger sign denotes the matrix adjoint. Thus, S is both symmetric and unitary.

The importance of these general properties cannot be overemphasized from the computational point of view. Equations (6.1) and (6.2) serve as a basic check for any numerical calculations involving non-spherical and non-circular scatterers.

Translation of the T-matrix

Especially in problems involving two or more scatterers, it is preferable to compute the T-matrix of each scatterer with respect to a coordinate system located inside that scatterer. In computing the T-matrix of the whole configuration, we may want to choose an arbitrary coordinate system. Then it becomes necessary to translate the T-matrix from one reference system to another (see, for example Peterson and Strom (33)). Even in problems involving just one scatterer in order to evaluate the accuracy of the numerical computations, it is sometimes useful to compute the T-matrix with respect to different coordinate systems centered at O and O′ (see Fig. 3). The transformation of the T-matrices under translation is governed by the transformation properties of the vector basis functions in which they are evaluated. Cruzan (34) has given the translation theorems for the vector spherical functions when they are defined in terms of a complex basis for the angular functions, i.e., the spherical harmonics are defined with exp ($im\phi$), $-n \le m \le n$. Peterson and Ström (35) have given the translation theorems for the transverse vector functions with real angular functions, i.e. in terms of cos $m\phi$ and sin $m\phi$, $o \le m \le n$.

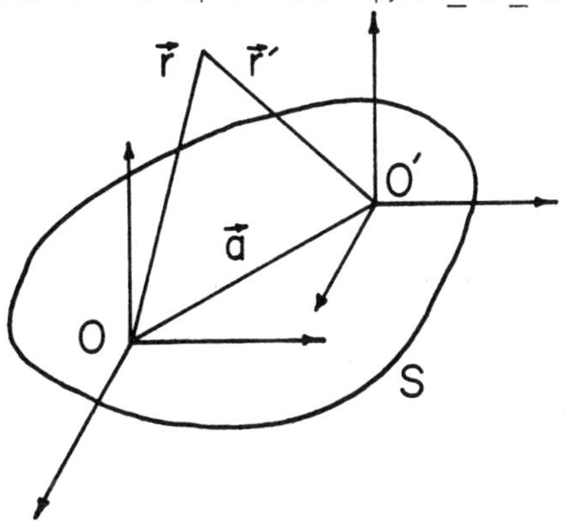

Fig. 3. Translation of the coordinate system

The translation of the longitudinal or P-wave function is the same as that for scalar waves. From the computational point of view it is preferable to use real spherical harmonics and in this case the translation theorem is given by

$$\mathrm{Re}\,\vec{\psi}_{\tau n}(\vec{r}+\vec{a}) = \sum_{\tau' n'} R_{\tau n,\tau' n'}(\vec{a})\,\mathrm{Re}\,\vec{\psi}_{\tau' n'}(\vec{r}) \tag{6.4}$$

$$\vec{\psi}_{\tau n}(\vec{r}+\vec{a}) = \sum_{\tau' n'} \sigma_{\tau n,\tau' n'}(\vec{a})\,\mathrm{Re}\,\vec{\psi}_{\tau' n'}(\vec{r});\ |\vec{a}| > |\vec{r}|, \tag{6.5}$$

$$\vec{\psi}_{\tau n}(\vec{r}+\vec{a}) = \sum_{\tau' n'} R_{\tau n,\tau' n'}(\vec{a})\,\vec{\psi}_{\tau' n'}(\vec{r});\ |\vec{a}| < |\vec{r}|. \tag{6.6}$$

In Eqs. (6.4) and (6.6), R and σ are the translation matrices. R can be obtained from σ by simply replacing the spherical Hankel functions that appear in σ by the spherical Bessel functions.

Explicit expressions for the translation have been given in (35) and since they are rather involved, they are not repeated here. The super matrix R has the following structure in the modal indices

$$R_{\tau\tau'} = \begin{bmatrix} R_{11} & 0 & 0 \\ 0 & R_{22} & R_{23} \\ 0 & R_{32} & R_{33} \end{bmatrix}. \tag{6.7}$$

There is no coupling of the longitudinal and transverse vector functions on translation. We further note that

$$R_{22} = R_{33}\ ;\ R_{23} = R_{32}\ ;\ R_{12} = R_{21} = R_{13} = R_{31} = 0. \tag{6.8}$$

One of the important properties of the σ and R matrices is that

$$\sigma^t(\vec{a}) = \sigma(-\vec{a})\ ;\ R^t(\vec{a}) = R(-\vec{a}) \tag{6.9}$$

Let the wave functions and coefficients defined with respect to the coordinate system at O′ be distinguished by a prime. The incident and scattered displacement field can be expanded in vector spherical functions with respect to the two origins. Thus,

$$\begin{aligned}\vec{u}^s &= \sum_{\tau n} f'_{\tau n}\,\vec{\psi}_{\tau n}(\vec{r}\,') = \sum_{\tau n} f'_{\tau n}\,\vec{\psi}'_{\tau n}(\vec{r}-\vec{a}) \\ &= \sum_{\tau n} f_{\tau n}\,\vec{\psi}_{\tau n}(\vec{r})\end{aligned} \tag{6.10}$$

where $\vec{a}$ is the vector drawn from O to O′ (that is restricted to be within the sphere circumscribing S with center at O. Similarly,

$$\begin{aligned}\vec{u}^0 &= \sum_{\tau n} a'_{\tau n}\,\mathrm{Re}\,\vec{\psi}'_{\tau n}(\vec{r}\,') = \sum_{\tau n} a'_{\tau n}\,\mathrm{Re}\,\vec{\psi}'_{\tau n}(\vec{r}-\vec{a}) \\ &= \sum_{\tau n} a_{\tau n}\,\mathrm{Re}\,\vec{\psi}_{\tau n}(\vec{r}).\end{aligned} \tag{6.11}$$

Using the translation theorems given in Eqs. (6.4) - (6.6) and the orthogonality of the vector spherical functions, we obtain in vector matrix notation

$$f = R(\vec{a})\ f' \tag{6.12}$$

$$a = R(\vec{a})\ a' \ . \tag{6.13}$$

Further, if T and T′ are the T-matrices relating the incident and scattered wave coefficients in the two coordinates systems, then

$$f = -Ta \tag{6.14}$$

$$f' = -T'a' \tag{6.15}$$

From Eqs. (6.12) - (6.15), we have

$$T' = [R(\vec{a})]^{-1}\ T\ R(\vec{a}) \tag{6.16}$$

which gives the transformation rule for the translation of the origin within the sphere circumscribing S.

Rotation of the T-matrix

In some problems, it is necessary to study how the T-matrix transforms under a rotation of the coordinate system. Let α, β, γ be the Euler angles of the primed coordinate system with respect to the unprimed system, see Fig. 4. The origin

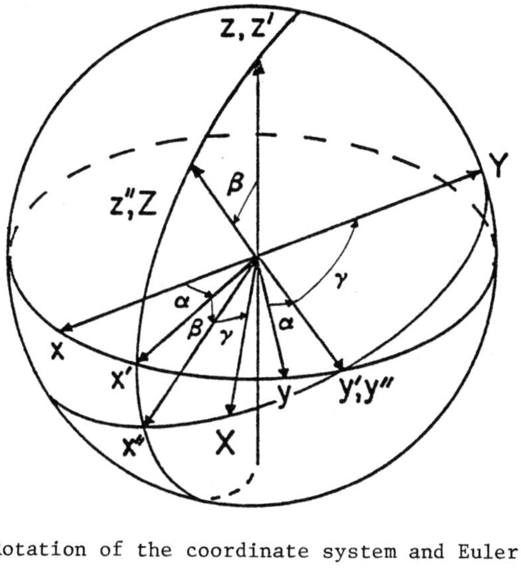

Fig. 4. Rotation of the coordinate system and Euler angles

of the two systems are at 0. The incident and scattered displacement fields can be expanded with respect to both systems. Since the length of the position vector $\vec{r}$ is invariant under rotation, the transformation of the vector spherical function is governed by the transformation of the spherical harmonics under finite rotations. Since the spherical harmonics are the eigen functions of the angular momentum operator, their transformation properties are well documented in the quantum mechanics literature, see Edmonds (36). Edmonds has given the transformation formulae for complex spherical harmonics from which is fairly easy to obtain the corresponding formula for real spherical harmonics. Let (θ',ϕ') and (θ,ϕ) be the angular coordinates of a certain point with respect to the primed and unprimed coordinate system. Then, we can write

$$Y_{nm\sigma}(\theta',\phi') = \sum_{m'\nu} D^{n}_{m'\nu,m\sigma}(\alpha,\beta,\gamma)\, Y_{nm'\nu}(\theta,\phi). \tag{6.17}$$

The index n is invariant under rotations and the rotation matrix D is a function of the Euler angles that define the relative orientation of the two coordinate systems. The expression for the rotation matrix is not repeated here and we refer the reader to Edmonds (36).

If $\vec{\psi}'_{\tau n}$ and $\vec{\psi}_{\tau n}$ are the vector spherical functions in the two coordinate systems, then one can write

$$\vec{\psi}'_{\tau nm\sigma} = \sum_{m'\nu} D^{n}_{m'\nu,m\sigma}(\alpha,\beta,\gamma)\, \vec{\psi}_{\tau nm'\nu}. \tag{6.18}$$

The incident and scattered displacement fields are once again expanded with respect to the two coordinate systems as in Eqs. (6.10) and (6.11). Using the orthogonality of the vector spherical functions, we can establish a relationship between the expansion coefficients for $\vec{u}$ and $\vec{u}^0$ in the two systems. Using the definition of the T-matrix, we finally get

$$T' = D^{-1}(\alpha,\beta,\gamma)\; T\; D(\alpha,\beta,\gamma). \tag{6.19}$$

Equation (6.19) is particularly important in comparing theoretical and experimental scattering data. Theoretically one can take advantage of any symmetry present in the geometry of the scatterer to choose a convenient coordinate system. The experimentalist on the other hand will want to choose a coordinate system convenient to the experimental set up. Equation (6.19) is then useful to transform from the coordinate system of the scatterer to the laboratory axes.

Another application of Eq. (6.19) is in arbitrary translation of the origin. Although the translation theorems in Eqs. (6.4) - (6.6) are for an arbitrary vector $\vec{a}$, numerically it may be more efficient to first translate along the z-axis of the original system and then perform a rotation to achieve the same end results. The transformation under rotations and translations is also very useful when a finite or even statistical ensemble of scatterers are involved.

VII. NUMERICAL APPLICATIONS

The greatest advantage of the null field method is that it is very well suited for numerical computations when applied to non-spherical and non-circular scatterers. Till a few years ago except for spheres and circular cylinders, no results were available for other shapes at wavelengths comparable to the size of the scatterer. Results for cylindrical cavities of elliptic cross section have been given by Tan (2) using the moment method and by Varadan (4) using the T-matrix method. In (4), a complete set of results have also been given for cylindrical inclusions. These results are for plane and anti-plane problems with the wave incident normal to the axis of the cylinder. The scattering of waves incident obliquely to the axis of infinite cylinders is yet to be studied using the null field method. Recently, Varadan and Varadan (5) have obtained extensive results for spheroidal inclusions and cavities. Visscher (37) has also studied the problem of the spheroidal cavity using a modified T-matrix approach. Preliminary results have also been obtained by Varadan and Varadan (38) for rough spheroidal cavities and for elliptical and penny shaped cracks.

In this section, the various aspects of the numerical calculation, the limitations and efficiency of each step for the various problems studied are discussed in detail. The basic steps involved in the computations for a single scatterer embedded in an infinite elastic solid are:

(1) Generation of the Q-matrix elements by numerical quadrature
(2) Inversion of the Q-matrix
(3) Calculation of scattered fields and scattering cross sections for various scattering geometries.

(1) Numerical Quadrature

It is efficient to generate the whole Q-matrix for a given incident wave frequency simultaneously, so the Bessel, Hankel and Legendre functions that have to be calculated for each point on the surface need not be computed unnecessarily for each matrix element. Since some of the special functions are calculated by backward recursion and some by forward recursion, this is especially efficient. At the present time, Gaussian quadrature using the Legendre polynomials P_n^0 seems most efficient, although for the elliptic cylinder Bodes fourth order formula for integration is equally good. When using the Gauss quadrature formula for spheroidal cavities and inclusions, difficulties were encountered for all values of the azimuthal index $m \neq 0$ especially for the higher order Q-matrix elements. The matrix elements tend to become very large and the integrations fail to converge even for a very large number of Gaussian points. This is of course more noticeable at higher frequencies when both the Hankel functions and the associated Legendre polynomials ($m > 0$) tend to oscillate very rapidly even for small changes in the argument.

(2) Inversion of the Q-matrix

The Q-matrix can be inverted by direct methods like the Gaussian elimination process. However, there are problems of overflow when the Q-matrices are ill-conditioned with very large imaginary parts for the higher order elements. Waterman (39) has suggested another approach in which the unitarity and symmetry properties of the S-matrix are incorporated in the inversion of the Q-matrix. There are advantages and disadvantages in using Waterman's program for the inversion of the Q-matrix. The disadvantages are that the symmetry and unitarity of the S-matrix cannot be used as checks to estimate the accuracy of the computations, since these properties are already forced on the T-matrix ; secondly the approach is not suited for lossy or dissipative scatterers because the S-matrix is no longer unitary. The advantages are that computation time is lower and the Q-matrix is conditioned and triangularized before the inversion procedure so that no overflow problems are encountered due to large imaginary parts. The inversion itself is achieved by Schmidt orthogonalization of the rows of the conditioned Q-matrix. For an otherwise well tested program, there are distinct advantages in using the Waterman procedure for inversion. Once the Q-matrix is inverted, it is a straight forward procedure to construct the T-matrix.

(3) Scattering Cross Sections

Since the T-matrix depends only on the frequency of the incident wave and is independent of the scattering geometry, it is advantageous to compute the scattering cross section for several different incident and scattered wave directions before going on to a different frequency. This is more efficient than storing the T-matrix on a disk or magnetic tape and reading it over and over again for different geometries. It is, however, practical to store the T-matrix on a storage device for future use.

Elliptic Cylinders

The equation to the boundary in cylindrical polar coordinates r,θ,z, see Fig. 1 b, is given by

$$r(\theta) = (\cos^2\theta/a^2 + \sin^2\theta/b^2)^{-\frac{1}{2}} \tag{7.1}$$

where a and b are the semi-major and semi-minor axes. For plane and anti-plane problems, i.e., for waves incident normal to the axis of the cylinder, the incident and scattered fields are independent of the z-coordinates so that the problem effectively reduces to two dimensions. For an elliptical boundary

$$\int_S \ldots \hat{n}\, dS = \int_0^{2\pi} \ldots [\hat{r} - \frac{1}{r} \frac{\partial r}{\partial \theta} \hat{\theta}]\, r(\theta)\, d\theta \tag{7.2}$$

where r is set equal to r(θ) in the vector cylindrical functions after all the derivatives have been taken.

Due to the symmetry of the ellipse, many of the elements of the Q-matrix for cavity, rigid, elastic solid and fluid inclusion are all zero. It can be shown that

$$Q_{\tau n\sigma,\tau' n'\nu} = 0 \text{ if } n+n' \text{ odd}$$

$$Q_{\tau n\sigma\ \tau' n'\nu} = 0 \text{ if } \tau = \tau' \text{ and } \sigma \neq \nu \tag{7.3}$$

$$Q_{\tau n\sigma},\ Q_{\tau' n'\nu} = 0 \text{ if } \tau \neq \tau' \text{ and } \sigma = \nu \; .$$

Further, the range of integration can be reduced to 0 - π/2 because

$$r(\theta) = r(\pi-\theta) = r(3\pi/2 - \theta) = r(2\pi - \theta) \cdot \tag{7.4}$$

The special properties for the ellipse can be first used to check the program and then incorporated into the program to reduce computation time by a significant amount.

If computations are required for the same embedding medium with different types of scatterers, it is economical to generate the different Q-matrices simultaneously provided storage requirements are not expensive.

In Ref. (4), the scattered stress field and the scattering cross section are presented as a function of $k_p a$ for aspect ratio b/a ranging from 0.25 to 1.0 for $0.05 < k_p a < 3.0$ for P-, SV- and SH-waves incident at 0°, 30°, 60° and 90° to the major axis of the ellipse. The host material is Aluminium and the scatterers are cavity and tungsten inclusion representatives of soft and hard scatterers. For the range of wavelengths considered it was sufficient to keep terms up to n = 10 resulting in a T-matrix size of 40 × 40 to obtain a 3 decimal place accuracy in the scattering cross section. The interested reader may refer to (4) for numerical results.

Spheroids

The equation to the surface of a spheroid with its axis of revolution parallel to the z-axis, see Fig. 5a, in spheroidal coordinates r,θ,ϕ is given by

$$r(\theta) = (\cos^2\theta/b^2 + \sin^2\theta/a^2)^{-\frac{1}{2}} \tag{7.5}$$

where 'a' is the radius in the x-y plane and b is the semi-axis in the z-direction. Since r is independent of ϕ, we have

$$\int_S \cdots \hat{n}\ dS = \int_0^{2\pi} d\phi \int_0^{\pi} d(\cos\theta) \cdots (\hat{r} - \frac{1}{r}\frac{\partial r}{\partial\theta}\hat{\theta})\ r^2(\theta) \tag{7.6}$$

and the ϕ integration can be performed directly. Due to the orthogonality of the angular functions cos $m\phi$ and sin $m\phi$, the Q-matrix and hence the T-matrix are diagonal in the azimuthal index m. In addition due to the fact that $r(\theta) = r(\pi-\theta)$, the range of integration on θ can be reduced to half and we note the following symmetries of the Q-matrix (see Ref.(5)) :

$$Q_{\tau nm\sigma,\ \tau' n' m' \nu} = 0 \quad \text{if } m \neq m'$$

$$Q_{\tau nm\sigma,\ \tau' n' m\ \nu} = 0 \quad \text{if } n + n' \text{ odd or } \sigma \neq \nu \text{ for } \tau = \tau' \text{ or } \tau,\tau' = 1,3 \tag{7.7}$$

$$Q_{\tau nm\sigma,\ \tau' n' m\ \nu} = 0 \quad \text{if } n+n' \text{ even or } \sigma = \nu \text{ or } m=0 \text{ for } \tau=2 \text{ or } \tau'= 2$$

Since Q and T are diagonal in the azimuthal index, it is efficient to compute the T-matrix separately for each azimuthal index. For P-waves incident along the z-axis only the $m = 0$ term contributes to the scattered field while for S-wave incidence along the z-axis only the $m = 1$ term contributes to the scattered field. For all other angles of incidence, there is a summation on the azimuthal modes. Many of the remarks made for elliptic cylinders apply to the computations for the spheroid also.

In (5), results are presented for titanium as the host material for spheroidal cavities and spheroidal tungsten carbide inclusions for aspect ratios ranging from 0.2 to 2.0. Note that for $b/a < 1$, the spheroid is oblate and for $b/a > 1$, the spheroid is prolate. We considered both P- and S- waves incident at various angles to the axis of revolution for $0.1 \leq k_p a \leq 4.0$. For the range of frequencies and aspect ratios considered, we begin with $n = 20$ for $m = 0$ and consider up to $n = 16$ for $m = 6$ recalling that $n \geq m$.

For $k_p a > 3.0$, numerical difficulties are already encountered for $m > 0$ both with the integration scheme as well as highly illconditioned Q-matrices for both cavities and inclusions especially for the higher values of m. Hence, with the integration and inversion routines presently available, $k_p a$ must be less than 4.0 to obtain stable results.

The results obtained using the T-matrix method have compared very favorably with experiments (40) and have also served as a data base for inverse methods like adaptive learning and Born inversion that are currently being tried for flaw detection in structural materials (41).

Rough Spheroidal Cavities

One of the major applications for the T-matrix approach to elastic wave scattering has been in flaw detection. Flaws do not occur as smooth spheroids. So an attempt was made to model the flaw more realistically by perturbing the smoothness with periodic corrugations. The equation to the surface of the rough spheroid is written as

$$r(\theta) \;=\; (\cos^2\theta/b^2 \;+\; \sin^2\theta/a^2)^{-\frac{1}{2}} + \delta\cos\ell\theta \tag{7.8}$$

where δ gives the amplitude of the roughness and ℓ defines the periodicity of the roughness, see Fig. 5b. Since $r(\theta)$ is still independent of ϕ, it has the symmetries of a smooth spheroid if ℓ is a multiple of four.

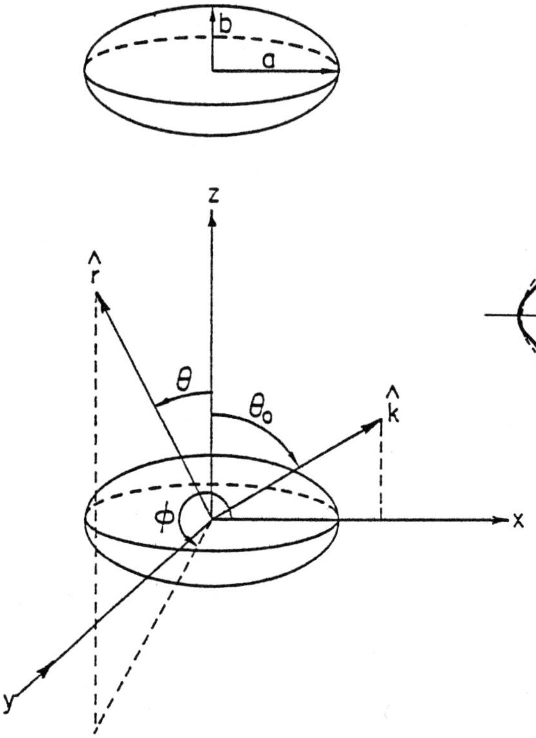

Fig. 5a. Spheroidal scatterer

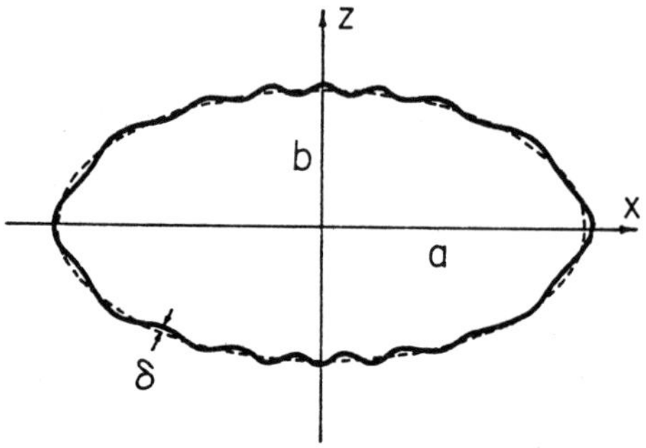

Fig. 5b. Rough spheroidal scatterer

In numerical computations, it was found that fairly stable Q- and T-matrices could be computed for $b/a > 0.7$, $\delta/a < 0.02$ and $\ell > 12$. If the amplitude of the roughness is increased beyond this limit and the wavelength decreased, the Q-matrix becomes highly illconditioned and the integration fails to converge. If the wavelength of the roughness is decreased beyond $\ell = 64$, there is not much difference between a smooth and rough spheroid up to $k_pa = 2.0$. The convergence of the integration was checked for all values of k_pa to ensure that this was not the cause of the agreement with the smooth spheroid. Thus, for wavenumber up to $k_pa = 2.0$, if the wavelength of the roughness is not too large, there is no appreciable change from a smooth spheroid. This suggests that modelling a flaw in a structural material as a oblate spheroid is not too far removed from reality. The reader may refer to (38) for the numerical results.

Penny Shaped Cracks and Strips

Although the oblate spheroid is a good model for a flaw in a structural material, the aspect ratio of the flaw is typically quite small making it resemble a penny shaped crack or an elliptical crack. These may be visualized as an oblate spheroid and an ellipsoid of zero aspect ratio. Although the null-field or T-matrix method is in general not well suited for numerical computations on long slender obstacles, the degenerate spheroid and the degenerate ellipse are special cases where the limit of zero aspect ratio can be taken by choosing suitable surface field expansions that incorporate the proper conditions at the crack edge. Waterman (23) in his original formulation of the T-matrix method for acoustic wave scattering has discussed the scattering from infinitely long strips for both Neumann and Dirichlet conditions where he has discussed the choice of proper weight factors that lead the correct edge conditions at the crack tip when the limit of zero aspect ratio is taken. Bates and Wall (30) have also dealt with the choice of proper surface field expressions for different types if sharp corners in some detail.

The scattering of normally incident acoustic waves by a strip on which Neumann boundary conditions are prescribed is mathematically the same as the scattering of SH-elastic waves by a stress free strip. The scattered ($f_{n\sigma}$) and incident ($a_{n\sigma}$) field coefficients are given by

$$f_{n\sigma} = \lim_{b/a\to 0} \frac{i}{4} \int_S w_+(\vec{r}) \quad \hat{n}\cdot\nabla \operatorname{Re}\phi_{n\sigma}(\vec{r}) \quad dS(\vec{r}) \tag{7.9}$$

$$a_{n\sigma} = -\lim_{b/a\to 0} \frac{i}{4} \int_S w_+(\vec{r}) \quad \hat{n}\cdot\nabla\phi_{n\sigma}(\vec{r}) \quad dS(\vec{r}) \tag{7.10}$$

where $\vec{r}$ on S is given by Eq. (7.1) and w_+ is the unknown anti-plane displacement field (in the z direction) on the surface of the strip.

For numerical computations, the limit of zero aspect ratio has to be taken before integration and this necessitates a careful choice for the expression of w_+. Since the range of integration now becomes $-a \le x \le a$ which includes the origin, the Hankel functions become singular. Further, the edge conditions at the crack tip require that the crack opening displacement should vanish as $\sqrt{1 - x^2/a^2}$ at the tips. Waterman suggested the following expression for w_+

$$w_+ = \frac{\sqrt{1 + [(\partial r/\partial\theta)/r]^2}}{k_s^2 r} \quad \hat{n}\cdot\nabla \sum_{n,\sigma} \alpha_{n\sigma} \operatorname{Re}\phi_{n\sigma} \tag{7.11}$$

This leads to a sy-metric Q-matrix for ellipses of arbitrary aspect ratio. When the limit is taken in Eqs. (7.9 and (7.10), the Q-matrix for the strip is given by

$$Q_{n\sigma,n'\nu} = \frac{\pi n}{b} \int_{-a}^{a} \frac{\sqrt{1-x^2/a^2}}{k^2x^2} H_n(k|x|)\ J_{n'}(k|x|)\ dx\ \delta_{\sigma\nu},\delta_{\nu,2}$$
$$= Q_{n'\nu,n\sigma} \qquad (7.12)$$

Since the Q-matrix is symmetric, only the elements above the diagonal need be evaluated numerically when the order of the Bessel function is greater than or equal to the order of the Hankel function. The only other troublesome term in the Q-matrix is the constant term in the expansion of $J_{n'}(k|x|)\ H_n(k|x|)$ due to the factors $1/x^2$ in the Q-matrix. This integral can, however, be evaluated analytically for arbitrary aspect ratio and the limit then taken yielding a finite contribution. All the other terms in the expansion of $J_{n'}H_n\ (n'\geq n)$ can be integrated numerically or given in the form of a finite series.

Numerical results were obtained using the Q-matrix given in Eq. (7.12) as well as the Q-matrix for an elliptical cavity for an aspect ratio of b/a = 0.05. The scattering cross section was obtained as a function of ka and compared with available results using Mathieu function (10). The agreement was excellent for both approaches. We have concluded that not much is gained by attempting to put b/a = 0 because values of b/a < 0.1 mimic the results for a crack very closely and not that much is gained by setting b/a exactly to zero. The additional advantage is that no new computer programs need be developed as long as stability of the numerical results can be ensured for b/a < 0.1.

For P- and SV-wave scattering from strips again we have used the programs developed for the elliptical cavity. Stable results can easily be obtained down to b/a values of 0.05. These results have been checked with Tan's results (2) for the strip using the moment method and again the agreement is very good. An important difference between SH-wave scattering and P-, SV-wave scattering is that even for incidence parallel to the strip there is a non-zero scattered field.

The penny shaped crack is an oblate spheroid of zero aspect ratio. Calculations for b/a = 0.05 are compared with those obtained by Mal (17) using integral equations and again the agreement was quite good. For waves incident normal to the crack, results have been obtained up to $k_pa = 2.0$ for which the symmetry properties of the T-matrix have been verified using direct inversion algorithms for the Q-matrix. All the numerical results for crack like scatterers are presented in (38).

ACKNOWLEDGEMENT

This research was sponsored by the Center for Advanced NDE operated by the Science Center, Rockwell International for the Advanced Research Projects Agency and the Air Force Materials Laboratory under contract F33615-74-C-5180

REFERENCES

1. ARPA/AFML Review on methods in QNDE, 1975, 1976, 1977, 1978, 1979, prepared by D.O. Thompson, Science Center, Rockwell International, Thousand Oaks, CA.

2. T.H. Tan, Diffraction theory for time harmonic elastic waves, doctoral thesis, Delft University of Technology, Delft, The Netherlands, (1975).

3. V. Varatharajulu (Varadan) and Y.H. Pao, Scattering matrix for elastic waves.

I. Theory, J. Accoust. Soc. Am. 60, 556, (1976).

4. V.V. Varadan, Scattering matrix for elastic waves. II. Application to elliptic cylinders, J. Accoust. Soc. Am. 63, 1014 (1978).

5. V.V. Varadan and V.K. Varadan, Scattering matrix for elastic waves. III. Application to spheroids, J. Acoust. Soc. Am. 65, 896, (1979).

6. V.K. Varadan, V.V. Varadan and Y.H. Pao, Multiple Scattering of elastic waves by cylinders of arbitrary cross section. I. SH waves, J. Acoust. Soc. Am. 63, 1310, (1978).

7. A. Boström, Multiple scattering of elastic waves by bounded obstacles, Report No.: 794, Inst. Theoretical Physics, S-41296, Göteborg, Sweden.

8. B. Peterson, V.V. Varadan and V.K. Varadan, On the multiple scattering of waves with solid-fluid interfaces, these proceedings.

9. J.D. Achenbach, A. Gautesen and H. McMaken, Application of ray theory to elastic wave scattering by cracks, these proceedings and the references therein.

10. J.J. Bowman, T.B.A. Senior and P.L.E. Uslenghi, Electromagnetic and acoustic scattering by simple shapes, North Holland, Amsterdam, (1969).

11. R.M. White, Elastic wave scattering at a cylindrical discontinuity in a solid, J. Acoust. Soc. Am. 30, 771, (1958).

12. C.F. Ying and R. Truell, Scattering of a plane longitudinal wave by a spherical obstacle in an isotropically elastic solid, J. Appl. Phys. 27, 1086, (1956).

13. A. Eringen and E. Suhubi, Elastodynamics, Volume II, Academic Press, New York, (1974).

14. P.M. Morse and H. Feshbach, Methods of Theoretical Physics, McGraw Hill, New York, (1953), p. 1408, Volume II.

15. D.D. Ang and L. Knopoff, Diffraction of scalar elastic waves by a clamped finite strip, Proc. Nat. Acad. Sci. U.S.A. 51, 471, (1964).

16. D.D. Ang and L. Knopoff, Diffraction of elastic waves by a finite crack, Proc. Nat. Acad. Sci. U.S.A. 51, 1075, (1964).

17. A.K. Mal, Interaction of elastic waves with a penny shaped crack, Int. J. Engng. Sci. 8, 381, (1970).

18. K. Harumi, Scattering of plane waves by a cavity ribbon in a solid, J. Appl. Phys. 33, 3588, (1962).

19. Chang Shih-Jung, Diffraction of plane dilatational waves by a finite crack, Quart. Journ. Mech. and Appl. Math. 24, 421, (1971).

20. S.K. Datta, Diffraction of plane elastic waves by ellipsoidal inclusions, J. Acoust. Soc. Am. 61, 1432, (1977).

21. J.E. Gubernatis, E. Domany, J.A. Krumhansl and M. Huberman, The Born approximation in the theory of the scattering of elastic waves by flaws, J. Appl. Phys. 48, 2812, (1977).

22. K.E. Newman and E. Domany, Calculation of scattering by the distorted wave Born approximation, ARPA/AFML progress on QNDE-1978, prepared by D.O.Thompson, Science Center, Rockwell International, Thousand Oaks, CA 91360.

23. P.C. Waterman, New formulation of acoustic scattering, J. Acoust. Soc. Am. 45, 1417, (1969).

24. P.C. Waterman, Symmetry, unitarity and geometry in electromagnetic scattering, Phys. Rev. D 3, 825, (1971).

25. P.C. Waterman, Matrix theory of elastic wave scattering. I., J. Accoust. Soc. Am. 60, 567, (1976).

26. Y.H. Pao and V. Varatharajulu (Varadan) Huygens' principle, radiation conditions and integral formulas for the scattering of elastic waves, J. Acous. Soc. Am. 59, 1361, (1976).

27. Ref. 14, p. 1865, Volume II.

28. D.J.N. Wall, Circularly symmetric Green tensors for the harmonic vector wave equation in spheroidal coordinate systems, J. Phys. A : Math. Gen. 11, 749, (1978).

29. R.F. Millar, The Rayleigh hypothesis and a related least squares solution to scattering problems for periodic surfaces and other scatterers, Radio Science 8, 785, (1973).

30. R.H.T. Bates and D.J.N. Wall, Null field Approach to scalar diffraction I. General method, II. Approximate methods, III. Inverse methods, Philos. Trans. Roy. Soc. Lon. 45, 287, (1977).

31. Y.H. Pao, The transition matrix for the scattering of acoustic waves and elastic waves, Modern Problems in elastic wave propagation edited by J. Miklowitz and J.D. Achenbach, Wiley Interscience, New York, (1978).

32. A. Boström, Scattering of stationary acoustic waves by an elastic obstacle immersed in a fluid, these proceedings.

33. B. Peterson and S. Ström, T-matrix for electromagnetic scattering from an arbitrary number of scatterers and representations of E(3), Phys. Rev. D. 8, 3661, (1973).

34. O.R. Cruzan, Translational addition theorems for spherical vector wave equations, Q. Appl. Math. 20, 33, (1962).

35. A. Boström, Multiple scattering of elastic waves by bounded obstacles, Report No.: 79-4, Inst. Theoretical Physics, S-41296 Göteborg, Sweden

36. A.R. Edmonds, Angular momentum in Quantum mechanics, Princeton Univ. Press, Princeton, New Jersey (1957).

37. W.M. Visscher, Method of optimal truncation, A new T-matrix approach to elastic wave scattering, these proceedings.

38. V.V. Varadan and V.K. Varadan, Elastic wave scattering from rough spheroidal cavities and cracks, ARPA/AF Review of progress in quantitative NDE, 1979, prepared by D.O. Thompson, Science Center, Rockwell International, Thousand Oaks, CA 91360.

39. P.C. Waterman, Computer Techniques for electromagnetics, volume 7, edited by R. Mittra, Pergamon Press, New York, (1973).

40. R.E. Elsley, J.M. Richardson, R.B. Thompson, and B.R. Tittman, Comparison between experimental and computational results for elastic wave scattering, these proceedings.

41. J.H. Rose, R.E. Elsley, B. Tittman, V.V. Varadan and V.K. Varadan, Inversion of ultrasonic scattering data, these proceedings.

SURVEY OF T-MATRIX METHODS AND SURFACE FIELD REPRESENTATIONS

P. C. Waterman
Center for Science and Technology, Limited
8 Baron Park Lane, Burlington, Massachusetts 01803, USA

ABSTRACT

The development of T-matrix methods is traced through the first applications in potential theory to scattering problems involving acoustic, electromagnetic and elastic waves. Throughout, the origins of the theory in Huygens' principle are stressed, and possible further developments are noted. Detailed numerical studies are also presented which point out the relative merits of various wave function representations for surface current in the electromagnetic case.

INTRODUCTION

Recent years have seen a significant growth rate in the literature employing T-matrix methods. We want to attempt a brief survey of this work, including speculation on possible further developments. With regard to future work, incidentally, we take the point of view that Huygens' principle is in fact the fundamental underlying concept. Whether or not the principle can then be reduced to a T-matrix formulation, or something analogous, in specific areas of interest (e.g., time-dependent problems) depends on whether one is clever or lucky enough to find appropriate sets of basis functions that will cast the problem in matrix form.

It should be stressed that the present discussion is by no means intended as a comprehensive review; apologies are offered in advance to those who have made important contributions which, through oversight, are not mentioned here.

One specific question which has remained unresolved for several years now is that of how best to represent surface fields that arise in connection with two- and three-dimensional scattering problems. In addition to our survey, it thus is appropriate to present some numerical results which may perhaps narrow the options in this area. In particular, we will look at truncation behavior for various choices of vector wave functions used to represent surface current. The manner in which convergence is affected by the choice of coordinate origin will also be examined, and finally the advantages of orthagonalization (i.e., constraining the S-matrix to be symmetric and unitary at finite truncation) will be demonstrated numerically.

THE T-MATRIX AND HUYGENS' PRINCIPLE

As already noted, we regard Huygens' principle as the fundamental physical concept underlying all of the T-matrix work. Specifically, when one writes down the Green's identity using the appropriate free-space Green's function, and integrates over only a portion of space, the result is quite different for points within and without the integration volume. The key to further progress then lies in recognizing that for those latter points, where the singularity of the Green's function does not come into force, the resulting statement, which one is easily tempted to pass over as a trivial and obvious consequence of the mathematics, is in fact a useful and apparently rather powerful computational tool.

The mathematical details of all this have been given many times; it would be superfluous to show them here. We could note, however, that Huygens' principle used in this way has been variously known as the extended boundary condition, Schelkunoff equivalent current method, Ewald-Oseen extinction theorem, null-field method, and perhaps other names. We will try to point out where these different names occur as we go along.

Smythe was one of the earliest to employ Huygens' principle this way. He treated various electrostatic problems, e.g., the charged right circular cylinder, by noting that the resulting charge distribution must be such as to give a constant potential, and thus no field, in the interior (1). Taylor made further extensions in both electrostatics (2) and magnetostatics (3). Fikioris employed similar techniques to show that the closed-form solution can be recovered for the conducting spheroid in a uniform applied field (4). In view of the analogy between magnetic polarizability and virtual mass (5), some of this work carries over immediately to the fluid dynamics case. Problems of incompressible and perhaps compressible fluid flow are presently being examined in this light by Eyges, who already has found new insights in the thin-body limit (6).

With regard to the classical scattering of time-harmonic waves, the acoustic case is by far the simplest to follow because of its scalar nature. Papers by Schenck (7) and the present writer (8) discussed this case from somewhat different points of view, in that the former did not employ a T-matrix formalism even though both began essentially from Huygens' principle. More recently, the problem was re-examined by Bates and Wall (9, null-field method). They have examined the consequences of using alternative expansion functions, and present many numerical results in support of their analysis.

Because of the wide ranging applications, the electromagnetic case has received the most attention. Earlier work considered only conducting objects (10, extended boundary condition), but the method was soon extended to dielectrics (11). Extensive numerical results for dielectrics were given by Barber and Yeh (12, Schelkunoff equivalent current method), and Barber went on to consider effects in lossy dielectric biological models (13). By this time, it had also been shown how to incorporate symmetry and unitary constraints, corresponding physically to reciprocity and energy conservation, into the numerical solutions (14). Whether or not such constraints should be included in <u>truncated</u> solutions remains controversial (15); some numerical results supporting the approach will be given below.

In parallel but independent work, Wolf observed that the Ewald-Oseen extinction theorem, which historically played such an important role in obtaining the Lorentz-Lorenz formula, could be used directly as a computational tool (16). Agarwal subsequently pointed out the equivalence of methods using the extinction theorem and the extended boundary condition (17). A least-squares approach has been advocated by Visscher which, as he points out, has the net effect of leading to almost the same system of equations as those obtained by other authors (15). The key difference lies in the choice of expansion functions for the surface field, which must be chosen as incoming waves in order to correspond to the least-squares result.

A further degree of complexity is introduced when one considers wave scattering in an elastic solid, because of the necessity of dealing simultaneously with longitudinal and transverse modes and their interactions. It turns out in this case that two different forms of Huygens' principle are available for deriving the T-matrix equations. The resulting equations are superficially somewhat different, one set involving the vector gradient operation $\nabla \underline{F}$ (18), the other the surface vector calculus of Weatherburn (19). The equivalence of the final results was later shown, however (20), and Varadan went on to give detailed numerical computations (21).

Two methods seen in the literature to date do not rely on Huygens' principle. The first is the least-squares formulation already mentioned (15), while the second involves a conserved flux which can only be identified with energy flux when there is no dissipation present. This conserved flux method has been described for all three cases, acoustic (22), electromagnetic (23), and elastic waves (20,24).

Under the direction of Strom, workers in Goteborg were among the earliest to recognize the single-object T-matrix as a fundamental building block that could be used to great advantage in compound problems, specifically those involving layered scatterers (25) and/or multiple scattering (26). As they observed, this extension is possible because the single-object T-matrix allows for the possibility of quite general incident waves, e.g., a wave impinging on the object from a second object in the vicinity. Additional multiple scattering computations have also been carried out by the Varadans at Ohio State, in conjunction with Pao at Cornell (27).

It will be interesting to observe the eventual impact of T-matrix methods in the field of multiple scattering, where a great many questions arise which do not occur in the single-object context. There is, for example, the behavior of periodic vs. irregular vs. random arrays of objects. One is also interested in the "medium" or "bulk" properties of such arrays, assuming of course that such properties exist in the first place. Many of these of course are old questions, but only a few have been resolved with any thoroughness.

The T-matrix approach should also prove useful for computing scattering from inhomogeneous objects that cannot be modeled by layered media, although this has not been tried, to our knowledge. The idea here would be to build up a set of field solutions within the object by solving appropriate partial difference equations numerically. Using a linear combination of these fields to represent the interior solution, the result is then substituted in the moment equations (e.g., Equation 7 of Reference 14 for the electromagnetic case), and coupled with boundary conditions leads in the usual way to a matrix equation for T.

Huygens' principle would also seem to present useful possibilities for general time-dependent problems. Here fields are governed by the full wave equations, and for acoustic problems, for example, we can write (u is the velocity potential, and underline indicates vector quantity)

$$\left.\begin{matrix} u(\underline{r},t) \\ 0 \end{matrix}\right\} = u^{inc}(\underline{r},t) + (1/4\pi)\int d\underline{\sigma}' \cdot [u(\underline{r}',t^*)\nabla'(1/R)$$

$$-(1/cR)\nabla' R\partial u(\underline{r}',t^*)/\partial t^* - (1/R)\nabla' u(\underline{r}',t^*)];$$

$$\underline{r}\begin{cases}\text{outside } \sigma \\ \text{inside } \sigma\end{cases} \qquad (1)$$

This equation describes scattering by an object bounded by the closed surface σ of integration. The quantity $R = |\underline{r}-\underline{r}'|$, c is the velocity of sound, and t* is set equal to t-R/c (retarded time) after the differentiations are carried out. One easily verifies that Eq. 1 is a slight generalization of the expression given by Morse and Feshbach (28).

Just as in the time-harmonic case, the right-hand side of Eq. 1 must vanish identically throughout the entire interior volume.

When supplemented with boundary conditions (e.g., for the acoustically "soft" object only the last term in the integrand is non-vanishing) this provides a prescription for finding the unknown surface fields. The incident wave u^{inc} is of course known.

The problem can be expressed in matrix form provided one can find complete sets of time-dependent wave functions to use as basis functions. Although we have not investigated this question in any detail, it appears that the spherical wave functions derived by Granzow may be appropriate here (29). Granzow gives both the acoustic and electromagnetic wave functions; the corresponding Huygens' principle for the time-dependent electromagnetic case can be constructed by extension of the equations given by Stickler (30).

WAVE FUNCTION REPRESENTATION OF SURFACE CURRENTS

For electromagnetic scattering by a perfectly conducting object, recall that the basic equations read as follows: For the incident and scattered waves, respectively, one has (14)

$$\underline{\psi}^{inc} = \Sigma a_n \mathrm{Re}\underline{\psi}_n, \tag{2}$$

$$\underline{\psi}^{scat} = \Sigma f_n \underline{\psi}_n, \tag{3}$$

where the $\underline{\psi}_n$ are the outgoing spherical vector partial waves, and the $\mathrm{Re}\underline{\psi}_n$ their regular parts (replace Hankel by Bessel functions). The incident wave coefficients a_n are of course known, and the scattered wave coefficients f_n are to be found. The moment relations are then

$$a_n = (k^2/\pi)\int d\sigma \underline{\psi}_n \cdot \underline{j} \tag{4}$$

$$f_n = -(k^2/\pi)\int d\sigma\, \mathrm{Re}(\underline{\psi}_n) \cdot \underline{j}, \quad n=1,2,\ldots \tag{5}$$

where the integration is over the surface of the object, which need not be smooth, and $\underline{j} \equiv n \times \underline{H}$ is the unknown surface current.

Now the T-matrix enables us to compute the scattering directly according to the prescription

$$f = Ta, \tag{6}$$

without the intermediate step of finding the surface currents. In order to get T, we expand $\underline{j}$ in some complete set of tangential vector functions, viz.

$$\underline{j} = \Sigma \alpha_n \underline{F}_n(\underline{r}), \quad \underline{r} \text{ on } \sigma . \tag{7}$$

Substituting this expansion in Eqs. 4 and 5, the α_n can be eliminated to give

$$QT = -\mathrm{Re}Q, \tag{8}$$

with elements of Q given by

$$Q_{mn} = (k^2/\pi)\int d\sigma\, \underline{F}_m \cdot \underline{\psi}_n$$
$$\text{(for ReQ, replace } \underline{\psi}_n \text{ by } \mathrm{Re}\underline{\psi}_n) \tag{9}$$

Equation 8 is now truncated and solved numerically for the T-matrix.

The crucial question at this point involves the choice of expansion functions $\underline{F}_n$ for the surface current. One possibility is to use the wave functions themselves. There are in this event several alternatives, the relative merits of which we want to examine.

Table 1 lists six possible choices for the $\underline{F}_n$. The electric and magnetic partial waves, which together make up the $\underline{\psi}_n$, are each other's curls, so that the table is unaffected should $\underline{\psi}_n$ be replaced by $\nabla \times \underline{\psi}_n$ anywhere. Based only on the table, case A, the outgoing waves, appears by far the most advantageous, and this was in fact our original choice in 1965 (10). Note first that an exact decoupling of the equations is possible, reducing the calculation from one N x N problem to two independent $\frac{1}{2}N \times \frac{1}{2}N$ problems. Furthermore, for a given truncation N, case A involves the smallest number of independent real matrix elements. This is important because the evaluation of these elements by numerical quadrature consumes probably 90% of the required computation time. Note, incidentally, that the skew-symmetry of Q for general bodies is fairly easily apparent, upon substituting $\underline{F}_n = \hat{n} \times \underline{\psi}_n$ in Eq. 9. The symmetry of Q for rotationally symmetric bodies is somewhat contrived, however, arising from further condensation due to azimuthal orthogonality (10,14).

Case B uses the regular wave functions, rotated 90 degrees about the unit normal. Because they are real, these functions, along with those of case D, allow constraints of symmetry and unitarity to be built into the solution, as discussed below. The case B functions also have a quasi-orthogonality property which makes them clearly the best choice for spheroids (14), and share the numerical advantage discussed for the analogous scalar functions, namely that the numerically dominant parts of Q-matrix integrands above the diagonal vanish identically upon integration (8).

The case C functions are of interest in that they represent the functions arising from a least-squares approach to the problem. This may be verified by applying Visscher's technique (15) to the electromagnetic case. Finally, for completeness we have indicated the general characteristics of the rotated incoming and unrotated outgoing wave choices E, F although we shall not look further at these latter two.

As shown in Fig. 1, the object we consider is a 15 degree half-angle sphere-cone-sphere, with variations in both shape and size, specified by b/a and $ka = 2\pi a/\lambda$, respectively. The location of the origin can also be varied over the points indicated in the figure.

Let us first look at truncation behavior for the first four cases of Table 1, for a fairly simple object, i.e., relatively small (ka=1) and not too different from spherical (b/a=0.8), with origin at the center of the large sphere (point 1). Results for the real and imaginary parts of T_{11} are shown in Figs. 2 and 3.

In full index notation (14), $T_{11} = T_{1e011e01}$, describing the coupling between rotationally symmetric magnetic dipole fields. Note that for all four cases, the results have settled down to the same value, to within fractional discrepancies of about 10^{-3}. The outgoing functions (case A) have the largest disparities, while the real function representations (cases B and D) perhaps show the best behavior.

The truncation behavior is also strongly influenced by the choice of coordinate origin. We can see this by repeating the computations for the different origins shown in Fig. 1, still with ka=1 but this time with b/a=0.6. Figure 4 shows the scattering cross section σ_s normalized by πa^2, plotted vs. truncation size (plane wave axially incident from the left). The case A outgoing functions are being employed here. The resulting curves all appear to be settling down to a common final value, but those representing origin choices 3,4,5,6 roughly "centered"

TABLE 1 Several choices of expansion functions are enumerated for the surface currents. Tangential functions must be employed, formed by rotating or not rotating (i.e., rotating 180°) about the unit normal. The last column shows the number of independent real matrix elements of Q which must be computed for truncation at N equations in N unknowns for bodies with an axis of rotational symmetry.

Case	Expansion Functions	Type	Significance	Q (General Body)	Q (Rot. Symmetry)	Elements of Q Indep., real ($m \neq 0$)
A	$\hat{n} \times \underline{\psi}_n$	Outgoing, rotated	Decoupled equations	Skew-symmetric	Symmetric	$(1/2)N^2$
B	$\hat{n} \times Re\underline{\psi}_n$	Regular, rotated	Orthogonalization, above-diagonal numerical advantage, best for spheroids	Real part skew symmetric	Real part symmetric	$(3/4)N^2$
C	$\hat{n} \times \hat{n} \times \underline{\psi}^*$	Incoming, unrotated	Least squares analysis	Hermitean	Hermitean	N^2
D	$\hat{n} \times \hat{n} \times Re\underline{\psi}_n$	Regular, unrotated	Orthogonalization	Real part symmetric	Real part symmetric	$(3/2)N^2$
E	$\hat{n} \times \underline{\psi}_n{}^*$	Incoming, rotated	?	Skew-hermitean	Hermitean	N^2
F	$\hat{n} \times \hat{n} \times \underline{\psi}_n$	Outgoing, unrotated	?	Symmetric	Symmetric	N^2

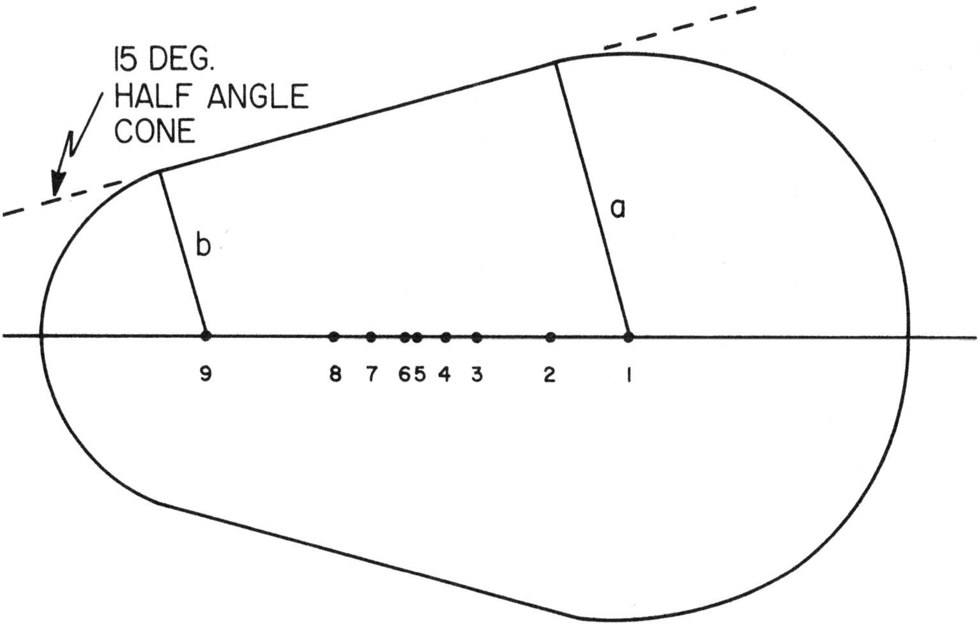

Fig. 1. The sphere-cone-sphere geometry.

in the object (see Fig. 1) are clearly superior. (The results from origin 9, incidentally, fall off scale and hence are not shown.) Quite analogous results are found for the radar cross section, as shown in Fig. 5. In Fig. 6 we have plotted the difference between scattering and total cross section, a quantity which must vanish if energy conservation requirements are met. Again origins 4,5,6 show the best truncation behavior. These three figures together offer strong evidence for choosing an origin centered to the best of one's ability, and also give a measure of the sensitivity of the computation when the origin is moved.

In order to see how things depend on the shape of the object we have varied the ratio of sphere radii over the range $0 \leq b/a \leq 0.8$, generating the various shapes shown in Fig. 7. In each case the coordinate origin is located at the midpoint of the axis of symmetry through the object, so the calculations should be near optimum from that point of view. Again ka=1, and we will compare results using the incoming and regular function representations, cases C and D respectively in Table 1.

The imaginary part of T_{11} (same full indices as earlier), which is the numerically dominant part, is shown in Fig. 8, while Fig. 9 gives the real part. Looking first at the curves for b/a=0.8 in Fig. 8, we see a remarkable improvement over those of Fig. 2, because of the new coordinate origin (the final values have also changed slightly for the same reason). There is no clear choice between the incoming and regular wave functions for this case.

As the object becomes more elongated, however, with decreasing values of b/a, the regular functions continue to show good convergence, at least by the time N=10, while the behavior using the incoming functions deteriorates fairly rapidly. An analogous result obtains for the real part of T_{11}, as can be seen from Fig. 9.

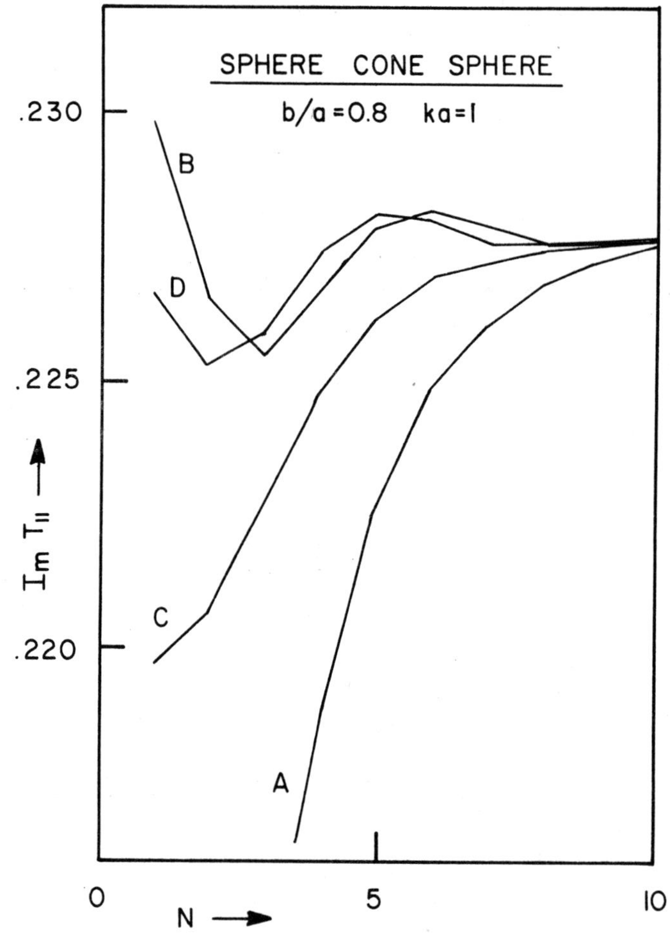

Fig. 2 The imaginary part of T_{11} vs. truncation size N.

From these two figures, we conclude that the incoming wave functions are perhaps not the best choice, especially when elongated shapes are involved.

The final question of interest in our numerical study involves the orthogonalization technique whereby the scattering matrix S is constrained by symmetry and unitarity (14). Salient features of the technique are as follows: with S=1+2T, Eq. 8 becomes

$$QS = -Q^*. \tag{10}$$

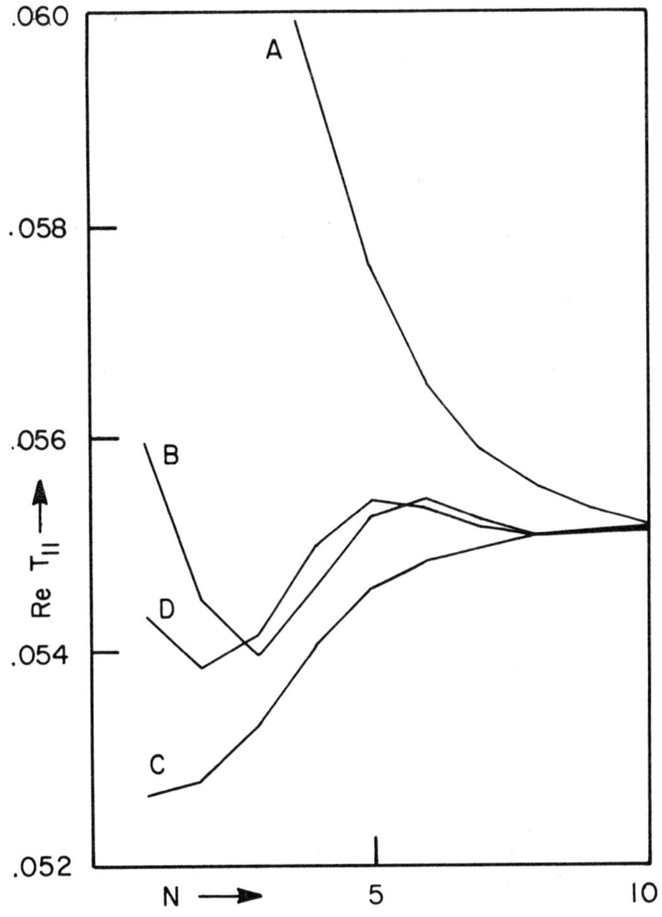

Fig. 3. The real part of T_{11} vs. truncation size N.

Now let M be the (upper triangular) matrix which Schmidt orthogonalizes A, i.e.,

$$\hat{Q} = MQ, \tag{11}$$

where Q is unitary. We then find straightforwardly that

$$S = -\hat{Q}'^*(MM^{*-1})\hat{Q}^*, \tag{12}$$

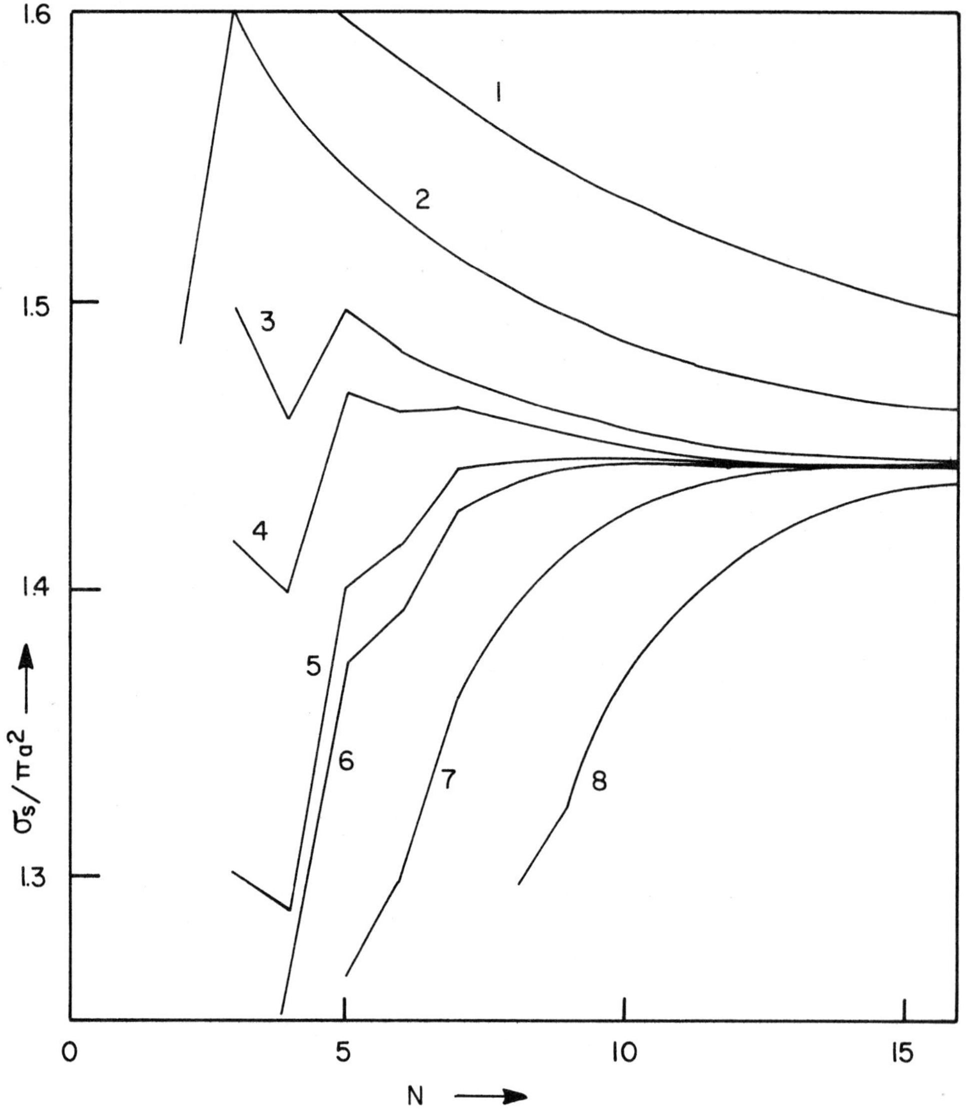

Fig. 4. The normalized scattering cross section vs. truncation size for various choices of coordinate origin.

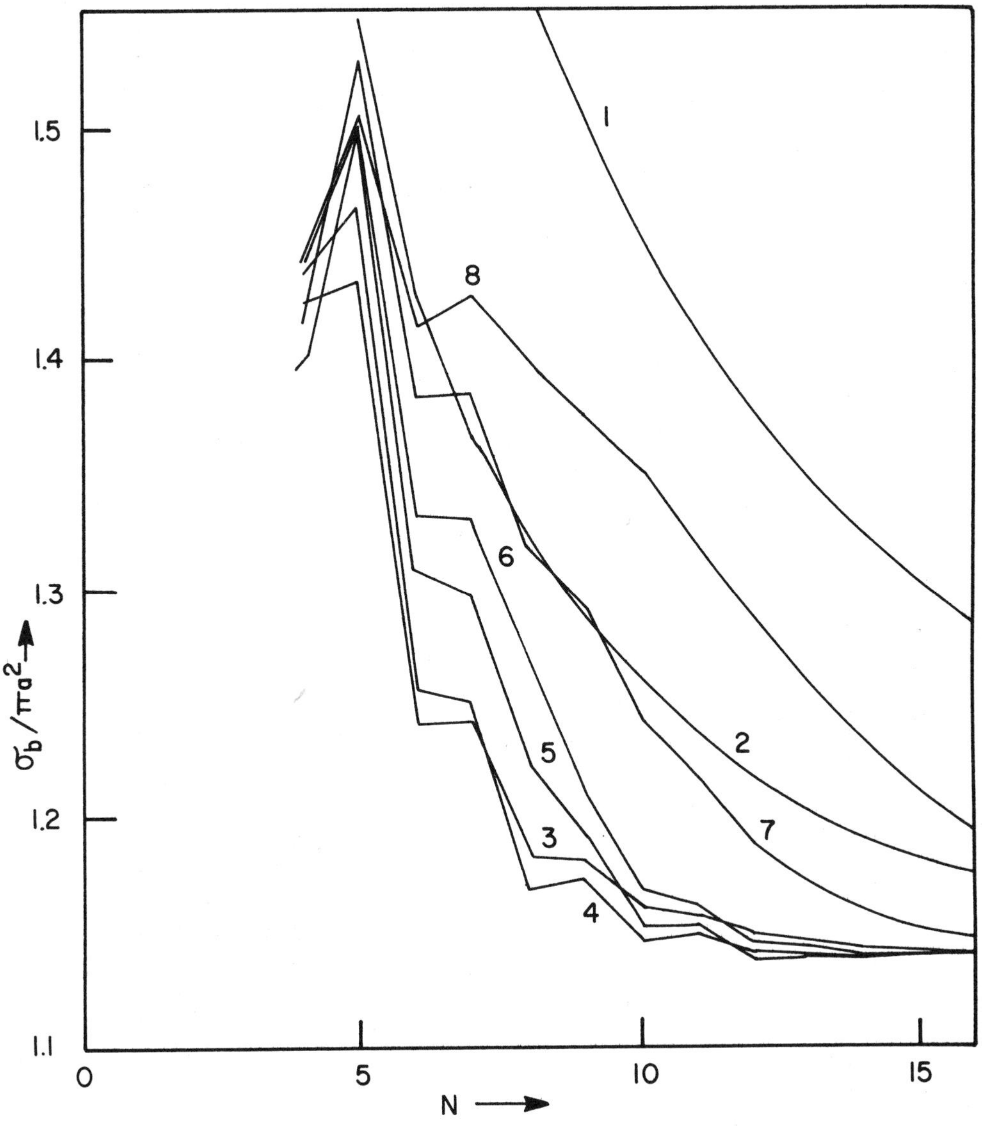

Fig. 5. The truncation behavior of the normalized radar cross section, for various choices of coordinate origin.

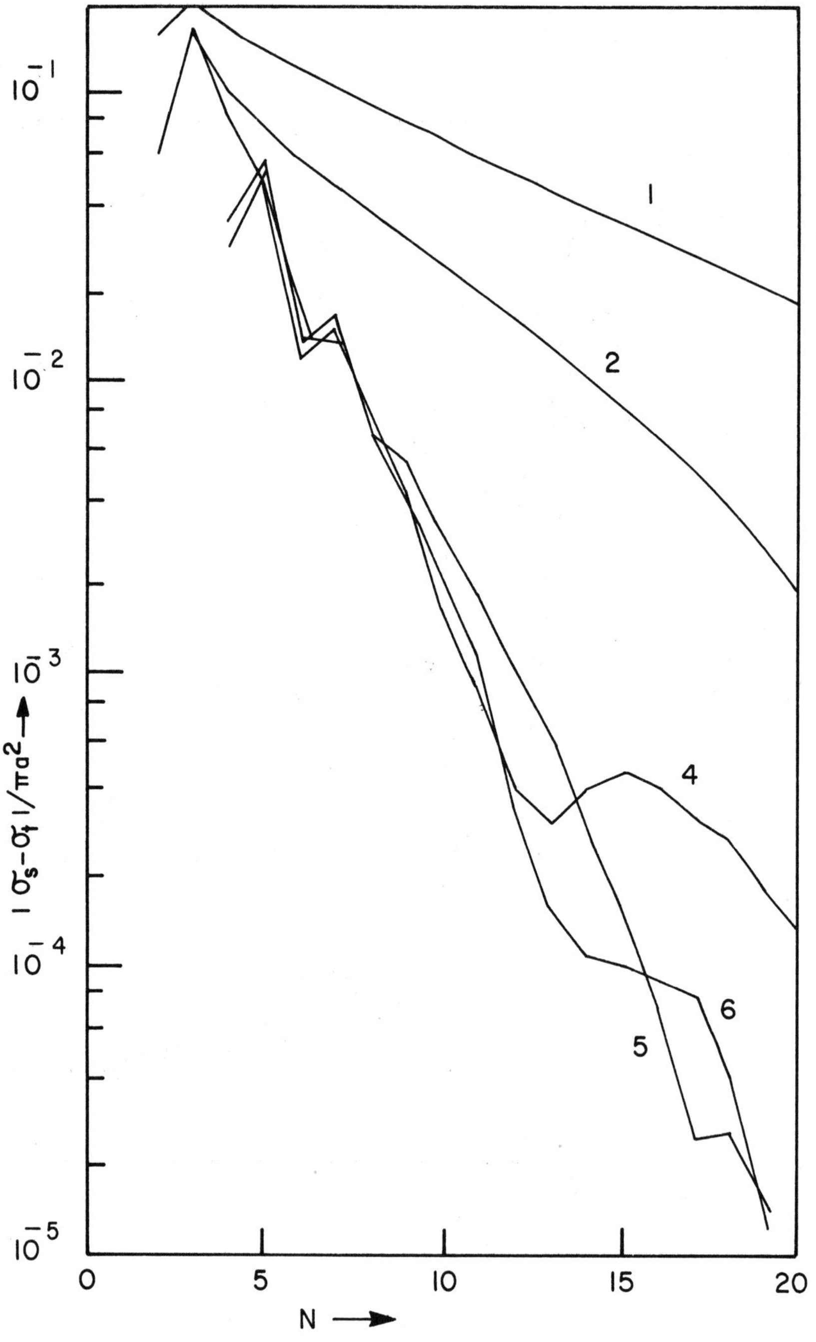

Fig. 6. Truncation behavior for the difference of scattering and total cross section, which theoretically should vanish, for various coordinate origins.

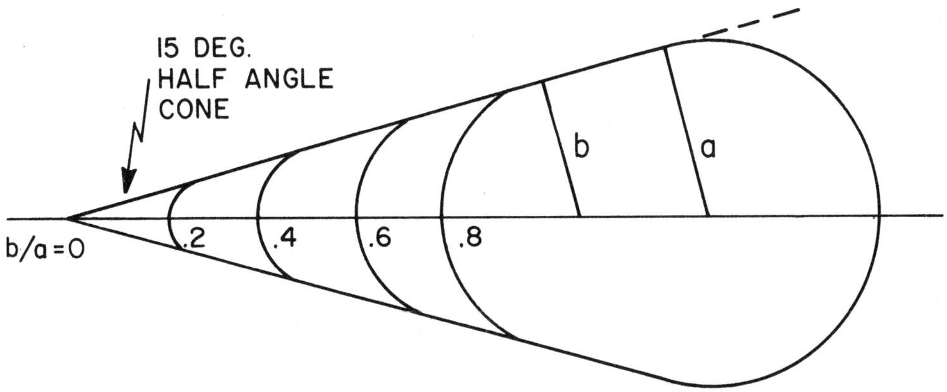

Fig. 7. Different shapes generated by varying the front sphere radius.

the prime indicating matrix transpose. This equation gives one possible procedure (although not necessarily the most efficient) for solving Eq. 10.

More information is available, however, We know that in the limit of infinite matrix size

$$S' = S, \tag{13}$$

$$S'^*S = \text{Identity}, \tag{14}$$

i.e., S is both symmetric and unitary (14).

Considering S as given in Eq. 12, the last two equations turn out to be necessary and sufficient conditions under which MM^{*-1} = Identity, so that M is real and

$$S = -\hat{Q}'^*\hat{Q}^* \quad (\text{or } T = -\hat{Q}'^*\text{Re}\hat{Q}). \tag{15}$$

Strictly speaking, Eq. 15 is valid only in the limit of infinite matrix size. It is nevertheless natural to consider using it in finite truncation. Notice that it prescribes a fairly efficient numerical approach. No record of M need be kept, nor must one carry along the right-hand side of Eq. 10 during the processing.

As a test case we consider a large sphere-cone-sphere, b/a=0.8, ka=10, with coordinate origin centered on the axis of symmetry. The regular wave functions (case B) are employed, and results are shown vs. truncation for some of the $T_{mn} = T_{1e0m1e0n}$ in Fig. 10. Solid curves show the truncation behavior of Eq. 15. The radius of the smallest sphere circumscribing the object is specified by kr_{min} = 12.8 (shown by the vertical line in Fig. 10). It is interesting to note that, with the exception of the lowest curve shown, the results have pretty much settled down to remain constant by the time N reaches this value. These curves are clearly superior to

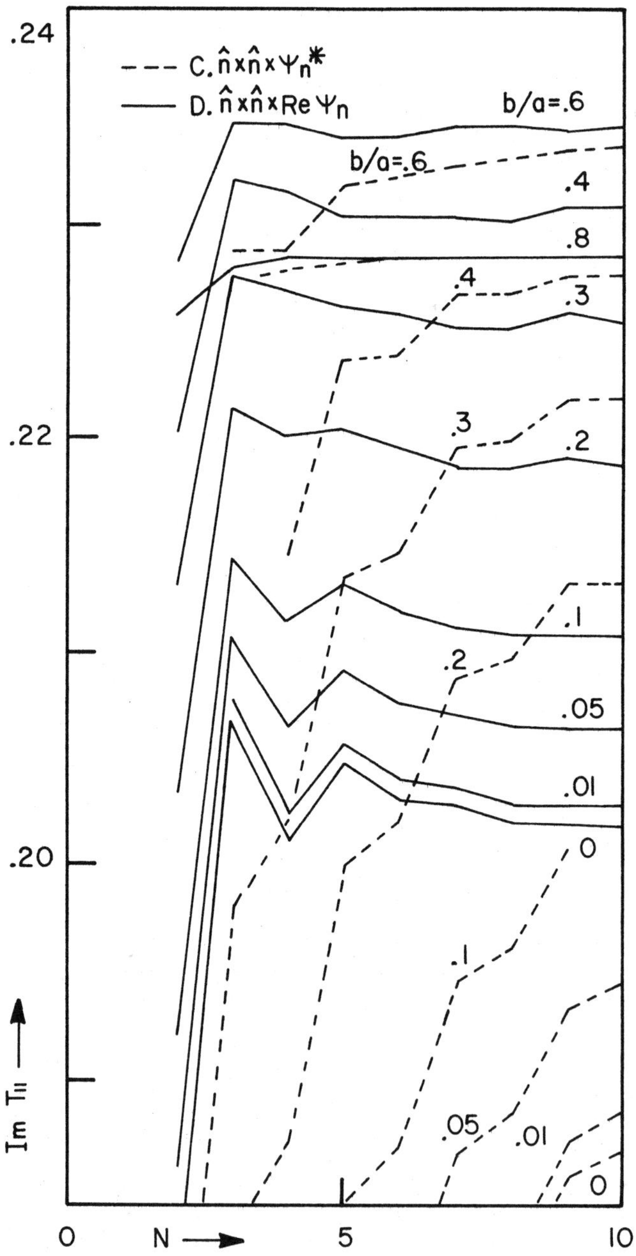

Fig. 8. Truncation behavior of the imaginary part of T_{11} for the shapes given in the preceding figure.

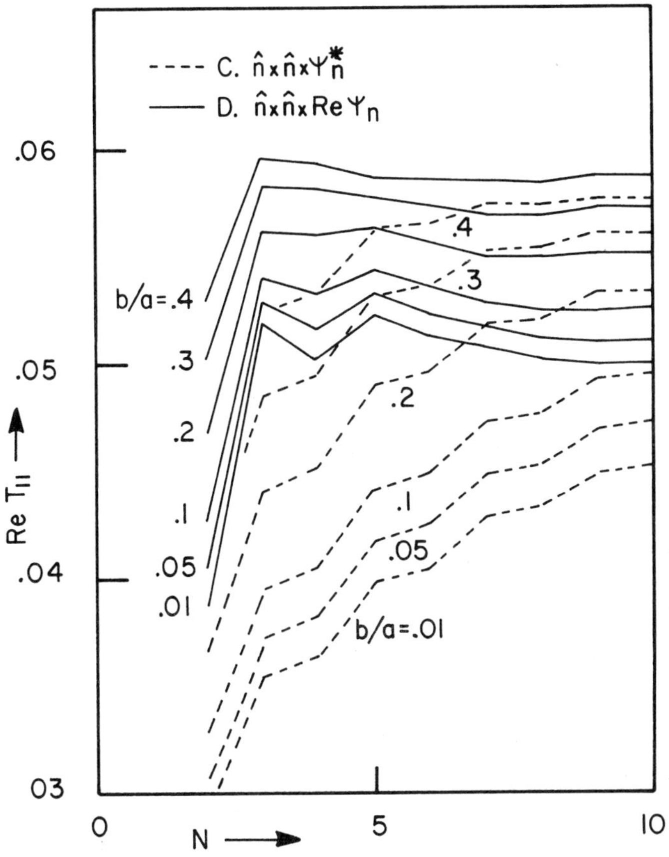

Fig. 9. Truncation behavior for the real part of T_{11} for the different geometrical shapes.

results obtained from Eq. 8, using full Gaussian elimination to convert Q on the left-hand side to the identity matrix. To confirm the extent to which the two calculations agree, the matrix inversion was carried out at N=28 (not shown), giving results that for the most part differed only in the fourth significant figure from the orthogonalization results at N=14.

From this and the preceding results, we conclude that the case B regular wave function representation of surface currents, coupled withthe orthogonalization procedure of Eq. 15, is probably the best approach of those studied here. This is not to say

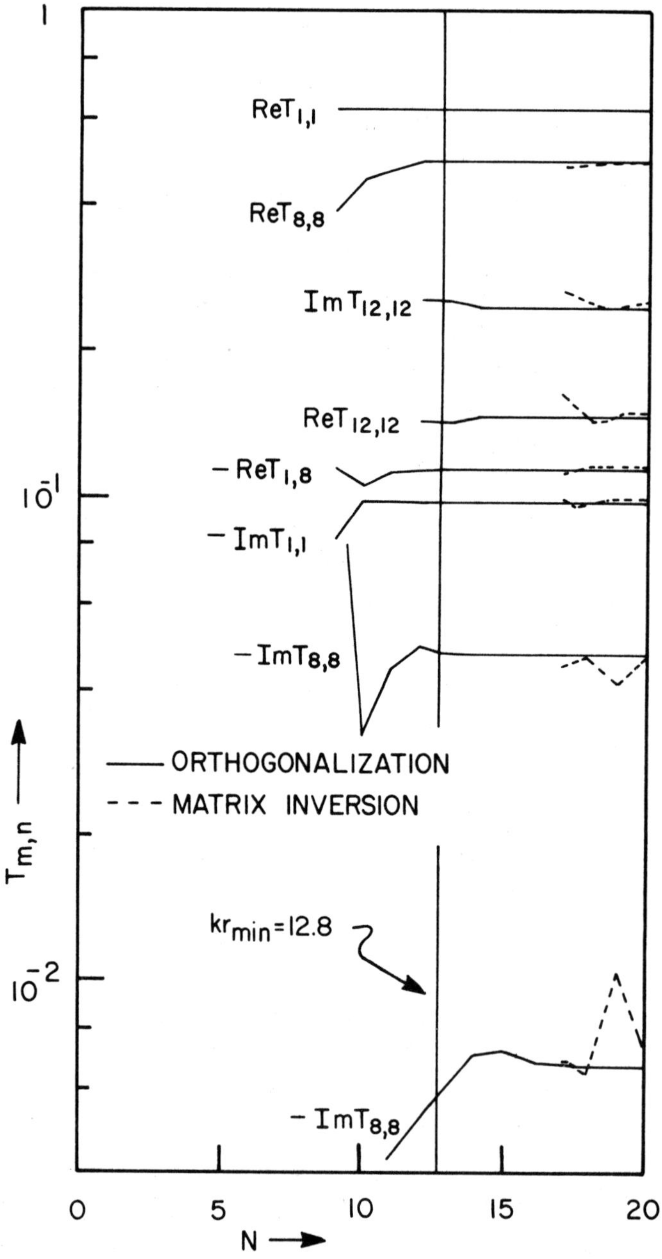

Fig. 10. Truncation behavior for various elements of T, showing the improvement obtained using orthogonalization.

that some other function choice may not be valuable, especially when particular shapes are involved; further study of this last question is warranted.

ACKNOWLEDGEMENT

The author is indebted to W. Millar, for his assistance in the work dealing with varied origin of coordinates.

REFERENCES

(1) W. R. Smythe, Charged right circular cylinder, J. Appl. Phys. 27, 917 (1956); Charged sphere in cylinder, J. Appl. Phys. 31, 553 (1960); Charged right circular cylinder, J. Appl. Phys. 33, 2966 (1962); charged spheroid in cylinder, J. Math. Phys. 4, 833 (1963).

(2) T. T. Taylor, Electric polarizability of a short, right circular conducting cylinder, J. Res. NBS 64B, 135 (1960).

(3) T. T. Taylor, Magnetic polarizability of a short, right circular conducting cylinder, J. Res. NBS 64B, 199 (1960).

(4) J. G. Fikioris, On the boundary-value problem of a spheroid, Quart. Appl. Math. 31, 143 (1973).

(5) J. B. Keller, R. E. Kleinman and T. B. A. Senior, Dipole movements in Rayleigh scattering, J. Inst. Maths. Applics. 9, 14 (1972).

(6) L. Eyges, private communication.

(7) H. A. Schenck, Improved integral formulation for acoustic radiation problems, J. Acoust. Soc. Am. 44, 41 (1968).

(8) P. C. Waterman, New formulation of acoustic scattering, J. Acoust. Soc. Am. 45, 1417 (1969).

(9) R. H. T. Bates and D. J. N. Wall, Null field approach to scalar diffraction, Phil. Trans. Roy. Soc. London 287, 45 (1977).

(10) P. C. Waterman, Matrix formulation of electromagnetic scattering, Proc. IEEE 53, 805 (1965).

(11) P. C. Waterman, Scattering by dielectric obstacles, Alta Freq. 38 (Speciale), 348 1969).

(12) P. Barber and C. Yeh, Scattering of electromagnetic waves by arbitrarily shaped dielectric bodies, Appl. Optics 14, 2864 (1975).

(13) P. W. Barber, Resonance electromagnetic absorption by nonspherical dielectric objects, IEEE Trans.Microwave Theory and Tech. 25. 373 (1977); Scattering and absorption efficiencies for nonspherical dielectric objects - biological models, IEEE Trans Biomed. Eng. 25, 155 (1978).

(14) P. C. Waterman, Symmetry, unitarity, and geometry in electromagnetic scattering, Phys. Rev. D 3, 825 (1971).

(15) W. M. Visscher, A new way to calculate scattering of acoustic and elastic waves, J. Appl. Phys. (to be published).

(16) E. Wolf (1973), A generalized extinction theorem and its role in scattering theory, in Coherence and Quantum Optics, ed. L. Mandel and E. Wolf, Plenum, New York.

(17) G. S. Agarwal, Relation between Waterman's extended boundary condition and the generalized extinction theorem, Phys. Rev. D 14, 1168 (1976).

(18) V. Varadan and Y.-H. Pao, Scattering matrix for elastic waves, I. Theory, J. Acoust. Soc. Am. 60, 556 (1976).

(19) P. C. Waterman, Matrix theory of elastic wave scattering, J. Acoust. Soc. Am. 60, 567 (1976).

(20) P. C. Waterman, Matrix theory of elastic wave scattering. II. A new conservation law, J. Acoust. Soc. Am. 63, 1320 (1977).

(21) V. V. Varadan, Scattering matrix for elastic waves. II. Application to elliptic cylinder, J. Acoust. Soc. Am. 63, 1014 (1978).

(22) Y.-H. Pao (1978) The transition matrix for the scattering of acoustic waves and for elastic waves, in Modern Problems in Elastic Wave Propagation, Wiley and Interscience, New York.

(23) P. C. Waterman, Matrix methods in potential theory and electromagnetic scattering J. Appl. Phys. to be published (1979).

(24) Y.-H. Pao, Betti's identity and transition matrix for elastic waves, J. Acoust. Soc. Am. 64, 302 (1978).

(25) B. Peterson and S. Strom, T-matrix formulation of electromagnetic scattering from multilayered scatterers, Phys. Rev. D 10, 2670 (1974); Matrix formulation of acoustic scattering from multilayered scatterers, J. Acoust. Soc. Am. 57, 2 (1975).

(26) B. Peterson and S. Strom, T-matrix for electromagnetic scattering from an arbitrary number of scatterers and representations of E(3), Phys. Rev. D 8, 3661 (1973); Matrix formulation of acoustic scattering from an arbitrary number of scatterers, J. Acoust. Soc. Am. 56, 771 (1974).

(27) V. S. Varadan, V.V. Varadan and Y.-H. Pao, Multiple scattering of elastic waves by cylinders of arbitrary cross section. I. SH waves, J. Acoust. Soc. Am. 63, 1310 (1978).

(28) P. M. Morse and H. Feshbach (1953) Methods of Theoretical Physics, McGraw-Hill, New York, p. 848.

(29) K. D. Granzow, Multipole theory in the time domain, J. Math. Phys. 7, 634 (1966); Time-domain treatment of a spherical boundary-value problem, J. Appl. Phys. 39, 3435 (1968).

(30) D. C. Stickler, Integral representation for Maxwell's equations with arbitrary time dependence, Proc. IEE 114, 169 (1967).

SURFACE CURRENTS AND 'NEAR' ZONE FIELDS

V. N. Bringi and T. A. Seliga
Atmospheric Sciences Program and Department of Electrical Engineering,
The Ohio State University, Columbus, Ohio 43210, U.S.A.

ABSTRACT

Waterman's method is used to compute the induced surface current on perfect conductors under plane wave illumination. The basic E-field integral equation is expanded into vector spherical harmonics and by using the extended boundary condition the coefficients of expansion of the surface current (in terms of regular type functions alone) are related to the incident plane wave expansion coefficients via the Q(Out, Re) matrix. Direct inversion of this Q-matrix yields the surface current expansion coefficients which are subsequently used to calculate the surface current on the scatterer. This procedure explicitly avoids consideration of fields in the region between the scatterer surface and the outer circumscribed sphere (the 'near' zone region). For comparison, the total H-field in this region is expanded using vector spherical harmonics of both the regular and outgoing types similar to the procedure of Hizal and Marincic (1970). The surface current is then expressed as the tangential component of the total H-field evaluated at the scatterer surface and the tangential component of the total E-field should vanish on the scatterer surface. Again, the extended boundary condition is used together with field matching at the outer circumscribed sphere to express the 'near' zone field expansion coefficients in terms of Q(Out, Re), Q(Out, Out) and the T-matrix of the scatterer. These coefficients are subsequently used to compute the tangential E and H fields on the scatterer surface. The magnitude of $\hat{n} \times \underline{E}$ on the scatterer surface was found to be -20dB to -80dB below the normalized surface current for a variety of scatterer configurations. Calculations of surface current, $\hat{n} \times \underline{H}$, are presented for a number of scatterer geometries including hemi-spherically capped cylinder, cone and spheroid configurations. These results are in excellent agreement with previous calculations for the same scatterer configurations by Andreasan (1965).

INTRODUCTION

The extended boundary condition (or T-matrix) method due to Waterman (1969, 1971) is being used by many workers in the analysis of acoustic, electromagnetic and elastic wave scattering; see, for example, Peterson and Ström (1973, 1974) and Varatharajulu and Pao (1976). From a computational viewpoint, the T-matrix approach has been used with success for the calculation of scattering properties of dielectric scatterers (Barber and Yeh, 1975; Warner and Hizal, 1976), layered scatterers (Bringi and Seliga, 1977) and elastic wave scattering (Varadan, 1978). Multiple scattering using the T-matrix approach has been analyzed for both deterministic scatterer configurations (Peterson, 1977) and random configurations (Varadan et al., 1978, 1979). Recently, Bates and Wall (1977) have proposed extensions of Waterman's method for calculating the induced surface fields on totally reflecting, arbitrarily shaped, two-dimensional scatterers under plane wave illumination; an extensive list of related references is also given in their paper.

This paper deals with the computation of the induced surface current on perfect conductors under plane wave excitation using Waterman's method (also termed the spherical null field method by Bates and Wall, 1977). Additionally, it is shown that the coefficients of the vector wave function expansion in the so-called 'near' zone region between the scatterer surface and the minimum circumscribing sphere (with origin inside the scatterer) can be determined by an extension of this method. These coefficients are subsequently used to compute the surface current, $\hat{n} \times \underline{H}$, on the scatterer surface and also to verify the boundary condition $\hat{n} \times \underline{E} = 0$ on the perfectly conducting surface. Numerical results are presented for cylinder, cone and spheroid configurations under plane wave incidence.

REVIEW OF T-MATRIX THEORY FOR PERFECT CONDUCTORS

Consider an incident wave, $\underline{E}^i, \underline{H}^i$, exciting a closed, perfectly conducting scatterer consisting of a smooth surface S as in Fig. 1. An origin O is selected inside S and two spheres ρ_1 and ρ_2 are selected such that ρ_1 is the maximum inscribed sphere and ρ_2 is the minimum circumscribed sphere. The region between S and ρ_2 is referred to as the 'near' zone region. The total electric field, $\underline{E}$, is expressed as the sum of the incident field, $\underline{E}^i$, and scattered field, $\underline{E}^s$; $\underline{E} = \underline{E}^i + \underline{E}^s$.

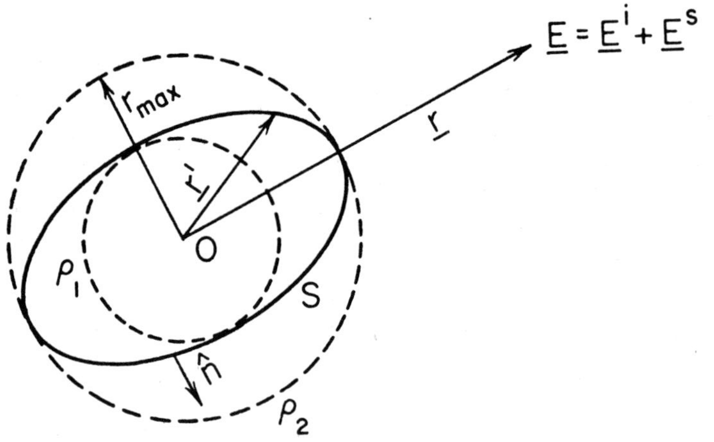

Fig. 1. Geometrical configuration of a perfectly conducting scatterer bounded by a closed surface S.

For field points $\underline{r}$ restricted to be wholly within S, the E-field integral equation is given by

$$-\underline{E}^i(\underline{r}) = j\omega\mu_0 \int_S \overline{\overline{G}}_0(\underline{r},\underline{r}') \cdot \underline{J}_s(\underline{r}')\, dS' \qquad ; r \in S \tag{1}$$

where $\underline{J}_s(\underline{r}')$ is the induced surface current density and $\overline{\overline{G}}_0(\underline{r},\underline{r}')$ is the free-space Green's dyadic. An exp(-jωt) time dependence is assumed and suppressed in all equations. Equation (1), which refers to the interior scattering problem, is expanded into spherical vector wave functions $\underline{M}^{1,3}, \underline{N}^{1,3}$ which satisfy the vector

Helmholtz equation (Stratton, 1941):

$$\underline{M}^3_{\substack{e\\o}mn} = \nabla \times [\underline{r}\, h_n(kr)\, P_n^m(\cos\theta) \begin{matrix}\cos m\phi\\ \sin m\phi\end{matrix}]$$

$$\underline{N}^3_{\substack{e\\o}mn} = \frac{1}{k} \nabla \times \underline{M}^3_{\substack{e\\o}mn}$$

where $\substack{e\\o}$ stands for even or odd term corresponding to the choice of the trignometric function and $\underline{M}^1$, $\underline{N}^1$ are obtained by taking the real part of $\underline{M}^3$, $\underline{N}^3$, respectively. The superscripts ^{1} or ^{3} on $\underline{M}$ and $\underline{N}$ refer to expansions in terms of regular or outgoing wave functions, respectively. The incident field $\underline{E}^i$ and the scattered field $\underline{E}^s$ can be expanded as

$$\underline{E}^i(\underline{r}) = \sum_\nu D_\nu [a_\nu \underline{M}^1_\nu(k\underline{r}) + b_\nu \underline{N}^1_\nu(k\underline{r})] \tag{2}$$

$$\underline{E}^s(\underline{r}) = \sum_\nu 4D_\nu [f_\nu \underline{M}^3_\nu(k\underline{r}) + g_\nu \underline{N}^3_\nu(k\underline{r})] \ ; \ r > r_{max} \tag{3}$$

where ν represents a combined index notation incorporating $\substack{e\\o}mn$ of the vector spherical harmonics, D_ν is a normalization constant, k represents the incident field wave number, r_{max} refers to the radius of the circumscribing sphere as shown in Fig. 1, (a_ν, b_ν) are the known incident field expansion coefficients and (f_ν, g_ν) are the unknown scattered field expansion coefficients. Note that the expansion of Eq. (3) in terms of outgoing wave functions is strictly valid for $\underline{r}$ outside the circumscribed sphere ρ_2.

The induced surface current, $\underline{J}_s(\underline{r}')$, is next expanded into a complete set of tangential vector functions $\hat{n} \times \underline{M}^1$ and $\hat{n} \times \underline{N}^1$ where $\hat{n}$ is the unit normal to S:

$$\underline{J}_s(\underline{r}') = -j\sqrt{\frac{\varepsilon_o}{\mu_o}} \sum_\mu [c_\mu \hat{n} \times \underline{N}^1_\mu(k\underline{r}') + d_\mu \hat{n} \times \underline{M}^1_\mu(k\underline{r}')]. \tag{4}$$

Caldéron (1954) has proven the completeness and mean square convergence properties of these tangential vector functions $\hat{n} \times \underline{M}^1$ and $\hat{n} \times \underline{N}^1$ for representing a purely tangential field $\underline{J}_s(\underline{r}')$ on the surface of a smooth, non-spherical scatterer.

Using the spherical vector wave function expansion of the free-space Green's dyadic $\overline{G}_o(\underline{r},\underline{r}')$ with $r < r'$ (Morse and Feshbach, 1953) and substituting Eqs. (2) and (4) in (1) results in the spherical null field equations. In compact matrix notation these are given by

$$[Q(\text{Out, Re})] \begin{bmatrix} c \\ d \end{bmatrix} = -j \begin{bmatrix} a \\ b \end{bmatrix} \tag{5}$$

where c, d, a, b are N-component column vectors and the 2N x 2N Q-matrix as defined by Waterman (1971) consists of four N x N submatrices. The arguments of Q stipulate the type (outgoing or regular) of wave functions used in arriving at the elements of the submatrices.

Bates and Wall (1977) have proposed two extensions with respect to the derivation of Eq. (5). One is the use of spheroidal wave functions in the expansion of the free-space Green's dyadic (Wall, 1978), and the second is in the choice of the form of the expansion for the induced surface current which is made in terms of functions natural to the surface of the scatterer. A computational approach using these two extensions of Waterman's method was performed by Bates and Wall (1977) on two-dimensional, totally reflecting scatterers, and they report increased numerical stability and lower resulting matrix orders than obtainable using a direct Waterman approach, especially for elliptical cylinders with large aspect ratios. However, such extensions are not considered here due to their complexity in the full three-dimensional vector case and because the scatterer configurations dealt with here are not grossly non-spherical.

Waterman (1971) has shown that the T-matrix, which relates the incident field expansion coefficients (a_ν, b_ν) to the scattered field expansion coefficients (f_ν, g_ν) is given by

$$\begin{bmatrix} f \\ g \end{bmatrix} = -\frac{1}{4}[Q(\mathrm{Re},\ \mathrm{Re})][Q(\mathrm{out},\ \mathrm{Re})]^{-1}\begin{bmatrix} a \\ b \end{bmatrix} = -\frac{1}{4}[T]\begin{bmatrix} a \\ b \end{bmatrix} \tag{6}$$

where (f, g) are again N-component column vectors. Also, Q(Re, Re) may be obtained from the real part of Q(Out, Re). Note that the expansion coefficients (f_ν, g_ν) are valid only outside the sphere ρ_2 (see Fig. 1). Hence, the 'near' zone region between S and ρ_2 is completely avoided in Waterman's formulation.

FIELD EXPANSION IN THE 'NEAR' ZONE REGION

It will now be shown that the coefficients of expansion of the total field in the region between the surface, S, of the scatterer and the outer circumscribed sphere, ρ_2, can be determined by a simple extension of Waterman's method.

The total $\underline{E}$ field in this 'near' zone region is expanded into spherical vector wave functions of both the regular type and the outgoing type:

$$\underline{E}(k\underline{r}) = \sum_\mu [\gamma_\mu \underline{M}^1_\mu + \alpha_\mu \underline{M}^3_\mu + \delta_\mu \underline{N}^1_\mu + \beta_\mu \underline{N}^3_\mu] \quad , \ \underline{r} \text{ between S and } \rho_2 \tag{7}$$

where γ_μ, α_μ, δ_μ, β_μ are unknown expansion coefficients. The induced surface current, $\underline{J}_S(\underline{r}')$, is now derived from the tangential component of the total magnetic field, $\underline{H}(\underline{r}) = \dfrac{\nabla \times \underline{E}(\underline{r})}{j\omega\mu_0}$ on the surface S:

$$\underline{J}_s(\underline{r}') = \hat{n} \times \underline{H}_+(\underline{r}') = -j\sqrt{\frac{\varepsilon_0}{\mu_0}}\ \hat{n} \times \sum_\mu [\gamma_\mu \underline{N}^1_\mu + \alpha_\mu \underline{N}^3_\mu + \delta_\mu \underline{M}^1_\mu + \beta_\mu \underline{M}^3_\mu] \tag{8}$$

where the subscript + on $\underline{H}(\underline{r}')$ refers to approaching the surface of the scatterer from the outside. Substituting Eqs. (8) and (2) in (1) together with the expansion of the free space Green's dyadic yields in compact matrix notation:

$$[Q(\mathrm{Out},\ \mathrm{Re})]\begin{bmatrix} \gamma \\ \delta \end{bmatrix} + [Q(\mathrm{Out},\ \mathrm{Out})]\begin{bmatrix} \alpha \\ \beta \end{bmatrix} = -j\begin{bmatrix} a \\ b \end{bmatrix} \tag{9}$$

where γ, δ, α, β are N-component column vectors and Q(Out, Out) is the Q-matrix constructed with outgoing wave functions. The form of Eq. (9) is similar to that derived previously for layered scatterers by Bringi and Seliga (1977). Continuity of the total field across the surface $r = r_{max}$ yields (see Fig. 1):

$$\underline{E}(k\underline{r}_{max}) = \underline{E}^{i}(k\underline{r}_{max}) + \underline{E}^{s}(k\underline{r}_{max}) \tag{10}$$

Substituting the respective vector wave function expansions in Eq. (10) and assuming their convergence on the surface $r = r_{max}$ yields a matrix equation between the various expansion coefficients:

$$\begin{bmatrix} 4\,D_n f_n \\ 4\,D_n g_n \end{bmatrix} + \begin{bmatrix} D_n a_n X \\ D_n b_n X \end{bmatrix} = \begin{bmatrix} \delta_{nn'}X & \delta_{nn'} & 0 & 0 \\ 0 & 0 & \delta_{nn'}X & \delta_{nn'} \end{bmatrix} \begin{bmatrix} \gamma_{n'} \\ \alpha_{n'} \\ \delta_{n'} \\ \beta_{n'} \end{bmatrix} \tag{11}$$

where $\delta_{nn'}$ is the Kronecker delta, and $X = j_q(kr_{max})/h_q^{(1)}(kr_{max})$, j_q and $h_q^{(1)}$ being the spherical bessel and hankel functions of order q, respectively.

By combining Eqs. (11) and (9), a matrix equation for determining the expansion coefficients γ, δ, α, β results.

$$\begin{pmatrix} \gamma \\ \delta \\ \alpha \\ \beta \end{pmatrix} = (A)^{-1} \begin{pmatrix} -ja \\ -jb \\ 4Df + DaX \\ 4Dg + DbX \end{pmatrix} \tag{12}$$

where the 4N x 4N matrix A is given by

$$(A) = \begin{pmatrix} Q(Out,\ Re) & Q(Out,\ Out) \\ \delta_{nn'}X & \delta_{nn'} \end{pmatrix} \tag{13}$$

and the coefficients (f_n, g_n) are given by Eq. (6).

Examining Eqs. (5), (6), (12), and (13) it is seen that the expansion coefficients of the surface current, far-zone scattered field and 'near' zone total field can be determined from the three Q-matrices, viz., Q(Out, Re), Q(Re, Re) and Q(Out, Out). These coefficients can be used to compute the induced surface current, $\underline{J}_s(\underline{r}')$, using Eq. (4) or (8); both should give identical results. Additionally, the same 'near' zone coefficients can be used to test via Eq. (7) if $\hat{n} \times \underline{E} = 0$ on the perfectly conducting scatterer surface. Note that such a test cannot be performed using the direct Waterman approach which completely avoids the 'near' zone region.

Expansions of the type given in Eq. (7) or (8) have been used by Hizal and Marincinc (1970) to calculate the scattering cross-sections of perfect conductors. However, we claim that expansions of the type given in Eqs. (7) or (8) are only useful in determining the field in the 'near' zone region and that both the induced surface current and scattered fields (outside the circumscribing sphere) are completely determined by the direct Waterman method. In other words, the Rayleigh hypothesis (Millar, 1973; Bates, 1975) is not invoked in the present formalism since the expansion coefficients (f_n, g_n) are not directly used to determine the field in the 'near' zone region between the surface, S, of the scatterer and the outer circumscribing sphere.

COMPUTATIONS

As mentioned previously, knowledge of the three Q-matrices, Q(Out, Re), Q(Re, Re), and Q(Out, Out) is sufficient to determine the expansion coefficients of the surface current, far-zone scattered field and the 'near' zone field via Eqs. (5), (6) and (12), respectively. This is convenient from a computational viewpoint since the different Q-matrices can be constructed within the same sub-program. Computations of the three Q-matrices and matrix solutions of (5), (6) and (12) were performed for plane wave excitation of perfectly conducting bodies. A sample calculation of the induced surface current, $\underline{J}_S(\underline{r}')$, on a hemispherically-capped cylinder as shown in Fig. 2a was performed using Eqs. (4) and (5). Note that the expansion for the surface current is in terms of regular wave functions only. A plane wave is incident endfire with polarizations $\hat{e}_{/\!/}$ and $\hat{e}_{\perp}$ which are parallel and perpendicular to the XZ-plane, respectively. For $\hat{e}_{/\!/}$, the surface current,

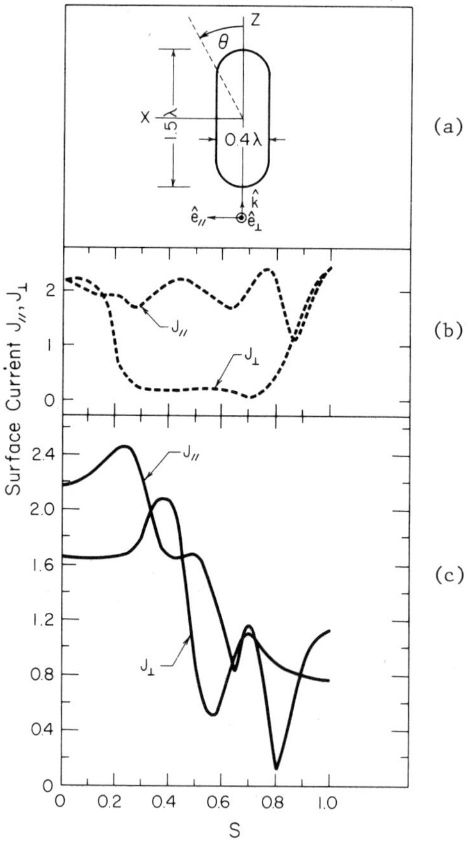

Fig. 2. (a) Hemi-spherically capped cylinder under plane wave excitation with polarizations $\hat{e}_{/\!/}$ or $\hat{e}_{\perp}$ which are parallel or perpendicular to the XZ-plane, respectively; (b) Surface current components $J_{/\!/}$ and $J_{\perp}$ as a function of normalized distance s along the scatterer surface for endfire incidence; (c) Surface current components $J_{/\!/}$ and $J_{\perp}$ as a function of s for broadside incidence.

$\underline{J}_s(\underline{r}') = J_{/\!/}\hat{t}$ where $\hat{t}$ is a local unit vector tangential to the surface in the XZ-plane; for $\hat{e}_\perp$, the surface current, $\underline{J}_s(\underline{r}') = J_\perp\hat{\phi}$, where $\hat{\phi}$ is a local unit vector tangential to the surface along the azimuthal direction. All calculations are performed over the XZ-plane but can be extended over other planes. Figure 2b shows $J_{/\!/}$ and $J_\perp$ as a function of normalized distance, s, measured along the surface of the scatterer. Note the standing wave pattern exhibited by $J_{/\!/}$ and the low magnitudes of $J_\perp$ in the center portion $0.2 < s < 0.8$. These results are in excellent agreement with point-matching solutions of Andreasan (1965) for the same body configuration. Figure 2c shows the surface current component $J_{/\!/}$ and $J_\perp$ for broadside incidence. Note the 'forced' behavior in the region $0 < s < 0.3$ and the 'resonant' behavior in the region $0.3 < s < 1.0$, which are consistent with the experimental results of Burton et al. (1976).

We next consider a spherical scatterer with $ka = \pi$, 'a' being the sphere radius, located with its center at the origin of a XYZ coordinate system and scattering geometry as explained above. The incident plane wave propagates along the positive Z-axis with polarizations $\hat{e}_{/\!/}$ and $\hat{e}_\perp$. In Fig. 3 we show $J_{/\!/}$ and $J_\perp$ as a function of normalized distance s along the scatterer surface using both Eqs. (4) and (8). Note that Eq. (8) uses the 'near' zone field coefficients where kr_{max} was chosen to be slightly larger than ka. Identical results were obtained for the surface current using the expansions of Eqs. (4) or (8), and are in excellent agreement with Andreasan (1965). The magnitude of $\hat{n} \times \underline{E}$ on the surface using the 'near' zone field coefficients (γ, α, δ, β) in Eq. (7) was of the order of 10^{-15}. Note that in this case the coefficients (γ_n, δ_n) are identical to ($D_n a_n$, $D_n b_n$) and (α_n, β_n) are identical to ($4D_n f_n$, $4D_n g_n$).

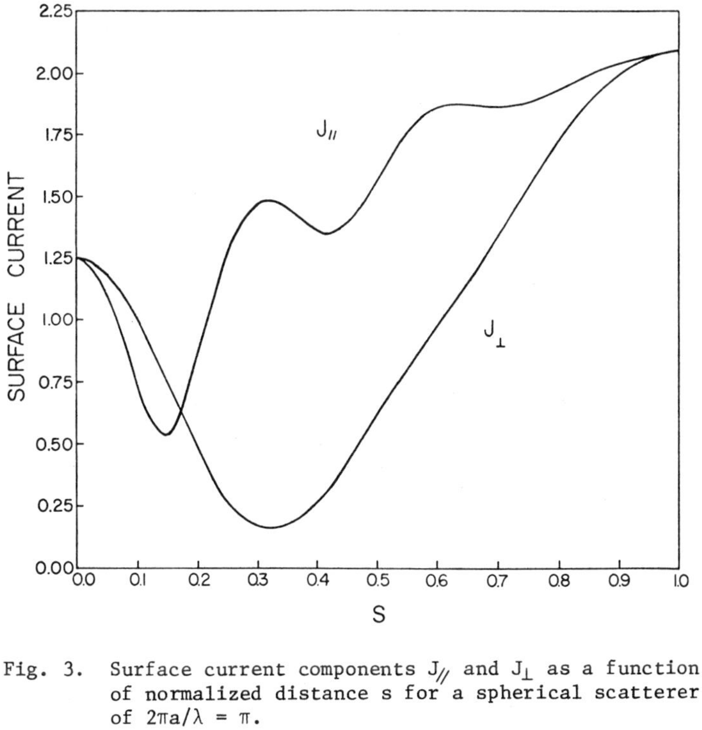

Fig. 3. Surface current components $J_{/\!/}$ and $J_\perp$ as a function of normalized distance s for a spherical scatterer of $2\pi a/\lambda = \pi$.

To consider the effects of non-sphericity we consider the scatterer configuration shown in Fig. 4a. Surface currents were calculated using both the full expansion of Eq. (8) as well as the regular wavefunction expansion of Eq. (4). As depicted in Fig. 4b, 4c, the two solutions agree remarkably well for both polarizations over most of the range $0 \leq s \leq 1$. The magnitude of $\hat{n} \times \underline{E}$ on the surface as compared to the magnitude of $\sqrt{\frac{\mu_0}{\varepsilon_0}}\, \hat{n} \times \underline{H}$ on the surface was of the order of -40dB to -80dB confirming adequate satisfaction of the boundary condition $\hat{n} \times \underline{E} = 0$ on the perfectly conducting surface. Note that the coefficients of expansion (γ, α, δ, β) of the near-zone field can, in principle, be used to compute the $\underline{E}$ and $\underline{H}$ fields in the region between the scatterer surface and the outer circumscribed sphere using Eq. (7); this is the region precisely avoided by the direct application of Waterman's method.

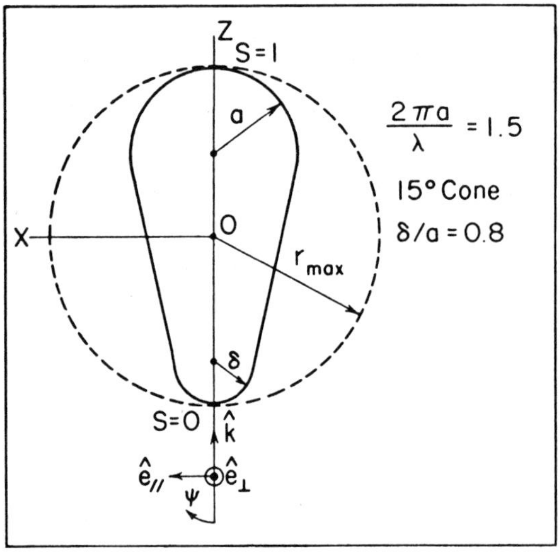

(a)

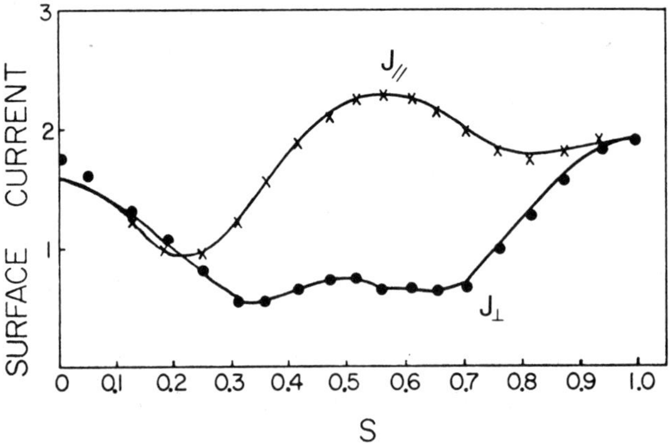

(b)

Fig. 4. (a) Hemi-spherically capped 15^{0} cone under plane wave illumination; (b) Surface current $J_{//}$ and $J_{\perp}$ as a function of s using both the 'near' zone field approach (—) and the direct approach (x,•).

The last example we consider is the oblate spheroid of kb = 1.0 and a/b = 0.75 where a and b are the semi-minor and semi-major axes, respectively. The spheroid is oriented with its axis of symmetry along the Z-axis and its center coincides with the origin of a XYZ coordinate system. The incident plane wave propagates along the positive Z-axis with polarizations $\hat{e}_{\parallel}$ and $\hat{e}_{\perp}$. In Fig. 5a, 5b we show the surface current as a function of normalized distance s along the scatterer surface using both Eqs. (4) and (8) for the two polarizations. Both solutions again agree remarkably well over most of the range $0 \leq s \leq 1$. The magnitude of $\hat{n} \times \underline{E}$ relative to $\sqrt{\frac{\mu_0}{\varepsilon_0}}|\hat{n} \times \underline{H}|$ on the surface was of the order of -20dB to -40dB over the range $0 \leq s \leq 1$.

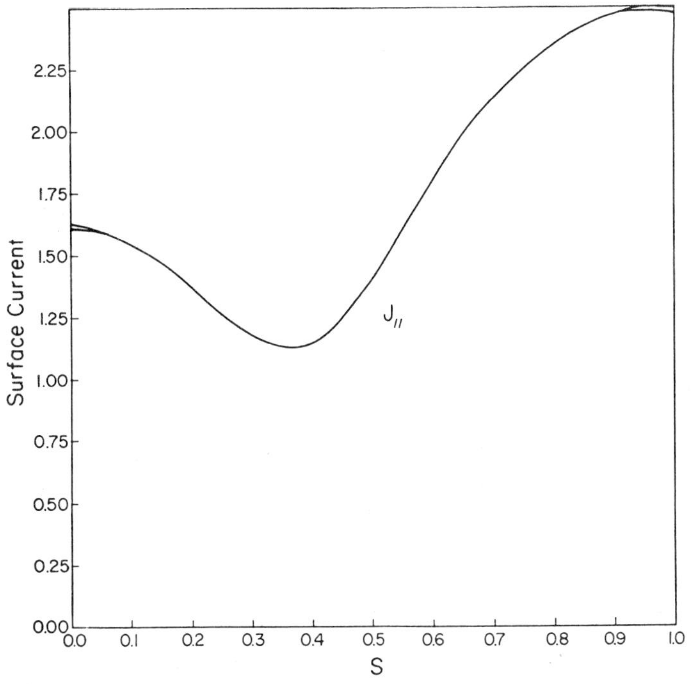

Fig. 5(a). Surface current $J_{\parallel}$ for oblate spheroid with $kb = 1$ and $\frac{a}{b} = .75$. Comparison using the 'near' zone field approach and the direct approach.

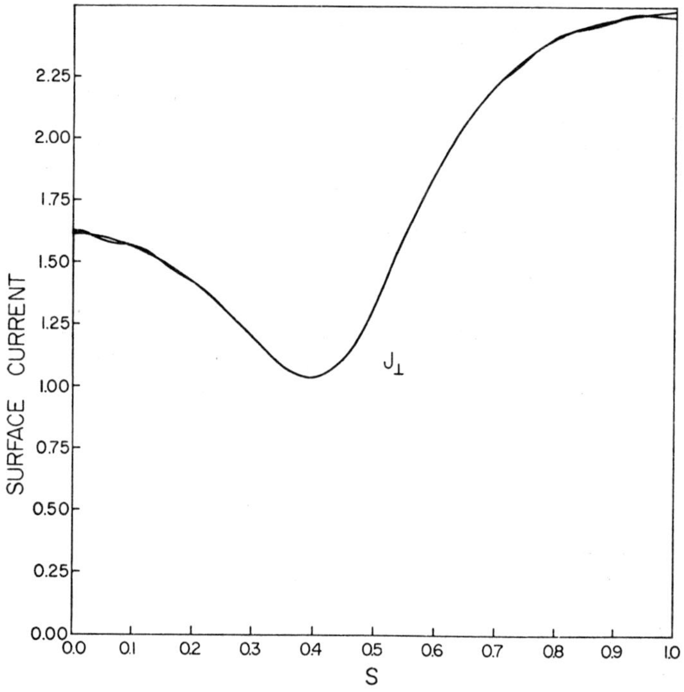

Fig. 5(b). Surface current $J_\perp$ for oblate spheroid with $kb = 1$ and $\frac{a}{b} = .75$. Comparison using the 'near' zone field approach and the direct approach.

CONCLUSIONS

Induced surface currents and 'near' zone fields were examined using Waterman's method for perfectly conducting scatterers under plane wave illumination. A spherical vector wave function expansion consisting of both outgoing and regular wave functions was used to represent the total field in the 'near' zone region between the surface, S, of the scatterer and the outer circumscribing sphere with center inside the scatterer. The coefficients of such an expansion were determined using an extension of Waterman's method together with field matching at the outer circumscribed sphere. These coefficients were subsequently used to compute $\hat{n} \times \underline{H}$ and $\hat{n} \times \underline{E}$ on the scatterer surface. A variety of scatterer configurations were considered including hemi-spherically capped cylinder, cone and spheroid; excellent agreement with past computations of surface current using point matching methods was demonstrated.

ACKNOWLEDGEMENTS

This research supported by the Atmospheric Research Section, National Science Foundation and the National Oceanic and Atmospheric Administration under Grant No. 04-78-B01-21.

REFERENCES

Andreasan, M. G., Scattering from bodies of revolution, Trans. IEEE, AP-13, 303-310 (1965).

Barber, P. and C. Yeh, Scattering of electromagnetic waves by arbitrarily shaped dielectric bodies, Applied Optics, 14, 2684-2872 (1975).

Bates, R.H.T., Analytic constraints on electromagnetic field computations, Trans. IEEE, MTT-23, 605-623 (1975).

Bates R.H.T. and D.J.N. Wall, Null field approach to scalar diffraction, I. General method, Phil. Trans. R. Soc. Lond. A, 287, 45-78 (1977).

Bringi, V. N. and T. A. Seliga, Scattering from axisymmetric dielectrics or perfect conductors imbedded in an axisymmetric dielectric, Trans. IEEE, AP-25, 575-580 (1977).

Burton, R. W., R.W.P. King, and D. Blejer, Surface currents and charges on an electrically thick conducting tube in an E-polarized, normally incident, plane-wave field, 2. Measurements, Radio Science, 11, 701-712 (1976).

Calderón, A. P., The multipole expansion of radiation fields, J. Rational Mech. Anal., 3, 523-537 (1954).

Hizal, A. and A. Marincic, New rigorous formulation of electromagnetic scattering from perfectly conducting bodies of arbitrary shape, Proc. IEE, 117, 1639-1647 (1970).

Millar, R. F., The Rayleigh hypothesis and a related least-squares solution to scattering problems for periodic surfaces and other scatterers, Radio Science, 8, 785-796 (1973).

Morse, P. M. and H. Feshbach (1953), Methods of Theoretical Physics, McGraw-Hill, New York.

Peterson, B., Multiple scattering of waves by an arbitrary lattice, Phys. Rev. A, 16, 1363-1370 (1977).

Peterson, B. and S. Ström, T-matrix for EM scattering from an arbitrary number of scatterers and representations of E(3), Phys. Rev. D, 8, 3661-3678 (1973).

Peterson, G. and S. Ström, T-matrix formulation of electromagnetic scattering from multilayered scatterers, Phys. Rev. D, 10, 2670-2684 (1974).

Stratton, J. A. (1941), Electromagnetic Theory, McGraw-Hill, New York.

Varadan, V. V., Scattering matrix for elastic waves, II. Application to elliptic cylinders, J. Acoust. Soc. Am., 63, 1014-1024 (1978).

Varadan, V. K., V. V. Varadan, and Y. H. Pao, Multiple scattering of elastic waves by cylinders of arbitrary cross section, I. SH waves, J. Acoust. Soc. Am., 63, 1310-1319 (1978).

Varadan, V. K., V. N. Bringi, and V. V. Varadan, Coherent electromagnetic wave propagation through randomly distributed dielectric scatterers, Phys. Rev. D, 19, 2480-2489 (1979).

Vartharajulu, V. and Y. H. Pao, Scattering matrix for elastic waves, I. Theory, J. Acoust. Soc. Am., 60, 556-566 (1976).

Wall, D.J.N., Circularly symmetric green tensors for the harmonic vector wave equation in spheroidal coordinate systems, J. Phys. A: Math. Gen., 11, 749-757 (1978).

Warner, C. and A. Hizal, Scattering and depolarization of microwaves by spheroidal raindrops, Radio Science, 11, 921-930 (1976).

Waterman, P. C., Scattering by dielectric obstacles, Alta Frequenza, 38, 348-352 (1969).

Waterman, P. C., Symmetry, unitarity, and geometry in electromagnetic scattering, Phys. Rev. D, 4, 825-839 (1971).

SCATTERING BY AN ELASTIC OBSTACLE IN A FLUID AND A SMOOTH ELASTIC OBSTACLE

Anders Boström
Institute of Theoretical Physics, S-412 96 Göteborg, Sweden

ABSTRACT

The transition matrix method for scattering is adapted to treat two new cases: an elastic obstacle in a fluid and a smooth elastic obstacle. These two cases are conveniently treated together as the main difference from previously treated cases is the same, namely that the boundary conditions must also be applied in the integral representation for the obstacle. Rather complicated expressions for the transition matrices are obtained, and some numerical results are given for spheroids.

I. INTRODUCTION

In the present paper we will adapt the transition matrix method for scattering of elastic waves (see Waterman (1) and Pao and Varatharajulu (2) and also Refs. (3)-(8)) to two other cases of interest. We will namely consider an elastic obstacle in a fluid (6) and a smooth elastic obstacle in an elastic medium (8). These cases require the same type of modifications in the theory and are thus conveniently treated together.

The usually employed boundary conditions (the "welded" contact) in elastodynamics require the displacements and surface tractions to be continuous across a boundary between two elastic bodies. The second of these conditions is necessary as long as the bodies keep contact, since otherwise infinite accelerations would arise at the boundary. However, the continuity of the displacements is a condition not always satisfied in practice. Friction, slip and cavitation are phenomena which at least to some degree are present in most cases. The problem with models of these phenomena is that they lead to non-linearities in general. There are two limiting cases, however, that give linear boundary conditions. If the coefficient of friction is infinite we have the welded contact and if it is zero we have a perfectly smooth boundary. The boundary conditions for the latter are continuity of the normal components of the displacements and the surface tractions and that the tangential surface tractions vanish:

$$\hat{n} \cdot \vec{u}^{0} = \hat{n} \cdot \vec{u}^{1} \tag{1}$$

$$\hat{n} \cdot \vec{t}^{0} = \hat{n} \cdot \vec{t}^{1} \tag{2}$$

$$\vec{t}^{0}_{tan} = 0 \tag{3}$$

$$\vec{t}^{1}_{tan} = 0 \tag{4}$$

where $\vec{u}^j$, j=0,1, are the displacements, $\vec{t}^j$ are the surface tractions and n is the normal to the boundary. If we compare these boundary conditions with those for the boundary between a fluid and an elastic medium we see that (taking medium 0 as the fluid) we still have Eqs. (1), (2) and (4). No further conditions are needed as the fluid has fewer degrees of freedom.

Taking medium 1 to be the bounded elastic obstacle we realize that it is the vanishing of the inner tangential surface traction, Eq. (4), that is the root of our difficulties if we try to use the usual approach of the T matrix method. As in this method all the boundary conditions have to be applied in the integral representations, the natural way to solve this difficulty is to modify the method and use also the inner integral representation (the integral representation for the obstacle).

In the following we will give methods to solve the two above-mentioned problems. The algebraic structures of these solutions are more complicated than those of the usual T matrix approach, as is apparent on comparison with the solution for the welded contact. Finally we will give a few numerical results comparing the welded and smooth obstacles and also an elastic body in a fluid.

II. THEORY

In this section the transition matrices for a smooth elastic obstacle in an elastic media and an elastic obstacle in a fluid will be developed. The two derivations will in fact be given in parallel as they closely resemble each other. As they can also be found elsewhere (Refs. (6) and (8), see also Ref. (2) for more details concerning some of the steps) only the main steps will be indicated.

We thus consider a bounded elastic obstacle with density ρ_1 and Lamé parameters λ_1 and μ_1. The surrounding medium is either a fluid (parameters ρ_0 and λ_0) or an elastic medium (parameters ρ_0, λ_0 and μ_0). The surface S of the obstacle is assumed to satisfy the requirements of the divergence theorem and to contain an origin such that the radius to a point on S is a continuous one-valued function of the spherical angles.

Only scattering of stationary waves is considered, so the common time factor $\exp(-i\omega t)$ can be suppressed. Introducing the longitudinal wave numbers $k_j^2=\rho_j\omega^2/(\lambda_j+2\mu_j)$, j=1,2, and the transverse wave numbers $\kappa_j^2=\rho_j\omega^2/\mu_j$ (κ_0 does not exist for the fluid, but should be put equal to infinity in the equation of motion) the equations of motion are

$$(1/k_j^2)\nabla\nabla\cdot\vec{u}^j - (1/\kappa_j^2)\nabla \times \nabla \times \vec{u}^j + \vec{u}^j = 0 \tag{5}$$

These equations will not be employed directly, but instead the following integral representations will be the starting point:

$$\vec{u}^i(\vec{r}) + \kappa_0/\mu_0 \int_S \{\vec{u}^o_+(\vec{r}') \cdot (n \cdot \bar{\Sigma}_o(\vec{r},\vec{r}')) -$$
$$\vec{t}^o_+(\vec{r}') \cdot \bar{G}_o(\vec{r},\vec{r}')\}dS'$$

$$= \begin{cases} \vec{u}\,(\vec{r}) & \vec{r} \text{ outside S} \\ 0 & \vec{r} \text{ inside S} \end{cases} \tag{6}$$

$$- \kappa_1/\mu_1 \int_S \{ \vec{u}_-^1(\vec{r}\,') \cdot (\hat{n} \cdot \overline{\Sigma}_1(\vec{r},\vec{r}\,')) - \vec{t}_-^1(r') \cdot \overline{G}_1(\vec{r},\vec{r}\,') \} \, ds^1 =$$

$$= \begin{cases} \vec{u}^1(\vec{r}) & \vec{r} \text{ inside S} \\ 0 & \vec{r} \text{ outside S.} \end{cases} \tag{7}$$

When the surrounding medium is a fluid the factor in front of the integral in Eq. (6) is k_0/λ_0 instead of κ_0/μ_0. $\hat{n}$ is the outward pointing normal to S, and a plus (minus) sign on $\vec{u}^j$ or $\vec{t}^j$ denotes the limiting value on S taken from the outside (inside). The surface tractions are

$$\vec{t}^j(\vec{u}^j(\vec{r})) = \lambda_j \hat{n} \nabla \cdot \vec{u}^j + 2\mu_j \frac{\partial}{\partial n} \vec{u}^j + \mu_j \hat{n} \times (\nabla \times \vec{u}^j) \qquad (j = 0,1) \tag{8}$$

where the last two terms vanish for the fluid. $\vec{u}^i$ is the (known) incoming wave and $\vec{u}^s = \vec{u}^o - \vec{u}^i$ is the (sought) scattered wave.

The Green's dyadics satisfy

$$(1/k_j^2) \nabla\nabla \cdot \overline{G}_j(\vec{r},\vec{r}\,') - (1/\kappa_j^2)\, \nabla \times \nabla \times \overline{G}_j(\vec{r},\vec{r}\,') + \overline{G}_j(\vec{r},\vec{r}\,') = \tag{9}$$

$$- (1/\kappa_j^3)\, \overline{I}\delta(\vec{r} - \vec{r}\,') \qquad (j = 0,1)$$

($\overline{I}$ being the unit dyadic) and are further required to satisfy radiation conditions. For the fluid κ_0^{-3} is changed to k_0^{-3}. The Green's stress triadics are

$$\hat{n} \cdot \overline{\Sigma}_j(\vec{r},\vec{r}\,') = \lambda_j \hat{n} \nabla \cdot \overline{G}_j(\vec{r},\vec{r}\,') + 2\mu_j \frac{\partial}{\partial n} \overline{G}_j(\vec{r},\vec{r}\,') \tag{10}$$

$$+ \mu_j \hat{n} \times (\nabla \times \overline{G}_j(\vec{r},\vec{r}\,')) \qquad (j = 0,1)$$

For a derivation and further discussion of the integral representations see Pao and Varatharajulu (9).

We introduce the outgoing spherical partial wave solutions of Eq. (5):

$$\vec{\psi}^j_{1\sigma m\ell}(\vec{r}) = (\ell(\ell+1))^{-1/2} \nabla \times [\vec{r} h_\ell(\kappa_j r) Y_{\sigma m\ell}(\theta,\phi)] \tag{11}$$

$$\vec{\psi}^j_{2\sigma m\ell}(\vec{r}) = (1/\kappa_j) \nabla \times \vec{\psi}^j_{1\sigma m\ell}(\vec{r})$$

$$\vec{\psi}^{j}_{3\sigma m\ell}(\vec{r}) = (k_j/\kappa_j)^{3/2}(1/k\ )\ \nabla\left[h_\ell(k_j r)\ Y_{\sigma m\ell}(\theta,\phi)\right]$$

for j=0,1. h_ℓ is the spherical Hankel function of the first kind and the normalized real spherical harmonics are

$$Y_{\sigma m\ell}(\theta,\phi) = \left[\varepsilon_m \frac{2\ell+1}{4\pi}\frac{(\ell-m)!}{(\ell+m)!}\right]^{1/2} P^m_\ell(\cos\theta) \begin{cases} \cos m\phi & \sigma = e \\ \sin m\phi & \sigma = o \end{cases} \tag{12}$$

where $\varepsilon_o=1$, $\varepsilon_m=2$, m=1,2,... . For the fluid only the last basis functions are needed and the normalizing factor $(k_o/\kappa_o)^{3/2}$ is then removed. In the following we abbreviate the indices: $\vec{\psi}^{j}_{n} \equiv \vec{\psi}^{j}_{\tau\sigma m\ell}$. Also the regular functions $\mathrm{Re}\vec{\psi}^{j}_{n}$, obtained by taking spherical Bessel rather than Hankel functions, are used in the following.

We expand the Green's dyadics

$$\bar{G}_j(\vec{r},\vec{r}\,') = i\sum_n \mathrm{Re}\vec{\psi}^{j}_{n}(\vec{r}_<)\vec{\psi}^{j}_{n}(\vec{r}_>) \qquad (j = 0,1) \tag{13}$$

From Eqs. (6) we then obtain expansions for the scattered and incoming fields

$$\vec{u}^{s} = \sum_n f_n \vec{\psi}^{o}_{n} \tag{14}$$

$$\vec{u}^{i} = \sum_n a_n \mathrm{Re}\vec{\psi}^{o}_{n} \tag{15}$$

valid outside the circumscribed and inside the inscribed sphere of S, respectively. For the fluid the summations are of course only for $\tau=3$. The expansion coefficients are

$$f_n = i\kappa_o/\mu_o \int_S \{\vec{u}^{o}_{+}\cdot \vec{t}^{o}(\mathrm{Re}\vec{\psi}^{o}_{n}) - \vec{t}^{o}_{+}\cdot \mathrm{Re}\vec{\psi}^{o}_{n}\}dS \tag{16}$$

$$a_n = -\ i\kappa_o/\mu_o \int_S \{\vec{u}^{o}_{+}\cdot \vec{t}^{o}(\vec{\psi}^{o}_{n}) - \vec{t}^{o}_{+}\cdot \vec{\psi}^{o}_{n}\}dS \tag{17}$$

(for the fluid κ_o/μ_o is replaced by k_o/λ_o). From the second of Eqs. (7) (the first only determines the interior field) we have in a similar manner

$$0 = i\kappa_1/\mu_1 \int_S \{\vec{u}^{1}_{-}\cdot \vec{t}^{1}(\mathrm{Re}\vec{\psi}^{1}_{n}) - \vec{t}^{1}_{-}\cdot \mathrm{Re}\vec{\psi}^{1}_{n}\}\ dS \tag{18}$$

Between Eqs. (16)-(18) we now have to eliminate the surface fields to obtain f_n in terms of a_n, i.e. to obtain the transition matrix T defined by

$$\vec{f} = T\vec{a} \tag{19}$$

in vector and matrix notation. To do this we apply the boundary conditions and expand both the exterior and interior surface fields

$$\vec{u}^{o}_{+} = \sum_n \beta_n \mathrm{Re}\vec{\psi}^{o}_{n} \tag{20}$$

$$\vec{u}^{1}_{-} = \sum_n \alpha_n \mathrm{Re}\vec{\psi}^{1}_{n} \tag{21}$$

The completeness of these kinds of expansions has been discussed by Millar (10) for the scalar case and by Müller (11) for the tangential vector case (and also originally by Waterman (12)). We note that the convergence is in a mean square sense on S and that the expansion coefficients depend on the truncation (because of the non-orthogonality of the basis functions on S). The inside expansion, Eq. (21), can be differentiated (as may be shown by using Eq. (18) without applying the boundary conditions), whereas the outside expansion, Eq. (20), can only be differentiated in the tangential direction. These facts about differentiability partly motivate the way in which we apply the boundary conditions.

We now consider the elastic body in a fluid and apply the boundary conditions Eqs. (1) and (2) in Eqs. (16) and (17) and Eqs. (1) and (4) in Eq. (18). Inserting Eqs. (20) and (21) we then get

$$\begin{aligned} \vec{f} &= i\mathrm{Re}Q^{1}\vec{\alpha} \\ \vec{a} &= -\,iQ^{1}\vec{\alpha} \\ 0 &= iQ^{2}\vec{\beta} + iQ^{3}\vec{\alpha} \end{aligned} \tag{22}$$

where the matrices are

$$Q^{1}_{nn'} = k_o/\lambda_o \int_S \{\hat{n} \cdot \mathrm{Re}\vec{\psi}^{1}_{n'} \lambda_o \nabla \cdot \vec{\psi}^{o}_{n} - \hat{n} \cdot \vec{t}^{1}(\mathrm{Re}\vec{\psi}^{1}_{n'})\hat{n} \cdot \vec{\psi}^{o}_{n}\} dS \tag{23}$$

$$Q^{2}_{nn'} = \kappa_1/\mu_1 \int_S \hat{n} \cdot \mathrm{Re}\vec{\psi}^{o}_{n'} \hat{n} \cdot \vec{t}^{1}(\mathrm{re}\vec{\psi}^{1}_{n}) dS \tag{24}$$

$$Q^{3}_{nn'} = \kappa_1 \mu_1 \int_S \{\mathrm{Re}\vec{\psi}^{1}_{n'\,\mathrm{tan}} \cdot \vec{t}^{1}(\mathrm{Re}\vec{\psi}^{1}_{n}) - \hat{n} \cdot \vec{t}^{1}(\mathrm{Re}\vec{\psi}^{1}_{n'})\hat{n} \cdot \mathrm{Re}\vec{\psi}^{1}_{n}\} dS \tag{25}$$

and $\mathrm{Re}Q^{1}$ is obtained from Q^{1} by taking regular functions in both places. We note that Q^{1} and Q^{2} are not "square", and hence the first two of Eqs. (22) are not enough to obtain the transition matrix. Solving Eqs. (22) we get

$$T = -\mathrm{Re}QQ^{-1} \tag{26}$$

where

$$Q = Q^1(Q^3)^{-1}Q^2 \tag{27}$$

Turning to the smooth elastic obstacle we apply the boundary conditions Eqs. (2) and (3) in Eqs. (16) and (17) and Eqs. (1) and (4) in Eq. (18). Inserting Eqs. (20) and (21) we get

$$\begin{aligned} \vec{f} &= i\mathrm{Re}Q^4\vec{\alpha} + i\mathrm{Re}Q^5\vec{\beta} \\ \vec{a} &= -iQ^4\vec{\alpha} - iQ^5\vec{\beta} \\ 0 &= iQ^2\vec{\beta} + iQ^3\vec{\alpha} \end{aligned} \tag{28}$$

where

$$Q^4_{nn'} = -\kappa_0/\mu_0 \int_S \hat{n}\cdot\vec{t}^1(\mathrm{Re}\vec{\psi}^1_{n'})\,\hat{n}\cdot\vec{\psi}^0_n dS \tag{29}$$

$$Q^5_{nn'} = \kappa_0/\mu_0 \int_S \mathrm{Re}\vec{\psi}^0_{n'}\cdot\vec{t}^0(\vec{\psi}^0_n)\,dS \tag{30}$$

and Q^2 and Q^3 are given by Eqs. (24) and (25) (but n' in Q^2 this time ranges over $\tau'=1,2,3$). Solving Eqs. (28) we still have Eq. (26) but with

$$Q = -Q^4(Q^3)^{-1}Q^2 + Q^5 \tag{31}$$

Comparing the Q matrices for an elastic body in a fluid, Eq. (27), with that for a smooth elastic obstacle, Eq. (31), we see that they have a similar structure. In fact we can obtain Q for an elastic body in a fluid in exactly the form of Eq. (31), see Ref. 6. We further note that it is important that we obtain the transition matrix in the form in Eq. (26), as we can then employ the method of Waterman (1) rather than invert Q directly (Q is ill-conditioned for direct inversion as it has elements in the lower triangular part increasing rapidly with order). The inversion of Q^3 in Eqs. (27) and (31), on the other hand, can be done directly, because Q^3 is a real and well-behaved matrix with the elements diminishing as the indices grow.

For a welded elastic obstacle the transition matrix is still given by Eq. (26), but for this case Q is given by (see Refs. 1 and 2)

$$Q_{nn'} = \kappa_0/\mu_0 \int_S \{\mathrm{Re}\vec{\psi}^1_{n'}\cdot\vec{t}^0(\vec{\psi}^0_n) - \vec{t}^1(\mathrm{Re}\vec{\psi}^1_{n'})\cdot\vec{\psi}^0_n\}dS \tag{32}$$

Compared with the two cases treated above this is thus a simpler case as we have both fewer integrations and fewer matrix manipulations to perform.

III. NUMERICAL APPLICATIONS AND DISCUSSION

In this section we will give a few applications of the foregoing formalism. As details of the numerical treatment can be found elsewhere (see Refs. (6) and (8))

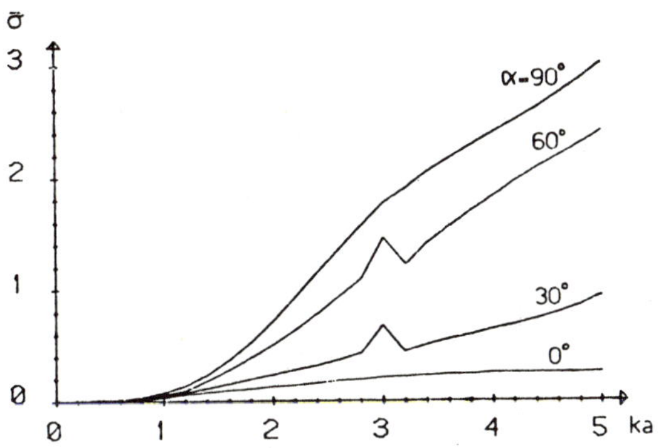

Fig. 1. The normalized total cross section vs. k_0a for an elastic spheroid in a fluid and four directions of incidence.

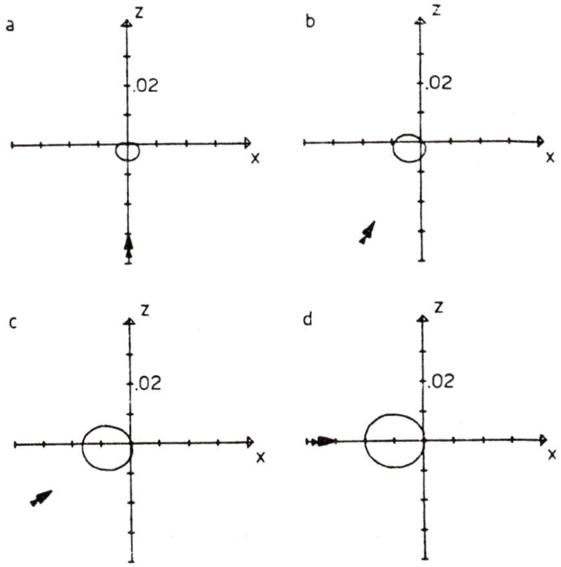

Fig. 2. The normalized differential cross section for an elastic spheroid in a fluid and four directions of incidence.

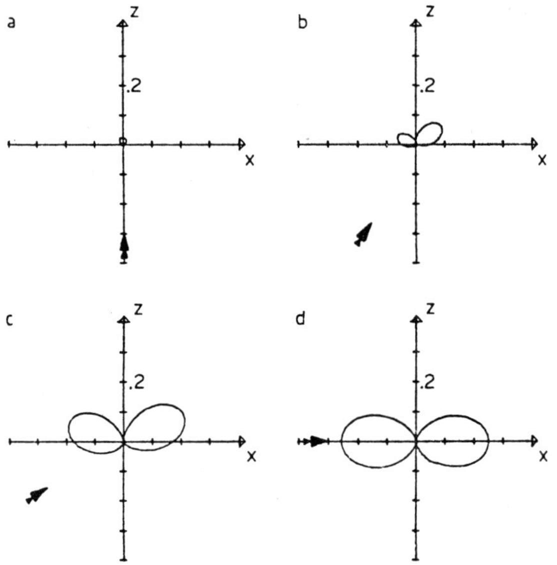

Fig. 3. The normalized longitudinal differential cross section for a smooth spheroid and four directions of incidence.

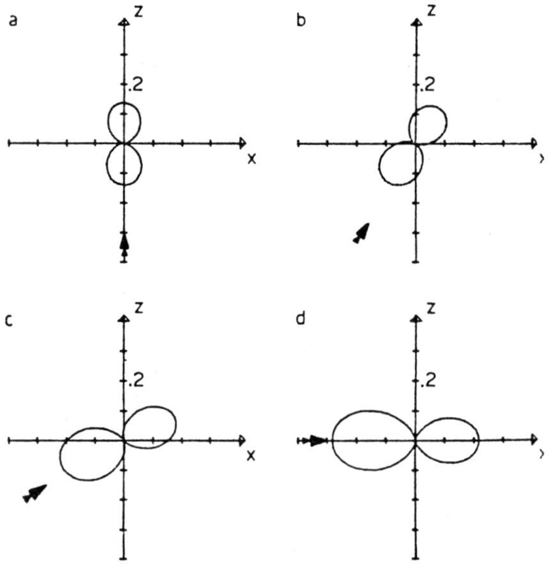

Fig. 4. The normalized longitudinal differential cross section for a welded spheroid and four directions of incidence.

we will be very brief.

We only consider an obstacle which is a prolate spheroid with main axis a along the z-axis and axis b in the xy-plane. The entities of physical interest we compute are cross sections and we normalize these with πb^2. The incoming wave is taken to be a longitudinal (P) plane wave propagating in the xz-plane making an angle α with the z-axis.

An elastic body in a fluid is first treated; we choose the material parameters to correspond to stainless steel in water, thus $k_1/k_o=0.3$, $\rho_1/\rho_o=7.9$ and $\nu_1=0.3$ (ν_1 is the Poisson ratio of the obstacle). The normalized total cross section $\bar{\sigma}$ for a prolate spheroid with b/a=0.333 is shown in Fig. 1 as a function of k_oa for $k_oa<5$, and for four different directions of the incoming wave, $\alpha=0^o$, 30°, 60° and 90°. In Fig. 2 we show the normalized differential cross section $\frac{d\bar{\sigma}}{d\Omega}$ for b/a=0.333 and $k_oa=1$ and four directions of incidence.

For a smooth obstacle we choose material parameters as close as possible to those above. The obstacle is still made of stainless steel and the surrounding medium is a fictitious elastic material with the same density and longitudinal wave speed as water. We thus choose $k_1/k_o=0.3$, $\rho_1/\rho_o=7.9$, $\nu_o=0.3$ and $\nu_1=0.3$, In Fig. 3 the normalized longitudinal differential cross section is shown for b/a=0.333 and $k_oa=1$ and four directions of incidence, and Fig. 4 is an analogous plot for the welded contact.

Comparing Figs. 2-4 we immediately notice that the elastic body in a fluid behaves in a markedly different manner compared to both the welded and smooth elastic obstacles. The two latter, on the other hand, show at least some qualitatively similar characteristics, see Ref. (8) for more examples of this. It seems, however, difficult to make some general remarks concerning the differences between the welded and smooth obstacles. Possibly, there is a tendency for the smooth obstacle to show more violent behaviour (thus the cross section in Fig. 3 is more dependent on the direction of incidence than in Fig. 4).

Finally we remark that more complex situations can also be treated. Using the results in Ref. (13) and (7) we can thus consider scattering by two or several obstacles, whereas the scattering from a layered obstacle (smooth or surrounded by a fluid) will require some modifications of the work in Ref. (7). It is also straightforward to treat the case with an elastic body in a fluid halfspace by employing the results of Kristensson and Ström (14).

ACKNOWLEDGEMENTS

The present work is supported by the National Swedish Board for Technical Development (STU) and this is gratefully acknowledged.

REFERENCES

1. P.C. Waterman, Matrix theory of elastic wave scattering, J. Acoust. Soc. Am. 60, 567-580 (1976).

2. V. Varatharajulu and Y.-H. Pao, Scattering matrix for elastic waves. I. Theory, J. Acoust. Soc. Am. 60, 556-566 (1976).

3. V.V. Varadan, Scattering matrix for elastic waves. II. Application to elliptic cylinders, J. Acoust. Soc. Am. 63, 1014-1024 (1978).

4. P.C. Waterman, Matrix theory of elastic wave scattering. II. A new conservation law, J. Acoust. Soc. Am. 63, 1320-1325 (1978).

5. Y-H. Pao, Bettis identity and transition matrix for elastic waves, J. Acoust. Soc. Am. 64, 302-310 (1978).

6. A. Boström, Scattering of stationary acoustic waves by an elastic obstacle immersed in a fluid, Rep. 78-43, Inst. Theoretical Physics, Göteborg (1978).

7. A. Boström, Multiple-scattering of elastic waves by bounded obstacles, Rep. 79-4, Inst. Theoretical Physics, Göteborg (1979).

8. A. Boström, Scattering by a smooth elastic obstacle, Rep. 79-14, Inst. Theoretical Physics, Göteborg (1979).

9. Y-H. Pao and V. Varatharajulu, Huygens' principle, radiation conditions, and integral formulas for scattering of elastic waves, J. Acoust. Soc. Am. 59, 1361-1371 (1976).

10. R.F. Millar, The Rayleigh hypothesis and a related least-squares solution to scattering problems for periodic surfaces and other scatterers, Radio Sci. 785-796 (1973).

11. C. Müller, Boundary values and diffraction problems, Symposia Mathematica XVIII, 353-367 (Academic Press, 1976).

12. P.C. Waterman, New Formulation of Acoustic Scattering, J. Acoust. Soc. Am. 45, 1417-1429 (1969).

13. B. Peterson and S. Ström, Matrix formulation of acoustic scattering from an arbitrary number of scatterers, J. Acoust. Soc. Am. 56, 771-780 (1974).

14. G. Kristensson and S. Ström, Scattering from buried inhomogeneities - a general three-dimensional formalism, J. Acoust. Soc. Am. 64, 917-936 (1978).

Part 2
T-Matrix Approach
Multiple Scattering

MULTIPLE SCATTERING OF ACOUSTIC, ELECTROMAGNETIC AND ELASTIC WAVES

V.K. Varadan
Wave Propagation Group
Department of Engineering Mechanics and Atmospheric Sciences Program
The Ohio State University, Columbus, Ohio 43210

ABSTRACT

In this article we present a multiple scattering analysis of the coherent wave propagation through an inhomogeneous medium consisting of either random or periodic distribution of scatterers of arbitrary shape. Both specific and random orientations of the scatterers are considered. The mathematical unity inherently present in the T-matrix formalism for the three wave fields, namely acoustic, electromagnetic and elastic, is employed in conjunction with suitable averaging procedures to formulate a self-consistent multiple scattering theory. For a random distribution of scatterers we use a configurational averaging procedure, while for a periodic distribution, we use a suitable lattice sum based on crystallographic theory. The information about the orientation of the scatterers has been incorporated into the T-matrix of the scatterer itself thus making formalism a convenient computational scheme to study the anisotropic effects in an inhomogeneous medium. The statistically averaged equations obtained by the analysis are then solved by using Lax's quasicrystalline approximation to obtain the bulk or effective properties of the medium. Numerical results are presented for propagation speeds, attenuation and frequency dependent elastic properties for a range of frequencies to demonstrate the broad applicability of the T-matrix method.

I. INTRODUCTION

The subject of wave propagation and scattering in inhomogeneous media has become increasingly important in many fields of engineering and science, for e.g., in studies of bubbles in a fluid, distribution of flaws in solids (NDE), ionospheric irregularities, geophysical exploration, artificial dielectrics, millimeter wave propagation in ocean fogs and mists, and through rain, cloud and hail particles, underwater signal transmission, porous media, composites, attenuation of waves in biological media, remote sensing, etc. All of these problems are characterized by a suitable statistical description of the media. The similarity in statistical descriptions and the mathematical unity present in the description of all three classical wave fields, namely acoustic, electromagnetic and elastic, may be taken into consideration in formulating a unifying approach to all these problems. In this article, we present one such unifying approach based on the T-matrix or the null field method that has been recently formulated for all three fields and promises to be an efficient computational scheme to analyze the scattering of waves from several different geometries.

At any point in a random medium, the total wave fields can be considered as a sum of two components, viz, a coherent or average wave (which is the statistical average over all possible configurations of the scatterers with regard to location and state) and an incoherent component due to changes in scatterer positions and states from configuration to configuration. The averages of the square of magnitude of the coherent and incoherent fields are called the coherent and incoherent intensities, respectively. The total intensity is equal to the sum of coherent intensity and incoherent intensity. For a plane wave incident on a medium containing random particles, the coherent intensity attenuates due to scattering and absorption. Incoherent scattering effects introduce 'noise' into the system and cause fluctuations in the coherent received amplitude and phase. In radar meteorological applications it is important to assess the incoherent scattered intensity relative to the total intensity in order to relate theoretical and experimental results (e.g., power law relations between attenuation and rainfall rate). Propagation of the coherent wave is generally expressed in terms of a bulk propagation coefficient characterizing the scatterer filled medium. Incoherent effects are usually determined by solving 'approximate' integral equations or by solving special forms of the radiative transfer equations. Such formulations are generally valid under conditions of sparse concentrations and/or weak multiple scattering for either Rayleigh scatterers or large scatterers which scatter primarily in the forward direction. Incoherent scattering is beyond the scope of this present paper, and those who are interested in such analysis may refer to Ishimaru (1). Here, our attention focuses on formulating a multiple scattering theory for studying coherent wave propagation through an inhomogeneous medium consisting of either random or periodic distribution of scatterers of arbitrary shape.

Several theories and models on diffraction and scattering of waves have been pursued ever since Rayleigh's (2) analytical treatment of scattering of waves by randomly distributed particle to study the color of the sky. We cite here papers that are related to our present analysis (3-21). The limitations, difficulties and advantages of various approaches are discussed in a recent review article (3). We also refer to (1) and the references cited therein. Since scattering theory starts with discrete ensemble of inhomogeneities before statistical averaging is carried out, the specific geometry and orientation of each scatterer can be easily incorporated into the formulation. This is not possible with continuum theories. The additional advantage of the scattering theory approach is that it is the counterpart of actual experiments performed for coherent wave propagation in heterogeneous media using elastic, electromagnetic or acoustic waves as appropriate.

In multiple scattering theories, approximations are usually made for the geometry and size of the scatterer relative to the wavelength of incident waves, and the distribution of scatterers in the medium. The approximations with respect to geometry and size of the scatterers are related. If the scatterers are small compared to the incident wavelength, one usually obtains the gross scattering properties of the medium. This is the so-called Rayleigh or low frequency limit, and yields corrections to point scatterers. As far as the distribution of the scatterers is concerned, we either have regular arrays of scatterers or a random distribution. In the first case, we employ suitable lattice sum while in the latter case, we use a configuration averaging procedure. If the scatterers are sparsely distributed, (i.e., the concentration is small) we may employ a single scattering or first Born approximation.

Most of the previous computational results using scattering theory are, however, often confined to the Rayleigh or low frequency limit, a specific angle of incidence, single scattering approximation and simple geometries such as circular cylinders and spheres. Our methodology which employs the T-matrix approach and suitable statistical averaging procedure, is completely general and can be used for multiple scattering analyses of acoustic, electromagnetic and elastic waves emphasizing the inherent unity of the approach. The analysis provides a powerful computational scheme for determining the coherent field characteristics of a medium containing arbitrary shaped scatterers over a wide range of frequencies.

We consider N number of discrete scatterers of arbitrary shape either randomly or periodically distributed. Both parallel and random orientations of the scatterers are considered. When N is very large (such that the number of scatterers per unit volume is finite), the T-matrix formalism is combined with a statistical analysis to provide a self-consistent multiple scattering theory. For a random distribution of scatterers we use a configurational averaging procedure, while for a periodic distribution we use a suitable lattice sum based on crystallographic theory. Inhomogeneous media in which scatterers have a specific orientation exhibit anisotropy and polarization effects. These effects are, however, nullified for the case of randomly oriented scatterers. The information about the orientation of the scatterers has been incorporated (in terms of a rotation matrix consisting of Euler angles) into the T-matrix of the scatterer itself thus making formalism a convenient computational scheme to study the anisotropic effects in an inhomogeneous medium. The statistically averaged equations are then solved using Lax's quasicrystalline approximation to yield the propagation characteristics of the coherent waves in the medium. Analytical results are obtained for the dispersion relation in the Rayleigh or low-frequency limit for both 2- and 3-dimensional scatterer geometries. Numerical results are presented for propagation speeds, attenuation and frequency dependent properties at higher frequencies to demonstrate the broad applicability of the T-matrix approach. The approach presented in this article has been successfully employed to many practical scattering problems on acoustic, electromagnetic and elastic wave fields (22-29).

II. T-MATRIX FORMULATION OF MULTIPLE SCATTERING

We consider N number of arbitrary shaped scatterers with a smooth surface S which are referred to a coordinate system as shown in Fig. 1. The scatterers may either be randomly distributed or periodically arranged. The orientation of the scatterers may be quite general. Here, we consider both specific and random orientations of the scatterers. In Fig. 1, O_i and O_j refer to the centers of the i-th and j-th scatterers, respectively and they are referred to the origin O by the spherical polar coordinates (r_i, θ_i, ϕ_i). For two dimensional problems, we use cylindrical polar coordinates (r_i, θ_i). $\vec{r}_{ij}$ refers to the vector connecting O_i and O_j. P is any point in the medium outside the scatterers (which we call the matrix medium).

We now describe the medium and the scatterers for all three wave fields. For an acoustic problem, we consider fluid scatterers immersed in another fluid, bubbles in a fluid, elastic or viscoelastic scatterers immersed in a fluid, etc. For an electromagnetic scattering problem, we consider dielectric scatterers in free space, dielectric scatterers embedded in a different dielectric medium, etc. For an elastic wave scattering problem, we consider elastic or viscoelastic inclusions embedded in another elastic or viscoelastic material, stress free or fluid filled cavities and cracks in an elastic or viscoelastic material, etc. The properties of the medium and the scatterers are given in terms of Lamé constants λ, μ and density ρ for an elastic material, compressibility λ_f and density ρ_f for non-viscous fluids

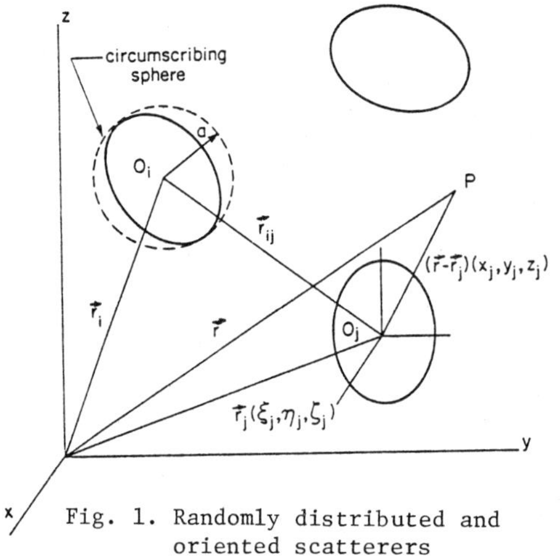

Fig. 1. Randomly distributed and oriented scatterers

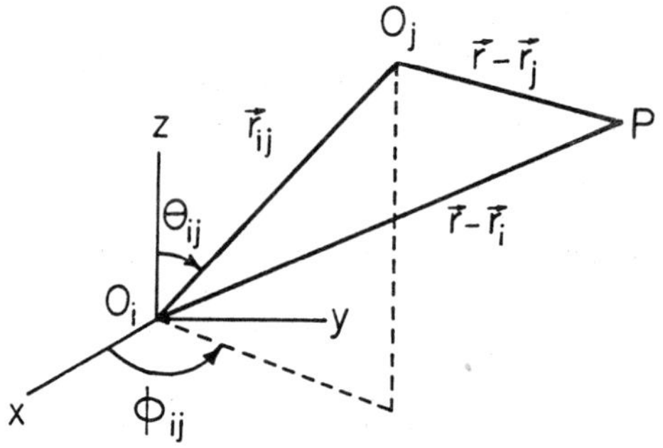

Fig. 2. Translation of the coordinate system of the j-th and i-th scatterers.

and relative dielectric constant ε_r and permeability μ_r with respect to free space describing dielectric medium. We use subscript 1 to denote these qualities inside the scatterers.

A time harmonic plane wave of unit amplitude and frequency ω is incident on the medium such that the direction of propagation of the incident waves is along the z-axis. The incident wave field may then be represented by

$$\vec{u}^{o} = e^{i(k_p z - \omega t)}\hat{z} + e^{i(k_s z - \omega t)}\hat{x} \tag{2.1}$$

where k_p and k_s are the compressional (longitudinal) and transverse (shear) wave numbers, respectively, and t is the time. The acoustic waves are purely compressional type and thus, the second term of Eq. (2.1) is set equal to zero; for electromagnetic waves which are transverse type, the first term on the right hand side of (2.1) is zero; for elastic waves which contain both compressional and transverse types, all the term of (2.1) are present. For acoustic and elastic wave problems, $\vec{u}^{o}$ refers to the incident displacement field vector, while for electromagnetic case, it refers to the incident electric field vector. In Eq. (2.1), we use the superscript (o) to indicate an incident wave. The wave numbers k_p and k_s are given by

$$k_p = \omega/c_p \quad ; \quad k_s = \omega/c_s \ , \tag{2.2}$$

respectively, where

$$c_p = \sqrt{(\lambda + 2\mu)/\rho} \qquad \text{(elastic waves)} \tag{2.3}$$

$$= \sqrt{\lambda_f/\rho_f} \qquad \text{(acoustic waves in fluids)} \tag{2.4}$$

$$c_s = \sqrt{\mu / \rho} \qquad \text{(elastic waves)} \tag{2.5}$$

$$= C/\sqrt{\mu_r \varepsilon_r} \qquad \text{(electromagnetic waves)} \tag{2.6}$$

In Eqs. (2.3) - (2.6), c_p and c_s refer to compressional and transverse wave velocities, respectively. In Eq.(6), C refers to the velocity of light in free space. The corresponding quantities inside the scatterers are differentiated by subscript 1. For brevity, we use the notation k_τ for the wave numbers ; $\tau = 1$ corresponds to compressional wave while $\tau = 2,3$ corresponds to shear wave.

When the wave impinges on a scatterer, part of the incident wave is scattered back into the matrix (medium outside the scatterer) and the rest is refracted into the scatterers and transmitted back into the matrix.We represent the scattered field by $\vec{u}^s(\vec{r})$ and the total field inside the scatterers by $\vec{u}^t(\vec{r})$. Since the incident wave has time dependence given by exp(-iωt), all these field quantities have the same time factor which will be suppressed henceforth. Both these fields and the incident wave field satisfy the vector Helmholtz equation, see for example (30). Even though for acoustic wave propagation in fluids one could work with velocity potential and scalar Helmholtz equation, we prefer to use the vector displacement vector field and the vector Helmholtz equation for the reasons mentioned in our previous papers (20,31,32). The problem at hand reduces to computing the total wave field at any point in the matrix satisfying the appropriate boundary condition on the surface of the scatterers and radiation conditions at infinity.

The total field at any point in the matrix (outside the scatterers) is the sum of the incident field and the fields scattered by all the scatterers. This is written as

$$\vec{u}(\vec{r}) = \vec{u}^o(\vec{r}) + \sum_{i=1}^{N} \vec{u}_i^s(\vec{r} - \vec{r}_i) \tag{2.7}$$

where $\vec{u}_i^s(\vec{r} - \vec{r}_i)$ is the field scattered by the i-th scatterer to the point of observation $\vec{r}$. The field that excites the i-th scatterer is the incident field $\vec{u}^o$ plus the fields scattered from all other scatterers. The exciting-field term $\vec{u}^e$ is used to distinguish between the field actually incident on a scatterer and the external incident field $\vec{u}^o$ produced by a source at infinity. Thus, at a point $\vec{r}$ in the vicinity of the i-th scatterer, we write

$$\vec{u}_i^e(\vec{r}) = \vec{u}^o(\vec{r}) + \sum_{j\neq i}^{N} \vec{u}_j^s(\vec{r} - \vec{r}_j) \quad , \quad a \leq |\vec{r} - \vec{r}_i| \leq 2a \tag{2.8}$$

where 'a' is the radius of the imaginary sphere circumscribing a scatterer. In this analysis, we have assumed that there is no interpenetration of the imaginary spheres of radius 'a' which circumscribe each scatterer.

The T-matrix formalism of scattering we employ here is based on the extended boundary condition method which has been discussed by many authors in this book (33). This formalism differs from the eigen function expansion technique in that the same basis sets namely, the vector spherical wave functions for 3-D problems and the vector cylindrical wave functions for 2-D problems, may be used for scatterers of any closed boundary S with a continuous turning normal. The vector fields are expanded in terms of a complete set of basis functions which form solutions to the vector Helmholtz equation and are given by

$$\vec{\psi}_{1\sigma mn}(\vec{r}) = \left(k_p/k_s\right)^{\frac{1}{2}} \xi_{mn} \nabla \left[h_n(k_p r) \, p_n^m(\cos\theta) \begin{pmatrix} \cos m\phi \, ; \sigma = 1 \\ \sin m\phi \, ; \sigma = 2 \end{pmatrix} \right] \tag{2.9}$$

$$\vec{\psi}_{2\sigma mn}(\vec{r}) = k_s \, \eta_{mn} \nabla \times \left[\vec{r} \; h_n(k_s r) \; p_n^m(\cos\theta) \begin{pmatrix} \cos m\phi \, ; \sigma = 1 \\ \sin m\phi \, ; \sigma = 2 \end{pmatrix} \right] \tag{2.10}$$

$$\vec{\psi}_{3\sigma mn}(\vec{r}) = (1/k_s) \, \nabla \times \vec{\psi}_{2\sigma mn}(\vec{r}) \tag{2.11}$$

where

$$\xi_{mn} = \left[\varepsilon_n \frac{(2n+1)\,(n-m)!}{4\pi(n+m)!} \right]^{\frac{1}{2}}$$

$$\eta_{mn} = \left[\frac{\varepsilon_n (2n+1)\,(n-m)!}{4\pi\, n(n+1)\,(n+m)!} \right]^{\frac{1}{2}} \tag{2.12}$$

with $\varepsilon_n = 1$ for $n = 0$ and $\varepsilon_n = 2$ for $n > 0$.

For brevity, we abbreviate these vector basis functions as $\vec{\psi}_{1\sigma mn}$; $\vec{\psi}_{2\sigma mn}$; $\vec{\psi}_{3\sigma mn} = \vec{\psi}_{\tau n}$; $\tau = 1,2,3$. For acoustic wave problems, we have only $\vec{\psi}_{1n}$; for electromagnetic waves, we have $\vec{\psi}_{2n}$ and $\vec{\psi}_{3n}$; for elastic waves, we have all three of them. In Eqs. (2.9 - 2.11), we have used spherical polar coordinates r,θ,ϕ with the origin of the coordinate system inside S, $h_n(1)$ are the spherical Hankel functions of order n, p_n^m are the associated Legendre polynomials and m is an integer that takes values $0, 1, 2, \ldots, n$; $n = 0, 1, 2, \ldots,\infty$ for $\sigma = 1$ and $n = 1,2,3,\ldots,\infty$ for $\sigma = 2$. Field quantities that are regular at the origin are expanded in terms of the regular (Re) basis set (Re $\vec{\psi}_{\tau mn}$) obtained by replacing h_n in the above equations by j_n, the spherical Bessel functions of the first kind of order n. For two dimensional problems, we employ cylindrical basis functions, see, for example (31).

We now expand the exciting and scattered fields in terms of basis functions Re $\vec{\psi}_{\tau n}$ and $\vec{\psi}_{\tau n}$, respectively :

$$u_i^e(\vec{r}) = \sum_\tau \sum_{n=0}^{\infty} \sum_{\ell=-n}^{n} \sum_{\sigma=1}^{2} b^i_{\tau n \ell \sigma} \, \mathrm{Re}\, \vec{\psi}^i_{\tau n \ell \sigma}(\vec{r} - \vec{r}_i)$$

$$= \sum_{\tau n} b^{\ell(i)}_{\tau n} \mathrm{Re}\, \vec{\psi}^i_{\tau n \ell} \equiv \sum_{\tau n} b^i_{\tau n} \, \mathrm{Re}\, \vec{\psi}^i_{\tau n} \tag{2.13}$$

$$\vec{u}^s_j(\vec{r}) = \sum_{\tau n \ell} B^{\ell(j)}_{\tau n} \vec{\psi}^j_{\tau n \ell} \equiv \sum_{\tau n} B^j_{\tau n} \vec{\psi}^j_{\tau n} \tag{2.14}$$

where the superscripts i or j on the basis functions refer to expansions with respect to O_i and O_j, respectively, and $B^j_{\tau n}$ and $b^i_{\tau n}$ are unknown coefficients to be evaluated. The choice of basis set in (2.14) satisfies the radiation condition at infinity for the scattered field, while the choice in (2.13) satisfies the regularity of the exciting field in the region $0<|\vec{r} - \vec{r}_i|<2a$.

Substituting (2.13) and (2.14) in (2.8), we obtain

$$\sum_{\tau n} b^i_{\tau n} \, \mathrm{Re}\, \vec{\psi}^i_{\tau n} = \vec{u}^o(\vec{r} - \vec{r}_i) + \sum_{j\neq i}^{N} \sum_{\tau n} B^j_{\tau n} \, \vec{\psi}^j_{\tau n} \tag{2.15}$$

It can be seen that the second series on the right-hand side of (2.15) are expressed with respect to the center of the j-th scatterer. In order to express them with respect to the i-th center, we employ the addition theorems of vector spherical harmonics for 3-D problems and of Bessel functions for 2-D problems. This translation has been discussed in detail in (34,35) and employed for acoustic, elctromagnetic and elastic wave problems (22-29). For the sake of uniformity, we reproduce the essential equations which translate the basis function from j-th center to the i-th center :

$$\begin{aligned}
\vec{\psi}^j_{1n} &= \vec{\psi}^j_{1n\ell} = \sum_{\nu=0}^{\infty} \sum_{\mu=-\nu}^{\nu} A^{n\ell}_{\mu\nu} \, \mathrm{Re}\, \vec{\psi}^i_{1\nu\mu} \\
\vec{\psi}^j_{2n} &= \vec{\psi}^j_{2n\ell} = \sum_{\nu=0}^{\infty} \sum_{\mu=-\nu}^{\nu} \left[B^{n\ell}_{\mu\nu} \, \mathrm{Re}\, \vec{\psi}^i_{2\mu\nu} + C^{n\ell}_{\mu\nu} \, \mathrm{Re}\, \vec{\psi}^i_{3\mu\nu} \right] \\
\vec{\psi}^j_{3n} &= \vec{\psi}^j_{3n\ell} = \sum_{\nu=0}^{\infty} \sum_{\mu=-\nu}^{\nu} \left[C^{n\ell}_{\mu\nu} \, \mathrm{Re}\, \vec{\psi}^i_{2\mu\nu} + B^{n\ell}_{\mu\nu} \, \mathrm{Re}\, \vec{\psi}^i_{3\mu\nu} \right]
\end{aligned} \tag{2.16}$$

where

$$\begin{aligned}
A^{n\ell}_{\mu\nu} &= \sum_q (-1)^{\mu} i^{\nu+q-\ell} (2\nu+1) a(n,\ell|-\mu,\nu|q) h_q(k_p r_{ij}) P^{n-\mu}_q (\cos\theta_{ij}) e^{i(n-\mu)\phi_{ij}} \\
B^{n\ell}_{\mu\nu} &= \sum_q (-1)^{\mu} i^{\nu+q-\ell} a(\ell,\nu,q) a(n,\ell|-\mu,\nu|q) h_q(k_s r_{ij}) P^{n-\mu}_q (\cos\theta_{ij}) e^{i(n-\mu)\phi_{ij}} \\
C^{n\ell}_{\mu\nu} &= \sum_q (-1)^{\mu+1} i^{\nu+q-\ell} b(\ell,\nu,q) a(n,\ell|-\mu,\nu|q,q-1) h_q(k_s r_{ij}) P^{n-\mu}_q (\cos\theta_{ij}) \\
&\qquad e^{i(n-\mu)\phi_{ij}}
\end{aligned} \tag{2.17}$$

Expressions for $a(\ell,\nu,q)$, $b(\ell,\nu,q)$, $a(n,\ell|-\mu,\nu|q)$ and $a(n,\ell|-\mu,\nu|q,q-1)$ are given by Cruzan (34) and Fig. 2 depicts the geometry of the translation between the j-th and i-th scatterers. In Eq. (2.16), for acoustic wave scattering, we keep only $\vec{\psi}^j_{1n}$; for electromagnetic wave scattering, we keep $\vec{\psi}^j_{2n}$ and $\vec{\psi}^j_{3n}$; for elastic wave scattering, we need all three of them. For two dimensional problems, the translation is given by

$$\psi^j_{\tau n} = \psi_{\tau n}(\vec{r}-\vec{r}_j) = (-1)^n \sum_m (-1)^m \operatorname{Re}\psi_{\tau m}(\vec{r}-\vec{r}_i)\,\psi_{\tau,n-m}(\vec{r}_i-\vec{r}_j) \quad (2.18)$$

which contain cylindrical Hankel and Bessel functions. For details, we refer to (22, 26, 36). For the sake of simplicity we use an abbreviation $\sigma_{\tau n\ \tau' n'}$ to represent the translation of the basis functions from the j-th center to the i-th center and then we write

$$\vec{\psi}^j_{\tau n} = \vec{\psi}_{\tau n}(\vec{r}-\vec{r}_j) = \sum_{\tau' n'} \sigma_{\tau n\ \tau' n'}(\vec{r}_i-\vec{r}_j)\ \operatorname{Re}\vec{\psi}^i_{\tau' n'} \quad (2.20)$$

It then remains to expand the incident wave also in the form of a series centered at the i-th scatterer. Referring to Fig. 1, we can rewrite Eq.(2.1) with exp(-iωt) term suppressed

$$\vec{u}^0 = e^{ik_p(\zeta_i + z_i)}\hat{z} + e^{ik_s(\zeta_i + z_i)}\hat{x}\,. \quad (2.20)$$

Expanding the terms $\exp(ik_p z_i)$ and $\exp(ik_s z_i)$ in (2.20) in terms of spherical Bessel functions and Legendre polynomials, and using the integral representation for spherical Bessel function of the first kind and orthogonality relations, we can express (2.20) in the following form, see for example Stratton (37) :

$$\vec{u}^0 = \frac{e^{i k_p \zeta_i}}{i k_p} \sum_{s=0}^{\infty} \sum_{t=-s}^{s} (2s+1)\, i^s \operatorname{Re}\vec{\psi}^i_{1ts}\, \delta_{t,0}$$

$$+ \frac{1}{2i} e^{i k_p \zeta_i} \sum_{s=1}^{\infty} \sum_{t=-s}^{s} \frac{2s+1}{s(s+1)}\, i^s \Bigg\{ \operatorname{Re}\vec{\psi}^i_{2ts}\left[\delta_{t,1} + s(s+1)\delta_{t,-1}\right]$$

$$+ \frac{1}{k_s} \operatorname{Re}\vec{\psi}^i_{3ts}\left[\delta_{t,1} - s(s+1)\,\delta_{t,-1}\right]\Bigg\} \quad (2.21)$$

where δ_{mn} is the Kronecker δ. For two dimensional problems, the incident waves are expanded by using Fourier series expansion in complex form, see for example (22,26). For the sake of simplicity, we write the incident wave field in terms of expansion co-efficients $a_{\tau n}$ as follows

$$\vec{u}^0 = \sum_{\tau n} a_{\tau n} \operatorname{Re}\vec{\psi}^i_{\tau n}\, e^{i k_\tau \cdot \vec{r}_i} \quad (2.22)$$

In writing (2.22), we have used the concept that for $\vec{k} = k\hat{z}$, $k\hat{z}\cdot\vec{r} = k\hat{z}\cdot(\vec{r} - \vec{r}_i) + k\hat{z}\cdot\vec{r}_i$ and $\hat{z}\cdot\vec{r}_i = \zeta_i$. For acoustic wave scattering ($\tau = 1$), $a_{\tau n}$ is equal to the first term of Eq. (2.21) multiplying exp $(i\, k_p\, \zeta_i)$ and Re $\vec{\psi}_{1ts}$; for electromagnetic waves ($\tau = 2,3$), $a_{\tau n}$ is equal to the second term of Eq. (2.21) multiplying exp $(i\, k_s \zeta_i)$ and Re $\vec{\psi}_{\tau ts}$. Since the incident wave field is a known field, $a_{\tau n}$ are, hence, known expansion field coefficients. Substituting Eqs. (2.19) and (2.21) in (2.15) and using orthogonality of basis functions, we obtain the following relation between the unknown expansion coefficients $b_{\tau n}$ of the exciting field and the unknown expansion coefficients $B_{\tau n}$ of the scattered field :

$$b^i_{\tau n} = a_{\tau n}\, e^{i\,\vec{k}_\tau\cdot\vec{r}_i} + \sum_{j\neq i}^{N} \sum_{\tau' n'} B^j_{\tau' n'}\, \sigma_{\tau' n';\, \tau n}\, (\vec{r}_i - \vec{r}_j) \qquad (2.23)$$

It has been shown by the previous papers on T-matrix approach that if the total field outside a scatterer is the sum of the incident and the scattered fields, the unknown scatentererd field expansion coefficients can be related to the incident field expansion coefficients through the transition or T-matrix. We extend this definition to our present problem. Since $(\vec{u}^e_i + \vec{u}^s_i)$ is the total field at any point in the matrix, the expansion coefficients of the field scattererd by the i-th scatterer may be formally related to the coefficients of the field exciting the i-th scatterer through the T-matrix :

$$B^i_{\tau n} = \sum_{\tau' n'} T^i_{\tau n, \tau' n'}\, b^i_{\tau' n'} \qquad (2.24)$$

In its expanded version, Eq. (2.24) can be written as

$$\begin{bmatrix} B^{\ell(i)}_{1n} \\ B^{\ell(i)}_{2n} \\ B^{\ell(i)}_{3n} \end{bmatrix} = \begin{bmatrix} (T^{11})^{\ell p(i)}_{nm} & (T^{12})^{\ell p(i)}_{nm} & (T^{13})^{\ell p(i)}_{nm} \\ (T^{21})^{\ell p(i)}_{nm} & (T^{22})^{\ell p(i)}_{nm} & (T^{23})^{\ell p(i)}_{nm} \\ (T^{31})^{\ell p(i)}_{nm} & (T^{32})^{\ell p(i)}_{nm} & (T^{33})^{\ell p(i)}_{nm} \end{bmatrix} \begin{bmatrix} b^{p(i)}_{1m} \\ b^{p(i)}_{2m} \\ b^{p(i)}_{3m} \end{bmatrix} \qquad (2.25)$$

The elements of the T-matrix, $T_{\tau n, \tau' n'}$, involve surface integrals which can be evaluated in closed form for cylindrical geometry (2-D) and spherical geometry (3-D), while for obstacles of arbitrary shape they can be evaluated numerically. The T-matrix for a single scatterer is of the form

$$T = -Q(Re, Re)\ \ Q^{-1}(Ou, Re) \qquad (2.26)$$

where Q(Re, Re) and Q(Ou, Re) are matrices which are functions of the surface S of the scatterer and of the nature of the boundary conditions. This form is quite common for acoustic, electromagnetic and elastic wave problems, except when there are solid-fluid interfaces. A detailed derivation of the T-matrix for a solid in a fluid and fluid in a solid can be found in the paper by Varadan (31).

With the scattered field coefficients $B^j_{\tau n}$ expressed in terms of exciting field coefficients $b^j_{\tau n}$ and the T-matrix as given by (2.24), Eq. (2.23) gives the exciting field formulation of the multiple scattering. If we multiply both sides of Eq. (2.23) by the T-matrix, then we obtain the scattered field formulation of multiple scattering which may be written as

$$B^i_{\tau n} \equiv B^{\ell(i)}_{\tau n} = \sum_{\tau''n''} T^i_{\tau n,\tau''n''} \left[a_{\tau''n''} \exp(i\, \vec{k}_\tau \cdot \vec{r}\,) + \sum_{j\neq i}^{N} \sum_{\tau'n'} B^j_{\tau'n'}\, \sigma_{\tau'n',\tau''n''}(\vec{r}_i - \vec{r}_j) \right] . \tag{2.27}$$

In its expanded version, Eq. (2.27) can be written as

$$\begin{bmatrix} B^{\ell(i)}_{1n} \\ B^{\ell(i)}_{2n} \\ B^{\ell(i)}_{3n} \end{bmatrix} = \begin{bmatrix} (T^{11})^{\ell p(i)}_{nm} & (T^{12})^{\ell p(i)}_{nm} & (T^{13})^{\ell p(i)}_{nm} \\ (T^{21})^{\ell p(i)}_{nm} & (T^{22})^{\ell p(i)}_{nm} & (T^{23})^{\ell p(i)}_{nm} \\ (T^{31})^{\ell p(i)}_{nm} & (T^{32})^{\ell p(i)}_{nm} & (T^{33})^{\ell p(i)}_{nm} \end{bmatrix} \begin{bmatrix} \Phi^i_{1mp} \\ \Phi^i_{2mp} \\ \Phi^i_{3mp} \end{bmatrix} \tag{2.28}$$

where

$$\Phi^i_{1pm} = (2m+1)\, i^m \frac{e^{i k_p \zeta_i}}{i k_p} \delta_{p,0} + \sum_{j\neq 1}^{N} \sum_{\nu=0}^{\infty} \sum_{\mu-\nu}^{\nu} B^\mu_{1\nu} A^{\nu\mu}_{mp}$$

$$\Phi^i_{2mp} = \frac{2m+1}{m(m+1)}\, i^m \frac{e^{i k_s \zeta_i}}{2i} \left[\delta_{p,1} + m(m+1)\, \delta_{p,-1} \right] + \sum_{j=1}^{N}{}' \sum_{\nu=0}^{\infty} \sum_{\mu=-\nu}^{\nu} \left[B^\mu_{2\nu} B^{\nu\mu}_{mp} + B^\mu_{3\nu} C^{\nu\mu}_{mp} \right] \tag{2.29}$$

$$\Phi^i_{3mp} = \frac{2m+1}{m(m+1)}\, i^m \frac{e^{i k_s \zeta_i}}{2i} \left[\delta_{p,1} - m(m+1)\, \delta_{p,-1} \right] + \sum_{j=1}^{N}{}' \sum_{\nu=0}^{\infty} \sum_{\mu=-\nu}^{\nu} \left[B^\mu_{2\nu} C^{\nu\mu}_{mp} + B^\mu_{3\nu} B^{\nu\mu}_{mp} \right]$$

In Eq. (2.29), $\sum'$ denotes the sum over all scatterers except the i-th. Thus, we have eliminated the unknown exciting field expansion coefficients through the use of the T-matrix resulting in a set of equations involving the expansion coefficients of the scattered field only. If the scatterers are all identical, then $T^i = T^j = T$. If they are of different sizes, we introduce a suitable size distribution for the scatterers and find average value of the T-matrix.

It can be seen from Eq. (2.27) that the scattererd field coefficients explicitly depend on the position and orientation of the scatterers. Depending on what information we choose to put into the distribution function, we can make our model more realistic but the analysis in turn gets more complicated. Here, we consider both random and periodic distributions with specific and random orientations.

III. RANDOM DISTRIBUTION OF SCATTERERS WITH SPECIFIC ORIENTATION

We consider N number of randomly distributed homogeneous scatterers. The scatterers are assumed to have a specific orientation (with major axis parallel to the y-axis) as shown in Fig. 3.

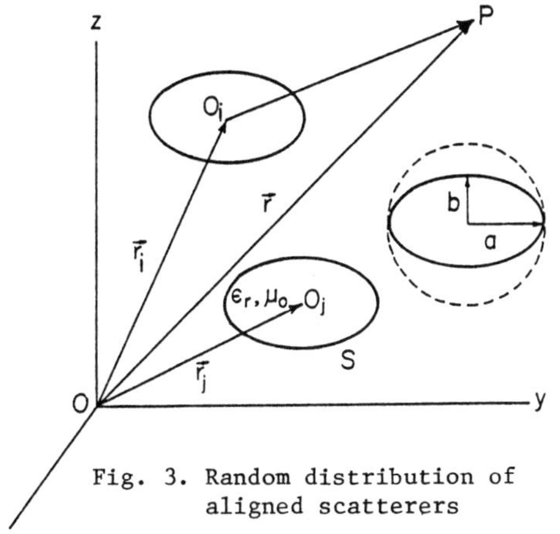

Fig. 3. Random distribution of aligned scatterers

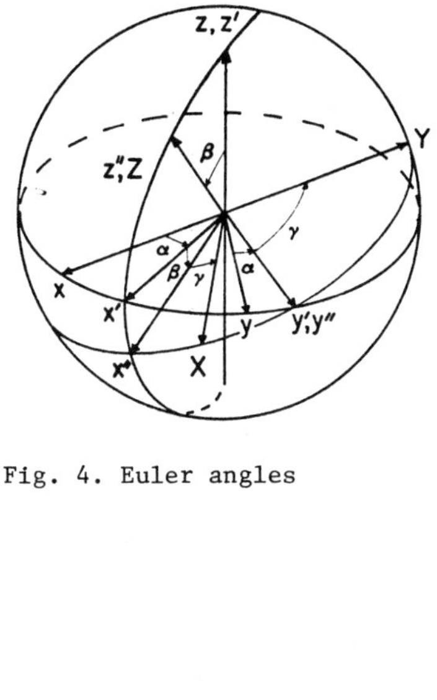

Fig. 4. Euler angles

Distribution functions are given by the total probability density of finding the scatterers at positions $\vec{r}_1, \vec{r}_2, \ldots, \vec{r}_N$ etc., which in turn can be written in terms of conditional probabilities :

$$p(\vec{r}_1, \vec{r}_2, \ldots, \vec{r}_N) = p(\vec{r}_i)\, p(\vec{r}_1\, \vec{r}_2, \ldots'. \ldots, \vec{r}_N|\vec{r}_i)$$

$$= p(\vec{r}_i)\quad p(\vec{r}_j|\vec{r}_i)\, p(\vec{r}_1, \vec{r}_2, \ldots', \ldots', \vec{r}_N|\vec{r}_j, \vec{r}_i) \tag{3.1}$$

In Eq. (3.1), $p(\vec{r}_i)$ denote the probability of finding a scatterer $\vec{r}_i$ and $p(\vec{r}_j|\vec{r}_i)$ denotes the conditional probability of locating a scatterer at $\vec{r}_j$ given that a scatterer is located at $\vec{r}$ etc. A prime in the first of Eq. (24) means $\vec{r}_i$ is absent while two primes in the second of Eq. (24) means both $\vec{r}_i$ and $\vec{r}_j$ are absent. We also denote the configurational averaging of a statistical quantity f as

$$<f>_i = \int_V \cdots \int_V f \; p(\vec{r}_1, \vec{r}_2, \ldots', \vec{r}_N \,|\, \vec{r}_i)\, d\vec{r}_1\, d\vec{r}_2 \ldots', \ldots, d\vec{r}_N$$

$$<f>_{ij} = \int_V \cdots \int_V f \; p(\vec{r}_1, \vec{r}_2, \ldots', \ldots', \ldots \vec{r}_N \,|\, \vec{r}_i, \vec{r}_j)\, d\vec{r}_1 d\vec{r}_2 \ldots', \ldots', \ldots d\vec{r}_N \qquad (3.2)$$

Equation (3.2a) implies that we have averaged over all scatterers except the i-th, while Eq. (3.2b) implies that we average over all scatterers except the i-th and j-th, and so on.

If the scatterers are randomly distributed, the positions of all scatterers are equally probable within volume V, and hence its distribution is uniform with probability density

$$p(\vec{r}_i) = \begin{cases} n_0/N, & \vec{r}_i \in V \\ 0, & \vec{r}_i \notin V \end{cases} \qquad (3.3)$$

where n_0 is the uniform number density of the scatterers and N is the total number of scatterers. In addition, for non-overlap of the imaginary spheres circumscribing each scatterer, we approximate the conditional probability as follows

$$p(\vec{r}_j|\vec{r}_i) \begin{cases} n_0/N, & |\vec{r}_i - \vec{r}_j| > 2a \\ 0, & |\vec{r}_i - \vec{r}_j| < 2a \end{cases} \qquad (3.4)$$

where 'a' is the radius of the circumscribing sphere. A suitable correlation in the position may also be added to (3.4) but is omitted here for simplicity. The form of pair correlation in Eq. (3.4) describes the usual radially symmetric distribution function with an exclusion surface or "hole" corresponding to a sphere of radius 2a. Other types of exclusion surfaces are discussed in recent papers by Twersky (8, 9). We refer the reader to Twersky's comments on choosing radially symmetric statistics for non-symmetrical scatterers, namely making the region of non-overlap of two scatterers to be a circle or sphere circumscribing the scatterers.

Multiplying both sides of Eq. (2.26) by the probability density given by Eq. (3.1) and using (3.2) - (3.4), we obtain the configurational average of the scattered field coefficients :

$$<B^i_{\tau n}>_i = \sum_{\tau''n''} T_{\tau n,\tau''n''} \Big[a_{\tau''n''}\, e^{i\vec{k}_\tau \cdot \vec{r}}$$

$$+ \frac{1}{V} \sum_{j \neq i}^{N} \sum_{\tau' n'} \int_{V'}^{j} \langle B_{\tau' n'} \rangle_{ij} \quad \sigma_{\tau' n', \tau'' n''} \quad d\vec{r}_j \Big] \tag{3.5}$$

where V´ denotes the volume of the medium excluding the volume of a sphere of radius 2a. For identical scatterers, $\sum_{j \neq i}^{N}$ can be replaced by (N-1). The above equation indicates that the configurational average with one scatterer fixed is given in terms of the configurational average with two scatterers fixed. We choose to solve these equations by using Lax's quasicrystalline approximation where we assume that there is no correlation between the scatterers except that of non-overlapping which can be written mathematically as

$$\langle B_{\tau n} \rangle_{ij} = \langle B_{\tau n} \rangle_{j} \tag{3.6}$$

Equation (3.6) also implies that the neighbourhood of every scatterer is the same. For identical scatterers, Eq. (3.5) may be rewritten using Eq. (3.6) as

$$\langle B^{i}_{\tau n} \rangle_{i} = \sum_{\tau'' n''} T_{\tau n, \tau'' n''} \Big[a_{\tau'' n''} \; e^{i \vec{k}_{\tau} \cdot \vec{r}_i} + \frac{N-1}{V} \sum_{\tau' n'} \int_{V'} \langle B^{j}_{\tau n} \rangle_{j} \; \sigma_{\tau' n', \tau'' n''} \; d\vec{r}_j \Big] \tag{3.7}$$

In its expanded version, Eq. (3.7) can be written as

$$\begin{bmatrix} \langle B^{\ell(i)}_{1n} \rangle_i \\ \langle B^{\ell(i)}_{2n} \rangle_i \\ \langle B^{\ell(i)}_{3n} \rangle_i \end{bmatrix} = \begin{bmatrix} (T^{11})^{\ell p}_{nm} & (T^{12})^{\ell p}_{nm} & (T^{13})^{\ell p}_{nm} \\ (T^{21})^{\ell p}_{nm} & (T^{22})^{\ell p}_{nm} & (T^{23})^{\ell p}_{nm} \\ (T^{31})^{\ell p}_{nm} & (T^{32})^{\ell p}_{nm} & (T^{33})^{\ell p}_{nm} \end{bmatrix} \begin{bmatrix} \langle \Phi^{i}_{1mp} \rangle \\ \langle \Phi^{i}_{2mp} \rangle \\ \langle \Phi^{i}_{3mp} \rangle \end{bmatrix} \tag{3.8}$$

where

$$\langle \Phi^{i}_{1mp} \rangle = (2m + 1) \, i^{m} \frac{e^{i k_p \zeta_i}}{i k_p} \delta_{p,0} + \frac{1}{V} \sum_{j=1}^{N'} \sum_{\nu=0}^{\infty} \sum_{\mu=-\nu}^{\nu} \int_{V'} \langle B^{\mu(j)}_{1\nu} \rangle_{ij} \; A^{\nu\mu}_{mp} \; d\vec{r}_j \; \ldots \tag{3.9}$$

This is a system of integral equations for the unknown coefficients $\langle B^{i}_{\tau n} \rangle_i$. Similar expressions for the average exciting field coefficients may be obtained from the exciting field formalism of multiple scattering.

IV. PROPAGATION CHARACTERISTICS OF THE AVERAGE WAVES IN THE MEDIUM

To solve the integral equations given by (3.7), we consider the inhomogeneous medium with discrete scatterers as a homogeneous continuum and assume that the average coherent wave is a plane wave propagating with an effective wave number K in the same direction as the incident plane wave. We can thus write

$$< B^{i}_{\tau n} > = X_{\tau n} \; e^{i \vec{K} \cdot \vec{r}_i} \tag{4.1}$$

where $X_{\tau n}$ is the amplitude of the coherent wave.

Substituting Eq. (4.1) in (3.7), employing divergence theorem to convert the volume integral in (3.7) to surface integrals and using the extinction theorem which cancels the incident wave, see for example (8,9,22), we obtain a set of simultaneous coupled homogeneous equations for the coefficients $X_{\tau n}$ given by

$$X_{\tau n} = c \sum_{\tau'' n''} \sum_{\tau' n'} \sum_{q=|n'-n''|}^{|n'+n''|} X_{\tau' n'} \; T_{\tau n, \tau'' n''} \; C^{q}_{\tau' n', \tau'' n''} \; \frac{JHq}{(k^2_{\tau'} - K^2)a^2} \tag{4.2}$$

where $c = 4\pi a^3 n_0/3$ is the effective spherical concentration of the scatterers per unit volume, C^q is an expression containing Wigner coefficients, and

$$JH_q = 2 k_{\tau'} a \;\; j(2 Ka) h'_q(2 k_{\tau'} a) - 2 K a \;\; h_q(2 k_{\tau'} a) \; j'_q (2 Ka) \tag{4.3}$$

Equation (4.2) can be written in its expanded form as follows :

$$X^{\ell}_{1n} = 6c \sum_{q=|n_1-m|}^{|n_1+m|} \sum_{m=0}^{\infty} \sum_{p=-m}^{m} \sum_{n_1=0}^{\infty} \sum_{m_1=-n_1}^{n_1} (-1)^p (-i)^q$$

$$i^{m_1+m+q-n_1} \; \delta_{m_1 p} \;\; JH_q \left\{ \frac{1}{(k^2_p - K^2)a^2} \; X^{m_1}_{1n_1} \; (T^{11})^{\ell p}_{nm} \right.$$

$$a(m_1, n_1 \; | -p, m \; | \; q) + \frac{1}{(k^2_s - K^2)a^2} \left\{ X^{m_1}_{2n_1} \left[(T^{12})^{\ell p}_{nm} \;\; a(n_1, m, q) \right. \right.$$

$$a(m_1, n_1 \; | -p, m \; | \; q) - (T^{13})^{\ell p}_{nm} \;\; b(n_1, m, q)$$

$$\left. a(m_1, n_1 \; | -p, m \; | \; q, q-1) \right] + X^{m_1}_{3n_1} \left[(T^{13})^{\ell p}_{nm} \;\; a(n_1, m, q) \right.$$

$$a(m_1,n_1|-p,m|q) - (T^{12})^{\ell p}_{nm} \quad b(n_1,m,q)$$

$$a(m_1,n_1|-p,m|q,\ q-1)\Big]\Big\}\Big\} \tag{4.4a}$$

$$X^{\ell}_{2n} = \ldots \tag{4.4b}$$

$$X^{\ell}_{3n} = \ldots \tag{4.4c}$$

Equation (4.4b) can be obtained from (4.4a) by replacing T^{11}, T^{12},T^{13} by T^{21}, T^{22}, T^{23}, while (4.4c) can be obtained by replacing T^{11}, T^{12}, T^{13} by T^{31}, T^{32}, and T^{33}. For acoustic wave problem, we get uncoupled equation for X^{ℓ}_{1n} ; for electromagnetic wave problem, we get coupled equations in terms of X^{ℓ}_{2n} and X^{ℓ}_{3n}; for elastic wave problem, we obtain coupled equations in terms of X^{ℓ}_{1n}, X^{ℓ}_{2n} and X^{ℓ}_{3n}. Similar equations can also be obtained from the average exciting field coefficients, if one chooses to use exciting field formalism.

Equation (4.2) is a system of simultaneous linear homogeneous equations for the unknown amplitudes $X_{\tau n}$. For a nontrivial solution, we require that the determinant of the truncated coefficient matrix vanishes, which yields an equation for the effective wave number K in terms of k_τ and the T-matrix of the scatterer. This is the dispersion relation for the scatterer filled medium. Equation (32) is a general expression valied for any arbitrary shaped scatterer, since the T-matrix is the only factor that contains information about the exact shape and boundary conditions at the scatterer. Thus the formalism presented here is valid for all the three wave fields. The effective wave number K obtained in the analysis is a complex quantity, the real part of which relates to the phase velocity, while the imaginary part relates to attenuation of coherent waves in the medium.

V. PERIODIC DISTRIBUTION WITH PARALLEL ORIENTATION

For periodic distribution the analysis will introduce the geometry of the lattice leading to different results for different packing, and is similar to the one employed in crystallographic theories, see for example (38,39). Here we apply lattice sum and include contributions from nearest neighbors. One couls use other refinements following the theories presented in (8,9,38,39). The averaging over the position of individual scatterers can be easily accomplished since there is no restriction on the position of one scatterer.

A procedure for evaluating a lattice sum over a simple lattice is to obtain by direct summation the contributions by the lattice points within a certain radius R_1 given by

$$\frac{4}{3}\ \frac{\pi R_1{}^3}{V} = N_0 \tag{5.1}$$

and to replace the summation over the points beyond this radius by an integral. In Eq. (5.1), N_0 is the number of scatterers counting the point $\vec{r}_j = 0$ and V is the volume of the unit cell, (see (40) for a given type of an array). The analysis is presented in (41,42) and the resulting scattered field coefficients are given by

$$B^i_{\tau n} = \sum_{\tau''n''} T^i_{\tau n,\tau''n''} \left[a_{\tau''n''} \; e^{i \vec{k}_\tau \cdot \vec{r}_i} + \sum_{j \neq i}^{N_0}{}' \sum_{\tau'n'} B^j_{\tau'n'} \; \sigma_{\tau'n',\tau''n''} (\vec{r}_i - \vec{r}_j) + \frac{1}{V} \int_{R_1}^{\infty} \sum_{\tau'n'} B^j_{\tau'n'} \; \sigma_{\tau'n',\tau''n''} (\vec{r}_i - \vec{r}_j) \, d\vec{r}_j \right] \quad (5.2)$$

To obtain a solution to the above equation, we assume a plane wave propagating with an effective wave number K in the same direction as the incident wave :

$$B^i_{\tau n} = X_{\tau n} \; e^{i \vec{K} \cdot \vec{r}_i} \quad (5.3)$$

Substituting Eq. (5.3) in Eq. (5.2) and following the procedure we had outlined in the previous section, we obtain a set of simultaneous coupled homogeneous equations for the coefficients $X_{\tau n}$ given by

$$X_{\tau n} = c \sum_{\tau''n''} \sum_{\tau'n'} \sum_{q=|n'-n''|}^{|n'+n''|} X_{\tau'n'} \; T_{\tau n,\tau''n''} \; C^q_{\tau'n',\tau''n''} \frac{1}{(k^2_\tau - K^2)a^2} \left[R_q + \frac{(k^2_\tau - K^2)a^2}{c} P_q \right] \quad (5.4)$$

where

$$R_q = k_\tau \; R_1 \; j_q(K R_1) \; h'_q(k_\tau R_1) - K R_1 \; h_n(k_\tau R_1) \; j'_n \; (K R_1)$$

$$P_q = \sum_j e^{i K \cdot \vec{r}_j} \; \sigma_{\tau'n', \tau''n''} \quad (5.5)$$

and c' is the concentration as defined before.

VI. ARBITRARY ANGLE OF INCIDENCE AND RANDOM ORIENTATION

Inhomogeneous media in which scatterers have a specific orientation will exhibit anisotropy and polarization effects when the waves are incident at arbitrary angles to the symmetry axes of the scatterers. When the incident wave direction is along the symmetry axis of the oblate spheroidal scatterers, we found that the effect on polarization is zero, and the coherent wave propagates with an effective wave number K as if it propagates in an effective homogeneous and isotropic continuum, see for example (25). When the scatterers are randomly oriented, the effects of anisotropy and polarization are also nullified (29).

For symmetrically oriented scatterers, we define the Euler's angles α,β,γ of the symmetry axes of the scatterers (x,y,z) with respect to the laboratory co-ordinate system (X,Y,Z), see Fig.4. All quantities that are referred to the x,y,z system are distinguished by a $\wedge$. Then, one could write the spherical harmonics using the relation operator as follows (for details, we refer to (43)) :

$$Y_{\ell m\sigma} = \sum_{m'\sigma'} D_{mm'\sigma\sigma'}(\alpha,\beta,\gamma)\ \hat{Y}_{\ell m'\sigma'} \tag{6.1}$$

where D is the rotation matrix associated with rotation operator. The rotation operator leaves the length of the position vector $|\vec{r}|$ invariant. The rotation matrix can be easily incorporated into the T-matrix to obtain a new T-matrix which can be written as

$$T = (D^t)^{-1}\ \hat{T}\ (D^t) \tag{6.2}$$

where D^t is the matrix transpose of D. Equation (6.2) gives the desired relation between the T-matrices evaluated with respect to the two set of coordinate axes. $\hat{T}$ is independent of position and orientation and is hence the same for identical scatterers. The matrices T, however, is different if the orientation of the scatterers is not the same. Substituting Eq. (6.2) in Eq. (4.2) and Eq. (5.4), we obtain the dispersion relation, phase velocity, coherent wave attenuation and polarization effects for an arbitrary angle of incidence for random and periodic distribution, respectively. This is one of the basic advantage of formulating the multiple scattering in terms of a T-matrix.

For random orientation, one has to average over all Euler angles α,β,γ, and the information can thus be incorporated into the T-matrix :

$$\langle T\rangle = \frac{1}{8\pi^2}\int_0^{2\pi}\int_0^{2\pi}\int_0^{2\pi} D(\alpha,\beta,\gamma)\ \hat{T}\ D^{-1}(\alpha,\beta,\gamma)\ \sin\beta\ d\beta\ d\gamma\ d\alpha\ . \tag{6.3}$$

With Eq. (6.3), Eqs. (4.2) and (5.4) yield the dispersion relation for both random and periodic distributions, respectively, with random orientation.

VII. RESULTS AND CONCLUSIONS

Using the theories outlines in previous sections, we present some analytical and numerical results for variety of 2-D and 3-D problems in all three wave fields to show the broad applicability of our multiple scattering approach. The sample examples given may find applications in many fields of engineering and science as outlined in the introduction. We present results for two dimensional parallel cylinders of elliptical cross section randomly and periodically (hexagonal array) distributed with specific and random orientation. The wave is incident normal to the cylinders. The polarization of the wave is along the axis of the cylinder (SH-waves). We also consider spherical, oblate spheroidal scatterers of various aspect ratios randomly distributed with specific and random orientations. At low frequencies, analytical expressions are derived for the effective wave number in the average medium as a function of the geometry, the material properties and the angle of orientation of the scatterers. The formulation is ideally suited for numerical computation of phase velocity and attenuation at higher frequencies as evidenced by the results presented here and in (22 - 29). In addition, we present numerical results for dynamic shear moduli for 2-D case as a function of frequency using the T-matrix approach and the work of Bedeux and Mazur (44) and Varadan and Vezzetti (45).

VIIa. Rayleigh or Low Frequency Limit

In the Rayleigh or low frequency limit, the size of the scatterers is considered to be small when compared to the incident wavelength. It is then sufficient to take only the lowest order coefficient in the expansion of the fields. In this limit, the elements of the T-matrix can be obtained in closed form for various simple shapes (46). It can be shown that at low frequencies, only $X_{\tau 0}$, $X_{\tau 1}$ and $X_{\tau,-1}$ of Eq. (4.2) or Eq. (5.4) make a contribution. After some manipulations of the resulting 3 × 3 determinant, we obtain the dispersion relations :

(i) Parallel Elliptic Cylinders (Elastic SH-Waves)

a) Specific Orientation of Cross Section (see Fig. 5 and paper (23))

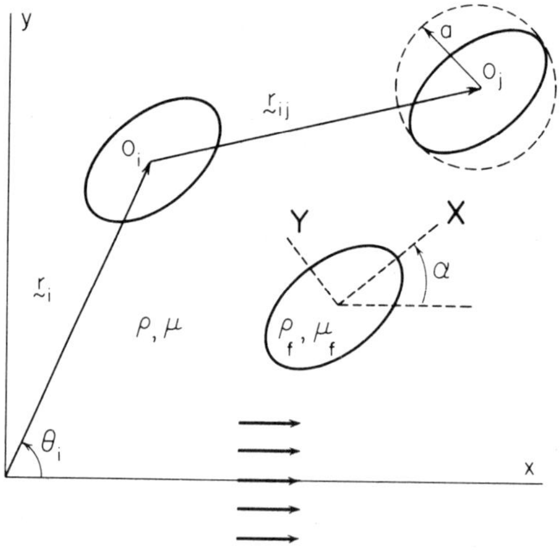

Fig. 5. Random distribution of aligned scatterers - oblique incidence

$$\frac{K^2}{k_s^2} = \frac{e_1 d_1 + e_2 d_2}{b_1 d_1 + b_2 d_2} \tag{7.1}$$

where

$$e_1 = (2ic/\pi^2)\ a_1\ e^{-i2\alpha} \left[1 - 2k_s^2a^2\ \ell n(k_sa)\right] \times$$
$$\times \left[\pi - 4ic\ S_0 + 8ic\ S_0\ k_s^2\ a^2\ \ell n(k_sa)\right]$$
$$e_2 = (1/\pi^2) \left[\pi - 4ic\ S_0 + 8ic\ S_0\ k_s^2\ a^2\ \ell n(k_sa)\right] \times$$
$$\times \left[\pi - 2ic \left\{1 - 2k_s^2a^2\ \ell n(k_sa)\right\} (S_1 + a_1\ e^{-i2\alpha})\right]$$
$$d_1 = (4c/\pi)\ a_1\ \sin 2\alpha \left[1 - 2k_s^2\ a^2\ \ell n(k_sa)\right]$$
$$d_2 = (1/\pi) \left[\pi - 2ic \left\{1 - 2k_s^2\ a^2\ \ell n(k_sa)\right\} (S_1 - a_1\ e^{-i2\alpha})\right]$$
$$b_1 = (2ic/\pi^2)\ S_1 \left[\pi + 8ic\ S_0\ k_s^2a^2\ \ell n(k_sa)\right]$$
$$b_2 = (1/\pi^2) \left[\pi + 8ic\ S_0\ k_s^2\ a^2\ \ell n(k_sa)\right] \left[\pi + 2ic\ (S_1 + a_1\ e^{i2\alpha})\right]$$
$$S_0 = i(\pi/8)\ (d - 1)\ (b/a) \left[2 - (d - 1)\ k_s^2 ab\ \ell n(k_sa)\right] \quad (7.2)$$

$$S_1 = \frac{i\pi}{8} \frac{(1 - m)(1 + b/a)^2(b/a)}{(b/a + m)(1 + mb/a)} \left\{1 + m - (\frac{1}{4})(1 - m)\left[(1+b^2/a^2)\times\right.\right.$$
$$\left.\left.\times\ (1 + m^2) + 4\ mb/a\right] k_s^2 ab\ \ell n(k_sa)/(1 + mb/a)(b/a + m)\right\}$$

$$a_1 = \frac{i\pi}{8} \frac{(1 - m)(1 - b^2/a^2)(b/a)}{(b/a + m)(1 + mb/a)} \left[1 - m - \frac{1}{4} \frac{(1 + b/a)^2(1 - m^2)}{(b/a + m)(1 + mb/a)} \times\right.$$
$$\left. k_s^2 ab\ \ell n(k_sa)\right]$$

and $c = \pi a^2 n_0$ is the concentration of the circumscribing circle.

b) Random Orientation

$$K^2/k_s^2 = (1 + c_1t_1) \left[1 + c_1(d - 1)\right] /(1 - c_1t_1) \quad (7.3)$$

where

$$t_1 = (1 - m^2)(a + b)^2 / 4(b + ma)\ (mb + a) \quad (7.4)$$

(ii) Random Distribution of 2-D cracks (Elastic SH - Waves)

a) Parallel Orientation (7.5)

$$\frac{K^2}{k_s^2} = \frac{(4 + \pi a^2 n_0)^2 - (\pi a^2 n_0)^2\ e^{-i4\alpha} - 2i(\pi a^2 n_0)^2\ e^{-i2\alpha} \sin 2\alpha}{\left[4 - \pi a^2 n_0(1 - e^{i2\alpha})\right] \left[4 + \pi a^2 n_0(1 + e^{-i2\alpha})\right] - 2i(\pi a^2 n_0)^2 \sin 2\alpha}$$

b) Random Orientation

$$K^2/k_s^2 = (1 + \pi a^2 n_0/4)(1 - \pi a^2 n_0/4) \quad (7.6)$$

where 2a is the length of the crack and n_0 is its number density, and α is the angle the incident wave makes with the major axis. In these equations, d = ratio of densities of the scatterer to that of the matrix = ρ_1/ρ and m = ratio of shear moduli of the scatterer to that of the matrix = μ_1/μ and a and b are semi major and minor axes of the elliptic cylinder, respectively. The analysis for both compressional and shear waves incident normal to cylinders is presented in (27).

(iii) Spheres (elastic waves)

$$\left(\frac{K_p}{k_s}\right)^2 = \frac{(1 + 9c\,E_1)(1 + 3c\,E_0)\left[1 + 3c\,\frac{E_2}{2}\,(2 + \frac{3k_s^2}{k_p^2})\right]}{1 - 15c\,E_2\left[1 + 3c\,E_0\right] + \frac{3}{2}\,c\,E_2\,(2 + \frac{3k_s^2}{k_p^2})} \quad (7.7)$$

$$\left(\frac{K_s}{k_s}\right)^2 = \frac{(1 + 9c\,E_1)(1 + \frac{3}{2}\,c\,E_2\left[2 + \frac{3k_s^2}{k_p^2}\right])}{1 + \frac{3}{4}\,c\,E_2(4 - \frac{9\,k_s^2}{k_p^2})} \quad (7.8)$$

where

$$E_0 = \frac{1}{3}\,\frac{3\lambda + 2\mu - (3\lambda_1 + 2\mu_1)}{4\mu + 3\lambda_1 + 2\mu_1}$$

$$E_1 = \frac{1}{9}\,(\frac{\rho_1}{\rho} - 1)$$

$$E_2 = -\frac{\frac{4}{3}\,\mu\,(\mu_1-\mu)\quad 24\mu_1(\mu_1-\mu) - (\lambda_1+2\mu_1)\,(19\mu_1+16\mu)}{24\mu_1(\mu_1-\mu) - (\lambda_1+2\mu_1)\,(19\mu_1+16\mu)\cdot} \quad (7.9)$$

$$\times\ \frac{1}{4\mu(\mu_1-\mu) + 3(\lambda + 2\mu)\,(2\mu_1 + 3\mu)}$$

and $c = 4\,\pi a^3 n_0/3$ is the concentration of spheres.

(iv) Spheres in Free Space (Electromagnetic Waves)

$$\left(\frac{K}{k_s}\right)^2 = \frac{1 + 2c\,\dfrac{\varepsilon_{r_1} - 1}{\varepsilon_{r_1} + 2}}{1 - c\,\dfrac{\varepsilon_{r_1} - 1}{\varepsilon_{r_1} + 2}} \quad (7.10)$$

(v) Spheroids in Free Space (Electromagnetic Waves)

Random Orientation

$$\left(\frac{K}{k_s}\right)^2 = \left[1 + \frac{\frac{3C}{2}\frac{\varepsilon_{r_1}-1}{\varepsilon_{r_1}+2}(f_2+f_3)}{1-\frac{3}{2}\frac{\varepsilon_{r_1}-1}{\varepsilon_{r_1}+2}f_1}\right]\left[1 - \frac{\frac{3c}{4}\frac{\varepsilon_{r_1}-1}{\varepsilon_{r_1}+2}(f_2+f_3)}{1-\frac{3}{2}\frac{\varepsilon_{r_1}-1}{\varepsilon_{r_1}+2}f_1}\right]^{-1} \tag{7.11}$$

where c is the concentration of the circumscribing spheres and f_1, f_2 and f_3 are functions of the eccentricity of the spheroids $e = [(a/b)^2 - 1]^{\frac{1}{2}}$ and are given in (25).

In the Rayleigh limit, the value of K as determined by the above dispersion relations is a real quantity for lossless material and a complex quantity for lossy material, and relates to phase velocity $V_p = \omega/K$. In this limit, we normally study the dependence of phase velocity on concentration, angle of incidence and aspect ratio of the scatterers. In Figs. 6 and 7, we have plotted the normalized phase velocity V_p/c_s versus concentration for various values of angle of incidence α for 2-D (SH-Wave) cylinders randomly distributed with parallel and random orientations. We assume the density ratio d = 2.53/2.72 and shear modulus ratio m = 25/3.87 (no loss in the material) for obtaining results in Fig. 6 while Fig. 7 gives the result for cracks. The general tendency of the phase velocity is to increase (for inclusions) and decrease (for cracks and cavities) as concentration increases. The results also indicate that phase velocity decreases as the angle of incidence α increases. For $\alpha = 0$ and $\alpha = \pi/2$, our results agree with well known results for cracks (3). It can easily be seen from Eq. (7.2) that for a given concentration and α, the phase velocity decreases with decreasing b/a. In Figs. 6 and 7, we have also plotted the corresponding results for random orientation.

Equation (7.10) is recognized as the dispersion relation of the Clausius-Mossotti form. The dispersion relation for spheroidal scatterers given by Eq. (7.11) appears to be new ; it reduces to Eq. (7.10) in the limit $e \to 0$. In Fig. 8, we have plotted the normalized phase velocity V_p/c_s for dielectric spheres and spheroidal scatterers in free space randomly distributed and oriented versus concentration. Both real and complex dielectric constants $(\varepsilon_r)_1$ of the scatterers are assumed. The values are taken for ice and water particles from (47).

The dispersion relation given in Eqs. (7.1) and (7.3) may also be useful in obtaining the effective shear modulus at low frequencies. Defining an average shear modulus $\langle\mu\rangle = \omega^2 \langle\rho\rangle K^2$ where $\langle\rho\rangle = c\,\rho_1 + (1 - c)\rho$ is the average density, we find the effective shear modulus of the composite medium containing cylindrical inclusions :

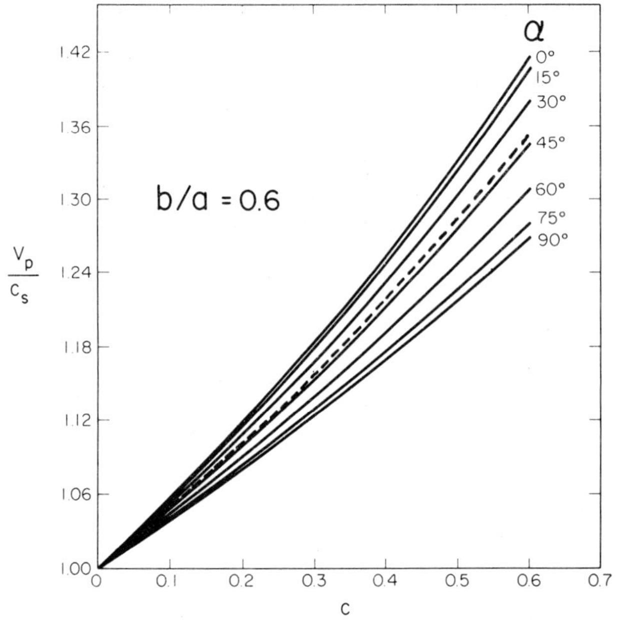

Fig. 6. Normalized phase velocity vs. concentration c for b/a = 0.6 and for various values of α; ——— aligned scatterers, ---randomly oriented scatterers

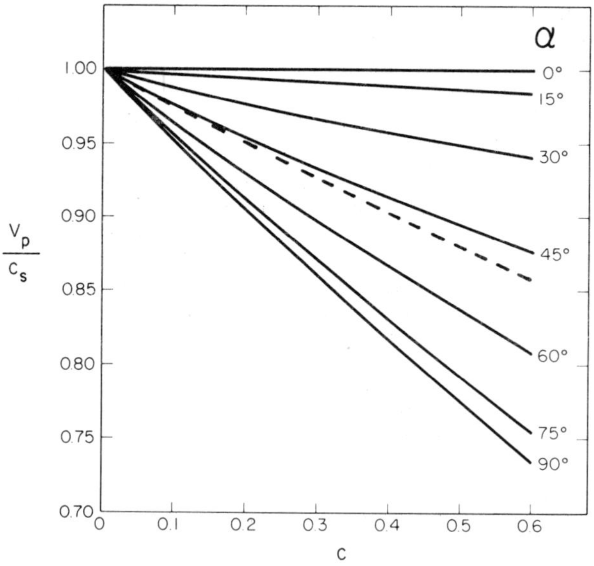

Fig. 7. Normalized phase velocity vs. concentration c for a random distribution of cracks; ——— aligned, --- random orientation

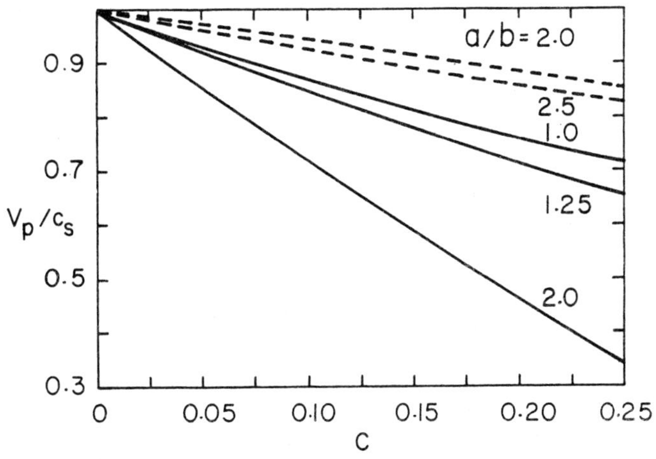

Fig. 8a. Normalized phase velocity vs. concentration for spherical and spheroidal dielectric scatterers in free space; —— lossy, --- lossless

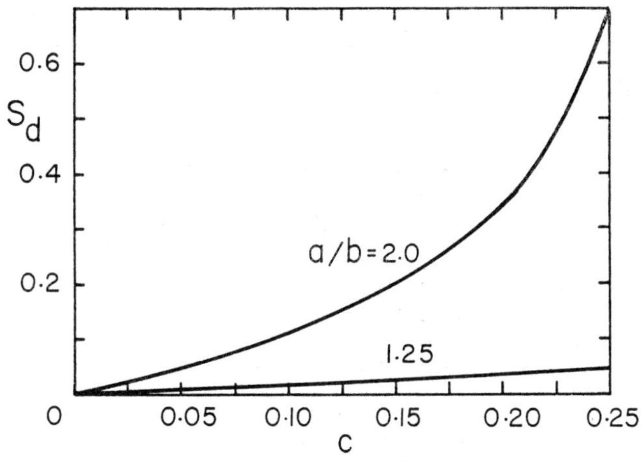

Fig. 8b. Attenuation S_d vs. concentration c for spheroidal lossy dielectric scatterers in free space

$$\langle\mu\rangle/\mu = (\langle\rho\rangle/\rho)\ (k_s/K)^2 \tag{7.12}$$

for more detail, we refer to (28).

Using the dispersion relation given in Eqs. (7.7) and (7.8), one could also compute the effective Shear and Bulk moduli of an elastic material containing a random distribution of stress free bubbles or cavities :

$$\frac{\langle\mu\rangle}{\mu} = \frac{4\mu - 3c\, E_2(9\lambda + 14\mu)}{4\mu + 6c\, E_2(3\lambda + 8\mu)} \;\; ; \;\; \frac{\langle B\rangle}{B} = \frac{(3\lambda + 2\mu)\,[1 - 6c\, E_0]}{(3\lambda + 2\mu)\,[1 + 3c\, E_0]} \qquad (7.13)$$

where $\langle\mu\rangle$ and $\langle B\rangle$ are the effective Shear and Bulk moduli of the inhomogeneous medium, and μ and B are the Shear and Bulk moduli of the matrix medium. This result might find applications in geophysics and material science, see for example Chaban (48).

VIIb. Phase Velocity and Attenuation at Higher Frequencies

To study the response at resonant and higher frequencies, we must consider higher powers of $k_\tau a$, and this implies that a larger number of terms ($X_{\tau n}$) must be kept in the expansion of the average field. This is best done numerically. For a given value of ka, the T-matrix for the scatterer is computed. Next, the coefficient matrix M corresponding to $X_{\tau n}$ (Eqs. (4.2) and (5.4)) is formed. The complex determinant of the coefficient matrix is computed using standard Gauss elimination techniques. For a given $k_\tau a$, the root of the equation det M = 0 is searched in the complex K plane ($K_1 + iK_2$) using Muller's method. Good initial guesses were provided by the Rayleigh limit solutions at low values of $k_\tau a$ and these could be used systematically to obtain convergence of roots at increasingly higher values of $k_\tau a$. The real part K_1 determines the phase velocity, while the imaginary part K_2 determines the coherent wave attenuation. Here, we present results for 2 - D SH-Wave and 3 - D electromagnetic problems. We define the normalized phase velocity as $V_p/c_s = k_s/K_1$ and the attenuation coefficient $S_d = 4\pi\, K_2/K_1$ so as to make it dimensionless.

For two dimensional elliptical cylinders (SH-Waves), phase velocity and attenuation are plotted versus $k_s a$ in Figs. 9-13 for various values of b/a, α and actual concentration $c_1 = \pi ab\, n_0$ for both random distribution with specific and random orientations and periodic distribution with specific orientation. We assume d = 2.53/2.72 and m = 25/3.87. The general behavior of the phase velocity is to increase at relatively low frequencies and then decrease rather sharply for higher frequencies. For b/a = 1.0, the results agree qualitatively with those obtained by Sutherland and Lingle (49). As α increases from 0° to 90°, it has been found that the normalized phase velocity is reduced as evidenced by Fig. 9. The anisotropic effect is greatly pronounced for lower values of b/a indicating that there is a substantial reduction in phase velocity for nearly flat oriented inclusions when the incident wave is in the direction perpendicular to the inclusions. This observation is of importance in geophysics. The results in Fig. 9 indicate that there is a significant difference in phase velocity for various orientations and that the corresponding values for random orientation lie in between 0° and 90° orientations.

In Figs. 10 and 12, we have plotted the coefficient of attenuation S_d versus $k_s a$. These results indicate that anisotropic effect is greatly pronounced ; substantial reduction of attenuation is observed when α = 90°. The general behavior of the damping or attenuation is to increase rapidly with frequency at low frequency range and decrease sharply or fluctuates slightly when frequency is increased. The results also indicate that there is a tendency for the attenuation to increase rapidly again at higher frequencies. The attenuation calculations for random orientation lie between 0° and 90° orientations, see Fig. 10. In Fig. 10, we have also plotted the coefficient of attenuation versus $k_s a$ for 2 - D cracks

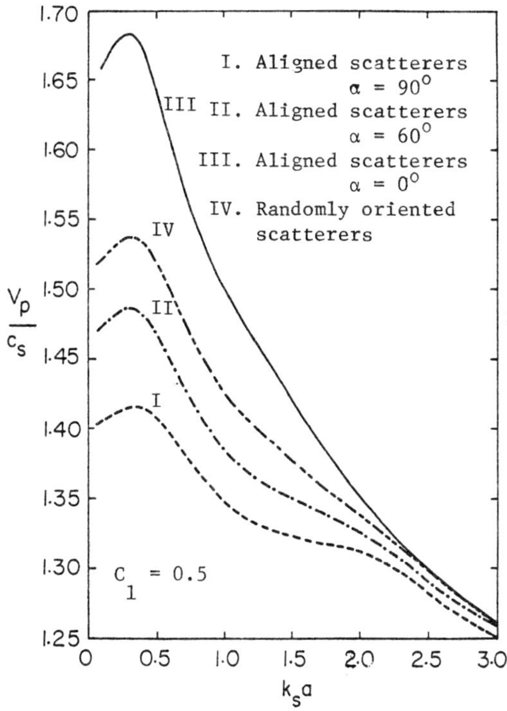

Fig. 9. Normalized phase velocity vs. $k_s a$ for a random distribution of aligned and randomly oriented scatterers

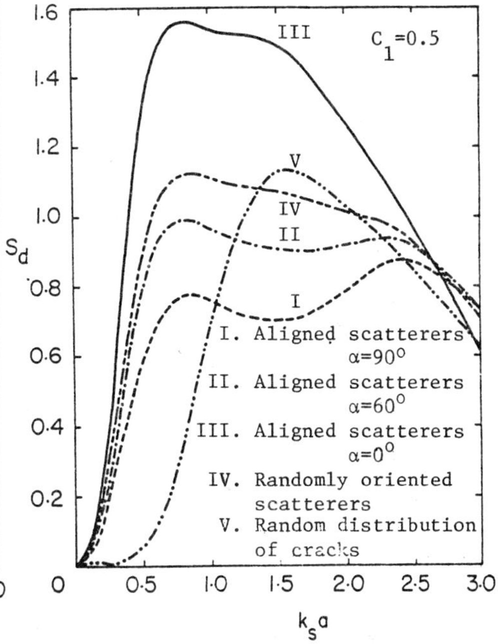

Fig. 10. Attenuation S_d vs. $k_s a$ for a random distribution of aligned and randomly oriented scatterers and cracks.

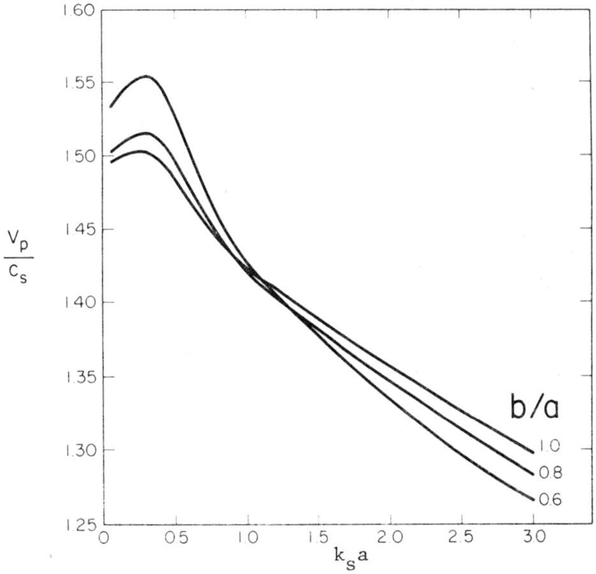

Fig. 11. Normalized phase velocity vs. $k_s a$ for $\alpha=45^0$ and $C_1=0.5$

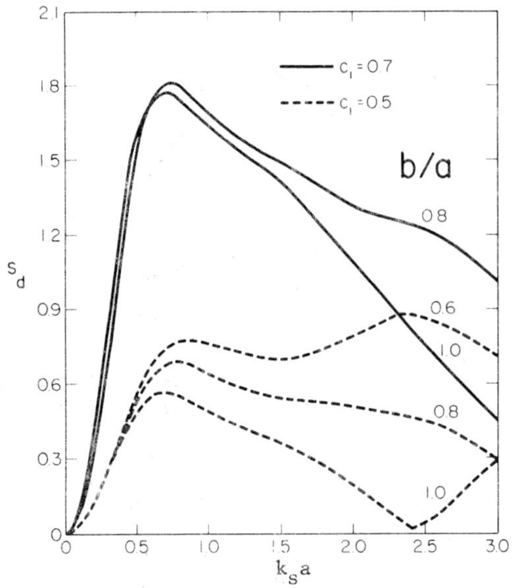

Fig.12. Attenuation S_d vs. $k_s a$ for various values of b/a and concentration C_1

Fig. 13a

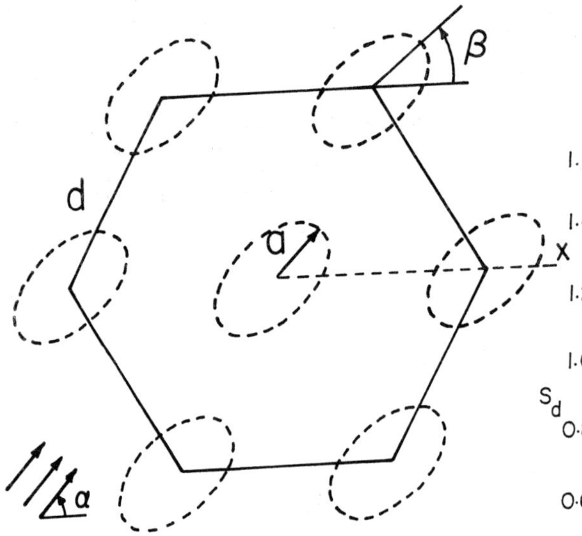

Fig. 13b

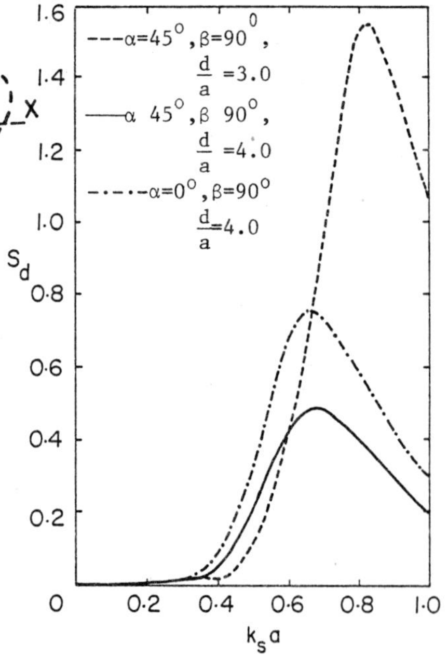

Fig. 13a. Hexagonally arranged scatterers

Fig. 13b. Attenuation S_d vs. $k_s a$ for periodically (hexagonal) arranged scatterers randomly distributed.

The frequency at which S_d begins to decrease is usually referred to as cut off frequency of the first pass band. At the cut off frequency, S_d decreases because energy begins to pass into the second pass band, see the explanations in (49,50). For periodic distribution, these cut off frequencies seem to move towards lower $k_s a$ as the intercylindrical spacing between the cylinders increase, see Fig. 13.

In Figs. 14 - 17, we have plotted phase velocity and attenuation coefficients for spherical and spheroidal dielectric scatterers in free space randomly distributed with parallel and random orientations. The calculations were performed for both lossless and lossy dielectric scatterers. For lossless scatterers, we assume dielectric constant ε_r = 3.168 which corresponds to ice at microwave frequencies. The imaginary part of the dielectric constant for ice is relatively small when compared to the real part, see Ray (47). For lossy case, we consider rain particles which have complex dielectric constants given as functions of temperature and frequency. The complex dielectric constants of rain particles are taken from the paper by Ray (47). For our calculations, we assume the temperature to be equal to 5°C.

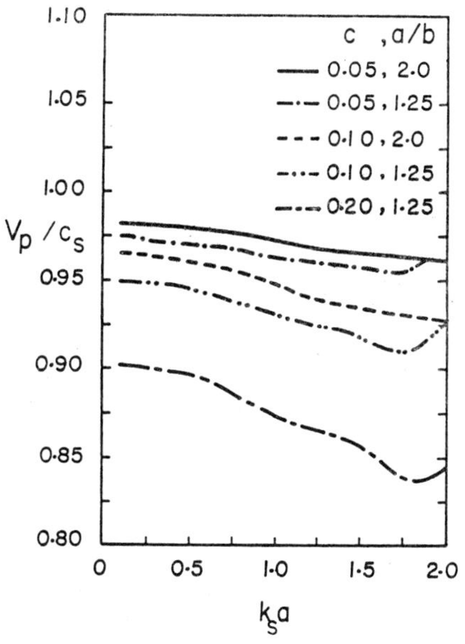

Fig. 14a. Normalized phase velocity vs $k_s a$ for aligned spheroidal dielectric scatterers

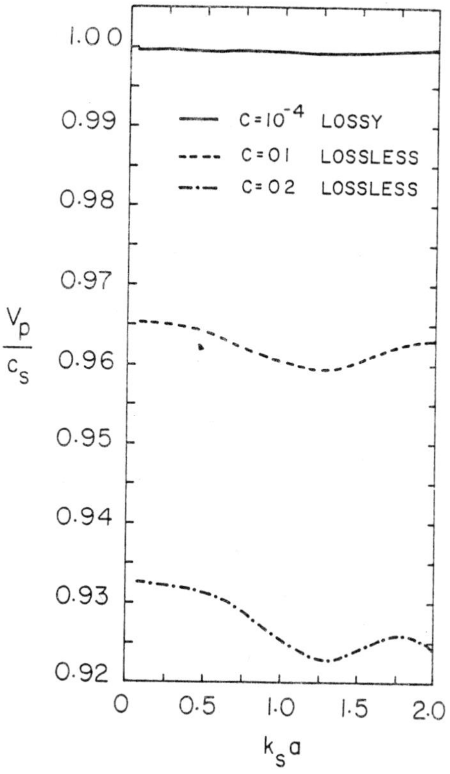

Fig. 14b. Normalized phase velocity vs $k_s a$ for randomly oriented spheroidal dielectric scatterers

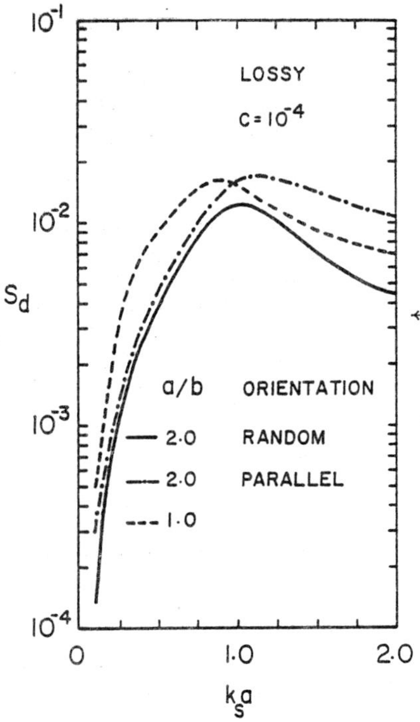

←Fig. 15

Fig. 16→

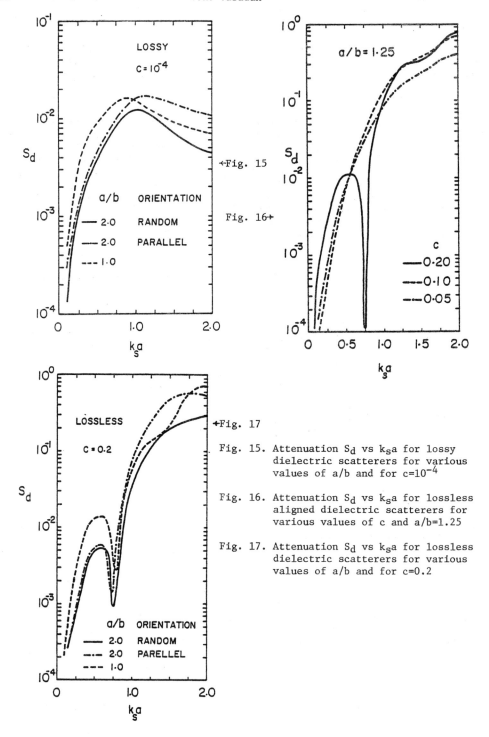

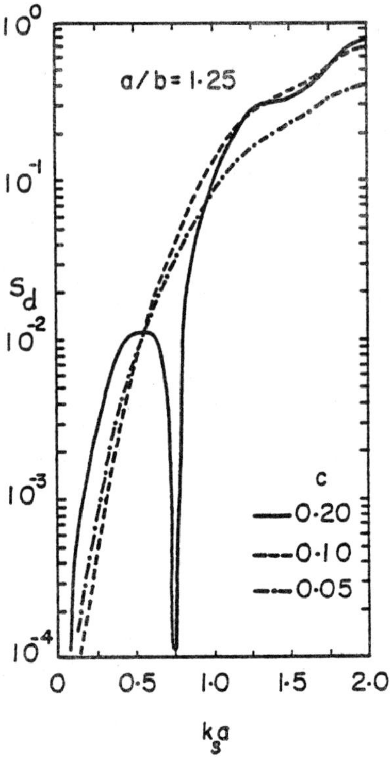

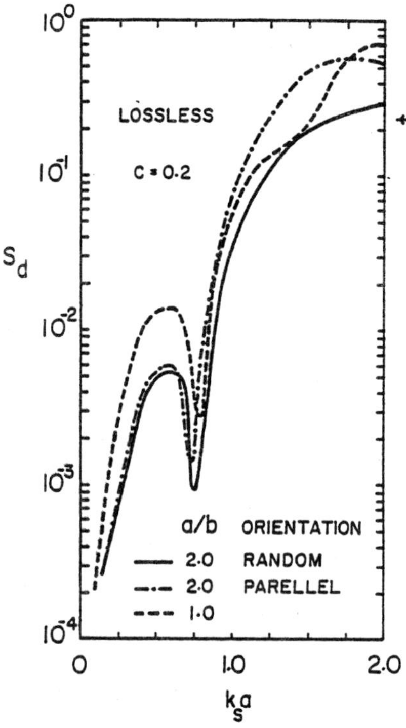

←Fig. 17

Fig. 15. Attenuation S_d vs $k_s a$ for lossy dielectric scatterers for various values of a/b and for c=10^{-4}

Fig. 16. Attenuation S_d vs $k_s a$ for lossless aligned dielectric scatterers for various values of c and a/b=1.25

Fig. 17. Attenuation S_d vs $k_s a$ for lossless dielectric scatterers for various values of a/b and for c=0.2

In Fig. 15, we show the attenuation coefficient S_d for lossy dielectric scatterers as a function of $k_s a$ for concentration $c = 10^{-4}$. The results are plotted for spherical and oblate spheroidal scatterers randomly distributed with both parallel and random orientations. For the case of parallel orientation, the spheroids are assumed to be oriented with their axes of revolution along the z-axis (see Fig. 3), and the incident wave is also assumed to propagate along the z-axis. In both specific and random orientation cases, it is well known that the bulk medium is not anisotropic and is characterized by a single wave number K. However, when the wave is incident obliquely to parallelly oriented scatterers, we obtain the anisotropic effects (just like the 2 - D case discussed above) and the bulk medium will then be characterized by an effective wave number with two different polarizations.

In Figs. 16 and 17, we have plotted S_d for lossless spheroidal dielectric scatterers for two different concentrations, c = 0.1 and 0.2, and for two different aspect ratios, a/b = 1.25, 2. The significant feature in the case of c = 0.2 is the presence of a sharp null at $k_s a \simeq 0.75$.

The multiple scattering analysis presented so far can also be extended to compute effective dynamic moduli of a composite medium (28). Bedeaux and Mazur (44) have obtained expressions for the average dielectric tensor in the medium and Varatharajulu (Varadan) and Vezzetti (45) have shown that for a given statistical model the zeros of the propagator in this model medium yield the dielectric tensor defined by Bedeaux and Mazur (44). In (28), we have extended this discussion to elastic wave propagation wherein we have given sample calculations for dynamic shear moduli as a function of $k_s a$.

In conclusion, we have presented a multiple scattering formalism for coherent wave propagation of acoustic, electromagnetic and elastic waves through an inhomogeneous medium composed of either random or periodic distribution of 2-dimensional and 3-dimensional scatterers. The scatterers are assumed to have either specific or random orientation. An important advantage is realized through the use of the T-matrix to characterize the scattering properties of any one scatterer. The methodology adapted in this analysis is general and can be used to include such effects as scatterer size distribution, depolarization, etc.

ACKNOWLEDGEMENTS

This work was supported in part by NOAA under Grant No: 04-78-B01-21 and in part by the U.S. Office of Naval Research under contract No: N00014-78-C-0559. The author would like to acknowledge helpful discussions with Professor V.V. Varadan and Dr. V.N. Bringi.

REFERENCES

(1) A. Ishimaru (1978), Wave Propagation and Scattering in Random Media, 2, Academic Press, New York.

(2) Lord Rayleigh, On the transmission of light through an atmosphere containing small particles in suspension, and on the origin of the blue of the sky, Philos. Mag. 47, 375 (1899).

(3) J.P. Watt, G.F. Davies and R.J. O'Connell, The elastic properties of composite materials, Rev. Geophys. Space Phys. 14, 541 (1976).

(4) L.L. Foldy, The multiple scattering of waves I. General theory of isotropic scattering by randomly distributed scatterers, Phys. Rev. 67, 107 (1945).

(5) M. Lax, Multiple scattering of waves II. Effective field in dense systems, Phys. Rev. 85, 621 (1952).

(6) V. Twersky, On scattering and reflection of sound by rough surfaces, J. Acoust. Soc. Am. 29, 209 (1957).

(7) V. Twersky, On scattering of waves by random distributions. I. Free-space scatterer formalism, J. Math. Phys. 3, 700 (1962).

(8) V. Twersky, Coherent scalar field in pair-correlated random distribution of aligned scatterers, J. Math. Phys. 18, 2468 (1977).

(9) V. Twersky, Coherent electromagnetic waves in pair-correlated random distribution of aligned scatterers, J. Math. Phys. 19, 215 (1978).

(10) F.C. Karal, Jr. and J.B. Keller, Elastic, electromagnetic and other waves in a random medium, J. Math. Phys. 5, 537 (1964).

(11) S.K. Bose and A.K. Mal, Longitudinal shear waves in a fiber-reinforced composite, Int. J. Solids and Structures 9, 1075 (1973).

(12) S.K. Bose and A.K. Mal, Elastic waves in a fiber-reinforced composite, J. Mech. Phys. Solids. 22, 217 (1974).

(13) A.K. Mal and S.K. Bose, Dynamic elastic moduli of a suspension of imperfectly bonded spheres, Proc. Camb. Phil. Soc. 76, 587 (1974).

(14) S.K. Datta, Scattering of a random distribution of inclusions and effective elastic properties, Continuum Models of Discrete Systems, University of Waterloo Press (1977).

(15) D.J. Vezzetti and J.B. Keller, Refractive index, attenuation, dielectric constant, and permeability for waves in a polarizable medium, J. Math. Phys. 8, 1861 (1967).

(16) P.C. Waterman and R. Truell, Multiple scattering of waves, J. Math. Phys. 2, 512 (1961).

(17) J.G. Fikioris and P.C. Waterman, Multiple scattering of waves. II. Hole corrections in the scalar case, J. Math. Phys. 5, 1413 (1964).

(18) B. Peterson and S. Ström, T-matrix for electromagnetic scattering from an arbitrary number of scatterers and representations of E(3), Phys. Rev. D. 8, 3661 (1973).

(19) B. Peterson and S. Ström, Matrix formulation of acoustic scattering from an arbitrary number of scatterers, J. Acoust. Soc. Am. 56, 771 (1974).

(20) B. Peterson, V.V. Varadan and V.K. Varadan, On the multiple scattering of waves from obstacles with solid-fluid interfaces, these proceedings.

(21) J.D. Achenbach, Waves and vibrations in directionally reinforced composites, Composite Materials (Edited by L.J. Broutman and R.H. Crock), Academic Press, New York, 2 (1972).

(22) V.K. Varadan, V.V. Varadan and Y.H. Pao, Multiple scattering of elastic waves by cylinders of arbitrary cross section. I. SH-waves, J. Acoust. Soc. Am. 63, 1310 (1978).

(23) V.K. Varadan and V.V. Varadan, Frequency dependence of elastic (SH-) wave velocity and attenuation in anisotropic two phase media, Int. J. Wave Motion. 1, 53 (1979).

(24) V.K. Varadan, Scattering of elastic waves by randomly distributed and oriented scatterers, J. Acoust. Soc. Am. 65, 655 (1979).

(25) V.K. Varadan, V.N. Bringi and V.V. Varadan, Coherent electromagnetic wave propagation through randomly distributed dielectric scatterers, Phys. Rev. D. 19, 2480 (1979).

(26) V.K. Varadan and V.V. Varadan, Dynamic elastic properties of a medium containing a random distribution of obstacles : Scattering matrix theory, The Materials Science Center Report # 2740, Cornell University, New York (1976).

(27) V.K. Varadan and V.V. Varadan, Multiple scattering of elastic waves by cylinders of arbitrary cross section. II. P- and SV- waves, The Materials Science Center Report # 2937, Cornell University, New York (1977).

(28) V.K. Varadan and V.V. Varadan, Characterization of dynamic shear modulus in inhomogeneous media using ultrasonic waves, Proceedings of the First International Symposium on Ultrasonic Materials Characterization, National Bureau of Standards, Washington, D.C., in press.

(29) V.V. Varadan and V.K. Varadan, Multiple scattering of electromagnetic waves by randomly distributed and oriented dielectric scatterers, submitted for publication.

(30) V.V. Varadan and V.K. Varadan, Acoustic, electromagnetic and elastic field equations pertinent to scattering problems, these proceedings.

(31) V.V. Varadan, Elastic wave scattering, these proceedings.

(32) B. Peterson, V.V. Varadan and V.K. Varadan, Scattering of acoustic waves by elastic and viscoelastic obstacles immersed in a fluid, Int. J. Wave Motion, in press.

(33) V.K. Varadan and V.V. Varadan, these proceedings.

(34) O.R. Cruzan, Translational addition theorems for spherical vector wave equations, Q. Appl. Math. 20, 33 (1962).

(35) B. Peterson and S. Ström, T-matrix formulation of electromagnetic scattering from multilayered scatterers, Phys. Rev. D. 10, 2670 (1974).

(36) G.N. Watson (1945), A Treatise on the Theory of Bessel Functions, The Macmillan Company, New York.

(37) J.A. Stratton (1941), Electromagnetic Theory, McGraw-Hill, New York.

(38) M.J.P. Musgrave (1970), Crystal Acoustics - Introduction to the Study of Elastic Waves and Vibrations in Crystals, Holden-Day.

(39) A.K. Ghatak and L.S. Kothari (1972), An Introduction to Lattice Dynamics, Addition-Wesley Publishing Company, Massachusetts.

(40) C. Kittel (1956), Introduction to Solid State Physics, John Wiley and Sons, Inc.

(41) V.K. Varadan and V.V. Varadan, Self-consistent approach to elastic wave propagation : application to periodic composite media, submitted for publication.

(42) V.K. Varadan, Dispersion of longitudinal shear waves in a medium containing periodic arrays of cylinders of arbitrary cross section Proc. 14th. Annual Meeting of the Soc. Eng. Sci., (1977).

(43) A.R. Edmonds (1957), Angular Momentum in Quantum Mechanics, Princetown University Press, Princetown, New Jersey.

(44) D. Bedeaux and P. Mazur, On the critical behavior of the dielectric constant for a nonpolar fluid, Physica, 67, 23 (1973).

(45) V.V. Varatharajulu (Varadan) and D.J. Vezzetti, Approach of the statistical theory of light scattering to the phenomenological theory, J. Math. Phys. 17, 232 (1976).

(46) V.V. Varadan and V.K. Varadan, Low Frequency expressions for acoustic wave scattering using Waterman's T-matrix method, J. Acoust. Soc. Am. in press.

(47) S.P. Ray, Broad band complex refractive index of ice and water, Appl. Optics. 11, 1836 (1972).

(48) I.A. Chaban, Calculation of effective parameters of micro inhomogeneous media by the self-consistent field method, Soviet Phys. Acoust. 11, 81 (1965).

(49) H.J. Sutherland and R. Lingle, Geometric dispersion of acoustic waves by a fiberous composite, J. Comp. Mater. 6, 490 (1972).

(50) F.C. Moon and C.C. Mow, Wave propagation in a composite material containing dispersed rigid spherical inclusions, The Rand Corporation Report, RM-6139-PR, Santa Monica, California (1970).

THE T MATRIX APPROACH TO SCATTERING FROM BURIED INHOMOGENEITIES

Gerhard Kristensson and Staffan Ström
Institute of Theoretical Physics, S-412 96 Göteborg, Sweden

ABSTRACT

The T matrix approach to acoustic and electromagnetic scattering from buried inhomogeneities is reviewed. The main structural features of the problem are outlined and the similarities with and differences from the layered scattered problem are pointed out. The general requirements on "useful" expansion systems are briefly discussed. The application of the T matrix approach is first described for the acoustic case and the analogous developments for the electromagnetic case are then indicated. Some general features of the results are discussed for a fairly general geometric shape of the ground surface. It is shown how an explicit solution is obtained when the ground surface is a plane. The case with the source above the ground is treated in some detail and it is shown how a source below the ground can be accommodated. Some further developments, such as the use of the flat surface case solution in an approximate treatment of a ground surface with a hill of finite extension, are also discussed briefly. A few examples of computations of the anomalous scattered field are given for unsymmetrically located, non-spherical inhomogeneities.

I. INTRODUCTION

In the present contribution we shall describe how the T matrix approach (1)-(4) can be used, with appropriate modifications, to determine the scattering from buried inhomogeneities. In order to make the presentation as simple and clear as possible we shall usually treat the acoustic case (5) first and then indicate what the generalizations to the electromagnetic case look like (6). As is often the case when one uses the T matrix approach, the similarities and analogies between the acoustic and electromagnetic cases will then become particularly transparent.

When presenting our formalism we shall first give the general analytical results which depend only on a minimum of assumptions concerning the interface and the inhomogeneity. In particular, it should be noted that our configuration is three-dimensional. After having displayed the general structure of the results, we shall demonstrate how more explicit useful results are obtained in special cases.

The interface above the inhomogeneity is an infinite surface and in the course of the development of our formalism we shall introduce certain regularity assumptions concerning that surface. Some of these will have to remain implicit. For instance, we shall always assume that the appropriate Green's representation of the field in terms of integrals over the interface exists. The weakest possible assumptions under which this is guaranteed do not seem to be known at present. The general features of the situation we shall consider as well as some relevant notations are given in Fig. 1

It is clear that the number and nature of the scattering surfaces which are present in the problem determine the essential features of the equations. It is therefore

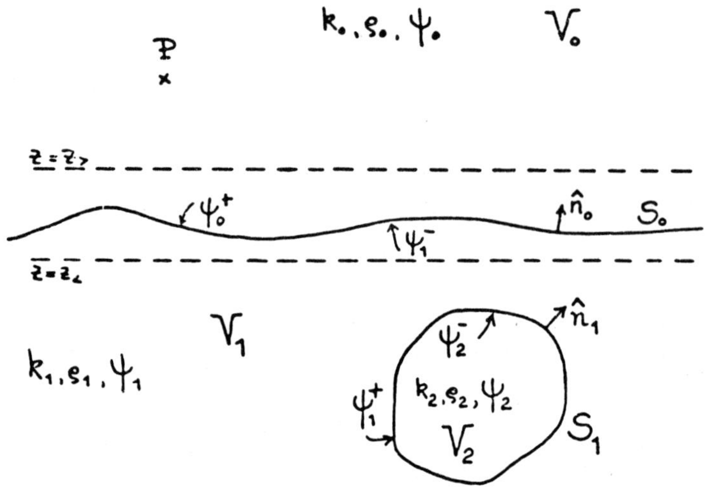

Fig. 1. Geometry and notation for the buried scatterer problem.

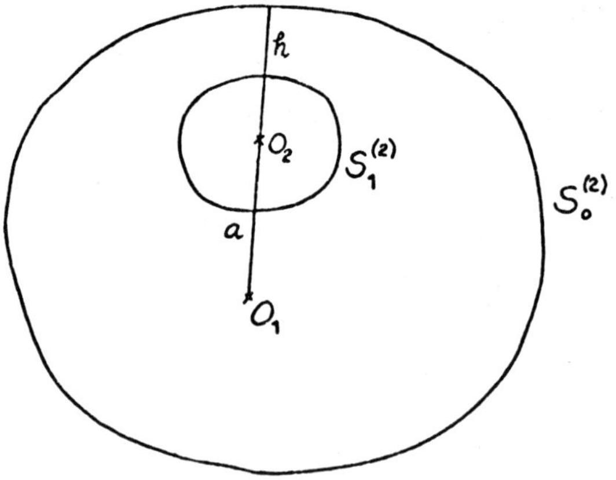

Fig. 2. Geometry and notation for the two-layered scatterer.

to be expected that some useful analogies may be made with the problem concerning a scatterer of the type depicted in Fig. 2. In particular, by letting O_1 in Fig. 2 recede to infinity (i.e. $a \to \infty$ with fixed O_2 and h) one approaches, in some sense, the situation in Fig. 1 in the vicinity of S_1. It turns out that an insight into the solution of the layered scatterer problem in Fig. 2 (7) is indeed very useful in constructing the solution to the buried scatterer problem.

In several instances one can in fact follow the details of the limiting procedure indicated above. The T matrix approach relies on repeated use of particular expansions for the free space Green's function $G(\vec{r}, \vec{r}'; k)$. It is therefore interesting to note that a fairly well-defined limiting procedure can be constructed which describes the transformation of the spherical wave expansion

$$G(\vec{r},\vec{r};k) = ik\sum_n \mathrm{Re}\psi_n(k\vec{r}_<)\ \psi_n(k\vec{r}_>) \tag{1}$$

in the limit $a \to \infty$ (the notation $\mathrm{Re}\psi_n$ and ψ_n for the spherical waves is explained in the appendix). By invoking the additional rules of contraction of group representations one is led to consider the limit

$$j_n(kr')\ h_n(kr)\ Y_{nm}(\theta,\phi)\ \bar{Y}_{nm}(\theta_1'\phi') \xrightarrow[(n\to\infty,\,r\to\infty,\,r^{-1}n=\lambda)]{}$$

$$\to (-i)\ \frac{\lambda d\lambda}{(\lambda^2-k^2)^{1/2}} \exp\left[-(\lambda^2-k^2)^{1/2}\cdot|z-z'|\right]\cdot J_m(\lambda\xi)\ J_m(\lambda\xi')e^{im(\phi-\phi')} \tag{2}$$

where (r,θ,ϕ) is a spherical coordinate system around O_1 and (ξ,ϕ,z) is a cylindrical system around O_2. The relations between the coordinates are: $r-a \to z$, $\theta r=\xi$, $\lambda=r^{-1}\cdot n$ (additional details are given in Ref. (5)). Thus in this limit, the spherical wave expansion (1) is transformed into a cylinder wave expansion:

$$G(r,r';k) = \sum_m \frac{2-\delta_{o,m}}{4\pi} \cos m(\phi-\phi')\ \cdot$$

$$\cdot\int_0^\infty \lambda d\lambda\cdot\ (\lambda^2-k^2)^{1/2}\ J_m(\lambda\xi)\ J_m(\lambda\xi')\ \exp\left[-(\lambda^2-k^2)^{1/2}|z-z'|\right]\cdot \tag{3}$$

An elaboration of other aspects of this limiting procedure yields further useful insights. However, here we content ourselves by pointing at these interesting relationships and we shall choose to develop our formalism directly for the configuration in Fig. 1. As we go along, we shall note numerous analogies with the T matrix approach to the layered scatterer problem of Fig. 2.

An important step in the T matrix approach is the expansion of the surface fields in a suitable system of functions which is complete on the surface. For a closed scattering surface the standard choice is the spherical waves. We note that in proving the completeness one makes use of the uniqueness properties of the inner and outer boundary value problems for Helmholtz' equation (1), (8). Our work on buried scatterers has led us to take a renewed interest in the uniqueness properties of solutions to Helmholtz' equation in the presence of infinite boundaries. The results of D.S. Jones (9) for Dirichlet's and Neumann's boundary conditions on an infinite surface which is a cone (with arbitrary cross section) outside a finite distance can be modified (10) so as to allow also finite inhomogeneities in the region considered. The opening angle of the cone may be arbitrary and therefore these results apply e.g. to a situation in which S_o in Fig. 1 is flat outside a finite circle. Furthermore, new results for the case of permeable media, separated by an interface which is a cone outside a finite distance have been obtained by G. Kristensson (10) (both media may contain a finite number of inhomogeneities).

The results which we have referred to above concern media without losses. If losses are introduced, it is usually substantially simpler to obtain e.g. uniqueness results. In the following we shall on occasion introduce a small imaginary part to the wave number k in intermediate stages. Thus we expect that if we allow this limiting absorbtion principle, several of the abovementioned results will in fact hold also for more general infinite surfaces.

Another key element in the T matrix approach is the extraction of equations for various expansion coefficients. This is done by means of a study of these expansions on the orthogonality surfaces associated with the particular functions which are used in the expansions of the fields. These orthogonality surfaces must not intersect the scattering surface (i.e. in the standard treatment of a closed scattering surface one considers the inscribed and circumscribed spheres). In the present problem natural choices of expansion systems lead to orthogonality surfaces which are planes. Consequently we shall furthermore assume that S_o is such that it lies between two parallel planes. In order to simplify the presentation we shall also assume that the region V_2 is homogeneous and S_1 such that the T matrix approach can be applied to compute the scattering from V_2. It will be seen that this assumption can be relaxed. The inhomogeneity V_2 enters in the final reresult only through its T matrix and it is irrelevant which particular method one uses in the computation of this T matrix.

II. DERIVATION OF THE BASIC EQUATIONS

We now turn to the derivation of a set of basic equations which will be used to compute the scattered field. To be specific, we assume that the source P of a scalar field ψ_o^{inc} lies in V_o. In V_o, V_1 and V_2 we then have fields ψ_r, r=0,1,2 which, except at P, satisfy (a time factor exp(-iωt) is suppressed throughout)

$$(\nabla^2+k^2)\psi_r = 0 \text{ in } V_r, \; r = 0,1,2. \tag{4}$$

In V_o we have $\psi_o=\psi_o^{inc}+\psi_o^{sc}$ and we shall concentrate on the calculation of ψ_o^{sc} in terms of ψ_o^{inc}. Furthermore, we would like to separate ψ_o^{sc} in a directly reflected component $\psi_o^{sc,dir.}$ and another component which depends on the presence of the inhomogeneity.

The starting point is Green's theorem applied to the field and the outgoing wave Green's function $G(\vec{r},\vec{r}; k)$. Consider first V_o. If we assume (cf. the introduction) that the integral over S exists, we have

$$\left.\psi_o(\vec{r})\right| = \psi_o^{inc}(\vec{r}) + \int_{S}\hat{n}_o \cdot \left[\psi_o^+(\vec{r}')\nabla'G(\vec{r},\vec{r}';k_o) - \right.$$

$$- G(\vec{r},\vec{r}';k_o)\ \nabla'\psi_o^+(\vec{r}')\Big]ds' \text{ for } \begin{cases}\vec{r} \text{ in } V_o \\ \vec{r} \text{ outside } V_o\end{cases} \tag{5}$$

(in general, the surface fields are denoted $\psi_r^{\pm}$ r=0,1,2 according to Fig.1). The next step is to introduce suitable expansions for G and ψ_o^+. The main consideration in choosing the expansion for G is that one should be able to use one and the same expansion in the whole integral over S_o in the right hand side of Eq. (5). It is further advantageous to be able to use the same system of functions for the expansion of G and ψ_o^+, $(\nabla\psi_o)^+$ (although this is by no means necessary). With the restrictions on S_o which were introduced in the introduction, plane wave or cylinder wave expansions are natural candidates. As was indicated previously, cylinder wave expansions can appear as limits of spherical wave expansions. Although much of our formalism was originally developed for cylinder waves we shall here use plane wave expansions in Eq. (4) since this is technically simpler and makes the structure of the equations somewhat more easily identified.

Let the z axis be perpendicular to the planes which enclose S_o. Among the available representations of G which involve plane waves we choose specifically the following expansion which contains both harmonic and evanescent waves:

$$G(\vec{r},\vec{r}';k_o) = i\ k_o(8\pi^2)^{-1}\int_0^{2\pi} d\beta \int_{C\pm} d\alpha\ \sin\alpha\ \exp\left[i\vec{k}_o \cdot (\vec{r}-\vec{r}')\right] \tag{6}$$

where $\vec{k}_o \equiv k_o$ (sinα cosβ, sinα sinβ , cosα) and where the integration is over C_+ if $z-z'>0$ and over C_- if $z-z'<0$. The contours $C_{\pm}$ are given in Fig. 3. The expansion (6) leads to a description of the fields in terms of waves which propagate and decrease in specific ways in half-spaces, a feature which makes it a natural choice in our problem. For instance, by considering an $\vec{r}$ with $z>z_>$ we get a representation of $G(\vec{r},\vec{r}'; k_o)$ as an integral over C_+ for all $\vec{r}'$ on S_o. Then the scattered field ψ_o^{sc} which is given by the integral over S_o in the right hand side of Eq. (5), will be given in terms of upwards-travelling plane harmonic waves and waves which decrease exponentially with increasing z. We recall that (6) is obtained from the standard representation of $G(\vec{r},\vec{r}'; k_o)$ as a three-dimensional fourier integral in k_{ox}, k_{oy} and k_{oz} by integrating over k_{oz} and introducing angular variables in the k_{ox}, k_{oy} plane (11). We note that similarly the cylinder wave expansion (3) contains both oscillating and evanescent contributions.

The prescribed incoming field ψ_o^{inc} is assumed to be generated at the point P with position vector $\vec{r}_P$ in V_o. We shall consider a ψ_o^{inc} which is a superposition of multipole fields emanating from P i.e.

$$\psi_o^{inc}(\vec{r}) = \sum_n a_n\psi_n(k_o(\vec{r}-\vec{r}_P)) \tag{7}$$

and in this way spherical waves are introduced into the problem. Furthermore, the scattering from the closed surface S_1 will be described in terms of spherical rather than plane waves. Consequently both spherical and plane waves will appear

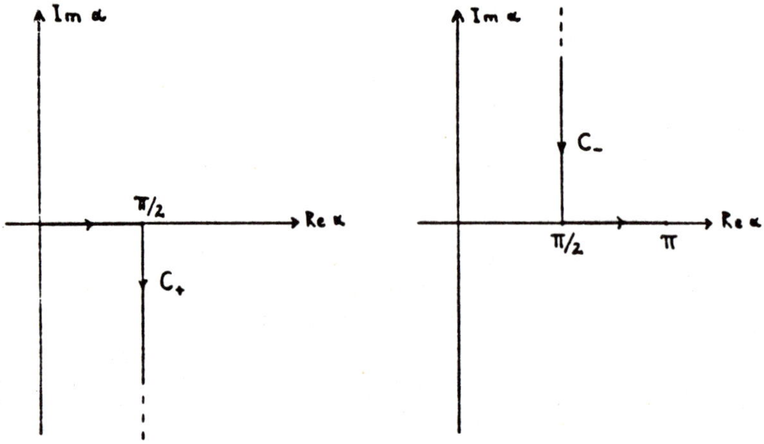

Fig. 3. The integration contours C_+ and C_-.

in several places in the formalism and we shall need the transformation formulas between these two sets of waves. For the scalar case they are (12)

$$\mathrm{Re}\,\psi_n(k\vec{r}) = (4\pi i^n)^{-1} \int_0^{2\pi} d\beta \cdot \int_0^{\pi} Y_n(\hat{k})\, e^{i\vec{k}\cdot\vec{r}} \sin\alpha d\alpha \tag{8}$$

$$\psi_n(k\vec{r}) = (2\pi i^n)^{-1} \int_0^{2\pi}\int_{C_\pm} Y_n(\hat{k}) e^{i\vec{k}\cdot\vec{r}} \cdot \sin\alpha d\alpha,\ z \gtrless 0 \tag{9}$$

$$d^{i\vec{k}\cdot\vec{r}} = \sum_n 4\pi i^n R_e\psi_n(k\vec{r})\ Y_n(\hat{k}) \tag{10}$$

Note that (6) is just a special case of (9). The explicit features of these relations limit the way in which they can be used and therefore they determine in part the structure of the equations we shall derive. For instance, there is no relation corresponding to (10) which contains $h_n(kr)$ on the right hand side. On the other hand, the two sides of (8) and (10) define analytic functions in the cosα plane (except for the branch points and cuts) i.e. they hold for all complex $\hat{k}$ in their common region of analyticity. Just as was the case with (6), we shall wish to arrange that when applying (9), we consider only z<0 or only z>0, so that a single C_- or C_+ integration will suffice.

It may be maintained that in particular cases the choice of plane and spherical waves is not an optimal one. However, we wish to emphasize that particularly in multiple scattering problems like the one considered here, there is another aspect which is of considerable importance for the choice of expansion systems, namely the rotation and translation properties of these systems. Rotations and translations in three-space togehter form the Euclidean (Lie) group E(3) and the natural language for describing rotations and translations of various function systems is that of the group representation theory for E(3) (in particular the theory of the induced unitary irreducible representations of E(3)).

In order for a system of functions to be useful, it is desirable that it exhibits a definite hierarchy of symmetries. This will then be manifested by the fact that there is a corresponding hierarchy of indices which characterize the functions so that to a specific subset of these indices there belongs a particular subset of transformations which affect only that particular subset of indices. This leads to the problem of identifying all the possible ways of decomposing E(3) transformations into (Lie) subgroups and of studying the corresponding decompositions of the representations of E(3). Here we will only limit ourselves to pointing out that these questions have been pursued vigorously during the latest decade and they are now essentially completely solved. Of relevance to us here is the fact (13), see also (14), that among these subgroup decompositions of E(3) there are only three inequivalent "maximal" ones which when applied to the representations, lead to maximal "subgroup-type" basis functions. These three subgroups are the three-dimensional rotation group, the three-dimensional translation group and the direct product of a two-dimensional Euclidean subgroup E(2) and a one-dimensional translation subgroup. The corresponding function systems are the spherical waves, the plane waves and the cylindrical waves where respectively a rotation, a translation and an E(2) transformation are expressed in a particularly simple fashion. The transformations between these systems have also been studied systematically within the group-theoretical context for the scalar case (see e.g. (13) and numerous references given there); considerable work remains to be done to obtain all the relations relevant to electromagnetic and elastic wave problems). Transformations which lie outside the subgroup in question will be described by more complicated relations but if the subgroup is "maximal" the number of such transformations is "minimal". However, also for these transformations, and especially for products of them, group theory provides a systematic approach. For instance, translations of spherical waves will not be a particularly simple operation, but the group-theoretical approach is still very useful since it provides integral representations, recursion relations as well as addition theorems for the translation matrices. To summarize, we see that the requirement that the function systems one uses should have reasonably simple transformation properties under E(3) transformations substantially reduces the number of function systems which are really useful.

By means of Eqs. (5), (6), (7) and (9) we now obtain specific representations for the incoming and scattered field in V_o. From Eq. (5), applied to an $\vec r$ with $z>z_>$ we get, after a change of the order of integration, the following representation of the scattered field

$$\psi_o^{sc}(\vec r) = \int_0^{2\pi} d\beta_o \int_{C_+} F(k_o)\, e^{i\vec k_o\cdot\vec r} \sin\alpha_o\, d\alpha_o \,, \quad z > z_> \tag{11}$$

where

$$F(k_o) = ik_o(8\pi^2)^{-1} \int_{S_o} \hat n_o \cdot \left[\psi_o^+(\vec r\,')\nabla'(e^{-i\vec k_o\cdot\vec r\,'}) - e^{-i\vec k_o\cdot\vec r\,'} \cdot \nabla'\psi_o^+(\vec r\,')\right] dS_o' \qquad \hat k_o \in C+ \tag{12}$$

For $z>z_P$ and $z<z_P$ ψ_o^{inc} can be expanded in terms of harmonic and evanescent plane waves by means of Eq. (9). Thus for $z<z_P$ we can also express ψ_o^{inc} as

$$\psi_o(\vec{r}) = \int_0^{2\pi} d\beta_o \int_{C_-} a(\vec{k}_o) e^{i\vec{k}_o\cdot\vec{r}} \cdot \sin\alpha_o d\alpha_o \tag{13}$$

where $a(\vec{k}_o)$ is a known function obtained from Eqs. (7) and (9). By introducing Eqs. (13) and (6) in the "extinction part" of Eq. (5) and by considering an $\vec{r}$ with $z<z_<$ we get

$$a(\vec{k}_o) = -ik_o(8\pi^2)^{-1} \cdot \int_{S_o} \hat{n}_o \cdot \left[\psi_o^+(\vec{r}')\nabla'(e^{-i\vec{k}_o\cdot\vec{r}'}) - e^{-i\vec{k}_o\cdot\vec{r}'}\nabla'\psi_o^+(\vec{r}')\right] dS_o' \Bigg|_{\hat{k}_o\in C_-} \tag{14}$$

Note that the integrals on the right hand side of Eq. (12) and Eq. (14) differ only in that $\hat{k}_o \in C_-$ in Eq. (14) whereas $\hat{k}_o \in C_+$ in Eq. (12).

In the T matrix approach the procedure is to eliminate the surface fields ψ_o^+ and $\nabla'\psi_o^+$ and this is done by means of the boundary conditions and an application of Green's theorem to the remaining parts of space which in the present case means one application to V_1 and another to V_2. From Green's theorem applied to V_1 we get

$$\left.\begin{matrix}\psi_1(\vec{r}) \\ 0\end{matrix}\right\} = -\int_{S_o} \hat{n}_o \cdot \left[\psi_1^-(\vec{r}')\nabla'G(\vec{r},\vec{r}';k_1) - G(\vec{r},\vec{r}';k_1)\nabla'\psi_1^-(\vec{r}')\right] dS_o' +$$

$$+\int_{S_1} \hat{n}_1 \cdot \left[\psi_1^+(\vec{r}')\nabla'G(\vec{r},\vec{r}';k_1) - G(\vec{r},\vec{r}';k_1)\nabla'\psi_1^+(\vec{r}')\right] dS_1' \quad \text{for} \begin{cases}\vec{r} \text{ in } V_1 \\ \vec{r} \text{ in } V_o \text{ or } 1/2\end{cases} \tag{15}$$

The next few steps can be patterned on the treatment of a layered finite scatterer (7). ψ_1^- and ψ_1^+ in Eq. (15) are related to ψ_o^+ and ψ_2^- by means of the boundary conditions. As in the case of a bounded two-layered scatterer we retain ψ_1^- and ψ_2^- as unknowns which are to be eliminated. The equations needed for this elimination are obtained from the two "extinction parts" of Eq. (15), namely $\vec{r}$ in V_o and $\vec{r}$ in V_2 (the case of $\vec{r}$ in V_1 will not be needed but it can of course be used to obtain various expressions for $\psi_1(\vec{r})$ in V_1).

The scattering from S_1 will be treated in the standard fashion by means of its T matrix referring to spherical waves. This description takes it simplest form when the origin O is chosen inside S_1 (cf. Fig. 1). The expansions of $\psi_2^-(\vec{r})$ is taken to be

$$\psi_2^-(\vec{r}) = \sum_n \alpha_n^{(2)} \mathrm{Re}\psi_n(k_2\vec{r}) \tag{16}$$

We note that the existence of an expansion of this type does not rely on the Rayleigh hypothesis. Since the spherical waves are not orthogonal on S_1, the coefficients $\alpha_n^{(2)}$ depend on the order of truncation used in Eq. (16). Furthermore, if we assume a corresponding expansion of $\hat{n}\cdot\nabla\psi_2^-$ in the complete system $\{\hat{n}\cdot\nabla\mathrm{Re}\psi_n\}$:

$$\hat{n}_1\cdot\nabla\Psi_2^- = \sum_n \beta_n^{(2)}\hat{n}_1\cdot\nabla\mathrm{Re}\psi_n(k_2\vec{r}) \tag{17}$$

it follows that $\alpha_n^{(2)}=\beta_n^{(2)}$ (2) (for non-resonant k_2 values). We also emphasize that if S_1 is characterized by a Dirichlet or Neumann boundary condition, we have a much greater freedom in the choice of expansions for $\hat{n}\nabla\psi_2^-$ and ψ_2^- respectively (15) (for instance, the expansion systems need not be solutions of Helmholtz' equation).

Consider now the corresponding question of finding suitable expansions for ψ_1^- and $\hat{n}\cdot\nabla\psi_1^-$ on S_o. Again it is instructive to compare with the situation for a layered scatterer. For two closed surfaces the set formed by the regular and outgoing waves is complete (5). By applying the same reasoning as is done in (2) for one closed surface, to two closed surfaces it follows that for non-resonant k values, the same expansion coefficients appear in the one of the field and its normal derivative. The modifications which occur in the presence of one finite and one infinite surface are discussed in (5). A complete set can be chosen as a combination of downgoing plane waves and outgoing spherical waves. On S_o this set can then be expressed in terms of downgoing and upgoing plane waves. In this way we are led to the ansatz

$$\psi_1^-(\vec{r}) = \int_0^{2\pi} d\beta_1\left[\int_{C_-}\alpha(\vec{k}_1) + \int_{C_+}\beta(\vec{k}_1)\right]e^{i\vec{k}_1\cdot\vec{r}}\cdot \sin\alpha_1 d\alpha_1 \tag{20}$$

We note that there are no nontrivial solutions to the homogeneous Neumann or Direchlet problem for V_1 when S_o fulfills the geometric condition given in (9) (for instance, when S_o is flat outside a finite region).

As indicated above the technique is now to use Eq. (15) for an $\vec{r}$ inside the inscribed sphere of S_1 (with origin in O) and for an $\vec{r}$ above $z=z_>$ in order to obtain two more relations involving $\alpha(\vec{k}_1)$, $\beta(\vec{k}_1)$ and $\alpha_n^{(2)}$. This is seen to be sufficient since by furthermore introducing the boundary conditions in Eqs. (12) and (14) and then expanding ψ_1^- and $\hat{n}\cdot\nabla\psi_1^-$, we obtain altogether four relations among the five coefficients $a(\vec{k}_o)$, $f(\vec{k}_o)$, $\alpha(\vec{k}_1)$, $\beta(\vec{k}_1)$ and $\alpha_n^{(2)}$. Thus we can calculate $f(\vec{k}_o)$ in terms of $a(\vec{k}_o)$ by means of elimination of $\alpha(\vec{k}_1)$, $\beta(\vec{k}_1)$ and $\alpha_n^{(2)}$. In the course of the derivation of these relations there appears the integral

$$I(\vec{k}_1,\vec{r}) \equiv \int_{S_o}\hat{n}_o\cdot\left[e^{i\vec{k}_1\cdot\vec{r}}\nabla' G(\vec{r},\vec{r}';k_1) - (\nabla' e^{i\vec{k}_1\cdot\vec{r}'})G(\vec{r},\vec{r}';k_1)\right]\cdot dS_o' \tag{21}$$

By appealing to the limiting absorbtion principle one finds that (5)

$$I(\vec{k}_1,\vec{r}) = \begin{cases} e^{i\vec{k}_1\cdot\vec{r}} & \text{for } k_1\varepsilon C_+ \\ 0 & \text{for } k_1\varepsilon C_- \end{cases} \Bigg\}\ \text{for } \vec{r} \text{ above } S_o \tag{22}$$

$$I(\vec{k}_1,\vec{r}) = \left\{\begin{matrix} 0 & \text{for } \hat{k}_1 \varepsilon C_+ \\ -e^{i\vec{k}_1\cdot\vec{r}} & \text{for } \hat{k}_1 \varepsilon C_- \end{matrix}\right\} \text{for } \vec{r} \text{ below } S_o \tag{23}$$

a result which is somewhat analogous to the orthogonality properties for various combinations of the spherical waves and their normal derivatives, which one makes use of in the treatment of the finite layered scatterer. In general, our aim is to extract coefficient relations from the equations one gets from Eqs. (5) and (15). This is particularly simple in the case of $f(\vec{k}_o)$ and $a(\vec{k}_o)$ since all the expansions which occur are in terms of plane waves and we can compare these expansions on a plane. However, the equations obtained from Eq. (15) contain both spherical and plane waves. The appropriate transformations between these systems must then be introduced before we can obtain coefficient relations by comparing expansions on planes or spheres. The explicit form of the transformations between spherical and plane waves in this way determines which of all the possible relations that can be obtained from Eq. (15), that are useful for our purposes. By carrying out the indicated technical steps one obtains the following four basic equations (and we refer to Ref. (5) for additional details of the derivation)

$$f(\vec{k}_o) = -i\int_0^{2\pi} d\beta_1\left[\int_{C_-}\alpha(\vec{k}_1) + \int_{C_+}\beta(k_1)\right] Q(\vec{k}_o,\vec{k}_1)\sin\alpha_1 d\alpha_1 \qquad \hat{k}_o\varepsilon C_+ \tag{24}$$

$$a(\vec{k}_o) = i\int_0^{2\pi} d\beta_1\left[\int_{C_-}\alpha(\vec{k}_1) + \int_{C_+}\beta(\vec{k}_1)\right] Q(\vec{k}_o,\vec{k}_1)\sin\alpha_1 d\alpha_1 \qquad \hat{k}_o\varepsilon C_- \tag{25}$$

$$4\pi i^{n-1}\int_0^{2\pi} d\beta_1\int_{C_-}\alpha(\vec{k}_1)\,Y_n(\hat{k}_1)\sin\alpha_1 d\alpha_1 = \sum_{n'} Q_{nn'}(\text{Out, Re})\alpha_{n'}^{(2)} \tag{26}$$

$$\beta(\vec{k}_1) = \sum_{nn'}(2\pi)^{-1}i^{-n-1}\,Y_n(\hat{k}_1)\,Q_{nn'}(\text{Re,Re})\alpha_{n'}^{(2)} \tag{27}$$

If the boundary conditions on S_o and S_1 are taken to be respectively $\psi_o^+=C_{o1}\psi_1^-$, $\hat{n}_o\cdot\nabla\psi_o^+=\hat{n}_o\cdot\nabla\psi_1^-$ and $\psi_1^+=C_{12}\psi_2^-$, $\hat{n}_1\cdot\nabla\psi_1^+=\hat{n}_1\cdot\nabla\psi_2^-$, the explicit form of the Q:s are

$$Q(\vec{k}_o,\vec{k}_1) \equiv k_o(8\pi^2)^{-1}\int_{S_o}\left[e^{-i\vec{k}_o\cdot\vec{r}'}\nabla'(e^{i\vec{k}_1\cdot\vec{r}'}) - C_{o1}(\nabla' e^{-i\vec{k}_o\cdot\vec{r}'})e^{i\vec{k}_1\cdot\vec{r}'}\right]\cdot\hat{n}_o dS_o' \tag{28}$$

$$Q_{nn'}(\text{Out,Re}) \equiv k_1\int_{S_1}\left[\psi_n(k_1\vec{r}')\nabla'\text{Re}\psi_{n'}(k_2\vec{r}') - C_{12}\,\nabla'\psi_n(k_1\vec{r}')\,\text{Re}\psi_n(k_2\vec{r}')\right]\cdot\hat{n}_1 dS_1' \tag{29}$$

(and analogously for $Q_{nn'}(\mathrm{Re},\mathrm{Re})$). The relations Eqs. (24)-(29) constitute the starting point in the present approach to the buried scatterer problem.

Before we describe how the scattered field is computed within this framwork we will show how a closely analogous set of relations is obtained in the electromagnetic case (6). Thus we consider the same geometry and we consider the equations for the electric field $\vec{E}$ and in analogy with the scalar case this is denoted $\vec{E}_o$, $\vec{E}_1$, $\vec{E}_2$, $\vec{E}_o^{\pm}$, $\vec{E}_1^{\pm}$ etc. in the different regions and on the different surfaces. Instead of Eqs. (4), (5) and (15) we now have

$$\nabla\times\left(\nabla\times\vec{E}_i(\vec{r})\right) - k_i^2\,\vec{E}_i(\vec{r}) = 0,\ i = 0,1,2 \tag{30}$$

$$\left.\begin{matrix}\vec{E}_o(\vec{r})\\ \vec{0}\end{matrix}\right\} = \vec{E}_o^{inc}(\vec{r}) + \nabla\times\int_{S_o}\hat{n}_o\times\vec{E}_o^{+}(\vec{r}\,')G(\vec{r},\vec{r}\,';k_o)dS_o +$$

$$+ k_o^{-2}\,\nabla\times\left\{\nabla\times\int_{S_o}\hat{n}_o\times\nabla'\times\vec{E}_o^{+}(\vec{r}\,')G(\vec{r},\vec{r}\,';k_o)dS_o'\right\}\quad\text{for}\begin{cases}\vec{r}\text{ in }V_o\\ \vec{r}\text{ outside }V_o\end{cases} \tag{31}$$

$$\left.\begin{matrix}\vec{E}_1(\vec{r})\\ \vec{0}\end{matrix}\right\} = -\nabla\times\int_{S_o}\hat{n}_o\times\vec{E}_1^{-}(\vec{r}\,')G(\vec{r},\vec{r}\,';k_1) -$$

$$-k_1^{-2}\,\nabla\times\left\{\nabla\times\int_{S_o}\hat{n}_o\times\left(\nabla'\times\vec{E}_1^{-}(\vec{r}\,')\right)G(\vec{r},\vec{r}\,';k_1)\,dS_o'\right\} +\nabla\times\int_{S_1}\hat{n}_1\times\vec{E}_1^{+}(\vec{r}\,')G(\vec{r},\vec{r}\,';k_1)dS_1'$$

$$+k_1^{-2}\nabla\times\left\{\nabla\times\int_{S_1}\hat{n}_1\times\left(\nabla'\times\vec{E}_1^{+}(\vec{r}\,')\right)G(\vec{r},\vec{r}\,';k_1)dS_1'\right\}\quad\text{for}\begin{cases}\vec{r}\text{ in }V_1\\ \vec{r}\text{ in }V_o\text{ or }V_2\end{cases} \tag{32}$$

In the electromagnetic case it is convenient to introduce Green's dyadic $\vec{J}G$ (2) ($\vec{J}$ denotes the unit dyadic). The expansion of $\vec{J}G$ in terms of vector spherical waves is wellknown. One has

$$\overleftrightarrow{J}\,G(\vec{r},\vec{r}\,';k) = ik\sum_{\substack{n,m,\delta\\ \tau=1,2}}\vec{\psi}_n(k\vec{r}_>)\,\mathrm{Re}\vec{\psi}_n(k\vec{r}_<) + \overleftrightarrow{J}_{irr} \tag{33}$$

where $\psi_n \equiv \psi_{nm\sigma\tau}$ etc. are the vector spherical waves (more details on the conventions for these functions are given in the appendix). $\vec{J}_{irr}$ is the irrotational part of $\vec{J}G$ and it contains the longitudinal vector waves (τ=3)

For the present problem one also needs the analogy of Eq. (33) in terms of plane waves. Since $\vec{J}$ can be written as $\sum_j \hat{a}_j\hat{a}_j$ where $\{\hat{a}_j\}$ is any righthanded orthogonal triplet of unit vectors, several choices are possible here. The choice of $\hat{a}_j$:s should preferrably meet the following requirements: i) it should lead to a separation of the longitudinal part of $\vec{J}G$. ii) it should be possible to combine it in a simple way with the transformations between spherical and plane vector waves. These transformations (12) are (cf. Eqs. (8) and (9)):

$$\vec{\psi}_n(k\vec{r}) = i(2\pi)^{-1}\cdot \int_0^{2\pi} d\beta \cdot \int_{C_\pm} \vec{B}_n(\hat{k})e^{i\vec{k}\cdot\vec{r}} \sin\alpha d\alpha, \; z \gtrless 0. \qquad (34)$$

$$\mathrm{Re}\vec{\psi}_n(k\vec{r}) = i(4\pi)^{-1} \int_0^{2\pi} d\beta \int_0^{\pi} \vec{B}_n(\hat{k})e^{i\vec{k}\cdot\vec{r}} \sin\alpha d\alpha \qquad (35)$$

where

$$\vec{B}_n(\hat{k}) \equiv (-i)^{n+2-\tau}\, \vec{A}_n(\hat{k}), \; \tau = 1,2.$$

(the $\vec{A}_n(\hat{k})$:s are defined in the appendix). Thus one finds (6) that a suitable choice is to take $\hat{a}_j$ to be the spherical unit vector of $\hat{k}\equiv\vec{k}/k$ and we use the notation $\hat{a}_1\equiv\hat{\alpha}$, $\hat{a}_2\equiv\hat{\beta}$, $\hat{a}_3\equiv\hat{k}$. Green's dyadic can then be written

$$\overleftrightarrow{J}G(\vec{r},\vec{r}';k) = \sum_{j=1}^{3} ik(8\pi^2)^{-1} \int_0^{2\pi} d\beta \cdot \int_{C_\pm} \hat{a}_j\cdot\hat{a}_j\cdot e^{i\vec{k}\cdot(\vec{r}-\vec{r}')} \sin\alpha d\alpha \qquad (36)$$

Using Eq. (31) we now get, corresponding to Eqs. (11)-(14)

$$\vec{E}_o^{sc}(\vec{r}) = \int_0^{2\pi} d\beta_o \int_{C_+} \vec{f}(\vec{k}_o)e^{i\vec{k}_o\cdot\vec{r}} \sin\alpha_o d\alpha_o \; , \; z > z \qquad (37)$$

where

$$\vec{f}(\vec{k}_o) = \sum_{j=1,2} f_j(\vec{k}_o)\,\hat{a}_j$$

and

$$f_j(\vec{k}_o) \equiv i\, k_o(8\pi^2)^{-1} \int_{S_o} \left\{(\hat{n}_o x\vec{E}_o^{\,+})\cdot(-i\vec{k}_o x\vec{a}_j) + \hat{n}_o x(\nabla' x\vec{E}_o^{\,+}) \cdot\hat{a}_j\right\} e^{-i\vec{k}_o\cdot\vec{r}'} dS_o'$$

$*$

$$\hat{k}_o \varepsilon C_+$$

$$\vec{E}_o^{inc}(\vec{r}) = \int_0^{2\pi} d\beta_o \int_{C_-} \vec{a}(\vec{k}_o)e^{i\vec{k}_o\cdot\vec{r}} \sin\alpha_o d\alpha_o \; , \; z < z_< \qquad (38)$$

where

$$\vec{a}(\vec{k}_o) = \sum_{j=1,2} a_j\cdot(\vec{k}_o)\,\hat{a}_j$$

and

$$a_j(\vec{k}_o) \equiv -i\, k_o(8\pi^2)^{-1} \int_{S_o} \left\{ (\hat{n}_o x\vec{E}_o^{\,+})\cdot(-i\,\vec{k}_o x\vec{a}_j) + \left[\hat{n}_o x(\nabla' x\vec{E}_o^{\,+}\right]\cdot\hat{a}_j \right\} e^{-i\vec{k}_o\cdot\vec{r}'} dS'_o \qquad \hat{k}_o \epsilon C_- \tag{39}$$

The tangential surface field expansions are now taken to be

$$\hat{n}_o x\vec{E}_1^{\,-}(\vec{r}') = \int_o^{2\pi} d\beta_1 \hat{n}_o x\left[\int_{C_-} \vec{\alpha}(\vec{k}_1) + \int_{C_+} \vec{\beta}(\vec{k}_1)\right] e^{i\vec{k}_1\cdot\vec{r}'} \sin\alpha_1 d\alpha_1 \tag{40}$$

$$\hat{n}_1 x\vec{E}_2^{\,-}(\vec{r}') = \sum_n \alpha_n\, \hat{n}_1 x\mathrm{Re}\vec{\psi}_n(k_2\vec{r}') \tag{41}$$

(the completeness of the function systems on the right hand sides of Eqs. (40) and (41) are discussed in Ref. (6)).

The continuity conditions for the tangential components on S_o and S_1 are

$$\hat{n}_o x\vec{E}_o^{\,+}(\vec{r}') = \hat{n}_o x\vec{E}_1^{\,-}(\vec{r}') \qquad \hat{n}_1 x\vec{E}_1^{\,+}(\vec{r}') = \hat{n}_1 x\vec{E}_2^{\,-}(\vec{r}')$$

$$\hat{n}_o x(\nabla' x\vec{E}_o^{\,+}(\vec{r}')) = C_{01}\, \hat{n}_o x(\nabla' x\vec{E}_1^{\,-}(\vec{r}')) \qquad \hat{n}_1 x(\nabla' x\vec{E}_1^{\,+}(\vec{r}')) = C_{12}\hat{n}_1 x(\nabla' x\vec{E}_2^{\,-}(\vec{r}')) \tag{42}$$

where $C_{01} = \frac{\mu_{0r}}{\mu_{1r}}$, $C_{12} = \frac{\mu_{1r}}{\mu_{2r}}$. By introducing Eqs. (42) and (40) into Eqs. (38) and (39) we get

$$f_j(\vec{k}_o) = i \sum_{j'=1,2} \int_o^{2\pi} d\beta_1 \left\{ \int_{C_-} \alpha_{j'}(\vec{k}_1) + \int_{C_+} \beta_{j'}(\vec{k}_1) \right\} Q_{jj'}(\vec{k}_o,\vec{k}_1) \sin\alpha_1 d\alpha_1, \qquad k_o \epsilon C_+ \tag{43}$$

$$a_j(\vec{k}_o) = i \sum_{j'=1,2} \int_o^{2\pi} d\beta_1 \left\{ \int_{C_-} \alpha_{j'}(\vec{k}_1) + \int_{C_+} \beta_{j'}(\vec{k}_1) \right\} Q_{jj'}(\vec{k}_o,\vec{k}_1) \sin\alpha_1 d\alpha_1, \quad \hat{k}_o \epsilon C_- \tag{44}$$

where

$$Q_{jj'}(\vec{k}_o,\vec{k}_1) \equiv -k_o(8\pi^2)^{-1} \int_{S_o} \left\{ (\hat{n}_o x\hat{a}_{j'})(-i\,\vec{k}_o x\hat{a}_j) + C_{01}\left[\hat{n}_o x(i\,\vec{k}_1 x\hat{a}_{j'})\right]\cdot\hat{a}_j \right\} e^{i(\vec{k}_1-\vec{k}_o)\cdot\vec{r}'} dS'_o \tag{45}$$

Analogously, we next consider Eq. (32) for $\vec{r}$ in V_o and V_2 respectively and introduce Eqs. (40)-(42). We then need the technical result, analogous to Eqs. (22) and (23) that $\vec{I}_j(\vec{k}_1,\vec{r}) \equiv \nabla\times \int_{S_o} \hat{n}_o\, \hat{a}_j G(\vec{r},\vec{r}';k_1)\exp(i\vec{k}_1\vec{r})dS'_o$ +

$$+\, k_1^{-2}\, \nabla\times\left\{ \nabla\times\int_{S_o} \hat{n}_o\, (ik_1\times\hat{a}_j)G(\vec{r},\vec{r}';k_1)\exp(i\vec{k}_1\cdot\vec{r};)dS'_o \right\}$$

simplifies according to (6)

$$\vec{I}_j(\vec{k}_1,\vec{r}) = \begin{cases} \hat{a}_j\, e^{i\,\vec{k}_1\cdot\vec{r}} \\ \vec{0} \end{cases}, \quad \hat{k}_1 \varepsilon C_{\pm} \text{ for } \vec{r} \text{ above } S_o \tag{47}$$

$$\vec{I}_j(\vec{k}_1,\vec{r}) = \begin{cases} \vec{0} \\ -\hat{a}_j e^{i\,\vec{k}_1\cdot\vec{r}} \end{cases}, \quad \hat{k}_1\ C_{\pm} \text{ for } \vec{r} \text{ below } S_o \tag{48}$$

At this stage the transformations between spherical and plane waves are needed. Taking these into ccount we obtain the following two relations from Eq. (32)

$$\int_0^{2\pi} d\beta_1 \int_{C_-} \vec{\alpha}(\vec{k}_1)\vec{B}_n^{\,+}(\hat{k}_1)\, \sin\alpha_1 d\alpha_1 = -(4\pi)^{-1} \sum_{n'} Q^{EM}_{nn'}\,(\text{Out, Re})\alpha_{n'} \tag{49}$$

$$\beta(\vec{k}_1) = (2\pi)^{-1} \sum_{nn'} \vec{B}_n(\hat{k}_1)\, Q^{EM}_{nn'}\,(\text{Re, Re})\alpha_{n'}, \quad k_1 \varepsilon C_+ \tag{50}$$

where

$$\vec{B}_n^{\,+}(\hat{k}) \equiv i^{n+2-\tau}\,\vec{A}_n(\hat{k})$$

and

$$Q^{EM}_{nn'}\,(\text{Out,Re}) \equiv k_1 \int_{S_1} \hat{n}_1 \cdot \Big\{ (\nabla' x \vec{\psi}_{n'}(k_1\vec{r}')\ x \text{Re}\vec{\psi}_{n'}(k_2\vec{r}') +$$

$$+ C_{12}\, \vec{\psi}_n(k_1\vec{r}')\, x\ \nabla' x \text{Re}\vec{\psi}_{n'}(k_2\vec{r}')) \Big\}\, dS_1' \tag{51}$$

and similarly for $Q^{EM}_{nn'}$(Re, Re). Consequently one obtains in the electromagnetic case, a set of basic equations (43), (44), (49) and (50) which are close analogue of the corresponding scalar equations (24)-(27).

It is clear from the treatment given above for the case of a source situated in V_o, that the same approach can be used when the source is situated e.g. in V_1. One gets the same number of equations and one can use the same surface fields as unknowns to be eliminated. The only difference is that the source terms i.e. the coefficients $\{a_n\}$ describing the incoming field appear in a different way in the equations. When the source lies in V_1 one needs an expansion of the incoming field inside the inscribed sphere of S_1 and also above $z=z_>$. In the first case one uses an expansion in terms of regular spherical waves and in the second case an expansion in terms of upgoing plane waves.

If we return to the scalar field case and specifically assume that a multipole source is situated at a point in V_1 with position vector $\vec{b}$ from O, we have

$$\psi_1^{inc}(\vec{r}) = \sum_n a_n\, \psi_n(\vec{r}-\vec{b}) \tag{52}$$

which then inside the inscribed sphere of S_1 is written

$$\psi_1^{inc}(\vec{r}) = \sum_n a'_n Re\psi_n(\vec{r}) \qquad (53)$$

whereas above $z=z_>$ we can write

$$\psi_1^{inc}(\vec{r}) = \int_0^{2\pi} d\beta_1 \int_{C_+} \hat{a}(\vec{k}_1) e^{i\vec{k}_1\cdot\vec{r}} \sin\alpha_1 d\alpha_1 \qquad (54)$$

Here a'_n and $a^1(\vec{k}_1)$ are known quantities, determined by a_n, by means of the translation properties of ψ_n and the transformation (9) of spherical waves into plane waves. By applying Green's theorem in the same way as before and by using the same surface field expansion one obtains now the following basic equations (5)

$$f(\vec{k}_o) = -i \int_0^{2\pi} d\beta_1 \left[\int_{C_-}\alpha(\vec{k}_1) + \int_{C_+} \beta(\vec{k}_1)\right] Q(\vec{k}_o,\vec{k}_1)\sin\alpha_1 d\alpha_1, \quad \hat{k}_o \varepsilon C_+ \qquad (55)$$

$$0 = \int_0^{2\pi} d\beta_1 \left[\int_{C_-} \alpha(\vec{k}_1) + \int_{C_+} \beta(\vec{k}_1)\right] Q(\vec{k}_o,\vec{k}_1)\sin\alpha_1 d\alpha_1, \quad \hat{k}_o \varepsilon C_- \qquad (56)$$

$$4\pi i^n \int_0^{2\pi} d\beta_1 \int_{C_-} \alpha(\vec{k}_1) Y_n(\vec{k}_1) \sin\alpha_1 d\alpha_1 =$$

$$= -a'_n + i \sum_{n'} Q_{nn'}(Out,Re)\alpha_{n'}^{(2)} \qquad (57)$$

$$\beta(\vec{k}_1) = a^1(\vec{k}_1) + \sum_{nn'} (2\pi)^{-1} i^{-n-1}\, Y_n(\hat{k}_1)\, Q_{nn'}(Re,Re)\alpha_{n'}(2) \qquad (58)$$

The complicated part of the solution of the buried scatterer problem, namely the multiple scatterings between S_o and S_1, is essentially independent of whether the source lies above or below S_o. The difference is confined to a factor multiplying an infinite series of terms corresponding to successive scatterings between S_o and S_1. This will be clear from the calculation of the scattered field, a subject to which we now address ourselves.

III. CALCULATION OF THE SCATTERED FIELD

We now have four relations among five quantities so that any of the coefficients $f(\vec{k}_o)$, $\alpha(\vec{k}_1)$, $\beta(\vec{k}_1)$ and $\alpha_n^{(2)}$ can be expressed in terms of $a(\vec{k}_o)$ (we consider the scalar case first). We shall concentrate on expressing $f(\vec{k}_o)$ in terms of $a(\vec{k}_o)$. When doing this we would like to separate that part, $\psi_o^{sc,dir}$, of ψ_o^{sc} which corresponds to a direct reflection at S_o i.e. we want to decompose ψ_o^{sc} as $\psi_o^{sc}=\psi_o^{sc,dir}+\psi_o^{sc,anom}$ where $\psi_o^{sc,anom}$ is different from zero only in the presence of an inhomogeneity.

The scattering from S_1 is naturally described in terms of its T matrix which is (1)

$$T_{nn'}(1) = - \sum_{n''} Q_{nn''}(Re,Re)\,[Q^{-1}(Out,Re)]_{n''n'} \tag{59}$$

This quantity appears when we eliminate $\alpha_n^{(2)}$ between Eqs. (26) and (27). We then get

$$\beta(\vec{k}_1) = 2\sum_{n'} i^{n'-n} Y_n(\hat{k}_1)\, T_{nn'}(1) \int_0^{2\pi} d\beta_1' \int_{C_-} \alpha(\vec{k}_1')\, Y_n(\hat{k}_1')\, \sin\alpha_1'\, d\alpha_1' \tag{60}$$

$$\hat{k}_1 \varepsilon C_+$$

This relation has the expected structure. It expresses how downgoing plane waves ($\alpha(\vec{k}_1')$) are scattered into plane upgoing plane waves ($\beta(\vec{k}_1)$). Since the scattering is described by a T matrix referring to spherical waves, the plane waves are first transformed into spherical waves ($Y_n(\hat{k}_1')$), scattered ($T_{nn'}(1)$) and then transformed back to plane waves ($Y_n(\hat{k}_1)$).

The three remaining relations i.e. Eqs. (24), (25) and (60) relate plane wave coefficients and these functional relations should in principle be solved for all relevant $\vec{k}$-values. In practice some form of discretization has to be introduced and this can conceivably be done in several ways. We introduce a partial descretization by defining the set $\{C_n\}$ according to

$$C_n \equiv \int_0^{2\pi} d\beta_1 \int_{C_-} \alpha(\vec{k}_1)\, Y_n(\hat{k}_1)\, \sin\alpha_1 d\alpha_1 \tag{61}$$

The C_n:s may be called the spherical wave projections of $\alpha(\vec{k}_1)$ (note that C_n appears on the righthand side of Eq. (60)). We note that $\alpha_1 \varepsilon C_-$, $\beta_1 \varepsilon$ 0,2π does not constitute an orthogonality region for $Y_n(\hat{k}_1)$ and therefore there is no simple direct way of inverting Eq. (61). However, if we insert Eq. (61) into Eq. (25) we get

$$a(\vec{k}_o) = i \int_0^{2\pi} d\beta_1 \left[\int_{C_-} \alpha(\vec{k}_1) + \int_{C_+} 2 \sum_{nn'} i^{n'-n}\, Y_n(\hat{k}_1) T_{nn'}(1) C_{n'} \right]$$

$$\cdot\, Q(\vec{k}_o,\vec{k}_1)\, \sin\alpha_1\, d\alpha_1,\ \hat{k}_o \varepsilon C_- \tag{62}$$

Eq. (62) may be said to be in a hybrid form since it contains the unknown $\alpha(\vec{k}_1)$ both in plane wave form and in terms of spherical wave projections.

In analogy with the $Q_{nn'}$ matrices for finite scatterers, $Q(\vec{k}_o,\vec{k}_1)$ can be interpreted to correspond to a passage of a plane wave through S_o from below. Its formal inverse $Q^{-1}(\vec{k}_o,\vec{k}_1)$ then corresponds to a passage through S_o from above. We emphasize, however, that $Q(\vec{k}_o,\vec{k}_1)$ is in general not a function and relations involving $Q(\vec{k}_o,\vec{k}_1)$ must be understood in a distribution sense. For instance if S_o is a plane $Q(\vec{k}_o,\vec{k}_1)$ is proportional to $\delta^{(2)}(\hat{n} \times (\vec{k}_1-\vec{k}_o))$. It is still suggestive to consider some strictly formal manipulations involving also the inverse of $Q(\vec{k}_o,\vec{k}_1)$. If we allow that operation, Eq. (62) can be written as follows

$$\alpha(\vec{k}_1) = -i \int_o^{2\pi} d\beta_o \int_{C_-} Q^{-1}(\vec{k}_o,\vec{k}_1)\ [a(\vec{k}_o) -$$

$$-2i \sum_{nn'} i^{n'-n}\ T_{nn'}(1)\ C_n \int_o^{2\pi} d\beta_1' \int_{C_+} Q(\vec{k}_1,\vec{k}_1')\ Y_n(\hat{k}_1)\ \sin\alpha_1'\ d\alpha_1']\sin\alpha_o d\alpha_o$$

$$\hat{k}_1 \varepsilon C_- \tag{63}$$

which is still in hybrid form. However, if Eq. (63) is multiplied by $Y_n(\vec{k}_1)$ and integrated, $\alpha(\vec{k}_1)$ will appear only in terms of its spherical wave projections. By means of these operations we get the matrix equation (Ref. (5))

$$C_n \equiv d_n - \sum_{n'} A_{nn'}\ d_{n'} \tag{64}$$

where

$$d_n \equiv -i \int_o^{2\pi} d\beta_1 \int_{C_-} Y_n(\hat{k}_1) \left[\int_o^{2\pi} d\beta_o \int_{C_-} Q^{-1}(\vec{k}_o,\vec{k}_1)\ a(\vec{k}_1)\ \sin\alpha_o d\alpha_o\right] \sin\alpha, d\alpha_1 \tag{65}$$

$$A_{nn'} \equiv -2 \sum_{n''} i^{n'-n''}\ T_{n''n'}(1) \int_o^{2\pi} d\beta_1 \int_{C_-} Y_n(\hat{k}_1)\ \cdot$$

$$\cdot \int_o^{2\pi} d\beta_1' \int_{C_-} R(\hat{k}_1,\vec{k}_1')\ Y_{n''}(\hat{k}_1')\sin\alpha_1'\ d\alpha_1'\ \sin\alpha_1\ d\alpha_1 \tag{66}$$

$$R(\vec{k}_1,\vec{k}_1') \equiv - \int_o^{2\pi} d\beta_o \int_{C_-} Q^{-1}(\vec{k}_1,\vec{k}_1)\ Q(\vec{k}_o,\vec{k}_1')\ \sin\alpha_o\ d\alpha_o$$

$$\hat{k}_1 \varepsilon C_-,\ \hat{k}_1' \varepsilon C_+ \tag{67}$$

In view of the interpretation of Q and Q^{-1} it is seen that $R(\vec{k}_1,\vec{k}_1^1)$ has the meaning of a reflection coefficient for the interface S_o for plane waves coming from below (note that there is a misprint in the corresponding Eq. (52) of Ref. (5) as well as in the lowest of the figures 5 there: the factors Q and Q^{-1} there should be commuted). We can also form

$$\tilde{R}(\vec{k}_o,\vec{k}_o') \equiv - \int_0^{2\pi} d\beta_1 \int_{C_-} Q(\vec{k}_o,\vec{k}_1)\, Q^{-1}(\vec{k}_o,\vec{k}_1) \sin\alpha_1\, d\alpha_1$$

$$\hat{k}_o \varepsilon\, C_+,\ \hat{k}_o' \varepsilon C_- \tag{68}$$

which is then the reflection coefficient of S_o for plane waves coming from above. Similarly, the interpretation of d_n is clear: it is the spherical wave projections $(Y_n(\vec{k}_1))$ of the incoming field $(a(\vec{k}_o))$ after it has passed through S_o from above $(Q^{-1}(\vec{k}_o,\vec{k}_1))$. Furthermore we note that $A_{nn'}$ is independent of the fields and is determined solely by the scattering properties of S_1 $(T_{nn'}(1))$ and S_o $(R(\vec{k}_o,\vec{k}_1))$. Consequently, d_n and $A_{nn'}$ are known quantities and thus Eq. (63) can be used to calculate C_n. In a vector and matrix notation the solution is

$$\vec{C} = (1+A)^{-1}\vec{d} \tag{69}$$

Solving Eq. (69) actually means solving the multiple scattering problem. A contains a product of $T_{nn'}$ and $R(\vec{k}_1,\vec{k}_1^1)$ and therefore describes a reflection from below at S_o and a consecutive scattering from S_1 and thus higher powers of A corresponds to repeated scatterings between S_o and S_1. By making an expansion of $(1+A)^{-1}$ in terms powers of A we thus obtain the solution in a form, which has a natural multiple scattering interpretation.

When the C_n:s have been determined (we refer to (5) for a more detailed discussion of the solution of Eq. (69)), the result can be introduced into Eqs. (63) and (60) to give $\alpha(\vec{k}_1)$ and $\beta(\vec{k}_1)$ and thus also $f(\vec{k}_o)$ (from Eq. (24)). In this way one obtains $\psi_o^{sc}(\vec{r})=\psi_o^{sc,dir}(\vec{r})+\psi_o^{sc,anom}(\vec{r})$ where

$$\psi_o^{sc,dir}(\vec{r}) \simeq \int_0^{2\pi} d\beta_o \int_{C_+} \left\{ \int_0^{2\pi} d\beta'_o \int_{C_-} \tilde{R}(\vec{k}_o,\vec{k}_o')\, a(\vec{k}_o') \sin\alpha_o'\, d\alpha_o' \right\} \cdot \tag{70}$$

$$\cdot e^{i\vec{k}_o\cdot\vec{r}} \cdot \sin\alpha_o\, d\alpha_o$$

$$\psi_o^{sc,anom}(\vec{r}) = \int_0^{2\pi} d\beta_o \int_{C_+} 2 \sum_{nn'} i^{n'-n-1}\, T_{nn'}(1)\, C_{n'} \cdot$$

$$\cdot \left\{ \int_0^{2\pi} d\beta_1 \int_{C_+} Y_n(\hat{k}_1) [Q(\vec{k}_1,\vec{k}_1) + \int_0^{2\pi} d\beta_1' \int_{C_-} R(\vec{k}_1',\vec{k}_1)\, Q(\vec{k}_o,\vec{k}_1') \sin\alpha_1'\, d\alpha_1'] \cdot \right.$$

$$\left. \sin\alpha_1\, d\alpha_1 \right\} e^{i\vec{k}_o\cdot\vec{r}} \sin\alpha_o\, d\alpha_o \tag{71}$$

Thus, although most of the above rearranging of the equations has been of a formal nature, it provides useful insights into the physical meaning of the various quantities which appear, as well as into the structure of the resulting equations and field representations. In the above formal development we have not introduced any additional assumptions on the interface S_o or the inhomogeneity S_1. Below we shall do this and consider the case of a plane surface S_o. Naturally, a much more explicit solution can be obtained in this case. However, we shall first give the results corresponding to Eqs. (59)-(71) for the electromagnetic case.

There are in fact close analogies in the electromagnetic case to all the steps leading to the final expressions (71) and (72) for the scattered field (Ref. (6)). As is well known, there is a corresponding T matrix $T^{EM}_{nn'}(1)$ for S_1 and if $\alpha_{n'}$ is eliminated between Eqs. (49) and (50), we get

$$\vec{\beta}\,(\vec{k}_1) = 2 \sum_{nn'} \vec{B}_n(\hat{k}_1)\; T^{EM}_{nn'}(1) \int_0^{2\pi} d\beta_1 \int_{C_-} \vec{\alpha}(\vec{k}_1)\vec{B}^{+}_n(\hat{k}_1)\sin\alpha_1 d\alpha_1 \tag{72}$$

Thus it is now useful to introduce the spherical vector wave projections

$$C^{EM}_n \equiv C^{EM}_n \equiv \int_o^{2\pi} d\beta_1 \int_{C_-} \vec{\alpha}(\vec{k}_1) \cdot \vec{B}_n^{\,+}(\hat{k}_1)\; \sin\alpha_1\; d\alpha_1 \tag{73}$$

and as a result Eq. (44) can be written (cf. Eq. (62))

$$a_j(\vec{k}_o) = i \sum_{j'} \int_0^{2\pi} d\beta_1 \left\{ \int_{C_-} \alpha_{j'}(\vec{k}_1) + 2 \sum_{nn'} \int_{C_+} B_{nj'}(\hat{k}_1)\; T^{EM}_{nn'}(1)\; C^{EM}_{n'} \right\} \cdot$$

$$\cdot\; Q_{jj'}(\vec{k}_o,\vec{k}_1)\; \sin\alpha_1\; d\alpha_1,\; \hat{k}_o \varepsilon C_- \tag{74}$$

where $B_{nj}(\hat{k}) \equiv \vec{B}_n(\hat{k})\cdot\hat{a}_j$ (and similarly we shall use $B^{\dagger}_{nj}(\hat{k}) \equiv \vec{B}^{\dagger}_n(\hat{k})\cdot\hat{a}_j$). The remarks made previously on the structure of the corresponding relation Eq. (62) and the meaning of $Q(\vec{k}_o,\vec{k}_1)$ (now $Q_{jj'}(\vec{k}_o,\vec{k}_1)$) still apply. We proceed formally as before and rearrange Eq. (73) (cf. Eq. (63))

$$\alpha_j(\vec{k}_1) = -i \sum_{j_1} \int_0^{2\pi} d\beta_o \int_{C_-} Q^{-1}_{jj_1}(\vec{k}_o,\vec{k}_1)\; [a_{j1}(\vec{k}_o) +$$

$$+2 \sum_{nn'j'} T^{EM}_{nn'}(1)\; C^{EM}_{n'} \int_o^{2\pi} d\beta_1' \int_{C_+} B_{nj'}(\vec{k}_1')\; Q_{j_1j'}(\vec{k}_o,\vec{k}_1')\; \sin\alpha_1'\; d\alpha_1'] \sin\alpha_o\; d\alpha_o \tag{75}$$

We perform those operations on Eq. (74) which are required in order to get C^{EM}_n on the left hand side i.e. we multiply by $B^{\dagger}_{nj}(\hat{k}_1)$, sum over j and integrate over $(0,2\pi)$ and C_-. In this way one obtains again

$$C^{EM}_n = d^{EM}_n - \sum_{n'} A^{EM}_{nn'} \cdot C^{EM}_{n'} \tag{76}$$

where

$$d_n^{EM} \equiv -i \sum_{jj'} \int_0^{2\pi} d\beta_1 \int_{C_-} \sin\alpha_1 \, d\alpha_1 \int_0^{2\pi} d\beta_0 \int_{C_-} a_j'(\vec{k}_0) \, Q^{-1}_{jj'}(\vec{k}_1,\vec{k}_0) \, B^+_{nj}(k_1) \sin\alpha_0 \, d\alpha_0 \tag{77}$$

$$A^{EM}_{nn'} \equiv -2 \sum_{n''jj'} \int_0^{2\pi} d\beta_1 \int_{C_-} \sin\alpha_1 d\alpha_1 \int_0^{2\pi} d\beta_0' \int_{C_-} \sin\alpha_0' \, d\alpha_0' B_{n''j'}(\hat{k}_1') \cdot T^{EM}_{n''n'}(1) \, R_{jj'}(\vec{k}_1,\vec{k}_1') \, B^+_{nj}(k_1) \tag{78}$$

Here $R_{jj'}(\vec{k}_1,\vec{k}_1')$ is the reflection coefficient at S_0 for plane vector waves coming from below:

$$R_{jj'}(\vec{k}_1,\vec{k}_1') \equiv - \sum_{j_1} \int_0^{2\pi} d\beta_0 \int_{C_-} Q^{-1}_{jj_1}(\vec{k}_0,\vec{k}_1) \, Q_{j_1j'}(\vec{k}_0,\vec{k}_1') \sin\alpha_0 d\alpha_0 \tag{79}$$

$$\hat{k}_1 \varepsilon C_-, \hat{k}_1 \varepsilon C_+$$

Comparing Eqs. (76)-(79) with the corresponding relations Eqs. (64)-(67), it is seen that on can proceed as in the scalar case and get $\vec{E}_0^{sc}=\vec{E}_0^{sc,dir}+\vec{E}_0^{sc,anom}$ where

$$\vec{E}_0^{sc,dir}(\vec{r}) = \int_0^{2\pi} d\beta_0 \int_{C_+} \sin\alpha_0 d\alpha_0 \Big\{\int_0^{2\pi} d\beta_0' \int_{C_-} \sin\alpha_0' d\alpha_0' \sum_{jj'} \tilde{R}_{jj'}(\vec{k}_0,\vec{k}_0) \cdot a_{j'}(\vec{k}_0') \cdot \hat{a}_j \cdot \exp(i\,\vec{k}_0\cdot\vec{r})\Big\} \tag{80}$$

$$E_0^{sc,anom}(r) = -2i \int_0^{2\pi} d\beta_0 \int_{C_+} \sin\alpha_0 d\alpha_0 \Big\{\int_0^{2\pi} d\beta_1 \int_{C_+} \sin\alpha_1 d\alpha_1 \cdot \sum_{nn'jj'} B_{nj}(\hat{k}_1) \, T^{EM}_{nn'}(1) \, C^{EM}_{n'} \cdot [Q_{jj'}(\vec{k}_0,\vec{k}_1) + \sum_{j'} \int_0^{2\pi} d\beta_1' \int_{C_-} \sin\alpha_1' d\alpha_1' \, Q_{jj''}(\vec{k}_0,\vec{k}_1') R_{j''j'}(\vec{k}_1,\vec{k}_1) \, \hat{a}_j \exp(i\,\vec{k}_0\cdot r)]\Big\} \tag{81}$$

In analogy with Eq. (68), $\tilde{R}_{jj'}$ is here the reflection coefficient at S_0 for plane vector waves coming from above:

$$\tilde{R}_{jj'}(\vec{k}_0,\vec{k}_0') \equiv - \sum_{j''} \int_0^{2\pi} d\beta_1 \int_{C_-} Q_{jj''}(\vec{k}_0,\vec{k}_1) \, Q^{-1}_{j''j'}(\vec{k}_0',\vec{k}_1) \sin\alpha_1 d\alpha 1$$

$$\vec{k}_o \varepsilon C_+ , \quad \vec{k}'_o \varepsilon C_- \tag{82}$$

Thus, as is amply illustrated by Eqs. (72)-(82), the solution for the electromagnetic case exhibits the same structure as was found in the scalar case.

In view of the formal nature of some of the steps in the above developments, we shall next demonstrate in greater detail how this scheme can be used to obtain much more explicit solutions in particular cases. One simplified case, which it is natural to treat first, is that of an inhomogeneity below a plane interface S_o. In the scalar case, the quantities which then simplify drastically are $Q(\vec{k}_o,\vec{k}_1)$, $R(\vec{k}_1,\vec{k}'_1)$ and $\tilde{R}(\vec{k}_o,\vec{k}'_o)$. $Q(\vec{k}_o,\vec{k}_1)$ can then be calculated explicitly from Eq. (28) and it is seen to be proportional to $\delta^{(2)}(\hat{n}_o \times (\vec{k}_o-\vec{k}_1))$, a fact which just expresses the refraction law for plane waves at S_o. Consequently, Eqs. (24) and (25) simplify into algebraic relations. One finds (S_o is given by $z=z_o$):

$$f(\vec{k}_o) = k_o(2k_1)^{-1} (k_1^2-\lambda^2)^{-1/2} e^{-i z_o (k_o^2-\lambda^2)^{1/2}} \{\alpha(\vec{k}_o^{*-}) e^{-i z_o (k_1^2-\lambda^2)^{1/2}}$$
$$[C_{01}(k_o^2-\lambda^2)^{1/2} - (k_1^2-\lambda^2)^{1/2}] + \beta(\vec{k}_o^{*+}) e^{i z_o (k_1^2-\lambda^2)^{1/2}} [C_{01}(k_o^2-\lambda^2)^{1/2} +$$
$$(k_1^2-\lambda^2)^{1/2}]\} \tag{83}$$

$$a(\vec{k}_o) = k_o(2k_1)^{-1} (k_1^2-\lambda^2)^{1/2} \cdot e^{i z_o (k_o^2-\lambda^2)^{1/2}} \cdot \{\alpha(\vec{k}_o^{*-}) e^{-i z_o (k_1^2-\lambda^2)} \cdot$$
$$\cdot [C_{01}(k_o^2-\lambda^2)^{1/2} + (k_1^2-\lambda^2)^{1/2}] + \beta(\vec{k}_o^{*+}) e^{i z_o (k_1^2-\lambda^2)} [C_{01}(k_o^2-\lambda^2)^{1/2} -$$
$$(k_1^2-\lambda^2)^{1/2}] \tag{84}$$

Here $\lambda \equiv k_o \sin\alpha_o = k_1 \sin\alpha_1$ (according to Snell's law) and $\lambda \varepsilon [0,\infty)$, $\hat{k}\varepsilon C_{\pm}$. The square roots are always chosen to have $\mathrm{Im}(\ldots)^{1/2}>0$. Since $\mathrm{Im}\cos\alpha_o>0$ on C_+ and <0 on C_- we write $(k_{ox},k_{oy}, \pm(k_o^2-\ ^2)^{1/2})$ for $\hat{k}_o$ on $C_{\pm}$. The vectors $\vec{k}_o^{\,-}$ and $\vec{k}_o^{\,+}$ are the Snell-transforms of $\vec{k}_o\varepsilon C_-$ where $\vec{k}_o^{\,-}\varepsilon C_-$ and $\vec{k}_o^{\,+}\varepsilon C_+$. Thus we have

$$\vec{k}_o^{*\pm} = (k_o\sin\alpha_o\cos\beta_o, k_o\sin\alpha_o\sin\beta_o, \pm(k_1^2-\lambda^2)^{1/2})$$

Proceeding as before, one uses Eq. (84) to express α in terms of a and β, where β is expressed by means of $T_{nn'}(1)$ and the spherical projections C_n of α(cf. Eq. (63) for the general case). This time Snell's law of refraction at S_o is expressed by $Q^{-1}(\vec{k}_o.\vec{k}_1)$ and in this connection we regard the argument of $a(\vec{k}_o)$ as Snell-transformed from $\vec{k}_1$ in $\alpha(\vec{k}_1)$. This transformed vector is then written

$$\vec{k}_1^{*-} \equiv (k_1\sin\alpha_1\cos\beta_1, k_1\sin\alpha_1\sin\beta_1, -(k_o^2-\lambda^2)^{1/2} \quad \varepsilon C_-$$

Similar considerations apply in analogous situations in several other instances. The resulting formulas for the flat interface case can be written

$$R(\vec{k}_1,\vec{k}_1') = -R(\lambda)\exp[2\,i\,z_o(k_1^2-\lambda^2)^{1/2}]\,(\sin\alpha_1)^{-1}\,\delta(\beta_1-\beta_1')\,\delta(\alpha+\alpha_1'+\pi)$$

$$\hat{k}_1\varepsilon C_-\,,\ \hat{k}_1'\varepsilon C_+ \tag{85}$$

$$\tilde{R}(\vec{k}_o,\vec{k}_o') = R(\lambda)\exp[-2\,i\,z_o(k_o^2-\lambda^2)^{1/2}]\,(\sin\alpha_o)^{-1}\delta(\beta_o-\beta_o')\delta(\alpha_o+\alpha_o'+\pi)$$

$$\hat{k}_o\varepsilon C_+\,,\ \hat{k}_o'\varepsilon C_- \tag{86}$$

Here

$$R(\lambda) = \frac{C_{o1}(k_o^2-\lambda^2)^{1/2} - (k_1^2-\lambda^2)^{1/2}}{C_{o1}(k_o^2-\lambda^2)^{1/2} + (k_1^2-\lambda^2)^{1/2}} \tag{87}$$

is the wellknown reflection coefficient of the plane (additional phase factors appear as a result of our choice of origin). $1+R(\lambda)$ and $1-R(\lambda)$ are both transmission coefficients through S_o. $1-R(\lambda)$ refers to transmission from V_o to V_1 and $1+R(\lambda)$ to transmission from V_1 to V_o. One finds the following simplified expressions for d_n, $A_{nn'}$ and ψ_o^{sc}

$$d_n = k_1(k_o)^{-1}\int_0^{2\pi} d\beta_1 \int_{C_-} Y_n(\hat{k}_1)\,[1-R(\lambda)]\,\cdot$$

$$\cdot\,\exp i\,z_o[(k_1^2-\lambda^2)^{1/2} - (k_o^2-\lambda^2)^{1/2}\,a(k_1^*-)\,\sin\alpha_1 d\alpha_1 \tag{88}$$

$$A_{nn'} = 2\sum_{n''} i^{n'-n''}\,T_{n''n'}(1)\int_0^{2\pi} d\beta_1 \int_{C_-} Y_n(\hat{k}_1)\,R(\lambda)\,Y_{n''}(\hat{k}_1^-)\,\cdot$$

$$\exp[2iz_o(k_1^2-\lambda^2)^{1/2}]\cdot\sin\alpha_1 d\alpha_1 \tag{89}$$

$$\psi_o^{sc,dir}(\vec{r}) = \int_0^{2\pi} d\beta_o \int_{C_+} R(\lambda)\exp[-2\,iz_o(k_o^2-\lambda^2)^{1/2}]\,a(\vec{k}_o-)\,e^{i\,\vec{k}_o\cdot\vec{r}}$$

$$\sin\alpha_o d\alpha_o \tag{90}$$

$$\psi_o^{sc,anon}(\vec{r}) = 2\, k_o k_1^{-1} \sum_{nn'} i^{n'-n}\, T_{nn'}(1) C_{n'} \int_o^{2\pi} d\beta_o \int_{C_+} Y_n(\tilde{\hat{k}}_1) \cdot$$

$$\cdot \exp i\, z_o [(k_1^2-\lambda^2)^{1/2} - (k_o^2-\lambda^2)^{1/2}] \cdot (1+R(\lambda)) e^{i\, \vec{k}_o \cdot \vec{r}} \sin\alpha_o d\alpha_o \qquad (91)$$

where

$$\vec{k}_o^{\,-} \equiv k_o(\sin\alpha_o \cos\beta_o, \sin\alpha_o \sin\beta_o, -\cos\alpha_o)$$

$$\hat{k}_1^{\,-} \equiv (\sin\alpha_1 \cos\beta_1, \sin\alpha_1 \sin\beta_1, -\cos\alpha_1)$$

$$\tilde{\hat{k}}_o \equiv k_1^{-1} \cdot \vec{k}_o^{*+} \equiv k_o \cdot k_1^{-1} \cdot (\sin\alpha_o \cos\beta_o, \sin\alpha_o \sin\beta_o, (k_1^2 k_1^{-2} - \sin^2\alpha_o)^{1/2})$$

Again, closely corresponding expressions are obtained in the electromagnetic case; the spherical harmonics Y_n are replaced by B_{nj}:s and $B_{nj}^{\dagger}$:s, C_n by C_n^{EM} etc. One has

$$d_n^{EM} = \sum_j \int_o^{2\pi} d\beta_o \int_{C_-} a_j(\vec{k}_o)\, B_{nj}^{+}(\hat{k}_o^*) \exp i\, z_o [(k_1^2-\lambda^2)^{1/2} - (k_o^2-\lambda^2)^{1/2}] \cdot$$

$$\cdot\, 2(k_o^2-\lambda^2)^{1/2} \cdot D_j^{-1}(\lambda) \cdot \sin\alpha_o d\alpha_o \qquad (92)$$

$$A_{nn'}^{EM} = 2 \sum_{n''j} \int_o^{2\pi} d\beta_1 \int_{C_-} B_{nj}^{+}(\hat{k}_1)\, B_{n''j}(\hat{k}_1^-)\, T_{n''n'}^{EM}(1) \cdot R_j(\lambda) \cdot$$

$$\cdot \exp[2\, i\, z_o (k_1^2-\lambda^2)^{1/2}] \cdot \sin\alpha_1 d\alpha_1 \qquad (93)$$

$$\vec{E}_o^{sc}(\vec{r}) = \vec{E}_o^{sc,dir}(\vec{r}) + \vec{E}_o^{sc,anon}(\vec{r}) =$$

$$= \sum_j \int_o^{2\pi} d\beta_o \int_{C_+} a_j(\vec{k}_o^-)\, R_j(\lambda) \exp[-2\, i\, z_o (k_o^2-\lambda^2)^{1/2} + i\, \vec{k}\cdot\vec{r}] \hat{a}_j \sin\alpha_o d\alpha_o$$

$$+ 4\, C_{o1} k_o k_1^{-1} \sum_{nn'j} \int_o^{2\pi} d\beta_o \int_{C_+} B_{nj}(\tilde{\hat{k}}_o)\, T_{nn'}^{EM}(1) \cdot C_{n'}^{EM} \exp[i\, z_o ((k_1^2-\lambda^2)^{1/2} - (k_o^2-\lambda^2)^{1/2})] \cdot$$

$$\cdot \exp(i\, \vec{k}_o \cdot \vec{r}) \cdot \hat{a}_j (k_o^2-\lambda^2)^{1/2} \quad D_j^{-1}(\lambda) \cdot \sin\alpha_o d\alpha_o \qquad (94)$$

Here

$$R_1(\lambda) \equiv \frac{N_1(\lambda)}{D_1(\lambda)} \equiv \frac{C_{o1}k_1(1-(\lambda\cdot k_o^{-1})^2)^{1/2} - k_o(1-(\lambda\cdot k_1^{-1})^2)^{1/2}}{C_{o1}k_1(1-(\lambda\cdot k_o^{-1})^2)^{1/2} + k_o(1-(\lambda\cdot k_1^{-1})^2)^{1/2}}$$

$$R_2(\lambda) \equiv \frac{N_2(\lambda)}{D_2(\lambda)} \equiv \frac{(k_o^2-\lambda^2)^{1/2} - C_{o1}(k_1^2-\lambda^2)^{1/2}}{(k_o^2-\lambda^2)^{1/2} + C_{o1}(k_1^2-\lambda^2)^{1/2}}$$

$$\hat{k}_o^* = k_1^{-1}\cdot \vec{k}_o^{*-}$$

Note that the integration in d_n^{EM} is chosen different from that in d_n. This is a matter of convenience, depending on the type of contrasts one usually considers in the acoustic and electromagnetic cases respectively.

IV. FURTHER DEVELOPMENTS

In the previous sections we have presented a general scheme for treating the buried scatterer problem and we have seen how it simplifies in the case of a flat interface. We shall briefly comment on some further developments of this approach which are studied presently. The T matrix approach emphasizes the essential similarities that exist between the linear theories of stationary scattering of acoustic, electromagnetic and elastic waves. The present approach can thus be expected to provide a possible way of attacking the corresponding elastic wave scattering problem. Inevitably this case will involve more complicated reflection and transmission coefficients, as well as T matrices, describing the mode conversion phenomena. A case of great practical importance is when V_o is a vacuum, since this case is a model for propagation and scattering of elastic waves in the ground. A case of particular interest is when the incoming field is a surface wave (i.e. a Rayleigh wave). The basic formalism for this case has been developed and some numerical results for scattering of a Rayleigh wave from a buried spherical inhomogeneity have been obtained (16). A related problem which can be treated in an analogous way is that of scattering from a buried inhomogeneity of a surface wave at an fluid-elastic interface (a Stoneley wave). As can be seen from the derivation of our formalism, the buried scatterer can equally well be a more complicated object than the homogeneous one considered so far (cf. Ref. (17)). Then its T matrix is not given by the simple expression Eq. (59) but must be calculated separately for the inhomogeneity under consideration (e.g. by T matrix methods for piecewise homogeneous objects, or by Fredholm integral equation methods for objects with arbitrarily varying material parameters).

We also remark that the recently developed T matrix formulation of scattering of acoustic waves from an elastic obstacle in a fluid (18) can be inserted directly into the scalar wave formalism given here, thereby extending the range of applications considerably.

Furthermore, it is of interest to extend this approach to the case of several interfaces. If both the source and the inhomogeneity are situated between the layers, one may introduce boundary conditions which confine the wave propagation to the region between the layers i.e. one then has a two-dimensional waveguide with a two- or three-dimensional obstacle (the details of the two-dimensional version of the formalism corresponding to Fig. 1 have also been worked out). Here we note that the use of the T matrix approach has recently been used to treat the scattering from an obstacle in a cylindrical waveguide (19). Also in this case the multiple scatterings are described by an equation which is analogous to Eq. (64).

In section III we have seen that a solution can be obtained for the flat interface case in a fairly explicit manner, by means of matrix operations. However, in order to solve a general case one has to invert a two-dimensional integral transfrom and this usually constitutes a formidable analytic and numerical problem. It is therefore of great interest to find special approximate methods which apply to specific classes of non-planar surfaces. One generalization of the planar case which immediately suggests itself, is when the deviation from a plane is confined to a finite region (20). It is then natural to try to use the planar case as a starting point in some approximate approach and we shall make a few remarks on that possibility. One might try to treat the case of a finite deviation from a plane by taking the configuration in Fig. 1 and let V_o and V_2 have the same propagation characteristics and by taking the source in V_1 and furthermore letting S_1 approach S_o so as to constitute the deviation from the plane. However, for several reasons such an approach is expected to be of limited value. The main reason for this is that one would then violate the geometrical constraints which are basic to the T matrix approach. Suitable modifications of the geometry could be considered but this would then jeopardize the simplicity and usefulness of the results. Furthermore, for a low-hill one would, in a direct application of the T matrix method above, want to compute the T matrix of a very oblate object which would increase the demand on the matrix size. Thus we conclude that it is desirable to develop approximate approaches of a different character.

One obvious way of making use of the plane interface solution is to write the integrals over S_o in the expression Eq. (28) for $Q(\vec{k}_o,\vec{k}_1)$ as one integral over the plane, cutting through the hill, plus the difference between the integrals over the hill and over the bottom of the hill. Let S_p denote the flat part of the interface S (cf. Fig. 6) and let ΔS be the hill so that $S=S_p+\Delta S$. Let furthermore S_o denote the whole plane and ΔS_o that part of S_o which lies underneath ΔS, so that $S_o=S_p+\Delta S_o$. As is easily seen, the treatment below applies to a depression as well as to a hill (or any arbitrary combination of these). However, for brevity we shall always refer to the finite deviation from the plane as the "hill". According to the division

$$\int_S [\ldots.] \cdot dS = \int_{S_o} [\ldots.] \cdot dS + \int_{\Delta S-\Delta S_o} [\ldots.]\, dS$$

of an integral over S, we have a corresponding division of $Q(\vec{k}_o\vec{k}_1)$:

$$Q(\vec{k}_o,\vec{k}_1) = Q_o(\vec{k}_o,\vec{k}_1) + \Delta Q(\vec{k}_o,\vec{k}_1)$$

where Q_o is the Q-function for the plane surface case (i.e. $Q_o \sim \delta^{(2)}(\hat{n}_o \times (\vec{k}_o,\vec{k}_1))$.

Consider first the case when the two halfspaces are homogeneous (no scatterer S_1, $\beta(\vec{k}_1)\equiv 0$). Then

$$a(\vec{k}_1) = i\int_0^{2\pi} d\beta_1 \int_{C_-} \alpha(\vec{k}_1)\, Q(\vec{k}_o,\vec{k}_1)\, \sin\alpha_1 d\alpha_1 =$$

$$= D(\lambda)\, \alpha(\vec{k}_o^{*-}) + i\int_0^{2\pi} d\beta_1 \int_{C_-} \alpha(\vec{k}_1)\Delta Q(\vec{k}_o,\vec{k}_1)\, \sin\alpha_1 d\alpha_1 \qquad (95)$$

where $D(\lambda)$ is determined by Eq. (84). So far no approximation has been made and Eq. (95) can be regarded as an integral equation of the second kind for α, with the driving term $D^{-1}(\lambda)\ a(\vec{k}_o)$, corresponding to the solution to the planar case.

One area of practical applications of the present buried scatterer formalism is in VLF prospecting and for long wavelengths ($2\pi k_i^{-1}$, $i=0,1$ greater than the main dimensions of ΔS) an iteration approach should be useful. The once-iterated solution is then obtained by introducing the plane surface solution on the right hand side of Eq. (95).

Since the solution for a plane surface plus inhomogeneity is also known we consider the possibility of using this solution as a starting point in a similar iteration process, which would then include both the influences of the inhomogeneity and of the hill, as well as interactions between these two structures. With $Q=Q_o+\Delta Q$, we get from Eq. (62)

$$a(\vec{k}_o) = \int_0^{2\pi} d\beta_1 \int_{C_-} \alpha(\vec{k}_1)\ Q_o(\vec{k}_o,\vec{k}_1)\ \sin\alpha_1 d\alpha_1 +$$

$$+ \int_0^{2\pi} d\beta_1 \int_{C_-} \alpha(\vec{k}_1)\ \Delta Q(\vec{k}_o,\vec{k}_1)\ \sin\alpha_1 d\alpha_1 +$$

$$+ \int_0^{2\pi} d\beta_1 \int_{C_+} \sum_{nn'} 2\ i^{n'-n}\ Y_n(\hat{k}_1)\ T_{nn'}(1) C_{n'}\ Q_o(\vec{k}_o,\vec{k}_1)\ \sin\alpha_1 d\alpha_1 +$$

$$+ \int_0^{2\pi} d\beta_1 \int_{C_+} \sum_{nn'} 2\ i^{n'-n}\ Y_n(\hat{k}_1)\ T_{nn'}(1) C_{n'}\ \Delta Q(\vec{k}_o,\vec{k}_1)\ \sin\alpha_1 d\alpha_1 \tag{96}$$

The first term on the right hand side reduces to an algebraic expression so that this can be extracted. If we then introduce the flat surface plus inhomogeneity solution in the second term, we obtain an equation which can be solved by the procedure used in section III. One finds

$$C_n = [(1+\tilde{A})^{-1}\ \tilde{d}]_n$$

where

$$\tilde{A}_{nn'} = A_{nn'} + (\delta A)_{nn'}$$

$$\tilde{d}_n = d_n + \delta_n$$

$A_{nn'}$ and d_n are the flat surface plus inhomogeneity expressions Eqs. (88) and (89) whereas $(\delta A)_{nn'}$ and δ_n involve integrals over $C_\pm$ of ΔQ. Thus $(\delta A)_{nn'}$ expresses the interaction between the inhomogeneity and the hill and this interaction is taken into account in each multiple scattering. Note that if we had introduced the flat surface plus inhomogeneity solution in both terms on the right hand side of Eq. (96) which contain ΔQ, the multiple scattering equation would be modified only in a trivial way. We would then get $C_n = (1+A)d_n$ where A is given by Eq. (88) and where $d_n = d_n + \delta'_n$ is a different modification of d_n in Eq. (88). Naturally, the above approximate schemes can be applied also to the electromagnetic case. Various aspects of the problems raised by these developments are being studied at present.

In view of the prospecting applications it is of great interest to extend all of the results in section II to lossy media. Because of the multitude of expansions and transformations which enter, the extension to complex k values involves considerable detailed work. This case has been treated and the results will appear in the near future (21).

V. SOME NUMERICAL RESULTS

In this section we give a few examples of calculations which have been done by means of the formalism presented in section III (we refer to (16) for some numerical results concerning the elastic case). The numerical results all refer to the flat interface case. Calculations have been performed in the acoustic and electromagnetic cases for several types of non-spherical buried inhomogeneities in three dimensions. In the acoustic case the source is a monopole, in the electromagnetic case it is a vertical electric dipole, situated on S_o in both cases. The scattered field is also computed on S_o.

We recall that the matrix $A_{nn'}$ depends only on the geometry and scattering characteristics of the media. Since it is independent of the fields one and the same matrix can be used in calculations where e.g. only the source position or source characteristics are varied. If one wants to study the influence of a given inhomogeneity as a function of its depth, it is advantageous to translate the T matrix of the inhomogeneity and let the origin be fixed, rather than to keep the origin inside the translated S_1. A change of origin means a change of z_o. Thus if the origin is changed, the integrals in the expressions Eqs. (88) and (89) for d_n and $A_{nn'}$ have to be recomputed for each depth. If the origin is kept fixed and the T matrix translated, d_n will be the same for all depths and the same applies to the integral in the expression for $A_{nn'}$. Of course, the full expression for $A_{nn'}$ depends on the depth, since it is obtained by summation over the elements of a translated T matrix, which varies with the depth. These remarks illustrate the importance of working with function systems which have reasonably simple translation properties as was emphasized in the introduction.

The results will be given in terms of $\psi_o^{sc,anom}$ and $\vec{\psi}_o^{sc,anom}$. Ratios like $|\psi_o^{sc,anom}/\psi_o^{sc,dir}|$ and $|\vec{\psi}_o^{sc,anom}/\vec{\psi}_o^{sc,dir}|$ as well as the corresponding phases are of course also of great interest and they can be obtained in a straightforward way within the present framework. The V_o-V_1 and V_1-V_2 contrasts have been chosen rather moderate. For more drastic contrasts further simplifications can conceivably be introduced into our general formulas. Further work is presently being done in this direction. However, in the computations presented here, we use the full expression for d_n, $A_{nn'}$, C_n etc. The numerical procedure can be divided into the following steps.

1) For the chosen source, compute the d_n-vector.
2) Compute the $T_{nn'}$ matrix of the inhomogeneity.
3) Compute the integrals in $A_{nn'}$ and generate the matrix $\{A_{nn'}\}$.
4) Solve for C_n by iteration of Eq. (64) or Eq. (75).
5) Evaluate $\psi_o^{sc,anom}(\vec{r})$ for a suitable set of measuring points $\vec{r}$.

In Figs. 4-6 we give a few examples of the kind of results which have been obtained by means of the present formalism. Further examples are found in (5) and (6) and others will appear shortly.

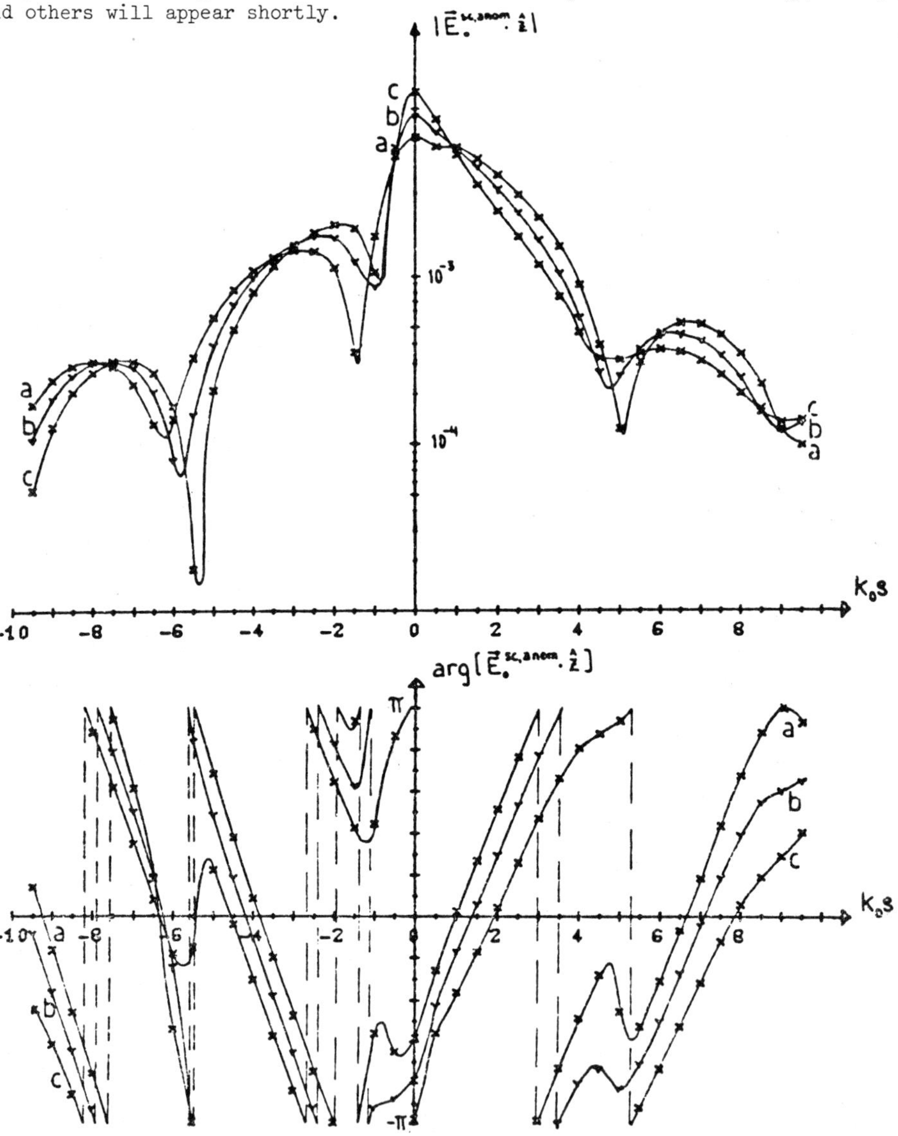

Fig. 4. The amplitude and phase variation of $\vec{E}_0^{sc,anom} \cdot \hat{z}$ along a ray $\phi=180^\circ$ and $\phi=0^\circ$ on the surface S_0 for a single buried sphere $k_1a_1=0.5$, $k_1/k_0=2.5$, $k_1/k_2=2$, $C_{01}=1$, $C_{12}=1$. The source is a vertical dipole located on the x axis at $k_0s_t=3$ as a variation of the distance h between the surface S_0 and the centre of the sphere. a) $k_1h=2.5$, b) $k_1h=2$, c) $k_1h=1.5$ (cf. Ref. (6)).

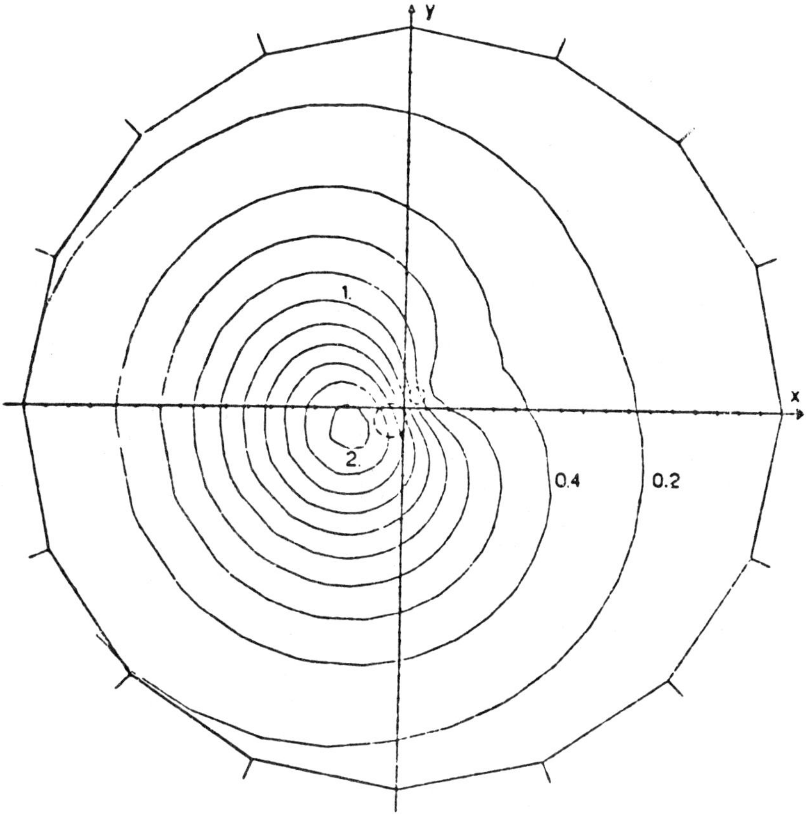

Fig. 5. The amplitude $|\psi_o^{sc,anom}|$ on the surface $S_o(k_1z_o=2)$ for two buried spheres of radii $k_1a_2=0.5$, $k_1a_3=0.25$, $\rho_1/\rho_o=2$, $\rho_2/\rho_1=\rho_3/\rho_1=2$, $k_o/k_1=1.5$, $k_2/k_1=2$. The separation distance is $k_1d=1.5$ and the symmetry axis has the polar angles $\theta=3\pi/4$, $\chi=5\pi/4$ and the point source is located at $k_o\rho_t=3$ on the x-axis (reprinted from Ref. (5)).

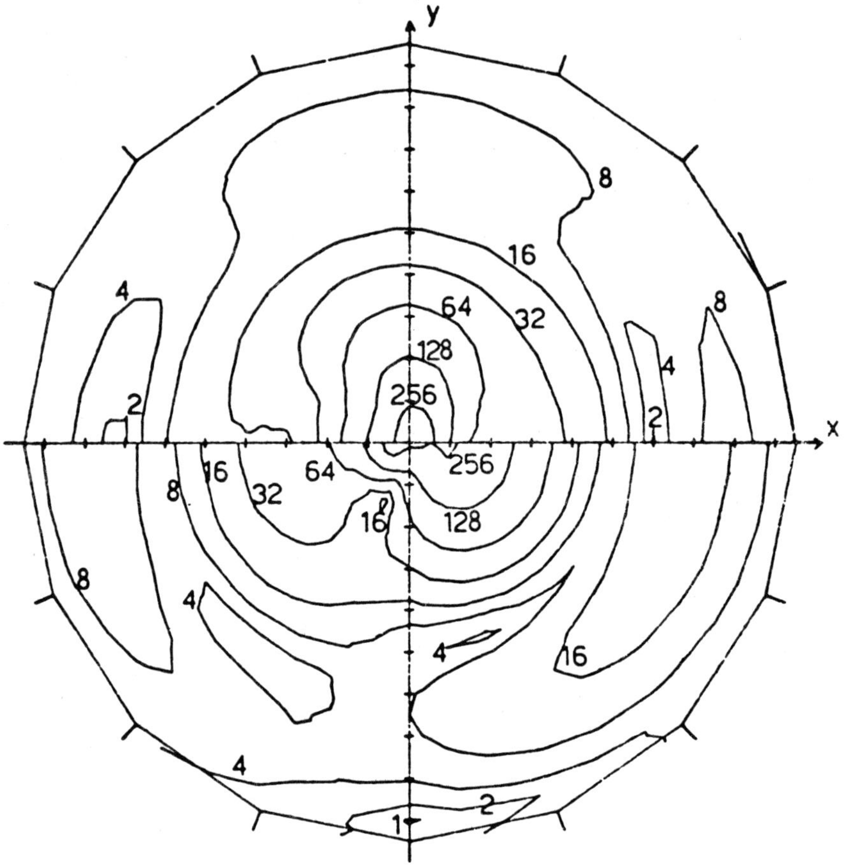

Fig. 6. The amplitude of the anomalous scattered field $\vec{E}_o^{sc,anom}$ on the surface S_o ($k_1z_o=2$) for a single buried perfectly conducting spheroid. The semi axis in the direction of rotational symmetry is $k_1a=0.2$ and the other semi axis is $k_1b=0.6$. The orientation of symmetry axis is $\theta=\pi/6$, $\chi=0$, the vertical dipole is located at $k_0\rho_t=3$ on the x-axis, $k_1/k_0=2.5$, $C_{o1}=1$ and the scale factor is 10^{-4}. Upper part: the x-component. Lower part: the z-component (Cf. Ref. (6).)

APPENDIX: THE SCALAR AND VECTOR SPHERICAL WAVES

The conventions for the scalar spherical waves are as follows.

$$\psi_n(k\vec{r}) \equiv \psi_{nm\sigma}(k\vec{r}) = \gamma_{nm}\, h_n^{(1)}(kr)\; P_n^m(\cos\theta) \begin{Bmatrix} \cos m\phi \\ \sin m\phi \end{Bmatrix}$$

$$\text{for} \begin{Bmatrix} \sigma = e \text{ (even)} \\ \sigma = o \text{ (odd)} \end{Bmatrix} \quad n = 1,2 \ldots\ldots,\ m = 0,1 \ldots n.$$

is the outgoing spherical wave. The corresponding regular wave $\mathrm{Re}\psi_n$ is obtained by replacing $h_n^{(1)}(kr)$ by $j_n(kr)$. $P_m^m(\cos\theta)$ is an associated Legendre polynomial as defined in Ref. (21) and

$$\gamma_{nm} = (-1)^m \left[\frac{\epsilon_m}{4\pi} \cdot \frac{(2n+1)\,(n-m)}{(n+m)\,!}\right]^{1/2},\quad \epsilon_m = 2 - \delta_{m,o}$$

The outgoing spherical vector waves are defined by

$$\psi_n(kr) \equiv \vec{\gamma}_{nm\sigma\tau}(k\vec{r}) \equiv \gamma_{nm}[n(n+1)]^{-1/2}(k^{-1}\nabla x)^{\tau}\,[k\vec{r}\; h_n^{(1)}(kr) P_n^m(\cos\theta) \begin{Bmatrix} \cos m\phi \\ \sin m\phi \end{Bmatrix}]$$

where τ=1,2. For τ=3 one has

$$\vec{\psi}_{nm\sigma 3} \simeq \gamma_{nm}(k^{-1}\nabla)\ [h^{(1)}(kr)\ P_n^m(\cos\theta) \begin{Bmatrix} \cos m\phi \\ \sin m\phi \end{Bmatrix}]$$

$\mathrm{Re}\vec{\psi}_n(k\vec{r})$ is defined analogously. If the angular part is expressed in terms of vector spherical harmonics $\vec{A}_n(\hat{r}) \equiv \vec{A}_{nm\sigma\tau}(\hat{r})$ one has

$$\vec{\psi}_{nm\sigma_1}(k\vec{r}) = \vec{A}_{nm\sigma_1}(\hat{r})\ h_n^{(1)}(kr)$$

$$\vec{\psi}_{nm\sigma 2}(k\vec{r}) = \vec{A}_{nm\sigma 2}(\hat{r})\,(kr)^{-1}\,\frac{d}{d(kr)}\,(kr\ h_n^{(1)}(kr)) \quad + [n(n+1)]^{1/2}\,\vec{A}_{nm\sigma 3}(kr)^{-1}\, h_n^{(1)}(kr)$$

where

$$\vec{A}_{nm\sigma 1} \equiv [n(n+1)]^{-1/2}\ \nabla x(\vec{r}\gamma_{nm}\ P_n^m(\cos\theta) \begin{Bmatrix} \cos m\phi \\ \sin m\phi \end{Bmatrix})$$

$$\vec{A}_{nm\sigma 2} \equiv \hat{r} \times \vec{A}_{nm\sigma 1},\quad \vec{A}_{nm\sigma 3} \equiv \hat{r}\ \gamma_{nm}\ P_n^m(\cos\theta) \begin{Bmatrix} \cos m\phi \\ \sin m\phi \end{Bmatrix}$$

REFERENCES

(1) P.C. Waterman, New formulation of acoustic scattering, J. Acoust. Soc. Am. 45, 1417-1429 (1969).

2. P.C. Waterman, Symmetry, Unitarity and Geometry in Electromagnetic Scattering, Phys. Rev. D 3, 825-839 (1971).

3. P.C. Waterman, Matrix theory of elastic wave scattering, J. Acoust. Soc. Am. 60, 567-580 (1976).

4. V. Varatharajulu and Y-H. Pao, Scattering matrix for elastic waves. I. Theory, J. Acoust. Soc. Am. 60, 556-566 (1976).

5. G. Kristensson and S. Ström, Scattering from buried inhomogeneities - a general three-dimensional formalism, J. Acoust. Soc. Am. 64, 917-936 (1978).

6. G. Kristensson, Electromagnetic scattering from buried inhomogeneities - a general three-dimensional formalism, Report 78-42, Institute of theoretical physics, S-412 96 Göteborg, November 1978.

7. B. Peterson and S. Ström, Matrix formulation of acoustic scattering from multilayered scatterers, J. Acoust. Soc. Am. 57, 2-13 (1975).

8. R.F. Millar, The Rayleigh hypothesis and a related least-squares solution to scattering problems for periodic surfaces and other scatterers, Radio Science 8, 785-796 (1973).

9. D.S. Jones, The eigenvalues of $\nabla^2 u+\lambda u=0$ when the boundary conditions are given on semi-infinite domains, Proc. Cambr. Phil. Soc. 49, 668-684 (1973).

10. G. Kristensson, A uniqueness theorem for the Helmholtz equation: Penetrable media with an infinite interface, Report 79-10, Institute of theoretical physics, S-412 96 Göteborg, March 1979.

11. See e.g. A. Baños, Dipole radiation in the presence of a conducting half-space, Pergamon Press, New York, 1966.

12. See e.g. A.J. Devaney and E. Wolf, Multipole expansions and plane wave representations of the electromagnetic field, J. Math. Phys. 15, 234-244 (1974).

13. E.G. Kalnins, J. Patera, R.T. Sharp and P. Winternitz, Two-variable Galilei group expansions of non-relativistic scattering amplitudes, Phys. Rev. D 8, 2552-2572 (1973).

14. C.P. Boyer, E.G. Kalnins and W. Miller, Symmetry and separation of variables for the Helmholtz and Laplace equations, Nagoya Math. J. 60, 35-80 (1976).

15. R.H.T. Bates and D.J.N. Wall, Null field approach to scalar diffraction A287, 45-114 (1977).

16. A. Boström and G. Kristensson, Contribution to the Tenth Nordic Seminar on Detection Seismology, Stockholm, May 21-23 1979.

17. S. Ström, T matrix for electromagnetic scattering from an arbitrary number of scatterers with continuously varying electromagnetic properties, Phys. Rev. D 10, 2685-2690 (1974).

18. A. Boström, Scattering of stationary acoustic waves by an elastic obstacle immersed in a fluid, Report 78-43, Institute of theoretical physics, S-412 96 Göteborg, November 1978.

19. A. Boström, The T matrix method for a scattering by an obstacle in a waveguide, these proceedings.

20. G. Kristensson and S. Ström, Scattering from inhomogeneities below a non-planar interface, Report TMF 79-1 Institute of theoretical physics, S-412 96 Göteborg, February 1979.

21. G. Kristensson, to appear.

22. A.R. Edmonds, Angular momentum in quantum mechanics, Princeton University Press, 1957.

SCATTERING FROM PERFECT CONDUCTORS AND LAYERED DIELECTRICS USING BOTH INCOMING AND OUTGOING WAVE FUNCTIONS

Altunkan Hızal
Middle East Technical University,
Department of Electrical Engineering, Ankara, Turkey

ABSTRACT

Vekua's theorems concerning the completeness of the ingoing metaharmonic spherical functions on a closed smooth surface S and the convergence properties of the expansions using these functions in and on S are generalized to exterior regions and to annular regions between non-intersecting concentric closed smooth surfaces for ingoing and/or outgoing metaharmonic spherical functions. The relevance of the Rayleigh hypothesis to Vekua's theorems for an exterior region is discussed.

T-matrix formulations of electromagnetic scattering for single homogeneous or layered dielectrics and perfect conductors are studied on the basis of the generalization of Vekua's theorems. Also, it is observed that there are two fundamental forms of the T-matrix which are numerically equivalent but may lead to two different multiple scattering interpretations.

The following numerical examples are considered: Perfectly conducting prolate spheroid and a dumbbell shaped body of revolution consisting of generatrix formed by three equi-radius circles with centers located on the vertices of a right isosceles triangle and two-layered dielectric bodies consisting of spherical and oblate spheroidal surfaces.

INTRODUCTION

Electromagnetic scattering from non-spherical perfectly conducting and homogeneous dielectric bodies using the spherical vector wave functions have been formulated in the form of T-matrices by Waterman (1965, 1969, 1971) and by other researchers. It can be shown that all these methods can be interpreted in terms of the method of moments introduced by Harrington (1968). In the method of moments applied to the present problem, the unknown surface fields can be expanded in terms of a large class of basis functions which may lead to different numerical properties of the solution. The use of spherical vector wave functions as the basis functions has certain advantages because of their unique analytical properties. Some of these are exploited by Waterman (1965, 1971) and by Hızal and Marinçiç (1970). Moreover, spherical wave functions are generally complete not only on spherical surfaces, but also on non-spherical surfaces. The completeness of the spherical wave functions is discussed by Vekua (1953), Calderon (1954) and by Waterman (1969a).

Vekua (1953) discussed the convergence properties of the expansions using the spherical wave functions for an interior region by formulating a number of theorems. It would be of both theoretical and practical interest to know the convergence properties of the expansions used in various T-matrix formulations in which ingoing and/ or outgoing spherical wave functions are used. For this purpose Vekua's theorems are interpreted and generalized to the exterior region of a closed smooth surface and to annular regions between non-intersecting concentric closed smooth surfaces.

To facilitate the assessment of various T-matrix formulations on the basis of the interpretation and the generalization of Vekua's theorems and to establish the relationships of various possible T-matrices, a compact T-matrix notation is used. Then it is shown that the existing forms of the T-matrices and some possible new ones can be cast into two fundamental forms whose convergence properties are known. Also, the two forms of the T-matrix, which are numerically equivalent, may be given two different multiple scattering interpretations.

Certain numerical examples are given. For perfectly conducting bodies, prolate spheroid and a dumbbell shaped body of revolution consisting of generatrix formed by three equi-radius circles with centers located on the vertices of a right isosceles triangle are considered. For dielectrics, two-layered bodies consisting of spherical and oblate spheroidal surfaces are considered. The above examples illustrate certain numerical aspects of the T-matrix formulations used in the computations.

EXTENSIONS AND INTERPRETATIONS OF VEKUA'S THEOREMS

Vekua's Theorems for an Interior Region

Vekua gave the following theorems which concern the spherical wave solutions of the metaharmonic equation $\Delta U + k^2 U = 0$; k=constant, in a closed region $D_i(S)$ bounded by a closed surface S which is smooth in Lyapunov's sense (Vladimir, 1971). Moreover, the origin is assumed to be in $D_i(S)$. The functions

$$U_{nm}^{(1)}(k\vec{r}) = j_n(kr)\, Y_{nm}(\theta,\phi) \tag{1}$$

are called regular (ingoing) metaharmonic spherical functions. Here, $n=0,1,2,\ldots,\infty$ $m=-n,\ldots,0,\ldots,n$, $j_n(kr)$ is the spherical Bessel function of order n and

$$Y_{nm}(\theta,\phi) = [\,(2n+1)(n-m)!\,/\,(n+m)!\,/\,(4\pi)\,]^{\frac{1}{2}}\; P_n^m(\cos\theta)\; e^{im\phi} \tag{2}$$

where $P_n^m(\cos\theta)$ is the associated Legendre polynomial.

Let $S(D_i)$ be the spectrum of eigenvalues of the homogeneous Dirichlet problem for the equation $\Delta U + k^2 U = 0$ in $D_i = D_i(S)$.

Theorem 1 The system of metaharmonic spherical functions (1) is complete with respect to surface S if and only if $k \notin S(D_i)$. If $k \in S(D_i)$ then the finite number of functions $\partial\psi_i / \partial n_s$; $i=1,2,\ldots$, where ψ_i; $i=1,2,\ldots$ are the corresponding eigenfunctions and $\vec{n}_s$ is normal to S, must be added to the system (1) to preserve the completeness.

Vekua outlines the steps of the proof of this theorem. Also Waterman (1969a) has shown the completeness of (1) when $k \notin S(D_i)$. Let the system (1) be orthonormalized with respect to the surface S to yield the set $\{V_n^{(1)}(k\vec{r})\}$; $n=1,2,\ldots,\infty$.

Theorem 3 If $k \notin S(D_i)$ then a metaharmonic function $U(k\vec{r})$ in D_i taking the continuous value $U(k\vec{r}_s)$ on S can be expanded in the form

$$U(k\vec{r}) = \sum_{n=1}^{\infty} c_n(1)\, V_n^{(1)}(k\vec{r}) \qquad \vec{r} \in D_i + S \tag{3}$$

and the series converges absolutely and uniformly in D_i. The Fourier coefficient $c_n(1)$ is expressed by

$$c_n(1) = \int_S U(k\vec{r}_s)\ V_n^{(1)}(k\vec{r}_s)\ ds$$

An outline of the proof of this theorem is given by Vekua. The expansion corresponding to (3) without orthogonalization can be obtained by using

$$V_n^{(1)}(k\vec{r}) = \sum_{v=1}^{n} b_v^{(n)}\ U_v^{(1)}(k\vec{r}) \tag{4}$$

where the double index nm in (1) is grouped in $v = 1,2,\ldots$ The constants $b_v^{(n)}$ are known and arise due to the orthogonalization. When (4) is substituted in (3),

$$U(k\vec{r}) = \lim_{N\to\infty} \sum_{n=1}^{N} a_n(1,N)\ U_n^{(1)}(k\vec{r}) \tag{5}$$

$$a_n(1,N) = \sum_{v=n}^{N} c_v(1)\ b_n^{(v)} \tag{6}$$

are obtained. The series (5) also converges uniformly in D_i. The important point to notice here is that the coefficients in (5) depend on N while in (3), $c_n(1)$'s are independent of the number of terms taken in the series expansion.

Lemma 1 Let G_o be a harmonic function in the exterior of S, which converges to zero on S and to unity at infinity. Let S_ρ be the outer surface of $G_o = \rho$, $0 \leq \rho < 1$. It is clear that $S_{\rho'}$ is completely inside $S_{\rho''}$ if $\rho' < \rho''$ where $S_o = S$. Let $U(k\vec{r})$ be a metaharmonic function in D_i, continuous in $D_i + S$. Then for every $\varepsilon > 0$ there exists an $\eta(\varepsilon) > 0$ such that to any outer surface S_ρ for $\rho < \eta(\varepsilon)$, a metaharmonic function $U_\rho(k\vec{r})$ in S_ρ can be related, which satisfies the inequality

$$|U(k\vec{r}) - U_\rho(k\vec{r})| < \varepsilon \qquad \text{for} \quad \vec{r} \in D_i + S\,, \quad \rho < \eta(\varepsilon) \tag{7}$$

Vekua does not give the proof of this lemma. A proof of the lemma is given by Hızal and Aydın (1979).

Remark In the lemma, $U(k\vec{r})$ is a metaharmonic function in D_i which has the continuous value $U(k\vec{r}_s)$ on S. The lemma is not concerned with the uniqueness of the solution of the interior Dirichlet problem in which $U(k\vec{r})$ is to be determined from a given boundary value $U(k\vec{r}_s)$. Thus, the lemma is valid whether $k \in S(D_i)$ or not.

Theorem 4 Every function $U(k\vec{r})$ which is metaharmonic in D_i and continuous in $D_i + S$, can be uniformly approximated in a closed region $D_i + S$ by the regular metaharmonic spherical polynomials.

Vekua proves this theorem using lemma 1 and theorem 3 : In the inequality (7), ρ can be so chosen that $k \notin S(D_\rho)$, where D_ρ is the region bounded by S_ρ such that $0 < \rho \leq \eta(\varepsilon)$. Then by theorem 3 the metaharmonic function $U_\rho(k\vec{r})$ can be represented by a uniformly converging expansion in $D_\rho \supset D_i + S$ in terms of the regular metaharmonic spherical functions. By use of the inequality (7), the theorem is proved.

Theorem 5 Any function $f(\vec{r}_s)$ continuous on S can be uniformly approximated by the regular metaharmonic spherical polynomials if and only if $k \notin S(D_i)$.

Vekua does not give a proof of this theorem. The following proof is based on theorms 4 and 1. If $k \notin S(D_i)$ then one can relate to S a metaharmonic function $U(k\vec{r})$ in D_i taking the continuous value $U(k\vec{r}_s) = f(\vec{r}_s)$ on S. This follows from the uniqueness of the solution of the interior Dirichlet problem. Then by theorem 4, $U(k\vec{r})$;

$\vec{r} \in D_i+S$ can be uniformly approximated by the regular metaharmonic spherical polynomials. Thus, $f(\vec{r}_s)$ is also approximated uniformly by the same polynomials. This is the sufficiency condition. To show the necessity condition one must use theorem 1: If $f(\vec{r}_s)$, which is an arbitrary continuous function on S, can be uniformly approximated, then the mean-square approximation also holds. This implies that the regular metaharmonic spherical functions used in the approximating polynomial are complete on S. Thus, by theorem 1 $k \notin S(D_i)$.

Remark This theorem does not contradict with theorem 4 in which if $k \in S(D_i)$, one can still uniformly approximate a $U(k\vec{r})$; $\vec{r} \in D_i+S$, which is metaharmonic in D_i. The approximation is such that $U(k\vec{r})$ is represented uniformly by a $U_\rho(k\vec{r})$ which is metaharmonic in $D_\rho+S_\rho \supset D_i+S$.

Extension of Vekua's Theorems to an Annular Region

Vekua's theorems are extended to an annular region E_i bounded by two closed Lyapunov surfaces S_1 and S_2, where S_2 encloses S_1. The proof of the following theorems are given by Hızal and Aydın (1979).

Theorem 1' The system of metaharmonic functions $\{U_{nm}^{(1)}(k\vec{r}),\ U_{nm}^{(2)}(k\vec{r})\}$, where $U_{nm}^{(2)}(k\vec{r})$ is obtained from (1) by the replacement of $j_n(kr)$ by the spherical Hankel function of the first kind $h_n^{(1)}(kr)$, is complete on S_1+S_2 if and only if $k \notin S(E_i)$ where $S(E_i)$ is the spectrum of eigenvalues of the homogeneous Dirichlet problem for the metaharmonic equation in the region E_i. If $k \in S(E_i)$, the normal derivatives of the corresponding eigenfunctions must be added to the above set to preserve the completeness.

Let the set in theorem 1' be orthonormalized on S_1+S_2 to yield the set $\{V_n(k\vec{r})\}$, $n=1,2,\ldots,\infty$.

Theorem 3' If $k \notin S(E_i)$, then a metaharmonic function $U(k\vec{r})$ in E_i, taking the continuous value $U(k\vec{r}_s)$ on S_1+S_2, can be expanded in $E_i+S_1+S_2$ as

$$U(k\vec{r}) = \sum_{n=1}^{\infty} c_n V_n(k\vec{r}) = \lim_{N\to\infty} \sum_{n=1}^{N} \left[a_n(1,N)\, U_n^{(1)}(k\vec{r}) + a_n(2,N)\, U_n^{(2)}(k\vec{r}) \right] \quad (8)$$

where the series converges absolutely and uniformly in E_i. The coefficients c_n, $a_n(1,N)$ and $a_n(2,N)$ are defined in a manner analogous to those in (3) and (6).

Lemma 1' Vekua's lemma 1 is extended to an annular region by considering an inner surface S_ζ^{in} in $D_i(S_1)$ which is defined by a harmonic function $G_0^{in}=\zeta$; $0 \leqslant \zeta \leqslant 1$ which converges to zero on S_1 and to unity on a closed surface S_1^{in} in $D_i(S_1)$ and an outer surface S_ρ^{out} in $D_e(S_2)$ which is defined by a harmonic function $G_0^{out}=\rho$; $0 \leqslant \rho < 1$, which converges to zero on S_2 and to unity at infinity, where $S_0^{in}=S_1$ and $S_0^{out}=S_2$.

Theorem 4' Every function $U(k\vec{r})$ which is metaharmonic in E_i and continuous in $E_i+S_1+S_2$ can be uniformly approximated by polynomials in terms of both the ingoing and the outgoing (those which contain the Hankels) metaharmonic spherical functions.

Theorem 5' Any function $f(\vec{r}_s)$ continuous on S_1+S_2 can be uniformly approximated by polynomials in terms of both the ingoing and the outgoing metaharmonic spherical functions if and only if $k \notin S(E_i)$.

Extension of Vekua's Theorems to an Exterior Region

Let $D_e=D_e(S)$ be the region outside a closed Lyapunov surface S. Then the following theorems hold.

Theorem 1" The system of the outgoing metaharmonic functions $\{U_{nm}^{(2)}(k\vec{r})\}$ is complete with respect to any Lyapunov surface S for any parameter value k.

This theorem is given as theorem 6 in Vekua's paper without proof. Following the lines of Vekua as in theorem 1, Millar (1973) gave a proof of this theorem for two dimensions. Calderon (1954) also proved the corresponding theorem for the vector case. Let the set in theorem 1" be orthonormalized on S to yield the set $\{V_n^{(2)}(k\vec{r})\}$.

Theorem 3" Every metaharmonic function $U(k\vec{r})$ in D_e, which satisfies the radiation condition at infinity and takes the continuous boundary value $U(k\vec{r}_s)$ on S can be expanded in the form of

$$U(k\vec{r}) = \sum_{n=1}^{\infty} c_n(2)\, V_n^{(2)}(k\vec{r}) = \lim_{N\to\infty} \sum_{n=1}^{N} a_n(2,N)\, U_n^{(2)}(k\vec{r}) \tag{9}$$

which converges in the mean-square sense on S and uniformly in D_e. The coefficients $c_n(2)$ and $a_n(2,N)$ are defined similarly to those in (3) and (6). This theorem can be proved by a procedure similar to that given by Millar (1973).

Lemma 1" Let G_o be a harmonic function in $D_i=D_i(S)$, which converges to zero on S and to unity on an inner closed surface S_1 in D_i. Let S_ρ be the inner surface of $G_o=\rho$ in D_i ; $0\leq\rho\leq 1$. It is clear that $S_{\rho'}$ encloses $S_{\rho''}$ completely if $\rho'<\rho''$, and $S_o=S$. Let $U(k\vec{r})$ be a metaharmonic function in D_e continuous in D_e+S and satisfy the radiation condition at infinity. If the singularities of the analytic continuation of $U(k\vec{r})$ in D_i are located inside the inner surface S_1, then for every $\varepsilon>0$ there exists an $\eta(\varepsilon)>0$ such that to any inner surface S_ρ; $\rho\leq\eta(\varepsilon)$ a metaharmonic function $U_\rho(k\vec{r})$ outside S_ρ can be related, which satisfies

$$|U(k\vec{r}) - U_\rho(k\vec{r})| < \varepsilon \quad , \quad \vec{r}\in D_e+S \quad , \quad \rho<\eta(\varepsilon) \tag{10}$$

The proof of this lemma is similar to that of lemma 1' for an annular region where the surface S_2 recedes to infinity. But in this case $U(k\vec{r})$ satisfying the radiation condition at infinity can be analytically continued into S up to S_1.

Theorem 4" Every function $U(k\vec{r})$, which is metaharmonic in D_e, satisfying the radiation condition at infinity and continuous in D_e+S, can be represented by a uniformly converging expansion in D_e+S using the outgoing metaharmonic spherical functions.

Proof : The singularities of the analytic continuation of $U(k\vec{r})$ accross S are located in D_i. Thus, an inner closed surface completely lying in D_i can be found which encloses all these singularities. Then in lemma 1", for a $\rho<1$ $U_\rho(k\vec{r})$ on S_ρ can be chosen to be equal to the analytic continuation of $U(k\vec{r})$ in D_i. Then, from the uniqueness of the exterior Dirichlet problem with respect to S_ρ it follows that $U_\rho(k\vec{r})=U(k\vec{r})$ in $D_e(S_\rho)+S_\rho$. By theorem 3" $U_\rho(k\vec{r})$ and hence $U(k\vec{r})$ can be represented by a uniformly convergent expansion of the form (9) with respect to S_ρ in $D_e(S_\rho)+S_\rho \supset D_e+S$. Thus the theorem follows.

Rayleigh hypothesis : If the inner surface S_1 lies inside $S_{ins}(S)$, the inscribing sphere of S, then S_1 can be chosen as $S_1= S_{ins}(S)$. Also, for $\vec{r}\in S_1$ if $U_1(k\vec{r})$ is chosen to be equal to $U(k\vec{r})$ which is the analytic continuation of the exterior field in D_e into S, then from the uniqueness of the exterior Dirichlet problem with respect to S_1, it follows that $U_1(k\vec{r})=U(k\vec{r})$ for $\vec{r}\in D_e(S_1)+S_1$. Then by theorem 3" $U_1(k\vec{r})$ and hence $U(k\vec{r})$ can be represented by (9) in $D_e(S_1)+S_1$ with respect to the surface S_1. As S_1 is spherical $V_n^{(2)}(k\vec{r})$ coincides with a normalized form of $U_n^{(2)}(k\vec{r})$, and $a_n(2,N)$, being proportional to $c_n(2)$, is independent of N. This is the sufficiency condition for the Rayleigh hypothesis, i.e. if all the singularities of the analytic continuation of $U(k\vec{r})$ in $D_i(S)$ lie in $S_{ins}(S)$, (9) converges uniformly in $D_e(S_1)+S_1$ with coefficients $a_n(2)$ independent of N. The necessary condition is evident. If (9) with $a_n(2,N)\equiv a_n(2)$ converges uniformly in $D_e(S_1)+S_1$, then $U(k\vec{r})$ must be analytic in $D_e(S_1)+S_1$ implying that the singularities of its analytic continuation must lie in $D_i(S_1)$.

Extension of Vekua's Theorems to Concentric Annular Regions

In the proof of theorems 1, 1' and 1" for the completeness, the wave number k appears as a parameter and one associates k with a region where the chosen metaharmonic spherical functions are analytic. This point may be exploited to generalize Vekua's theorems to an arbitrary number of non-intersecting concentric closed smooth surfaces $S_1,\dots,S_q,\dots,S_M$, where S_q encloses S_{q-1}. Then the following theorems hold.

Theorem 1a The system $\{U_{nm}^{(1)}(k\vec{r})\}$ is complete on S_q if and only if $k\notin S[D_i(S_q)]$, where $D_i(S_q)$ is the interior region bounded by the surface S_q. if $k\in S[D_i(S_q)]$, then the normal derivatives of the corresponding eigenfunctions must be added to the above set to preserve the completeness.

Theorem 3a If $k\notin S[D_i(S_q)]$ then a metaharmonic function $U(k\vec{r})$ in $D_i(S_q)$ taking the continuous value $U(k\vec{r}_s)$ on S_q can be expanded as in (3) or (5) with respect to the surface S_q using the regular metaharmonic spherical functions, which converges uniformly in $D_i(S_q)$.

Let k belong to the closed homogeneous region between S_{q-1} and S_q. Then this theorem shows that $U(k\vec{r})$ as given by (3) or (5) converges to $U(k\vec{r})$ uniformly in the annular region between S_{q-1} and S_q, including the surface S_{q-1}, and to the analytic continuation of $U(k\vec{r})$ in $D_i(S_{q-1})$. Note that the series converges in the mean-square sense on S_q.

Remark If $k\in S[D_i(S_{q-1})]$ then $\{U_{nm}^{(1)}(k\vec{r})\}$ is not complete on S_{q-1} and one may doubt the validity of the uniform convergence of the series (3) or (5) on S_{q-1} according to theorem 5. There is no ambiguity here as the expansions are written with respect to the surface S_q. They converge uniformly to the analytic continuation of $U(k\vec{r})$ in $D_i(S_{q-1})+S_{q-1}$, which is metaharmonic everywhere in $D_i(S_q)$ including the surface S_{q-1}. See also the remark following theorem 5.

Theorem 4a Every function $U(k\vec{r})$, which is metaharmonic in $D_i(S_q)$ and continuous in $D_i(S_q)+S_q$ can be uniformly approximated in $D_i(S_q)+S_q$ by the regular metaharmonic spherical polynomials.

The proof of this theorem is based on theorems 3a and a corresponding lemma for S_q which is analogous to lemma 1.

Theorem 5a Any function $f(\vec{r}_s)$; $\vec{r}_s\in S_q$ continuous on S_q can be uniformly approximated by the regular metaharmonic spherical polynomials if and only if $k\notin S[D_i(S_q)]$.

Theorem 1"a The set of the outgoing metaharmonic spherical functions $\{U_{nm}^{(2)}(k\vec{r})\}$ is complete on S_q for any parameter value k.

Theorem 3"a Every metaharmonic function $U(k\vec{r})$ in the annular region between S_q and S_{q+1} which takes the continuous boundary value $U(k\vec{r}_s)$ on S_q can be expanded in the form of (9) with respect to the surface S_q which converges in the mean-square sense on S_q and uniformly in $D_e(S_q)$. For $\vec{r} \in D_e(S_{q+1})$ the expansion converges to the analytic continuation of $U(k\vec{r})$ in $D_e(S_{q+1})$, which satisfies the radiation condition at infinity. Note that the expansion converges uniformly on S_{q+1}.

Theorem 4"a Every metaharmonic function $U(k\vec{r})$ in the annular region $E_{q,q+1}$ between S_q and S_{q+1} continuous in $E_{q,q+1}+S_q+S_{q+1}$ can be represented by a uniformly converging expansion in the closure of the above annular region in terms of the outgoing metaharmonic spherical functions.

The proof of this theorem is similar to that of theorem 4". The expansion converges to the analytic continuation of $U(k\vec{r})$ in $D_e(S_{q+1})$. It should be remarked here that the expansion, when evaluated in $D_e(S_{q+1})$ satisfies the radiation condition at infinity. Consequently the expansion, when evaluated in $D_i(S_q)$ must diverge inside an inner closed surface S_1 in $D_i(S_q)$. If S_1 lies in $S_{ins}(S_q)$ then the corresponding Rayleigh hypothesis becomes valid.

T-MATRIX FORMULATIONS AND VEKUA'S THEOREMS

Matrix formulation of electromagnetic scattering from single scatterers was presented previously by Waterman (1965, 1971) and Hızal and Marinçiç (1970) for conductors and by Waterman (1969b) for homogeneous dielectric bodies. The formulations of Barber and Yeh (1975) and Warner and Hızal (1976) for dielectric bodies are identical to that of Waterman (1969b). The corresponding formulations for multilayered scatterers are presented by Peterson and Ström (1974), Aydın (1976) and Bringi and Seliga (1977). In these formulations, Spherical Vector Wave Functions (SVWF) of the ingoing and/or the outgoing types on the boundary surfaces or in the regions between them are used as the basis functions for the expansion of the unknown fields. The convergence properties of these expansions on the surfaces and in the regions between them are not discussed sufficiently, except in the cases concerning the validity of the Rayleigh hypothesis (Millar, 1973 and Bates, 1975).

It would be of both theoretical and practical interest to know the convergence properties of various T-matrix formulations. For this purpose the corresponding theorems of the preceding section for the SVWF are needed. These theorems can be extended in an analogous manner for the ingoing and/or the outgoing SVWF. Extensions of some of the theorems to SVWF with complex arguments are given by Aydın and Hızal (1979) and Aydın (1978). In this section various T-matrix formulations and some possible new ones will be analyzed using the corresponding theorems for the SVWF. Moreover, a compact notation will be used, which makes the analysis easier. Also, it will be shown that the T-matrices can be expressed in either of the two forms which have different multiple scattering interpretations.

Definition of the SVWF

An outgoing SVWF at a point $\vec{r} : (r,\theta,\phi)$ is defined below in the form given by Jackson (1962).

$$\vec{F}_{nm1}^{(2)}(k\vec{r}) = -i\, h_n^{(1)}(kr)\, [n(n+1)]^{-\frac{1}{2}}\, (\vec{r}\times\vec{\nabla})\, Y_{nm}(\theta,\phi) \tag{11}$$

$$\vec{F}_{nm\tau}^{(2)}(k\vec{r}) = \frac{1}{k}\, \vec{\nabla}\times \vec{F}_{nm,3-\tau}^{(2)}(k\vec{r}) \tag{12}$$

where $n=1,2,\ldots,\infty$, $m=-n,\ldots,n$, $\tau=1$ or 2 and $Y_{nm}(\theta,\phi)$ is the spherical harmonic defined by (2). The spherical Hankel function of the first kind in (11), which was also introduced in theorem 1', is appropriate for an outgoing wave having a time dependence of $\exp(-i\omega t)$. An ingoing (regular) SVWF will be denoted by $\vec{F}_{nm\tau}^{(1)}(k\vec{r})$ which contains a spherical Bessel function $j_n(kr)$ instead of $h_n^{(1)}(kr)$.

Definition of the Radiation Matrix $Q^q_{nn'}(k,k',p,p')$

Consider a sufficiently smooth closed surface S_q having an outward unit normal $\vec{n}_q$. Let $\vec{J}_q(k'\vec{r}_s)$ and $\vec{M}_q(k'\vec{r}_s)$ be, respectively the electric and magnetic surface current densities on S_q; $\vec{r}_s\in S_q$. The wave number k' is associated with either $D_i(S_q)$ or $D_e(S_q)$. The surface current densities will be defined by

$$\vec{J}_q(k'\vec{r}_s) = \vec{n}_q\times \vec{H}(k'\vec{r}_s) \tag{13}$$

$$\vec{M}_q(k'\vec{r}_s) = -\vec{n}_q\times \vec{E}(k'\vec{r}_s) = \vec{n}_q\times[\vec{\nabla}\times \vec{H}(k'\vec{r}_s)]/(i\omega\varepsilon') \tag{14}$$

where ε' is the electrical permittivity in k' and $\vec{H}(k'\vec{r}_s)$ and $\vec{E}(k'\vec{r}_s)$ are, respectively the total magnetic and electric fields on S_q as one approaches S_q from the region where the wave number is k'. The SVWF expansion of the radiation from the given surface source densities will be considered in an unbounded homogeneous region which has the wave number k. For this purpose the vector potential formulation and the SVWF expansion of the free-space dyadic Green's function will be used. If $\overleftrightarrow{I}$ is the unity dyadic, it can be shown that

$$\overleftrightarrow{I}\exp(ik|\vec{r}-\vec{r}_s|)/(4\pi|\vec{r}-\vec{r}_s|) = ik\sum_n \vec{F}_n^{(3-p)}(k\vec{r})\; \vec{F}_n^{(p)*}(k\vec{r}_s) \tag{15}$$

where n stands for the triple index $nm\tau$ and $p=1$ or $p=2$. The asterix * implies complex conjugation in the sense that it does not act on the spherical Bessel and Hankel functions and the wave number k. For $p=1$ and $p=2$, (15) converges uniformly for $\vec{r}\in D_e[S_{cir}(S_q)]$ and $\vec{r}\in D_i[S_{ins}(S_q)]$, respectively. Here $S_{cir}(S_q)$ is the circumscribing sphere of S_q and $S_{ins}(S_q)$ is the inscribing sphere of S_q.
Using the vector potential formulation and (11), (12) and (15) one obtains

$$\vec{H}^{rad}(k\vec{r}) = \sum_n a_n^{rad}(3-p)\; \vec{F}_n^{(3-p)}(k\vec{r}) \tag{16}$$

where

$$a_n^{rad}(3-p) = ik\int_{S_q}[\vec{J}_q(k'\vec{r}_s)\cdot \vec{\nabla}\times\vec{F}_n^{(p)*}(k\vec{r}_s)+i\omega\varepsilon\vec{M}_q(k'\vec{r}_s)\cdot\vec{F}_n^{(p)*}(k\vec{r}_s)]\, ds_q \tag{17}$$

If the total magnetic field on S_q is represented by

$$\vec{H}(k'\vec{r}_s) = \sum_{n'} a_{n'}(p')\, \vec{F}_{n'}^{(p')}(k'\vec{r}_s) \quad ;\ p'=1 \text{ or } 2\ ,\quad \vec{r}_s\in S_q \tag{18}$$

which will be assumed to be convergent in the mean-square sense, and substituted in (17) the following matrix equation results.

$$a_n^{rad}(3-p) = Q^q_{nn'}(k,k',p,p')\ a_{n'}(p') \tag{19}$$

where summation over n' is implied, and

$$Q^q_{nn'}(k,k',p,p') = -ik \int_{S_q} \{[\vec{\nabla}\times\vec{F}_n^{(p)*}(k\vec{r}_s)]\times\vec{F}_{n'}^{(p')}(k'\vec{r}_s) + \frac{\varepsilon}{\varepsilon'}\ \vec{F}_n^{(p)*}(k\vec{r}_s)\times[\vec{\nabla}\times\vec{F}_{n'}^{(p')}(k'\vec{r}_s)]\}\cdot\vec{n}_q(\vec{r}_s)\ ds_q \tag{20}$$

The matrix $Q^q_{nn'}(k,k',p,p')$ will be referred to as the radiation matrix or the radiation operator for it yields the coefficients of the radiated field when operated on the coefficients of the source field. It should be noted that $a_n^{rad}(1)$ and $a_n^{rad}(2)$ are the coefficients of the radiated field in $D_i[S_{ins}(S_q)]$ and in $D_e[S_{cir}(S_q)]$, respectively. In the above formulation the unbounded medium having the wave number k will be referred to as the host medium.

Properties of the Radiation Matrix $Q^q_{nn'}(k,k',p,p')$

From (20) it readily folows that

$$Q^{q*}_{n'n}(k,k',p,p') = \frac{k\ \varepsilon}{k'\varepsilon'}\ Q^q_{nn'}(k',k,p',p) \tag{21}$$

where * has the same meaning as that in (15). This property is analogous to the symmetry property of the free-space Green's function. As an example consider a surface S_q whose interior and exterior regions have the wave numbers k' and k, respectively. In the equation $a_n^{rad}(2) = Q^q_{nn'}(k,k',1,2)\ a_{n'}(2)$, the radiation operator causes a passage of an outgoing wave out through the surface S_q, while in the equation $a_n^{rad}(1) = Q^q_{nn'}(k',k,2,1)\ a_{n'}(1)$ the radiation operator causes a passage of an ingoing wave in through the surface S_q. It is seen that, according to (21), the two radiation operators are, apart from a factor, the conjugate transpose of each other.

If the host medium k and the medium k' where the fields are expanded are the same, the radiation matrix becomes diagonal. Applying the Green's second vector identity to an annular region between S_q and a sphere and using the Wronskian relationships of the spherical Bessel and Hankel functions one can show that

$$Q^q_{nn'}(k,k,p,p') = (1-\delta_{pp'})\ (1-2\delta_{p2})\ \delta_{nn'} \tag{22}$$

where the Kronecker deltas are used. This relationship is given also by Peterson and Ström (1974). The radiation matrix is also diagonal on a spherical surface :

$$Q^q_{nn'}(k,k',p,p') = ikr[(A_n - \frac{\varepsilon}{\varepsilon'}B_n)\delta_{\tau 1} + (\frac{\varepsilon}{\varepsilon'}\ \frac{k'}{k}A_n - \frac{k}{k'}B_n)\delta_{\tau 2}]\delta_{nn'} \tag{23}$$

$$A_n = \frac{d}{dr}[r\ z_n^p(kr)]\ z_n^{p'}(k'r) \tag{24}$$

$$B_n = z_n^p(kr)\ \frac{d}{dr}[r\ z_n^{p'}(k'r)] \tag{25}$$

where $z_n^1(kr) = j_n(kr)$ and $z_n^2(kr) = h_n^{(1)}(kr)$.

Radiation Matrix for a Perfectly Conducting Surface : $G^{q}_{nn'}(k,p,p')$

On a perfectly conducting surface $\vec{M}_q(k\vec{r}_s)=0$. Thus, the radiation matrix becomes

$$G^{q}_{nn'}(k,p,p') = -ik \int_{S_q}\{[\vec{\nabla}\times\vec{F}_n^{(p)*}(k\vec{r}_s)]\times\vec{F}_{n'}^{(p')}(k\vec{r}_s)\}\cdot\vec{n}_q(\vec{r}_s)\, ds_q \tag{26}$$

which has the following property

$$G^{q*}_{n'n}(k,p,p') = -G^{q}_{nn'}(k,p',p)+(1-\delta_{pp'})\,\delta_{nn'}\,(-1)^{p+\tau} \tag{27}$$

On a spherical surface (26) reduces to

$$G^{q}_{nn'}(k,p,p') = ikr\,(A_n\delta_{\tau 1}-B_n\delta_{\tau 2})\,\delta_{nn'} \tag{28}$$

where A_n and B_n are given by (24) and (25) for k=k'.

T-Matrix Formulations for Dielectric Bodies

Electromagnetic Scattering from single homogeneous or multilayered scatteres can be formulated in the form of T-matrices by using the generalized Extended Boundary Condition (EBC) integral equations of Al-Badwaihy and Yen (1975), which are originally used by Waterman (1965,1969). Al-Badwaihy and Yen also proved the uniqueness of the coupled integral equations for the unknown electric and magnetic surface currents which are considered as independent unknowns. However, it is proved by Aydın (1978) that using only the electric or magnetic field integral equation, uniqueness is preserved if the surface currents are expressed in terms of the fields which satisfy Maxwell's equations in appropriate regions.

In terms of the SVWF expansions, the EBC integral equations are reduced to equations for the coefficients such that for an interior null field condition the coefficient $a_n^{rad}(1)$, with p=2 vanishes while for an exterior null field condition the coefficient $a_n^{rad}(2)$, with p=1 vanishes. Note that if the sources of the incident radiation are located in the host medium, then the coefficients of the total radiated field should be equated to the negative of the incident field's coefficients. Moreover, in writing out the matrix equations for a surface S_q corresponding to a null field condition or to a radiation from S_q a negative sign should precede the radiation matrix if the sources $\vec{J}_q$ and $\vec{M}_q$ defined by (13) and (14) with $\vec{n}_q$ directed outward with respect to S_q, are placed on the inner side of S_q. Note that $\vec{J}_q$ and $\vec{M}_q$ are to be placed on that side of S_q which faces the host medium k.

After the above preliminary work, one can now discuss various T-matrix formulations in the light of the theorems given previously.

Single surface scattering (Waterman, 1969b) : Form I Let a closed surface S separate the inner and the outer regions having the parameters k,ε and k_o, ε_o, respectively (Fig. 1). The total field in $D_i=D_i(S)$ can be expanded in terms of the ingoing SVWF with coefficient $a_n(1)^i$. The EBC to be used is that $\vec{J}^+$, $\vec{M}^+$, which are obtained from the total field in D_i yield $-a_n^{inc}(1)$ in D_i in the host medium k_o in accordance with (17). The folded sections on the vectors of $\vec{J}^+$, $\vec{M}^+$ indicate that they are obtained from the total field in D_i in accordance with (13) and (14). Thus in the radiation matrix notation the EBC stated above is expressed by $Q_{nn'}(k_o,k,2,1)\ a_{n'}(1) = -a_n^{inc}(1)$ for p=2. Applying (19) for p=1 one obtains $Q_{nn'}(k_o,k,1,1)\ a_{n'}(1) = a_n^{sc}(2)$. From these two equations the T-matrix defined by $a_n^{sc}(2) = T_{nn'}\ a_{n'}^{inc}(1)$ is obtained in Form I as

$$T_{nn'} = -\,Q_{nn'}(k_o,k,1,1)\,[\,Q_{nn'}(k_o,k,2,1)]^{-1} \tag{29}$$

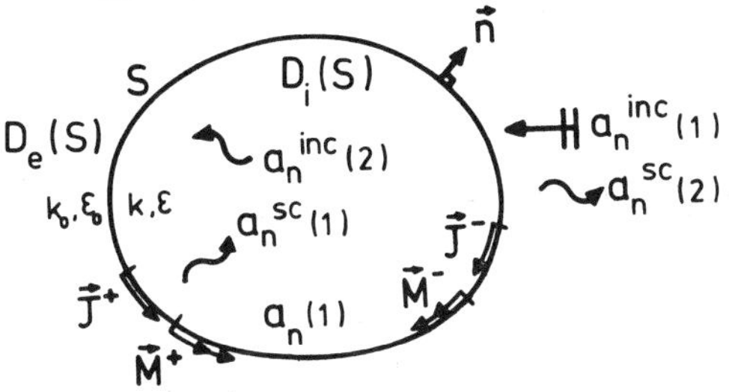

Fig. 1. Scattering from a homogeneous dielectric body

In compact matrix notation one may write (29) as $T=-Q_{11}Q_{21}^{-1}$ where Q_{21}^{-1} is associated with a passage of a wave in through S while Q_{11} is associated with a passage of a wave out through S. Thus, T is obviously associated with a reflection at the outer side of S. According to theorem 3 the expansion of the total field in D_i converges uniformly in D_i if $k \notin S(D_i)$ while on S the convergence is in the mean-square sense by theorem 1. See also the section on T-matrix formulations and theorem 4.

Single surface scattering (proposed) : Form II The EBC that $\vec{J}^-$, $\vec{M}^-$, which are obtained from the sum of the ingoing incident and the outgoing scattered fields in $D_e(S)$, yield null field in $D_e(S)$ in the host medium k leads to

$$-Q_{nn'}(k,k_o,1,1)\ a_{n'}^{inc}(1)\ -Q_{nn'}(k,k_o,1,2)\ a_{n'}^{sc}(2) = 0$$

where the negative signs precede the radiation matrices because $\vec{J}^-$, $\vec{M}^-$ are placed on the inner side of S. Thus Form II of the T-matrix is given by

$$T_{nn'} = -[Q_{nn'}(k,k_o,1,2)]^{-1}\ Q_{nn'}(k,k_o,1,1) \tag{30}$$

In this formulation the outgoing scattered field is used on S. This is allowed by theorems 1", 3" and 4", where the Rayleigh hypothesis may not necessarily be valid. See also the section on T-matrix formulations and theorem 4.

The two T-matrix forms given in (29) and (30) are equivalent. If the above formulations were carried out with the SVWF 's which are orthogonalized on S, then using the corresponding equation (21) in (29) and (30), it would be possible to show that (29) and (30) are identical. Nevertheless, for a spherical surface, the analytic equivalence can be shown. For a non-spherical surface, without the orthogonalization of the SVWF 's on S, the radiation matrices are not diagonal, so that only the numerical equivalence can be demonstrated. This is done for a two-layered dielectric scatterer, whose T-matrix in Form I and Form II will be given.

Single surface scattering in a cavity (proposed) Let $a_n^{inc}(2)$ be the coefficient of an outgoing incident field in D_i which scatters (reflects) from the inner side (Fig.1) of S to yield the ingoing scattered field having the coefficient $a_n^{sc}(1)$. The EBC that

$\vec{J}^+$, $\vec{M}^+$, which are obtained from the sum of the outgoing incident and the ingoing scattered fields in D_i, yield null field in D_i in the host medium k_o leads to

$$Q_{nn'}(k_o,k,2,1)\, a_{n'}^{sc}(1) + Q_{nn'}(k_o,k,2,2)\, a_{n'}^{inc}(2) = 0\ .$$

Thus the reflection matrix (internal T-matrix) defined by $a_n^{sc}(1)=R_{nn'}\, a_{n'}^{inc}(2)$ is expressed by

$$R_{nn'} = -[Q_{nn'}(k_o,k,2,1)]^{-1}\, Q_{nn'}(k_o,k,2,2) \qquad (31)$$

The use of the ingoing SVWF expansion for the scattered field in D_i can be justified by theorems 1, 3 and 4.

Scattering from two-layered dielectric body (Aydın, 1978) : Form I In the two-layered dielectric body shown in Fig. 2, the total field in E_i, the region between the surfaces S_1 and S_2, is expanded in terms of both the ingoing and the outgoing SVWF 's with coefficients $a_n(1)$ and $a_n(2)$, respectively. Thus, theorems 1', 3' and 4' are relevant in this case. The EBC that $\vec{J}_1^-$, $\vec{M}_1^-$ yield null field in $D_e(S_1)$ in the host medium k_1, leads to the equation

$$-Q^1_{nn'}(k_1,k_2,1,1)\, a_{n'}(1) - Q^1_{nn'}(k_1,k_2,1,2)\, a_{n'}(2) = 0\ .$$

Also the EBC that $\vec{J}_2^+$, $\vec{M}_2^+$ yield $-a_n^{inc}(1)$ in $D_i(S_2)$ in the host medium k_o leads to

$$Q^2_{nn'}(k_o,k_2,2,1)\, a_{n'}(1) + Q^2_{nn'}(k_o,k_2,2,2)\, a_{n'}(2) = -a_n^{inc}(1)\ .$$

The coefficients of the scattered field are given by

$$Q^2_{nn'}(k_o,k_2,1,1)\, a_{n'}(1) + Q^2_{nn'}(k_o,k_2,1,2)\, a_{n'}(2) = a_n^{sc}(2)\ .$$

Solving these matrix equations for the scatterer's T-matrix defined by $a_n^{sc}(2)=T_{nn'}(1,2)\, a_{n'}^{inc}(1)$ one obtains, in compact matrix notation Form I, namely

$$T(1,2) = -[Q^2_{11}+Q^2_{12}\, T(1)][Q^2_{21}+Q^2_{22}\, T(1)]^{-1} \qquad (32)$$

where the arguments of the radiation matricies for S_2 are evident. T(1) is the T-matrix of the inner surface and $T(1)=-[Q^1_{12}]^{-1}\, Q^1_{11}$ which is in Form II. T(1,2) can be expanded as an infinite series and a multiple scattering interpretation may be given by following the lines of Peterson and Ström (1974). This is shown in Fig. 3, where $T(2)=-Q^2_{11}[Q^2_{21}]^{-1}$ for S_2 which is in Form I and $R(2)=-[Q^2_{21}]^{-1}\, Q^2_{22}$ which is the reflection matrix for a cavity scattering problem for the surface S_2.

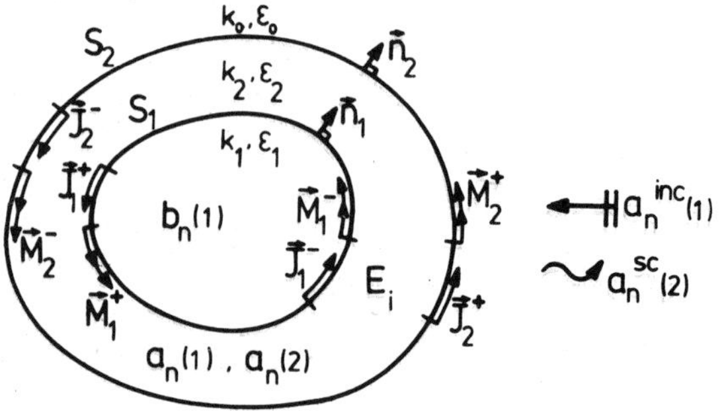

Fig. 2. Scattering from a two-layered dielectric body

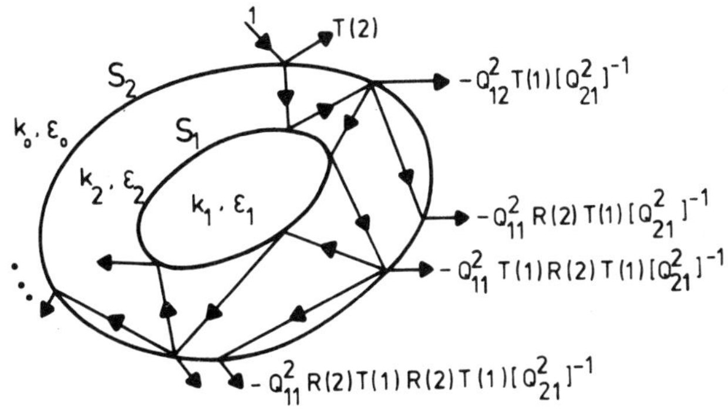

Fig. 3. Multiple scattering interpretation of Form I

Scattering from two-layered dielectric body (Aydın, 1978) : Form II An alternative formulation of the scattering problem cited in Fig. 2 would be to expand the total field in $D_i(S_i)$ in regular SVWF 's having the coefficients $b_n(1)$. With E_i as the host medium, the EBC that $\vec{J}_1^{\,+}$, $\vec{M}_1^{\,+}$ and $\vec{J}_2^{\,-}$, $\vec{M}_2^{\,-}$, which are respectively obtained from the total field in $D_i(S_1)$ and the sum of the ingoing incident and the outgoing scattered fields in $D_e(S_2)$ yield null field in $D_e(S_2)$ and $D_i(S_1)$ lead to

$$-Q^2_{nn'}(k_2,k_o,1,1)\, a_{n'}^{inc}(1) + Q^2_{nn'}(k_2,k_o,1,2)\, a_{n'}^{sc}(2) + Q^1_{nn'}(k_2,k_1,1,1)\, b_{n'}(1) = 0$$

$$-Q^2_{nn'}(k_2,k_o,2,1)\, a_{n'}^{inc}(1) + Q^2_{nn'}(k_2,k_o,2,2)\, a_{n'}^{sc}(2) + Q^1_{nn'}(k_2,k_1,2,1)\, b_{n'}(1) = 0$$

respectively. Upon solving these equations one obtains, in compact matrix notation, Form II of $T(1,2)$, namely

$$T(1,2) = -[Q^2_{12} + T(1)\, Q^2_{22}]^{-1} [\, Q^2_{11} + T(1)\, Q^2_{21}] \qquad (33)$$

where the arguments of the radiation matrices are evident from the EBC equations above and $T(1) = -Q^1_{11}\, [\, Q^1_{21}]^{-1}$ for the inner surface S_1 which is in Form I. In this formulation theorems 1, 3 and 4 for $D_i(S_1)+S_1$ and theorems 1", 3" and 4" for $D_e(S_2)+S_2$ apply. A multiple scattering interpretation of Form II for $T(1,2)$ may be developed by expanding (33) as an infinite series as in Fig. 4, where $T(2) = -[Q^2_{12}]^{-1}\, Q^2_{11}$ for the surface S_2 which is in Form II.

The two formulations given above are equivalent. This can be verified analytically for a spherical surface. For a non-spherical surface numerical equivalence can be demonstrated. This is done by Aydın (1978) for a number of two-layered oblate and prolate spheroidal shapes.

Extensions of Form I and Form II to multiyared scatterers Peterson and Ström (1974) have formulated the scattering from multilayered scatterers in the T-matrix form by obtaining an iterative formula which is a generalization of the T-matrix for a two-layered scatterer. The latter problem is formulated by a procedure similar to that of Form I. If one applies the present notation to the surface sources and the EBC used by Peterson and Ström (1974) for a two-layered scaterer, it can be readily

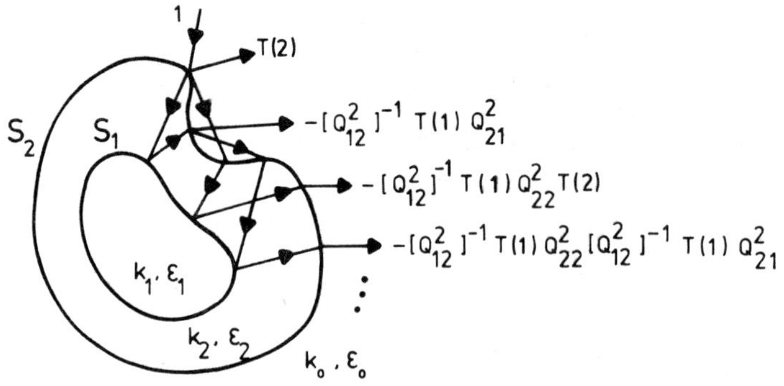

Fig. 4. Multiple scattering interpretation of Form II

verified that their formulation is exactly in Form I given by (32), except that T(1) is in Form I instead of being in Form II.

Formulations I and II can be generalized to a multilayered scatterer in a straightforward manner. Let a scatterer be composed of M layers bounded by surfaces S_1, S_2, ...,S_M, where S_1 and S_M are the innermost and the outermost surfaces, respectively. In the regions facing the inner side of these surfaces let the wave numbers be k_1, k_2,...,k_M and that in $D_e(S_M)$ the wave number is k_o. It can be shown that T-matrix of the multilayered scatterer described above may be expressed either in Form I as

$$T(1,2,...,M) = -[Q^M_{11}+Q^M_{12}\;T(1,2,...,M-1)][Q^M_{21}+Q^M_{22}\;T(1,2,...,M-1)]^{-1} \tag{34}$$

or in Form II as

$$T(1,2,...,M) = -[Q^M_{12}+T(1,2,...,M-1)\;Q^M_{22}]^{-1}[Q^M_{11}+T(1,2,...,M-1)\;Q^M_{21}] \tag{35}$$

where the arguments of the radiation matrices can be readily deduced from (32) and (33).

In the above formulations the property of the radiation matrix in (22) for k=k' is used. This property has a precise meaning in terms of Vekua's theorems. Consider a closed smooth surface S and let the wave number in $D_e(S)$ be $k \neq k'$. Also let $\vec{J}^+$ and $\vec{M}^+$ be the sources on the outer side of S expressed as in (13) and (14). If these sources are expanded in terms of the ingoing and the outgoing SVWF 's with coefficients $a_n(1)$ and $a_n(2)$, respectively, then the radiation operator yields $a_n^{rad}(1)=0$ and $a_n^{rad}(2)=a_n(2)$ in $D_e[S_{cir}(S)]$. This result is in agreement with theorems 1" and 3" for the exterior region. A similar application of the radiation operator in $D_i(S)$ reveals that $a_n^{rad}(1)=a_n(1)$ and $a_n^{rad}(2)=0$ in $D_i[S_{ins}(S)]$, which is in agreement with theorems 1 and 3 for the interior region. This property has been checked numerically for an exterior region using Form II applied to a two-layered dielectric body having rotational symmetry.

T-Matrix Formulations for Conducting and Coated Conducting Bodies

SVWF formulation of electromagnetic scattering from a conducting body is described by Waterman (1965, 1971) and Hızal and Marinçiç (1970). Waterman's (1965) formulation can be expressed as

$$T_{nn'} = -G_{nn'}(k_o,1,2)[G_{nn'}(k_o,2,2)]^{-1} \tag{36}$$

where the radiation matrix for a conducting surface defined by (26) is used. In this formulation the total magnetic field on the surface is expanded in terms of the outgoing SVWF 's. Then, by theorem 1" the series converges in the mean-square sense whether the Rayleigh hypothesis is valid or not. Equation (36) is the corresponding Form I for conducting bodies. Waterman's (1971) formulation reduces to

$$T_{nn'} = -G_{nn'}(k_o,1,1)[G_{nn'}(k_o,2,1)]^{-1} \tag{37}$$

which is in Form I. In this formulation the ingoing SVWF expansion of the total field on S converges in the mean-square sense if and only if $k_o \notin S\ D_i(S)$ by theorem 1. See also the section on T-matrix formulations and theorem 4. Waterman's (1971) observation of comparatively faster convergence of this formulation may be also explained by the results of the section cited above.

If the scattered field on a perfectly conducting surface S is represented by the outgoing SVWF 's by theorems 1", 3" and 4" one obtains

$$G_{nn'}(k_o,2,1)\ a_{n'}^{inc}(1) + G_{nn'}(k_o,2,2)\ a_{n'}^{sc}(2) = -a_n^{inc}(1)$$

which leads to

$$T_{nn'} = -[G_{nn'}(k_o,2,2)]^{-1}[G_{nn'}(k_o,2,1) + \delta_{nn'}] \tag{38}$$

It can be shown that the formulation of Hızal and Marinçiç (1970) can be reduced to the one expressed in (38).

T-matrix description of scattering from a conducting body ($k_1=\infty$) coated with a dielectric material is given by Bringi and Seliga (1977) and Aydın (1976, 1978).

In Bringi and Seliga's formulation the total field on S_1 (Fig. 2) is expanded using the ingoing SVWF while on the inner side of S_2 both the ingoing and the outgoing SVWF 's are used to express the total field. Using the EBC that $\vec{J}_1^{\,+}$ and $\{\vec{J}_2^{\,-}, \vec{M}_2^{\,-}\}$ yield zero field in $D_i(S_1)$ and in $D_e(S_2)$ in the host medium k_2 together with the property (22) and that $\vec{J}_2^{\,+}$, $\vec{M}_2^{\,+}$ yield $-a_n^{inc}(1)$ in $D_i(S_2)$, one obtains Form I of the T-matrix given by (32) where T(1) is that of Waterman (1971) in (37) with the replacement of k_o by k_2. In this formulation the total field on S_1 converges in the mean-square sense if and only if $k_2 \notin S[D_i(S_1)]$. The convergence properties of the SVWF expansion of the total field in the region between the surfaces S_1 and S_2 are in accordance with theorems 1', 3' and 4'.

Aydın's (1978) Form I for the present problem is given by (32) where T(1) is either that given by (37) with the replacement of k_o by k_2 or by

$$T_{nn'}(1) = -[G^1_{nn'}(k_2,1,2) - \delta_{nn'}]^{-1}\ G^1_{nn'}(k_2,1,1)$$

This new Form II of T(1) arises from the EBC in $D_e(S_2)$. For a single conducting body this Form II can be obtained by considering an annular region between the surface of the body and another closed surface enclosing the body and applying the radiation matrix notation in this region after expressing the total field as the sum of the ingoing incident and the outgoing scattered fields.

In Aydın's (1978) Form II, applied to the present problem, the total field on S_1 is expanded using the outgoing SVWF 's while $\vec{J}_2^{\,-}$, $\vec{M}_2^{\,-}$ are obtained from the sum of the ingoing incident and the outgoing scattered fields in $D_e(S_2)$. The EBC that $\vec{J}_1^+$ and $\{\vec{J}_2^-, \vec{M}_2^-\}$ yield zero field in $D_i(S_1)$ and $D_e(S_2)$ in the host medium k_2 leads to Form II as in (33), where T(1) is that in (36) with the replacement of k_o by k_2. The convergence properties of this formulation are in accordance with theorems 1"a, 3"a and 4"a for S_1 and theorems 1", 3" and 4" for S_2.

In Aydın's Form II, if one expands the total field on S_1 using the ingoing SVWF 's instead of the outgoing ones and follows the same lines as in the rest of the formulation one obtains (33), where T(1) is now given by (37) with the replacement of k_o by k_2. Here the convergence property on S_1 is governed by theorems 1a, 3a and 4a with respect to the surface S_2.

T-Matrix Formulations and Theorem 4

The role of theorem 4 in T-matrix formulations are kept out of the attention so far, to avoid any ambiguous interpretation. However, this theorem plays a significant role in T-matrix formulations. The following conclusions are derived by Hızal and Aydın (1979).

From theorem 3, lemma 1 and theorem 4 for an interior region, it follows that if $k \notin S(D_i)$, the computed SVWF expansion of the interior field in terms of the ingoing SVWF 's can converge uniformly in D_i+S. On the other hand if $k \in S(D_i)$, the computed expansion can only be a uniform approximation. This conclusion can be extended to an annular region also.

In a T-matrix formulation in which the outgoing SVWF expansion of the scattered field is incorporated, the computed expansion can converge uniformly in D_e+S if the scattered field has no singularities on S.

NUMERICAL EXAMPLES

Perfectly Conducting Rotationally Symmetrical Bodies

A computer program based on the method of Hızal and Marinçiç (1970), which is equivalent to the formulation in (38) is used to solve the two examples of the scattering problem shown in Fig. 5 and Fig. 6. The above formulation and the computer program used were checked previously with other numerical and experimental results (Hızal and Yasa, 1973).

The back-scattering cross-section of a prolate spheroid at oblique incidence is investigated as a function of its elongation (Fig. 5). The accuracy by which the numerical integrations are performed is observed by the use of the properties in (27). It is observed that as the body is elongated, more Gauss-Legendre points G are needed to maintain a fixed accuracy by which these properties are satisfied. In Fig. 5, 20 to 32 points are used. The number of elevation modes N, which is the maximum order of the spherical Hankel functions in the SVWF expansion of the scattered field, is increased up to 10. The results became stable at about N=8. Increasing N beyond 10 caused numerical difficulties. It is observed that the convergence depends not only on the optic size ka, but also on the elongation, the steepness of the curve and the polarization of the incident field.

The second example for the perfectly conducting bodies is shown in Fig. 6. The generatrix of the body consists of three equi-radius circles tangent to each other such that at the point of contact the derivative of the surface equation in spherical co-ordinates becomes infinite. Thus this shape is difficult to handle numerically with respect to an origin at its center. Waterman (1971) faced a similar

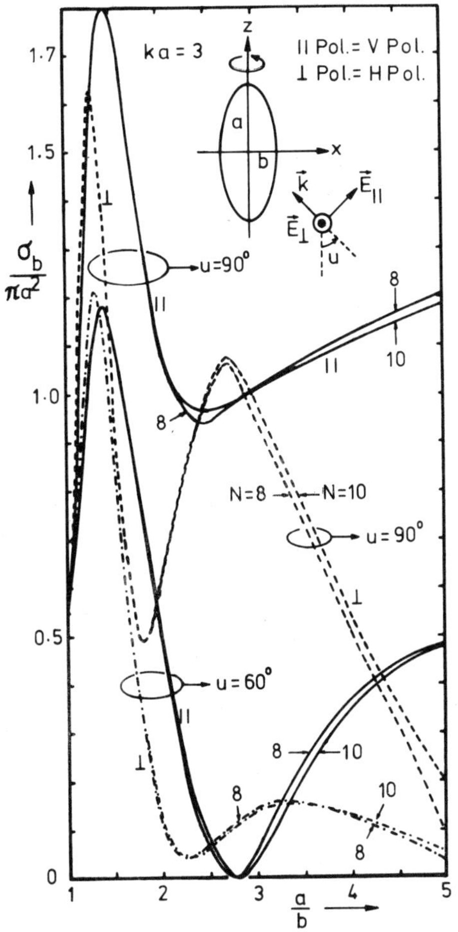

Fig. 5. The back-scattering cross-section of a conducting prolate spheroid

difficulty for spherically capped cylinders and cones. Here a simple method is used to avoid the singularity of the surface equation at the points of contact. The method basically is to divide the range of integration into three parts, namely AB, BC and CD. Then the expressions of the integrals with respect to 0 are transformed to the expressions containing surface equations with respect to the points 0_1, 0_2 and 0_3 using the Jacobian relationships of the transformations of the coordinates. It is observed that this method gives accurate results for the integrations, which are checked by the matrix properties (27). In Fig. 6 the choice of G=32 for the sections AB and CD and G=64 for the section BC are found to be sufficient. The variation of the back-scattering cross-section with the angle of incidence is seen to be sensitive to the optic size of the body. The convergence is checked by changing the number of elevation modes N. It is observed that the sensitivity of the results with respect to N is greater in the steep portion of the curves in comparison to the extremum points. Too much increase in N for a fixed optic size kr_m results in deteriorations instead of better results because of the numerical difficulties encountered. Thus, there is an optimum range of values for N which

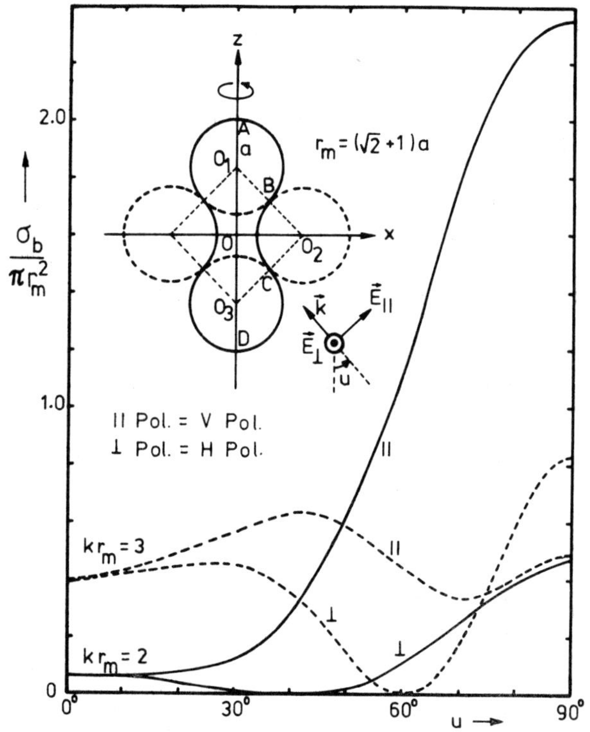

Fig. 6 The back-scattering cross-section of a dumbbell shaped conducting body

enables one to obtain satisfactory results for a given optic size, shape, polarization and angle of incidence.

Two-Layered Rotationally Symmetrical Dielectric Bodies

Two computer programs, which are developed by Aydın (1978) and based on Formulations I and II are used to solve the problems given below. The dependability of these programs were determined by comparisons of the results with those in the literature and with the exact Mie solutions for concentric spheres.

The first example consists of a teflon sphere having a dielectric constant ε_{r1}=2.1+i0.000315 and radius r=5.5 mm which is symmetrically coated by a melamine layer of oblate spheroidal shape with dielectric constant ε_{r2}=4.93+i0.50779 and major to minor axis ratio a/b. A plane wave whose magnetic field has unity amplitude is incident along the major axis which corresponds to an angle of incidence of 90° as measured from the axis of rotation. The optic size of the body is $k_o a$=1 where k_o is that of free-space. At the frequency of 3 GHz the normalized back-scattering cross-sections σ_{BV}, σ_{BH} and the normalized extinction cross-sections σ_{EV}, σ_{EH} for vertical (V) and horizontal (H) polarizations are computed. For V (H) polarization, the incident electric field is in (perpendicular to) the plane formed by the axis of rotation and the direction of incidence. The results are displayed in Table 1. The number of Gauss-points for the integration over the inner surface is chosen as G=10 while for the outer surface G=32. The number of the azimuthal modes in the computations, max.(m) is 5. The number of the elevation modes is indicated by N in the Table. The cross-sections presented are normalized with the area of a circle whose diameter is equal to the equi-volume spherical diameter of the object. For a/b=1 the results obtained from the exact Mie solutions are :

$$\sigma_{BV}=\sigma_{BH}=0.45940709 \quad \text{and} \quad \sigma_{EV}=\sigma_{EH}=1.3879080$$

TABLE 1 Results for the Cross-Sections of an Oblate Spheroid With a Spherical Core

$\frac{a}{b}$	N	σ_{BV}	σ_{BH}	σ_{EV}	σ_{EH}
1.0	4	0.4594068	0.4594064	1.3879079	1.3879079
	5	0.45940704	0.45940716	1.3879080	1.3879080
1.5	6	0.2001	0.409	0.5214	1.0387
	7	0.20000	0.40886935	0.521254	1.0383836
	8	0.20001	0.40886939	0.521256	1.0383837
2.0	11	0.1094	0.338	0.271	0.830
	13	0.1099	0.339	0.272	0.833
2.5	13	0.09	0.41	0.24	0.95
	15	0.04	0.14	0.06	0.37
2.5*	13	0.035	0.084	0.052	0.111
	15	0.039	0.088	0.051	0.116

* ε_{r1} and ε_{r2} are interchanged

It is seen that the convergence occurs for a/b up to 2.5, at which the convergence fails. However, it is observed that when the dielectric constants of the two layers are interchanged, i.e. when a melamine sphere is coated with a teflon oblate spheroid, the convergence is achieved. This can be due to the decrease of the optic size k_2a. The results of Table 1 are obtained by Form II and some of them are also checked with those obtained by Form I, which yield almost exactly the same results.

The second example consists of a melamine oblate spheroid symmetrically coated with a teflon sphere. At 3 GHz the optic radius of the sphere is $k_0r=1$, while the major axis of the oblate spheroid is a=8 mm. The number of Gauss-points over the inner and outer surfaces are G=32 and G=10, respectively and the number of azimuthal modes used is 5. The number of elevation modes N are indicated in Table 2. The angle of incidence is 90°. The normalized back-scattering and extinction cross-sections, with N as a parameter, are computed by Form II to observe the convergence. It is seen that satisfactory convergence is achieved with a/b up to 3. For a/b=1 the results obtained from the exact Mie solutions are $\sigma_{BV}=\sigma_{BH}=0.26796331$ and $\sigma_{EV}=\sigma_{EH}=0.32163494$.

It may be also of interest to calculate the total field within the inner surface. In such applications Form II is suitable. The non-zero field components at the spherical coordinates r=3 mm, $\theta=90^\circ$ and $\phi=0^\circ$ (3 mm from the center along the direction of incidence) are computed and displayed in Table 3. It is seen that, as expected by theorem 3, the convergence occurs even for a considerably deformed inner surface. For a/b=1 the exact Mie solutions are $H_{V\phi}=0.12498926-i0.010862079$ and $H_{H\theta}=0.12498926-i0.010862079$.

The number of the elevation modes required for a given optic size, dielectric constant, shape and polarization cannot be determined or known apriori. This point, which is discussed by Bates (1975) is a major drawback of such formulations. The numerical experience for two-layered dielectric bodies shows that $N \geqslant |k|_{max}\, r_{max}$.

TABLE 2 Results for the Cross-Sections of a Sphere with an Oblate Spheroidal Core

a/b	N	σ_{BV}	σ_{BH}	σ_{EV}	σ_{EH}
1.0	5	0·2679632	0.26796334	0.32163491	0.32163492
	7	0.2679632	0.26796334	0.32163491	0.32163492
	8	0.2679632	0.26796334	0.32163491	0.32163492
1.5	6	0.21808343	0.23137256	0.25627275	0.27748866
	7	0.21808343	0.23137256	0.25627275	0.27748866
2.0	9	0.19724157	0.21277474	0.22916697	0.25431524
	10	0.19724148	0.21277474	0.22916700	0.25431523
2.5	13	0.1861	0.2014	0.214928	0.2399
	15	0.1862	0.2015	0.214933	0.2400
3.0	13	0.179	0.194	0.2064	0.229
	15	0.181	0.197	0.2062	0.235

TABLE 3 Results for the Total Magnetic Field in the Oblate Spheroidal core in a sphere

a/b	N	$H_{V\phi}$	$H_{H\theta}$
1.0	5	0.12498926-i0.010862071	0.12498926-i0.010862075
	7	0.12498926-i0.010862073	0.12498926-i0.010862083
	8	0.12498926-i0.010862073	0.12498926-i0.010862083
1.5	6	0.12003323-i0.010680693	0.12132825-i0.0093603548
	7	0.12003323-i0.010680693	0.12132825-i0.0093603549
2.0	9	0.11716896-i0.010455263	0.11881981-i0.008050622
	10	0.11716899-i0.010455255	0.11881982-i0.008050628
2.5	13	0.11535 -i0.010252	0.116989 -i0.006957
	15	0.11532 -i0.010257	0.116990 -i0.006959
3.0	13	0.114 -i0.0101	0.11558 -i0.0060
	15	0.113 -i0.0102	0.11563 -i0.0061

CONCLUSIONS

On the basis of the generalizations of Vekua's theorems it is shown that the T-matrix formulations are well founded. Thus, the convergence properties of various T-matrix formulations are known and this is important in numerical applications. The extensions of the theorems to an exterior region elucidates the controversial aspects of the Rayleigh hypothesis.

The unified radiation matrix and T-matrix notations presented make the formulations and physical interpretations of various scattering problems easier. Moreover, two fundamental T-matrix forms, which essentially arise from the use of the interior and exterior null field conditions, are observed. Without the orthogonalization of the SVWF 's on the surface of a scatterer only the numerical equivalence of the two forms of the T-matrix can be demonstrated, except in the case of a spherical surface on which the SVWF's are already orthogonal and the analytic equivalence can be readily established. Depending upon the problem, the first type of formulation may be more convenient compared to the other. For instance in a biomedical application if the field between two surfaces of a composite body is required, formulation I is suitable, whereas if the field within the inner surface is required formulation II is more suitable. Also, different multiple scattering interpretations of the two formulations might be useful in analysing the scattering properties of a composite scatterer such as a hydrometeor.

The numerical examples considered reveal certain computational aspects of the formulations used in the computations regarding the convergence. It is observed that the convergence depends on various factors such as optic size, shape and polarization of the incident field. For a two-layered dielectric body uniform convergence of the total interior field is demonstrated.

Although T-matrix formulations are capable of solving a large class of problems they are limited by the numerical difficulties encountered when the parameters of the given scattering problem are not favorable for precise computations. These may be (i) large optic sizes requiring large matrices which may be ill-conditioned, (ii) shapes severely deformed from a sphere, (iii) increased number of layers, (iv) steep variation of the scattering parameters for a given polarization of the incident field and (v) given wave numbers close to the interior resonance wave numbers.

ACKNOWLEDGMENTS

The author is indebted to Kültegin Aydın for the great assistance he has given in the preparation of this work, who obtained the numerical results for the dielectric bodies and through continual discussions increased the author's understanding of many of the concepts discussed in this paper. The numerical results were obtained using the IBM 370/145 computer of the Department of Computer Engineering of METU.

REFERENCES

Al-Badwaihy, K.A., and Yen, J.L. (1975) Extended boundary condition integral equations for perfectly conducting and dielectric bodies : Formulation and uniqueness, IEEE Trans. Antennas Propagat., AP-23, 546.

Aydın, K. (1976) Spherical wave formulation of electromagnetic scattering from composite dielectric bodies of arbitrary shape, Departmental Memorandum No.3, Elect. Eng. Dept., Middle East Technical University, Ankara, Turkey.

Aydın, K. (1978) Electromagnetic scattering from two-layered dielectric bodies, Ph.D. Dissertation, Elect. Eng. Dept., Middle East Technical University, Ankara Turkey.

Aydın, K. and Hızal, A. (1979) On the convergence of the spherical vector wave function expansions used in electromagnetic scattering problems, International IEEE/APS Symposium, University of Washington, Seattle, Washington, U.S.A.

Bates, R.H.T. (1975) Analytic constraints on electromagnetic field computations, IEEE Trans. Microwave Theory and Techniques, MTT-23, 605.

Barber, P., and Yeh, C. (1975) Scattering of electromagnetic waves by arbitrarily shaped dielectric bodies, Applied Optics, 14, 2864.

Bringi, V.N., and Seliga, T.A. (1977) Scattering from axisymmetric dielectric or perfect conductors imbedded in an axisymmetric dielectric, IEEE Trans. Antennas Propagat., AP-25, 575.

Calderon, A.P. (1954) The multipole expansion of radiation fields, J. Ration. Mech. Analy., 3, 523.

Harrington, R.F. (1968) Field Computation by Moment Methods, Macmillam, New York.

Hızal, A., and Aydın, K. (1979) On the generalization and interpretation of Vekua's theorems, Departmental Memorandum No.8, Elect. Eng. Dept., Middle East Technical

University, Ankara, Turkey.

Hızal, A., and Marinçiç, A. (1970) New rigorous formulation of electromagnetic scattering from perfectly conducting bodies of arbitrary shape, Proc. Inst. Elect. Eng., 117, 1639.

Hızal, A., and Yasa, Z. (1973) Scattering by perfectly conducting rotational bodies of arbitrary form excited by an obliquely incident plane wave or by a linear antenna, Proc. Inst. Elect. Eng., 120, 181.

Jackson, J.D. (1962) Classical Electrodynamics, Wiley, New York.

Millar, R.F. (1973) The Rayleigh hypothesis and a related least-squares solution to scattering problems for periodic surfaces and other scatterers, Radio Science, 8, 785.

Peterson, B., and Ström, S. (1974) T-matrix formulation of electromagnetic scattering from multilayered scatterers, Physical Review D, 10, 2684.

Vekua, I.N. (1953) About the completeness of the system of metaharmonic functions, Doklady Akademii Nauk., USSR, 90, 715.

Vladimir, V.S. (1971) Equations of Mathematical Physics, Marcel Dekker, New York.

Waterman, P.C. (1965) Matrix formulation of electromagnetic scattering, Proc. IEEE, 53, 805.

Waterman, P.C. (1969a) New formulation of acoustic scattering, The J. of the Acoustical Society of America, 45, 1417.

Waterman, P.C. (1969b) Scattering by dielectric obstacles, Alta Frequenza, 38, 348.

Waterman, P.C. (1971) Symmetry, unitarity and geometry in electromagnetic scattering, Physical Review D, 3, 825.

Warner, C., and Hızal, A. (1976) Scattering and depolarization of microwaves by spheroidal raindrops, Radio Science, 11, 921.

SCATTERING AND ABSORPTION BY HOMOGENEOUS AND LAYERED DIELECTRICS

Peter W. Barber
Department of Electrical Engineering
University of Utah
Salt Lake City, Utah 84112

ABSTRACT

The extended boundary condition method is used to make scattering and absorption calculations for a variety of nonspherical dielectric objects. Both the general electromagnetic interaction and specific problems are investigated.

INTRODUCTION

There is currently a great interest in the scattering and absorption characteristics of dielectric objects. This interest is the result of a variety of new and diverse studies involving the interaction of electromagnetic (EM) waves with closed dielectric bodies. These studies include a determination of the power-absorption characteristics of man due to exposure to EM waves (1), the absorption and scattering of microwaves by raindrops (2), and continuing investigations to apply laser light-scattering techniques to problems in microbiology (3). In all of these studies, a theoretical solution which describes the EM interaction is crucial. For spherical objects, the Lorenz-Mie theory is used (4), while for other geometries it is necessary to employ alternate techniques, and these generally involve approximations with limited range of applicability, e.g., low ka, small deviation from spherical shape, etc. The need for new methods of analysis which will provide quantitative results and also increase our understanding of the interaction mechanisms, especially techniques suitable for nonspherical objects on the order of a wavelength in size (resonance-sized objects), is clearly indicated.

In the last few years, a number of numerical techniques have been developed for solving EM problems involving nonspherical (and in some cases inhomogeneous) dielectric objects. These methods include the extended boundary condition or T-matrix method (5-8), the spheroidal separation of variables method (9), the finite-element method (10), and the method of moments(11, 12). All of these techniques have been implemented on the digital computer, although only for axisymmetric bodies.

The extended boundary condition method (EBCM) is particularly versatile and is ideally suited for performing the types of calculations indicated in the first paragraph. In this chapter, some of the problems to which the EBCM has been applied are reviewed. The goal is to give the reader a clear understanding of the capabilities of the EBCM in solving 3-D scattering and absorption problems.

As far as the numerical implementation of the EBCM is concerned, a family of computer programs has been written -- each program an optimized version to solve specific problems, e.g., absorption by homogeneous objects and scattering by lossless layered objects. One of the most important activities has been validity testing. Results generated by the EBCM have been checked against almost every known method with total success. These other methods include all the other numerical techniques listed above as well as the Lorenz-Mie theory (spheres), and the Stevenson and Rayleigh-Gans-Debye approximations. Furthermore, conformance with the laws of reciprocity and conservation of energy has been verified. One of the most useful tests involves making calculations for a sphere with offset origin. The EBCM treats the sphere as a nonspherical object, but the results for absorption and scattering must be identical to the usual sphere results obtained by the Lorenz-Mie theory. The EBCM computer programs have been successfully run on a number of computers, including the IBM 360/91 and 370/165, the UNIVAC 1108, the CDC 6600 and 7600, and the CRAY 1.

In what follows, the theory is briefly reviewed and those quantities which will be used later are defined. Then a variety of absorption and scattering problems that have been investigated with the aid of the EBCM will be reviewed.

THEORY

The EBCM utilizes spherical harmonic expansions of the incident and scattered fields in conjunction with the boundary conditions at the surface to obtain a system of linear equations relating the unknown expansion coefficients of the scattered field to the known coefficients of the incident field. The nonsphericity of the object is handled by an analytic continuation process. Details of the development can be found elsewhere (8) and we will only summarize those specific equations which are used here.

The incident field expansion is given by

$$\bar{E}^i(\bar{r}) = \sum_{\nu=1}^{\infty} D_\nu \left[a_\nu \bar{M}_\nu^1(k\bar{r}) + b_\nu \bar{N}_\nu^1(k\bar{r}) \right] \tag{1}$$

where $\bar{M}$ and $\bar{N}$ are the vector spherical wave functions (13), $k = 2\pi/\lambda$, ν is a combined index incorporating the spherical harmonic indices, D_ν is a normalization constant, and the expansion coefficients a and b are assumed known for a specified incident field. The superscript 1 on $\bar{M}$ and $\bar{N}$ indicates that these functions are of the type which are finite at the origin (Bessel function radial dependence). The scattered field has a similar form

$$\bar{E}^s(\bar{r}) = \sum_{\nu=1}^{\infty} 4D_\nu \left[f_\nu \bar{M}_\nu^3(k\bar{r}) + g_\nu \bar{N}_\nu^3(k\bar{r}) \right] \tag{2}$$

where the f and g coefficients are unknown and the superscript 3 on $\bar{M}$ and $\bar{N}$ indicates that these functions are of the type suitable for radiation fields (Hankel function radial dependence). Applying the EBCM and truncating the expansions results in a linear system of equations relating the unknown scattered field coefficients to the known incident field coefficients. For homogeneous dielectric objects, we obtain,

$$\begin{bmatrix} f_\nu \\ \\ g_\nu \end{bmatrix} = - \begin{bmatrix} T_1 \end{bmatrix} \begin{bmatrix} \frac{a_\nu}{4} \\ \\ \frac{b_\nu}{4} \end{bmatrix} \tag{3}$$

where

$$\begin{bmatrix} T_1 \end{bmatrix} = \begin{bmatrix} Q_1^{11} \end{bmatrix} \begin{bmatrix} Q_1^{31} \end{bmatrix}^{-1} \tag{4}$$

and the [Q] matrices, which are defined below, contain all the information about the dielectric body, such as its size, shape, and permittivity. The matrix $[T_1]$ transforms the incident field coefficients to the scattered field coefficients and hence is called the transition matrix.

For an n-layered object, the transition matrix $[T_n]$ can be found by a recursion relation working from the inside out (14).

$$[T_n] = \left\{[Q_n^{11}] - [Q_n^{13}][D][T_{n-1}][D]^{-1}\right\} \cdot \left\{[Q_n^{31}] - [Q_n^{33}][D][T_{n-1}][D]^{-1}\right\}^{-1} \quad (5)$$

By calculating $[T_1]$ for the most inner layer using Eq. 4, Eq. 5 is applied repeatedly to calculate $[T_2]$, $[T_3]$,..., until the final $[T_n]$ matrix is obtained. Replacing $[T_1]$ by $[T_n]$, Eq. 3 is used to calculate f_ν and g_ν.

The $[Q_i]$ matrices in Eqs. 4 and 5 are $2N \times 2N$ matrices, each consisting of four $N \times N$ submatrices. The quantity N is the truncation size (determined by convergence requirements) on the expansion in Eqs. 1 and 2. For example, the $[Q_i^{13}]$ matrix is given by

$$[Q_i^{13}] = \begin{bmatrix} K_i^{13} + (\varepsilon_{r_i})^{1/2} J_i^{13} & L_i^{13} + (\varepsilon_{r_i})^{1/2} I_i^{13} \\ I_i^{13} + (\varepsilon_{r_i})^{1/2} L_i^{13} & J_i^{13} + (\varepsilon_{r_i})^{1/2} K_i^{13} \end{bmatrix} \quad (6)$$

where i = the *i*th layer numbered from the inside out, and $(\varepsilon_{r_i})^{1/2} = (\varepsilon_i)^{1/2}/(\varepsilon_{i+1})^{1/2}$ is the dielectric constant of the *i*th layer. The elements of the $N \times N$ matrices I, J, K, and L are surface integrals. For example, the I elements are given by

$$[I_i^{13}]_{\mu\nu} = \frac{k_{i+1}^2}{\pi} \int_{S_i} \bar{n}_i \cdot \bar{M}_\mu^1(k_{i+1}\bar{r}) \times \bar{M}_\nu^3(k_i\bar{r})\,ds$$

where $\bar{n}_i$ = the outward directed surface normal on S_i. The J, K, and L terms are defined similarly except for cross products, which are given by

$$\bar{M}_\mu^1 \times \bar{N}_\nu^3,\ \bar{N}_\mu^1 \times \bar{M}_\nu^3,$$

and

$$\bar{N}_\mu^1 \times \bar{N}_\nu^3,$$

respectively.

A useful definition where far-zone scattered fields are of interest is the vector far field amplitude $\bar{F}$ defined by

$$\bar{E}^S(k\bar{r}) = \bar{F}(\bar{o}, \bar{i}) \frac{e^{+jkr}}{r}, \quad kr \to \infty \quad (7)$$

where $\bar{i}$ is a unit vector in the direction of the incident field and $\bar{o}$ is a unit vector in the direction of the scattered field. The differential scattering cross section is the scattered power per unit solid angle in the direction $\bar{o}$ divided by the incident power density, which for unit incident field is given by

$$\sigma_d(\bar{o}, \bar{i}) = |\bar{F}(\bar{o}, \bar{i})|^2 \tag{8}$$

The scattering cross section, which is the total scattered power divided by the incident power density, is then

$$\sigma_s = \int_{4\pi} \sigma_d \, d\Omega = \int_{4\pi} |\bar{F}(\bar{o}, \bar{i})|^2 \, d\Omega \tag{9}$$

where $d\Omega$ is a differential solid angle. In a similar manner, the absorption cross section is the total absorbed power divided by the incident power density. The sum of the scattering and absorption cross sections is the total, or extinction cross section:

$$\sigma_e = \sigma_s + \sigma_a \tag{10}$$

The Forward Amplitude Theorem permits solution for the extinction cross section from the forward scattered field:

$$\sigma_e = \frac{4\pi}{k} \mathrm{Im}\left[\bar{e}_i \cdot \bar{F}(\bar{i}, \bar{i})\right] \tag{11}$$

where $\bar{e}_i$ is a unit vector which defines the polarization of the incident wave.

The EBCM is used to find the vector far field amplitude $\bar{F}(\bar{o}, \bar{i})$. Substitution of the scattered field into Eqs. 9 and 11 gives the scattering and extinction cross sections, respectively. These are then used in Eq. 10 to find the absorption cross section. The cross section quantities are divided by the shadow area of the object under illumination to obtain the scattering, absorption, and extinction efficiencies, which are then a measure of the fraction of the power incident on the *geometrical* cross section of an object which is reradiated or absorbed or reradiated and absorbed, respectively.

Absorption Calculations

A number of studies have been made to determine the absorption characteristics of nonspherical objects. Two types of calculations have been made. First, calculations have been made to determine the general absorption behavior of nonspherical objects and the dependence of the absorption on size, shape, and complex dielectric constant. Secondly, calculations have been made to determine radio frequency absorption by specific man and animal models as an aid to setting a radio frequency population exposure standard.

As an example of the first type of results (15), Fig. 1 compares the absorption characteristics of a finite circular cylinder and a prolate spheroid, both objects having identical overall dimensions. The absorption characteristics of the two shapes are very similar and it would appear that the details of the shape are not critical in determining the absorption. Figure 2 gives the internal-field distributions at $ka = 1.5$ (the approximate resonant point) for both the 3:1 prolate spheroid and the 3:1 circular cylinder. The internal-field distribution for the spheroid and cylinder are also very similar. One interesting feature here is that the field is greater on the backside of the cylinder than on the side upon which the incident wave strikes.

Figure 3 gives the magnitude of the surface field around the major and minor dimension for the 3:1 prolate spheroid shown in Fig. 1. The capability of being able to

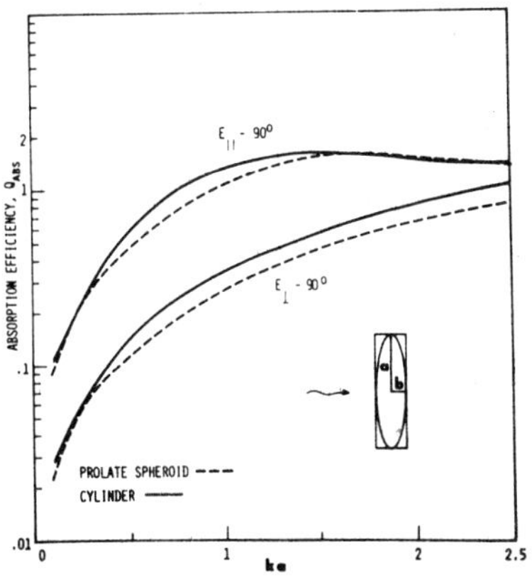

Fig. 1. Absorption efficiency versus *ka* for a lossy dielectric 3:1 finite cylinder and prolate spheroid with ε_r = 3.536 - j3.536.

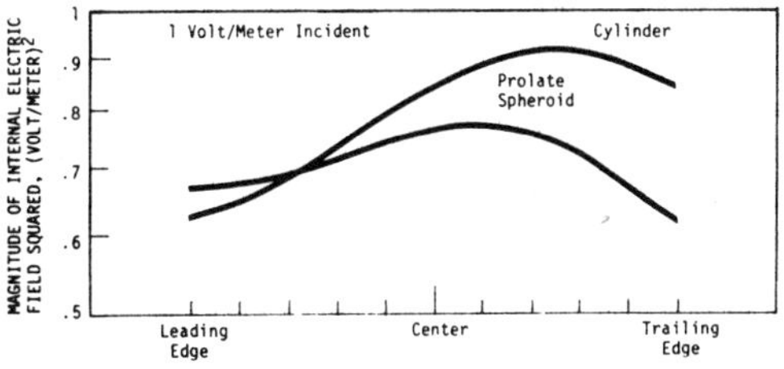

Fig. 2. Internal field $|E|^2$ distributions within a 3:1 prolate spheroid and cylinder at the approximate resonant point (*ka* = 1.5). Broadside incident wave, $E_{\parallel}$ polarization.

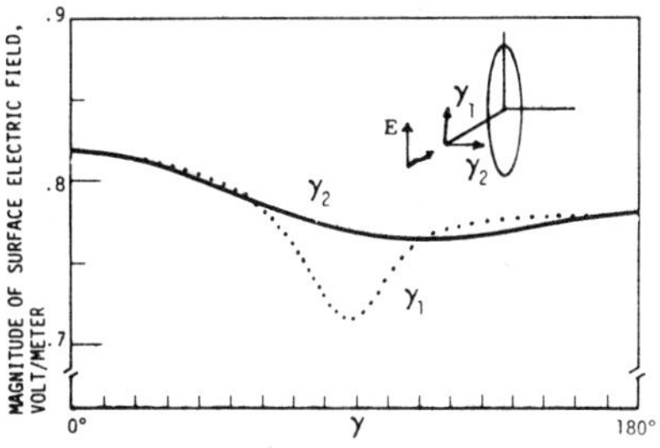

Fig. 3. Surface field $|E|$ along the major and minor surfaces for the 3:1 prolate spheroid. ε_r = 3.536 - j3.536 and *ka* = 1.5.

determine surface fields should permit further understanding of the reactive-field energy-storage characteristics of dielectric objects, which is important in microwave-resonator applications. Although these curves simply show distributions along the surface in two directions, it would be quite easy to generate contour plots for a detailed study of surface-field behavior.

The interaction of electromagnetic waves with biological systems is currently a subject of intense research effort by many investigators. This activity is a result of recent concerns regarding the potential hazards of this radiation to man. The overall need is to provide a scientific basis for the establishment of an electromagnetic radiation safety standard. To attain this goal will require a multitude of experimental tests to quantify physiological and psychological effects. Human experimentation is not possible and irradiation experiments must therefore be performed on animals.

The intent is to obtain biological effects data in animal experiments and then establish a radiation safety standard to insure that the tissue fields and internal power absorption levels which cause an observed effect in animals are never reached in humans. Direct extrapolation is impossible because scaling the frequency (to maintain a given size/wavelength ratio for both experimental animal and human) sets certain requirements on the constitutive parameters -- requirements which cannot be satisfied due to the frequency dependence of the complex dielectric constant of tissue. Therefore, theoretical solutions for power absorption by both man and animals are required to develop an extrapolation process. Only when tissue field and power absorption characteristics can be predicted for both man and experimental animals will it be possible to interpret and confirm experimentally derived biological effects in animals and thereby develop a radiation safety standard for man.

The EBCM has been used to calculate resonance region power deposition in prolate spheroidal models of animals and man (16). The mathematical models consist of homogeneous distributions of muscle tissue. Calculations are made for bodies isolated in free space.

The electromagnetics problem to be solved is illustrated schematically in Fig. 4.

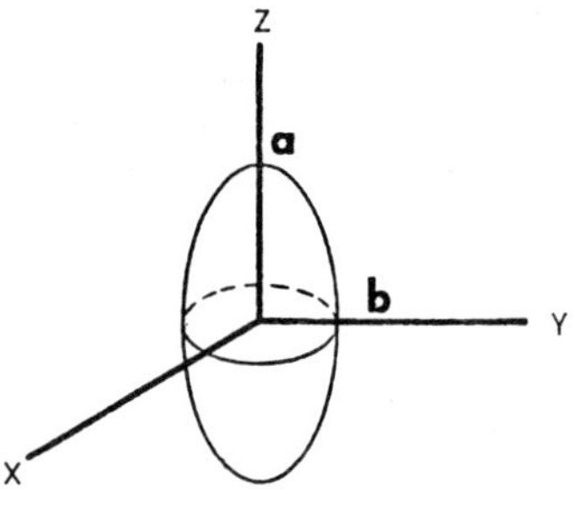

Fig. 4. The prolate spheroidal model centered at the origin of a spherical coordinate system.

The biological tissue body, assumed homogeneous and isotropic, is characterized by the constitutive parameters $\mu = \mu_0$ and $\varepsilon = \varepsilon_r\varepsilon_0$, where $\varepsilon_r = \varepsilon_r' - j\varepsilon_r''$ is the relative dielectric constant of the tissue and $\varepsilon_r'' = \sigma/\omega\varepsilon_0$, where σ is the tissue conductivity. The tissue body is illuminated by a uniform plane wave. In general, we consider three particular incident waves, one from the end-on direction and two from the broadside direction. The end-on wave we denote as E - 0° and the broadside waves we denote as $E_{||}$ - 90° when the wave is polarized in the plane defined by the incident wave and the major axis of the spheroid, and $E_{\perp}$ - 90° when it is polarized at right angles to this plane. Waves incident from directions other than end-on or broadside also use the $E_{||}$, $E_{\perp}$ designation with the particular angle of incidence specified. The angle of incidence corresponds to the polar angle in the spherical coordinate system.

A typical result, as shown in Fig. 5, gives the average specific absorbed power for

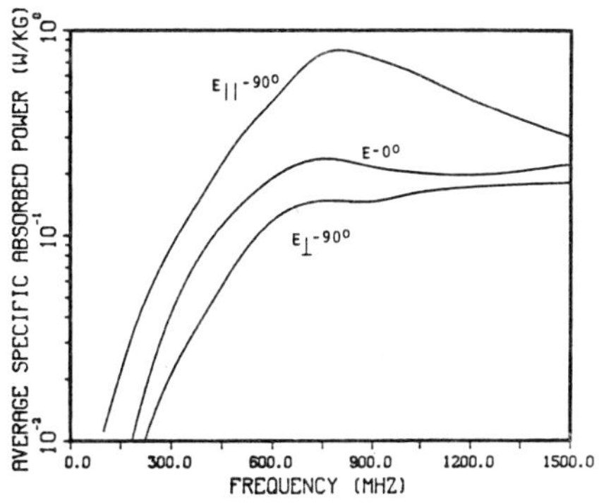

Fig. 5. Average absorbed power density in a prolate spheroidal model of a 200 g rat. a/b = 3.0, a = .075 m. Incident power density = 1 mW/cm^2.

a prolate spheroidal model of a 200 g rat. The resonant frequency of maximum absorption occurs for the E-parallel broadside wave at 800 MHz giving an absorption efficiency of 2.7. The average specific absorbed power is defined as the total absorbed power divided by the volume. An assumed tissue density of 1 g/cc permits presentation of the power deposition in units of watts/kilogram. Numerical results of the type shown in Fig. 5 have been found to agree quite closely with experimental results obtained from live animals.

Internal standing waves of internal power density $1/2\ \sigma\ |E|^2$ are shown in Fig. 6 for the 200 g rat model at 800 MHz.

Calculations similar to those shown in Fig. 5 have found that a 1.75 m tall, 70 kg, 6.34:1 prolate spheroidal model of man absorbs maximum power at a frequency of 70 MHz and with the incident wave in the $E_{||}$ - 90° configuration.

The EBCM has also been useful in quantifying the scattering by biological bodies (17). In many biological effects experiments, it is important to know what fraction of the incident power is absorbed and what fraction is scattered. Depending on the specific laboratory set-up, scattering effects can introduce experimental artifacts. Figure 7 shows the scattering, absorption, and extinction efficiencies for a prolate spheroidal model of a 320 g rat, a commonly used animal in microwave biological effects research.

Figure 7 indicates that scattering is less than absorption for the incident wave polarized perpendicular to the long axis of the spheroid (although they are coming together close to ka = 2.5), but that the scattering can be significantly greater than the absorption when the incident wave is polarized parallel to the axis of the spheroid. This latter polarization is very commonly used in experimental work to most rapidly heat up animals (due to the higher absorption that can be obtained than with other incident wave configurations). Note that the resonances or peak of the

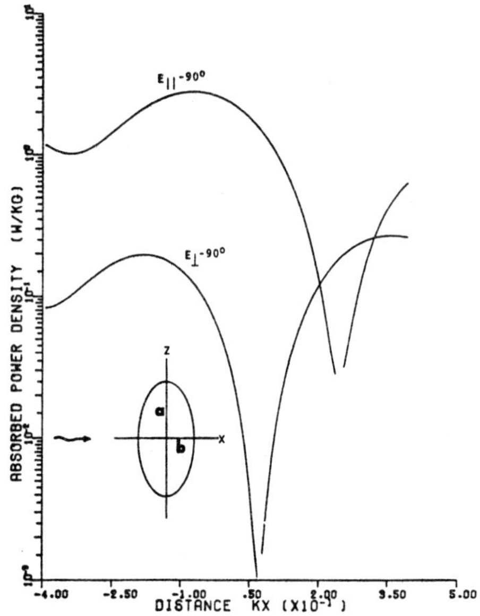

Fig. 6. Internal power density distribution in the 200 g rat model of Fig. 5. Broadside incidence at 800 MHz.

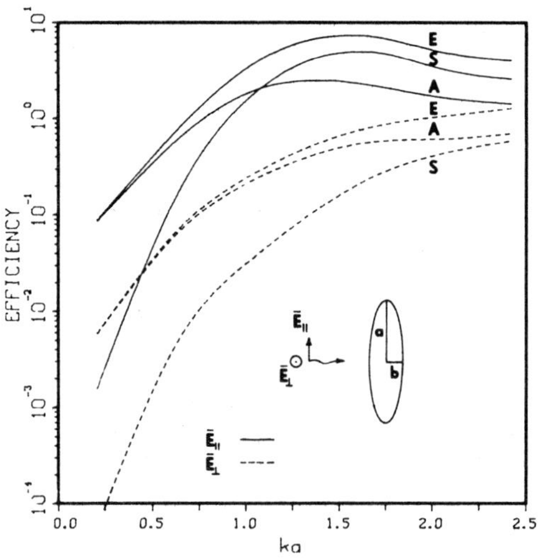

Fig. 7. Scattering (S), absorption (A), and extinction (E) efficiency as a function of electrical size for broadside incidence. The dielectric constant varies from $\varepsilon_r = 45.3 - j106.7$ at $ka = 0.21$ to $\varepsilon_r = 33.3 - j17.7$ at $ka = 2.42$. $a/b = 3.62$.

efficiency curves occur at different *ka* values for scattering and absorption. Figure 7 indicates that for broadside incidence, radiating this particular model at a maximum frequency corresponding to $ka = 1.25$ would be best. Radiating at higher frequencies (corresponding to a higher *ka*) gives no increase in absorption, but a considerable increase in scattering.

Figure 8 shows the dependence of the cross sections on the angle of incidence for

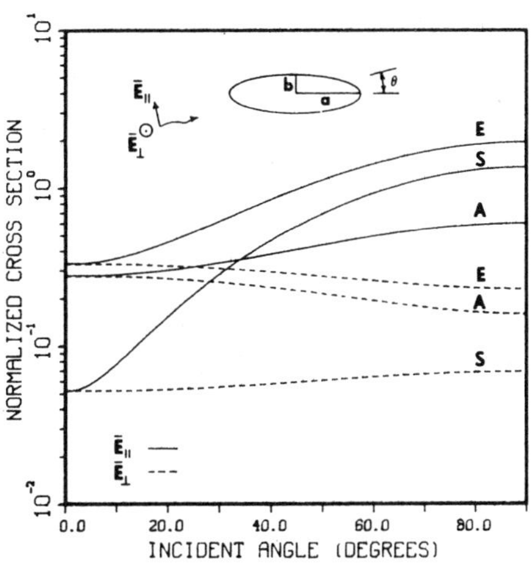

Fig. 8. Scattering (S), absorption (A), and extinction (E) cross sections normalized to πa^2 as a function of angle incidence. $ka = 1.68$, $a/b = 3.62$, and $\varepsilon_r = 34.7 - j23.3$.

the case $ka = 1.68$ (corresponding to a frequency of 800 MHz for this model). For the parallel-polarized case, the scattering exceeds the absorption as soon as the angle of incidence exceeds about thirty-five degrees. For the perpendicular-polarized case, the absorption decreases relative to the scattering as the incident angle increases from zero. It is clear that radiating the model in an end-on fashion gives the greatest absorption for the least amount of scattering.

Scattering Calculations

As with the absorption studies, two types of scattering calculations have been made. First, calculations have been made to determine the general scattering behavior of nonspherical objects and the dependence of the scattering on size, shape, and dielectric constant (or index of refraction). This type of calculation has also been useful in determining the range of validity of the Rayleigh-Gans-Debye approximation. Secondly, calculations have been made to determine the scattering characteristics of specific objects, namely the scattering of microwaves by meteorological particles and the scattering of light by chemical and biological particles, such as bacteria.

As an example of the first type of calculation (18), Figure 9 shows the differential scattering in the equatorial plane for 2:1 and 3:1 capped cylinders with the same minor diameters, but different lengths. It can be seen that both vertical and horizontal curves are almost identical in shape. Vertical means that the incident wave

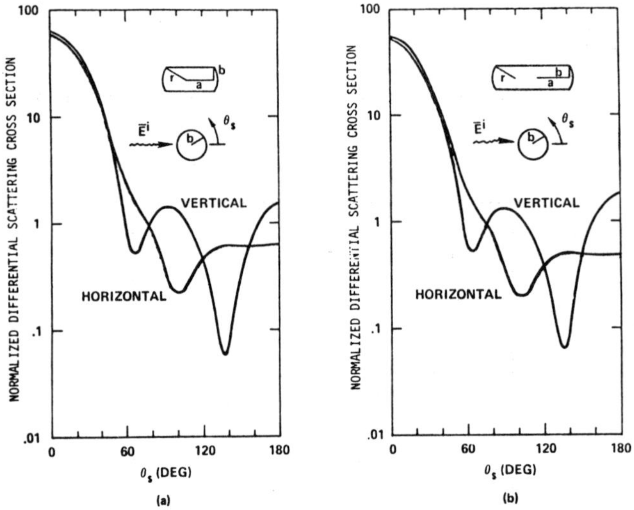

Fig. 9. Cylinder scattering, equatorial plane, $\varepsilon_r = 1.96$, $r = 2b$: (a) 2:1, $ka = 5.288$, $b/a = 0.577$; (b) 3:1, $ka = 8.341$, $b/a = 0.366$. The curves are normalized to the a dimension of each cylinder.

is polarized perpendicular to the scattering plane (the plane of the paper in Fig. 9), while horizontal means that the incident wave is polarized parallel to the scattering plane. Calculations were also made for a shorter cylinder (1.5:1) with the same minor diameter, and the scattering curves, although similar to the 2:1 curves in Fig. 9(a), did not have the almost exact correlation that is evident for the 2:1 and 3:1 cylinders. The conclusion is that the shape of the scattering curve in the equatorial plane becomes independent of length for longer cylinders.

Figure 10 compares the vertical polarization differential scattering curves for a lossless cylinder (real index of refraction) and for the same cylinder with a small amount of loss added (complex index of refraction). The behavior of the curves is very similar to that which is observed for spherical scatterers. It can be seen that the absorption has caused a change in the differential scattering characteristic that is manifested primarily by a reduction in the scattered energy all the way around the cylinder, although the effect of the internal loss is much more noticeable in the backscatter direction than in the forward direction. The reason for this is probably attributable in part to a diffraction of a large percentage of the incident radiation around the edge of the cylinder. This component is unaffected by the internal features and therefore shows no reduction when the interior becomes absorbing. The small reduction in the forward direction is due to that portion of the forward scattered wave that is refracted through the cylinder, while a major portion of the backscattered energy is due to that portion of the incident wave that is refracted into the interior region and then exits in the backward direction after multiple internal reflections. This component is directly dependent on the internal characteristics of the scattering object and therefore is attenuated when loss is added. The net effect of all this is that the effect of loss in the interior is much more noticeable in the backward than in the forward directions. This is

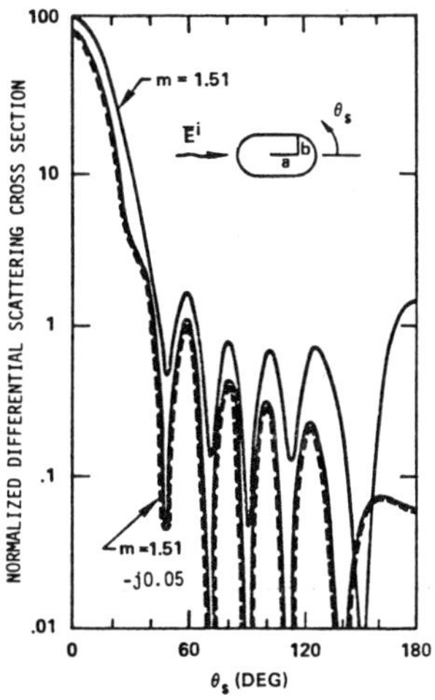

Fig. 10. Cylinder scattering, azimuthal plane, $ka = 4.765$, $a/b = 1.0$, $m = 1.51$, $m = 1.51 - j0.05$. The curves are normalized to the a dimension of the cylinder.

illustrated very graphically by looking at the ratio of forward to backscatter in the two cases. That ratio is 80 in the lossless case and almost 1500 in the lossy case.

The EBCM has been quite useful in testing the validity of the Rayleigh-Gans-Debye (RGD) approximation (19). The RGD method applies to particles satisfying $|m - 1| << 1$ and $2ka|m - 1| << 1$. The refractive index of the particle (relative to the surrounding medium) is m, $k = 2\pi/\lambda$, where λ is the wavelength in the medium, and a is a characteristic dimension of the particle, usually taken as one-half of the major dimension, e.g., the radius for a sphere. The fundamental RGD assumption is that the phase shift of the wave traversing the particle is negligibly different from the phase shift experienced by the same wave traveling in the surrounding medium; i.e., each small volume element in the scatterer is assumed to be excited by the incident field. This implies that neither the particle size nor m can be too large. Although the range of validity of the RGD approximation can easily be tested for homogeneous or radially inhomogeneous spheres and infinite cylinders for which exact solutions are available(20, 21), the method is increasingly being applied to relatively large nonspherical particles where its validity is uncertain. For example, it is often used to calculate the light-scattering characteristics of bacteria, e.g., the rodlike *Escherichia coli* (22). Although the refractive index of this bacterium relative to water is assumed to be about 1.05, which probably satisfies $|m - 1| << 1$, its size is such that $2ka|m - 1|$ is unity or greater, e.g., for a 2-μm long microbe illuminated by an argon-ion laser $\left(\lambda_{air} = 0.5145\ \mu m,\ \lambda_{water} = 0.3868\ \mu m\right)$, $2ka|m - 1|$ is 1.62.

The range of validity of the RGD approximation as applied to homogeneous nonspherical particles has been investigated by comparing computed scattering results to those obtained from the EBCM. Numerical calculations were made for a set of nonabsorbing homogeneous prolate spheroids with a relative refractive index of $m = 1.05$. This particular m was chosen because many biological specimens are characterized by a refractive index of 1.04-1.05. Furthermore, previous validity testing for homogeneous spheres (21) indicates that this value of m easily satisfies the first RGD criteria $|m - 1| \ll 1$, and therefore independent testing of the second criteria $2ka|m - 1| \ll 1$ can be accomplished by increasing particle size (ka) only. The prolate spheroid is a particularly convenient model because its shape can easily be altered by varying the axial ratio.

It has been found that the greatest error in using the RGD approximation occurs when the incident wave is parallel to the major axis of the prolate spheroid. The greater error in RGD for end-on incidence is indicative of the relatively large phase difference between the internal and external fields for this orientation and thereby less conformance with the basic premise of the RGD approximation.

Figure 11 shows prolate spheroid scattering cross sections as a function of ka and

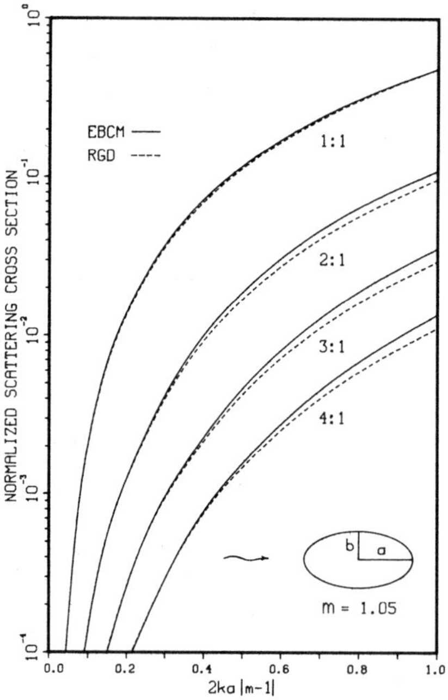

Fig. 11. Scattering cross section (normalized to πa^2) as a function of ka and axial ratio $a{:}b$.

axial ratio for the worst-error case of end-on incidence. For small particles, RGD gives uniformly good results for all spheroids as expected, but with the exception of the sphere, the discrepancy in the RGD approximation becomes greater with increasing ka. The dependence of the error on axial ratio is not as clear. At $2ka|m - 1| = 0.3$ ($ka = 3$), the error is greatest for the sphere and decreases as the axial ratio increases. However, for larger particles, the error increases with increasing axial ratio. At $2ka|m - 1| = 1$, the error is less than 1 percent for the sphere, but is 12 percent, 17 percent, and 18 percent for the 2:1, 3:1, and 4:1 spheroids, respectively. For this worst-error case of end-on incidence, it is clear that if $2ka|m - 1| \ll 1$ for a particular particle, the greatest error in using the RGD approximation occurs if the particle is a sphere, but as $2ka|m - 1| \rightarrow 1$, the least error occurs if the particle is a sphere.

Figure 12 shows the angular distribution for a 4:1 prolate spheroid for ka = 10 and

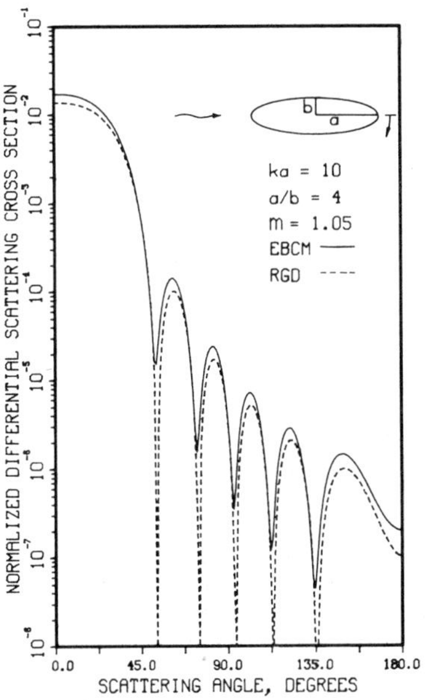

Fig. 12. Differential scattering cross section (normalized to πa^2) for a 4:1 prolate spheroid, end-on incidence. Vertical polarization.

end-on incidence. It can be seen that RGD gives essentially the same angular distribution, but the amplitude is uniformly reduced, resulting in a significant reduction in the calculated total scattering (an error of 18 percent for this model) as indicated in Fig. 11.

The scattering cross section for a randomly oriented 4:1 prolate spheroid is shown in Fig. 13. Less error is anticipated in applying the RGD approximation to randomly oriented particles than to the worst-case fixed orientations in Fig. 11. This expectation is borne out in Fig. 13 in that little difference in the two methods occurs up to about $2ka|m - 1| = 0.5$, and the maximum difference at $2ka|m - 1| = 1$ is 6.2 percent, which is significantly less than the 18 percent error which occurred for the oriented particle in Fig. 11.

An interesting problem for which specific calculations have been made concerns the scattering of microwaves by melting hailstones, a problem of considerable importance in radar studies of precipitation processes (23). This problem has been studied extensively using a coated spherical model (24). The EBCM provides a means to

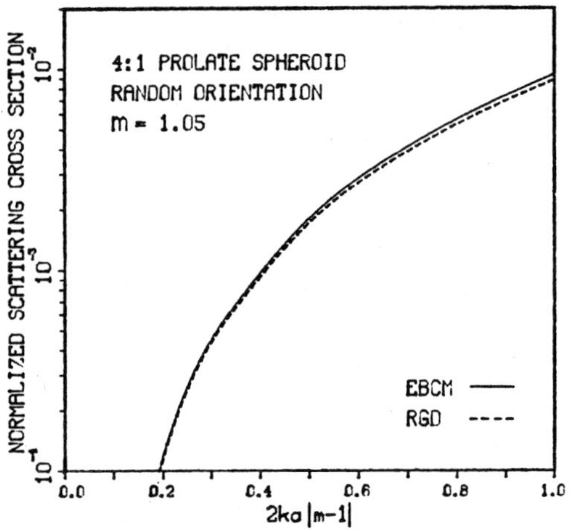

Fig. 13. Scattering cross section normalized to πa^2 as a function of ka for a randomly oriented 4:1 prolate spheroid.

investigate the dependence of the various efficiencies on the size and shape of the melting hailstone. In Fig. 14(a), the variation of the extinction, scattering, and absorption efficiencies as a function of the axial ratio of an oblate spheroidal hailstone is shown. The total volume and the volume of the ice layer are kept constant while the axial ratio changes from 1 to 1.5. The sphere (a/b = 1) is 4 cm in radius with a water layer 1 cm thick. The incident wave is at a frequency of 3 GHz and propagates along the axis of symmetry (minor axis). The efficiencies are defined as the cross sections divided by the geometrical cross section as seen by the incident wave. The curves show that the extinction, scattering, and absorption efficiencies are almost constant as the axial ratio increases, although the corresponding cross sections will of course increase. (The cross sections are πa^2 greater than the efficiencies.) On the other hand, the backscattering efficiency increases significantly for axial incidence as the axial ratio increases, as shown in Fig. 14(b) Figure 14(b) also shows that the backscattering efficiencies of these particular oblate spheroids are lower than that of a sphere with equal volume when the direction of the incident wave is along the major axis of the oblate spheroid.

The EBCM has been quite useful in investigating the light scattering characteristics of bacteria (23). An example of the type of calculations is shown in Fig. 15. Assuming the incident wave is vertically polarized with a wavelength of 0.3868 μm (one of the emitting wavelengths of the ArI laser, measured in water), the prolate spheroidal models in Fig. 15 represent 1.6 μm by 0.85 μm bacteria. In Fig. 15(a), comparison is made between the differential scattering curves of the object with and without a 0.025 μm thick cell wall. The indices of refraction used for the object are estimated values for the cell wall and protoplasm. It is clear that the cell wall plays a major role in the electromagnetic interaction. Figure 15(b) shows the effect on the differential scattering curve when a small sphere 0.08 μm in diameter is embedded at the center of the layered spheroid of Fig. 15(a). The solid line is for the three-layered model, and the dotted line for the two-layered model.

The results shown in Fig. 15 indicate that the differential scattering cross section changes significantly when the internal structure of the object changes. Hence, in theoretical calculations of light scattering by biological cells, it is probably more appropriate to use layered models instead of homogeneous models in order to yield accurate predictions of the angular scattering.

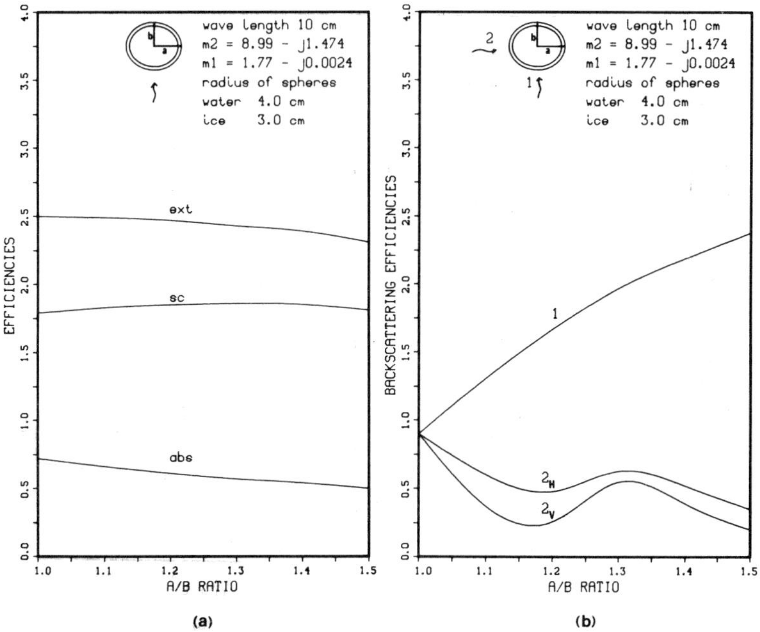

Fig. 14. (a) The variations of the scattering (sc), extinction (ext), and absorption (abs) efficiency as the shape of the hailstone changes from spherical to oblate spheroidal. (b) The backscattering efficiency as the angle of incidence changes from end-on (1) to broadside with horizontal polarization (2_H), and vertical polarization (2_V).

Light scattering calculations have also been made for the more realistic case of polydisperse suspensions of randomly oriented bacteria (25). In particular, calculations have been made for suspensions of two-layered prolate spheroidal models. Assuming a vertically polarized incident wave with a wavelength of 0.4757 μm (one of the emitting wavelengths of the HeNe laser, measured in water), the models have a minor dimension of 0.8 μm and a major dimension varying from 0.8 to 1.6 μm according to the quasi-Gaussian distribution shown in Fig. 16. The thickness of the cell wall is kept constant at 0.08 μm for all particles in the suspension. The refractive indices are 1.045 for the core (cytoplasm) and 1.1 for the coating (cell wall).

The differential scattering cross section over the distribution at a given angle is given by:

$$\sigma_d(\theta) = \frac{1}{A}\int_{D_L}^{D_R} \sigma_d\,(\theta,\, D)\; P(D)\; dD \tag{12}$$

where $\sigma_d(\theta, D)$ is the differential scattering cross section of particle size D at a given scattering angle θ.

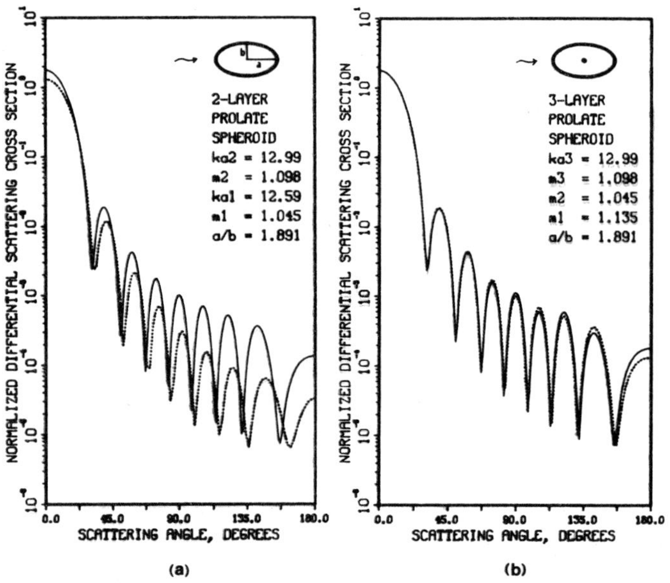

Fig. 15. Differential scattering cross section for a prolate spheroid. Vertical polarization. (a) With (——) and without (- - -) a thin surface layer. Normalized to πa_2^2. (b) The layered prolate spheroid in (a) with (——) and without (- - -) a spherical inclusion. Normalized to πa_3^2. The layers are numbered from the inside out.

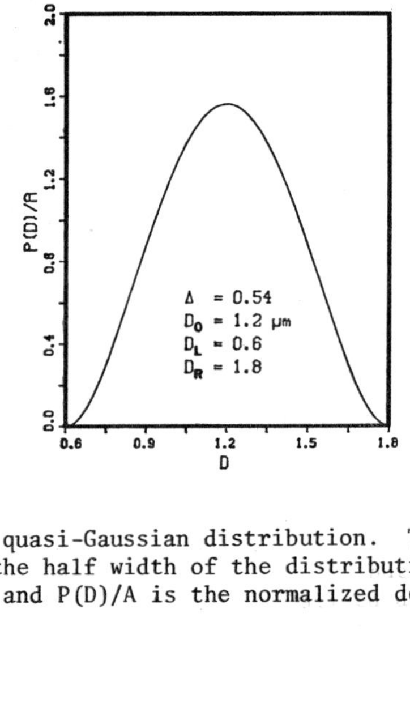

Fig. 16. The quasi-Gaussian distribution. The quantity Δ is the half width of the distribution at half maximum and P(D)/A is the normalized density function.

Figure 17 shows the angular scattering characteristics of the suspension as calcu-

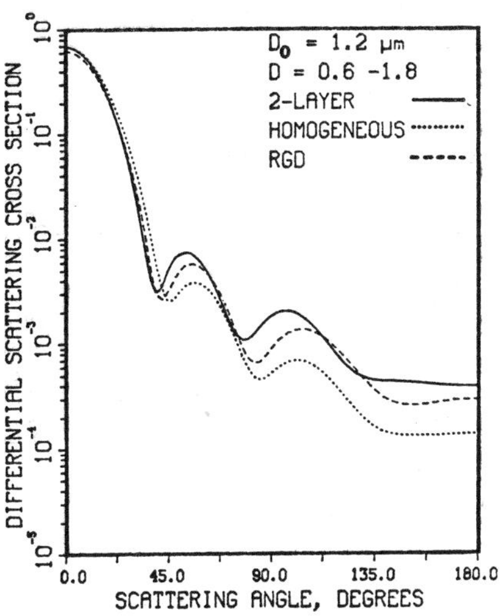

Fig. 17. The scattering curve of a polydispersion of randomly oriented prolate spheroids with the distribution shown in Fig. 16. The size parameter *ka* of the particles in the suspension varies from 5.282 to 10.565.

lated by the EBCM (solid line). Also shown for comparison is the scattering result for the same distribution except that the particles are homogeneous with an index of refraction of 1.0692 (the volume averaged index of refraction of the two-layered models). The Rayleigh-Gans-Debye (RGD) result for two-layered particles is also shown.

The results in Fig. 17 are useful in showing the limitations of homogeneous models and the RGD approximation in predicting the scattering by suspensions of bacteria-like particles.

CONCLUSION

The extended boundary condition method has been shown to be a useful tool for determining the electromagnetic scattering and absorption characteristics of resonance-sized nonspherical dielectric objects. The method has been used to determine the general electromagnetic interaction as well as to make calculations for specific problems. Specific areas which have been investigated include the radio frequency absorption behavior of homogeneous prolate spheroidal models of man and animals, the microwave absorption and scattering characteristics of oblate spheroidal layered models of hailstones, and the light scattering characteristics of prolate spheroidal layered models of bacteria.

The contributions of Dau-Sing Wang of Clarkson College of Technology and Harry C. H. Chen of the University of Utah are gratefully acknowledged.

REFERENCES

1. C. H. Durney, et al., Radiofrequency Radiation Dosimetry Handbook, Second Edition, USAF School of Aerospace Medicine, Brooks Air Force Base, Texas, Report SAM-TR-78-22 (1978).

2. J. A. Morrison and M. J. Cross, Scattering of a plane electromagnetic wave by axisymmetric raindrops, Bell System Technical Journal 53, 955 (1974).

3. P. J. Wyatt (1973) in Methods in Microbiology, J. R. Norris and D. W. Robbins, Eds., Academic Press, New York.

4. M. Kerker (1969), The Scattering of Light and Other Electromagnetic Radiation, Academic Press, New York.

5. P. C. Waterman, Matrix formulation of electromagnetic scattering, Proceedings of the IEEE 53, 805 (1965).

6. P. C. Waterman and C. V. McCarthy, Numerical Solution of Electromagnetic Scattering Problems, Mitre Corporation, Bedford, Massachusetts, Report MTP-74, N69-31912 (1968).

7. P. C. Waterman, Scattering by dielectric obstacles, Alta Frequenza 38 (Speciale), 348 (1969).

8. P. C. Waterman, Symmetry, unitarity, and geometry in electromagnetic scattering, Physical Review D 3, 825 (1971).

9. S. Asano, Light scattering properties of spheroidal particles, Applied Optics 18, 712 (1979).

10. M. A. Morgan and K. K. Mei, Finite-element computation of scattering by inhomogeneous penetrable bodies of revolution, IEEE Transactions on Antennas and Propagation 27, 202 (1979).

11. J. R. Mautz and R. F. Harrington, Electromagnetic Scattering from a Homogeneous Body of Revolution, Electrical and Computer Engineering, Syracuse University, Syracuse, New York, Report TR-77-10 (1977).

12. L. N. Medgyesi-Mitschang and C. Eftimiu, Scattering from axisymmetric obstacles embedded in axisymmetric dielectrics: the method of moments solution, Applied Physics in press (1979).

13. P. M. Morse and H. Feshbach (1953) Methods of Theoretical Physics, McGraw-Hill, New York.

14. B. Peterson and S. Ström, T-matrix formulation of electromagnetic scattering from multilayered scatterers, Physics Review D 10, 2670 (1974).

15. P. W. Barber, Resonance electromagnetic absorption by nonspherical dielectric objects, IEEE Transactions on Microwave Theory and Techniques 25, 373 (1977).

16. P. W. Barber, Electromagnetic power deposition in prolate spheroid models of man and animals at resonance, IEEE Transactions on Biomedical Engineering 24, 513 (1977).

17. P. W. Barber, Scattering and absorption efficiencies for nonspherical dielectric objects -- biological models, IEEE Transactions on Biomedical Engineering 25,

155 (1978).

18. P. W. Barber and C. Yeh, Scattering of electromagnetic waves by arbitrarily shaped dielectric bodies, Applied Optics 14, 2864 (1975).

19. P. W. Barber and D-S. Wang, Rayleigh-Gans-Debye applicability to scattering by nonspherical particles, Applied Optics 17 797 (1978) and 18, 962 (1979).

20. W. A. Farone, M. Kerker, and E. Matijevic (1963) in Interdisciplinary Conference on Electromagnetic Scattering, M. Kerker, Ed., Pergamon Press, New York.

21. M. Kerker, W. A. Farone, and E. Matijevic, Applicability of Rayleigh-Gans scattering to spherical particles, Journal of the Optical Society of America 53, 758 (1963).

22. R. E. Buchanan and N. E. Gibbons, Eds. (1974) Bergey's Manual of Determinative Bacteriology, Williams and Wilkins, Baltimore, Maryland.

23. D-S. Wang and P. W. Barber, Scattering by inhomogeneous nonspherical objects, Applied Optics 18, 1190 (1979).

24. B. M. Herman and L. J. Battan, Calculations of Mie back-scattering from melting ice spheres, Journal of Meteorology 18, 468 (1961).

25. D-S. Wang, C. H. Chen, P. W. Barber, and P. J. Wyatt, Light scattering by polydisperse suspensions of inhomogeneous nonspherical particles, Applied Optics in press (1979).

THE T MATRIX METHOD FOR SCATTERING BY AN OBSTACLE IN A WAVEGUIDE

Anders Boström
Institute of Theoretical Physics, S-412 96 Göteborg, Sweden

ABSTRACT

In the present paper the transition matrix method (see P.C. Waterman, J. Acoust. Soc. Am. 45, 1417-1429 (1969)) is extended to the case of acoustic scattering by an obstacle in a waveguide. The theory is developed for a general waveguide and the effect of the obstacle is thus seen to be completely determined by its transition matrix. Expressions for the transmission and reflection coefficients for a cylindrical waveguide are given, and for a spherical obstacle plots are shown for various values of the radii. For long wavelengths very simple approximate expressions are obtained.

I. INTRODUCTION

The propagation of waves in waveguides finds many important applications and has thus been studied extensively for various cases, see e.g. Refs. 1-6 for general surveys. The problems considered include e.g. the transmission and reflection from different kinds of inhomogeneities, such as a change in cross section or boundary condition, for acoustic, electromagnetic and elastic wave propagation. Often some kind of approximation scheme is employed, for instance perturbation methods. However, the treatments of the transmission and reflection from an obstacle inside a waveguide seem to be few; this will be the subject of the present paper.

We thus consider the propagation of an acoustic wave in a waveguide with an obstacle. We will assume stationary conditions with the time factor $\exp(-i\omega t)$ omitted, so the velocity potential ψ will be taken to satisfy the Helmholtz' equation

$$\nabla^2\psi + k^2\psi = 0 \tag{1}$$

Here $k=\omega/c$ is the wave number and c is the sound velocity in the waveguide. Rather than work directly with the differential equation we will let a surface integral representation be the starting point. Contrary to common practise we will, however, employ the free space Green's function in the integral representation and not a problem-related one.

Both regular and irregular ("outgoing") cylindrical and spherical basis functions will be needed and thus also the transformations between these. As an important step the Green's function, the incoming and scattered fields and also the surface fields on the waveguide and the obstacle will be expanded in one or both of these basis sets.

We will discover that the influence of the obstacle can be expressed solely by its transition matrix. In fact the present approach is an adaption of the transition matrix method originally introduced by Waterman for one bounded obstacle, see Ref. 7 for the acoustic case. We should also note the similarities in structure between the present case and the case of a buried obstacle, see Kristensson and Ström (8).

First the problem will be formally solved for a general waveguide, and later we will specialize to a cylindrical waveguide. It is then possible to obtain relatively explicit expressions for the transmission and reflection coefficients. If we let the obstacle be spherical and symmetrically placed in the waveguide some further simplifications occur. If the radius of the sphere is much less than that of the cylinder we can give simple approximate expressions for the transmission and reflection coefficients. Finally we will give some numerical data for a sphere for various values of the sphere and cylinder radii.

II. BASIS FUNCTIONS

In this preparatory section we will define the cylindrical and spherical basis functions and also give the transformations between these two sets.

The elementary solutions of Helmholtz' equation can in cylindrical coordinates be written

$$\chi_m(\alpha) \equiv \chi_{\sigma m}(\vec{r};\alpha) = (\varepsilon_m/8\pi)^{1/2} H_m^{(1)}(k\rho\sin\alpha) \begin{Bmatrix} \cos m\phi \\ \sin m\phi \end{Bmatrix} e^{ikz\cos\alpha} \tag{2}$$

where $\varepsilon_0=1$, $\varepsilon_m=2$, $m=1,2,\ldots$ and $H_m^{(1)}$ is the Hankel function of the first kind. σ=even, odd determines the azimuthal parity, $m=0,1,2,\ldots$ and α belongs to C, C_+ or C_-, see Fig. 1. When $\alpha \in C$ we have the usual Fourier transform in $h=k\cos\alpha$ and when $\alpha \in C_\pm$ we have the Fourier-Hankel transform in $q=k\sin\alpha$. We will further need the regular functions $\mathrm{Re}\chi_m(\alpha)$ obtained by taking Bessel rather than Hankel functions and also the functions $\bar{\chi}_m(\alpha)$ and $\mathrm{Re}\bar{\chi}_m(\alpha)$ which differ by a sign change in the exponential factor which thus is $\exp(-ikz\cos\alpha)$.

The spherical basis functions are

$$\psi_n \equiv \psi_{\sigma m \ell}(\vec{r}) = \left(\varepsilon_m \frac{2\ell+1}{4\pi} \frac{(\ell-m)!}{(\ell+m)!}\right)^{1/2} h_\ell^{(1)}(kr) P_\ell^m(\cos\theta) \begin{Bmatrix} \cos m\phi \\ \sin m\phi \end{Bmatrix} \tag{3}$$

where $h_\ell^{(1)}$ is the spherical Hankel function, σ=even, odd, $m=0,\ldots,\ell$ and $\ell=0,1,\ldots$. The corresponding regular functions are $\mathrm{Re}\psi_n$, which have spherical Bessel instead of Hankel functions.

The transformations between the two regular sets are wellknown, see e.g. Stratton (9),

$$\mathrm{Re}\chi_m(\alpha) = \sum_\ell B_n(\alpha)\,\mathrm{Re}\psi_n \tag{4}$$

$$\mathrm{Re}\psi_n = \int_{C_0} d\alpha\; B_n^\dagger(\alpha)\,\mathrm{Re}\chi_m(\alpha) \tag{5}$$

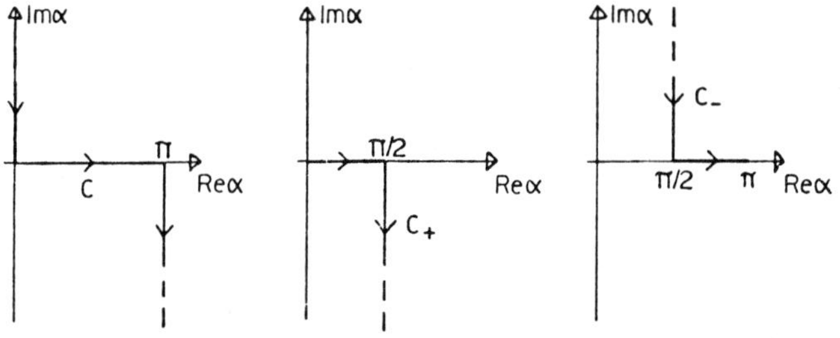

Fig. 1. Integration paths in the complex α plane.

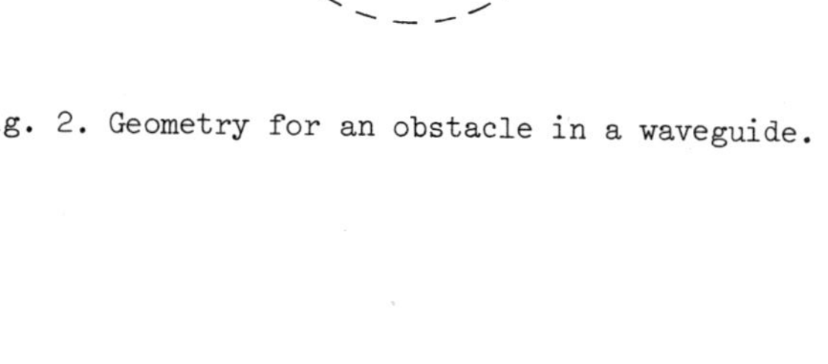

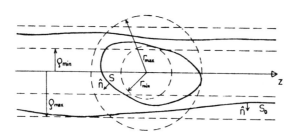

Fig. 2. Geometry for an obstacle in a waveguide.

C_0 is the real interval $(0,\pi)$ and as in the following we have omitted the factor $\sin\alpha$ from the integral. The transformation functions are essentially the normalized Legendre functions

$$B_n(\alpha) \equiv B_{m\ell}(\alpha) = i^{\ell-m}\left(\frac{2\ell+1}{2}\frac{(\ell-m)!}{(\ell+m)!}\right)^{1/2} P_\ell^m(\cos\alpha) \tag{6}$$

and $B_n^\dagger(\alpha)$ has the first factor changed to $(-i)^{\ell-m}$. We can also express the irregular spherical basis functions in the regular or irregular cylindrical basis functions

$$\psi_n = 2\int_{C\pm} d\alpha B_n^\dagger(\alpha)\,\mathrm{Re}\chi_m(\alpha) \qquad (z \gtrless 0) \tag{7}$$

$$\psi_n = \int_C d\alpha B_n^\dagger(\alpha)\chi_m(\alpha) \qquad (\rho>0) \tag{8}$$

The first of these can be obtained as a modification of expressions given by Danos and Maximon (10) and Eq. (8) can be obtained from Eq. (7) (and vice versa) by contour deformations, but we leave out all details.

III. GENERAL FORMULATION

Consider now the waveguide with an obstacle as depicted in Fig. 2. To begin with we shall let the geometry be quite general. Thus the waveguide wall S_0 is assumed to satisfy the requirements of the divergence theorem and the radius to a point on S_0 from a z-axis inside S_0 to be a one-valued function of the cylindrical coordinates ϕ and z. We also assume that there are an inscribed cylinder $\rho=\rho_{min}$ and a circumscribed cylinder $\rho=\rho_{max}$ to S_0. For the obstacle we similarly assume that its surface S satisfy the requirements of the divergence theorem, that the radius from an origin on the z-axis to a point on S is a one-valued function of the spherical angles and that there are an inscribed sphere $r=r_{min}$ and a circumscribed sphere $r=r_{max}$.

A monochromatic wave ψ^i is incident from the region $z<0$ and upon scattering by the obstacle (and the walls) gives the scattered wave ψ^s. Later we will give the more conventional division in incident, transmitted and reflected waves. The following integral representation for the total field $\psi=\psi^i+\psi^s$, which is an easy application of the divergence theorem, will be our starting point:

$$\psi^i(\vec{r}) + k\int_{S-S_0}\{\psi(\vec{r}\,')\frac{\partial}{\partial n'}G(\vec{r},\vec{r}\,') - G(\vec{r},\vec{r}\,')\frac{\partial}{\partial n'}\psi(\vec{r}\,')\}dS' = \begin{cases}\psi(\vec{r}) & \vec{r} \text{ between } S \text{ and } S_0\\ 0 & \vec{r} \text{ inside } S \text{ or outside } S_0\end{cases} \tag{9}$$

For simplicity we will here assume that the boundary conditions on both the wall and the obstacle are those for a hard boundary

$$\frac{\partial\psi}{\partial n} = 0 \tag{10}$$

The second term in the integrals thus disappear. The extension to other boundary conditions is straightforward and for the obstacle this will be further commented on below.

As stressed in the introduction we use the free-space Green's function

$$G(\vec{r},\vec{r}') = e^{ik|\vec{r}-\vec{r}'|}/(4\pi k|\vec{r}-\vec{r}'|) \tag{11}$$

The expansions of G in the basis functions are then straightforward. In cylindrical coordinates we have two different expansions, one as a Fourier integral

$$G(\vec{r},\vec{r}') = (i/8\pi k)\sum_{m=0}^{\infty}\varepsilon_m \cos m(\phi-\phi')\int_{-\infty}^{\infty} J_m(q\rho_<)H_m^{(1)}(q\rho_>)e^{ih(z-z')}dh =$$
$$i\sum_m\int_C d\alpha \operatorname{Re}\chi_{\sigma m}(\vec{r}_<;\alpha)\overline{\chi}_{\sigma m}(\vec{r}_>;\alpha) \tag{12}$$

and one as a Fourier-Hankel integral

$$G(\vec{r},\vec{r}') = (i/4\pi k)\sum_{m=0}^{\infty}\varepsilon_m \cos m(\phi-\phi')\int_0^{\infty} J_m(q\rho)J_m(q\rho')e^{ih|z-z'|}(q/h)dq=$$
$$2i\sum_m\int_{C\pm} d\alpha \operatorname{Re}\chi_{\sigma m}(\vec{r};\alpha)\operatorname{Re}\overline{\chi}_{\sigma m}(\vec{r}';\alpha) \qquad (z \gtrless z') \tag{13}$$

Here $q^2+h^2=k^2$ with all roots defined with non-negative imaginary parts. In all summations over m there is implicitly understood that there is also a σ-summation. In Eq. (12) we by $\vec{r}_<(\vec{r}_>)$ mean the radius vector with the smallest (greatest) value of ρ. The bar in Eq. (12) can be moved to the other function.

The expansion of the Green's function in spherical waves is

$$G(\vec{r},\vec{r}') = i\sum_{\ell=0}^{\infty}\sum_{m=0}^{\ell}\varepsilon_m \frac{2\ell+1}{4\pi}\frac{(\ell-m)!}{(\ell+m)!}\cos m(\phi-\phi') \times$$
$$\times P_\ell^m(\cos\theta)P_\ell^m(\cos\theta')j_\ell(kr_<)h_\ell^{(1)}(kr_>) = \tag{14}$$
$$i\sum_n \operatorname{Re}\psi_n(\vec{r}_<)\psi_n(\vec{r}_>)$$

where $\vec{r}_<(\vec{r}_>)$ means the radius vector with the smallest (greatest) value of r. For a discussion of Green's functions and their expansions see e.g. Morse and Feshbach (1) and Felsen and Markuvitz (11).

In Eq. (9) we now assume that $r<r_{min}$ and $\rho<\rho_{min}$. In the integral over S we use Eq. (14) and in the integral over S_0 Eq. (12) thus obtaining

$$\psi^i = i\sum_m \int_C d\alpha \text{Re}\chi_m(\alpha) k \int_{S_o} \psi \frac{\partial}{\partial n}\bar{\chi}_m(\alpha) dS - i\sum_n \text{Re}\psi_n k \int_S \psi \frac{\partial}{\partial n}\psi_n dS$$

Employing Eq. (4) to express $\text{Re}\chi_m(\alpha)$ in $\text{Re}\psi_n$ we have an expansion for the incoming field

$$\psi^i = \sum_n a_n \text{Re}\psi_n \qquad (r<r_{min}, \rho<\rho_{min}) \tag{15}$$

where

$$a_n = i\int_C d\alpha B_n(\alpha) k \int_{S_o} \psi \frac{\partial}{\partial n}\bar{\chi}_m(\alpha) dS - ik\int_S \psi \frac{\partial}{\partial n}\psi_n dS \tag{16}$$

Instead assuming that $\rho>\rho_{max}$ and $r>r_{max}$ and still using Eqs. (14) and (12) but also Eq. (8) we get

$$\psi^i = \sum_m \int_C d\alpha a_m(\alpha)\chi_m(\alpha) \qquad (\rho>\rho_{max}, \ r>r_{max}) \tag{17}$$

where

$$a_m(\alpha) = ik\int_{S_o} \psi \frac{\partial}{\partial n} \text{Re}\bar{\chi}_m(\alpha) - i\sum_\ell B_n^\dagger(\alpha) k \int_S \psi \frac{\partial}{\partial n} \text{Re}\psi_n dS \tag{18}$$

Further taking $\rho<\rho_{min}$ and $r>r_{max}$ in Eq. (9) we have an expansion for the scattered field

$$\psi^s = -i\sum_m \int_C d\alpha \text{Re}\chi_m(\alpha) k \int_{S_o} \psi \frac{\partial}{\partial n}\bar{\chi}_m(\alpha) dS + i\sum_n \psi_n \int_S \psi \frac{\partial}{\partial n} \text{Re}\psi_n dS \tag{19}$$

As the expansion coefficients a_n and $a_m(\alpha)$ for the incoming field are assumed to be known we only have to eliminate the surface fields with the help of Eqs. (16) and (18) and insert the result in Eq. (19) to complete the solution. To this end we assume that the surface fields have the following expansions

$$\psi = \sum_n \beta_n \text{Re}\psi_n \qquad (\text{on } S) \tag{20}$$

$$\psi = \sum_m \int_c d\alpha \gamma_m(\alpha) \text{Re}\chi_m(\alpha) \qquad (\text{on } S_o) \tag{21}$$

and define

$$Q_{nn'} = k\int_S \text{Re}\psi_{n'} \frac{\partial}{\partial n}\psi_n dS \tag{22}$$

$$Q_{nn'}(\alpha,\alpha') = k\int_{S_o} \text{Re}\chi_m'(\alpha') \frac{\partial}{\partial n}\bar{\chi}_m(\alpha) dS \tag{23}$$

Eqs. (16), (18) and (19) then become

$$a_n = i \int_C d\alpha B_n(\alpha) \Sigma_{m'} \int_C d\alpha' Q_{mm'}(\alpha,\alpha')\gamma_{m'}(\alpha') - i\Sigma_{n'} Q_{nn'}\beta_{n'} \tag{24}$$

$$a_m(\alpha) = i\Sigma_{m'} \int_C d\alpha' \,\mathrm{Re}Q_{mm'}(\alpha,\alpha')\gamma_{m'}(\alpha') - i\Sigma_\ell B^+_n(\alpha)\Sigma_{n'} \mathrm{Re}Q_{nn'}\beta_{n'} \tag{25}$$

$$\psi^s = - i\Sigma_m \int_C d\alpha \mathrm{Re}\chi_m(\alpha)\Sigma_{m'} \int_C d\alpha' Q_{mm'}(\alpha,\alpha')\gamma_{m'}(\alpha') + i\Sigma_n \psi_n \Sigma_{n'} \mathrm{Re}Q_{nn'}\beta_{n'} \tag{26}$$

where $\mathrm{Re}Q_{nn'}$ and $\mathrm{Re}Q_{mm'}(\alpha,\alpha')$ are obtained from Eqs. (22) and (23) by taking regular functions in both places.

Formally Eq. (25) can be solved to obtain $\gamma_m(\alpha)$ in terms of β_n which is then introduced into Eqs. (24) and (26). Then β_n is obtained from Eq. (24) and the scattered field is thus obtained in known quantities. However, the inversion of the integral operator in Eq. (25) is in general not an easy task. But if the waveguide has constant cross section $Q_{mm'}(\alpha,\alpha')$ (and $\mathrm{Re}Q_{mm'}(\alpha,\alpha')$) will be diagonal in α and α' (i.e. containing the factor $\delta(\alpha-\alpha')$), and the procedure for solving Eqs. (24)-(26) will only involve matrix inverses.

IV. CYLINDRICAL WAVEGUIDE

For simplicity we now consider a waveguide with a constant circular cross section with radius a. The integral in Eq. (23) can then be calculated

$$Q_{mm'}(\alpha,\alpha') = (\pi/2)ka \sin\alpha\, J_m(ka \sin \alpha)\, H_m^{(1)'}(ka\sin\alpha)\delta_{mm'}\delta(\cos\alpha - \cos\alpha')$$

$$\equiv Q_m(\alpha)\delta_{mm'}\delta(\cos\alpha-\cos\alpha') \tag{27}$$

We further define

$$R_m(\alpha) = - Q_m(\alpha)/\mathrm{Re}Q_m(\alpha) = - H_m^{(1)'}(ka\sin\alpha)/J'_m(ka\sin\alpha) \tag{28}$$

$$A_n = \int_C d\alpha B_n(\alpha) R_m(\alpha) a_m(\alpha) \tag{29}$$

$$R_{nn'} \equiv R_{\ell m \ell'} = \int_C d\alpha B_{m\ell}(\alpha) R_m(\alpha) B^\dagger_{m\ell'}(\alpha) \tag{30}$$

$$T_{nn'} = - \Sigma_{n''} \mathrm{Re}Q_{nn''}(Q^{-1})_{n''n'} \tag{31}$$

which are in turn the reflection coefficient for the wall, the spherical projection of the once reflected incoming wave (roughly speaking), the spherical projection of the reflection coefficient and the transition matrix for the obstacle.

We eliminate $\gamma_m(\alpha)$ with the help of Eq. (25) and thereafter β_n with the help of Eq. (24). Finally we obtain the scattered field from Eq. (26)

$$\psi^s = \sum_m \int_C d\alpha \mathrm{Re}\chi_m(\alpha) R_m(\alpha) \{a_m(\alpha) + \sum_{\ell n'} B_n^\dagger(\alpha) T_{nn'} c_{n'}\} + \sum_{nn'} \psi_n T_{nn'} c_{n'} \tag{32}$$

where c_n is the solution of

$$c_n - \sum_{n'n''} R_{nn'} T_{n'n''} c_{n''} = a_n + A_n \tag{33}$$

This solution for the scattered field does not have the appearance we would expect, primarily because Eq. (32) seem to contain a continuous spectrum, whereas we know that it can be written as a sum over the waveguide modes. We are now going to transform Eq. (32) to such a sum.

The eigenfunctions for a hard cylindrical waveguide are of course well-known. Apart from normalization they are just $\mathrm{Re}\chi_m(\alpha)$ with α belonging to a discrete point spectrum on C_+ (waves going in the positive direction) or C_- (waves going in the negative direction). The spectrum is determined by

$$J'_m(q_{ms}a) = J'_m(ka\sin\alpha_{ms}) = 0 \qquad s = 0,1, \ldots$$

We now assume that the incoming field is a waveguide mode (by superposition we can then treat a mode sum). We then note that its sources are a surface distribution on the wall and not a source inside the waveguide as we more or less assumed when writing the integral representation Eq. (9). But Eq. (9) is still valid, but ψ^i is now a part of the surface integral over S_0 (which upon integration yields the waveguide mode). The surface field on S_0 in the integral is then not the total surface field but only the scattered part.

For an incoming waveguide mode we see that the expansion coefficients $a_m(\alpha)$ (Eq. (17)) will be just a delta function and so A_n, Eq. (29), will vanish as the integral along C in Eq. (29) must be interpreted as a contour integral that should avoid the singularities (cuts and poles) of $R_m(\alpha)$ (one pole coincides with the argument of the delta function). This can also be understood physically; for an incoming wave satisfying the boundary conditions on the wall there will be no wave first reflected by the wall (roughly the meaning of A_n as mentioned above). The first term in the scattered field, Eq. (32), will vanish for exactly the same reason.

In the last term in Eq. (32) we now employ Eq. (7) to transform the spherical basis functions to cylindrical ones. Taking into consideration that the first term vanishes we get

$$\psi^s = \sum_{nn'} \{\int_C d\alpha R_m(\alpha) + 2\int_{C\pm} d\alpha\} \mathrm{Re}\chi_m(\alpha) B_n^\dagger(\alpha) T_{nn'} c_{n'} \tag{34}$$

for $z \gtrless r_{max}$. It is here advantageous to change integration variable to $h=k\cos\alpha$. The first integral is closed in the upper or lower halfplane and on taking the difference between the values on different sides of C_+ or C_- (where the first integrand has both a root and a logarithmic cut) will just cancel the second integral. Only the simple poles from $R_m(\alpha)$ are left. Leaving all details we eventually obtain

$$\psi^s = \sum_{snn'} 4\{ka^2(k^2 - {}_{ms}^{2})^{1/2}(1-(m/{}_{ms}a)^2)\}^{-1}\mathrm{Re}\chi_m(\alpha_{ms})B_n^{\dagger}(\alpha_{ms})T_{nn'}c_{n'} \tag{35}$$

for $z>r_{max}$ and α_{ms} changed to the corresponding value $(\pi-\alpha_{ms})$ on C_- for $z<r_{max}$.

We want now to express this result in terms of reflection and transmission coefficients. Though this can easily be done in general we for simplicity restrict ourselves to the case m=0 as we can then discard the σ index (which we remember is implicitly present together with m; thus there are two σ-summations in Eq. (35) in general). Thus we assume that the obstacle is rotationally symmetric and therefore its transition matrix has only two indices: $T_{\ell\ell'}$. We also assume the incoming wave to be a waveguide mode propagating in the positive z direction. This mode can be specified with the index s, say s=s'. The transmitted and reflected waves can in the same manner be specified with the index s. Switching to the usual waveguide eigenfunctions $J_0(q_s\rho)\exp(\pm iz(k^2-q_s^2)^{1/2})$ we from Eq. (35) then obtain the transmission and reflection coefficients

$$T^w_{s's} = \delta_{s's} + \{ka^2(k^2-q_s^2)^{1/2}J_0^2(q_sa)\}^{-1}\sum_{\ell=0}^{\infty}\sum_{\ell'=0}^{\infty} i^{-\ell}\{(2\ell+1)/\pi\}^{1/2} \times P_\ell(\{1-q_s^2/k^2\}^{1/2})T_{\ell\ell'}c_{\ell'} \tag{36}$$

$$R^w_{s's} = \{ka^2)k^2 - {}_s^2)^{1/2}J_0^2({}_sa)\}^{-1}\sum_{\ell=0}^{\infty}\sum_{\ell'=0}^{\infty} i^{-\ell}\{(2\ell+1)/\pi\}^{1/2})T_{\ell\ell'}c_{\ell'} \tag{37}$$

where c_ℓ is the solution of

$$c_\ell - \sum_{\ell'=0}^{\infty}\sum_{\ell''=0}^{\infty} R_{\ell 0\ell'}T_{\ell'\ell''}c_{\ell''} = 0 \tag{38}$$

In Eqs. (36) and (37) we have written q_s instead of q_{os} and thus $J_1(q_sa)=0$, s=0,1,... . We remember that the square roots in Eqs. (36) and (37) are defined with non-negative imaginary parts. Usually one only defines the transmission and reflection coefficients for modes below cut-off, and so s and s' will belong to the integers less than or equal to N where $q_N<k$ but $q_{N+1}>k$. For ka<3.83 only the fundamental mode $q_0=0$ will propagate and we will just have one transmission and one reflection coefficent.

a_ℓ is defined by Eq. (15) and employing Eq. (4) we get

$$a_\ell = 2i^\ell\{\pi(2\ell+1)\}^{1/2}P_\ell(\{1-q_{s'}^2/k^2\}^{1/2}) \tag{39}$$

$q_{s'}$ being the propagation constant of the incoming wave. $R_{\ell 0\ell'}$, defined by Eq.

(39), can be written in a more suitable form. First we change variable to $t=\cos\alpha$ and then rotate the integration path $\pi/4$ in the negative direction to avoid the cuts and poles of $R_0(\alpha)$. For $\ell+\ell'$ even we obtain

$$R_{\ell 0\ell'} = e^{-i\pi/4} i^{\ell-\ell'}\{(2\ell+1)(2\ell'+1)\}^{1/2}\int_0^\infty P_\ell(te^{-i\pi/4})P_{\ell'}(te^{-i\pi/4})\times \\ \times H_1^{(1)}(ka\{1+it^2\}^{1/2})/J_1(ka\{1+it^2\}^{1/2})dt \tag{40}$$

and for $\ell+\ell'$ odd $R_{\ell O\ell'}$ vanishes.

The transition matrix (Eqs. (31) and (22)) is for a soundhard rotationally symmetric obstacle and m=0:

$$T_{\ell\ell'} = -\sum_{\ell''=0}^{\infty} \mathrm{Re}Q_{\ell\ell''}(Q^{-1})_{\ell''\ell'} \tag{41}$$

$$Q_{\ell\ell'} = (k/2)\{(2\ell+1)(2\ell'+1)\}^{1/2}\int_0^\infty P_{\ell'}(\cos\theta)j_{\ell'}(kr)\frac{\partial}{\partial n}\{P_\ell(\cos\theta)h_\ell^{(1)}(kr)\}\times \\ \times r^2\sin\theta d\theta \tag{42}$$

where $r=r(\theta)$ defines the surface of the obstacle. For a hard sphere with radius b we have

$$T_{\ell\ell'} = -\delta_{\ell\ell'}j'_{\ell'}(kb)/h_\ell^{(1)\prime}(kb) \tag{43}$$

and is thus diagonal in ℓ and ℓ'. We remark that we could as well have used other boundary conditions on the obstacle, it could for instance be penetrable and with losses, see Waterman (7) for a thorough discussion of the transition matrix for a homogeneous obstacle. However, more complex cases can also be treated. Thus the obstacle can consist of homogeneous layers or we could even have two or several obstacles, see Peterson and Ström (12-13). We just have to put in the appropriate transition matrix instead of the ones above.

If the sphere radius is much less than the cylinder radius the product RT in Eq. (38) will be small compared with unity, and we will thus have $c_\ell \sim a_\ell$. If we further assume that kb<<1 we only need to take the terms for $\ell=0$ and $\ell=1$ in the sums and we can also expand the spherical Bessel and Hankel functions in the transition matrix. We then get

$$T^W_{oo} = 1 + \frac{1}{3}ikb(b/a)^2 \tag{44}$$

$$R^W_{oo} = -\frac{5}{3}ikb(b/a)^2 \tag{45}$$

For e.g. ka=1 and b/a=0.2 we from these formulas obtain an error that is less than 2% and for b/a=0.5 the error is still only about 10%. We may note that Eqs. (44) and (45) seem to contradict the energy conservation that demands that $|T^W_{00}|^2 + |R^W_{00}|^2 = 1$ (when only the first mode propagate). This contradiction is only apparent, however, as there is a negative real part of T^W_{00} of the next order.

We now turn to some simple numerical applications for a hard sphere in a cylinder. We take the incoming wave to be the first mode exp(ikz). We will give a plot for the transmission and reflection coefficients but we will mainly plot the energy coefficients. Taking the incoming wave to have unit energy these are

$$E^{T}_{os} = (1 - q_s^2/k^2)^{1/2} J_0^2(q_s a)\,|T^{W}_{os}|^2 \tag{46}$$

$$E^{R}_{os} = (1 - q_s^2/k^2)^{1/2} J_0^2(q_s a)\,|R^{W}_{os}| \tag{47}$$

if the mode s is propagating.

In Figs. 3 and 4 we show E_{00} and E_{01} as a function of cylinder radius (or frequency), $ka \leq 5$, and sphere-to-cylinder ratio $b/a=0.5$ and 0.75, respectively. E_{01} is only shown above the cut-off at $ka=3.83$. For $b/a=0.25$ above 99% of the energy is transmitted below cut-off and about 95% above. In Fig. 5 we show the absolute value of the reflection and transmission coefficients and in Fig. 6 their phases, all for $ka \leq 5$ and $b/a=0.5$. These figures should be compared with Fig. 3.

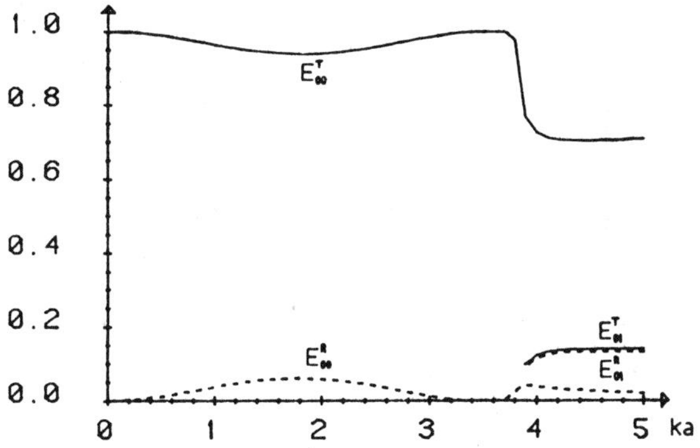

Fig. 3. Energy transmission and reflection vs. ka for b/a=0.5

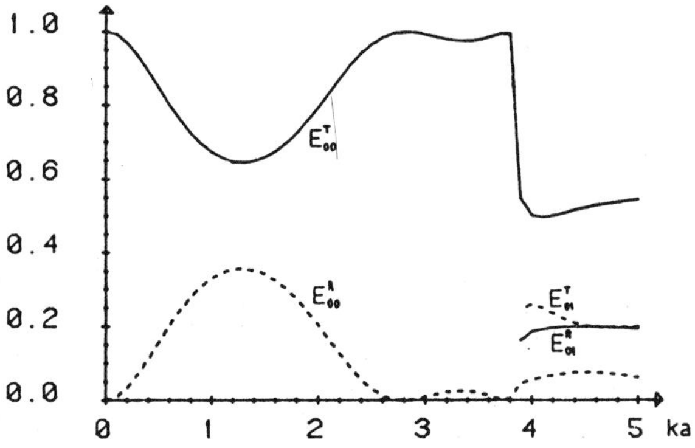

Fig. 4. Energy transmission and reflection vs. ka for b/a=0.75.

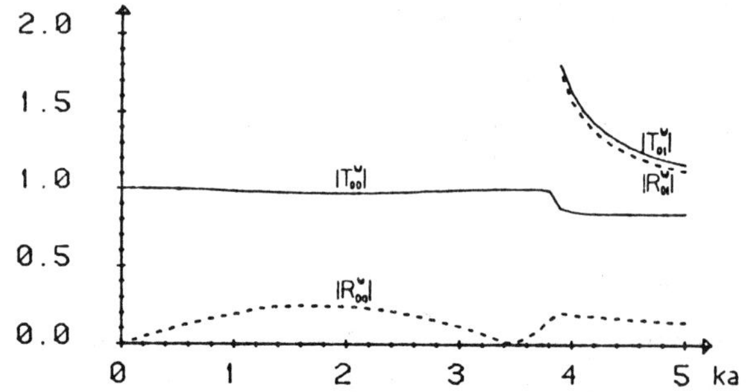

Fig. 5. Absolute value of the transmission and reflection coefficients vs. ka for b/a=0.5.

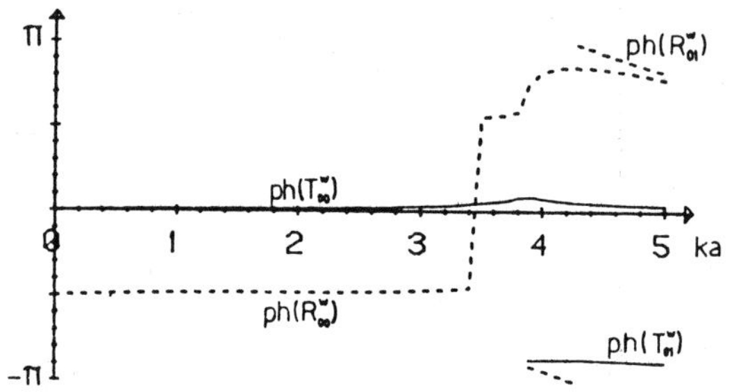

Fig. 6. Phase of the transmission and reflection coefficients vs. ka for b/a=0.5.

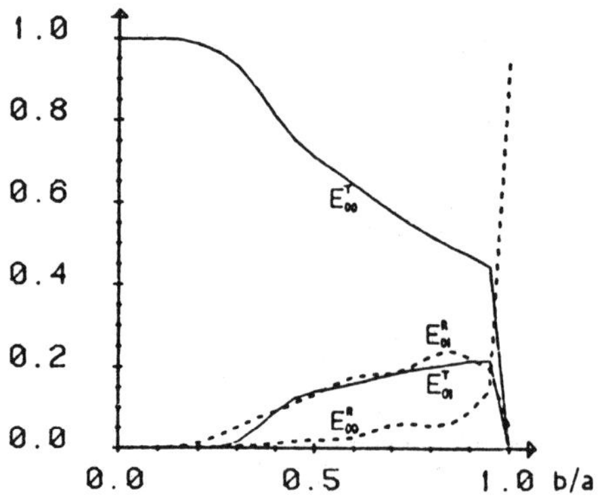

Fig. 7. Energy transmission and reflection vs. b/a for ka=1.

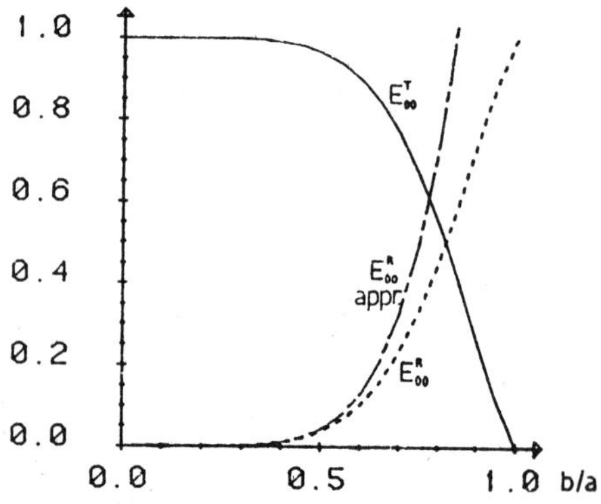

Fig. 8. Energy transmission and reflection vs. b/a for ka=5.

E_{00} and E_{01} are shown as a function of b/a in Figs. 7 and 8 for ka=1 and 5, respectively. In Fig. 7 we have also plotted the approximate curve for the reflected energy obtained from Eq. (45). We note that for b/a close to unity (b/a=0.95 and above) the results are somewhat uncertain. For e.g. ka=1 and b/a=1 we have obtained E^R_{00}=0.989 instead of 1, but by computing the integrals in $R_{\ell 0 \ell'}$ better and by using a higher truncation this could probably be improved.

V. CONCLUDING REMARKS

We have in this paper considered acoustic scattering by an obstacle in a waveguide and given relatively explicit expressions when the waveguide is a circular cylinder. It is of course of interest to study also other cross sections of the waveguide, say a rectangular one. One should then start with Eqs. (24)-(26) and solve these in the same manner as for the circular waveguide. The structure of Eqs. (32) and (33) will be the same but the expressions for $R_m(\alpha)$ and $R_{nn'}$ will be considerably more complex. The problem will be to recover the waveguide modes but this could possibly be done.

Eqs. (24)-(26) also hold for waveguides with non-constant cross section, but their solution requires the inversion of an integral operator. Therefore the present approach do not seem of so much value for such problems, but it could possibly be combined with other methods to treat say an obstacle in a circular waveguide with constant cross section except for a jump in the radius (see e.g. Brander and Nilsson (14)).

We can also "turn the problem inside-out" and thus consider the scattering by a cylinder (of constant but not necessarily circular cross section) in the presence of a bounded obstacle. If the obstacle is situated inside the cylinder no other complications arise then that we of course get a continuous spectrum. But if the obstacle is outside the cylinder we must further introduce the translation properties of the basis functions as we can no longer have a common origin for the cylinder and the obstacle. This will further complicate the solution to some extent.

Finally we can also turn to the more complicated cases of electromagnetics and elastodynamics. The present approach applies also to these cases; formally everything will look much the same with the exception of some vector notations and an extra index. However, the last step (i.e. the transformation to waveguide modes) will be more complicated especially in the elastic case.

ACKNOWLEDGEMENTS

The author wishes to thank Dr. Staffan Ström for carefully reading the entire manuscript, and Börje Nilsson for helpful discussions and for putting one of his computer programmes at my disposal. The present work is supported by the National Swedish Board for Technical Development (STU) and this is gratefully acknowledged.

REFERENCES

1. P.M. Morse and H. Feshbach, Methods of Theoretical Physics (McGraw-Hill, New York 1953).

2. R.E. Collin, Field Theory of Guided Waves (McGraw-Hill, New York 1960).

3. P.M. Morse and K.U. Ingard, Theoretical Acoustics (McGraw-Hill, New York 1968).

4. R. Mittra and S.W. Lee, Analytic Techniques in the Theory of Guided Waves (MacMillan, New York 1971).

5. L. Lewin, Theory of Waveguides (Newnes-Butterworths, London 1975).

6. I. Miklowitz, The Theory of Elastic Waves and Waveguides (North-Holland, Amsterdam 1978).

7. P.C. Waterman, New Formulation of Acoustic Scattering, J. Acoust. Soc. Am. 45, 1417-1429 (1969).

8. G. Kristensson and S. Ström, Scattering from Buried inhomogeneities - a general three-dimensional formalism, J. Acoust. Soc. Am. 64, 917-936 (1978).

9. I.A. Stratton, Electromagnetic Theory (McGraw-Hill, New York 1941).

10. M. Danos and L.C. Maximon, Multipole Matrix Elements of the Translation Operator, J. Math. Phys. 6, 766-778 (1965).

11. L.B. Felsen and N. Markuvitz, Radiation and Scattering of Waves (Prentice-Hall, Englewood Cliffs 1973).

12. B. Peterson and S. Ström, Matrix formulation of acoustic scattering from an arbitrary number of scatterers, J. Acoust. Soc. Am. 56, 771-780 (1974).

13. B. Peterson and S. Ström, Matrix formulation of acoustic scattering from multilayered scatterers, J. Acoust. Soc. Am. 57, 2-13 (1975).

14. O. Brander and B. Nilsson, A method for calculating the sound attenuation in a silencer, consisting of finite sections of bulkreacting lining in a cylindrical duct with mean flow, Rep. 77-52, Inst. Theoretical Physics, Göteborg (1977).

ON THE MULTIPLE SCATTERING OF WAVES FROM OBSTACLES WITH SOLID-FLUID INTERFACES

Bo Peterson, V.V. Varadan and V.K. Varadan
Wave Propagation Group
Department of Engineering Mechanics
Ohio State University, Columbus, Ohio 43210

I. INTRODUCTION

The T-matrix or null field method has proved to be an efficient computational scheme for analyzing the scattering of waves for several difficult geometries. Its broad applicability to many fields of engineering and science in handling acoustic, quantum mechanical, electro-magnetic or -static, and elastic problems has been realized only recently, see (1-16). The T-matrix approach has also been extended to treat increasingly complicated situations such as multilayered scatterers (17-25), multiple scattering by a finite number of scatterers (19-25), multiple scattering in lattices (26), and multiple scattering by statistical distribution of scatterers (27). However, in the theory of elasticity the analysis of structures involving several layers with solid-fluid interfaces is still lacking. The purpose of this article is to present a T-matrix formalism for elastic wave scattering by multilayered regions with solid-fluid interfaces. Here the layers are not necessarily consecutively enclosing and thus in some regions there may be several different scatterers. The basic methods for this extension are given in (21-25). We are using the Green's dyadic in an integral representation although the T-matrix can be obtained without using any Green's dyadic (8).

This article consists of 9 parts. Here, in part I, we introduce the basic problems to be treated. In part II we obtain the integral representations for a displacement field in the case of layered regions. Then, we introduce the spherical wave functions and use them to expand the Green's dyadic. In part III we expand the various fields in spherical wave functions. Using the integral representations derived in part II we obtain the Q-matrix for the outermost surface of a body separating two elastic materials. This matrix and the T-matrix associated with the body inside the next outermost surface form the total T-matrix for the body. The procedure in part IV is similar to part III; however, we must now deal with a fluid outside and an elastic material inside the outermost surface. This new situation requires the defining of three matrices (the P, Q, and R matrices) for the outermost surface, in contrast to the simpler, "single surface matrix" case of part III. These three matrices and the T-matrix associated with the body inside the next outermost surface form the total T-matrix for the body. In part V we have an elastic material outside and a fluid inside the outermost surface of the body. Here we have to introduce the U, V, W and Y matrices associated with the outermost surface. These four matrices, together with the T-matrix associated with the next outermost surface of the two layered body, form the total T-matrix for the body.

In part VI we study scattering from several nonoverlapping bodies. In part VII we use the results from the parts II-VI to get a general procedure for obtaining the T-matrix associated with a multilayered, finite body. These layers may be either elastic materials or fluids, and need not be successively enclosing. In part VIII we review the terminology for scattering data. In part IX we give numerical results for rotationally symmetric bodies in a fluid. The wavelengths under consideration are in the longwave- and resonance-regions. We also consider complex wave numbers, the moduli of which are less than the inverse power of some characteristic body dimension. The results are presented in the form of frequency dependent plots for the bistatic differential scattering cross section, the total scattering cross section, and the absorption cross section. We also present some polar plots of the differential scattering cross section.

II. INTEGRAL REPRESENTATIONS AND SPHERICAL WAVE FUNCTIONS

Consider an infinite, homogeneous elastic material with density ρ_0 and Lamé parameters λ_0 and μ_0. A finite layered body is introduced in this material. The body is such that between the outer bounding surface S_1 and an inner surface S_2 the density is ρ_1 and the Lamé parameters λ_1 and μ_1. The material parameters inside the surface S_2 can be varying with the space coordinates in an arbitrary way. The region outside the surface S_1 and the region between the surfaces S_1 and S_2 are called region 0 and region 1, respectively. We assume that an origin can be chosen inside the two surfaces in such a way that, for each surface, the radius (from the origin) to that surface is a continuous single-valued function of the spherical angles. Furthermore, we assume that both surfaces satisfy the requirements of the divergence theorem. Let a monochromatic displacement field $\vec{u}_0^i$, with time dependence $\exp(-i\omega t)$, impinge on the body. The scattered field $\vec{u}_0^s$ and the incoming field $\vec{u}_0^i$ give the total field in region 0 as:

$$\vec{u}_0 = \vec{u}_0^i + \vec{u}_0^s \tag{2.1}$$

The total field between the surface S_1 and S_2 (region 1) is called $\vec{u}_1$. Then with the exception of source points we have the following equations of motion for $\vec{u}_i(\vec{r})$:

$$\frac{\kappa_i^3}{k_i^2} \nabla\nabla\cdot\vec{u}_i - \kappa_i \nabla x(\nabla x\vec{u}_i) + \kappa_i^3\vec{u}_i = 0 \tag{2.2}$$

$$\text{where } i = 0 \text{ or } 1 \text{ and } k_i^2 = \frac{\rho_i\omega^2}{\lambda_i + 2\mu_i} \text{ and } \kappa_i^2 = \frac{\rho_i\omega^2}{\mu_i} .$$

For the same regions we define the Green's dyadic $\overline{G}_i(k_i,\kappa_i,|\vec{r}-\vec{r}'|)$ as the outgoing solution of

$$\frac{\kappa_i^3}{k_i^2} \nabla\nabla\cdot\overline{G}_i - \kappa_i \nabla x(\nabla x\overline{G}_i) + \kappa_i^3\overline{G}_i = - \overline{I}\delta(\vec{r}-\vec{r}') , \tag{2.3}$$

where $i = 0$ or 1 and $\overline{I}$ is the unit dyadic and δ is the Dirac distribution. The stress-dyadics and triadics for the same regions are given by (2.4) and (2.5), respectively.

$$\bar{\sigma}_i(\vec{u}) = \lambda_i \bar{I}\nabla\cdot\vec{u} + \mu_i(\nabla\vec{u} + \vec{u}\nabla) \tag{2.4}$$

$$\bar{\Sigma}_i(\bar{G}) = \lambda_i \bar{I}\nabla\cdot\bar{G} + \mu_i(\nabla\bar{G} + \bar{G}\nabla) \tag{2.5}$$

where i = 0 or 1, and $\vec{u}$ and $\bar{G}$ are the arbitrary displacement field and Green's dyadic, respectively. The surface tension $\vec{t}_i$ is related to the stress dyadic by the surface normal $\hat{n}$.

$$\vec{t}_i = \hat{n}\cdot\bar{\sigma}_i \tag{2.6}$$

The equations (2.2) and (2.3) can be rewritten in terms of the stress-dyadic and triadic, respectively.

$$\nabla\cdot\bar{\sigma}_i(\vec{u}_i) + \rho_i\omega^2\vec{u}_i = 0 \tag{2.2'}$$

$$\nabla\cdot\bar{\Sigma}_i(\bar{G}_i) + \rho_i\omega^2\bar{G}_i = -\bar{I}\,\frac{\rho_i\omega^2}{\kappa_i^3}\,\delta(\vec{r}-\vec{r}\,') \tag{2.3'}$$

These latter forms may be better suited for the derivation of the integral representation. Here equation (2.2') is expressed in the same terms, which appear in the boundary conditions of the integral representation. However, the earlier forms ((2.2) and (2.3)) may be better suited for obtaining the basis functions or Green's dyadic. Here we can clearly see how to separate the curl free and divergence free parts. By using equations (2.2), (2.3), and the divergence theorem we can obtain the integral representations for the various regions (9).

$$\vec{u}_0^{\,i}(\vec{r}) + \frac{\kappa_0^3}{\rho_0\omega^2}\int_{S_1}[\vec{u}_{0+}\cdot(\hat{n}_1\cdot\bar{\Sigma}_0) - \hat{n}_1\cdot\bar{\sigma}_{0+}\cdot\bar{G}_0]ds' = \begin{cases} \vec{u}_0(\vec{r}), & \vec{r} \text{ outside } S_1. \\ 0, & \vec{r} \text{ inside } S_1. \end{cases} \tag{2.7}$$

$$\begin{aligned} -\frac{\kappa_1^3}{\rho_1\omega^2}\int_{S_1}[\vec{u}_{1-}\cdot(\hat{n}_1\cdot\bar{\Sigma}_1) - \hat{n}_1\cdot\bar{\sigma}_{1+}\cdot\bar{G}_1]ds' \,+ \\ +\frac{\kappa_1^3}{\rho_1\omega^2}\int_{S_2}[\vec{u}_{1+}\cdot(\hat{n}_2\cdot\bar{\Sigma}_1) - \hat{n}_2\cdot\bar{\sigma}_{1+}\cdot\bar{G}_1]ds' = \begin{cases} \vec{u}_1(\vec{r}), & \left\{\begin{array}{l}\vec{r} \text{ between} \\ S_1 \text{ and } S_2.\end{array}\right. \\ 0, & \left\{\begin{array}{l}\vec{r} \text{ outside } S_1 \\ \text{or} \\ \text{inside } S_2.\end{array}\right. \end{cases} \end{aligned} \tag{2.8}$$

Here, the + and - subscripts stand for the limit values calculated from the outside and the inside of the surface, respectively. If the material in one of the regions was a fluid ($\mu_i = 0$ for some i), we would get (2.9) and (2.10) instead of (2.2) and (2.3).

$$\nabla\nabla\cdot\vec{u}_i + k_i^2\vec{u}_i = 0 \tag{2.9}$$

$$\nabla\nabla\cdot\overline{G}_i + k_i^2\overline{G}_i = -\frac{1}{k_i}\,\overline{I}\delta(\vec{r}-\vec{r}\,') \tag{2.10}$$

The only change in the integral representations (2.7) and (2.8) is that we have to change the κ_i to a k_i.

Alternatively, when dealing with a fluid we can work with a scalar potential ϕ_i, which is related to the displacement $\vec{u}_i$ by:

$$\vec{u}_i = \nabla\phi_i \; . \tag{2.11}$$

The potential $\phi_i(\vec{r})$ and the corresponding Green's function $g_i(k_i,|\vec{r}-\vec{r}\,'|)$ have the following equations of motion:

$$(\nabla^2+k_i^2)\phi_i = 0 \tag{2.12}$$

$$(\nabla^2+k_i^2)g_i = -\,\delta(\vec{r}-\vec{r}\,') \; . \tag{2.13}$$

From the equations (2.12) and (2.13) we can derive the following integral representations valid only for fluids.

$$\phi_0^i(\vec{r}) + \int_{S_1}[\phi_{0+}\nabla' g_0 - (\nabla'\phi_{0+})g_0]\cdot d\vec{s}\,' = \begin{cases} \phi_0(\vec{r}), & \vec{r} \text{ outside } S_1. \\ 0, & \vec{r} \text{ inside } S_1. \end{cases} \tag{2.14}$$

$$- \int_{S_1}[\phi_{1-}\nabla' g_1 - (\nabla'\phi_{1-})g_1]\cdot d\vec{s}\,' +$$

$$+ \int_{S_2}[\phi_{1+}\nabla' g_1 - (\nabla'\phi_{1+})g_1]\cdot d\vec{s}\,' = \begin{cases} \phi_1(\vec{r}), & \vec{r} \text{ between } S_1 \text{ and } S_2. \\ 0, & \vec{r} \text{ outside } S_1 \text{ or inside } S_2. \end{cases} \tag{2.15}$$

Here ϕ_0, the total field in region 0, is given in terms of the incoming field ϕ_0^i and scattered field ϕ_0^s by:

$$\phi_0 = \phi_0^i + \phi_0^s \; . \tag{2.16}$$

To use the above integral representations (2.7), (2.8), (2.14) or (2.15) one must introduce boundary conditions. These may be expressed in terms of physical observables such as displacement and tension at the surface S_i. The surface tension $\vec{t}_i$ is expressable in a general material as a function of the displacement $\vec{u}$ as seen from equations (2.4) and (2.6). In the particular case of a fluid the surface tension $\vec{t}_{0+}$ can be expressed as a function of the potential ϕ_0 as:

$$\vec{t}_{0+} = -\,k_0^2\lambda_0\phi_{0+}\hat{n}_1 = -\,\rho_0\omega^2\phi_{0+}\hat{n}_1 \; . \tag{2.17}$$

In the case of a fluid, either of the equations (2.7) or (2.14) as well as either of the equations (2.8) or (2.15) can be derived from the other.

It is our aim to develop a matrix representation of the fields. The spherical wave solutions (2.18)-(2.20) to eq. (2.2) are well suited as a basis in such a representation.

$$\vec{\psi}^i_{1\sigma mn}(\vec{r}) = \frac{1}{\sqrt{n(n+1)}} \nabla x[\vec{r}Y_{\sigma mn}(\theta,\phi)h_n(\kappa_i r)] \tag{2.18}$$

$$\vec{\psi}^i_{2\sigma mn}(\vec{r}) = \frac{1}{\kappa_i} \nabla x\vec{\psi}^i_{1\sigma mn}(\vec{r}) \tag{2.19}$$

$$\vec{\psi}^i_{3\sigma mn}(\vec{r}) = \left(\frac{k_i}{\kappa_i}\right)^{3/2} \frac{1}{k_i} \nabla[Y_{\sigma mn}(\theta,\phi)h_n(k_i r)] \tag{2.20}$$

$$\begin{aligned} Y_{\sigma mn}(\theta,\phi) &= \left[\varepsilon_m \frac{(2n+1)(n-m)!}{4\pi(n+m)!}\right]^{1/2} P_n^m(\cos\theta)\begin{cases}\cos m\phi, & \sigma = e \\ \sin m\phi, & \sigma = o\end{cases} \\ &= \gamma_{mn} P_n^m(\cos\theta)\begin{cases}\cos m\phi \\ \sin m\phi\end{cases}, \end{aligned} \tag{2.21}$$

where $\varepsilon_0 = 1$ and $\varepsilon_m = 2$ for $m > 0$, e and o stands for even resp. odd, h_n is the (outgoing) spherical Hankel function, P_n^m is the associated Legendre polynomial. We are going to use various abbreviations $\vec{\psi}^i_{\tau\sigma mn} \equiv \vec{\psi}^i_{\tau p} \equiv \vec{\psi}^i_q$. Further, the symbols Ou and Re will represent the outgoing and regular functions, respectively. For example, $\mathrm{Ou}\vec{\psi}^i_q = \vec{\psi}^i_q$, but, in contrast, $\mathrm{Re}\vec{\psi}^i_q$ means that, instead of using h_n, we use the regular function j_n (at the origin). This is not necessarily the same as taking the real part of $\vec{\psi}_q$. The spherical wave solution to eq. (2.12) is:

$$\phi^i_{\sigma mn}(\vec{r}) = Y_{\sigma mn}(\theta,\phi)h_n(k_i r) . \tag{2.22}$$

When dealing with a fluid we let $\vec{\psi}^i_{1p} = \vec{\psi}^i_{2p} = 0$ and remove the factor $(k_i/\kappa_i)^{3/2}$ in $\vec{\psi}^i_{3p}$. Given the above special conditions we can interpret the expansion of the Green's dyadic $\bar{G}_i$ in spherical wave functions for all cases as:

$$\bar{G}_i(k_i,\kappa_i,|\vec{r}-\vec{r}'|) = i\sum_q \vec{\psi}^i_q(\vec{r}_>)\mathrm{Re}\vec{\psi}^i_q(\vec{r}_<) , \tag{2.23}$$

where $\vec{r}_>$ and $\vec{r}_<$ stands for the greater resp. smaller of $\vec{r}$ and $\vec{r}'$. The Green's function g_i can be expanded as:

$$g_i(k_i,|\vec{r}-\vec{r}'|) = ik_i\sum_p \phi^i_p(\vec{r}_>)\mathrm{Re}\phi^i_p(\vec{r}_<) . \tag{2.24}$$

III. <u>LAYERED BODIES WITH ELASTIC MATERIAL ON BOTH SIDES OF THE BOUNDING SURFACE</u>

Consider an elastic body, with an inclusion, situated in an infinite elastic material. Outside the bounding surface S_1 as well as between S_1 and an inner surface S_2 the material parameters are constant and given in the notations from section II. In section II we introduced the spherical vector wave functions as a basis and expanded the Green's dyadic in this basis. Here we expand the displacement fields in the same basis. The expansion of the incoming and scattered field (in the infinite material) is given by (3.1) and (3.2), respectively.

$$\vec{u}_0^i(\vec{r}) = \sum_q a_q \mathrm{Re}\vec{\psi}_q^0(\vec{r}) \text{ for } r < r^{so} \tag{3.1}$$

$$\vec{u}_0^s(\vec{r}) = \sum_q f_q \vec{\psi}_q^0(\vec{r}) \text{ for } r > r^1_{max} \tag{3.2}$$

Here, r^{so} is the length of the radius vector from origin to the closest singularity of the sources of the incoming field. The symbol r^1_{max} is the radius of the sphere X_1, with center at the origin, circumscribing S_1 (see Fig. 1). By using the integral representation (2.7) and considering $\vec{r}$ outside the sphere X_1, applying the expansions (3.2) and (2.23), we get (3.3), after identifying coefficients.

$$f_q = i \frac{\kappa_0^3}{\rho_0 \omega^2} \int_{S_1} [\vec{t}_0(\mathrm{Re}\vec{\psi}_q^0)\cdot\vec{u}_{0+} - \mathrm{Re}\vec{\psi}_q^0\cdot\vec{t}_{0+}(\vec{u}_0)]ds \tag{3.3}$$

Let Z_1 be the sphere, with center at the origin, inscribed in S_1 (see Fig. 1). Considering $\vec{r}$ inside Z_1 we get (3.4), in a similar way as above, from (2.7).

$$a_q = - i \frac{\kappa_0^3}{\rho_0 \omega^2} \int_{S_1} [\vec{t}_0(\vec{\psi}_q^0)\cdot\vec{u}_{0+} - \vec{\psi}_q^0\cdot t_{0+}(\vec{u}_0)]ds \tag{3.4}$$

The total T-matrix (32-34) for the layered body, defined by (3.5), can not yet be obtained because we have not yet specified the properties of the body.

$$f_q = \sum_{q'} T_{qq'} a_{q'} \tag{3.5}$$

The relations (3.3) and (3.4) together constitute a formal definition of the T-matrix through the two mapings from the surface field to the scattered field and the incoming field, respectively. To realize these mappings we introduce the boundary conditions at S_1 and expand the field on the inside of S_1. The welded boundary conditions (3.6) and (3.7) relate the limit values of the field quantities from each side of the surface S_1.

$$\vec{u}_{0+} = \vec{u}_{1-} \text{ (for } \vec{r}\varepsilon S_1) \tag{3.6}$$

$$\vec{t}_{0+} = \vec{t}_{1-} \text{ (for } \vec{r}\varepsilon S_1) \tag{3.7}$$

The expansion of the field on the inside of S_1 is given by:

$$\vec{u}_{1-} = \sum_q (\alpha_q^1 \mathrm{Re}\vec{\psi}_q^1 + \beta_q^1 \vec{\psi}_q^1) \ . \tag{3.8}$$

The completeness of expansions of this kind has been treated in (28-31). Using (3.6), (3.7), and (3.8) in (3.3) and (3.4), we get:

$$f_q = i\sum_{q'} (\mathrm{ReRe}Q^1_{qq'}\alpha^1_{q'} + \mathrm{ReOu}Q^1_{qq'}\beta^1_{q'}) \tag{3.9}$$

and

$$a_q = - i\Sigma_{q'}(\text{OuRe}Q^1_{qq'}\alpha^1_{q'} + \text{OuOu}Q^1_{qq'}\beta^1_{q'}) \ , \tag{3.10}$$

where the matrix Q^1 is given by

$$\begin{Bmatrix}\text{Ou}\\\text{Re}\end{Bmatrix}\begin{Bmatrix}\text{Ou}\\\text{Re}\end{Bmatrix}Q^{i+1}_{qq'} = \frac{\kappa_i^3}{\rho_i\omega^2}\int_{S_{i+1}}\Big[\vec{t}_i\Big(\begin{Bmatrix}\text{Ou}\\\text{Re}\end{Bmatrix}\vec{\psi}^i_q\Big)\cdot\begin{Bmatrix}\text{Ou}\\\text{Re}\end{Bmatrix}\vec{\psi}^{i+1}_{q'} - \begin{Bmatrix}\text{Ou}\\\text{Re}\end{Bmatrix}\vec{\psi}^i_q\cdot\vec{t}_{i+1}\Big(\begin{Bmatrix}\text{Ou}\\\text{Re}\end{Bmatrix}\vec{\psi}^{i+1}_{q'}\Big)\Big]ds \quad \text{for} \quad i = 0. \tag{3.11}$$

Here, the first $\begin{Bmatrix}\text{Ou}\\\text{Re}\end{Bmatrix}$ in the left member of equation (3.11) corresponds to the selection of either the regular or outgoing functions associated with the first index in Q^i. Similarly, the second $\begin{Bmatrix}\text{Ou}\\\text{Re}\end{Bmatrix}$ is related to the second index. In vector matrix notation we rewrite (3.9) and (3.10) as:

$$\vec{f} = i\text{ReRe}Q^1\vec{\alpha}^1 + i\text{ReOu}Q^1\vec{\beta}^1 \tag{3.12}$$

$$\vec{a} = - i\text{OuRe}Q^1\vec{\alpha}^1 - i\text{OuOu}Q^1\vec{\beta}^1 \tag{3.13}$$

If the body under consideration is homogeneous (no inclusion) which, in turn, implies that $\vec{\beta}^1 \equiv 0$ we can solve the equations (3.12) and (3.13) by elimination of $\vec{\alpha}^1$. In this special case we get the T-matrix for a homogeneous elastic body as:

$$T = - \text{ReRe}Q^1(\text{OuRe}Q^1)^{-1} \ . \tag{3.14}$$

However, if the body has an inclusion, with bounding surface S_2, we have to introduce the integral representation (2.8). By using the integral representation (2.8) and considering $\vec{r}$ outside the sphere X_1 we get (3.15). Further, by using (2.8) and considering $\vec{r}$ inside the sphere Z_2 we get (3.16). The sphere Z_2 is defined for the surface S_2 in a way similar to the defining of the sphere Z_1 with respect to the surface S_1 (see Fig. 1).

$$\Sigma_{q'}(\text{ReRe}X^1_{qq'}\alpha^1_{q'} + \text{ReOu}X^1_{qq'}\beta^1_{q'}) - \frac{\kappa_1^3}{\rho_1\omega^2}\int_{S_2}[\vec{t}_1(\text{Re}\vec{\psi}^1_q)\cdot\vec{u}_{1+} - \text{Re}\vec{\psi}^1_q\cdot\vec{t}_{1+}(\vec{u}_1)]ds = 0 \ , \tag{3.15}$$

$$\Sigma_{q'}(\text{OuRe}X^1_{qq'}\alpha^1_{q'} + \text{OuOu}X^1_{qq'}\beta^1_{q'}) - \frac{\kappa_1^3}{\rho_1\omega^2}\int_{S_2}[\vec{t}_1(\vec{\psi}^1_q)\cdot\vec{u}_{1+} - \vec{\psi}^1_q\cdot\vec{t}_{1+}(\vec{u}_1)]ds = 0 \ , \tag{3.16}$$

where

$$\begin{Bmatrix}\text{Ou}\\\text{Re}\end{Bmatrix}\begin{Bmatrix}\text{Ou}\\\text{Re}\end{Bmatrix}X^i_{qq'} = \frac{\kappa_i^3}{\rho_i\omega^2}\int_{S_i}\left[\vec{t}_i\left(\begin{Bmatrix}\text{Ou}\\\text{Re}\end{Bmatrix}\vec{\psi}^i_q\right)\cdot\begin{Bmatrix}\text{Ou}\\\text{Re}\end{Bmatrix}\vec{\psi}^i_{q'} - \begin{Bmatrix}\text{Ou}\\\text{Re}\end{Bmatrix}\vec{\psi}^i_q\cdot\vec{t}_i\left(\begin{Bmatrix}\text{Ou}\\\text{Re}\end{Bmatrix}\vec{\psi}^i_{q'}\right)\right]ds \quad \text{for} \quad i = 1. \tag{3.17}$$

It is a straight forward matter to show (3.18).

$$\begin{Bmatrix}\text{Ou}\\\text{Re}\end{Bmatrix}\begin{Bmatrix}\text{Ou}\\\text{Re}\end{Bmatrix}X^i_{qq'} = \begin{cases} 0 \text{ for OuOu or ReRe.} \\ i\delta_{qq'}\times\begin{cases}(+1) \text{ for OuRe.}\\(-1) \text{ for ReOu.}\end{cases}\end{cases} \tag{3.18}$$

Expressions similar to (3.18) has been studied for the acoustic and electromagnetic cases in (21) and (22), respectively. In (7) X^1 is expressed as the difference between the Q-matrices for a cavity and a rigid body and the result (3.18) is obtained there. More interesting is perhaps the possibility to develop matrix formulations without using Green's functions or dyadics. This is done by constructing, for each kind of problem (acoustic, electromagnetic or elastic), the quantity corresponding to the integrand in (3.17), see (8). Using the remark about the T-matrix, just after equation (3.5), and the relation (3.18) we get (3.19) from (3.15) and (3.16).

$$\vec{\beta}^1 = T^2\vec{\alpha}^1 \tag{3.19}$$

Here T^2 is the T-matrix for the body inside the surface S_2 (i.e. the inclusion). We notice that the material parameters inside the surface S_2 can be dependent on the space coordinates. Furthermore, the surface S_2 might be interpreted solely as a mathematical abstraction (i.e., it doesn't have to be the boundary between different materials). Except for possible separated inclusions within S_1 the material contained in S_1 and S_2 may be identical. From (3.12), (3.13) and (3.19) we derive the T-matrix for the entire layered elastic body as:

$$T = -\,[\text{ReRe}Q^1 + \text{ReOu}Q^1T^2][\text{OuRe}Q^1 + \text{OuOu}Q^1T^2]^{-1}\,. \tag{3.20}$$

We will end this section by deriving the field between the surfaces S_1 and S_2 in the body. By using the integral representation (2.8) and considering $\vec{r}$ in the region between the sphere Z_1 and the sphere X_2, applying the expansions (3.8) and (2.23), we get:

$$\vec{u}_1 = -\,i\sum_{q,q'}(\text{OuRe}X^1_{qq'}\text{Re}\vec{\psi}^1_q\alpha^1_{q'} + \text{OuOu}X^1_{qq'}\text{Re}\vec{\psi}^1_q\beta^1_{q'}) + \sum_q\left(i\,\frac{\kappa_1^3}{\rho_1\omega^2}\int_{S_2}[\vec{t}_1(\text{Re}\vec{\psi}^1_q)\cdot\vec{u}_{1+} - \text{Re}\vec{\psi}^1_q\cdot\vec{t}_{1+}(\vec{u}_1)]ds\right)\vec{\psi}^1_q \tag{3.21}$$

Using the remark about the T-matrix, just after equation (3.5), and the relations (3.18) and (3.19), we get (3.22) from (3.21).

$$\vec{u}_1 = \sum_q(\alpha^1_q\text{Re}\vec{\psi}^1_q + \beta^1_q\vec{\psi}^1_q) \tag{3.22}$$

This relation is of course only valid when the sphere X_2 is truly inside the sphere Z_1.

IV. LAYERED BODIES WITH AN ELASTIC MATERIAL INSIDE THE BOUNDING SURFACE AND A FLUID OUTSIDE

Consider an elastic body, with an inclusion, situated in an infinite fluid. Outside the bounding surface S_1 as well as between S_1 and an inner surface S_2 the material parameters are constant and given in the notations from section II. (notice that $\mu_0 = 0$). The main difference from section III is that in this section we only have to deal with the curl free basis functions outside S_1 (in the fluid). This, in turn, will lead to some minor complications. The expansions of the incoming and scattered field (in the fluid) is given by (4.1) and (4.2), respectively.

$$\vec{u}_0^i(\vec{r}) = \sum_p a_p \mathrm{Re}\vec{\psi}_{3p}^0(\vec{r}) \quad \text{for} \quad r < r^{so} \tag{4.1}$$

$$\vec{u}_0^s(\vec{r}) = \sum_p f_p \vec{\psi}_{3p}^0(\vec{r}) \quad \text{for} \quad r > r_{max}^1 \tag{4.2}$$

Here, r^{so} and r_{max}^1 has the same meaning as in section III. By a procedure similar to the one used in section III we get (4.3) and (4.4). Notice that we have k_0 here instead of κ_0 as in section III.

$$f_p = i\frac{k_0^3}{\rho_0\omega^2}\int_{S_1}[\vec{t}_0(\mathrm{Re}\vec{\psi}_{3p}^0)\cdot\vec{u}_{0+} - \mathrm{Re}\vec{\psi}_{3p}^0\cdot\vec{t}_{0+}(\vec{u}_0)]ds \tag{4.3}$$

$$a_p = -i\frac{k_0^3}{\rho_0\omega^2}\int_{S_1}[\vec{t}_0(\vec{\psi}_{3p}^0)\cdot\vec{u}_{0+} - \vec{\psi}_{3p}^0\cdot\vec{t}_{0+}(\vec{u}_0)]ds \tag{4.4}$$

The total T-matrix for the layered body is defined directly by (4.5) and indirectly by (4.3) and (4.4).

$$f_p = \sum_{p'} T_{pp'} a_{p'} \tag{4.5}$$

The boundary conditions at the surface S_1 is given by (4.6), (4.7) and (4.8)

$$\hat{n}_1\cdot\vec{u}_{0+} = \hat{n}_1\cdot\vec{u}_{1-} \quad (\text{for } \vec{r}\varepsilon S_1) \tag{4.6}$$

$$\hat{n}_1\cdot\vec{t}_{0+} = \hat{n}_1\cdot\vec{t}_{1-} \quad (\text{for } \vec{r}\varepsilon S_1) \tag{4.7}$$

$$[\vec{t}_{0+}]_{tan} = [\vec{t}_{1-}]_{tan} = 0 \quad (\text{for } \vec{r}\varepsilon S_1) \tag{4.8}$$

Here, $[\ ]_{tan}$ stands for the tangential component. The tangential components of $\vec{u}_{0+}$ and $\vec{u}_{1-}$ are unrelated. However, because of the relation (4.9) we notice that no information about the tangential part of $\vec{u}_{0+}$ or $\vec{u}_{1-}$ is needed.

$$\vec{t}_0(\vec{v}_0) = \hat{n}_1\cdot\vec{\sigma}_0(\vec{v}_0) = \lambda_0(\nabla\cdot\vec{v}_0)\hat{n}_1 = (\hat{n}_1\cdot\vec{t}_0(\vec{v}_0))\hat{n}_1 \tag{4.9}$$

Using (4.6), (4.7), (4.9) and the expansion (3.8) in (4.3) and (4.4), we get:

$$f_p = i\Sigma_{q'}(\mathrm{ReRe}Q_{pq'}\alpha^1_{q'} + \mathrm{ReOu}Q_{pq'}\beta^1_{q'}) \tag{4.10}$$

$$a_p = -i\Sigma_{q'}(\mathrm{OuRe}Q_{pq'}\alpha^1_{q'} + \mathrm{OuOu}Q_{pq'}\beta^1_{q'}) , \tag{4.11}$$

where the matrix Q is given by

$$\begin{Bmatrix}\mathrm{Ou}\\ \mathrm{Re}\end{Bmatrix}\begin{Bmatrix}\mathrm{Ou}\\ \mathrm{Re}\end{Bmatrix}Q_{pq'} = \frac{k_0^3}{\rho_0\omega^2}\int_{S_1}\Big[\lambda_0\nabla\cdot\begin{Bmatrix}\mathrm{Ou}\\ \mathrm{Re}\end{Bmatrix}\vec{\psi}^0_{3p}\hat{n}_1\cdot\begin{Bmatrix}\mathrm{Ou}\\ \mathrm{Re}\end{Bmatrix}\vec{\psi}^1_{q'} - \hat{n}_1\cdot\begin{Bmatrix}\mathrm{Ou}\\ \mathrm{Re}\end{Bmatrix}\vec{\psi}^0_{3p}\hat{n}_1\cdot\vec{t}_1\left(\begin{Bmatrix}\mathrm{Ou}\\ \mathrm{Re}\end{Bmatrix}\vec{\psi}^1_{q'}\right)\Big]ds . \tag{4.12}$$

In vector matrix notations we rewrite (4.10) and (4.11) as:

$$\vec{f} = i\mathrm{ReRe}Q\vec{\alpha}^1 + i\mathrm{ReOu}Q\vec{\beta}^1 \tag{4.13}$$

$$\vec{a} = -i\mathrm{OuRe}Q\vec{\alpha}^1 - i\mathrm{OuOu}Q\vec{\beta}^1 . \tag{4.14}$$

Notice that the indices in the matrix Q belong to different sets (i.e., it can not be made square by choosing different finite truncations in the two indices n and n'. The most obvious counterexample is a sphere.) This means that even if the body under consideration is homogeneous (no inclusion) which, in turn, implies that $\vec{\beta}^1 \equiv 0$ we can not yet solve the equations (4.13) and (4.14) by elimination of $\vec{\alpha}^1$. To overcome this we introduce the expansion (4.15) of the field on the outside of S_1.

$$\vec{u}_{0+} = \Sigma_p\gamma_p\mathrm{Re}\vec{\psi}^0_{3p} \tag{4.15}$$

By using the integral representation (2.8) and considering $\vec{r}$ outside the sphere X_1, applying the expansion (4.15), and applying the boundary conditions (4.6), (4.7), and (4.8), we get (4.16), after identifying coefficients.

$$\Sigma_{p'}\mathrm{ReRe}P_{qp'}\gamma_{p'} + \Sigma_{q'}(\mathrm{ReRe}R_{qq'}\alpha^1_{q'} + \mathrm{ReOu}R_{qq'}\beta^1_{q'}) - \frac{\kappa_1^3}{\rho_1\omega^2}\int_{S_2}[\vec{t}_1(\mathrm{Re}\vec{\psi}^1_{q})\vec{u}_{1+} - \mathrm{Re}\vec{\psi}^1_q\cdot\vec{t}_{1+}(\vec{u}_1)]ds = 0 \tag{4.16}$$

where

$$\begin{Bmatrix}\mathrm{Ou}\\ \mathrm{Re}\end{Bmatrix}\begin{Bmatrix}\mathrm{Ou}\\ \mathrm{Re}\end{Bmatrix}P_{qp'} = \frac{\kappa_1^3}{\rho_1\omega^2}\int_{S_1}\vec{t}_1\left(\begin{Bmatrix}\mathrm{Ou}\\ \mathrm{Re}\end{Bmatrix}\vec{\psi}^1_q\right)\cdot\hat{n}_1\hat{n}_1\cdot\begin{Bmatrix}\mathrm{Ou}\\ \mathrm{Re}\end{Bmatrix}\vec{\psi}^0_{3p'}ds \tag{4.17}$$

and

$$\begin{Bmatrix}Ou\\Re\end{Bmatrix}\begin{Bmatrix}Ou\\Re\end{Bmatrix}R_{qq'} = \frac{\kappa_1^3}{\rho_1\omega^2}\int_{S_1}\left[\vec{t}_1\left(\begin{Bmatrix}Ou\\Re\end{Bmatrix}\vec{\psi}_q^1\right)\cdot\left[\begin{Bmatrix}Ou\\Re\end{Bmatrix}\vec{\psi}_{q'}^1\right]_{tan} - \begin{Bmatrix}Ou\\Re\end{Bmatrix}\vec{\psi}_q^1\cdot\hat{n}_1\hat{n}_1\cdot\vec{t}_1\left(\begin{Bmatrix}Ou\\Re\end{Bmatrix}\vec{\psi}_{q'}^1\right)\right]ds \,. \tag{4.18}$$

We rewrite (4.16) in a vector matrix notation using (3.19) and the remark about the T-matrix, just after equation (3.5).

$$\mathrm{ReReP}\vec{\gamma} + \mathrm{ReReR}\vec{\alpha}^1 + \mathrm{ReOuRT}^2\vec{\alpha}^1 + iT^2\vec{\alpha}^1 = 0 \tag{4.19}$$

We also rewrite (4.13) and (4.14) after using (3.19).

$$\vec{f} = i\mathrm{ReReQ}\vec{\alpha}^1 + i\mathrm{ReOuQT}^2\vec{\alpha}^1 \tag{4.20}$$

$$\vec{a} = - i\mathrm{OuReQ}\vec{\alpha}^1 - i\mathrm{OuOuQT}^2\vec{\alpha}^1 \tag{4.21}$$

If the body under consideration is homogeneous (no inclusion) we have $T^2 \equiv 0$. The T-matrix for the homogeneous body with no inclusion is given by (4.22).

$$T = - \mathrm{ReReQ}(\mathrm{ReReR})^{-1}\mathrm{ReReP}[\mathrm{OuReQ}(\mathrm{ReReR})^{-1}\mathrm{ReReP}]^{-1} \tag{4.22}$$

The T-matrix for the entire layered elastic body with an inclusion possibly as general as described in section III, is given by (4.23).

$$T = - (\mathrm{ReReQ} + \mathrm{ReOuQT}^2)(\mathrm{ReReR} + \mathrm{ReOuRT}^2 + iT^2)^{-1}\mathrm{ReReP}\times \times[(\mathrm{OuReQ} + \mathrm{OuOuQT}^2)(\mathrm{ReReR} + \mathrm{ReOuRT}^2 + iT^2)^{-1}\mathrm{ReReP}]^{-1} \tag{4.23}$$

V. LAYERED BODIES WITH A FLUID INSIDE THE BOUNDING SURFACE AND AN ELASTIC MATERIAL OUTSIDE

Consider a bounded volume of fluid, with an inclusion, situated in an infinite elastic material. Outside the bounding surface S_1 as well as between S_1 and an inner surface S_2 the material parameters are constant and given in the notations from section II (notice that $\mu_1 = 0$). We can proceed as in section III using the integral representation (2.7), the expansions (2.23), (3.1), and (3.2) to obtain the formulas (3.3) and (3.4), which are applicable here also. The boundary conditions are given by (4.6), (4.7), and (4.8) in section IV. We now introduce the expansions (5.1) and (5.2) for the field on the outside of S_1 and on the inside of S_1, respectively.

$$\vec{u}_{0+} = \sum_q \gamma_q \mathrm{Re}\vec{\psi}_q^0 \tag{5.1}$$

$$\vec{u}_{1-} = \sum_p (\alpha_p^1 \mathrm{Re}\vec{\psi}_{3p}^1 + \beta_p^1\vec{\psi}_{3p}^1) \tag{5.2}$$

Notice, that outside S_1 (in the elastic material) we need both curl and divergence free basis functions. But, inside S_1 (in the fluid) we need only the curl free basis functions. Using the boundary conditions (4.6), (4.7), and (4.8) together with the expansions (5.1) and (5.2) in (3.3) and (3.4), we get:

$$\vec{f} = i\mathrm{ReRe}U\vec{\gamma} + i\mathrm{ReRe}V\vec{\alpha}^1 + i\mathrm{ReOu}V\vec{\beta}^1 \tag{5.3}$$

$$\vec{a} = -\, i\mathrm{OuRe}U\vec{\gamma} - i\mathrm{OuRe}V\vec{\alpha}^1 - i\mathrm{OuOu}V\vec{\beta}^1 , \tag{5.4}$$

where the matrices U and V are given by:

$$\begin{Bmatrix}\mathrm{Ou}\\ \mathrm{Re}\end{Bmatrix}\mathrm{Re}U_{qq'} = \frac{\kappa_0^3}{\rho_0\omega^2}\int_{S_1}\vec{t}_0\left(\begin{Bmatrix}\mathrm{Ou}\\ \mathrm{Re}\end{Bmatrix}\vec{\psi}^0_q\right)\cdot\left[\mathrm{Re}\vec{\psi}^0_{q'}\right]_{\tan} ds \tag{5.5}$$

$$\begin{Bmatrix}\mathrm{Ou}\\ \mathrm{Re}\end{Bmatrix}\begin{Bmatrix}\mathrm{Ou}\\ \mathrm{Re}\end{Bmatrix}V_{qp'} = \frac{\kappa_0^3}{\rho_0\omega^2}\int_{S_1}\left[\vec{t}_0\left(\begin{Bmatrix}\mathrm{Ou}\\ \mathrm{Re}\end{Bmatrix}\vec{\psi}^0_q\right)\cdot\hat{n}_1\hat{n}_1\cdot\begin{Bmatrix}\mathrm{Ou}\\ \mathrm{Re}\end{Bmatrix}\vec{\psi}^1_{3p'} - \begin{Bmatrix}\mathrm{Ou}\\ \mathrm{Re}\end{Bmatrix}\vec{\psi}^0_q\cdot\hat{n}_1\lambda_1\nabla\cdot\begin{Bmatrix}\mathrm{Ou}\\ \mathrm{Re}\end{Bmatrix}\vec{\psi}^1_{3p'}\right]ds . \tag{5.6}$$

Notice that U is "square" but V is not. Here as in section IV we can see that even if the fluid is homogeneous (no inclusions) which, in turn, implies that $\vec{B}^1 \equiv 0$ we can not yet solve the equations (5.3) and (5.4). We proceed in a way similar to section III with two exceptions. First, unlike section III where κ_1 was used, we shall substitute k_1 in the integral representation (2.8). Second, unlike section III where both curl and divergence free basis functions were used we shall expand the various fields between S_1 and S_2 using curl free basis functions alone. By using the integral representation (2.8) (with k_1 instead of κ_1) and considering $\vec{r}$ outside the sphere X_1, applying the expansions (5.2), we get (5.7), after identifying coefficients. Similarly, considering $\vec{r}$ inside the sphere Z_2 we get (5.8).

$$\Sigma_{p'}(\mathrm{ReRe}X^1_{3p3p'}\alpha^1_{p'} + \mathrm{ReOu}X^1_{3p3p'}\beta^1_{p'}) - \frac{k_1^3}{\rho_1\omega^2}\int_{S_2}[\vec{t}_1(\mathrm{Re}\vec{\psi}^1_{3p})\cdot\vec{u}_{1+} - \mathrm{Re}\vec{\psi}^1_{3p}\cdot\vec{t}_{1+}(\vec{u}_1)]ds = 0 \tag{5.7}$$

$$\Sigma_{p'}(\mathrm{OuRe}X^1_{3p3p'}\alpha^1_{p'} + \mathrm{OuOu}X^1_{3p3p'}\beta^1_{p'}) - \frac{k_1^3}{\rho_1\omega^2}\int_{S_2}[\vec{t}_1(\vec{\psi}^1_{3p})\cdot\vec{u}_{1+} - \vec{\psi}^1_{3p}\cdot\vec{t}_{1+}(\vec{u}_1)]ds = 0 \tag{5.8}$$

Notice that two modifications have been made when changing from an elastic medium as in section III to a fluid. The first is the change from κ_1 to k_1 in the integral representation (2.8). The second modification is the removal of the factor $(k_1/\kappa_1)^{3/2}$ in the curl free basis function. These two changes cancel each other out. For this reason we obtain a subpart of the matrix X^1 as defined in (3.17). By using the properties of the matrix X^1 given by (3.18) and the remark about the T-matrix after equation (3.5), we get:

$$\vec{\beta}^1 = T^2\vec{\alpha}^1 , \tag{5.9}$$

where T^2 is the T-matrix for the body inside the surface S_2 (i.e. the inclusion). The equations (5.3), (5.4), and (5.9) are still not sufficient to give us the T-matrix for the total body. We have to use the integral representation (2.8), but now with the expansion (5.1) introduced through the boundary conditions at S_1. By using the integral representation (2.8) (with k_1 instead of κ_1) and considering $\vec{r}$ outside the sphere X_1, applying the expansions (5.1) and (5.2), and applying the boundary conditions (4.6) we get (5.10).

$$\sum_{q'} \mathrm{ReReW}_{3pq'}\gamma_{q'} - \sum_{p'}(\mathrm{ReReY}_{3p3p'}\alpha^1_{p'} + \mathrm{ReOuY}_{3p3p'}\beta^1_{p'}) - \frac{k_1^3}{\rho_1\omega^2}\int_{S_2}[\vec{t}_1(\mathrm{Re}\vec{\psi}^1_{3p})\cdot\vec{u}_{1+} - \mathrm{Re}\vec{\psi}^1_{3p}\cdot\vec{t}_{1+}(\vec{u}_1)]ds = 0 \tag{5.10}$$

where

$$\left\{\begin{matrix}\mathrm{Ou}\\ \mathrm{Re}\end{matrix}\right\}\mathrm{ReW}_{3pq'} = \frac{k_1^3}{\rho_1\omega^2}\int_{S_1}\lambda_1\nabla\cdot\left\{\begin{matrix}\mathrm{Ou}\\ \mathrm{Re}\end{matrix}\right\}\vec{\psi}^1_{3p}\hat{n}_1\cdot\mathrm{Re}\vec{\psi}^0_{q'}ds \tag{5.11}$$

$$\left\{\begin{matrix}\mathrm{Ou}\\ \mathrm{Re}\end{matrix}\right\}\left\{\begin{matrix}\mathrm{Ou}\\ \mathrm{Re}\end{matrix}\right\}Y_{3p3p'} = \frac{k_1^3}{\rho_1\omega^2}\int_{S_1}\left\{\begin{matrix}\mathrm{Ou}\\ \mathrm{Re}\end{matrix}\right\}\vec{\psi}^1_{3p}\cdot\hat{n}_1\lambda_1\nabla\cdot\left\{\begin{matrix}\mathrm{Ou}\\ \mathrm{Re}\end{matrix}\right\}\vec{\psi}^1_{3p'}ds \tag{5.12}$$

Using (3.19) and the remark about the T-matrix, just after equation (3.5) we get (5.13) from (5.10).

$$\mathrm{ReReW}\vec{\gamma} - \mathrm{ReReY}\vec{\alpha}^1 - \mathrm{ReOuY}T^2\vec{\alpha}^1 + iT^2\vec{\alpha}^1 = 0 \tag{5.13}$$

We also rewrite (5.3) and (5.4) after using (3.19).

$$\vec{f} = i\mathrm{ReReU}\vec{\gamma} + i\mathrm{ReReV}\vec{\alpha}^1 + i\mathrm{ReOuV}T^2\vec{\alpha}^1 \tag{5.14}$$

$$\vec{a} = -i\mathrm{OuReU}\vec{\gamma} - i\mathrm{OuReV}\vec{\alpha}^1 - i\mathrm{OuOuV}T^2\vec{\alpha}^1 \tag{5.15}$$

If the body under consideration is homogeneous (no inclusion) we have $T^2 \equiv 0$. The T-matrix for the homogeneous "fluid-body" with no inclusion is given by (5.16).

$$T = -(\mathrm{ReReU} + \mathrm{ReReV}(\mathrm{ReReY})^{-1}\mathrm{ReReW})(\mathrm{OuReU} + \mathrm{OuReV}(\mathrm{ReReY})^{-1}\mathrm{ReReW})^{-1} \tag{5.16}$$

The T-matrix for the entire layered "fluid-body", with an inclusion possibly as general as described in section III, is given by (5.17).

$$T = -[\mathrm{ReReU} + (\mathrm{ReReV} + \mathrm{ReOuV}T^2)(\mathrm{ReReY} + \mathrm{ReOuY}T^2 - iT^2)^{-1}\mathrm{ReReW}]\times[\mathrm{OuReU} + (\mathrm{OuReV} + \mathrm{OuOuV}T^2)(\mathrm{ReReY} + \mathrm{ReOuY}T^2 - iT^2)^{-1}\mathrm{ReReW}]^{-1} \tag{5.17}$$

VI. MULTIPLE SCATTERING

Consider N elastic bodies situated in an infinite elastic material. Outside the bounding surfaces of the N bodies the material parameters are constant and given in the notations from section II. The integral representation (2.7) can be used after changing the region of integration. We then have the following expression:

$$\vec{u}_0^i(\vec{r}) + \sum_i \frac{\kappa_0^3}{\rho_0\omega^2}\int_{S_i}[\vec{u}_{0+}\cdot(\hat{n}_i\cdot\bar{\Sigma}_0) - \hat{n}_i\cdot\bar{\sigma}_{0+}\cdot\bar{G}_0]ds' = \tag{6.1}$$

$$= \begin{cases} \vec{u}_0(\vec{r}), & \vec{r} \text{ outside } S_i \text{ for all } i \le N. \\ 0 & , \vec{r} \text{ inside } S_i \text{ for a specific } i \le N. \end{cases}$$

For the expansion of the Green's dyadic and the incoming field we use (2.23) and (3.1), respectively. However, for the scattered field we now must deal with two different regions. One region is the volume outside of the smallest sphere circumscribing all the surfaces S_i (see Fig. 2). The other region is the volume inside the largest sphere inscribed within all the surfaces S_i. Both spheres have center at the origin 0. The radii of these spheres are called r^{out} and r^{reg}, respectively (see Fig. 2). We now expand the scattered fields as follows:

$$\vec{u}_0^s = \sum_q f_q^{reg}\text{Re}\vec{\psi}_q^0(\vec{r}) \quad \text{for} \quad r < r^{reg} \tag{6.2}$$

$$\vec{u}_0^s = \sum_q f_q^{out}\vec{\psi}_q^0(\vec{r}) \quad \text{for} \quad r > r^{out} \tag{6.3}$$

Before proceeding further we require the translation properties for the basis functions. These are given by the following formulas:

$$\vec{\psi}_q^i(\vec{r}+\vec{a}) = \begin{cases} \sum_{q'} R(\vec{a})_{qq'}\vec{\psi}_{q'}^i(\vec{r}) & \text{for} \quad a < r. \\ \sum_{q'} \sigma(\vec{a})_{qq'}\text{Re}\vec{\psi}_{q'}^i(\vec{r}) & \text{for} \quad a > r, \end{cases} \tag{6.4}$$

$$\text{Re}\vec{\psi}_q^i(\vec{r}+\vec{a}) = \sum_{q'} R(\vec{a})_{qq'}\text{Re}\vec{\psi}_{q'}^i(\vec{r}) \quad \text{for all } a \text{ and } r. \tag{6.5}$$

The properties of the matrices R and σ for the acoustic and electromagnetic problems are treated in (24) and (23). Here we simply need the exterior product of these two kinds of representations. The idea is to express the surface integrals in terms of the radius vectors $\vec{r}_i$ extending from the origins 0_i. These coordinate systems are pure translations of each other (see Fig. 2). By considering $\vec{r}$ inside the inscribed sphere of surface S_i, with center at the origin 0_i, we obtain (6.6) by using the expansions (3.1), (2.23), (6.4), and (6.5) and by identifying the coefficients of the regular functions.

$$R^t(\vec{a}_i)\vec{a} = \vec{a}^i - \sum_{j\neq i}\sigma(-\vec{a}_i+\vec{a}_j)\vec{f}^j \quad \text{for} \quad i,j = 1,2,3...N, \tag{6.6}$$

with the restriction:

$$|-\vec{a}_i+\vec{a}_j| > r_i'', r_j'' . \tag{6.7}$$

The superscript t in (6.6) stands for the transpose of R (which in turn is equal to the inverse of R). The double prime on $\vec{r}_i$ in (6.7) indicates that $\vec{r}_i$ is on the surface S_i. Here we have used the following expressions for the components of the vectors $\vec{a}^i$ and $\vec{f}^i$.

$$f_q^i = i \frac{\kappa_0^3}{\rho_0\omega^2} \int_{S_i} [\vec{t}_0(\mathrm{Re}\vec{\psi}_q^0(\vec{r}_i''))\cdot\vec{u}_{0+} - \mathrm{Re}\vec{\psi}_q^0(\vec{r}_i'')\cdot\vec{t}_{0+}(\vec{u}_0)]ds'' \tag{6.8}$$

$$a_q^i = -i \frac{\kappa_0^3}{\rho_0\omega^2} \int_{S_i} [\vec{t}_0(\vec{\psi}_q^0(\vec{r}_i''))\cdot\vec{u}_{0+} - \vec{\psi}_q^0(\vec{r}_i'')\cdot\vec{t}_{0+}(\vec{u}_0)]ds'' \tag{6.9}$$

By considering $\vec{r}$ outside the sphere, with radius r^{out}, mentioned above, we obtain (6.10) by using the expansions (6.3), (2.23), (6.5) and by identifying coefficients of the outgoing functions.

$$\vec{f}^{out} = \sum_i R(\vec{a}_i)\vec{f}^i \tag{6.10}$$

Further, by considering $\vec{r}$ inside the sphere with radius r^{reg}, mentioned above, we obtain (6.11) by using the expansions (6.2), (2.23), (6.4) and by identifying coefficients of the regular functions.

$$\vec{f}^{reg} = \sum_i \sigma(\vec{a}_i)\vec{f}^i \tag{6.11}$$

We observe that the T-matrix T_i for the body bounded by the surface S_i gives the relation between $\vec{f}^i$ and $\vec{a}^i$ as follows:

$$\vec{f}^i = T_i\vec{a}^i . \tag{6.12}$$

We thus obtain the following system of algebraic equations with unknown coefficients $\vec{a}^i$, which can be eliminated:

$$R^t(\vec{a}_i)\vec{a} = \vec{a}^i - \sum_{j\neq i} \sigma(-\vec{a}_i+\vec{a}_j)T_j\vec{a}^j \quad \text{for} \quad i,j = 1,2,3\ldots N , \tag{6.13}$$

with the restriction: $|-\vec{a}_i+\vec{a}_j| > r_i''$,

$$\vec{f}^{out} = \sum_i R(\vec{a}_i)T_i\vec{a}^i , \tag{6.14}$$

$$\vec{f}^{reg} = \sum_i \sigma(\vec{a}_i)T_i\vec{a}^i . \tag{6.15}$$

The separation requirement is weaker than separability by planes or nonoverlapping spheres. We note that the result (6.13)-(6.15) is independent of whether the infinite medium is a fluid (with $\mu_0 = 0$) or a fully elastic material (with $\mu_0 \neq 0$). The differences in the derivation when the infinite material is a fluid is that κ_0 has to be changed to k_0 and only the curl free basis function is needed (i.e. $\tau = 3$). Furthermore, it is clear the kind of material inside S_i is irrelevant as pointed out before. Equation (6.16) gives the T-matrix for all the N bodies as a formal solution of (6.13) and (6.14).

$$T = \sum_{i,j} R(\vec{a}_i)T_i\{\delta_{mn} - \sum_{n\neq m} \sigma(-\vec{a}_m+\vec{a}_n)T_n\}^{-1}_{ij}R^t(\vec{a}_j) \tag{6.16}$$

Here, the symbol { } stands for a matrix of the space point indices. The elements of this matrix don't commute which makes the inversion more difficult than in the case of matrices with elements which are complex numbers. A procedure to obtain the T-matrix in a maximally symmetric form is described in (23) and (24). The T-matrix for two and three scatterers is also obtained and interpreted in physical terms in these references.

VII. EXTENSIONS

Let us consider a body with N consecutively enclosing surfaces S_i (i = 1,2...N) separating different elastic materials (see Fig. 3). We associate the Q-matrix Q^i, given by (3.11), with surface S_i. The T-matrix T^i is associated with the body inside the surface S_i. It is clear that the formula (3.21) can be generalized to (7.1) relating T^{i-1} with T^i by means of Q^{i-1}.

$$T^{i-1} = -[\mathrm{ReRe}Q^{i-1} + \mathrm{ReOu}Q^{i-1}T^i][\mathrm{OuRe}Q^{i-1} + \mathrm{OuOu}Q^{i-1}T^i]^{-1} \tag{7.1}$$

Starting with the T-matrix T^N for the innermost surface S_N and the Q-matrix Q^{N-1} for the next innermost surface S_{N-1} we get, using (7.1), the T-matrix T^{N-1} for the body inside the surface S^{N-1}. Repeated use of (7.1) finally gives us the T-matrix T^1 for the entire body.

As mentioned before in section III the innermost surface S_N may be interpreted solely as a mathematical abstraction and, hence, may have no physical interpretation. The surface S_N may contain a set of nonintersecting, nonembedded surfaces S^i (i = 1,2...M). These surfaces (S^i), in turn, may enclose separate regions having different material properties (see Fig. 4). In this case we get the T-matrix T^N for the M bodies inside S_N by the methods in section VI.

The body with the N surfaces S_i, with which we started this section, can be considered together with other neighboring bodies. These neighboring bodies can be enclosed by different layers of material. Again, the total T-matrix can be computed as outlined above.

We can generalize equation (7.1) producing:

$$T^{i-1} = f(Q^{i-1},T^i) \ . \tag{7.2}$$

Equation (4.23) for the body in a fluid, and equation (5.17) for the "fluid-body" in an elastic material can be generalized to (7.3) and (7.4), respectively.

$$T^{i-1} = g(P^{i-1},Q^{i-1},R^{i-1},T^i) \tag{7.3}$$

$$T^{i-1} = h(U^{i-1},V^{i-1},W^{i-1},Y^{i-1},T^i) \tag{7.4}$$

The structure of equations (7.2)-(7.4) implies we can relax our initial restriction to elastic inclusions. Starting from some innermost surface and using either equation (7.2), (7.3), (7.4), or (6.16) we can get the T-matrix for most combinations of fluid and elastic regions. Consider, for example, the three regions in

Fig. 5 where S_{01}, S_{02} and S_{12} are separating different materials. This configuration can, as shown in (21) and (22), be treated as multiple scattering from two bodies. However, the origins of the various coordinate systems involved must be chosen so that the geometrical constraint in (6.13) is satisfied. The radius vector describing the surfaces also must be single valued.

We can also study multiple scattering in lattices, with unit cells composed of more than one scatterer, as in (26). In this case we need only plug in the proper matrices pertinent to our elastic-fluid problem in the formulas, as given in (26).

VIII. CROSS SECTIONS

In this section we begin with wave scattering in fluids. As mentioned earlier this can be described by a scalar potential. The expansion $\phi_0^s(\vec{r}) = \sum_p f_p^s \phi_p^0(\vec{r})$ (ϕ_p^0 is given by (2.22)) of the scattered scalar field ϕ_0^s gives us the following asymptotic behavior of the scattered field:

$$\phi_0^s(\vec{r}) = \frac{1}{k_0} \sum_p (-i)^{n+1} f_p^s Y_p(\theta,\phi) \frac{e^{ik_0 r}}{r} \quad \text{for large r.} \tag{8.1}$$

This leads us to the following definition of the scattering amplitude A^s, which is independent of the radius r:

$$A^s(\theta,\phi) = \lim_{r\to\infty} re^{-ik_0 r} \phi_0^s(\vec{r}) \; . \tag{8.2}$$

We are going to define the various cross sections (32,33,34) in terms of incoming plane waves. For this case the cross sections can be interpreted physically. This physical interpretation will be treated later.

The incoming plane wave ϕ_0^i is given by:

$$\phi_0^i(\vec{r}) = Ce^{i\vec{k}_0\cdot\vec{r}} = \sum_p a_p^{pl}(\alpha,\beta)\mathrm{Re}\phi_p^0(\vec{r}) = 4\pi C \sum_p i^n Y_p(\alpha,\beta)\mathrm{Re}\phi_p^0(\vec{r}) \; . \tag{8.3}$$

Here, the angles α and β are the spherical angles for the incoming wave vector $\vec{k}_0$ (see Fig. 6). This wave vector is given by:

$$\vec{k}_0 = k_0(\sin\alpha\,\cos\beta,\ \sin\alpha\,\sin\beta,\ \cos\alpha) \; . \tag{8.4}$$

We define the differential scattering cross section $d\sigma/d\Omega$ by:

$$\frac{d\sigma(\alpha,\beta,\theta,\phi)}{d\Omega(\theta,\phi)} \equiv \lim_{r\to\infty} r^2 \left|\frac{\phi_0^s}{\phi_0^i}\right|^2 = \left|\frac{A^s(\alpha,\beta,\theta,\phi)}{C}\right|^2 , \tag{8.5}$$

where $d\Omega(\theta,\phi) = \sin\theta d\theta d\phi$. Again, α and β are the spherical angles for the incoming wave. Observe that both the scattered field ϕ_0^s and the scattering amplitude A^s are functions of these angles, as shown in equation (8.5). An incoming plane wave, with direction defined by α and β, produces wave scattering in the direction given by θ and ϕ. The amount of energy scattered is proportional to the differential scattering cross section $d\sigma/d\Omega$.

The total scattering cross section σ is defined by:

$$\sigma(\alpha,\beta) \equiv \int_{\phi=0}^{2\pi} \int_{\theta=0}^{\pi} \frac{d\sigma}{d\Omega}\, d\Omega = \frac{1}{|C|^2 k_0^2} \sum_p |f_p^s(\alpha,\beta)|^2 \; . \tag{8.6}$$

Here, the coefficients f_p^s in the expansion of the scattered field depend on the spherical angles α and β, which, in turn, define the direction of the incoming plane wave. The total scattering cross section is proportional to the total amount of energy scattered in all directions.

The extinction (or total) cross section σ_e is defined by:

$$\sigma_e(\alpha,\beta) \equiv \frac{4\pi}{k_0} \, \mathrm{Im} \, \frac{1}{C} \, A^s(\alpha,\beta,\alpha,\beta) \; . \tag{8.7}$$

Here, Im stands for "the imaginary part of", and C is the amplitude of the incoming plane wave. The extinction cross section is proportional to the total amount of energy lost from an incoming plane wave with direction given by α and β.

Finally, we define the absorption cross section σ_a by:

$$\sigma_a(\alpha,\beta) \equiv \sigma_e(\alpha,\beta) - \sigma(\alpha,\beta) \tag{8.8}$$

This cross section is proportional to the amount of energy that is absorbed by the body from the incoming plane wave. The angles α and β give the direction of this incoming wave.

Instead of using the potential to describe the scattering in a fluid we can use the displacement field. The expansion $\vec{u}_0^s(\vec{r}) = \sum_p f_p \vec{\psi}_{3p}^0(\vec{r})$ of the scattered displacement field $\vec{u}_0^s$ gives us the following asymptotic behavior of the scattered field:

$$\vec{u}_0^s(\vec{r}) = \frac{\hat{r}}{k_0} \sum_p (-i)^n f_p Y_p(\theta,\phi) \, \frac{e^{ik_0 r}}{r} \quad \text{for large r.} \tag{8.9}$$

We define the scattering amplitude $\vec{A}$, which now is a vector in the $\hat{r}$ direction and independent of the radius r, by:

$$\vec{A}(\theta,\phi) = \lim_{r\to\infty} r e^{-ik_0 r} \vec{u}_0^s(\vec{r}) \; . \tag{8.10}$$

As before we are only going to define the cross sections for incoming plane waves. For a fluid the incoming plane wave $\vec{u}_0^i$ is given by:

$$\vec{u}_0^i(\vec{r}) = \hat{k}_0 D e^{i\vec{k}_0 \cdot \vec{r}} = \frac{D}{iC} \sum_p a_p^{pl}(\alpha,\beta) \mathrm{Re}\vec{\psi}_{3p}^0(\vec{r}) = \sum_p a_p(\alpha,\beta) \mathrm{Re}\vec{\psi}_{3p}^0(\vec{r}) \; . \tag{8.11}$$

Here a_p^{pl} is the same as in (8.3). We now define, for displacement fields, the differential scattering cross section $d\sigma/d\Omega$, the total scattering cross section σ, the extinction (or total) cross section σ_e, and the absorption cross section σ_a by:

$$\frac{d\sigma(\alpha,\beta,\theta,\phi)}{d\Omega(\theta,\phi)} \equiv \lim_{r\to\infty} r^2 \frac{|\vec{u}_0^s|^2}{|\vec{u}_0^i|^2} \tag{8.12}$$

$$\sigma(\alpha,\beta) \equiv \int_{\phi=0}^{2\pi} \int_{\theta=0}^{\pi} \frac{d\sigma}{d\Omega} d\Omega \tag{8.13}$$

$$\sigma_e(\alpha,\beta) \equiv \frac{4\pi}{k_0} \text{Im} \frac{1}{D} A(\alpha,\beta,\alpha,\beta) \tag{8.14}$$

$$\sigma_a(\alpha,\beta) \equiv \sigma_e(\alpha,\beta) - \sigma(\alpha,\beta) \ . \tag{8.15}$$

Here D is the amplitude of the incoming plane wave. In section II we defined the relation $\vec{u}_0 = \nabla\phi_0$. From this relation and by comparison of equations (8.3) and (8.11) we get the relation $D = ik_0C$ between the amplitudes for the displacement field and scalar field. We notice that the same T-matrix relates the incoming field to the scattered field in the potential and displacement field descriptions. This fact, together with equation (8.11) and the relation $D = ik_0C$, leads us to the following two relations: $f_p = k_0 f_p^s$ and $A = ik_0A^s$. From the above we can see that the cross sections defined by (8.5)-(8.8) are the same as those defined by (8.12)-(8.15).

We now turn to the case where the unbounded medium is an elastic material. In this case we have to deal with both the divergence free and curl free basis functions. (i.e. τ = 1, 2, and 3). The expansion $\vec{u}_0^s(\vec{r}) = \sum_q f_q \vec{\psi}_q^0(\vec{r})$ (where $q \equiv \tau\sigma mn$) of the scattered displacement field $\vec{u}_0^s$ gives us the following asymptotic behavior of the scattered field:

$$\vec{u}_0^s(\vec{r}) = \sum_p \frac{(-i)^n}{\kappa_0}\left[\left(- if_{1p}\vec{A}_{1p}(\theta,\phi) + f_{2p}\vec{A}_{2p}(\theta,\phi)\right)\frac{e^{i\kappa_0 r}}{r} + \left(\frac{k_0}{\kappa_0}\right)^{1/2} f_{3p}\vec{A}_{3p}(\theta,\phi)\frac{e^{ik_0r}}{r}\right] \text{ for large r.} \tag{8.16}$$

The functions, $\vec{A}_{\tau p}$ for τ = 1, 2 are purely transversal. The function $\vec{A}_{3p}$ is purely longitudinal. In the case, with an unbounded elastic medium we have two kinds of incoming displacement fields. The first kind is the purely longitudinal wave given by (8.11) with a slight modification. Here, in the case with the elastic medium, we must introduce a factor $(\kappa_0/k_0)^{3/2}$ before the expansion coefficient a_p because of a change in the normalization of the curl free basis function $\vec{\psi}_{3p}^0$ when shifting from a fluid to an elastic medium. The second kind of incoming displacement field is the purely transversal one given by:

$$\vec{u}_0^i(\vec{r}) = \vec{E}e^{i\vec{k}_0\cdot\vec{r}} = \sum_{\substack{\tau=1,2\\p}} a_{\tau p}(\alpha,\beta)\vec{\psi}_{\tau p}^0(\vec{r}). \tag{8.17}$$

Here $\vec{E}$ is the amplitude vector, which is orthogonal to $\hat{r}$. The angles α and β are the spherical angles for the incoming wave vector. We define, for the displacement fields, the differential and total scattering cross sections by (8.12) and (8.13), respectively. Notice that we can separate the longitudinal and the transversal parts of the different cross sections. The total scattering cross section can, by using the relation $\int \vec{A}_q \cdot \vec{A}_{q'} d\Omega = \delta_{qq'}$, be expressed as:

$$\sigma(\alpha,\beta) = \frac{1}{|\vec{u}_0^i|^2 \kappa_0^2} \sum_p (|f_{1p}|^2 + |f_{2p}|^2 + \frac{k_0}{\kappa_0} |f_{3p}|^2) \tag{8.18}$$

IX. NUMERICAL RESULTS

Elastic wave scattering (single scattering) by homogeneous bodies of elastic materials or cavities on an infinite elastic material is studied numerically in (12,14). Multiple scattering of elastic waves by layered bodies (several bodies) in an infinite elastic material is studied numerically in (19). Numerical results for the scattering of waves by a homogeneous body in a fluid are given in (15). For more wave number dependent data see (16). In this article we shall give some numerical results for homogeneous and layered elastic bodies in a fluid. The surfaces of the layers in the bodies are prolate spheroids with axis ratios $a_i/b_i = 2.0$ (see Fig. 7). The ratios a_2/a_1 between the semiaxes of the inner and outer surfaces are: 0.0, 0.25, and 0.9. The inner surface always encloses a cavity. The material parameters of the fluid and of the elastic solid are given in Table 1.

We choose the z-axis as the axis of rotational symmetry. Without any loss of generality we set the spherical angle β for the incoming wave (see Fig. 6) equal to zero. This choice reduces the number of matrix elements needed in the finite truncation by a factor of 0.5. In the diagrams studied here we shall consider only incoming waves with the spherical angle $\alpha = 0.0$. In the polar plots we shall study the differential scattering cross section $d\sigma/d\Omega$ as a function of the scattered angle θ. In this context the other scattered angle ϕ will take on the values 0.0 or 180.0. In the Cartesian plots we shall study the differential and total scattering cross sections as a function of k_0a_1. We only consider the scattering angles $\theta = 90$ and $\phi = 0$ in these plots. For both the polar and Cartesian diagrams we shall consider the influence of losses in the elastic material. The complex wave numbers associated with these losses are given in Table 1.

Gauss-Legendre quadrature formulas were used to generate the matrix elements of Q, P, and R defined in Eqs. (4.12), (4.17), and (4.18). In the homogeneous case the matrix R was inverted by Gaussian elimination. However, the next inversion in Eq. (4.22) was done by Schmidt orthogonalization for the case of loss less materials. Here we used a computer routine developed by P. C. Waterman (35) wherein the symmetric and unitarity properties of the S-matrix (S = 1-2T) have been used to optimize the inversion procedure. In the layered, loss less case we could use only the Schmidt orthogonalization process for the T-matrix of the cavity. The other inversions in this case were done by using Gaussian elimination. In the cases dealing with losses the S-matrix is not unitary, and we have to use Gaussian elimination.

The T-matrix for rotationally symmetric bodies is diagonal in the azimuth index. The two submatrices (of the T-matrix) which corresponds to the two azimuth indices 0 and 1 were given the same dimension in the finite truncation. However, for each unit increase in the value of the azimuth index beyond 1 we decreased the dimension of the associated submatrix by one. The dimension of the two biggest submatrices was progressively increased from 8 at $k_0a_1 = 0.1$ to 12 for $k_0a_1 = 3.2$.

TABLE 1

	Water	Fict. Solid	Lossy Fict. Solid
Density ρ_i kg/dm^3	1.0	1.7	1.7
Comp. wave speed c_p m/s	1493	2000	2000
shear wave speed c_s m/s	zero	500	500
Comp. wave number k_i	k_0	$k_1 = 0.7465k_0$	$k_1 = 0.7465k_0[1-ik_0a_1 0.1]^{-1/2}$
Shear wave number κ_i	0	$\kappa_1 = 4.0k_0$	$\kappa_1 = 4.0k_0[1-ik_0a_1 0.1]^{-1/2}$

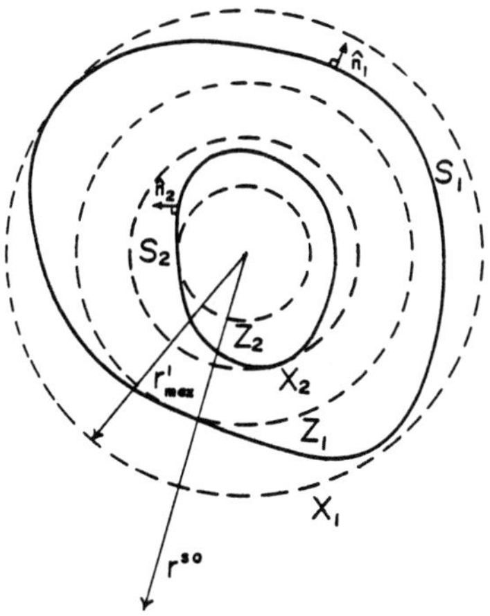

Fig. 1. Geometry of two layered body.

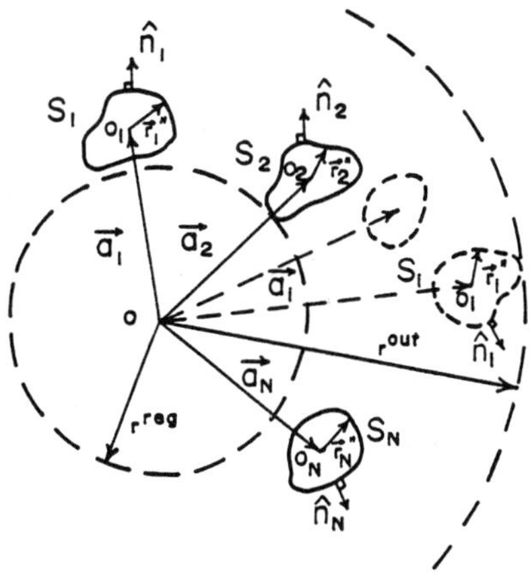

Fig. 2. Geometry of N scattering regions bounded by the closed surfaces S_i for i = 1,2,3...N.

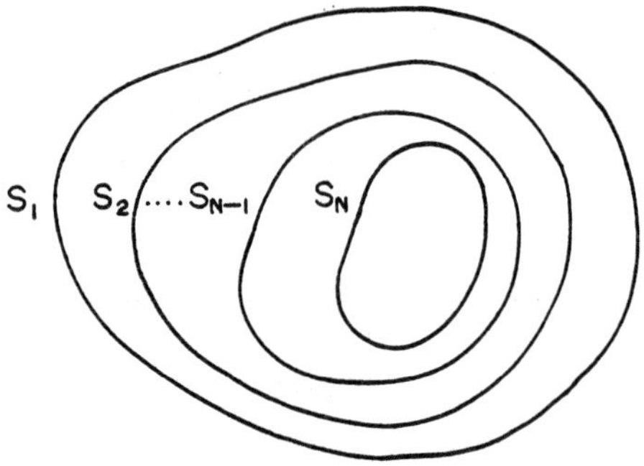

Fig. 3. Geometry of N layered body.

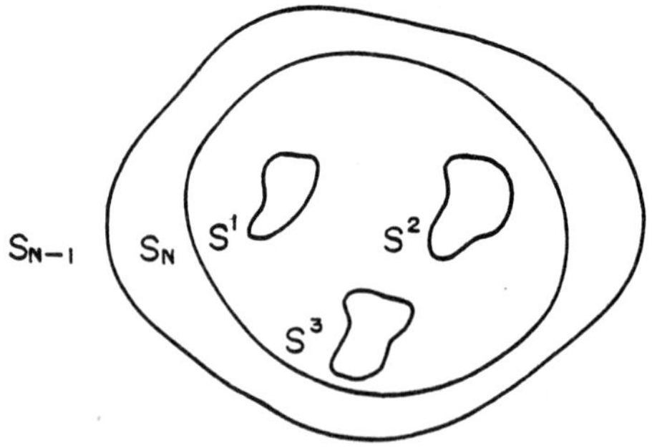

Fig. 4. Geometry of layered body with 3 scattering regions, bounded by the closed surfaces S_1, S_2, and S_3, inside the N'th surface.

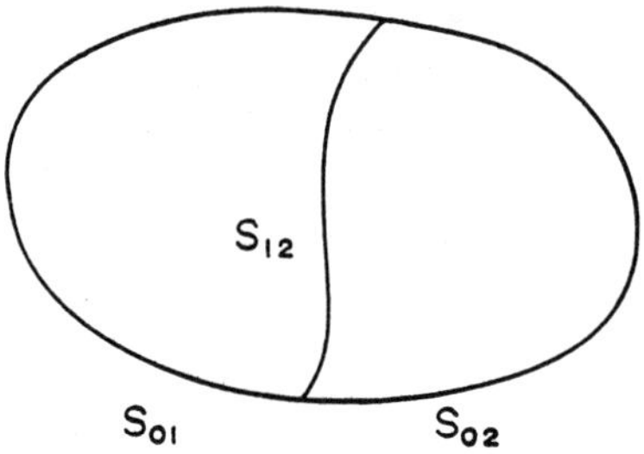

Fig. 5. Geometry of a scatterer consisting of two nonenclosing parts.

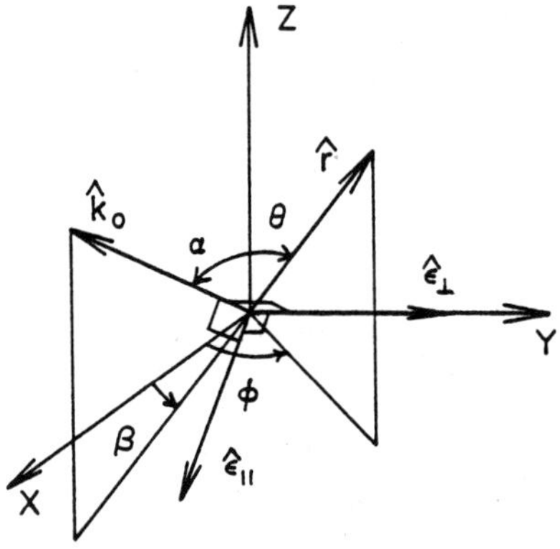

Fig. 6. Notations for incoming plane wave and scattered wave.

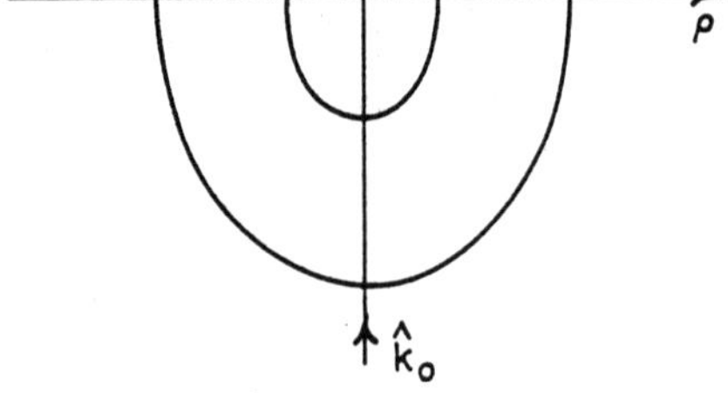

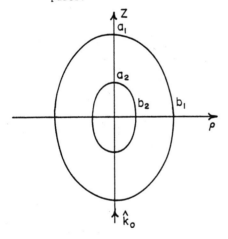

Fig. 7. Geometry of the two layered body, with two prolate spheroid surfaces, for which we give numerical results in this paper.

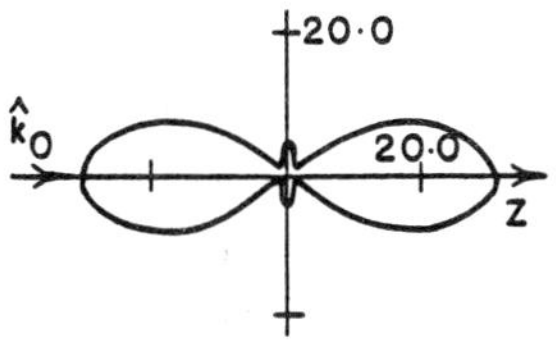

Fig. 8.a. Loss less solid.

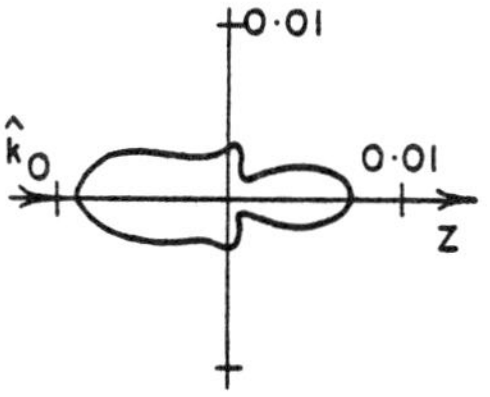

Fig. 8.b. Lossy solid.

Fig. 8. Polar plots of the differential scattering cross sections for homogeneous spheroids with $k_0a_1 = 1.0$.

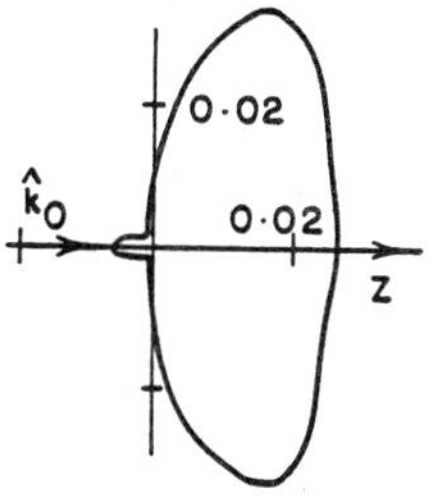

Fig. 9.a. Loss less solid.

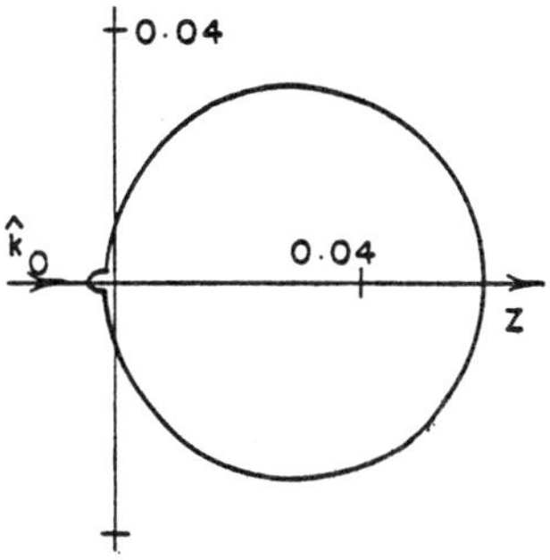

Fig. 9.b. Lossy solid.

Fig. 9. Polar plots of the differential scattering cross sections for homogeneous spheroids with $k_0a_1 = 3.0$.

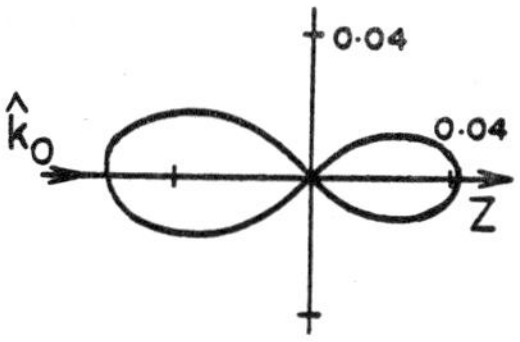

Fig. 10. Polar plot of differential scattering cross section for a layered spheroid with $a_2/a_1 = 0.25$ and $k_0a_1 = 1.0$.

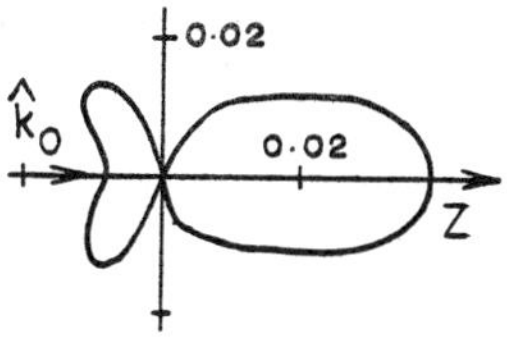

Fig. 11. Similar to Fig. 10 but here $k_0a_1 = 3.0$.

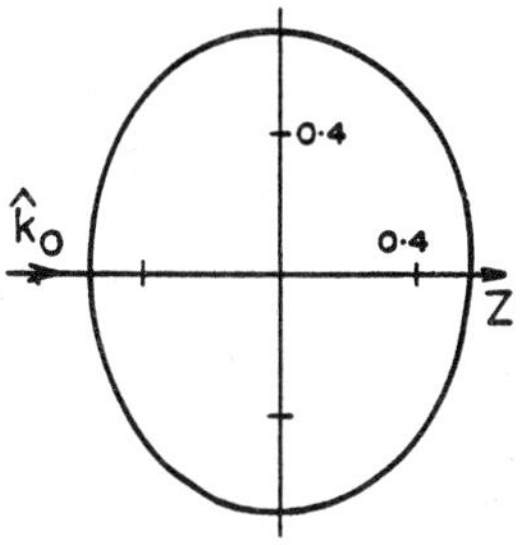

Fig. 12. Polar plot of differential scattering cross section for a layered spheroid with $a_2/a_1 = 0.90$ and $k_0a_1 = 1.0$.

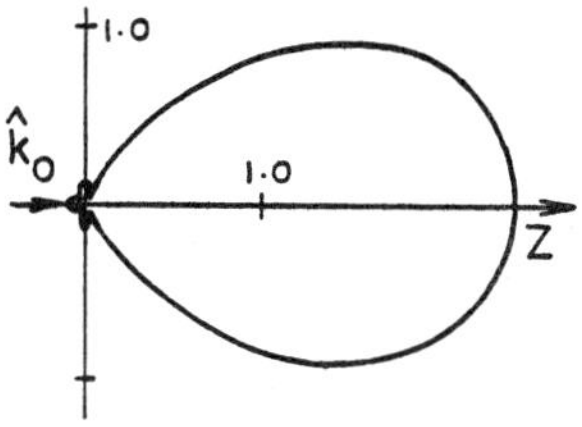

Fig. 13. Similar to Fig. 12 but here $k_0a_1 = 3.0$.

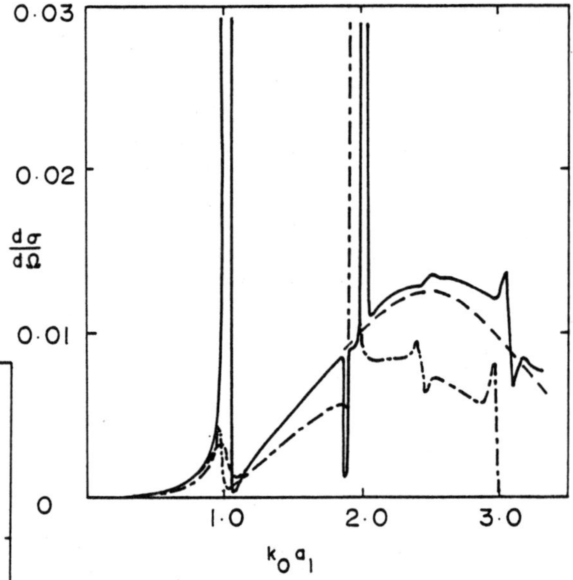

Fig. 14. Differential scattering cross section for spheroids for $\alpha = 0$, $\beta = 0$, $\theta = 90$ and $\phi = 0$; ——— for $a_2/a_1 = 0.0$ loss less solid; - - - for $a_2/a_1 = 0.0$ lossy solid; -•-•- for $a_2/a_1 = 0.25$ loss less solid.

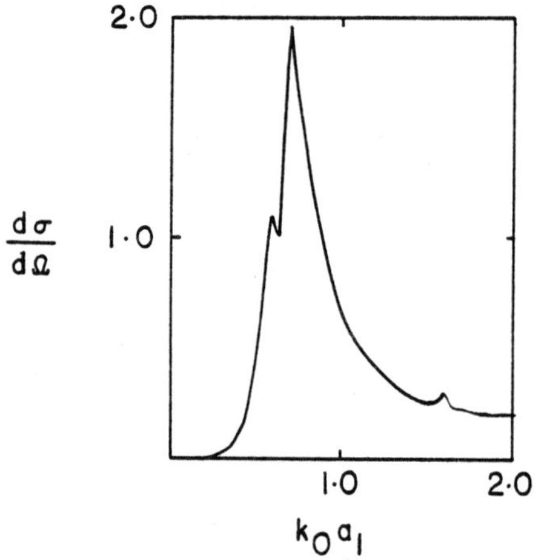

Fig. 15. Similar to Fig. 14 but here only for $a_2/a_1 = 0.90$ loss less solid.

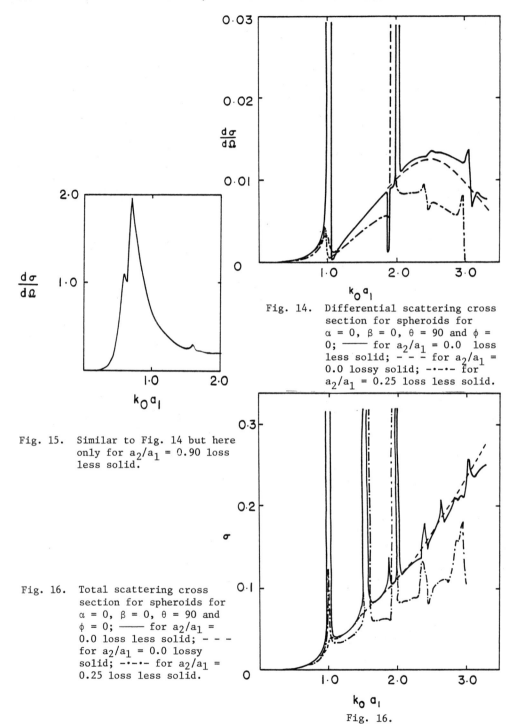

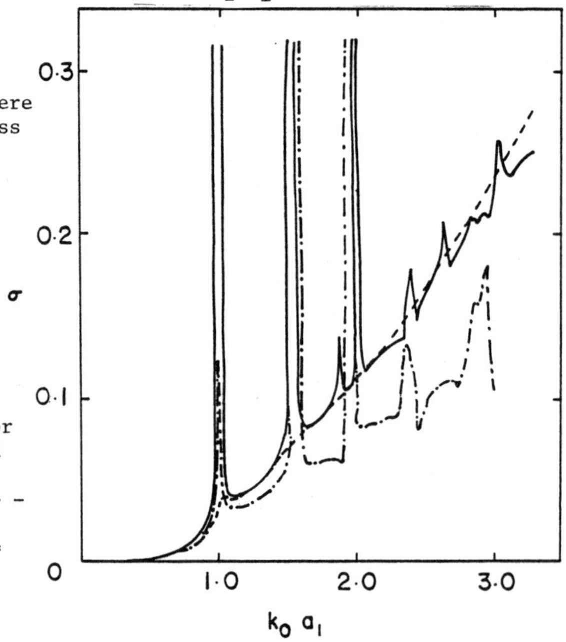

Fig. 16.

Fig. 16. Total scattering cross section for spheroids for $\alpha = 0$, $\beta = 0$, $\theta = 90$ and $\phi = 0$; ——— for $a_2/a_1 =$ 0.0 loss less solid; - - - for $a_2/a_1 = 0.0$ lossy solid; -•-•- for $a_2/a_1 =$ 0.25 loss less solid.

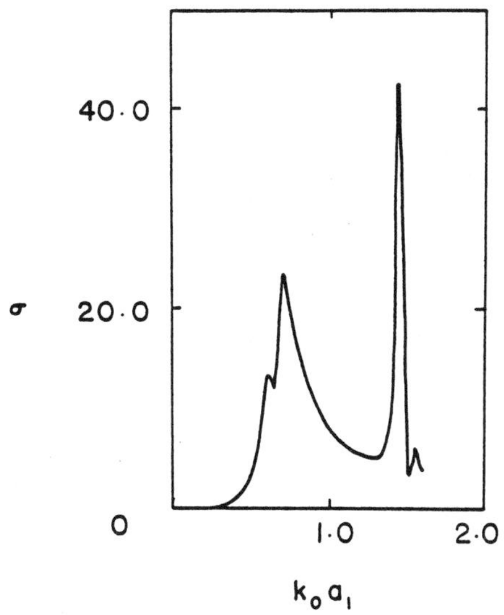

Fig. 17. Similar to Fig. 16 but here only for a_2/a_1 = 0.90 and a loss less solid.

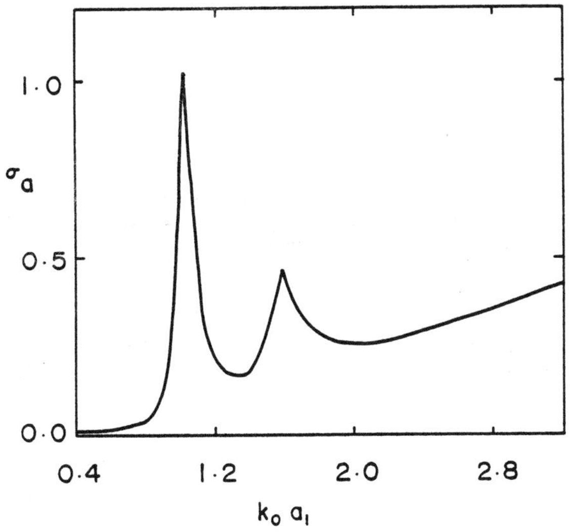

Fig. 18. Absorption cross section for a homogeneous spheroid for α = 0, β = 0 and a lossy solid.

Figures 8-18 are plots of the cross sections, all normalized with respect to a_1^2, for a fictitious solid. The properties of this solid are given in Table 1. Figures 8-13 are polar plots. Figures 14-18 give the cross sections as functions of k_0a_1. Figure captions give details of the individual plots. The spectra indicate that the very sharp resonances for a loss less solid almost completely disappear with the introduction of frequency dependent complex wave numbers in the solid. To obtain these sharp peaks we used a step size of 0.025 in k_0a_1.

The method used in this paper is to our knowledge the only one working for the problems treated here. This method shares with many other methods (in other areas) the lack of a rigorous convergence proof. The parameters we have to play with in order to obtain numerical convergence are: the number of integration points and the dimension of the matrices in the finite truncation. We have made computations using the Gauss elimination and numerically tested the relations (9.1) and (9.2) for a loss less solid.

$$T^{\dagger} = T \tag{9.1}$$

$$T^{\dagger}T = - \text{Real part of } T \tag{9.2}$$

Here, $\dagger$ means the Hermite conjugate. The discrepancy from these relations was measured relative to the maximum element of the T-matrix. We also use the ratio between the absorption cross section and the total scattering cross section as a convergence measure for the loss less solid. For homogeneous bodies our measures were usually better than 10^{-3}. However, for the layered bodies they were about $2\cdot10^{-2}$. In one exceptional case, near a peak, we got as bad a measure as 0.2. It is clear that some matrices will become more illconditioned for wave numbers close to peaks. It is also clear that matrices with the combination OuOu are especially difficult to obtain with good numerical accuracy.

ACKNOWLEDGEMENTS

This research was supported in part by the Center for Advanced NDE operated by the Science Center, Rockwell International for the Advanced Research Projects Agency and the Air Force Materials Laboratory under Contract No. F3361-74-C-5180. V.V. Varadan and V.K. Varadan were also supported by the U.S. Office of Naval Research under Contract No. N00014-78-C-0559. B.A. Peterson was supported in part by a post doctoral fellowship from the Graduate School and the Department of Engineering Mechanics of The Ohio State University. Use of the Instructional and Research Computer Center at OSU are gratefully acknowledged. Helpful discussions with Professor Staffan Ström and Mr. Anders Boström are gratefully acknowledged. We are also grateful to Mr. Howard Tamashiro for careful reading of the manuscript.

REFERENCES

(1) P.C. Waterman, New formulation of acoustic scattering, J. Acoust. Soc. Am. 45, 1417 (1969).

(2) D.N. Pattanyak and E. Wolf, Scattering states and bound states as solutions of the Schrödinger equation with nonlocal boundary conditions, Phys. Rev. D13, 913 (1976).

(3) D.N. Pattanyak and E. Wolf, Resonance states as solutions of the Schrödinger equation with a nonlocal boundary condition, Phys. Rev. D13, 2287 (1976).

(4) S. Ström, Quantum-mechanical scattering from an assembly of nonoverlapping potentials, Phys. Rev. D13, 3485 (1976).

(5) P.C. Waterman, Symmetry, unitarity, and geometry in electromagnetic scattering, Phys. Rev. D3, 825 (1971).

(6) B. Peterson, Matrix formulation of static field problems involving an arbitrary number of bodies, Report no. 75-12, Inst. Theoretical Physics, S-41296, Göteborg.

(7) P.C. Waterman, Matrix theory of elastic wave scattering. I, J. Acoust. Soc. Am. 60, 567 (1976).

(8) P.C. Waterman, Matrix theory of elastic wave scattering. II. A new conservation law, J. Acoust. Soc. Am., 63, 1320 (1978).

(9) Y-H. Pao and V. Varatharajulu (Varadan), Huygens' principle, radiation conditions, and integral formulas for the scattering of elastic waves, J. Acoust. Soc. Am. 59, 1361 (1976).

(10) V. Varatharajulu (Varadan) and Y-H. Pao, Scattering matrix for elastic waves. I. Theory, J. Acoust. Soc. Am., 60, 556 (1976).

(11) V. Varatharajulu (Varadan), Reciprocity relations and forward amplitude theorems for elastic waves, J. Math. Phys. 18, 537 (1977).

(12) V.V. Varadan and V.K. Varadan, Scattering matrix for elastic waves. III. Application to spheroids, J. Acoust. Soc. Am. 65, 896 (1979).

(13) V.V. Varadan and V.K. Varadan, Low-frequency expansions for acoustic wave scattering using Waterman's T-matrix method, J. Acoust. Soc. Am. To appear.

(14) A. Boström, Scattering by a smooth elastic obstacle, Report no. 79-14, Inst. Theoretical Physics, S-41296, Göteborg.

(15) A. Boström, Scattering of stationary acoustic waves by an elastic obstacle immersed in a fluid, Report no. 78-43, Inst. Theoretical Physics, S-41296, Göteborg.

(16) B. Peterson, V.V. Varadan and V.K. Varadan, Scattering of acoustic waves by elastic and viscoelastic obstacles immersed in a fluid, Int. J. Wave Motion, in press.

(17) P. Barber, these proceedings.

(18) V.N. Bringi and T.A. Seliga, Scattering from axisymmetric dielectrics or perfect conductors imbedded in an axisymmetric dielectric, IEEE Trans. Ant. Prop. 25, 575 (1977).

(19) A. Boström, Multiple-scattering of elastic waves by bounded obstacles, Report no. 79-4, Inst. Theoretical Physics, S-41296 Göteborg.

(20) K. Aydin, Electromagnetic scattering from two layered dielectrics, Ph.D. Thesis (1978) Middle East Technical University, Turkey.

(21) B. Peterson and S. Ström, Matrix formulation of acoustic scattering from multilayered scatterers, J. Acoust. Soc. Am. 57, 2 (1975).

(22) B. Peterson and S. Ström, T-matrix formulation of electromagnetic scattering from multilayered scatterers, Phys. Rev. D10, 2670 (1974).

(23) B. Peterson and S. Ström, T-matrix for electromagnetic scattering from an arbitrary number of scatterers and representations of E(3), Phys. Rev. D8, 3661 (1973).

(24) B. Peterson and S. Ström, Matrix formulation of acoustic scattering from an arbitrary number of scatterers, J. Acoust. Soc. Am. 56, 771 (1974).

(25) S. Ström, T-matrix for electromagnetic scattering from an arbitrary number of scatterers with continuously varying electromagnetic properties, Phys. Rev. D10, 2685 (1974).

(26) B. Peterson, Multiple scattering of waves by an arbitrary lattice, Phys. Rev. A16, 1363 (1977).

(27) V.K. Varadan, V.N. Bringi and V.V. Varadan, Coherent electromagnetic wave propagation through randomly distributed dielectric scatterers, Phys. Rev. D19, 2480 (1979).

(28) R.F. Millar, The Rayleigh hypothesis and a related least-squares solution to scattering problems for periodic surfaces and other scatterers, Radio. Sci. 8, 785 (1973).

(29) C. Müller, Boundary values and diffraction problems, Symposia Mathematica, XVIII, 353, Academic Press (1976).

(30) A.P. Calderon, The multipole expansion of radiation fields, J. Rat. Mech. Anal. 3, 523 (1954).

(31) A. Hizal, these proceedings.

(32) E. Gerjuoy and D.S. Saxon, Variational principles for acoustic field, Phys. Rev. 94, 1445 (1954).

(33) E. Gerjuoy and D.S. Saxon, Tensor Scattering matrix for the electromagnetic field, Phys. Rev. 100, 1771 (1955).

(34) R.G. Newton, Scattering theory of waves and particles, McGraw-Hill, New York, (1966).

(35) P.C. Waterman, Computer techniques for electromagnetics Vol. 7. Edited by R. Mittra, Pergamon Press, Oxford & New York (1973).

Part 3

Computational Aspects of T-Matrix
Comparison with Other Methods

THE FREDHOLM INTEGRAL EQUATION METHOD AND COMPARISON WITH THE T-MATRIX APPROACH

A.R. Holt
Department of Mathematics, University of Essex,
Wivenhoe Park, Colchester, CO4 3SQ, U.K.

ABSTRACT

The development and basic theory of the Fredholm integral equation approach to the solution of scattering problems is outlined and compared with the T-matrix method. The specific application to the scattering of electromagnetic waves by dielectric spheroids is considered in detail in both its theoretical and computational aspects.

INTRODUCTION

In this paper we are reviewing the Fredholm integral equation method (FIM) and attempting to bring into focus the differences and similarities between it and the T matrix method. In many ways they will be seen to be complementary, though in other ways they have similar problems.

It is appropriate firstly to consider the way in which the FIM was developed. It was developed in the context of atomic and molecular collision processes. After an extensive study of the second Born approximation it was realised from a paper of Reinhardt (Ref. 1) that such a method might be used to obtain accurate scattering cross-sections at intermediate impact energies. As a first step in this process the method was applied to the problem of obtaining phase-shifts for scattering by a (scalar) central potential - an integral equation in one variable (Ref. 2). In the next stage we considered the integral equation for the scattering amplitude for the same problem - an integral equation in one vector variable (Ref. 3). Again the method was successful. Next it was applied to the simple model of a collinear atom-molecule inelastic collision (Ref. 4) This results in an integral equation in two variables in which an infinite set of coupled channels occurs, some being "open", but most being "closed". Once again the method was successful though it was clear that computer core size was a real constraint.

The application to electromagnetic scattering by dielectric particles came about almost by accident, and was the result of an enquiry from the experimental propagation group in the University of Essex. On investigation the method seemed appropriate to apply since it had already been successfully applied to the inelastic scattering problem involving an infinite set of coupled equations, and only three coupled equations occur for electromagnetic wave scattering.

It is obvious then that the context of the development of FIM was entirely different to that of the T-matrix method which was developed totally within the context of electromagnetic wave scattering.

This paper is divided into three sections. In the first we shall look at the basic FIM theory and compare its salient points with those of the T-matrix theory. In the second section we look at the specific application of the FIM to scattering by spheroids. In the last section we discuss the computational aspects of FIM applied to scattering by spheroids and compare them with the T-matrix application.

Information on the T-matrix theory has been drawn in the main from articles by Barber and Yeh (Ref. 5), Barber (Ref. 6), Peterson (Ref. 7) and Warner (Ref. 8).

BASIC THEORY

Our starting point in the Fredholm Integral equation method is the (volume) integral equation for the electromagnetic field describing the scattering of a plane wave, of wave vector $\underline{k}_o$, by a scatterer of dielectric $\varepsilon(\underline{r})$ and volume V. We shall use dyadic notation (distinguished by script type). Then (cf Ref. 9 P. 102)

$$\mathcal{E}(\underline{r}) = \mathcal{J}_o \exp(i\,\underline{k}_o.\underline{r}) + \int_V \mathcal{G}(\underline{r},\underline{r}')\cdot\gamma(\underline{r}')\,\mathcal{E}(\underline{r}')\,d\underline{r}' \tag{1}$$

$$\text{where } \mathcal{G}(\underline{r},\underline{r}') \doteq \left(\mathcal{I} + \frac{1}{k_o^2}\,\underline{\nabla}\,\underline{\nabla}\right)\frac{\exp(i\,k_o|\underline{r}-\underline{r}'|)}{4\pi|\underline{r}-\underline{r}'|} \tag{2}$$

$$\gamma(\underline{r}) = k_o^2\,(\varepsilon(\underline{r}) - 1) \tag{3}$$

$$\mathcal{J}_\lambda = \mathcal{I} - \hat{\underline{k}}_\lambda\hat{\underline{k}}_\lambda \quad \text{(for any subscript } \lambda) \tag{4}$$

and $\mathcal{I}$ denotes the unit tensor

For simplicity we have assumed the dielectric constant is a scalar, though this is not necessary.

The dyadic scattering amplitude $\mathit{f}(\underline{k}_s, \underline{k}_o)$ for scattering into direction $\underline{k}_s$ is defined by

$$\mathcal{E}(\underline{r}) \underset{r\to\infty}{\sim} \mathcal{J}_i \exp(i\,\underline{k}_o.\underline{r}) + \frac{\exp(ik_o r)}{r}\,\mathit{f}(\underline{k}_s,\underline{k}_o) + O(r^{-2}) \tag{5}$$

Considering the asymptotic form of (1) gives

$$\mathit{f}(\underline{k}_s,\underline{k}_o) = \frac{1}{4\pi}\,\mathcal{J}_s\cdot\int_V \exp(-i\,\underline{k}_s.\underline{r})\,\gamma(\underline{r})\,\mathcal{E}(\underline{r})d\underline{r} \tag{6}$$

We thus see that the scattering parameters can be determined from knowledge of the field inside the scatterer only. However, the field integral equation has a singular Green's function which makes calculations on equation (1) extremely difficult to perform. To circumvent this singularity we iterate the integral equation and integrate, dealing with the singularity analytically.

Hence, we premultiply (1) by $\exp(-i\underline{\kappa}.\underline{r})\,\gamma(\underline{r})$, (where $\underline{\kappa}$ is presently arbitrary) and integrate throughout the scatterer, formally obtaining

$$<\kappa|\gamma|\mathcal{E}> = <\kappa|\gamma|\mathcal{E}_o> + <\kappa|\gamma\,\mathcal{G}\,\gamma|\mathcal{E}> \tag{7}$$

where $\mathcal{E}_o$ denotes the incident field. Since the integrations in (7) are volume integrations throughout the scatterer, only the interior field is involved in (7).

All information on the exterior field has been lost (though it may be regained by substituting the interior field in (1)). A solution to (7) may therefore not be unique as far as the exterior region is concerned, but it will be unique inside the scatterer. We may thus assume that the interior field is Fourier transformable and expand the interior field as

$$E_i(\underline{r}) = \int C(\underline{k},\underline{k}_o)\ \exp(i\ \underline{k}.\underline{r})\ d\underline{k} \tag{8}$$

noting that this will coincide with $E(\underline{r})$ inside the scatterer, but not outside the scatterer.

Substituting (8) into (7) and (6) gives

$$\int d\underline{k}\ K\ (\underline{\kappa},\underline{k})\cdot C(\underline{k},\underline{k}_o) = J_o\ U(\underline{\kappa},\underline{k}) \tag{9}$$

$$\delta(\underline{k}_s,\underline{k}_o) = \frac{1}{4\pi} J_s\cdot \int d\underline{k}\ U(\underline{k}_s,\underline{k})C(\underline{k},\underline{k}_o) \tag{10}$$

$$\text{where } U(\underline{\kappa},\underline{k}) = \langle\ \kappa|\gamma|k_o\rangle \tag{11}$$

$$\text{and } K\ (\underline{\kappa},\underline{k}) = \langle\ \kappa|\gamma\ 1 - \gamma\ G\ \gamma|k\ \rangle \tag{12}$$

Equation (9) is thus a Fredholm integral equation of the first kind for the transform of the interior field.

We solve this by a Galerkin procedure - that is, we approximate the integrations by numerical quadrature, which converts the integral equation (9) into a matrix equation. The arbitrary $\underline{\kappa}$ is now restricted to having as many values as there are integration pivots, and indeed we choose those values of $\underline{\kappa}$ to be identical with the pivots. Treating (10) in the same way, with the same pivots, results in the algebraic linear equations

$$K\ C = b$$

$$\delta = d^T\ C \tag{13}$$

where T denotes matrix transpose. The determination of the scattering amplitude is thus straight forward.

There are two important points to note. Firstly, K is a non-singular kernel. Secondly, although the solutions to Fredholm equations of the first kind can be unstable, it can be shown (Ref. 10) that the scattering amplitude obtained from solving equations (13) satisfies the Schwinger variational principle - hence first order error in the transform results only in second order error in the amplitude. Furthermore the Schwinger method is known to be convergent (with the number of quadrature pivots) to the exact solution (Ref. 11), and hence the FIM is a convergent method, converging to the exact results, and is numerically stable.

We now need to compare this basic theory with that of the T-matrix method. We summarise this comparison in Table 1. It is clear that the methods are entirely different and in many ways complementary; they are certainly independent of each other. If the FIM provides nothing else, it certainly provides a completely independent check on T-matrix calculations.

The main, and very real, limitation on the FIM is the calculation of the K matrix elements (via equation (12)). The U matrix elements (which are first Born elements, as appear in Rayleigh-Gans theory) need to be relatively simple if the K matrix elements are to be evaluated analytically. This limits the application to a limited number of bodies - infinite cylinders of elliptic or rectangular

TABLE 1 Comparison of basic FIM and T-Matrix methods

Aspect of Method	FIM	T-Matrix
Formulation	Volume integral equation for electric field	Surface integral equation and Huygens principle
Scattering Parameters determined from	Internal Field via integration	External Field via asymptotic form
Part of Field removed from calculation	External Field	Internal Field *
Surface enters calculation	Implicitly through volume integrals	Explicitly via surface integrals - implicitly matching occurs on surface
Expansion in terms of	Fourier transform variable	Position space variable
Numerical Stability	Theoretically stable practically : instabilities have not revealed themselves	Instability occurs when ka is increased too far (Ref. 6)
Easily Adaptable to various shapes	No	Yes (?)

* Alternative derivation in which the external field is removed has recently been given by Morita (Ref. 12) for cylinders.

cross-section, spheroids and ellipsoids, and finite circular cylinders. On the other hand it would seem probable that at least for spheroids,(of maximum dimension Δ) the FIM can be used at values of $k_o\Delta$ greater than those the T-matrix method can treat.

The application of the FIM is dependent somewhat on the shape. In the basic theory given above we have dealt with all integrations by numerical quadrature, and this was the way in which the method was first applied to spheroids and ellipsoids (Ref. 13). However in a later modification of the method we have shown (Ref. 14) how a great improvement may be effected by expressing the azimuthal dependence in terms of an exponential Fourier series when dealing with scattering by spheroids. This leads to a dramatic reduction in computer time and in core requirements. It is this modification which we now describe in detail.

FIM APPLIED TO SCATTERING BY HOMOGENEOUS SPHEROIDS

We shall assume the axis of symmetry of the spheroid to be the z axis, its principal semi-axes to be of length a,a,c, and the incident wave to lie in the $\phi = 0$ plane. Thus the incident wave vector is

$$\underline{k}_o = k_o\,(\sin\theta,\ 0,\ \cos\theta) \tag{14}$$

We define vertical and horizontal polarisations as being in, and perpendicular to, the plane formed by the incident direction and the scatterer axis respectively.

Thus $\hat{\underline{e}}_V = (\cos\theta,\ 0,\ -\sin\theta)$

$$\hat{\underline{e}}_H = (0,\ -1,\ 0) \tag{15}$$

Note that $J_i \cdot \hat{\underline{e}}_{\substack{V\\H}} = \hat{\underline{e}}_{\substack{V\\H}}$ (16)

For a homogeneous body, the wave velocity in the scatterer will be constant and hence the three-fold integration in (8) reduces to a two-fold integral over the angle variables θ_k, ϕ_k ; only $k = k_o n_o$ contributes to the radial integration, where

$$n_o^2 = \varepsilon - 1 \tag{17}$$

For the sake of simplicity, we shall assume an incident polarisation $\hat{\underline{e}}$ and write

$$\underline{c}\ (x,\ \phi) = C(\underline{k},\ \underline{k}_o)\cdot \hat{\underline{e}} \tag{18}$$

where $x = \cos\theta_k$

On reducing the x integration by N-point quadrature (pivots and weights $\{x_i, W_i \mid i = 1,\dots,N\}$), equations (9) and (10) become

$$\sum_{j=1}^{N} W_j \int_0^{2\pi} d\phi_2\ K(\underline{k}_{i1},\ \underline{k}_{j2})\underline{c}(x_j,\ \phi_2) = U(\underline{k}_{i1},\ \underline{k}_o)\hat{\underline{e}} \quad (i=1,\ \dots .N) \tag{19}$$

$$\underline{f}(\underline{k}_s,\ \underline{k}_o) = \frac{1}{4\pi} \sum_{j=1}^{N} W_j\ J_s \cdot \int_0^{2\pi} d\phi_2\ U(\underline{k}_s,\ \underline{k}_{j2})c(x_j,\ \phi_2) \tag{20}$$

where $\underline{k}_{i\ell}$ denotes a vector of magnitude $k_o n_o$ in the direction with polar angles $(\arccos x_i,\ \phi_\ell)$, and $\underline{f} = \mathcal{f}\cdot\hat{\underline{e}}$

$\underline{c}$ is a vector whose components describe the x,y,z components of the field. It transpires, however, that there is a great advantage in considering, rather, the combinations $\frac{1}{\sqrt{2}}\ (E_x \pm iE_y)$. Thus writing

$$A = \frac{1}{\sqrt{2}} \begin{pmatrix} 1 & 1 & 0 \\ -i & i & 0 \\ 0 & 0 & \sqrt{2} \end{pmatrix} \tag{21}$$

we write $\underline{c} = A\ \underline{d}$

$$L = A^{-1}\ K\ A$$

$$\underline{h} = A^{-1}\ \underline{e} \tag{22}$$

We now expand $\underline{d}$ as an exponential Fourier series in ϕ

$$W_j\underline{d}\ (x_j, \phi) = \sum_{s=-S}^{+S} \underline{d}_s(x_j)\ \exp\ (si\phi) \tag{23}$$

Transforming (19) according to (22), premultiplying by $\exp(-ir\phi_1)$ and integrating with respect to both ϕ_1 and ϕ_2 over $[0,2\pi]$ yields

$$\sum_{j=1}^{N} \sum_{s=-S}^{+S} L_{rs}(x_i,x_j)\underline{d}_s(x_j) = U_r(x_i,x_o)\underline{\hat{h}} \tag{24}$$

$$\text{where } U_r(x_i,x_o) = \int_0^{2\pi} d\phi \exp(-ir\phi)\, U(k_o n_o, x_i, \phi), \underline{k}_o) \tag{25}$$

The calculation of the L_{rs} matrix element is a central point in the method. It is a heavy piece of analysis, and can only be performed for a few scatterers. For spheroids the calculation follows the lines given in Reference (10) and uses the result

$$\int_0^{2\pi} d\phi \int_0^{2\pi} d\phi_1 \int_0^{2\pi} d\phi_2 \exp[i[n\phi - r\phi_1 + s\phi_2]]\, C_t^1 [\cos(\phi_1-\phi)]\, C_u^1 [\cos(\phi_2-\phi)]$$

$$= \begin{cases} 8\pi^3 \delta_{r,n+s} & |r|+t \text{ even}, |r| \le t \;;\; |s|+u \text{ even}, |s| \le u \\ 0 & \text{otherwise} \end{cases} \tag{26}$$

where C_t^1 denotes the Gegenbauer polynomial.

In fact only three values of s contribute to the sum in (24), and in such a way that the L matrix can be rearranged into block diagonal form, each block being of dimension 3N. The reason for using the transformations (22) is that it minimises both the number of terms contributing to the summation in (24) and also the coupling.

The resultant equations can be arranged in the form

$$\begin{pmatrix} L_{11}^{r+2} & L_{12}^{r+2} & L_{13}^{r+2} \\ (L_{12}^{r+2})^T & L_{11}^{r} & (L_{13}^{r+1})^T \\ (L_{13}^{r+2})^T & L_{13}^{r+1} & L_{33}^{r+1} \end{pmatrix} \begin{pmatrix} d_{1,r+2} \\ d_{2,r} \\ d_{3,r+1} \end{pmatrix} = \begin{pmatrix} U_{r+2}\, h_1 \\ U_r\, h_2 \\ U_{r+1}\, h_3 \end{pmatrix} \tag{27}$$

$$\text{where } \underline{d}_r = (d_{1,r}\,,\, d_{2,r}\,,\, d_{3,r})^T \tag{28}$$

$$U_r = \int_0^{2\pi} d\phi_1\, U(\underline{k}_{i1}, \underline{k}_o) \exp(-ir\phi_1) \tag{29}$$

and $\{L_{pq}^s\}$ are expressible in terms of a set of integrals

$$I_{\substack{1\\2\\3}}(m,n,s,r) = ai \int_0^1 \frac{dx}{X^2}\, j_{m>+1}(k_oX)\, h^{(1)}_{m<+1}(k_oX)\, Q^{t+1}_{m+1}(X)\, Q^{u+1}_{n+1}(X) \begin{bmatrix} 1 \\ x\sqrt{1-x^2} \\ 1-x^2 \end{bmatrix} \tag{30}$$

$$\text{where } X = \sqrt{(a^2 + (c^2 - a^2)x^2)} \tag{31}$$

$$\chi = cx/X \tag{32}$$

and $$Q_{m+1}^{t+1}(\chi) = \left(\frac{(m-t)!(t+1)}{(m+t+2)!}\right)^{\frac{1}{2}} \frac{P_{m+1}^{t+1}(\chi)}{\sqrt{(1-\chi^2)}} \tag{33}$$

by summations such as

$$L_{11}^{r}(x_1,x_2) = \frac{256a^3c^2\gamma^2\pi^3}{k_o} \sum_{\substack{m=|r| \\ m+n \text{ even}}}^{\infty} \sum_{n=|r|}^{\infty} \sum_{\substack{t=|r| \\ t+|r| \text{even}}}^{\infty} \sum_{\substack{u=|r| \\ u+|r| \text{even}}}^{\infty}$$

$$\tilde{j}_{m+1}(K_1)\tilde{j}_{n+1}(K_2)Q_{m+1}^{t+1}(\chi_1)Q_{n+1}^{u+1}(\chi_2)\{I_1(m,n,t,u) - \tfrac{1}{2} I_3(m,n,t,u)\} \tag{34}$$

where $$\tilde{j}_n(K) = (n+\tfrac{1}{2})j_n(K)/K \tag{35}$$

$$K_i = k_o n_o \sqrt{(a^2 + (c^2 - a^2)\chi_i^2)} \qquad (i=1,2) \tag{36}$$

and χ_i is defined similarly to equation (32).

In the above, $j_n, h_n^{(1)}$ denote the spherical Bessel and Hankel functions of order n, and P_m^n the associated Legendre polynomial.

It should be noted that the m,n summations in (34) must be truncated at some value N_o which will be determined by requiring the matrix elements to have no significant contribution from terms with m or n > N_o. Since the spherical Bessel functions diminish rapidly once the argument significantly exceeds the order (Ref. 9 p.70), we may presume that for large size parameters, $N_o \sim k_o n_R \Delta + 3$, where $n_R = \mathrm{Re}(n_o)$.

Notice that the matrix in (27) is symmetric. This is a consequence of the symmetry of the body. Furthermore it can be shown that

$$\begin{aligned} d_{1,-r} &= \pm d_{2,r} \\ d_{3,-r} &= \pm d_{3,r} \end{aligned} \qquad \text{(for all r)} , \tag{37}$$

the + sign relating to V polarisation and the - sign to H polarisation. Consequently we need only solve (27) for r=-1,....S, the cases r=-1, S-1,S being special cases; r=-1 is degenerate, whereas r=S-1,S are of lower dimension since we ignore all values of r > S, and hence ignore some of the equations in (27). It should be noted that increasing the size of S just requires some extra equations to be solved, and some extra matrix elements to be evaluated, rather than a completely new calculation, and hence within one calculation the convergence of the Fourier series can be displayed at minimal cost. This is a significant improvement on the scheme given in reference(10).

The integrals in (29) can be seen to be independent of the refractive index, and only dependent on the two parameters $k_o a$ and c/a.

Applying (22), (23), (37) to (20) gives

(i) Incident V Polarisation

$$\underline{f}(\underline{k}_s,\underline{k}_0) = \frac{1}{4\pi}\sum_{j=1}^{N} J_s \cdot \left[V_0 \begin{pmatrix} \sqrt{2}\, d_{2,0}(x_j) \\ 0 \\ d_{3,0}(x_j) \end{pmatrix} + \sum_{r=1}^{S} V_r \begin{pmatrix} \sqrt{2}[d_{1,r}(x_j)+d_{2,r}(x_j)] \\ 0 \\ 2d_{3,r}(x_j) \end{pmatrix} \right] \tag{38}$$

(ii) Incident H Polarisation

$$\underline{f}(\underline{k}_s,\underline{k}_0) = \frac{i\sqrt{2}}{4\pi}\sum_{j=1}^{N} J_s \cdot \left[\begin{pmatrix} 0 \\ V_0 d_{2,0}(x_j) \\ 0 \end{pmatrix} + \sum_{r=1}^{S} V_r(x_j) \begin{pmatrix} 0 \\ d_{2,r}(x_j)-d_{1,r}(x_j) \\ 0 \end{pmatrix} \right] \tag{39}$$

where $V_r(x_j) = \int_0^{2\pi} \exp(ri\phi_1)\; U(\underline{k}_s,\underline{k}_{ji})d\phi_1$ (40)

COMPUTER IMPLEMENTATION

To implement the theory for spheroids given in the previous section there are three stages in the calculation:-

1. Calculation of the integrals required in evaluating the matrix elements.
2. Calculation of the matrix elements required.
3. Solution of the linear equations and calculation of the scattering amplitudes.

These three stages are performed in separate programs, the intermediate results being written to disk. This organisation has been developed since

(a) If the integrals are written carefully to disk, then increasing N_0 does not require complete recalculation of all the integrals - just the extra integrals have to be evaluated.

(b) If results are required for a particular scatterer and wavelength, but for a range of refractive index, there is no need to calculate the integrals more than once.

(c) Since a number of matrix equations must be solved, there is a large saving in core if the equations are solved one at a time (the elements being read in when needed from the disk.)

(d) One calculation of the elements of the L matrix in (27) suffices for any incident direction and/or polarisation.
There is also a balance required between core size and CPU time, but since a major constraint is that a problem be tractable, it is more important to keep core requirements low. Therefore not all incident directions should be considered simultaneously.

All these arguments have favoured the division of obtaining the solution into three stages. The details of these stages are as follows

1. For a given m,n, the integrals $\{I_{1,2,3}(m,n,s,r)\}$ can mainly be evaluated by recurrence relations based on the recurrence relation for the associated Legendre polynomials

$$2sx\, p_m^s(x) = (1-x^2)^{\frac{1}{2}}[p_m^{s+1}(x) + (m+1-s)(m+s)p_m^{s-1}(x)] \quad (41)$$

The only integrals requiring numerical integration are

$\{I_3(m,n,s,0),\ s=0....m\ ;\ s \text{ even}\}$

$\{I_3(m,n,m,r),\ r=1....n\ ;\ r+m \text{ even}\}$ and $I_1(m,n,0,0)$

For given N_0, there are

$$Y(\ (3Z^2+3Z+2)Y+4Z+2\ )/2 \quad (42)$$

integrals to be evaluated, where $Y = [\frac{N_0}{2}] + 1$, $Z = [\frac{N_0+1}{2}]$, and [t] denotes the largest integer less than or equal to t.

The Spherical Bessel functions are evaluated by a routine which uses either a series expansion, a forward recursion relation, or the Miller algorithm (Ref. 15) depending on the values of the order and the argument. Numerical integration is performed by a variable step-length routine using a Clenshaw-Curtis quadrature and an inbuilt estimate of the absolute error obtained by comparing Newton-Cotes and Romberg estimates (Ref. 16). The program is written in double precision arithmetic and requires 23K (words) core for $N_0 \lesssim 30$. The disk area required for a given N_0 can be calculated from (42) - each integral is a complex number.

2. The calculation of the matrix elements will depend on the values of N_0, N and S. The number of matrix elements to be calculated is $N\{27S+4\cdot5)N +3(S+0\cdot5)\}$, where N will be roughly proportional to $k_0\Delta$ and S will depend both on $k_0\Delta$ and c/a. For example, for large raindrops at 30GHz (a^2c = 0.027, c/a = 0.7), S = 7, N = 15. The time for this stage will depend not only on the number of matrix elements to be calculated, but also on N_0. The core required depends on how much reading is done from the disk. If a third of the integrals are read in at a time, then for $N=N_0=16$, 40K words store is required, whereas for $N=N_0=20$ about 70K words are required. However these figures could be dramatically reduced by only reading the integrals for a given m,n at one time.

3. The third program constructs the matrix equations to be solved from the matrix elements on disk, and by calculating the appropriate right-hand side of (27), and then solves the linear equations one at a time. The scattering amplitudes are then calculated, using the solutions, from (38) or (39). The convergence of the scattering amplitude for different S is automatically contained within the program. For any incident direction, both V and H polarisations are included. Core requirement again depends on N,N_0. For $N=N_0=20$, the core required is 50K.

At the time of writing, program development is not yet completed, but a sample of program run times is given in Table 2. It is anticipated that further development should enable the time required for stage 2 to be halved.

TABLE 2 Examples of FIM spheroid program run lines

Parameters			CPU secs			
N_o	N	S	Stage 1	Stage 2	Stage 3	Total
6	3	4	17	2	4	23
9	5	5	63	14	11	88
11	7	5	124	52	17	193
13	8	6	223	130	28	381
16	14	8	450	283	73	806

The computing limitations on this application appear to be as follows

a) Stage 1. The main limitation is disk storage of the integrals. As will be seen from Table 2, program run times do not increase rapidly with N_o, but disk storage will increase approximately as the fourth power of N_o (cf.42), and for $N_o = 16$ already requires about 450 blocks. Core storage is not a problem, since the integrals are calculated and dumped.

b) Stage 2. As currently written, about one-third of the integrals are required in core and since the number of integrals increase rapidly with N_o, there is clearly a definite limit to the size parameter which can be treated. However this restriction can be lifted easily, at the cost of frequent reading from the disk. Since N is slowly increasing with scatterer size, the restrictions are likely to be more from CPU time than from core size, once the integrals are read from the disk as required.

c) Stage 3. The matrix equation to be solved is of dimension 3N (and being symmetric, not all matrix elements are required). N_o still enters the calculation at present since U_r is calculated from (29) by expanding the integrand in a partial wave expansion. Further development work should improve the situation and N should eventually be able to be increased to around 50 without exceeding 70K words core. The solutions are currently written to disk and read back when the scattering amplitudes are calculated. This stage should be the least demanding as far as CPU time and core storage are concerned.

The implementation uses double-precision arithmetic, but since complex double precision is not available in FORTRAN on the PDP-10 (KL10) machine used, complex variables have had to be treated as pairs of real variables. Consequently the program code is somewhat larger than would otherwise be necessary.

Comparison with T-matrix method implementation

Direct comparison is not easy since different machines have been used for the implementations, and this makes CPU time and even core size requirements impossible to compare. Moreover there have been at least three different implementations of the T-matrix method (Refs. 5,7,8) and the implementations are constantly being improved (cf. Refs. 8,17). However some useful comparisons can be made.

Points of similarity Both implementations consist of three stages: for the T-matrix method these are (Ref. 6) calculation of integrals, solution of linear equations, and evaluations of scattering parameters. Spherical Bessel and Hankel functions, and associated Legendre polynomials have to be calculated in both schemes. Both schemes allow the separation into azimuthmal modes (Ref. 6) and in both an expansion has to be truncated. In both methods the computational complexity increases as $k_o\Delta$ increases and eventually limits their applicability.

Points of difference It should be noted that since the FIM expands in the transform variable whereas T-matrix expands in the position space variable, there is no direct comparison between the "azimuthal" and "elevation" modes of the T-matrix method, and similar quantities in the FIM. In the latter the various azimuthal modes are decoupled and therefore one can examine the convergence of the scattering parameters, with respect to the number of azimuthal modes, within a single calculation. Whether this also holds in the T-matrix method is unclear in the literature; the decoupling is clearly indicated in reference (6) but in reference (7) it would appear that separate calculations were necessary for different numbers of azimuthal modes. The elevation modes are dealt with rather differently. In the T-matrix method the size of the matrix to be solved depends explicitly on the number of modes to be included. In the FIM modes could be said to enter in two ways; firstly through the partial wave expansions, such as occur in equation (34), which are truncated at some upper value N_o, and secondly through the quadrature pivots. The first of these two ways does not affect the number of equations, just the amount of work involved in calculating the matrix elements. The quadrature pivots are not strictly modes, but their number does directly affect the number of equations to be solved. It should be noted that a change of refractive index does require a complete new calculation in the T-matrix method, since the integrals depend on the refractive index (Ref. 6). In the FIM, the most time-consuming part of the calculation does not depend on the refractive index.

One other point to note, since it appears to be important for increasing the size parameter (Ref. 8), is that in the FIM we need the Hankel function of argument $k_o\Delta/2$, not of argument $k_o n_o\Delta/2$ as is needed by the T-matrix method.

Results

To demonstrate the convergence of the scattering amplitudes for a particular case, we give in Table 3 the forward and backward scattering amplitudes for a raindrop at 94GHz. For this scatterer $k_o n_R\Delta = 13.7$ and $f(0)$, $f(\pi/2)$ denote the scattering amplitude for incidence along and perpendicular to the axis of symmetry. For this example, $N_o = 10$.

Certain general comments can be made (i) For incidence along the z axis, the non-zero azimuthmal modes make no contribution to the amplitudes - only the zero-order mode contributes. (ii) In the forward direction, the H polarisation amplitudes converge more slowly than do the V polarisation amplitudes (iii) As should be expected, the forward amplitudes converge more quickly than do the backward amplitudes.

It is worthwhile pointing out, finally, that for scattering by spheroids for $k_o n_R\Delta \lesssim 5$, the forward amplitude at any incident angle θ may be accurately expressed in terms of those for incidence along the principal axes as

$$f(\theta) = f(0)\cos^2\theta + f(\pi/2)\sin^2\theta \qquad (43)$$

This rule, which was empirically deduced from results obtained using the FIM, has been shown to have some theoretical foundation in reference (18). A similar rule also applies to back-scattering, but for a more restricted range of size-parameters.

TABLE 3 Convergence of scattering amplitudes for scattering of electromagnetic waves by a dielectric spheroid at 94GHz

$$(a^2c)^{1/3} = 0.1\text{cm} \quad c/a = 0.9 \quad n_0 = 3.359 + i1.930$$

a) FORWARD Scattering Amplitudes

N	S	f(0)	$f_V(\pi/2)$	$f_H(\pi/2)$
9	3	$1.33^{-2} + i1.51^{-1}$	$2.03^{-2} + i1.39^{-1}$	$1.00^{-2} + i1.47^{-1}$
	4		$2.01^{-2} + i1.39^{-1}$	$9.62^{-3} + i1.48^{-1}$
	5		$2.01^{-2} + i1.39^{-1}$	$9.61^{-3} + i1.48^{-1}$
10	3	$1.32^{-2} + i1.51^{-1}$	$2.03^{-2} + i1.39^{-1}$	$1.00^{-2} + i1.47^{-1}$
	4		$2.01^{-2} + i1.39^{-1}$	$9.52^{-3} + i1.48^{-1}$
	5		$2.01^{-2} + i1.39^{-1}$	$9.51^{-3} + i1.48^{-1}$

b) BACKWARD Scattering amplitudes

N	S	f(0)	$f_V(\pi/2)$	$f_H(\pi/2)$
9	3	$2.10^{-2} - i3.13^{-2}$	$5.73^{-3} - i3.77^{-2}$	$1.20^{-2} - i3.97^{-2}$
	4		$5.55^{-3} - i3.73^{-2}$	$1.15^{-2} - i3.94^{-2}$
	5		$5.55^{-3} - i3.74^{-2}$	$1.15^{-2} - i3.94^{-2}$
10	3	$2.10^{-2} - i3.14^{-2}$	$5.71^{-3} - i3.77^{-2}$	$1.20^{-2} - i3.97^{-2}$
	4		$5.53^{-3} - i3.74^{-2}$	$1.15^{-2} - i3.94^{-2}$
	5		$5.53^{-3} - i3.74^{-2}$	$1.15^{-2} - i3.94^{-2}$

The index gives the power of ten by which the entry is to be multiplied.

CONCLUSION

One point that must be underlined about the FIM is that it is not easily adaptable to different shapes of scatterer. This is because of the complexity of the K matrix elements (equ. (12)). Calculations have been performed on infinite cylinders of elliptic cross-section (Ref. 19) and on general ellipsoids (Ref. 10, 20), and are at present in progress on finite cylinders, and infinite cylinders of rectangular cross-section. It should be noted that the case of scattering from thin finite cylinders has already been treated (Ref. 21). The FIM therefore is suitable for treating a limited number of model shapes, and because of its numerical stability, is likely to be able to deal with larger size parameters than is the T-matrix method. The latter will, however, be very much more suitable for treating bodies of arbitrary shape - such as, for example (Ref. 17), the Pruppacher and Pitter raindrop model (Ref. 22). The study of shape effects in the resonance region for scattering at microwave frequencies does seem a worthwhile study, and one to which simple scatterer models can contribute. There is evidence that in the resonance region shape may not be critical - the Pruppacher and Pitter raindrops scatter very similarly to spheroids (Ref. 17). However, interesting resonance effects have been found (Ref. 23) and there is evidence that slight changes in refractive index can be significant in this region (Ref. 24). This introduces another possible use of the FIM. All the quantum mechanical applications dealt with scattering interactions which were distance dependant in the interaction region, whereas so far the electromagnetic applications have dealt with homogeneous scatterers. The FIM could be extended to deal with Scatterers whose refractive index was either anisotropic and/or a smoothly varying function of position. Such an extension does not appear to be readily available in the T-matrix method.

Acknowledgements

The author wishes to thank Prof. B. L. Moiseiwitsch for a conversation which first suggested the FIM, Dr. Barry Evans for his initial enquiry and subsequent constant encouragement, and Dr. Nikos Uzunoglu with whom all the initial work on electromagnetic wave scattering was developed. All the computational work has been done on the PDP-10 installation at the University of Essex, and the help of the Computer service staff is gratefully acknowledged.

REFERENCES

(1) W. P. Reinhardt and A. Szabo, Fredholm Method I: Numerical Procedure for elastic scattering, Phys Rev A. 1, 1162 (1970).

(2) A. R. Holt and B. Santoso, A Fredholm integral equation method for scattering phase shifts, J. Phys B (Atom. Molec. Phys.) 5, 497 (1972).

(3) A. R. Holt and B. Santoso, The Fredholm integral method II. The calculation of scattering amplitudes for potential scattering J. Phys. B (Atom. Molec. Phys.) 6, 2010 (1973).

(4) G.W.F. Drake and A.R. Holt, Improved quantum calculation of the vibrational excitation of H_2 in collinear collisions with helium, J. Phys. B (Atom. Molec. Phys.) 8, 494 (1975).

(5) P. Barber and C. Yeh, Scattering of electromagnetic waves by arbitrarily shaped dielectric bodies Applied Optics. 14, 2864 (1975).

(6) P. Barber, Resonance electromagnetic absorption by non-spherical dielectric objects IEEE Trans on Microwave Theory and Techniques, MTT-25, 373 (1977).

(7) B. Peterson, Numerical computation of electromagnetic scattering by raindrops. Internal report TMF 76-1, Institute of theoretical physics, Fack, Sweden, (1976).

(8) C. Warner, Calculated scattering characteristics of hailstones at weather radar wavelengths. Internal report, Department of Environmental sciences, University of Virginia, Charlottesville, Va. (1978).

(9) R. G. Newton, Scattering theory of waves and particles (McGraw-Hill, New York), (1966).

(10) A.R. Holt, N. K. Uzunoglu, and B.G. Evans, An integral equation solution to the scattering of electromagnetic radiation by dielectric spheroids and ellipsoids IEEE Trans. Antennas and Propag. AP-26, 706 (1978)

(11) S. R. Singh and A. D. Stauffer, A convergence proof for the Schwinger variational method for the scattering amplitude J. Phys. A, (Math:Gen) 8, 1379 (1975).

(12) N. Morita, Another method of extending the boundary condition for the problem of scattering by dielectric cylinders IEEE Trans. Antennas and Propag AP-27 97 (1979).

(13) N.K. Uzunoglu, B.G. Evans and A. R. Holt, Evaluation of the scattering of an electromagnetic wave from precipitation particles by the use of Fredholm integral equations, Electronic Letters, 12, 312 (1976).

(14) N.K. Uzunoglu, A. R. Holt and B.G. Evans, The calculation of scattering from precipitation particles at millimetre wave frequencies. Abstracts IEE Conference on Antennas and Propagation (IEE, London) 2, 114 (1978).

(15) M. Abramowitz and I. A. Stegun, Handbook of Mathematical Functions (Dover, New York) (1965).

(16) H. O'Hara and F. J. Smith, The evaluation of definite integrals by interval subdivision Comput J 12, 179 (1969).

(17) C. Warner and A. Hizal, Scattering and depolarisation of microwaves by spheroidal raindrops Radio Science, 11, 921 (1976).

(18) A.R. Holt and J.W. Shepherd, Electromagnetic scattering by dielectric spheroids in the forward and backward directions J Phys.A.(Math.Gen) 12, 159 (1979).

(19) N.K. Uzunoglu and A. R. Holt, The scattering of electromagnetic radiation from dielectric scatterers, J. Phys. A.(Math. Gen), 413 (1977).

(20) A. R. Holt, N. K. Uzunoglu, and B.G. Evans, An integral equation solution to the scattering of electromagnetic radiation by dielectric spheroids and ellipsoids Abstracts IEEE/AP-S Symposium (IEEE, New York), 424 (1976).

(21) N. K. Uzunoglu, N. G. Alexopoulos and J. G. Fikioris, Scattering from thin and finite dielectric fibers J. Opt. Soc. Am, 68, 194 (1978).

(22) H. R. Pruppacher and R. L. Pitter, A semi-empirical determination of the shapes of cloud and rain drops J.Atmos. Sci, 28, 86 (1971).

(23) A. R. Holt and B. G. Evans, Some resonance effects in scattering of microwaves by hydrometeors, Proc IEE 124, 1114 (1977).

(24) L. E. Allan and G. C. McCormick, Measurements of the backscatter matrix of dielectric spheroids, IEEE Trans. Antennas and Propag, AP-26, 579 (1978).

METHODS OF OVERCOMING NUMERICAL INSTABILITIES ASSOCIATED WITH THE T-MATRIX METHOD

D. J. N. Wall
Department of Mathematics, University of Dundee, Scotland

§1 INTRODUCTION

In this paper we consider, from a computational viewpoint, a technique applied to the classical direct scattering problem which has been variously called the transition matrix formulation (Ref. 1 - 4), and the null field method (Ref. 5 - 6). We will henceforth call the method the null field method. The problem involves calculating the field scattered from a body of known constitution and location, given the incident field.

In order to clarify the exposition the analysis is restricted to the scattering of scalar waves from a totally reflecting single body. The analysis is therefore applicable to all scalar fields which behave in the same manner as acoustic fields of small amplitude. By this restriction we are able to illustrate all the points we wish to, without any of the analytic complications that occur in other scattering problems. In particular for application of the null field method to scattering problems involving vector waves, penetrable bodies or multiple bodies the reader is referred to other papers in this symposium.

In developing the null field method there are two important steps. The first is to obtain the 'extended integral equation' (e.i.e.); the second is the use of the bilinear expansion, in an appropriate coordinate system, of the free space Green's function. In the methods used prior to Ref. 6 the e.i.e. were satisfied explicitly within a sphere (for three-dimensional problems) or a circle (for two-dimensional problems) inscribed by the body. This was done by utilizing the bilinear expansions appropriate to the spherical polar and the circular cylinder coordinates respectively. It has been found that the system of equations derived from these methods tends to be numerically ill-conditioned when the body has a large aspect ratio. By the term 'aspect ratio' we mean the ratio of the largest dimension to the smallest dimension of the body. By examining the coordinate systems in which the free space Green's function possesses a suitable bilinear expansion, Bates and Wall (Ref. 6) were able to satisfy the e.i.e. explicitly within the spheroid (for three-dimensional problems) or the ellipse (for two-dimensional problems) inscribed by the body, thus overcoming these numerical problems. In §4 we follow the analysis of Bates and Wall (Ref. 6) in deriving the 'general' null field method. We do not, however, give detailed formulae for the various methods - the interested reader is referred to the original paper for these. Wall (Ref. 7) has extended these methods to the electromagnetic case.

One of the desirable features of the null field methods is that they are unique at all frequencies; in §2 we set up the necessary preliminaries and demonstrate this fact. The theoretical explanation for the numerical instability experienced with the 'spherical' and 'circular' null field methods for some body shapes is developed in §3. If the size of the linear system of equations which results from a particular null field method can be kept small, the onset of ill-conditioning for certain scattering body shapes can be reduced. We discuss in §5 ways in which this may be achieved by suitable choice of basis functions representing the unknown function in the equations.

In §6 we illustrate by numerical example how the region in which the e.i.e.'s are explicitly satisfied should be chosen. Although the general null field outlined in §4 enables many bodies to be analysed satisfactorily, one can devise shapes for which none of these methods is particularly suitable. In §6 we illustrate how the method of regularisation may be employed with any of the null field methods to overcome the numerical instability problems. Our approach to the use of the regularisation methods is heuristic; for more rigorous and detailed accounts of these methods refer to Refs. 8 - 10.

§2 THE EXTENDED INTEGRAL EQUATIONS

We let D denote the infinite region lying outside the simple closed surface ∂D of the scattering body. D_- will then be taken to denote the complement of $D \cup \partial D$. Upper case letters P, Q will denote points of D; P_-, Q_- will denote points of D_-, and lower case letters p, q will denote points of ∂D. The unit normal of the surface ∂D at a point P directed from ∂D towards D will be denoted by $\hat{\underline{n}}(p)$. The origin of coordinates 0 is taken at an arbitrary point of D_-. The position vector from 0 to the point P is denoted by $\underline{r}_P$ and its scalar length by $r_P = |\underline{r}_P|$.

The exterior boundary-value problem considered here consists of finding the total field u; the solution of the scalar Helmholtz equation

$$(\nabla^2+k^2)u(P) = -\chi(P) \quad \text{in } D, \tag{1}$$

subject to the radiation condition

$$r_p\left(\frac{\partial u}{\partial r_p} + iku\right) \to 0 \quad \text{as} \quad r_p \to \infty \tag{2}$$

for three-dimensional problems, and boundary conditions on ∂D to be prescribed later. For two-dimensional problems the r_p quantity outside the parenthesis in (2) is replaced by $r_p^{1/2}$. All sources and fields are complex functions of space with the time factor $\exp(i\omega t)$ suppressed.

Application of Green's theorem to (1) together with the corresponding partial differential equation satisfied by the free space Green's function, denoted by g, yields the well known identity

$$\int_D \chi(Q)g(P,Q)d\tau_Q + \int_{\partial D} \left[u(q)\frac{\partial g}{\partial n_q} - g(P,q)\frac{\partial u}{\partial n_q}\right]ds_q = \begin{cases} u(P) & P \in D \\ 0 & P \in D_- \end{cases} \tag{3}$$

Here $d\tau_q$ and ds_q denote, respectively, the volume and surface variables for the integrations. The surface ∂D need not possess a tangent which is a differentiable function of position at all points on ∂D: i.e., ∂D can include edges. The free space Green's function is

$$g(P,Q) = \exp(-ikR)/4\pi R \text{ , in three dimensions}$$

$$= -(i/4)H_0^{(2)}(kR) \text{ , in two dimensions}$$

where $R = |\underline{r}_Q - \underline{r}_P|$ and $H_0^{(2)}$ denotes the Hankel function of the second kind of order zero.

It is convenient for subsequent discussion to split the field into the sum of an incident wave u_{inc} and a scattered wave u_s , where

$$u_{inc} = \int_D \chi(Q)g(P,Q)d\tau_Q \text{ ,}$$

and u_s is given by the remaining integral on the left hand side of (3). Henceforth for ease of presentation we shall discuss only the Neumann and Dirichlet scattering problems, which occur in, for example, acoustic problems involving the scattering from sound-hard or sound-soft bodies respectively.

In the Dirichlet boundary condition case, $u(q) = 0$; by restricting P to lie in D_- , (3) then becomes

$$\int_{\partial D} \frac{\partial u}{\partial n_q} g(P_-,q)ds_q = u_{inc}(P_-) \text{ .} \tag{4}$$

The corresponding equation for the Neumann boundary condition $\frac{\partial u}{\partial n_q} = 0$,is

$$\int_{\partial D} u(q)\frac{\partial g}{\partial n_q}(P_-,q)ds_q = -u_{inc}(P_-) \text{ .} \tag{5}$$

These two integral equations for the unknown surface densities $\partial u/\partial n$ and u have come to be known as the 'extended integral equations' after Waterman (Ref. 1) who derived them via what he called the 'extended boundary condition'. This boundary condition has also been called the 'extinction theorem' by various authors (see Ref. 6).

We will now show that the two equations (4) and (5) are unique for all positive values of the wavenumber k . We note that this is in contrast with the 'conventional' surface integral equations, which can also be obtained from (3) by careful consideration as $P \to \partial D$. These conventional equations can be shown to be non-unique at wavenumbers corresponding to internal resonances of the complementary problem (Refs. 11 - 12).

We consider here the Dirichlet boundary condition; the proof for the Neumann boundary condition follows in a similar manner. As (4) is an integral equation of the first kind the Hilbert-Schmidt theory (Ref. 13) ensures that it has a unique solution provided that zero is not in the spectrum of the integral operator. We can show that this is not the case by the following device.

Suppose the homogeneous form of (4) has a non-trivial solution $\partial u_0/\partial n_q$ so that

$$\int_{\partial D} \frac{\partial u_0}{\partial n_q} g(P_-,q)ds_q = 0 \text{ .} \tag{6}$$

The integral on the left hand side of (6) can then be associated with the complementary problem, i.e. with the solution $U(P_-)$ of the homogeneous Helmholtz equation in D_- with Dirichlet boundary conditions. Therefore if (6) is satisfied $\forall\ P_- \in D_{null}$, where $D_{null} \in D_-$ and D_{null} has non-zero volume, it follows that

$$U(P_-) \equiv 0 \quad \forall \quad P_- \in D_- ,$$

and hence $\partial U/\partial n_q = 0$. This contradicts the original assumption; hence (4) is unique. Except for a few special geometries and incident field configurations D_{null} must have a non-zero volume to ensure uniqueness (Ref. 14).

§3 SINGULAR FUNCTION ANALYSIS OF THE E.I.E.

The e.i.e.'s are integral equations of the first kind which are well known to be ill-posed. By this we mean that the solution of the equation does not depend continuously on the right hand side u_{inc}. We should therefore expect ill-conditioned linear systems from any numerical method of solution of these equations. However this is not always the case and, as shown in §6, provided D_{null} is chosen to be 'near' ∂D, the linear system is well-conditioned.

To explain the numerical ill-conditioning of our equations and to elucidate one of the methods which we propose for overcoming this problem, we present some of the relevant theory of singular functions. More detailed accounts can be found in Refs. 15, 13, and 10.

For the purposes of this section we find it convenient to rewrite equations (4) and (5) as

$$\int_{\partial D} K(P_-,q)f(q)ds_q = h(P_-) \tag{8}$$

Here the respective kernels g and $\partial g/\partial n$ in (4) and (5) are denoted by $K(P_-,q)$ and the right hand side in both equations is denoted by h. The unknown quantity f denotes the surface source density in (4) and (5). We use K to denote the Hilbert-Schmidt integral operator in (8), which may then be written in the shorthand form $Kf = h$. Let us consider the self adjoint operators KK^* and K^*K where K^* is the adjoint operator of K defined by

$$K^*h = \int_{D_{null}} \overline{K(P,q)}h(P)d\tau_P$$

and the superbar denotes the complex conjugate. Then orthogonal sequences $\{u_j\}$ and $\{v_j\}$ of eigenfunctions can be defined for these self adjoint operators; viz.

$$KK^*u_j = \kappa_j^2 u_j \ , \quad K^*Kv_j = \kappa_j^2 v_j \ ,$$

where κ_j^2 are the eigenvalues. The u_j and v_j are called the singular functions, and the κ_j the singular values of the operator K. They are useful in the solution of (8) because of the properties

$$Kv_j = \kappa_j u_j \ , \quad K^* u_j = \kappa_j v_j \ .$$

Use of this result suggests that a formal solution of (8) is

$$f(q) = \sum_{j=1}^{\infty} \left(\frac{h_j}{\kappa_j}\right) v_j(q) \ , \tag{9}$$

where

$$h_j = \langle h, u_j \rangle = \int_{D_{null}} h(P)\overline{u_j(P)}\, d\tau_P \tag{10}$$

and $\langle .,. \rangle$ is used to denote the inner product. Necessary and sufficient conditions for (9) to be a solution can be found in Ref. 15. The uniqueness of the solution as asserted in §2 is assured, provided that $Kf = 0$ has no non-trivial solution.

The instability of the integral equations of the first kind can now be demonstrated by the following argument. As the sequence of singular values $\{\kappa_j\}$ has zero as its only limit point we may choose a j so that κ_j is as small as we like. Therefore if h_j is perturbed by δh_j then f is changed by $\delta f = (\delta h_j/\kappa_j)v_j$; hence $\|\delta f\| / \|\delta h\|$ can be made arbitrarily large. We can look upon δh_j as an error term resulting from the numerical process of 'solving' the e.i.e.

§4 THE NULL FIELD EQUATIONS

To convert the e.i.e.'s into an algebraic system of equations we must try to satisfy them throughout the volume D_{null} . Also, in the light of the analysis in §3, we might ask 'What is the best choice of D_{null} to offset the effects of numerical ill-conditioning?'. Before attempting to answer these questions we need to have available partial wave solutions of the homogeneous Helmholtz equation in various coordinate systems. These are: for three dimensional problems, the prolate and oblate spheroidal coordinates together with the spherical polar coordinates, and for two dimensional problems, the elliptic and circular polar cylinder coordinates. Following Waterman's notation we choose to denote the set of outgoing partial wave solutions in the aforementioned coordinate systems by

$$\{\psi_n(P,k) \ , n = 1, 2, \ldots\}$$

for an arbitrary point P . The various indices associated with the wave functions have been reordered into a single index for simplicity. The detailed functional form of these wave functions may be found in Refs. 16 and 6. For the special case of spherical polar coordinates we shall examine the detailed form subsequently. Together with the outgoing partial waves, which must be singular at the origin, we require the set of partial wave solutions $\{\mathrm{Reg}\,\psi_n(P,k)\}$ which are regular at the origin. The other property of these wave functions necessary for our later development is their orthogonality on the closed surface formed when the radial type coordinate, of the appropriate coordinate system, is kept constant while the other two coordinates vary over their ranges. For example, in the spherical polar coordinate system these surfaces are spheres and in the spheroidal coordinate systems these surfaces are spheroids. One such surface is crucial to our development. This surface is inscribed by ∂D and as D_{null} is to be defined as its interior, we will denote it by ∂D_{null} .

The method we choose for solution of the e.i.e.'s is the general Galerkin method (method of moments). In order to ensure the uniqueness of the e.i.e.'s verified in §2 Waterman introduced a systematic way of reducing them to an infinite set of algebraic equations. However to relate development of the null field equations to the conventional approach taken with the numerical solution of integral equations we modify Waterman's procedure slightly.

We briefly review here Galerkin's method as it applies to the e.i.e.'s. The operator form of (8) is used to denote both (4) and (5). We attempt to find an approximation to f of the form

$$f^N = \sum_{n=1}^{N} \beta_n f_n$$

where the f_n form a spanning set in the Hilbert space $L_2(\partial D)$, the domain of K . We use this expansion in (8) and demand that both sides of this equation have the same projection on the n-dimensional subspace spanned by the test functions ω_m . This yields the linear system of equations $A\underline{\beta} = \underline{h}$, solution of which determines the coefficients β_n where $A_{mn} = \langle\omega_m, Kf_n\rangle$, $\underline{h} = \langle\omega_m, h\rangle$. The ω_m should span the Hilbert space $L_2(D_{null})$ with the inner product as defined in (10).

For the testing functions we choose the set $\{\text{Reg}\,\psi_n\}$. By noting that for all $P \in D_{null}$ the free space Green's function g has a bilinear expansion of the form $g(P_-,q) = \sum_n c_n \psi_n(q,k)\,\text{Reg}\,\psi_n(P_-,k)$, where the c_n are normalising constants, we observe that because of the orthogonality properties of wave functions the testing inner products defined on D_{null} can be performed analytically. The detailed form of the bilinear expansion for g in the various coordinate systems may be found in Ref. 16. The equation (8) then becomes

$$\sum_{n=1}^{N} \beta_n \int_{\partial D} f_n(q)\Psi_m(q,k)ds_q = a_m , \quad m = 0, 1, \ldots \tag{10}$$

where $\Psi_m = \psi_m$ for the Dirichlet boundary condition and $\Psi_m = -\partial\psi_m/\partial n_q$ for the Neumann boundary condition. In (10) the

$$a_m = \int_D \chi(Q)\psi_m(Q,k)d\tau_Q \tag{11}$$

are known functions. The equations (10) are what we call the general null field equations. Substitution of the specific form of the wavefunctions ψ_m in (10) and (11) results in a particular null field method: e.g. by use of the oblate spheroidal wavefunctions, (10) becomes what we call the oblate spheroidal null field method. When specific forms are used for the wavefunctions the e.i.e.'s are satisfied in a D_{null} appropriate for the coordinates chosen: e.g. in the above example D_{null} is the oblate spheroid inscribed by ∂D .

Various authors (Refs. 11, 17, 18) have noticed that when the aspect ratio of ∂D becomes large, numerical instabilities become apparent when solving the null field equations appropriate when ∂D_{null} is either a sphere or a circle. Using a similar analysis to that presented above, Bates and Wall (Ref. 6) were able to reduce this tendency towards numerical instability by decreasing the part of D_- not included in D_{null}. Thus by use of the spheroidal wavefunctions D_{null} becomes the inscribed spheroid, or by the use of the elliptic cylinder wavefunctions D_{null} becomes the inscribed ellipse. We illustrate the improvements that can be obtained by these methods in §6.

Equations (10) and (11) in their present form are not ideally suited for direct use on a digital computer. This is because the part of the wavefunction ψ_m that depends upon the radial coordinate grows very rapidly in magnitude as m increases. Renormalisation of the equations is then necessary if the exponent overflow which will occur on most computers is to be prevented. Although this renormalisation can be performed numerically, it is preferable to renormalise analytically to prevent any loss of accuracy. We will illustrate how this must be done in the case of the spherical wavefunctions.

Examination of the detailed functional form of the ψ_m in the spherical coordinate system (r,θ,ϕ) shows that

$$\psi_m = h_n^{(2)}(kr)P_n^j(\cos\theta)\exp(ij\phi)$$

where P_n^j is the Legendre polynomial of order j and degree n and $h_n^{(2)}$ is the spherical Hankel function of the second kind of order n. We assume that there is a suitable mapping from m to n, j. The Hankel function can be written in terms of the Bessel functions j_n and y_n as

$$h_n^{(2)}(x) = j_n(x) - iy_n(x) .$$

It is the Hankel function that causes the problems in any numerical calculation. Recently in a problem involving the summation of many terms using these Hankel functions O'Brien and Wall (Ref. 19) suggested the definition of a new function to overcome these problems. The Bessel functions are unsuited to computers because for large n and fixed argument j_n and y_n will respectively underflow and overflow the exponent range of the computer. To circumvent this difficulty a factor which decays with n can be extracted from j_n and a factor which grows with n can be extracted from y_n; the residue in each case is then comparable with unity. Functions $u_n^{(j)}(x)$ can then be defined by

$$j_n(x) = \frac{\sqrt{\pi}}{2}\left(\frac{x}{2}\right)^n \frac{1}{\Gamma(n+\frac{3}{2})} u_n^{(1)}(x),$$

$$y_n(x) = -\frac{1}{\sqrt{\pi x}}\left(\frac{2}{x}\right)^n \Gamma(n+\tfrac{1}{2})u_n^{(2)}(x),$$

$$h_n^{(2)}(x) = \frac{i}{\sqrt{\pi x}}\left(\frac{x}{2}\right)^{-n} \Gamma(n+\tfrac{1}{2})u_n^{(4)}(x).$$

We note

$$u_n^{(1)}(x) = \sum_{m=0}^{\infty} \frac{(-x^2/4)^m}{m!\,\Gamma(n+m+\frac{3}{2})} \quad , \tag{12}$$

$$u_n^{(1)}(x) \to 1 \quad \text{as} \quad |n| \to \infty \quad ,$$

$$u_n^{(2)}(x) = u_{-1-n}^{(1)}(x) \quad ,$$

$$u_n^{(4)}(x) = u_n^{(2)}(x) + i\left(\frac{2n+1}{x}\right)\left(\frac{x}{2n+1}\right)^2\left(\frac{x}{2n-1}\right)^2 \cdots \left(\frac{x}{1}\right)^2 u_n^{(1)}(x) .$$

Since $u_n^{(j)}$, $0 \le n \le m$, can be calculated from $u_n^{(1)}$, $-m-1 \le n \le m$, it is only necessary to see how this latter quantity can be computed. As $u_n^{(1)}$ satisfies the recurrence relation

$$u_{n-1}^{(1)}(x) = u_n^{(1)}(x) - \left(\frac{x}{2}\right)^2 \frac{1}{(n+\frac{3}{2})(n+\frac{1}{2})}\, u_{n+1}^{(1)}(x) \ ,$$

which is stable for the direction of decreasing n , $u_n^{(1)}$ can be rapidly computed provided starting values can be found. These values can easily be found as series (12) converges rapidly for large n . Ref. 19 also defines new functions suitable for replacing the derivative of the Hankel function. These functions are useful in the Neumann boundary condition case. From the function $u_n^{(4)}$ we define

$$U_n^{(4)}(x) = \left(\frac{a}{x}\right)^{n+1} u_n^{(4)}(x)$$

where a is an arbitrary constant which is chosen to suit the particular problem being considered; often the maximum radius of the scattering body. In terms of these new functions the null field equations for the Dirichlet boundary condition become

$$\sum_{\ell=1}^{N} \beta_n \int_{\partial D} f_\ell(q) U_n^{(4)}(kr_q) P_n^j(\cos\theta_q)\exp(ij\phi_q)\,ds_q$$

$$= \int_D \chi(Q) U_n^{(4)}(kr_Q) P_n^j(\cos\theta_Q)\exp(ij\phi_Q)\,d\tau_Q \ . \tag{13}$$

§5 BASIS FUNCTION SELECTION

One way of looking at the numerical solution of integral equations is to consider the problem as one of approximation theory in which one desires to approximate some function by polynomials, spline functions, rational functions, special functions and other useful, easily evaluated functions. The main difference between the solution of an integral equation and the determination of a function approximation is that the function we are trying to approximate is defined via an integral equation rather than by functional or tabulated form. Approximation theory can give us useful results on convergence rates of an approximating sequence.

It is well known when solving integral equations of the first kind that besides the obvious advantages of smaller linear systems to solve, the judicious choice of the basis functions f_n can often offset the inherent instability in these equations.

The moment method gives very little in the way of guidance in the choice of basis functions other than that they should be linearly independent. As has been pointed out by Jones (Ref. 20), if the solution being sought from the integral equation does not lie in, or 'near', the space spanned by the basis functions, no amount of computational sophistication will generate a numerical solution near the answer.

We can divide the types of basis function into two possible classes. In the first class the basis functions are defined as non-zero, except for sets of measure zero, over the whole of ∂D. This class we call global bases. In the other, the bases are again defined on ∂D but they are zero over part of the domain. This class we call local bases (sometimes called sub-domain or sub-sectional bases).

Global basis functions have found the greatest use in the null field equations. Waterman (Refs. 1 - 2), and other workers following his method closely, approximate f by a finite number of the $\{\text{Reg}\,\psi_n\}$. Although these wavefunctions appear to provide a good approximation to f with few terms when ∂D is smooth and 'near' ∂D_{null}, they are disadvantageous when this is not the case. Also in common use as global bases are the trigonometric and polynomial functions. When f has discontinuities in its continuity or any of its derivatives, straightforward application of all these functions provides poor point-wise convergence to f. More general global basis representations may take the form

$$f^N = w\sum_{n=1}^{N}\beta_n f_n \tag{14}$$

where the f_n are again 'smooth' functions but the function w has the singularity properties of f; note that w may have local support. Use of a basis approximation of this form requires some a priori knowledge of f.

We illustrate in Fig. 1 results obtained by Bates and Wall (Ref. 6) with the circular null field method, for the surface source density f on a sound-soft infinite cylinder of square cross section when illuminated by a normally incident plane wave. In obtaining these results they have used a basis function representation of the form (14). This has enabled them to approximate f well with only the small values of N noted in the figure caption. If local coordinates are set up at a corner of the square and x denotes the distance measured along either face from the corner then the w chosen has the form $x^{-1/3}$; see Ref. 6 for more computational details.

Local bases can be chosen from a wide variety of functions. However for our purposes we will discuss only the spline functions. We briefly define spline functions (for detail see Refs. 21 - 22). If $a = x_0, x_1, \ldots, x_m = b$ is a subdivision of the interval $[a,b]$ then a spline function, denoted by s, of degree p, is a polynomial of degree p on each of the sub-intervals and its first $(p-1)$ derivatives are continuous on $[a,b]$. In the simplest case, $p = 1$, s is the usual piecewise linear triangle function. Another commonly used spline, obtained from the above definition when $p = 3$, is the B-spline. This spline has up to second derivative continuity between the nodes.

In general from our computational experience the local bases, when used with the null field equations, result in slightly larger condition numbers for the linear system than would be obtained using global bases. To illustrate this point we show in Table 1 the condition number obtained from the spherical null field method applied to a scattering problem when (a) global basis and (b) local bases are used. The problem is the scattering of a plane wave incident along the minor axis of a sound-soft oblate spheroid with a major to minor axis ratio of 2.0 . The origin of coordinates is located at the intersection of the major and minor axes. The global bases chosen are the Legendre polynomials and the local bases are splines of degree one, two and three. The asterisk indicates the value of N for which the solution is adjudged to have converged.

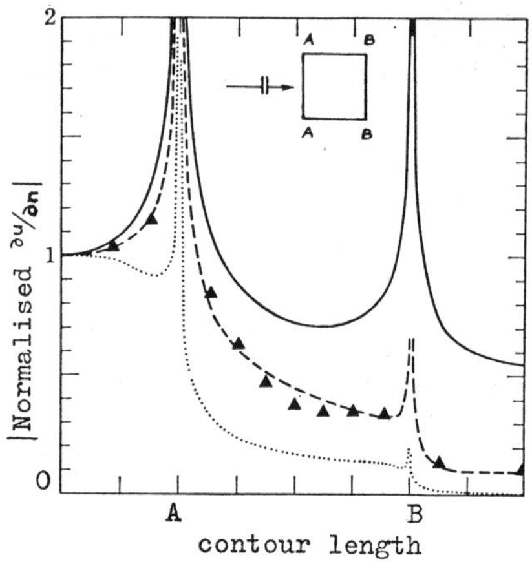

Fig. 1. Normalised surface density on a sound-soft square cylinder.
...... k.AB = 10 , N = 14 ; ------ k.AB = 2 , N = 10 ; ——— k.AB = 0.2 , N = 5 ; ▲ k.AB = 0.2 from Ref. 30.

TABLE 1 Condition numbers $\kappa(A)$ of the linear system obtained from the use of the spherical null field method with various bases

(a) Legendre polynomial basis	N		11	13	15
	$\kappa(A)$		$*0.23 \times 10^2$	0.55×10^2	0.13×10^3
(b) Spline basis	N		11	13	15
		degree			
	$\kappa(A)$	1	$0.52\ 10^3$	$0.83\ 10^4$	$0.35\ 10^6$
		2	$0.17\ 10^3$	$0.18\ 10^3$	$0.30\ 10^5$
		3	$*0.76\ 10^2$	$0.41\ 10^3$	$0.39\ 10^4$

Obtained from the plane wave scattering from an oblate spheroid with an aspect ratio of 2 and a ratio of minor axis to wavelength of 0.4 .

We show in Figs. 2(a) and (b) results obtained with the two different classes of basis when f has a singularity in its derivative. The results shown are from application of the spherical null field method to the electromagnetic antenna boundary-value problem. The curves depict the total current $I = (\hat{n} \times \underline{H}) 2\pi\rho$, where $\underline{H}$ denotes the magnetic field intensity on a perfectly conducting dipole. ρ denotes the radius of the monopole at any point. The hemispherically capped, axially symmetric dipole which is located on an infinite plane, is excited from a coaxial line (Refs. 7, 18).

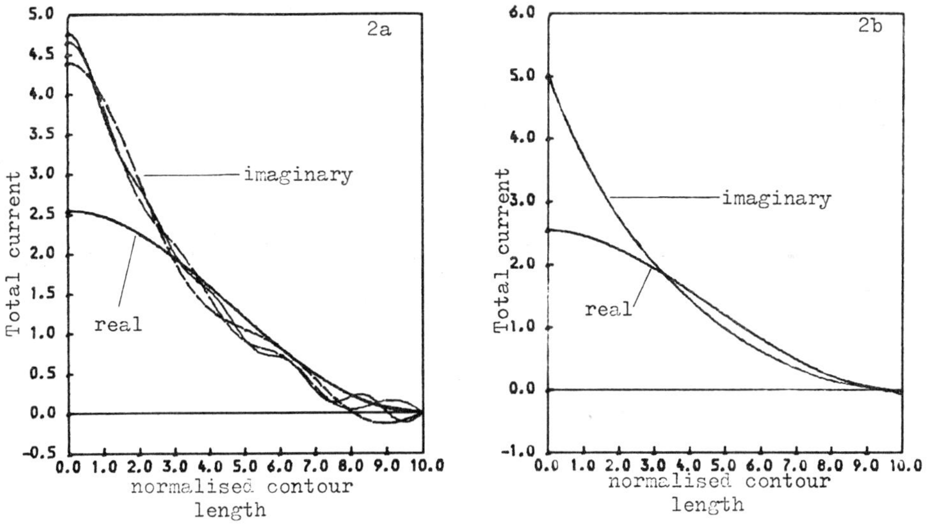

Figs. 2(a) and 2(b). Real and imaginary parts of the total current, in amperes $\times 10^{-2}$, on a hemispherically capped monopole antenna. Height to wavelength ratio = 0.15 , radius of monopole to wavelength ratio = 0.127 .
(a) Legendre polynomial basis, —— N = 10 , —·— N = 8 , - - - N = 4 .
(b) Spline basis of degree 3, —— N = 6 , ---- N = 4 .

It is well known that the imaginary part of the surface field has a singularity in its derivative at the junction of the coaxial line with the antenna (Ref. 23). We may therefore expect a simple global basis expansion to have difficulty at this point. This is shown clearly in Fig. 2(a), whereas the use of a local basis overcomes this problem, as shown in Fig. 2(b).

§6 NUMERICAL SOLUTION OF THE NULL FIELD EQUATIONS

Having obtained the system of equations (10) one may obtain a solution to them by Gaussian elimination, although it is perhaps naive to do so in the light of the results of §3. Because of the properties of the integral operator discussed at some length in §3, we are led to suspect that for some scattering problems and choice of basis functions the coefficient matrix A will become ill-conditioned. Iterative refinement of the solution obtained from the Gaussian elimination method can be employed to alleviate the effects of ill-conditioning. However, this will be of help only if the matrix elements of A have been evaluated to machine accuracy and the condition number is not too large. When the condition number of A is large, in order to obtain a sensible solution one must either change the null field equations or else utilize the concept of regularisation (Ref. 8).

We present results showing how the change of null field method can overcome the effects of ill-conditioning. As mentioned in §4, when the null field method based on either the sphere or the circle is used for bodies of large aspect ratio, the resulting system of equations becomes ill-conditioned. In Table 2 we illustrate this by listing the spectral condition number (in the L_2 norm), denoted by $\kappa(A)$, against aspect ratio of an oblate spheroid for the acoustic scattering problem mentioned in §5. Note that for these results D_{null} is the inscribed sphere.

TABLE 2 Condition numbers of the linear system obtained from use of the spherical null field method with Legendre polynomial basis

Aspect ratio	8	10	11	13	15
2	0.65×10^1	0.15×10^2	$*0.23 \times 10^2$	0.55×10^2	0.13×10^3
4	0.11×10^2	0.34×10^2	0.73×10^2	0.28×10^3	0.11×10^4
6	0.24×10^2	0.51×10^2	0.79×10^2	0.23×10^4	0.13×10^6
10	-	0.23×10^4	0.19×10^5	0.60×10^7	0.12×10^{10}

Obtained from the plane wave scattering from a sound-soft oblate spheroid with a ratio of minor axis to wavelength of 0.4 . * denotes the value of N for which the solution is adjudged to have converged.

We now present some results from Refs. 6 and 7 illustrating how for bodies of large aspect ratio the elliptic and spheroidal null field methods can overcome the ill-conditioning effects. Examination of the detailed form of the wavefunctions of the spheroidal and elliptic cylinder coordinates shows that they are dependent upon the semi-focal distance, denoted by d, of the coordinate system. Within the constraints allowed by the analysis of §4 we are free to choose $0 \le d \le \check{d}$. When $d = \check{d}$ the semi-focal distance is taken to be the largest value possible for the spheroid or ellipse which can still be inscribed by ∂D. When $d = 0$ the null field method reverts to either the spherical or circular null field method. We note that as d varies between 0 and $\check{d}$ the part of D_- spanned by D_{null} increases.

The results in the aforementioned papers utilize as a measure of the numerical condition of the linear system the normalised determinant; each element of the matrix is normalised by the Euclidean length of the row vector prior to evaluation of the determinant. (Note the typographical error in equation 6.12 of Ref. 6.) Denoting this condition number by Z we can say in general that the smaller Z is, the greater will be the error in the computed solution for a given error in the coefficients of A . Although the more usual spectral condition number enables an estimate of the relative error in the solution to be made, given the error in the matrix coefficients, Z is easier to calculate.

To illustrate how the use of the prolate spheroidal null field method can result in a numerically well-conditioned set of equations we show in Table 3 results obtained by Wall (Ref. 7). These are taken from application of this null field method to the aforementioned antenna boundary value problem. Table 3 shows how the order of Z , denoted by O(Z) , increases markedly as d increases from zero to $\check{d}$ for the monopole antenna.

TABLE 3 Prolate spheroidal null field method applied to the cylindrical monopole antenna with a hemispherical endcap

$d/\check{d}$	0	0.2	0.4	0.8	1.0
O(Z)	10^{-13}	10^{-11}	10^{-10}	10^{-6}	10^{-1}

Height of monopole to wavelength = 0.25 , radius of monopole to wavelength = 0.007 . Sine function basis, N = 7 .

We also illustrate in Table 4 (from the results of Ref. 6) how the correct choice of d , in the elliptic null field method, can improve the numerical condition of the resulting equations. In this case the method is applied to the problem of acoustic scattering from a sound-soft cylinder of rectangular cross-section.

TABLE 4 Elliptic null field method applied to the plane wave scattering of a sound-soft rectangular cylinder with an aspect ratio of 10
(a) major side to wavelength ratio = $1/\pi$, N = 10
(b) major side to wavelength ratio = 1 , N = 14

	d/d	0	0.25	0.5	0.75	1.0
(a)	O(Z)	10^{-10}	10^{-4}	10^{-4}	10^{-2}	10^{0}
(b)	O(Z)	10^{-20}	10^{-11}	10^{-9}	10^{-5}	10^{-1}

Obtained with basis functions of the form of (14).

We now discuss how the method of regularisation can be used to obtain a 'satisfactory' solution to a particular set of null field equations. As mentioned previously when the condition number of the matrix A is large the solution is strongly dependent upon any numerical errors made in forming the matrix A and $\underline{h}$. One way of overcoming this difficulty is to restrict the range of admissable functions by invoking some property of the solution which is not contained in the original equation. Smoothness of the solution or its first derivative is often used. This restriction on the domain of admissable functions provides a filter process.

The regularisation method consists of replacing this ill-posed problem by a stable minimisation problem involving a small positive parameter α ; i.e. instead of attempting to solve $Kf = h$ directly, we seek instead to minimise

$$||KF - h|| + \alpha||Lf|| , \tag{15}$$

where L is some linear operator chosen so that the second term has a stabilising effect on the solution. The regularised solution f_α to this minimisation problem has the desirable property that if $Kf = h$ possesses a unique solution f then the norm $||f - f_\alpha|| \to 0$ as $\alpha \to 0$. For certain choices of L it can be shown that (15) can be reduced to an integral equation of the second kind dependent on the parameter α . This approach is not suitable if we wish to retain the essential features of the null field method. However further analysis shows that a solution of (15) can be expressed in terms of the singular functions used in §3. The regularised solution f_α has an analytic solution similar to (9) except that each term is multiplied by the filter factor term $\kappa_j^r/(\kappa_j^r+\alpha^r)$, where r is a constant dependent upon the linear operator L . Examination of the filter factor shows that its value is unity so long as κ_j is large compared with α , but it tends to zero as $\kappa_j \to 0$, the rate of transition depending upon r and α . The choice $r = \infty$ produces the square filter where for all $\kappa_j > \alpha$ the filter factor has a value of one, and for all $\kappa_j < \alpha$ the filter factor has a value of zero. When this square filter is inserted into the series (9) the expansion is truncated when $\kappa_j < \alpha$ which provides a smoothing effect by removing the higher order singular functions.

By use of the singular-value decomposition of the matrix A (Refs. 25 - 26) we emulate the singular-value decomposition of the infinite-dimensional integral operator. If we assume A is an $m \times n$ matrix of rank r , then there exists an $m \times m$ unitary matrix U , an $n \times n$ unitary matrix V , and an $r \times r$ diagonal matrix D containing the singular values s_j of A , such that

$$A = U\,\Sigma V^* , \quad \Sigma = \begin{pmatrix} D & 0 \\ 0 & 0 \end{pmatrix},$$

where $*$ denotes the Hermitian operator and Σ is an $m \times n$ matrix. We assume for application of the regularisation technique that the null field equations may be overdetermined so that $m \geq n$. The singular value decomposition of A can be accurately evaluated using the algorithm of Golub and Reinsch (Ref. 26) or Ref. 25. The solution of the null field equations using this decomposition can be written

$$\underline{\beta} = \underline{A}^+\underline{h} , \tag{16}$$

where A^+ is the Moore-Penrose pseudo-inverse of A (Ref. 9). When $m = n = r$, A^+ becomes the more familiar inverse A^{-1} . A^+ can be expressed in terms of the unitary matrices as

$$A^+ = V\,\Sigma^+\,U^* ,$$

where Σ^+ is of the same form as Σ but the leading submatrix D is replaced by D^{-1} ; i.e. the only non-zero terms of Σ^+ are the diagonal ones $1/s_j$.

When A is severely ill-conditioned because of the near dependence of the adjacent row and column vectors, several of the singular values (normalised by the largest singular value) are less than or close to the relative arithmetic precision (denoted by ϵ) of the digital computer used in the calculations. If the criterion used for deciding on the rank of A is that all normalised singular values smaller than ϵ are zero, then $r < n$ and the Moore-Penrose inverse provides the unique least squares solution with the minimum solution norm. However the remaining normalised singular values that are much smaller than the accuracy to which the elements of A and $\underline{h}$ have been evaluated will cause amplification of the higher order column vectors of V and produce oscillatory components in the solution. This numerical instability can be overcome by the use of a technique similar to the one suggested in the second to last paragraph. We redefine A^+ to be the general pseudo-inverse (the regularised pseudo-inverse), namely $A^+ = V F \Sigma^+ U^*$, where F , the <u>filter matrix</u>, is of block diagonal form with the only non-zero terms $s_j^r/(s_j^r+\alpha^r)$. The filtered solution f_α can now be obtained by use of this pseudo-inverse in equation (16). For appropriate choices of α and r we will obtain a stable regularised solution f_α . There appears to be no simple rule as to the optimum choice of α and r for any given problem.

Various applications of this method may be found in the literature for numerical solution of integral equations of the first kind (e.g. Refs. 27 - 28). Zuckerman and Diament (Ref. 29) have applied this method, but using only a square filter to a waveguide problem. The linear system of equations was obtained by using a technique that utilised the extended boundary condition.

We will now illustrate how the regularisation method can be used to obtain a filtered solution from the spherical null field equations when the condition number is large. The scattering problem is the one previously considered in §5 and to which Table 1 applies; namely the plane wave scattering from an oblate spheroid. The basis functions chosen to represent the surface density were the quadratic splines and the elements of A we evaluated to a relative accuracy of approximately 0.01 . Examination of Table 1 shows that for this choice of basis function and with $N = 15$ the condition number of the linear system is 0.30×10^5 . The rapidly oscillating solution, shown in Fig. 3(a), which we get when a direct solution of the equations is attempted, is therefore not unexpected. Also shown for comparison is the 'true' solution obtained by solving the problem with a Legendre polynomial basis, $N = 13$. The condition number for this case can also be found in Table 1.

Application of the regularised pseudo-inverse to the linear system relevant to the quadratic spline basis yields many filtered solutions. In fact one solution results from each choice of r and α . To justify the r and α values chosen, we list in Table 5 the last five normalised and ordered singular values. Examination of Table 5 shows that the last two singular values are much smaller and more separated than the previous ones. Past experience leads us to expect that the removal of these two singular values by means of a square filter will remove the spurious oscillation observed in Fig. 3(a). This is verified in Fig. 3(b). Also displayed in Fig. 3(b) are filtered solutions procured by filters with less sharp cutoff characteristics. All these f_α provide an 'acceptable' solution to the problem.

TABLE 5 The last five of the ordered and normalised s_j obtained from the singular value decomposition of the matrix A

$0.50 \; 10^{-2}$
$0.32 \; 10^{-2}$
$0.23 \; 10^{-2}$
$0.44 \; 10^{-3}$
$0.34 \; 10^{-4}$

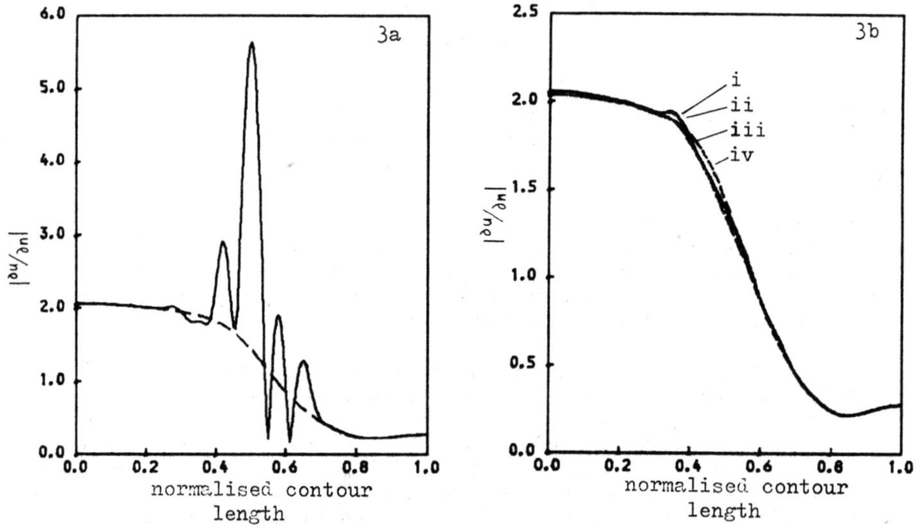

Figs. 3(a) and 3(b). Modulus of the surface density on a sound-soft oblate spheroid, showing the solution with and without regularisation.
(a) —— direct solution of the linear system,
---- correct solution.
(b) i square filter removing last two s_j,
ii $\alpha = 1 \times 10^{-3}$, $r = 4$,
iii $\alpha = 1 \times 10^{-3}$, $r = 2$,
iv correct solution.

§7 SUMMARY

We have demonstrated that for some scattering body shapes and basis function selections the null field equations are numerically ill-conditioned. The three procedures suggested for overcoming these numerical difficulties are the use of:

(1) the appropriate null field equations for the scattering body being considered,

(2) the appropriate basis function choice,

(3) the regularisation method.

The author thanks D. M. O'Brien for many helpful discussions concerning the regularisation method. This work was financed by the U.K. Ministry of Defence with an R.S.R.E. Research Fellowship.

REFERENCES

(1) P. C. Waterman, Matrix formulation of electromagnetic scattering, Proc. I.E.E.E. 53, 805 (1965).

(2) P. C. Waterman, New formulation of acoustic scattering, J. Acoust. Soc. Am. 45, 1417 (1969).

(3) B. Peterson and S. Strom, T-matrix formulation for electromagnetic scattering from an arbitrary number of scatterers and representations of E(3), Phys. Rev. D8, 3661 (1973).

(4) B. Peterson and S. Strom, Matrix formulation of acoustic scattering from an arbitrary number of scatterers, J. Acoust. Soc. Am. 56, 771 (1974).

(5) R. H. T. Bates, Modal expansions for electromagnetic scattering from perfectly conducting cylinders of arbitrary cross-section, Proc. I.E.E. 115, 1443 (1968).

(6) R. H. T. Bates and D. J. N. Wall, Null field approach to scalar diffraction, Phil. Trans. R. Soc. Lond. A287, 45 (1977).

(7) D. J. N. Wall, The null field approach to the antenna boundary value problem, I.E.E. International Conf. on Antennas and Prop. Part 1, 174 (1979).

(8) A. N. Tihonov, Solution of incorrectly formulated problems and the regularisation method, Soviet Math. 4, 1035 (1963).

(9) Nashed, M. Z. (1975) ed. Generalised Inverses and Applications, Academic Press, New York.

(10) G. F. Miller, Fredholm equations of the first kind, in Numerical Solution of Integral Equations ed. Delves, L. M. and Walsh, J. (1974) Chap. 13, Clarendon Press, Oxford.

(11) J. C. Bolomey and W. Tabbara, Numerical aspects on coupling between complementary value problems, I.E.E.E. Trans. Antennas Propag. AP-21, 356 (1973).

(12) A. J. Burton, Numerical solution of acoustic radiation problems, National Physical Laboratory Report (1976).

(13) Stakgold, I. (1967) Boundary Value Problems of Mathematical Physics, Chap. 3, MacMillan Co., New York.

(14) Al-Badwaihy and J. L. Yen, Extended boundary condition integral equations for perfectly conducting and dielectric bodies: formulation and uniqueness, I.E.E.E. Trans. Antennas Propag. AP-23, 546 (1975).

(15) Smithies, F. (1958) Integral Equations, Cambridge University Press, Cambridge.

(16) Morse, P. M. and Feshbach, H. (1953) Methods of Theoretical Physics, Chap. 11, McGraw-Hill, New York.

(17) J. C. Bolomey and A. Wirgin, Numerical comparison of the Green's function and the Waterman theories of scattering from a cylinder of arbitrary cross-section, Proc. I.E.E. 121, 794 (1974).

(18) R. H. T. Bates and C. T. Wong, The extended boundary condition and thick axially symmetric antennas, Appl. Sci. Res. 29, 19 (1974).

(19) D. M. O'Brien and D. J. N. Wall, Radiation from a whip antenna mounted on a sphere (to be published).

(20) D. S. Jones, Numerical methods for antenna problems, Proc. I.E.E. 121, 573 (1974).

(21) Schultz, M. H. (1973) Spline Analysis, Prentice Hall, Englewood Cliffs, N.J.

(22) Prenter, P. M. (1975) Splines and Variational Methods, Wiley, New York.

(23) R. H. Duncan, Theory of the infinite cylindrical antenna including the feed-point singularity in antenna current, J. Research N.B.S. Radio Prop. 66D, 181 (1962).

(24) Westlake, J. R. (1968) A Handbook of Numerical Matrix Inversion and Solution of Linear Equations, Wiley, New York.

(25) Lawson, C. L. and Hanson, R. J. (1974) Solving Least Squares Problems, Prentice Hall, Englewood Cliffs, N.J.

(26) G. H. Golub and C. Reinsch, Singular value decomposition and least squares, in Wilkinson, J. H. and Reinsch, C. (1971) ed. Handbook for Automatic Computation 2, 134, Springer, Berlin.

(27) R. J. Hanson, A numerical method for solving Fredholm integral equations of the first kind using singular values, S.I.A.M. J. Numer. Anal. 8, 616 (1971).

(28) D. B. L. Jupp and K. Vozoff, Stable iterative methods for the inversion of geophysical data, Geophys. J. R. Astro. Soc. 42, 957 (1975).

(29) D. N. Zuckerman and P. Diament, Rank reduction of ill-conditioned matrices in waveguide junction problems, I.E.E.E. Trans. Microwave Theory Tech. MTT-25, 613 (1977).

(30) K. Iizuka and J. L. Yen, Surface currents on triangular and square metal cylinders, I.E.E.E. Trans. Antennas Propag. AP-15, 795 (1967).

METHOD OF OPTIMAL TRUNCATION: A NEW T-MATRIX APPROACH TO ELASTIC WAVE SCATTERING

William M. Visscher*
Theoretical Division, Los Alamos Scientific Laboratory
Los Alamos, New Mexico 87545

ABSTRACT

A family of matrix theories of elastic wave scattering is derived, and one, which is in a certain sense optimal, is developed. Called the method of optimal truncation (MOOT), it results from a minimum principle and can be shown to yield a convergent sequence of approximations. Numerical results for scattering cross-sections for longitudinal incident waves with $ka \leq 10$ from fixed rigid obstacles and voids with axial symmetry are obtained using MOOT, and are compared with results of other matrix theories. Shapes considered include spheres, oblate and prolate spheroids, pillboxes, and cones. Convergence is demonstrated. Extension of the method to elastic and fluid inclusions is discussed, as is its application to cracks, which may be accomplished by simulating the crack with an incompletely bonded identical inclusion. Implications of reciprocity and time-reversal invariance are discussed.

I. INTRODUCTION

Proliferation of important scientific and technological applications in geology, materials science, and nondestructive testing, coupled with the current ubiquity of large computers, has spurred the development of methods for calculating the scattering of elastic waves from flaws. In this paper we will consider some of these methods, in particular the so-called T-matrix theories (1,2,3). We derive a family of these theories in Section II, and consider the problem of picking out an optimal one in Section III. These methods usually involve an expansion of the scattered displacement field in a finite (truncated) sum of partial-waves, and they result in sets of linear equations for the partial-wave amplitudes.

A particular choice, resulting from imposing the requirement that it minimize the mean-square deviance from the boundary conditions, is shown in Section IV to yield a sequence of approximations which converges as the truncation limit L increases. We call this matrix theory MOOT, the method of optimal truncation. Numerical results from MOOT and other methods are presented and compared in Section V, for some examples wherein the flaw is an axially symmetric void or a fixed rigid obstacle of spheroidal, cylindrical, or conical shape, and the incident longitudinal wave has $ka \leq 10$. In Section VI the extension of MOOT to elastic and fluid inclusions, and to cracks, is discussed.

*Supported by U.S. Department of Energy

II. FAMILY OF MATRIX METHODS

Because they are simpler than the equations of elasticity (4), we will illustrate the methods with the equations of the incompressible irrotational fluid (5). In that case, if $\phi(\vec{r},t)$ is the velocity potential

$$\vec{v} = - \vec{\nabla} \phi \quad , \tag{1}$$

then the linearized equation of motion for a homogeneous system is

$$\left(\nabla^2 - \frac{1}{c^2} \frac{\partial^2}{\partial t^2} \right) \phi(\vec{r},t) = 0 \ . \tag{2}$$

If harmonic time dependence is assumed, this becomes the Helmholtz equation

$$(\nabla^2 + k^2)\ \phi = 0 \ , \tag{3}$$

where the wavenumber k is given by

$$k^2 = \omega^2/c^2 = \omega^2 \rho/\kappa \ . \tag{4}$$

κ is the compressibility modulus, ρ is the equilibrium density, and the pressure is

$$P = \frac{\partial \phi}{\partial t} = i\omega\rho\phi \quad . \tag{5}$$

If the fluid has within it an inclusion with different density and compressibility modulus ρ' and κ', with volume V bounded by a surface Σ (see Fig. 1), then Eq. (3) with appropriately modified k will again be satisfied inside Σ. Continuity of pressure across the boundary requires

$$\rho' \phi(\vec{r}_-) = \rho\phi(\vec{r}_+) \tag{6}$$

(where $\vec{r}_\pm$ are positions just outside and inside Σ, respectively), and conservation of matter demands

$$\hat{n}\cdot\vec{\nabla}\phi(\vec{r}_-) = \hat{n}\cdot\vec{\nabla}\phi(\vec{r}_+) \ , \tag{7}$$

where $\hat{n}$ is the unit outward normal to Σ .

The scattering solution of Eq. (3) which we seek has the form

$$\phi(\vec{r}) = \phi_{inc}(\vec{r}) + \sum_s a_s \phi_s^{(+)}(\vec{r}) \ , \tag{8}$$

where ϕ_{inc} is an incident plane wave, $\phi_s^{(+)}$ is an outgoing spherical wave specified by eigenvalues s = (ℓ,m), and a_s is the corresponding complex amplitude. ϕ_{inc} and $\phi_s^{(+)}$ are solutions of Eq. (3). In detail,

$$\phi_{inc}(\vec{r}) = e^{i\vec{k}_0\cdot\vec{r}} = \sum_s d_s \tilde{\phi}_s(\vec{r}) = 4\pi \sum_{\ell,m} i^\ell Y^*_{\ell m}(\theta_o,\phi_o) Y_{\ell m}(\theta,\phi) j_\ell(kr) , \tag{9}$$

$$\phi_s^{(+)} = \phi_{\ell m}^{(+)} = h_\ell^{(1)}(kr) Y_{\ell m}(\theta,\phi) , \tag{10}$$

where the spherical Hankel function $h_\ell^{(1)} = j_\ell + iy_\ell$ in terms of the spherical Bessel and Neumann functions, and

$$\tilde{\phi}_s = \tilde{\phi}_{\ell m} = j_\ell(kr)Y_{\ell m}(\theta,\phi) \tag{11}$$

is the part of $\phi_s^{(+)}$ which is regular at the origin. $Y_{\ell m}$ is the usual spherical harmonic (6).

The partial-wave amplitudes a_s in Eq. (8) specify the outgoing wave from which the scattering cross-sections may be calculated. We will now derive, in a nearly trivial way, a family of sets of linear equations for the a's, the T-matrix theories.

Equations (6) and (7) specify the boundary conditions at the surface of a scatterer which is a homogeneous inclusion. For the moment we will further simplify the problem by restricting the inclusion to be one of two kinds, either a void (cavity), for which the Dirichlet boundary condition of vanishing pressure

$$\phi(\vec{r}_+) = 0 \qquad \text{void} \tag{12}$$

is satisfied, or a rigid fixed obstacle, for which the Neumann boundary condition of vanishing normal velocity

$$\hat{n}\cdot\vec{\nabla}\phi\ (\vec{r}_+) = 0 \qquad \text{obstacle} \tag{13}$$

holds. The more general case is discussed in Section VI.

If Eq. (8) is substituted into Eqs. (12) and (13), the result is, where the arguments of the wavefunctions ϕ are understood to be $\vec{r}_+$,

$$\sum_s (d_s\tilde{\phi}_s + a_s\phi_s^{(+)}) = 0 \qquad \text{void} \tag{14}$$

$$\sum_s (d_s\hat{n}\cdot\vec{\nabla}\tilde{\phi}_s + a_s\hat{n}\cdot\vec{\nabla}\phi_s^{(+)}) = 0 \qquad \text{obstacle.} \tag{15}$$

Now we introduce a set of functions $\{f_j\}$, $j = 1,2,\ldots\infty$ which is complete on Σ. Then if the notation

$$\int_\Sigma d\sigma\ u^*v = (u,v) \tag{16}$$

is introduced, Eqs. (14) and (15) when multiplied by $f_j^*(\vec{r}_+)$ and integrated over Σ become

$$\sum_s \left[(f_j,\tilde{\phi}_s)d_s + (f_j,\phi_s^{(+)})a_s\right] = 0 \qquad \text{void} \tag{17}$$

$$\sum_s \left[(f_j,\hat{n}\cdot\vec{\nabla}\tilde{\phi}_s)d_s + (f_j,\hat{n}\cdot\vec{\nabla}\phi_s^{(+)})a_s\right] = 0 \qquad \text{obstacle} \tag{18}$$

Equations (14), (15) hold for every r_+ on the surface Σ; because $\{f_j\}$ is postulated to be complete on Σ, they are wholly equivalent to Eqs. (17), (18) for $j = 1,2,\ldots\infty$.

These equations can be written in compact form

$$\tilde{Q}d + Qa = 0 \tag{19}$$

where

$$\tilde{Q}_{js} = \begin{cases} (f_j, \tilde{\phi}_s) & \text{void} \quad (20) \\ (f_j, \hat{n}\cdot\vec{\nabla}\tilde{\phi}_s) & \text{obstacle} \quad (21) \end{cases}$$

and

$$Q_{js} = \begin{cases} (f_j, \phi_s^{(+)}) & \text{void} \quad (22) \\ (f_j, \hat{n}\cdot\vec{\nabla}\phi_s^{(+)}) & \text{obstacle} \quad (23) \end{cases}$$

One can formally solve Eq. (19) for the vector a,

$$a = -Q^{-1}\tilde{Q}d = Td. \tag{24}$$

This defines the T-matrix which linearly transforms the incident wave amplitudes d_s into the outgoing wave amplitudes a_s.

The set of functions $\{f_j\}$ with which one works in a practical calculation is, however, never complete. One has a basis set $\{f_j\}$, $j = 1,2,...L$, where L, the truncation limit, is almost always significantly less than infinity. So although if $L \to \infty$ the matrix equations are completely equivalent to the boundary conditions and T will not depend on what complete set $\{f_j\}$ we choose, in practice L is rather small and the set must be chosen carefully. The sets $\{\phi_s\}$ will also be truncated at $s = L$. This insures that the matrices Q are square, and we assume that Q^{-1} exists.

Conversely, our choice of the truncated set $\{f_j\}_L$ will affect our calculated results, and we need to find some criterion to tell us which sets are better than others, and ideally to pick out an optimum one.

III. MINIMUM PRINCIPLE

We would like to choose $\{f_j\}_L$ so that the error in the results for physical observables is minimized. Because we do not know the exact values for cross-sections, we cannot formulate this condition. What we do know exactly are the boundary conditions, namely Eqs. (12) and (13), which the wavefunction ϕ must satisfy. This leads us to consider the absolute squares of the deviance from the boundary conditions, integrated over the surface of the scatterer.

$$I = \begin{cases} (\phi, \phi) & \text{void} \quad (25) \\ (\hat{n}\cdot\vec{\nabla}\phi, \hat{n}\cdot\vec{\nabla}\phi) & \text{obstacle} \quad (26) \end{cases}$$

I vanishes if and only if ϕ satisfies the boundary conditions exactly, which happens in fact only if $L \to \infty$. For finite L, $I > 0$, and we require that I be minimized with respect to variations in the coefficients a_s. We substitute Eq. (8) into Eqs. (25) and (26). If the resultant bilinear expression in the a's has a minimum, then the derivatives

$$\frac{\partial I}{\partial a_s^*} = 0 \qquad s = 1, 2, \ldots L, \tag{27}$$

must vanish. Computing them we find

$$\frac{\partial I}{\partial a_s^*} = \frac{\partial}{\partial a_s^*}\int_{\Sigma_1} d\sigma \begin{cases} \left|\phi_{inc} + \sum_{s'=1}^{L} a_{s'}\phi_{s'}^{(+)}\right|^2 & \text{void} \quad (28) \\ \left|\hat{n}\cdot\vec{\nabla}\phi_{inc} + \sum_{s'=1}^{L} a_{s'}\hat{n}\cdot\vec{\nabla}\phi_{s'}^{(+)}\right|^2 & \text{obstacle} \quad (29) \end{cases}$$

which are Eqs. (17), (18) with

$$f_s = \begin{cases} \phi_s^{(+)} & \text{void} \quad (30) \\ \hat{n}\cdot\vec{\nabla}\phi_s^{(+)} & \text{obstacle} \quad (31) \end{cases}$$

Equation (30), (31) specify one choice out of an infinite number which could be made for $\{f_s\}$. Waterman (1,2) using a very different approach, would prescribe

$$f_s = \begin{cases} \hat{n}\cdot\vec{\nabla}\tilde{\phi}_s & \text{void} \quad (32) \\ \tilde{\phi}_s & \text{obstacle} \quad (33) \end{cases}$$

for this scalar example. One needs to ask now: which of Eqs. (30), (31) or (32), (33) , or of an infinite variety of others, will give the most accurate and reliable answers with least labor? This question can really only be satisfactorily answered by trying them and comparing the numbers obtained with different choices of $\{f_s\}_L$.

We will do this for a few choices in Section V, but one fact, a priori, does favor Eqs. (30), (31). Namely, this choice, because it results from a minimum principle, yields a convergent sequence of approximations, as we now prove.

IV. CONVERGENCE AND OTHER CRITERIA OF CHOICE

Consider the surface integral Eqs. (25), (26)

$$I_L = I(a_1, a_2, \ldots a_L, 0, 0 \ldots) \tag{34}$$

to be a function of an infinite number of amplitudes a_s, with constraints

$$a_{L+1} = a_{L+2} = \ldots = 0 \tag{35}$$

For $L < \infty$, $I_L \geqslant 0$, and we know that if an exact solution exists, then the minimum value of I is zero for $L \to \infty$. Now define I_L^{min} to be the minimum value of I_L which is attained by variation of its L complex arguments a_s, that is

$$I_L^{min} = \min\left[I(a_1, a_2, \ldots a_L, 0, 0 \ldots)\right] \tag{36}$$

Call the values of $a_1 \ldots a_L$ for which the minimum is attained $a_1^{(L)}, a_2^{(L)}, \ldots a_L^{(L)}$; i.e.

$$I_L^{min} = I\left(a_1^{(L)}, a_2^{(L)} \ldots a_L^{(L)}, 0, 0 \ldots\right). \tag{37}$$

Now consider

$$I_{L+1}^{min} = \min\left[I(a_1, a_2 \ldots a_L, a_{L+1}, 0, \ldots)\right] \tag{38}$$

then, because $\left(a_1^{(L)}, a_2^{(L)}, \ldots a_L^{(L)}, 0\right)$ is a possible set of values of $(a_1, a_2, \ldots a_{L+1})$ in Eq. (38), it follows that

$$I_{L+1}^{min} \leqslant I_L^{min}, \tag{39}$$

and I_L^{min} forms a sequence which converges monotonically to zero as $L \to \infty$.

From the fact that a particular sequence of bilinear forms in the a_s's converges monotonically to the exact answer, it does not necessarily follow that other bilinear forms, such as the cross-sections, are also convergent. But it is reasonable that they are. The coefficients of the bilinear form I_L are Q-matrix elements, whose magnitudes become very large when partial waves with high radial eigenvalue ℓ are involved. On the other hand, the coefficients in the bilinear form for the cross-section σ do not increase rapidly with ℓ. Therefore I_L is more sensitive than σ to changes in a_s for large s, and the latter should converge faster as $L \to \infty$. That it does so will be illustrated for a particular case in Section V.

The minimization of I [Eqs. (25)(26)] leads uniquely to a set $\{f_s\}_L$ given by Eqs. (30), (31), which in turn leads to a monotonically convergent sequence $I_L, I_{L+1}, \ldots$. But one could use other criteria for choosing $\{f_s\}$. Examples are energy conservation and satisfaction of reciprocity (7,2).

In any non-dissipative system energy is conserved. In a scattering process this implies the optical theorem, which is a proportionality between the imaginary part of the forward scattering amplitude and the total cross-section. In different terms it implies unitarity of the S-matrix (8).

The optical theorem imposes a constraint on the truncated amplitudes, which will thereby be overdetermined, because they are already uniquely specified by the matrix equation [Eq. (24)]. Alternatively, one can try to choose $\{f_s\}_L$ so as to yield a unitary truncated S-matrix. This can be done (8), but not consistently with the minimization of I, which already uniquely determines $\{f_s\}_L$.

One manifestation of reciprocity is, in the scattering problem, that if the directions of observation and of the incoming wave are reversed and interchanged (i.e. $\theta \leftrightarrow \pi - \theta_0$, $\phi \leftrightarrow \pi + \phi_0$), then the cross-section is unchanged. Reciprocity is true more generally than unitarity; the system need not, for example, be conservative. Reciprocity is guaranteed if the T-matrix (or the S-matrix) is symmetric.

Reciprocity imposes another condition on the truncated amplitudes; again it cannot be satisfied simulatneously with minimization of I.

One might choose to satisfy the minimization principle exactly and use the other symmetries to check the accuracy of the results. Or one might choose to satisfy the symmetries exactly and the matrix equations approximately. A third alternative is to solve the overdetermined system in a least-squares sense, satisfying everything only approximately.

We take the first course. Waterman (2) effectively took the seond. Possibly the third course would be advantageous, but it is more complicated and as yet untried.

V. NUMERICAL EXAMPLES AND COMPARISONS

The T-matrix [Eq. (24)] is given by $T = -Q^{-1}\tilde{Q}$, where

$$Q_{ss'} = \left(\phi_s^{(+)}, \phi_{s'}^{(+)}\right)$$
$$\tilde{Q}_{ss'} = \left(\phi_s^{(+)}, \tilde{\phi}_{s'}\right) \tag{40}$$

and $s = (\ell,m)$; $\ell = 0,1, \ldots \ell_{max}$; $m = -\ell,-\ell+1, \ldots \ell$. For nonspherical shapes these matrix elements must be calculated numerically, which comprises most of the computational labor in a scattering calculation. An important simplification is obtained if the shape Σ is constrained to be axially symmetric. Then it is clear from Eq. (10) that the Q and $\tilde{Q}$ matrices are diagonal in m;

$$Q_{\ell m,\ell' m'} = \delta_{mm'} Q_{\ell m,\ell' m} = \delta_{mm'} Q_{\ell -m,\ell' -m} \tag{41}$$

where the second equality follows from

$$Y_{\ell -m} = (-1)^m Y^*_{\ell m}. \tag{42}$$

Thus the Q-matrices can be rearranged as follows,

$$Q = \begin{pmatrix} Q^{(0)} & 0 & 0 \\ 0 & Q^{(1)} & \cdot \\ 0 & \cdot & 0 \\ 0 & \cdot \quad \cdot & Q^{(\ell_{max})} \end{pmatrix} \tag{43}$$

which represents a block-diagonal matrix, wherein matrices along the diagonal are

$$Q^{(m)} = \begin{pmatrix} Q_{mm,mm} & Q_{mm,m+1\,m} & \cdots & Q_{mm,\ell_{max}m} \\ \cdot & & & \cdot \\ \cdot & & & \cdot \\ \cdot & & & \cdot \\ Q_{\ell_{max}m,mm} & \cdots & \cdots & Q_{\ell_{max}m,\ell_{max}m} \end{pmatrix} \tag{44}$$

with $Q^{(0)}$ being $(\ell_{max} + 1)x(\ell_{max} + 1)$ and $Q^{(\ell_{max})}$ being (1 x 1).

This is an important simplification because the inverses and products of block-diagonal matrices are again block-diagonal. Thus

$$Q^{-1} = \begin{pmatrix} Q^{(0)^{-1}} & 0 & 0 & 0 \\ 0 & \ddots & & 0 \\ 0 & & \ddots & 0 \\ 0 & 0 & 0 & Q^{(\ell_{max})^{-1}} \end{pmatrix} \tag{45}$$

and

$$T = \begin{pmatrix} T^{(0)} & 0 & 0 & 0 \\ 0 & \ddots & & 0 \\ 0 & & \ddots & 0 \\ 0 & 0 & 0 & T^{(\ell_{max})} \end{pmatrix} \tag{46}$$

where

$$T^{(m)} = - Q^{(m)^{-1}} \tilde{Q}^{(m)} \tag{47}$$

Therefore the T-matrix, and hence the amplitudes $a_{\ell m}$, can be calculated separately for each m. The largest matrix we ever need to invert or multiply has rank $(\ell_{max} + 1)$, which is the square root of the rank of the T-matrix for a shape with no symmetries.

It should be emphasized that although we have taken the axis of symmetry of the scatterer to be in the z-direction neither the incoming plane wave nor the direction of observation are constrained. The T-matrix does not depend on θ_0, ϕ_0 ; once we have calculated T we can immediately get the scattered amplitudes for any incident direction, viz, from Eq. (9),

$$a_{\ell m} = \sum_{\ell'=m}^{\ell_{max}} T^{(m)}_{\ell m,\ell' m}\, d_{\ell' m} = 4\pi \sum_{\ell'=m}^{\ell_{max}} T^{(m)}_{\ell m,\ell' m}\, i^{\ell'}\, Y^{*}_{\ell' m}(\theta_o,\phi_o) \tag{48}$$

The differential cross-section $\frac{d\sigma}{d\Omega}$ is the scattered power per unit solid angle divided by the incident power per unit area: it is easy to show that it is

$$\frac{d\sigma}{d\Omega} = k^{-2}\left|\sum_{\ell m} i^{-\ell-1} a_{\ell m} Y_{\ell m}(\theta,\phi)\right|^2 \tag{49}$$

Although the equations we have written have been exclusively for the scalar case, similar ones, with more components, can be written for elastic waves (4). The numerical results that we now present are for the latter case, which is more complicated because the elastic displacement is a vector. Thus the scalars in the surface integrals Eqs. (20)-(23) are replaced by the vectors

$$\rho\phi_{\ell m} \to \vec{S}_{p\ell m} \tag{50}$$

$$\hat{n} \cdot \vec{\nabla}\, \phi_{\ell m} \to \vec{t}_{p\ell m}, \tag{51}$$

and scalar products are taken in the integrals. $\vec{S}$ and $\vec{t}$ are the displacement and surface traction (stress tensor contracted with $\hat{n}$) vectors respectively. The radial and azimuthal eigenvalues have been supplemented by a polarization $p = 1$ (longitudinal), 2,3 (transverse).

Comparison Between Different Basis Sets $\{f_s\}_L$.

The case we consider here is the scattering of a longitudinally polarized elastic wave incident at 45° on an oblate spheroidal void. The prescription of MOOT for this case is

$$f_s = \vec{t}^{(+)}_{p\ell m}, \tag{52}$$

where, again, the (+) means that the surface traction is constructed from outgoing waves. But several other choices besides Eq. (52) suggest themselves as being just as easy to calculate with, to wit,

$$f_s = \begin{cases} \tilde{S}_{p\ell m} & (53a) \\ \tilde{t}_{p\ell m} & (53b) \\ S^{(o)}_{p\ell m} & (53c) \\ t^{(o)}_{p\ell m} & (53d) \\ S^{(t)}_{p\ell m} & (53e) \\ t^{(+)}_{p\ell m}\,. & (53f) \end{cases}$$

The first two are constructed from regular Bessel functions, the last two from outgoing waves (Hankel functions) and the middle two have no r dependence, but are merely linear combinations of vector spherical harmonics, obtained by setting r = a in Eqs. (53a) and (53b).

Figure 2 shows computed results for the total longitudinal cross-section for all of these choices for ka = 1 (a is the radius of the oblate spheroid, which in this figure has aspect ratio 2/3). These results support the view that for this scattering situation, at least, the outgoing waves (e) and (f) are to be preferred. This choice is reinforced by the results shown in Fig. 3, which is the same as the preceding one except that the oblate spheroidal void now has aspect ratio 1/2. Whether or not the surface traction (f) is superior to the displacement (e) is not decided by this data. On subjective esthetic grounds we prefer (f); it is prescribed by MOOT.

Convergence of Truncation Sequence

As discussed above, the sequence of surface integrals I_L, $I_{L+1}\cdots$ converges monotonically. It is of interest to see for a particular case just how fast the convergence is, and how it is correlated with the convergence of an observable cross-section. Numerical results are presented in Fig. 4 for scattering of an incident longitudinal wave from a sphere and from a prolate spheroidal fixed rigid obstacle. They confirm the monotonic convergence of I, and indicate that the sequence of approximations to the cross-section converge much faster, albeit not monotonically.

Examples of Differential Cross-Sections

Scattering of an incident longitudinal wave with ka = 10 from a conical void in titanium is shown in Fig. 5. This cone has radius equal to its height; the incident direction is $(\theta_o, \phi_o) = (135^o, 0^o)$. Peaks appear in the longitudinal cross-section both in the forward and the specularly reflected directions.

Mode-converted scattering of an incident longitudinal wave from an oblate spheroidal void in titanium is shown in Fig. 6. Again ka = 10, but $(\theta_o, \phi_o) = (90^o, 0)$. The incident wave is longitudinally polarized here; the boundary produces a transversely polarized component in the scattered wave (mode conversion). Symmetry requires that the cross-section vanish in the forward and backward directions; these points are plotted at -100 db.

Figure 7 has geometry similar to that of Fig. 6, but the target is now a pillbox with aspect ratio 1/2 (height = radius). The forward peak has diffraction minima surrounding it, and there is evidence of specular reflection from the pillbox side. The longitudinal cross-section is the ordinate.

Spot checks on the consistency of our results with the optical and reciprocity theorems have been made. Both agree within about 1/2 db for oblate spheroidal voids with aspect ratio 1/2 and ka $\leq$ 2. Agreement deteriorates for larger ka and more extreme shapes because the angular variation becomes more rapid.

VI. INCLUSIONS AND CRACKS

Referring again to Fig. 1, we imagine that Σ is filled with a compressible irrotational fluid with density ρ' and compressibility κ', and the conditions that must be satisfied at the boundary are given by Eqs. (6) and (7). In analogy with Eqs. (25), (26) we form the surface integrals

$$I = \int_\Sigma |\rho'\phi(\vec{r}_-) - \rho\phi(\vec{r}_+)|^2 d\sigma \tag{54}$$

$$J = \int_\Sigma |\hat{n}\cdot\vec{\nabla}\phi(\vec{r}_-) - \hat{n}\cdot\vec{\nabla}\phi(\vec{r}_+)|^2 d\sigma, \tag{55}$$

and we need an additional expansion for the wave function inside the defect. So for $\vec{r}$ inside Σ we put

$$\phi(\vec{r}) = \sum_s a'_s \tilde{\psi}_s(\vec{r}), \tag{56}$$

where only the regular solutions of the inside Helmholtz equation

$$(\nabla^2 + k'^2)\,\psi_s(\vec{r}) = 0 \tag{57}$$

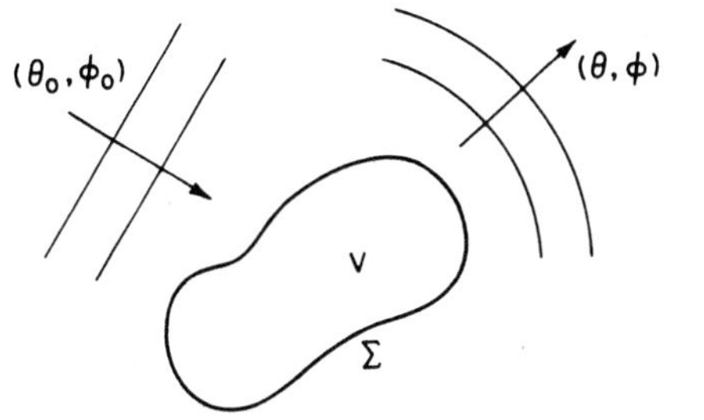

Fig. 1. Scattering geometry. A homogeneous flaw of volume V with a surface Σ is imbedded in the host medium. A wave is incident with wavenumber k at polar and azimuthal angles (θ_0, ϕ_0). Spherical outgoing scattered waves emanate from the flaw: the direction of observation is (θ, ϕ).

contribute, because the origin is always assumed to be inside Σ. Now both I and J are bilinear forms in a_s and a'_s, a total of 2L amplitudes. We can solve for them as follows. Form a positive definite linear combination of I and J:

$$K = \alpha I + (1 - \alpha)J, \tag{58}$$

with $0 \leqslant \alpha \leqslant 1$ and minimize K with respect to variations in a_s and a_s'.

$$\frac{\partial K}{\partial a_s} = 0 \tag{59}$$

$$\frac{\partial K}{\partial a'_s} = 0 \tag{60}$$

Equations (59) and (60) are both matrix equations for the amplitude vectors $\vec{a}$ and $\vec{a}'$ in L dimensions; the matrices occurring in them are surface integrals like Q and $\tilde{Q}$ (Eqs. (20)-(23)).

With the definitions

$$
\begin{aligned}
Q &= \left(\rho\phi^{(+)},\ \rho\phi^{(+)}\right) \\
\tilde{Q} &= \left(\rho\phi^{(+)},\ \rho\tilde{\phi}\right) \\
Q' &= \left(\rho\phi^{(+)},\ \rho'\tilde{\psi}\right) \\
Q'' &= \left(\rho'\tilde{\psi},\ \rho'\tilde{\psi}\right) \\
\tilde{Q}' &= \left(\rho'\tilde{\psi},\ \rho\tilde{\phi}\right),
\end{aligned}
\tag{61}
$$

and a set R, $\tilde{R}$, . . . , which is the same as Q, $\tilde{Q}$, . . . except ρ (and ρ') is replaced with $\hat{n}\cdot\vec{\nabla}$, we construct a set of matrices P, $\tilde{P}$, . . . according to

$$P = \alpha Q + (1 - \alpha)R . \tag{62}$$

Then Eqs. (59) and (60) can be written

$$P'\vec{a}' - \tilde{P}\vec{d} - P\vec{a} = 0 \tag{63}$$

$$P''\vec{a}' - \tilde{P}'\vec{d} - P'^{\dagger}\vec{a} = 0, \tag{64}$$

and the unobserved interior amplitudes $\vec{a}'$ eliminated;

$$
\begin{aligned}
\vec{a} &= -\left[P - P'P''^{-1}P'^{\dagger}\right]^{-1}\left[\tilde{P} - P'P''^{-1}\tilde{P}'\right]\vec{d}. \\
&= T\,\vec{d}
\end{aligned}
\tag{65}
$$

This equation still contains the parameter α, which can be chosen to affect one or more of a number of things: 1) the matrices which must be inverted may be ill-conditioned for some values of α, 2) the rate of convergence of the truncation sequence will depend on α, 3) the accuracy as reflected by how well the optical theorem and/or reciprocity are satisfied will depend on α.

Cracklike defects present a special problem in a scattering calculation. One might hope that a crack could be considered to be the limit of a void as opposite sides squeeze together and the volume goes to zero. This introduces serious problems in the partial wave expansion, because the origin must be inside the crack, and the outgoing partial waves contain irregular Bessel functions. So we would like to consider the crack to be part of a surface which has a mathematical interior. Thus consider Fig. (8), which depicts our view of a plane circular crack. It is simulated by a truncated spherical inclusion, for which the included material is identical to that outside, and the boundary conditions imposed are free-surface both inside and outside on the plane circle, and continuity of both displacement and surface traction is required over the spherical surface.

A straightforward treatment of the crack as a special case of the inclusion is foiled by a well-known feature of displacements and stresses in the neighborhood of a crack edge. Namely, they are singular, the displacements behaving like $\varepsilon^{\frac{1}{2}}$, the stresses like $\varepsilon^{-\frac{1}{2}}$, where ε is the distance from the field point to the edge of the crack. The futility of attempting to describe these with a partial-wave expansion is manifest; fortunately exact solutions in the neighborhood of the edge are known in terms of a small number of parameters. Using them, work on the application of MOOT to cracks is in progress and will be reported elsewhere.

Acknowledgements

Continued interactions with Dr. James E. Gubernatis have been generally informative and often useful.

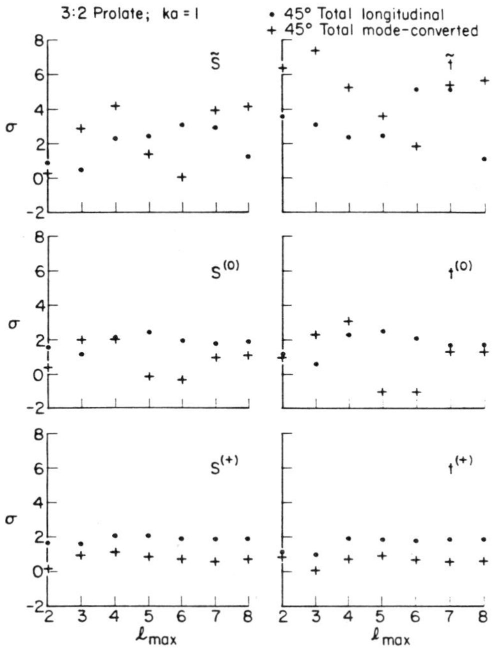

Fig. 2. Total longitudinal (•) and total mode-converted (+) cross-sections for scattering of an incident longitudinal wave with ka = 1 at $(\theta_0, \phi_0) = (45°, 0°)$ from a prolate spheroidal void with aspect ratio 2/3. Six different choices are made for f_s as discussed in the text, the abscissa is the truncation limit ℓ_{max}, and the cross-sections, divided by πa^2, are plotted in db.

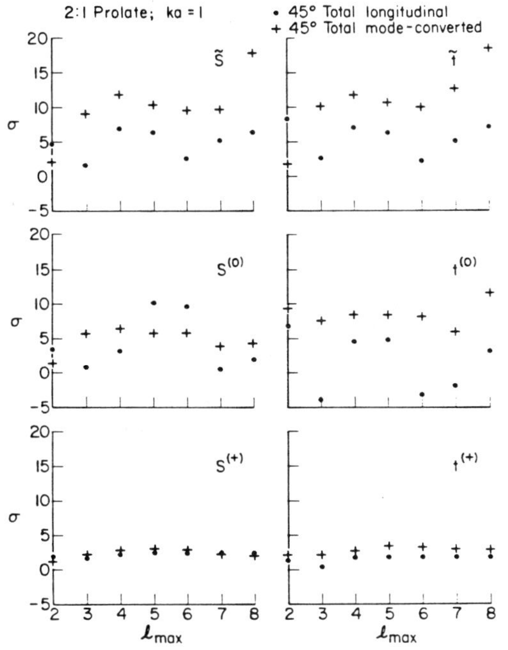

Fig. 3. Same as Fig. 2, but the void now has aspect ratio 1/2. The accuracy of most of the results here has deteriorated; note the change of scale in the ordinate.

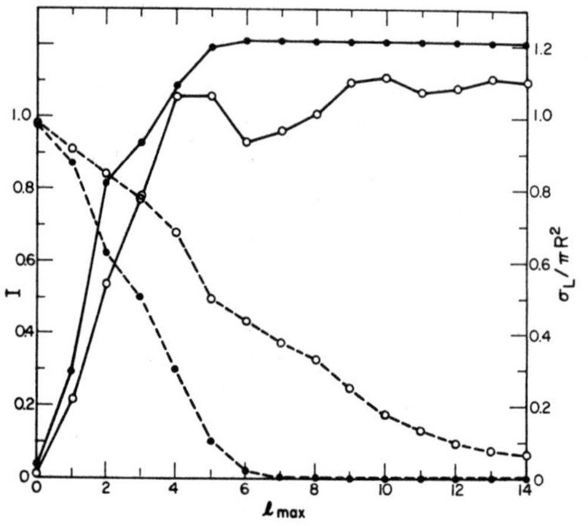

Fig. 4. Normalized surface integrals I (dashed lines) of the square of the vector displacement for scattering from spherical (•) and prolate spheroidal (ȯ) rigid obstacles. Also shown are calculated total longitudinal cross-sections for $(\theta_0, \phi_0) = (45°, 0°)$. The incident wave is longitudinal with ka = 5 (a = radius of the sphere); the spheroid has aspect ratio 2/1 and has the same volume as the sphere.

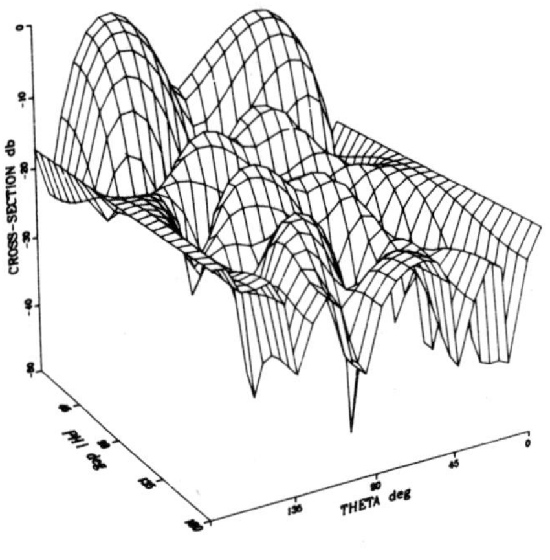

Fig. 5. Differential longitudinal cross-section for a right circular conical void with aspect ratio 1/2. The cone has its flat side up and $(\theta_0, \phi_0) = (135°, 0°)$; thus a specular reflection would be expected at $(\theta, \phi) = (45°, 0°)$. The medium is titanium; the incident wave is longitudinally polarized and has ka = 10.

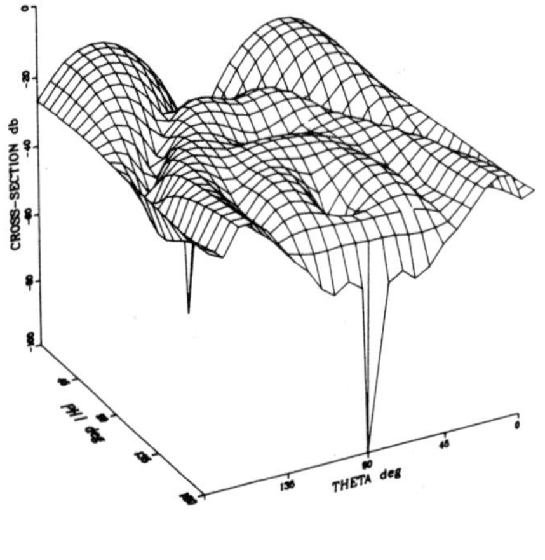

Fig. 6. Differential mode-converted cross-section for scattering of a longitudinal wave incident at $(\theta_0, \phi_0) = (90°, 0°)$ with ka = 10 from an oblate spheroidal void (aspect ratio 1/7) in titanium. Symmetry requires the cross-section to vanish in the forward and backward directions; it is plotted at -100 db.

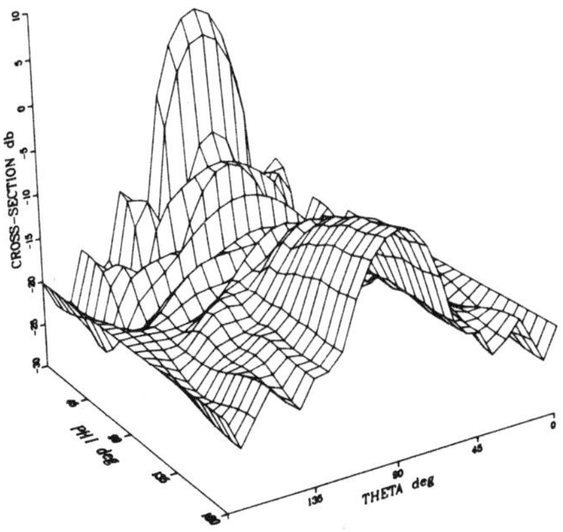

Fig. 7. Longitudinal scattering from a pillbox-shaped void in titanium. The incident wave has ka = 10 and $(\theta_0, \phi_0) = (90°, 0°)$; the aspect ratio is 1/2.

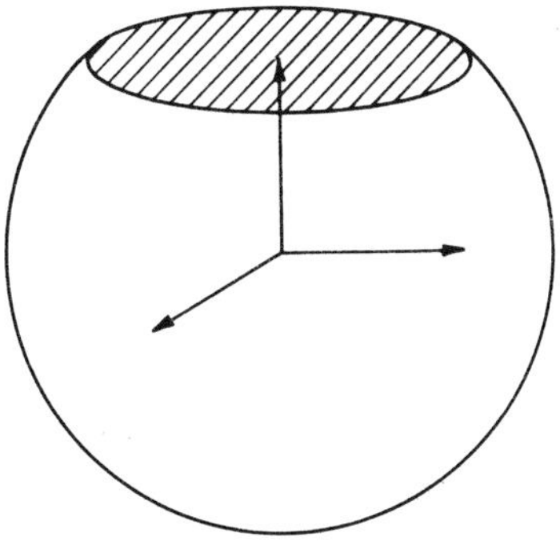

Fig. 8. Simulation of a circular crack by an identical inclusion. Continuous boundary conditions are imposed everywhere except on the cross-hatched circle where free-surface boundary conditions are imposed both from above and from below.

REFERENCES

1. P. C. Waterman, J. Acoust. Soc. Am., 45, 1418 (1968).
2. P. C. Waterman, J. Acoust. Soc. Am., 60, 567 (1976).
3. V. Varatharajulu and Y.-H. Pao, J. Acoust. Soc. Am., 60, 556 (1976).
4. See William M. Visscher, J. Appl. Phys. (in press) or Los Alamos Preprint LA-UR-79-399 for details of the formulation for elastic waves.
5. E. U. Condon and H. Odishaw, eds., "Handbook of Physics" (McGraw-Hill, N.Y. 1967), Part 3, Chapter 7 by P. M. Morse.
6. J. D. Jackson, "Classical Electrodynamics" (Wiley, N.Y., 1962), Chapter 3.
7. V. Varatharajulu, J. Math. Phys. 18, 537 (1977).
8. William M. Visscher, loc. cit. and Los Alamos Preprint LA-UR-78-3008.

Part 4

Moment Methods

APPLICATION OF THE COMBINED-SOURCE FORMULATION TO A CONDUCTING BODY OF REVOLUTION

Roger F. Harrington and Joseph R. Mautz
Department of Electrical and Computer Engineering
Syracuse University, Syracuse, New York 13210, U.S.A.

ABSTRACT

The combined-source integral equation for electromagnetic scattering and radiation is applied to a perfectly conducting body of revolution of arbitrary generating contour. For the scattering problem, the conducting body is assumed to be in a known incident electromagnetic field. For the radiation problem, the tangential component of the electric field is specified over the aperture (a portion of the surface of the body) and is zero elsewhere on the surface. The solutions remain valid at frequencies for which the surface forms a resonant cavity. This is in contrast to the usual H-field and E-field solutions which are not valid at such frequencies. Formulas for the electric current, scattering cross section, and gain are given.

I. INTRODUCTION

There are many integral equation formulations for the problem of electromagnetic scattering and radiation from a conducting body of revolution. All of these formulations are equivalent whenever the operators that must be inverted are nonsingular. (Two formulations are equivalent when the first one implies the second, and the second one implies the first.) Some of the various formulations that have been used extensively for conducting bodies are the T-matrix approach (Ref. 1), the H-field integral equation (Refs. 2,3,8,9), the E-field integral equation (Refs. 4,5,6,8,9), the combined-field integral equation (Refs. 7,8,9), and the combined-source formulation (Refs. 10,11). In this paper we specialize the combined-source formulation to a perfectly conducting body of revolution.

The general theory and some numerical examples for the combined-source solution for conducting bodies are given in (Ref. 10). The combined-source solution applies to a conducting body bounded by a closed surface S. The combined source consists of an a priori unknown electric current $\underline{J}$ and a magnetic current $\underline{M} = \eta \underline{n} \times \underline{J}$ placed simultaneously on S. Here, η is the intrinsic impedance and $\underline{n}$ is the unit normal vector which points outward from S. This combined source is an equivalent source which is assumed to radiate the correct scattered field outside S. An operator equation for $\underline{J}$ is obtained from the E-field boundary value equation which states that the tangential component of the scattered electric field cancels that of the incident electric field on S. The combined-source solution is the method of moments solution of this operator equation for $\underline{J}$.

The combined-source solution for aperture radiation is similar to that for scattering. In the aperture radiation problem, the combined source is an equivalent source assumed to produce the correct radiated field outside S. In this case, the tangential component of the radiated electric field is specified on a portion of S known as the aperture and is zero elsewhere on S.

This paper uses the general formulation for the combined-source solution given in (Ref. 10). Equation numbers drawn from (Ref. 10) are preceded by 10-. For instance, (10-3) denotes equation three of (Ref. 10). The notation and nomenclature used in the present paper are the same as those of (Ref. 10) insofar as possible.

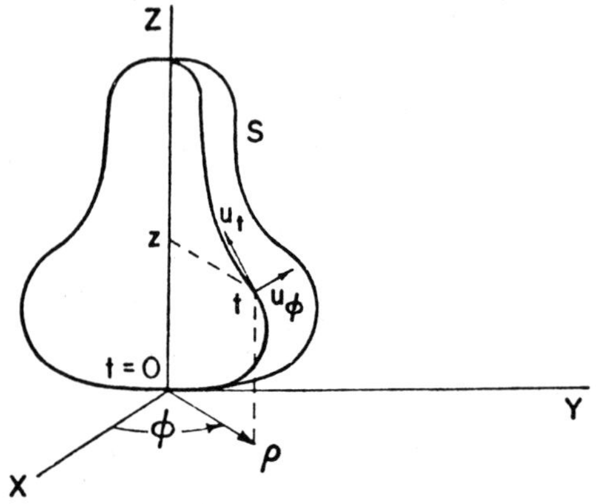

Fig. 1. Body of revolution and coordinate system.

II. COMBINED-SOURCE ELECTRIC CURRENT

The combined-source electric current $\underline{J}$ of (10-21) will be specialized to the body of revolution shown in Fig. 1. In Fig. 1, ρ is the distance from the z axis and (t,ϕ) is an orthogonal coordinate system with unit vectors $\underline{u}_t$ and $\underline{u}_\phi$ on the surface S of the body of revolution.

The expansion functions $\underline{J}_j$ appearing in (10-21) are specialized to

$$\left.\begin{aligned} \underline{J}^t_{nj} &= \underline{u}_t f_j(t)\exp(jn\phi) \\ \underline{J}^\phi_{nj} &= \underline{u}_\phi f_j(t)\exp(jn\phi) \end{aligned}\right\} \begin{aligned} j &= 1,2,\dots N \\ n &= 0, \pm 1, \pm 2\dots \end{aligned} \tag{1}$$

Here, $f_j(t)$ is an arbitrary real function of t. For computations, we took $\rho f_j(t)$ to be the triangle functions as defined by equation (31) of (Ref. 6). The testing functions $\underline{W}_i$ are specialized to

$$\left.\begin{aligned} \underline{W}^t_{ni} &= \underline{u}_t f_i(t)\exp(-jn\phi) \\ \underline{W}^\phi_{ni} &= \underline{u}_\phi f_i(t)\exp(-jn\phi) \end{aligned}\right\} \begin{aligned} i &= 1,2,\dots N \\ n &= 0, \pm 1, \pm 2\dots \end{aligned} \tag{2}$$

In view of (1), (10-21) specializes to

$$\underline{J} = \sum_{n=-\infty}^{\infty} \sum_{j=1}^{N} (I^t_{nj}\underline{J}^t_{nj} + I^\phi_{nj}\underline{J}^\phi_{nj}) \tag{3}$$

The unknown coefficients I^t_{nj} and I^ϕ_{nj} will be obtained by solving (10-23) for $\vec{I}$.

The matrix equation (10-23) can be written as

$$\tilde{Z}^{cf''}\vec{I} = \vec{V} \tag{4}$$

where $Z^{cf''}$ is the combined-field moment matrix which would result if the sets of expansion functions and testing functions were interchanged. The symbol ~ denotes transpose. Because, as shown in (Ref. 9) each element of Z^{cf} which is an interaction of an exp(jnφ) expansion function with an exp(-jmφ) testing function is zero for m ≠ n, the single matrix equation (4) decomposes into several matrix equations given by

$$\tilde{Z}^{cf}_{-n}\vec{I}_n = \vec{V}_n, \quad n = 0, \pm 1, \pm 2, \ldots \tag{5}$$

The subscript n in (5) denotes the exp(jnφ) group of expansion functions (1) and the exp(-jnφ) group of testing functions (2). In (5), the subscript on Z^{cf} is -n because interchanging the sets of expansion and testing functions (1) and (2) is equivalent to replacing n by -n.

Because of the t and φ types of expansion and testing functions, the matrices in (5) partition as

$$Z^{cf}_n = T_n = \begin{bmatrix} T^{tt}_n & T^{t\phi}_n \\ T^{\phi t}_n & T^{\phi\phi}_n \end{bmatrix} \tag{6}$$

$$\vec{V}_n = \begin{bmatrix} \vec{V}^t_n \\ \vec{V}^\phi_n \end{bmatrix} \tag{7}$$

$$\vec{I}_n = \begin{bmatrix} \vec{I}^t_n \\ \vec{I}^\phi_n \end{bmatrix} \tag{8}$$

Here, the ijth elements of T^{tt}_n, $T^{\phi t}_n$, $T^{t\phi}_n$, and $T^{\phi\phi}_n$ are given by

$$(T^{pq}_n)_{ij} = \iint_S \underline{W}^p_{ni} \cdot L^{cf}(\underline{J}^q_{nj})dS \tag{9}$$

where p is either t or φ and q is either t or φ. Also, L^{cf} is the combined-field operator defined by (10-16). The ith elements of $\vec{V}^t_n$ and $\vec{V}^\phi_n$ are given by

$$V^p_{ni} = \frac{1}{\eta}\iint_S \underline{W}^p_{ni} \cdot \underline{E}^i dS \tag{10}$$

The jth element of $\vec{I}^t_n$ is the coefficient I^t_{nj} appearing in (3). Similarly, the jth element of $\vec{I}^\phi_n$ is I^ϕ_{nj}. In (10), $\underline{E}^i$ is the incident electric field.

Now,

$$Z^{cf}_n = Y_n + Z_n \tag{11}$$

where Y_n represents the contribution due to the $-\underline{n} \times \underline{H}(\underline{J})$ term in (10-16) and Z_n represents the contribution due to the $-(1/\eta)\underline{E}_{tan}(\underline{J})$ term in (10-16) for exp(jnφ) expansion functions. Note that Y_n is the H-field moment matrix in (Ref. 9) and that Z_n is the E-field moment matrix in (Ref. 9). It can be shown that

$$\begin{bmatrix} T_{-n}^{tt} & T_{-n}^{t\phi} \\ T_{-n}^{\phi t} & T_{-n}^{\phi\phi} \end{bmatrix} = \begin{bmatrix} T_{n}^{tt} & -T_{n}^{t\phi} \\ -T_{n}^{\phi t} & T_{n}^{\phi\phi} \end{bmatrix} \tag{12}$$

where, as in (6), T_n is short for $\widetilde{Z_n^{cf}}$. Substitution of (12) for Z_{-n}^{cf} in (5) gives

$$\begin{bmatrix} T_{n}^{tt} & -T_{n}^{t\phi} \\ -T_{n}^{\phi t} & T_{n}^{\phi\phi} \end{bmatrix} \begin{bmatrix} \vec{I}_n^{t} \\ \vec{I}_n^{\phi} \end{bmatrix} = \begin{bmatrix} \vec{V}_n^{t} \\ \vec{V}_n^{\phi} \end{bmatrix} \tag{13}$$

A more convenient form of (13) is

$$\tilde{T}_n \begin{bmatrix} \vec{I}_n^{t} \\ -\vec{I}_n^{\phi} \end{bmatrix} = \begin{bmatrix} \vec{V}_n^{t} \\ -\vec{V}_n^{\phi} \end{bmatrix} \tag{14}$$

In view of (6) and (11), (14) becomes

$$\widetilde{[Y_n + Z_n]} \begin{bmatrix} \vec{I}_n^{t} \\ -\vec{I}_n^{\phi} \end{bmatrix} = \begin{bmatrix} \vec{V}_n^{t} \\ -\vec{V}_n^{\phi} \end{bmatrix} \tag{15}$$

Equation (15) determines the column vectors $\vec{I}_n^t$ and $\vec{I}_n^{\phi}$ of coefficients $I_{nj}^{t\phi}$ and I_{nj} in (3). The solution to (15) by LU decomposition of $Y_n + Z_n$ is given in (refs. 12,13).

III. SURFACE MAGNETIC FIELD DUE TO THE COMBINED SOURCE

The quantity $\underline{n} \times \underline{H}(\underline{J},\underline{M})$ of (10-30) will be specialized to the body of revolution. In view of (1), (10-30) becomes

$$\underline{n} \times \underline{H}(\underline{J},\underline{M}) = \sum_{n=-\infty}^{\infty} \sum_{j=1}^{N} (C_{nj}^{t} \underline{J}_{nj}^{t} + C_{nj}^{\phi} \underline{J}_{nj}^{\phi}) \tag{16}$$

The unknown coefficients C_{nj}^t and C_{nj}^{ϕ} will be obtained by solving (10-36) for $\vec{C}$.

The column vector on the left-hand side of (10-36) will be calculated next. Let

$$\vec{V}' = Z^{cs'}\vec{I} \tag{17}$$

Because the combined-source moment matrix is the transpose of the combined-field matrix which would result if the set of expansion functions and the set of testing functions were interchanged, (17) becomes

$$\vec{V}' = \tilde{Z}^{cf''}\vec{I} \tag{18}$$

where $Z^{cf''}$ is the combined-field moment matrix which would result if the jth expansion function were $\underline{n} \times \underline{W}_j$ and if the ith testing function were $\underline{n} \times \underline{J}_i$. This $Z^{cf''}$ is not to be confused with that in (4). In view of

$$\underline{n} \times \underline{u}_t = -\underline{u}_\phi$$
$$\underline{n} \times \underline{u}_\phi = \underline{u}_t \qquad (19)$$

and the fact that interchanging the sets of expansion and testing functions is equivalent to replacing n by -n, substitution of (6) into (18) gives

$$\begin{bmatrix} \vec{V}_n^{t\prime} \\ \vec{V}_n^{\phi\prime} \end{bmatrix} = \widetilde{\begin{bmatrix} T_{-n}^{\phi\phi} & -T_{-n}^{\phi t} \\ -T_{-n}^{t\phi} & T_{-n}^{tt} \end{bmatrix}} \begin{bmatrix} \vec{I}_n^{t} \\ \vec{I}_n^{\phi} \end{bmatrix}, \quad n=0, \pm 1, \pm 2 \ldots \qquad (20)$$

where the left-hand side of (20) is the portion of $\vec{V}'$ due to the portion of $\vec{I}$ given by (8). In view of (12), all minus signs in (20) disappear. A more convenient form of (20) is

$$\begin{bmatrix} \vec{V}_n^{\phi\prime} \\ \vec{V}_n^{t\prime} \end{bmatrix} = \tilde{Z}_n^{cf} \begin{bmatrix} \vec{I}_n^{\phi} \\ \vec{I}_n^{t} \end{bmatrix}, \quad n = 0, \pm 1, \pm 2 \qquad (21)$$

In view of (1) and (2), the matrix D appearing in (10-36) specializes to

$$\begin{bmatrix} D & 0 \\ 0 & D \end{bmatrix}, \quad n = 0, \pm 1, \pm 2, \ldots \qquad (22)$$

where D is an N × N matrix whose ijth element is given by

$$D_{ij} = \iint_S f_i(t)\, f_j(t) dS \qquad (23)$$

Consequently, (10-36) becomes

$$D\vec{C}_n^{t} = \vec{V}_n^{t\prime}, \quad n = 0, \pm 1, \pm 2, \ldots \qquad (24)$$

$$D\vec{C}_n^{\phi} = \vec{V}_n^{\phi\prime}, \quad n = 0, \pm 1, \pm 2, \ldots \qquad (25)$$

where $\vec{V}_n^{t\prime}$ and $\vec{V}_n^{\phi\prime}$ are given by (21). Equations (24) and (25) determine $\vec{C}_n^t$ and $\vec{C}_n^\phi$. The jth element of $\vec{C}_n^t$ is the coefficient C_{nj}^t appearing in (16). Similarly, the jth element of $\vec{C}_n^\phi$ is C_{nj}^ϕ. The solution to a system of equations with tridiagonal matrix such as (24) or (25) is described in Appendix B of (Ref. 13).

IV. ELECTRIC FIELD MEASUREMENT

The $\underline{u}^r$ component of the electric field $\underline{E}(\underline{J},\underline{M})$ due to the combined source $(\underline{J},\underline{M})$ is given by (10-62). If, instead of (10-60) and (10-61), $\underline{E}(\underline{u}^r/G_r)$ and $\underline{H}(\underline{u}^r/G_r)$ are substituted into (10-41) and (10-42) where $\underline{u}^r/G_r$ is a current element located at an arbitrary point $\underline{r}_r$ not necessarily distant from S, then (10-62) will give $\underline{E}(\underline{J},\underline{M}) \cdot \underline{u}^r$ at $\underline{r}_r$.

In view of (10-23), (10-62) can be written as

$$\underline{E}(\underline{J},\underline{M}) \cdot \underline{u}^r = \eta G_r [\tilde{R} + \tilde{R}'] \vec{I} \qquad (26)$$

With the expansion and testing functions (1) and (2), $\vec{I}$ goes over into $\vec{I}_n^t$ and $\vec{I}_n^\phi$ which depend on $\vec{V}_n^t$ and $\vec{V}_n^\phi$ according to (15). Furthermore, $\vec{R}$ and $\vec{R}'$ with elements (10-41) and (10-42) go over into vectors $\vec{R}_n^p \exp(jn\phi_r)$ and $\vec{R}_n^{p'} \exp(jn\phi_r)$ where the jth elements of $\vec{R}_n^p$ and $\vec{R}_n^{p'}$ are given by

$$R_{nj}^p = \frac{e^{-jn\phi_r}}{\eta} \iint_S \underline{E}(\underline{u}^r/G_r) \cdot \underline{J}_{nj}^p \, dS \tag{27}$$

$$R_{nj}^{p'} = e^{-jn\phi_r} \iint_S \underline{n} \times \underline{H}(\underline{u}^r/G_r) \cdot \underline{J}_{nj}^p \, dS \tag{28}$$

where p may be either t or ϕ. If ϕ_r is the azimuth of the receiving current element $\underline{u}^r/G_r$, then expressions (27) and (28) do not depend on ϕ_r. In view of the above mentioned forms of $\vec{I}$, $\vec{R}$, and $\vec{R}'$, (26) becomes

$$\underline{E}(\underline{J},\underline{M}) \cdot \underline{u}^r = \eta G_r \sum_{n=-\infty}^{\infty} [\tilde{R}_n^t + \tilde{R}_n^{t'} \quad \tilde{R}_n^\phi + \tilde{R}_n^{\phi'}] \begin{bmatrix} \vec{I}_n^t \\ \vec{I}_n^\phi \end{bmatrix} e^{jn\phi_r} \tag{29}$$

In (29), G_r is given by (10-57), the elements of $\vec{R}_n^p$ and $\vec{R}_n^{p'}$ are given by (27) and (28) and $\vec{I}_n^p$ depends on $\vec{V}_n^p$ according to (15). The point $\underline{r}_r$ at which the electric field is evaluated in (29) is arbitrary.

For large r_r, the fields $\underline{E}$ and $\underline{H}$ in (27) and (28) approach (10-60) and (10-61) when $\underline{u}^r$ is tangent to the radiation sphere and zero when $\underline{u}^r$ is radial. If $\underline{u}_\theta^r$ is first substituted for $\underline{u}^r$ in (29) and then $\underline{u}_\phi^r$ is substituted for $\underline{u}^r$ in (29), then

$$\begin{bmatrix} \underline{E}(\underline{J},\underline{M}) \cdot \underline{u}_\theta^r \\ \underline{E}(\underline{J},\underline{M}) \cdot \underline{u}_\phi^r \end{bmatrix} = \eta G_r \sum_{n=-\infty}^{\infty} \begin{bmatrix} \tilde{R}_n^{t\theta} + \tilde{R}_n^{t\theta'} & \tilde{R}_n^{\phi\theta} + \tilde{R}_n^{\phi\theta'} \\ \tilde{R}_n^{t\phi} + \tilde{R}_n^{t\phi'} & \tilde{R}_n^{\phi\phi} + \tilde{R}_n^{\phi\phi'} \end{bmatrix} \begin{bmatrix} \vec{I}_n^t \\ \vec{I}_n^\phi \end{bmatrix} e^{jn\phi_r} \tag{30}$$

where the jth elements of the vectors $\vec{R}_n^{pq}$ and $\vec{R}_n^{pq'}$ are given by

$$R_{nj}^{pq} = k e^{-jn\phi_r} \iint_S \underline{J}_{nj}^p \cdot \underline{u}_q^r \, e^{-j\underline{k}_r \cdot \underline{r}} \, dS \tag{31}$$

$$R_{nj}^{pq'} = - e^{-jn\phi_r} \iint_S (\underline{n} \times \underline{J}_{nj}^p) \cdot (\underline{k}_r \times \underline{u}_q^r) e^{-j\underline{k}_r \cdot \underline{r}} \, dS \tag{32}$$

Here, p may be either t or ϕ and q may be either θ or ϕ. The first superscript on $\vec{R}_n$ and $\vec{R}_n'$ in (30) is the same as the superscript in (29) and denotes the polarization of the expansion function. The second superscript on $\vec{R}_n$ and $\vec{R}_n'$ in (30) distinguishes the $\underline{u}_\theta^r$ polarized measurement from the $\underline{u}_\phi^r$ polarized measurement. Here, $\underline{u}_\theta^r$ and $\underline{u}_\phi^r$ are respectively the θ and ϕ directed unit vectors at the measurement point. In (31), k is the propagation constant, $\underline{k}_r$ is the propagation vector of the plane wave coming from the measurement point, and $\underline{r}$ is the position vector of dS. The vectors $\underline{u}_\theta^r$, $\underline{u}_\phi^r$, and $\underline{k}_r$ are shown along with the bearing (θ_r, ϕ_r) of the measurement point in Fig. 2.

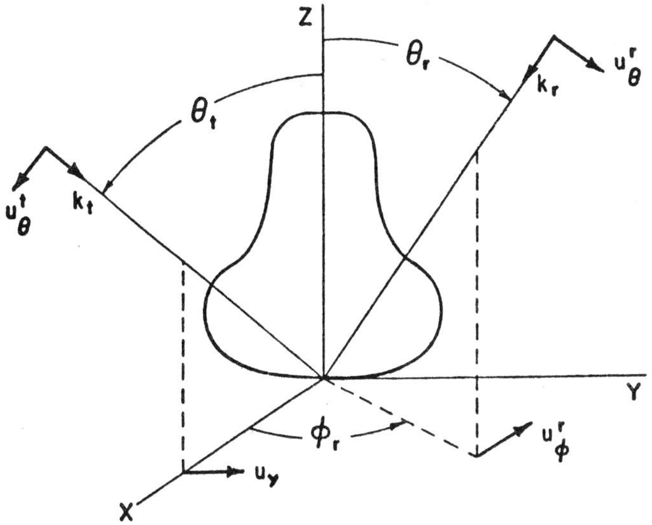

Fig. 2. Plane wave scattering by a conducting body of revolution.

It is evident from (1), (19), and

$$\begin{aligned} \underline{k}_r \times \underline{u}^r_\theta &= -\,k\,\underline{u}^r_\phi \\ \underline{k}_r \times \underline{u}^r_\phi &= \,k\,\underline{u}^r_\theta \end{aligned} \tag{33}$$

that the column vectors generated by (32) are related to those generated by (31) according to

$$\begin{bmatrix} \vec{R}_n^{t\theta'} & \vec{R}_n^{t\phi'} \\ \vec{R}_n^{\phi\theta'} & \vec{R}_n^{\phi\phi'} \end{bmatrix} = \begin{bmatrix} -\vec{R}_n^{\phi\phi} & \vec{R}_n^{\phi\theta} \\ \vec{R}_n^{t\phi} & -\vec{R}_n^{t\theta} \end{bmatrix} \tag{34}$$

Substitution of (34) into (30) gives

$$\begin{bmatrix} \underline{E}(\underline{J},\underline{M}) \cdot \underline{u}^r_\theta \\ \underline{E}(\underline{J},\underline{M}) \cdot \underline{u}^r_\phi \end{bmatrix} = \eta G_r \sum_{n=-\infty}^{\infty} \begin{bmatrix} \tilde{R}_n^{t\theta} - \tilde{R}_n^{\phi\phi} & \tilde{R}_n^{\phi\theta} + \tilde{R}_n^{t\phi} \\ \tilde{R}_n^{t\phi} + \tilde{R}_n^{\phi\theta} & \tilde{R}_n^{\phi\phi} - \tilde{R}_n^{t\theta} \end{bmatrix} \begin{bmatrix} \vec{I}_n^{t} \\ \vec{I}_n^{\phi} \end{bmatrix} e^{jn\phi_r} \tag{35}$$

The column vectors generated by (31) can be shown to satisfy

$$\begin{bmatrix} \vec{R}_{-n}^{t\theta} & \vec{R}_{-n}^{t\phi} \\ \vec{R}_{-n}^{\phi\theta} & \vec{R}_{-n}^{\phi\phi} \end{bmatrix} = \begin{bmatrix} \vec{R}_n^{t\theta} & -\vec{R}_n^{t\phi} \\ -\vec{R}_n^{\phi\theta} & \vec{R}_n^{\phi\phi} \end{bmatrix} \tag{36}$$

It follows from (36) and (34) that the measurement matrix in (35) satisfies

$$\begin{bmatrix} \tilde{R}^{t\theta}_{-n} - \tilde{R}^{\phi\phi}_{-n} & \tilde{R}^{\phi\theta}_{-n} + \tilde{R}^{t\phi}_{-n} \\ \\ \tilde{R}^{t\phi}_{-n} + \tilde{R}^{\phi\theta}_{-n} & \tilde{R}^{\phi\phi}_{-n} - \tilde{R}^{t\theta}_{-n} \end{bmatrix} = \begin{bmatrix} \tilde{R}^{t\theta}_{n} - \tilde{R}^{\phi\phi}_{n} & -\tilde{R}^{\phi\theta}_{n} - \tilde{R}^{t\phi}_{n} \\ \\ -\tilde{R}^{t\phi}_{n} - \tilde{R}^{\phi\theta}_{n} & \tilde{R}^{\phi\phi}_{n} - \tilde{R}^{t\theta}_{n} \end{bmatrix} . \tag{37}$$

V. PLANE WAVE SCATTERING

We consider separately a θ polarized incident plane wave defined by

$$\begin{aligned} \underline{E}^i &= k\eta\, \underline{u}^t_\theta\, e^{-j\underline{k}_t \cdot \underline{r}} \\ \\ \underline{H}^i &= (\underline{k}_t \times \underline{u}^t_\theta) e^{-j\underline{k}_t \cdot \underline{r}} \end{aligned} \tag{38}$$

and a ϕ polarized incident plane wave defined by

$$\begin{aligned} \underline{E}^i &= k\eta\, \underline{u}^t_\phi\, e^{-j\underline{k}_t \cdot \underline{r}} \\ \\ \underline{H}^i &= (\underline{k}_t \times \underline{u}^t_\phi) e^{-j\underline{k}_t \cdot \underline{r}} \end{aligned} \tag{39}$$

The above fields have been obtained by replacing $\underline{u}^t$ in (10-58) and (10-59) first by $\underline{u}^t_\theta$ and then by $\underline{u}^t_\phi$. Here, $\underline{u}^t_\theta$ and $\underline{u}^t_\phi$ are, respectively, the unit vectors in the θ and ϕ directions at the transmitter bearing $(\theta_t, 0)$. Since the azimuth of the transmitter is zero, the propagation vector $\underline{k}_t$ lies in the xz plane and $\underline{u}^t_\phi$ is actually the unit vector $\underline{u}_y$ in the y direction. See Fig. 2. In (38) and (39), $\underline{r}$ is the radius vector from the origin. The origin is on the axis of the body of revolution, but not necessarily at the lower pole as shown in Fig. 2.

The column vectors $\vec{V}^p_n$ from which $\vec{I}^p_n$ in (29) is determined according to (15) are obtained by substituting (38) and (39) into (10). As a result, column vectors $\vec{V}^{pq}_n$ whose ith elements are given by

$$V^{pq}_{ni} = k \iint_S \underline{W}^p_{ni} \cdot \underline{u}^t_q\, e^{-j\underline{k}_t \cdot \underline{r}}\, dS \tag{40}$$

are generated. Here, p may be either t or ϕ and q may be either θ or ϕ. The superscript p is the same as in (10) and denotes the polarization of the testing function. The superscript q denotes the polarization of the incident plane wave.

In view of (1), (2), and (36), comparison of (40) and (31) shows that

$$\begin{bmatrix} \vec{V}^{t\theta}_n & \vec{V}^{t\phi}_n \\ \\ \vec{V}^{\phi\theta}_n & \vec{V}^{\phi\phi}_n \end{bmatrix} = \begin{bmatrix} \vec{R}^{t\theta}_n & -\vec{R}^{t\phi}_n \\ \\ -\vec{R}^{\phi\theta}_n & \vec{R}^{\phi\phi}_n \end{bmatrix} \tag{41}$$

where the right-hand side of (41) is evaluated at the transmitter angle θ_t instead of the receiver angle θ_r. From (41) and (36),

$$\begin{bmatrix} \vec{V}_{-n}^{t\theta} & \vec{V}_{-n}^{t\phi} \\ \vec{V}_{-n}^{\phi\theta} & \vec{V}_{-n}^{\phi\phi} \end{bmatrix} = \begin{bmatrix} \vec{V}_{n}^{t\theta} & -\vec{V}_{n}^{t\phi} \\ -\vec{V}_{n}^{\phi\theta} & \vec{V}_{n}^{\phi\phi} \end{bmatrix} \tag{42}$$

If each column vector in (15) is replaced by two column vectors, one for the θ polarized incident plane wave and the other for the ϕ polarized incident wave, the result is

$$\widetilde{[Y_n + Z_n]} \begin{bmatrix} \vec{I}_{n}^{t\theta} & \vec{I}_{n}^{t\phi} \\ -\vec{I}_{n}^{\phi\theta} & -\vec{I}_{n}^{\phi\phi} \end{bmatrix} = \begin{bmatrix} \vec{V}_{n}^{t\theta} & \vec{V}_{n}^{t\phi} \\ -\vec{V}_{n}^{\phi\theta} & -\vec{V}_{n}^{\phi\phi} \end{bmatrix} \tag{43}$$

The first superscript on each column vector in (43) is the same as in (15). The second superscript denotes the polarization of the incident plane wave. From (12) in which T_n is a shorthand notation for $Y_n + Z_n$ and from (42), it is apparent that the solutions to (43) with n replaced by minus n are related to the solutions of the original (43) by

$$\begin{bmatrix} \vec{I}_{-n}^{t\theta} & \vec{I}_{-n}^{t\phi} \\ \vec{I}_{-n}^{\phi\theta} & \vec{I}_{-n}^{\phi\phi} \end{bmatrix} = \begin{bmatrix} \vec{I}_{n}^{t\theta} & -\vec{I}_{n}^{t\phi} \\ -\vec{I}_{n}^{\phi\theta} & \vec{I}_{n}^{\phi\phi} \end{bmatrix} \tag{44}$$

In view of (1) and (44), the combined source electric current (3) becomes

$$\underline{J}^{\theta} = (\tilde{f}\ \vec{I}_{o}^{t\theta})\underline{u}_t + \sum_{n=1}^{\infty} \{2(\tilde{f}\ \vec{I}_{n}^{t\theta})\underline{u}_t \cos(n\phi) + 2j(\tilde{f}\ \vec{I}_{n}^{\phi\theta})\underline{u}_{\phi} \sin(n\phi)\} \tag{45}$$

for the θ polarized incident plane wave and

$$\underline{J}^{\phi} = (\tilde{f}\ \vec{I}_{o}^{\phi\phi})\underline{u}_{\phi} + \sum_{n=1}^{\infty} \{2j(\tilde{f}\ \vec{I}_{n}^{t\phi})\underline{u}_t \sin(n\phi) + 2(\tilde{f}\ \vec{I}_{n}^{\phi\phi})\underline{u}_{\phi} \cos(n\phi)\} \tag{46}$$

for the ϕ polarized incident plane wave. The superscript on $\underline{J}$ denotes the polarization of the incident plane wave. In (45) and (46), $\tilde{f}$ is a row vector of the functions $f_j(t)$ appearing in (1). For axial incidence, the only non-zero excitation vectors on the right-hand side of (43) are those for which $n = \underline{+}1$ and, consequently, (45) reduces to

$$\underline{J}^{\theta} = 2(\tilde{f}\ \vec{I}_{1}^{t\theta})\underline{u}_t \cos\phi + 2j(\tilde{f}\ \vec{I}_{1}^{\phi\theta})\underline{u}_{\phi} \sin\phi \tag{47}$$

In this case $\underline{J}^{\phi}$ contains no new information because $\underline{J}^{\phi}$ is $\underline{J}^{\theta}$ with ϕ replaced by either $(\phi - 90°)$ or $(\phi + 90°)$ depending on whether θ_t is either 0° or 180°.

For plane wave incidence, (24) and (25) become

$$\begin{bmatrix} D & 0 \\ 0 & D \end{bmatrix} \begin{bmatrix} \vec{C}_n^{t\theta} & \vec{C}_n^{t\phi} \\ \vec{C}_n^{\phi\theta} & \vec{C}_n^{\phi\phi} \end{bmatrix} = \begin{bmatrix} \vec{V}_n^{t\theta'} & \vec{V}_n^{t\phi'} \\ \vec{V}_n^{\phi\theta'} & \vec{V}_n^{\phi\phi'} \end{bmatrix} \tag{48}$$

where

$$\begin{bmatrix} \vec{V}_n^{\phi\theta'} & \vec{V}_n^{\phi\phi'} \\ \vec{V}_n^{t\theta'} & \vec{V}_n^{t\phi'} \end{bmatrix} = \tilde{Z}_n^{cf} \begin{bmatrix} \vec{I}_n^{\phi\theta} & \vec{I}_n^{\phi\phi} \\ \vec{I}_n^{t\theta} & \vec{I}_n^{t\phi} \end{bmatrix} \tag{49}$$

The second subscript on the column vectors in (48) and (49) denotes the polarization of the incident plane wave. In view of (12) in which T_n is short for Z_n^{cf} and (44), the column vectors $\vec{V}_n^{pq'}$ given by (49) satisfy

$$\begin{bmatrix} \vec{V}_{-n}^{t\theta'} & \vec{V}_{-n}^{t\phi'} \\ \vec{V}_{-n}^{\phi\theta'} & \vec{V}_{-n}^{\phi\phi'} \end{bmatrix} = \begin{bmatrix} \vec{V}_n^{t\theta'} & -\vec{V}_n^{t\phi'} \\ -\vec{V}_n^{\phi\theta'} & \vec{V}_n^{\phi\phi'} \end{bmatrix} \tag{50}$$

It follows from (48) and (50) that

$$\begin{bmatrix} \vec{C}_{-n}^{t\theta} & \vec{C}_{-n}^{t\phi} \\ \vec{C}_{-n}^{\phi\theta} & \vec{C}_{-n}^{\phi\phi} \end{bmatrix} = \begin{bmatrix} \vec{C}_n^{t\theta} & -\vec{C}_n^{t\phi} \\ -\vec{C}_n^{\phi\theta} & \vec{C}_n^{\phi\phi} \end{bmatrix} \tag{51}$$

In view of (51), substitution of (1) into (16) gives

$$\underline{n} \times \underline{H}^{\theta}(\underline{J},\underline{M}) = (\tilde{f}\,\vec{C}_0^{t\theta})\underline{u}_t + \sum_{n=1}^{\infty} \{2(\tilde{f}\,\vec{C}_n^{t\theta})\underline{u}_t \cos(n\phi) + 2j(\tilde{f}\,\vec{C}_n^{\phi\theta})\underline{u}_\phi \sin(n\phi)\} \tag{52}$$

for the θ polarized incident plane wave and

$$\underline{n} \times \underline{H}^{\phi}(\underline{J},\underline{M}) = (\tilde{f}\,\vec{C}_0^{\phi\phi})\underline{u}_\phi + \sum_{n=1}^{\infty} \{2j(\tilde{f}\,\vec{C}_n^{t\phi})\underline{u}_t \sin(n\phi) + 2(\tilde{f}\,\vec{C}_n^{\phi\phi})\underline{u}_\phi \cos(n\phi)\} \tag{53}$$

for the ϕ polarized incident plane wave. The superscript on $\underline{H}(\underline{J},\underline{M})$ denotes the polarization of the incident plane wave. For axial incidence, (52) reduces to

$$\underline{n} \times \underline{H}^{\theta}(\underline{J},\underline{M}) = 2(\tilde{f}\,\vec{C}_1^{t\theta})\underline{u}_t \cos\phi + 2j(\tilde{f}\,\vec{C}_1^{\phi\theta})\underline{u}_\phi \sin\phi \tag{54}$$

The scattered electric field $\underline{E}(\underline{J},\underline{M})$ at an arbitrary point $\underline{r}_r$ is given by (29). However, $\vec{I}_n^p$ therein is for arbitrary excitation. Now, $\vec{I}_n^{p\theta}$ is for the θ polarized incident plane wave (38) and $\vec{I}_n^{p\phi}$ is for the ϕ polarized incident plane wave (39). Hence,

$$\underline{E}(\underline{J},\underline{M}) \cdot \underline{u}^r = \eta G_r \sum_{n=-\infty}^{\infty} [\tilde{R}_n^t + \tilde{R}_n^{t'} \quad \tilde{R}_n^{\phi} + \tilde{R}_n^{\phi'}] \begin{bmatrix} \vec{I}_n^{t\theta} \\ \vec{I}_n^{\phi\theta} \end{bmatrix} e^{jn\phi_r} \tag{55}$$

for the θ polarized incident plane wave and

$$\underline{E}(\underline{J},\underline{M}) \cdot \underline{u}^r = \eta G_r \sum_{n=-\infty}^{\infty} [\tilde{R}_n^t + \tilde{R}_n^{t'} \quad \tilde{R}_n^{\phi} + \tilde{R}_n^{\phi'}] \begin{bmatrix} \vec{I}_n^{t\phi} \\ \vec{I}_n^{\phi\phi} \end{bmatrix} e^{jn\phi_r} \tag{56}$$

for the ϕ polarized incident plane wave where, as given by (10-57),

$$G_r = \frac{-je^{-jkr_r}}{4\pi r_r} \tag{57}$$

A similar specialization of the far field result (35) to the θ and ϕ polarized plane waves (38) and (39) gives

$$E^{\theta\theta} = \eta G_r (S_o^{\theta\theta} + 2 \sum_{n=1}^{\infty} S_n^{\theta\theta} \cos(n\phi_r)) \tag{58}$$

$$E^{\phi\theta} = 2j \eta G_r \sum_{n=1}^{\infty} S_n^{\phi\theta} \sin(n\phi_r) \tag{59}$$

$$E^{\theta\phi} = 2j \eta G_r \sum_{n=1}^{\infty} S_n^{\theta\phi} \sin(n\phi_r) \tag{60}$$

$$E^{\phi\phi} = \eta G_r (S_o^{\phi\phi} + 2 \sum_{n=1}^{\infty} S_n^{\phi\phi} \cos(n\phi_r)) \tag{61}$$

where

$$\begin{bmatrix} S_n^{\theta\theta} & S_n^{\theta\phi} \\ S_n^{\phi\theta} & S_n^{\phi\phi} \end{bmatrix} = \begin{bmatrix} \tilde{R}_n^{t\theta} - \tilde{R}_n^{\phi\phi} & \tilde{R}_n^{\phi\theta} + \tilde{R}_n^{t\phi} \\ \tilde{R}_n^{t\phi} + \tilde{R}_n^{\phi\theta} & \tilde{R}_n^{\phi\phi} - \tilde{R}_n^{t\theta} \end{bmatrix} \begin{bmatrix} \vec{I}_n^{t\theta} & \vec{I}_n^{t\phi} \\ \vec{I}_n^{\phi\theta} & \vec{I}_n^{\phi\phi} \end{bmatrix} \tag{62}$$

The left-hand sides of (58) to (61) are components of the far scattered electric fields for plane wave incidences. The first superscript on E denotes the receiver polarization and the second superscript on E denotes the polarization of the incident plane wave.

The scattering cross section per square wavelength σ/λ^2 is given by (10-64) in which, from (10-62), the quantity enclosed by the magnitude signs is the appropriate component of the scattered field stripped of its ηG_r factor. The scattered field components (58) to (61) give rise to

$$\frac{\sigma^{\theta\theta}}{\lambda^2} = \frac{1}{4\pi^3} \left| \frac{1}{2} S_o^{\theta\theta} + \sum_{n=1}^{\infty} S_n^{\theta\theta} \cos(n\phi_r) \right|^2 \tag{63}$$

$$\frac{\sigma^{\phi\theta}}{\lambda^2} = \frac{1}{4\pi^3} \left| \sum_{n=1}^{\infty} S_n^{\phi\theta} \sin(n\phi_r) \right|^2 \tag{64}$$

$$\frac{\sigma^{\theta\phi}}{\lambda^2} = \frac{1}{4\pi^3}\left| \sum_{n=1}^{\infty} S_n^{\theta\phi} \sin(n\phi_r) \right|^2 \tag{65}$$

$$\frac{\sigma^{\phi\phi}}{\lambda^2} = \frac{1}{4\pi^3}\left| \frac{1}{2} S_o^{\phi\phi} + \sum_{n=1}^{\infty} S_n^{\phi\phi} \cos(n\phi_r) \right|^2 \tag{66}$$

where the superscripts on σ/λ^2 coincide with those on the left-hand sides of (58) to (61). For axial incidence, only the S_1^{pq} terms are present in (58) to (61) and (63) to (66).

VI. APERTURE RADIATION

For aperture radiation, the ith elements of the excitation vectors $\vec{V}^p$ are obtained by replacing $\underline{E}^i$ in (10) by $-\underline{E}^a$, where $\underline{E}^a$ is the aperture electric field. Hence,

$$V_{ni}^p = -\frac{1}{\eta}\iint_S \underline{W}_{ni}^p \cdot \underline{E}^a \, dS \tag{67}$$

If one agrees to replace (10) by (67), then the formulas for the combined-source electric current $\underline{J}$ in Section II, the surface field $\underline{n} \times \underline{H}(\underline{J},\underline{M})$ in Section III, and the measurements of the electric field $\underline{E}(\underline{J},\underline{M})$ in Section IV are valid for aperture radiation.

The gain is given by (10-69) in which, from (10-62), the quantity enclosed by the magnitude signs is the appropriate component of $\underline{E}(\underline{J},\underline{M})$ stripped of its ηG_r factor. The electric field components (35) give rise to

$$G^\theta = \frac{\eta}{4\pi P}\left| \sum_{n=-\infty}^{\infty} G_n^\theta e^{jn\phi_r} \right|^2 \tag{68}$$

$$G^\phi = \frac{\eta}{4\pi P}\left| \sum_{n=-\infty}^{\infty} G_n^\phi e^{jn\phi_r} \right|^2 \tag{69}$$

where

$$\begin{bmatrix} G_n^\theta \\ \\ G_n^\phi \end{bmatrix} = \begin{bmatrix} \tilde{R}_n^{t\theta} - \tilde{R}_n^{\phi\phi} & \tilde{R}_n^{\phi\theta} + \tilde{R}_n^{t\phi} \\ \\ \tilde{R}_n^{t\phi} + \tilde{R}_n^{\phi\theta} & \tilde{R}_n^{\phi\phi} - \tilde{R}_n^{t\theta} \end{bmatrix} \begin{bmatrix} \vec{I}_n^t \\ \\ \vec{I}_n^\phi \end{bmatrix} \tag{70}$$

The superscript on G on the left-hand sides of (68) and (69) denotes the receiver polarization. In (70), $\vec{I}_n^t$ and $\vec{I}_n^\phi$ are obtained by solving (15) when the elements of $\vec{V}_n^t$ and $\vec{V}_n^\phi$ are given by (67).

If the aperture electric field is axially symmetric, then only the n = 0 terms are present in (68) and (69). The coefficients G_o^θ and G_o^ϕ are especially easy to calculate because

$$T_o^{\phi t} = T_o^{t\phi} = 0 \tag{71}$$

$$R_o^{\phi\theta} = R_o^{t\phi} = 0 \tag{72}$$

In (71), T_o is short for $Y_o + Z_o$ which appears in (15).

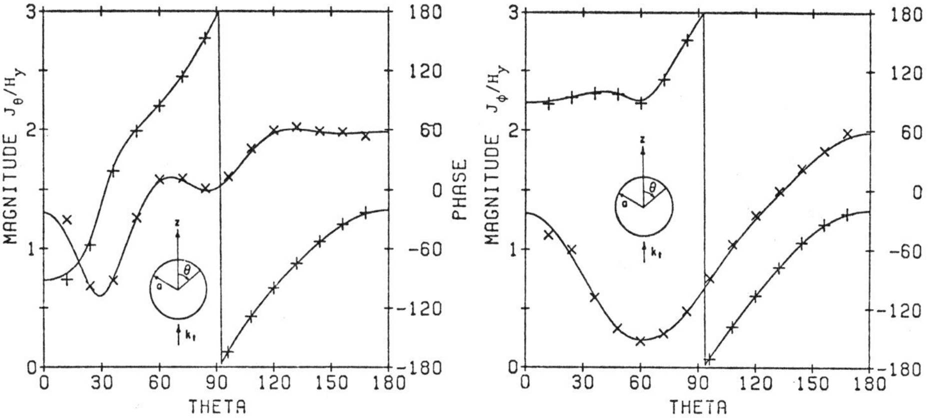

(a) (b)

Fig. 3. Electric current J_θ/H_y and J_ϕ/H_y induced on a conducting sphere, ka = 2.75, by an axially incident plane wave. Solid curves represent the exact solution. Symbols × and + denote magnitude and phase, respectively, of the combined-source solution.

VII. EXAMPLES

Computer programs to implement the formulas derived in this paper are available in a research report (Ref. 13). Some computed results are shown in this section.

Figure 3 shows the electric current $\underline{J}^c$ induced by a plane wave axially incident on a conducting sphere for which ka = 2.75, where a is the radius of the sphere. The first resonance of the spherical cavity is at ka = 2.744. If the incident magnetic field is in the y direction, the induced electric current is of the form

$$\underline{J}^c = \underline{u}_\theta J_\theta \cos\phi + \underline{u}_\phi J_\phi \sin\phi \qquad (73)$$

where neither J_θ nor J_ϕ depend on ϕ. In (73), $\underline{u}_\theta$ and $\underline{u}_\phi$ are unit vectors in the θ and ϕ directions respectively. Figure 3a shows the combined-source solution for J_θ/H_y versus θ. Figure 3b shows the combined-source solution for J_ϕ/H_y versus θ. Here, H_y is the y component of the incident magnetic field at the center of the sphere. In Figs. 3a and 3b, the label THETA on the horizontal axis stands for θ and $\theta = 0$ is the forward scattering direction. In Figs. 3a and 3b, the symbols × and + denote magnitude and phase respectively of the combined-source solution. The solid curves represent the exact Mie series solution.

The combined-source solution gives currents which differ only slightly from those computed by the combined-field solution (Ref. 9). In Fig. 5a of (Ref. 9), the electric current error Δ of the combined-field solution for the conducting sphere is linear from ka = 2.70 to ka = 2.80. Similarly, the electric current error Δ of the combined-source solution for the conducting sphere is also linear from ka = 2.70 to ka = 2.80. The values of Δ at ka = 2.70 and ka = 2.80 are given in the table below. Although not quite as accurate as the combined-field solution for ka between 2.70 and 2.80, the combined-source result for $\underline{J}^c$ is likewise unaffected by the first resonance at ka = 2.744. Both combined-source and combined-field results for J_θ and J_ϕ were obtained by covering the generating curve with 14 equally

Table 1. Electric current error Δ.

ka	Δ combined field	Δ combined source
2.70	0.0245	0.0347
2.80	0.0263	0.0358

spaced triangular functions. For integrations in t, each triangular function was sampled at 4 points. A 20-point Gaussian quadrature formula was used for integrations with respect to ϕ between the limits $\phi = 0°$ and $\phi = 180°$.

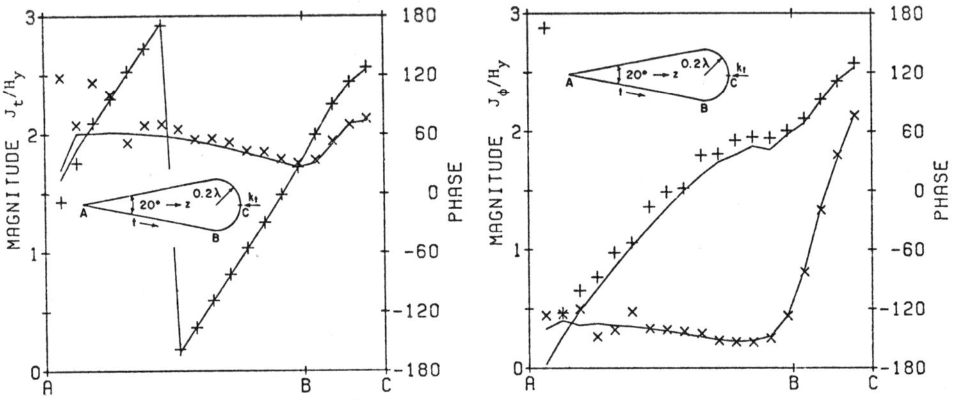

(a) (b)

Fig. 4. Electric current J_t/H_y and J_ϕ/H_y induced on a conducting cone-sphere, cone angle = 20°, sphere radius = 0.2λ, by a plane wave axially incident on the sphere. Solid curves show the combined-source solution. Symbols × and + denote magnitude and phase, respectively, of the E-field solution.

Figure 4 shows the electric current $\underline{J}^c$ induced on a conducting cone-sphere by a plane wave axially incident on the sphere end. As shown in the inserts of Fig. 4, the cone angle of this cone-sphere is 20° and the sphere radius is 0.2 wavelength. The cone and sphere are joined such that the tangent to the generating curve is continuous at the cone to sphere junction. If the incident magnetic field is in the y direction, the induced electric current is of the form

$$\underline{J}^c = \underline{u}_t J_t \cos\phi + \underline{u}_\phi J_\phi \sin\phi \tag{74}$$

Both J_t/H_y and J_ϕ/H_y are plotted versus t where, as shown in the inserts, t is the arc length along the generating curve. Here, H_y is the y component of the incident magnetic field at the tip of the cone. The solid curves in Fig. 4 represent the combined-source solution, and the symbols × and + denote magnitude and phase respectively of the E-field solution of (Ref. 6). An H-field solution for $|J_\phi|$ appears on page 218 of (Ref. 8).

The combined-source results for J_t and J_ϕ in Fig. 4 were obtained by covering the generating curve with 19 equally spaced triangular functions. For integrations in t, each triangular function was sampled at 4 points. A 20 point Gaussian quadrature formula was used for the integrations with respect to ϕ.

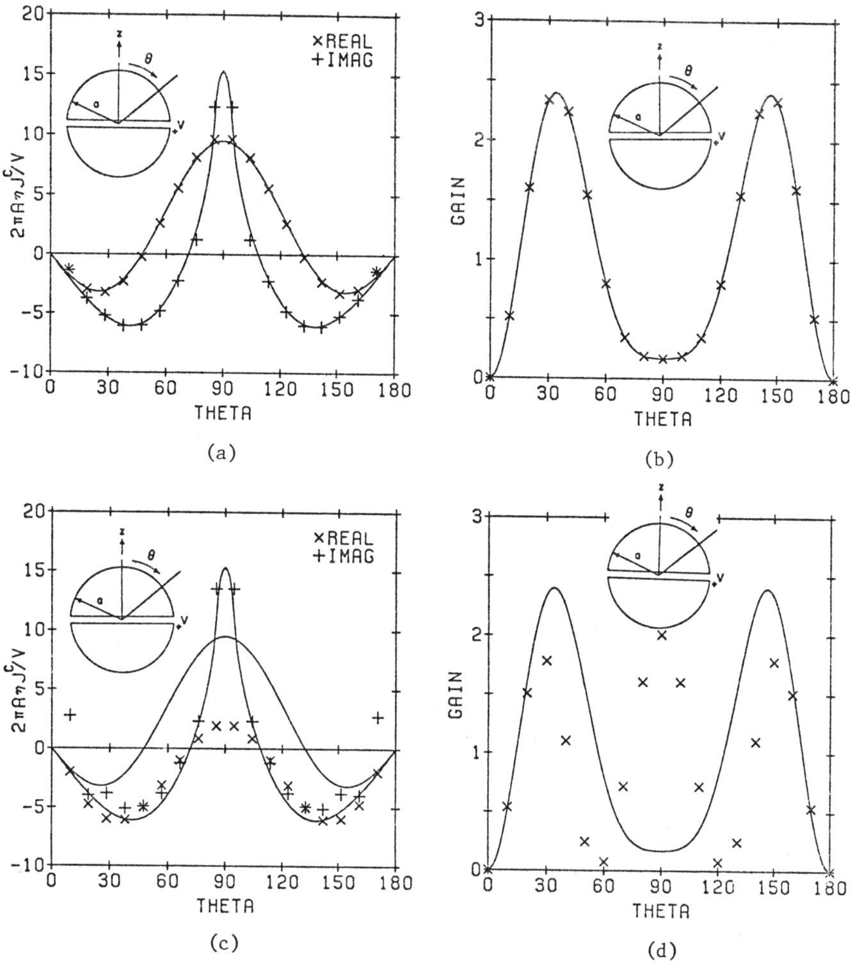

Fig. 5. Electric current J_θ^c and gain for a rotationally symmetric aperture at the equator of a conducting sphere, ka = 2.75. Solid curves represent the exact solution. Symbols × and + denote the combined-source solution in (a) and (b), and the E-field solution in (c) and (d). The first resonance of the spherical cavity is ka = 2.744.

The induced electric current and gain for a thin rotationally symmetric aperture at the equator of a conducting sphere of size ka = 2.75 driven by a minus θ directed electric field are shown in Figs. 5a and 5b. In Fig. 5a, the symbols × and + denote the real and imaginary parts of $2\pi a\eta J_\theta^c/V$ versus θ as obtained from the combined-source solution. Here, J_θ^c is the θ component of $\underline{J}^c$ and V is the voltage across the aperture. At θ = 90°, $2\pi a\eta J^c/V$ reduces to the input admittance normalized by dividing by the admittance 1/η of free space. In Fig. 5b, the symbol × denotes the gain pattern for the aperture problem of Fig. 5a as obtained from the combined-source solution. The solid curves in Figs. 5a and 5b represent the exact Mie series solution. For comparison, the E-field solution for the same sphere and same aperture as in Figs. 5a and 5b is shown in Figs. 5c and 5d. In Figs. 5c and

5d, the symbols × denote the E-field solution and the solid curves denote the Mie series solution. The E-field solution in Figs. 5c and 5d is poor because ka = 2.75 is very close to the first resonance which occurs at ka = 2.744.

The moment solutions of Fig. 5 use 18 expansion functions equally spaced on the generating curve such that the 9th and 10th expansion functions straddle the aperture. Both the impulsive aperture field at the equator and the aperture field

$$\underline{E}^a = \frac{-V}{a\Delta\theta}\,\underline{u}_\theta \quad , \qquad 90° - \frac{\Delta\theta}{2} \leq \theta \leq 90° + \frac{\Delta\theta}{2} \tag{75}$$

where $0 \leq \Delta\theta \leq 180°/19$ give exactly the same moment solution. For convenience, the aperture field (75) with $\Delta\theta = 180°/19$ was used to calculate the Mie series solution.

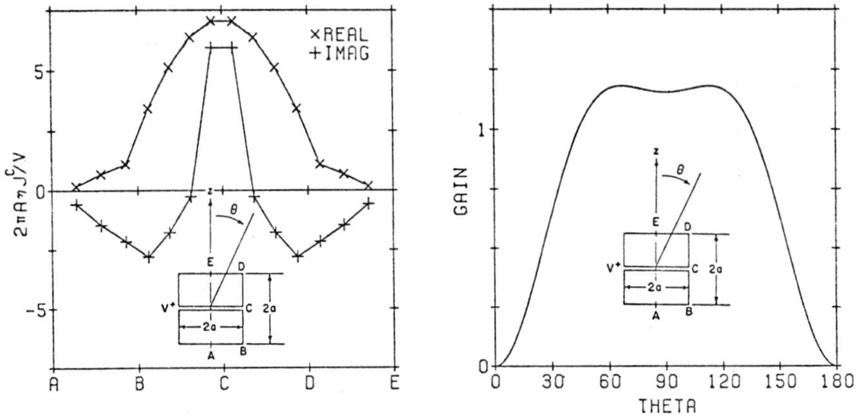

(a) (b)

Fig. 6. Electric current J_t^c and gain for a rotationally symmetric aperture at the center of a conducting cylinder of length 2a and radius $a = \lambda/4$, combined-source solution.

Figure 6 shows the combined-source solution for the induced electric current and for the gain of an axially symmetric aperture at the center of a conducting cylinder whose length is twice its radius a. Here, $a = \lambda/4$. In Fig. 6a, the symbols × and + denote the real and imaginary parts of $2\pi a\eta J_t^c/V$ versus t, where J_t^c is the t component of $\underline{J}^c$ and V is the voltage across the aperture. These symbols have been connected by straight lines in Fig. 6a to improve readability. Figure 6b shows the gain pattern for the aperture problem of Fig. 6a as obtained from the combined-source solution. There are 14 triangular expansion functions on the generating curve of the finite cylinder in Fig. 6. These triangular functions are not equally spaced. The generating curve consists of 7 equal length intervals on each end cap of the cylinder and 16 equal length intervals on the lateral side. The domain of each expansion function consists of 4 of these intervals. For the moment solutions of Figs. 5 and 6, the integrations with respect to t were approximated by sampling each triangular function at 4 points and a 20-point Gaussian quadrature formula was used for the integrations with respect to ϕ. The approximate (moment solution) gain patterns in Figs. 5 and 6 were calculated from (68) with P obtained from numerical integration of the far field power pattern.

VIII. DISCUSSION

A combined-source solution has been developed for electromagnetic radiation and scattering problems involving only the region external to a closed surface S. As such, this solution is not affected by resonances of the region internal to S. Because the combined source contains a magnetic current which implies a discontinuity of tangential electric field, the combined-source formulation does not apply to scattering from "zero thickness" bodies, like plates or washers, nor to aperture radiation from a conducting surface which was not closed before the aperture was cut out of it. The combined-source formulation could be adapted to electromagnetic problems involving only the region interior to a closed conducting surface, but then the solution would deteriorate near internal resonances.

The combined-source formulation is closely related to the combined-field formulation (Ref. 9). For instance, the combined-source operator is the adjoint of the combined-field operator. From this it follows that the combined-source moment matrix is the transpose of the combined-field matrix which would result if the set of expansion functions and the set of testing functions were interchanged. For dipole scattering, the combined-source excitation vector is the combined-field measurement vector which would result if the set of expansion functions were replaced by the set of testing functions and if the receiving current element were replaced by the transmitting current element. The combined-source measurement vector is the combined-field excitation vector which would result if the set of testing functions were replaced by the set of expansion functions and if the transmitting current element were replaced by the receiving current element. It follows from the above three statements that the combined-source solution for dipole scattering is the combined-field solution which would result if the sets of expansion and testing functions were interchanged and if the transmitting and receiving current elements were interchanged.

One disadvantage of the combined-source solution in comparison with the combined-field solution is that the combined source $(\underline{J},\underline{M})$ has no physical significance. The physically significant electric current requires calculation of $\underline{n} \times \underline{H}(\underline{J},\underline{M})$ from $(\underline{J},\underline{M})$. This calculation is especially simple for the body of revolution expansion functions because the operation $n \times$ is closed with respect to those expansion functions. However, the approximation (16) is still present. This could explain why the electric current error for the combined-source solution is slightly larger than that for the combined-field solution in Table 1.

One advantage of the combined-source solution over the combined-field solution is that the combined-source formulation is directly applicable to the aperture radiation problem in which the aperture electric field is specified. One must recast the aperture radiation problem into an equivalent scattering problem in order to apply the combined-field solution. In this equivalent scattering problem, the surface S is perfectly conducting everywhere and the incident field is the field radiated by the magnetic current $\underline{E}^a \times \underline{n}$ in free space, where $\underline{E}^a$ is the specified electric field in the aperture.

REFERENCES

(1) P. C. Waterman, Matrix formulation of electromagnetic scattering, Proc. IEEE 53, 805 (1965).

(2) F. K. Oshiro and K. M. Mitzner, Digital computer solution of three-dimensional scattering problems, Digest IEEE Symp. Ant. and Prop., Ann Arbor, Mich. 257 (1967).

(3) P.L.E. Uslenghi, Computation of surface currents on bodies of revolution, Alta Frequenza 39, 1 (1970).

(4) K. K. Mei and J. Van Bladel, Scattering by perfectly conducting rectangular cylinders, IEEE Trans. AP-11, 185 (1963).

(5) M. G. Andreasen, Scattering from parallel metallic cylinders with arbitrary cross sections, IEEE Trans. AP-12, 746 (1964).

(6) J. R. Mautz and R. F. Harrington, Radiation and scattering from bodies of revolution, Appl. Sci. Res. 20, 405 (1969).

(7) F. K. Oshiro et. al., Calculation of radar cross section, A. F. Avionics Lab., Tech. Rept. AFAL-TR-70-21, Part II (1970).

(8) A. J. Poggio and E. K. Miller, Integral equation solutions of three-dimensional scattering problems, Computer Techniques for Electromagnetics, editor R. Mittra, Pergamon Press, Oxford (1973).

(9) J. R. Mautz and R. F. Harrington, H-field, E-field, and combined-field solutions for conducting bodies of revolution, A.E.Ü. 32, 159 (1978).

(10) J. R. Mautz and R. F. Harrington, A combined-source solution for radiation and scattering from a perfectly conducting body, IEEE Trans. AP-27, July (1979).

(11) H. Brakhage and P. Werner, Über das Dirichletsche Aussenraumproblem für die Helmholtzsche Schwingungsgleichung, Archiv d. Math. 16, 325 (1965).

(12) G. Forsythe and C. Moler, Computer Solution of Linear Algebraic Systems, Prentice-Hall, Englewood Cliffs, N.J. (1967).

(13) J. R. Mautz and R. F. Harrington, Application of the combined-source solution to a conducting body of revolution, Report TR-78-6, Dept. of Electrical and Computer Engineering, Syracuse University (1978).

ACCURACY TESTS AND ITERATIVE PROCEDURES FOR HIGH FREQUENCY ASYMPTOTIC SOLUTIONS - A SPECTRAL DOMAIN APPROACH

Raj Mittra and Mark Tew
University of Illinois, Urbana, Illinois

ABSTRACT

In this paper we describe three procedures for evaluating the accuracy of high frequency asymptotic solutions. Two of the tests are based on the spectral domain approach while the third can be implemented either in the space domain or in the spectral domain. A method for improving the solution via an iterative procedure is also presented.

INTRODUCTION

High frequency asymptotic solutions of the wave equation play a very important role in electromagnetics and acoustics. For low frequencies, the integral equation formulation combined with the method of moments provides a convenient approach for solving radiation and scattering problems. However, because the matrix size required to handle such problems becomes too large above the resonance region, one is forced to seek alternate means, such as ray-optical techniques, for deriving asymptotic solutions in the high frequency range. Unlike moment method solutions which are numerically vigorous, asymptotic solutions are approximations, and as such, pose problems in evaluating their accuracy. The problem of assessing the accuracy of asymptotic solutions has come under investigation only in recent years [1] - [4]. Presented here are three procedures for evaluating the accuracy of an asymptotic solution by examining its effect on satisfaction of boundary conditions. In addition, two of the proposed tests lend themselves for use as an iterative equation, offering the possibility of systematic improvement of a proposed solution.

ACCURACY TESTS FOR ASYMPTOTIC SURFACE FIELDS

Two of the accuracy tests to be discussed deal with the problem of a magnetic dipole radiating in the presence of an infinitely long, perfectly conducting, circular cylinder. This problem has received increased attention in recent years because of the insights it provides into the performance of slot antennas on curved surfaces. Knowledge of the induced currents, for example, allows the engineer to calculate mutual coupling between slot antennas in a conformal array or to accurately compute far-field patterns from a single slot or slot array. The exact modal solution to this problem [5] - [6], which is in the form of an infinite series of infinite integrals, converges so slowly as to make its use impractical for numerical calculation. Various approximate solutions, more suitable for numerical computation, have been proposed [7] - [11]. Two of the solutions, [7] - [9], are derived from manipulation of the modal solution. These solutions are denoted as Asymptotic Solution - 1 (AS-1) [7] - [8], and Asymptotic Solution - 2 (AS-2) [9], based on their chronological order of publication. The third approximate solution [10] - [11] is based on a modification of the work of V. A. Fock, which addresses the problem of radiation on a sphere and is denoted Asymptotic Solution - 3, or AS-3.

The next section describes in detail the three published solutions and gives the appropriate formulae.

Proposed Solutions to Cylinder Problem

Figure 1a presents the geometry of the problem. An infinitely long, perfectly conducting, circular cylinder of radius R is located with the cylinder axis coinciding with the z-axis of a standard ρ, ϕ, z cylindrical coordinate system. An infinitesimal, phi-directed, magnetic dipole is located on the cylinder surface at Q' given by the coordinates $\rho = R$, $\phi = 0$, $z = 0$. The H-field on the cylinder surface is observed at a point Q located at $\rho = R$, $\phi = \phi_T$, $z = z_T$. The proposed solutions are ray-type solutions, and the surface fields are dependent on the geodesic path between Q and Q' defined by the surface path length, s, and ray angle, θ, measured from the ϕ-axis to the surface ray.

The cylinder is a developable surface and a geodesic path on the cylinder surface becomes a straight line on the infinite strip that makes up a developed cylinder. Figure 1b shows the developed cylinder and introduces the local $\hat{n}'$, $\hat{b}'$, $\hat{t}'$ and $\hat{n}$, $\hat{b}$, $\hat{t}$ coordinate systems, where $\hat{n}'$, $\hat{n}$ are the outward normal to the surface, and $\hat{t}'$, $\hat{t}$ are tangent to the surface path at the source and observation points respectively ($\hat{b}' = \hat{t}' \times \hat{n}'$, $\hat{b} = \hat{t} \times \hat{n}$). Both the AS-1 and AS-3 solutions give the surface field in terms of fields parallel to $\hat{b}$ and $\hat{t}$ as

$$\overline{H}(Q) = \overline{M} \cdot (\hat{b}'\hat{b}H_b + \hat{t}'\hat{t}H_t) \qquad (1)$$

where $\overline{M}$ is the magnetic dipole moment. In this section a circumferentially oriented dipole is treated, i.e., $\overline{M} = \hat{\phi}$. For this case, conventional H_z and H_ϕ fields can be found from H_b and H_t using the relationships

$$H_\phi = \cos^2 \Theta \; H_t + \sin^2 \Theta \; H_b \qquad (2a)$$

$$H_z = \sin \Theta \cos \Theta (H_t - H_b) \quad . \qquad (2b)$$

Each of the proposed solutions gives the surface H-fields in terms of a combination of "Fock functions," $u(\xi)$, $v(\xi)$, and $v_1(\xi)$ and their derivatives $u'(\xi)$, $v'(\xi)$, and $v_1'(\xi)$, respectively. ξ is a normalized distance parameter given by $\xi = \left(\frac{k}{2R_{\hat{t}}^2}\right)^{1/3} s$ where k is the wavenumber, R_t is the radius of curvature in the direction of $\hat{t}$ given by $R_t = R/\cos^2 \Theta$ and s is the path length, $s = \sqrt{(R\phi)^2 + z^2}$. The radius of curvature in the direction of $\hat{b}$ is also employed and is given by $R_b = R/\sin^2 \Theta$. The AS-1 solution as tested gives the surface fields as

$$H_b(Q) \sim v(\xi)G(s) \qquad (3a)$$

$$H_t(Q) \sim \left(\frac{2j}{ks}\right)u(\xi)G(s) \quad , \qquad (3b)$$

where H_t of (3b) differs from the H_t given in [7]-[8] by a factor of 2 (this is done so that as $k \to \infty$ the H_t of (3b) recovers identically the $(ks)^{-2}$ term of the known exact solution). The AS-2 solution is given by

$$H_\phi(Q) \sim \{v(\xi)[\sin^2\Theta + \frac{j}{ks}(1 - 3\sin^2\Theta)] + (\frac{j}{ks})\sec^2\Theta[u(\xi) - \sin^2\Theta v_1(\xi)]\, G(s) \tag{4a}$$

$$H_z(Q) \sim -\sin\Theta\cos\Theta\, v(\xi)[1 - \frac{3j}{ks}]G(s) \tag{4b}$$

for the case of a circumferentially oriented dipole. The AS-3 solution gives the surface fields as

$$H_b(Q) \sim \{(1 - \frac{j}{ks})v(\xi) - (\frac{1}{ks})^2 u(\xi) + j(\sqrt{2}\, kR_t)^{-2/3}[v'(\xi) + (R_t/R_b)u'(\xi)]\}G(s) \tag{5a}$$

$$H_t(Q) \sim (\frac{j}{ks})\{v(\xi) + (1 - \frac{2j}{ks})u(\xi) + j(\sqrt{2}\, kR_t)^{-2/3}u'(\xi)\}G(s)\ . \tag{5b}$$

In Equations (3)-(5), $G(s) = k^2/(2\pi j\eta)\cdot e^{-jks}/ks$, where η is the wave impedance of free space.

In addition to the three published solutions, a fourth solution has been constructed, which is a modified AS-1 solution and is denoted AS-4. The AS-4 solution is given by

$$H_b(Q) \sim [v(\xi) - j/ks - (1/ks)^2]G(s) \tag{6a}$$

$$H_t(Q) \sim (2j/ks)[u(\xi) - j/ks]G(s)\quad . \tag{6b}$$

As Θ goes to ninety degrees (a ray propagating down the cylinder axis), H_b becomes identical to H_ϕ and the solutions reduce to

$$H_\phi(Q) \sim G(s) \text{ for AS-1} \tag{7a}$$

$$H_\phi(Q) \sim [1 - \frac{j}{ks} + (\frac{\pi}{2})^{1/2}e^{-j\pi/4}\frac{(ks)^{1/2}}{kR}]G(s) \text{ for AS-2} \tag{7b}$$

$$H_\phi(Q) \quad [1 - \frac{j}{ks} - (\frac{1}{ks})^2 + \frac{3}{4}(\frac{\pi}{2})^{1/2}e^{-j\pi/4}\frac{(ks)^{1/2}}{kR}]G(s) \text{ for AS-3}\ . \tag{7c}$$

and

$$H_\phi(Q) = [1 - \frac{j}{ks} - (\frac{1}{ks})^2]G(s) \text{ for AS-4}\quad . \tag{7d}$$

The four solutions presented here embody some important differences. For example, as $kR \to \infty$, both the AS-3 and AS-4 solutions recover the exact solution to the problem of a magnetic dipole radiating over a ground plane. The AS-1 solution recovers the ks^{-1} terms; while the AS-2 solution recovers both the $(ks)^{-1}$ and $(ks)^{-2}$ terms, missing only the $(ks)^{-3}$ term. Since these differences involve terms that have significant contribution only for very short path lengths ($s < 1\lambda$), these differences are referred to as "source-region" differences.

Long-path length differences are most apparent along theta equal to ninety degrees. Here, for large s, both the AS-2 and AS-3 solutions vary asymptotically as $(ks)^{-1/2}$, while the AS-1 and AS-4 solutions show a free-space-like attenuation of $(ks)^{-1}$.

The accuracy tests presented are closely examined to determine if they can discriminate between the various solutions. An accuracy test which relies on the entire solution being correct is termed a global test. That is, a global test is not associated with any particular region of the proposed solution, but tests the entire solution. A test which is associated with a region or a point is termed a local test. A local test, then, would be able to discriminate between the AS-2 and AS-3 solutions even though their source region forms are identical, because of the differences in the long path length behavior.

THE E-FIELD TEST

In implementing an accuracy test, the best approach is usually to compare the approximate solution to the exact solution. As stated before, computational difficulties with the exact solution preclude that option in this case. If the exact solution is unavailable for comparison, then an accuracy test can be formulated by determining how well the approximate solution satisfies boundary conditions for the problem - in this case, the electric field boundary condition at the perfectly conducting surface.

The most direct and attractive method of determining the surface E-field would be to compute the normal derivative of the H-field. Unfortunately, the asymptotic solutions are valid only for points on the cylinder surface, so that such a direct computation is unfeasible. Accordingly, an indirect procedure must be used. An indirect evaluation of the surface E-field is conducted in the following manner: Each of the asymptotic solutions predicts the H-field on the cylinder surface. Through the use of Maxwell's equations, the surface H-field can be related to the surface E-field and the boundary condition checked. Use of the spectrum of the H-field instead of the direct surface field makes analysis straightforward and allows use of a Fast-Fourier Transform (FFT) algorithm for efficient numerical calculation.

The test proceeds as follows:

1. A cylindrical transform is defined

$$\tilde{H}_{\substack{z\\ \phi}}(n,k_z) = 1/2\pi \int_0^{2\pi} d\phi \int_{-\infty}^{\infty} dz\, H_{\substack{z\\ \phi}}(\phi,z)e^{-jn\phi}\, e^{-jk_z z} \quad . \tag{8}$$

2. Electric and magnetic vector potentials are expanded with unknown coefficients

$$\left.\begin{matrix} A_z \\ \\ F_z \end{matrix}\right\} = \frac{1}{2\pi} \sum_{n=-\infty}^{\infty} e^{jn\phi} \int_{-\infty}^{\infty} \begin{matrix} f_n(k_z) \\ \\ g_n(k_z) \end{matrix}\; H_n^{(2)}(\rho\sqrt{k^2-k_z^2})\, e^{jk_z z}\, dz \quad . \tag{9}$$

Observe that n and k_z are "transform variables," k is the wavenumber, and $H_n^{(2)}$ is the nth order Hankel function of the second kind, representing an outward-traveling cylindrical wave. For future notation, the complex variable γ is used to replace the radical in the argument of the Hankel function, $\gamma = \sqrt{k^2-k_z^2}$.

3. Through the use of

$$\overline{H} = \nabla x\overline{A} - j\omega\varepsilon_o\overline{F} + 1/j\omega\mu_o \quad \nabla\nabla\cdot\overline{F} \tag{10}$$

one may determine the unknown coefficients f_n and g_n in terms of the transform of the surface H-field as

$$f_n(k_z) = \frac{-1}{\gamma H_n'^{(2)}(\gamma R)} [\tilde{H}_\phi(n,k_z) + \frac{nk_z}{\gamma^2 R} \tilde{H}_z(n,k_z)]$$

$$g_n(k_z) = \frac{j\omega\mu_o}{\gamma^2 H_n^{(2)}(\gamma R)} \tilde{H}_z(n,k_z) \qquad (11)$$

where $H_n'^{(2)}$ is the derivative of the $H_n^{(2)}$ Hankel function.

4. Applying

$$\overline{E} = -\nabla x\overline{F} - j\omega\mu_o\overline{A} + 1/j\omega\varepsilon_o \nabla\nabla\cdot\overline{A} \qquad (12)$$

permits computation of the surface E-field, accomplishing the desired test.

The above procedure can be condensed into two steps of actual computation by combining Equations (9), (10), (11), and (12). While the resulting expressions appear to be complex, this two-step procedure is significant because it essentially involves only a two-dimensional Fourier transform, modification and combination of the transformed fields, and then inverse Fourier transformation and summation of Fourier coefficients. The analysis is computationally efficient because the FFT can be used to evaluate all the integrals involved.

Practically speaking, the greatest difficulty in the above procedure comes in accurately determining the spectrum of the surface H_ϕ field. Both the AS-3 and AS-4 solutions have $1/s^2$ and $1/s^3$ singularities, while the AS-1 solution has a 1/s singularity. The "peakiness" of these H_ϕ fields means that special care must be taken in using the FFT to determine the spectrum.

One attempt to overcome this problem involved raising the magnetic dipole slightly above the cylinder surface so that the field was no longer singular, but had a finite peak. After this step had been implemented, a convergence check of the FFT integral showed that the FFT was able to handle the 1/s peak correctly with reasonable sampling rates, but the $1/s^2$ and $1/s^3$ peaks yielded erroneous results. Further measures were necessary to achieve a reliable test of the AS-3 and AS-4 solutions (for a reasonable computer size).

The key to achieving a reliable check of the AS-3 and AS-4 solutions lies in recognizing that the singular form of the source region is that of a planar case, and that the singularity has an analytic transform. Specifically, the planar singularity can be expressed as

$$H_\phi|_s = \frac{1}{2\pi\eta jk} (\frac{\partial^2}{\partial^2(R\phi)^2} + k^2) \frac{e^{-jks}}{s}$$

$$H_z|_s = \frac{1}{2\pi\eta jk} (\frac{\partial^2}{\partial(R\phi)\partial z} \frac{e^{-jks}}{s}) \qquad (13)$$

where s is the path length given by $s = \sqrt{\Delta^2+(R\phi)^2+z^2}$, and Δ is the height of the dipole above the cylinder. Because the singularities can be expressed as derivatives, their analytic transform is obtainable and is of the form

$$\tilde{H}_z|_s = \frac{Q}{2\pi\eta jk} (\frac{-nk_z}{R}) \frac{1}{4\pi jR} \frac{e^{-j\Delta\sqrt{k^2-(n/R)^2-k_z^2}}}{\sqrt{k^2-(n/R)^2-k_z^2}} \qquad (14a)$$

$$\tilde{H}_\phi|_s = \frac{Q}{2\pi\eta jk}\left(-\frac{n^2}{R^2}+k^2\right)\frac{1}{4\pi jR}\frac{e^{-j\Delta\sqrt{k^2-(n/R)^2-k_z^2}}}{\sqrt{k^2-(n/R)^2-k_z^2}} \tag{14b}$$

where Q is a constant.

The total field on the cylinder can then be expressed as

$$\overline{H}(Q) = \overline{H}_{PLANAR}(Q) + \overline{H}_{DIFFERENCE}(Q) \tag{15}$$

where $\overline{H}(Q)$ is the total field as predicted by an asymptotic solution, and $\overline{H}_{PLANAR}(Q)$ is the field that would exist on a flat, infinite ground plane. (This division may be thought of as taking the planar field, "wrapping" it around the cylinder, and subtracting it from $\overline{H}(Q)$). The transform of the surface fields is given by

$$\overline{\tilde{H}}(n,k_z) = \overline{\tilde{H}}_{PLANAR}(n,k_z) + \overline{\tilde{H}}_{DIFFERENCE}(n,k_z) \tag{16}$$

and $\overline{\tilde{H}}_{PLANAR}(n,k_z)$ is given analytically in Equation (14). $\overline{H}_{DIFFERENCE}(Q)$ is at most on the order of $1/\sqrt{s}$, so that evaluation of $\overline{\tilde{H}}_{DIFFERENCE}(n,k_z)$ can be reliably obtained from application of the FFT. Any test which involves breaking the fields up into planar and difference fields will be termed a <u>hybrid</u> computation, because it combines analytic and numerical techniques. The only difference between a "hybrid computation" and a "direct computation" is in the method of obtaining the spectrum. After the spectrum is found, both tests proceed identically. Figure 2 compares the phase of $\tilde{H}_\phi$ for a fixed value of k_z, when the transform was derived from hybrid and direct computations. Comparison with the phase of the modal transform reveals the increased accuracy of the hybrid method.

To utilize the E-Field Test the three asymptotic solutions are compared to an exact modal solution [6]. The procedure used in deriving the exact modal solution is essentially the same as that used to perform the E-Field Test. The difference lies in the fact that while the test begins with the asymptotic H-field, the modal solution begins with the known E-field (known for an elemental source). In a manner similar to the test, the H-spectrum can be found from the known E-fields and be expressed in terms of the E-spectrum. This provides analytic H_z and H_ϕ spectrums that can be compared to those resulting from the asymptotic solutions. After the H-spectrum is obtained from the E-field, it can be tested just like any other spectrum. This "check" that begins with a surface E-field, finds the H-spectrum and then returns to the surface E-field is also valuable inassuring that the FFT sampling of the spectrum is sufficient.

During actual application of the test, the source used was a slot radiator instead of an elemental dipole. This was necessary because the surface E-field of the slot is finite, although discontinuous, while the surface E-field from the elemental source is singular. Sampling the E-spectrum sufficiently well to represent the singular surface field would make the computer requirements prohibitively large, while the more regular slot is readily handled.

Representation of the slot spectrum was achieved by first determining the H-field spectrum due to an elemental dipole source (the direct asymptotic solutions or, for the analytic case, $\overline{E}_{SURFACE} = \hat{z}(\delta(\rho-R)\delta(\phi)\delta(z))$. The H-spectrum was then multiplied by the transform of the slot distribution, which is equivalent to convolving the elemental source with a distribution in the space domain. For a finite slot as shown in Fig. 1a, the transforms used were of the form

$[\sin(k_z b/2)]/(k_z b/2)$ to represent a uniform E_z field, and of the form $\cos(\alpha n)/[\pi^2/4 - (\alpha n)^2]$, $\alpha = \arcsin(a/2R)$ to represent a half-period cosine spreading in ϕ.

The totality of the E-field check has now been determined. The components of the spectrum of the H-field that result from an elemental source are determined. For the asymptotic solutions the spectrum is determined either by direct application of the FFT or by use of the hybrid technique; for the exact case, the H-spectrum may be found analytically. Regardless of its source, the H-spectrum is then multiplied by the transform of the assumed slot distribution; thus, the convolution that is necessary to represent the slot distribution is accomplished. The E-field spectrum is then found from modification and combination of the H-field spectrum components. Finally, inverse Fourier transformation and summation of the Fourier coefficients give the surface E-field.

Results of the E-Field Test

Figures 3 and 4 give representative results after the E-Field Test has been implemented. The test was first applied to the analytic H-spectrum. This was the standard to which all other solutions were compared. The analytic solution resulted in an E_ϕ that was essentially zero (a totally flat curve) and an E_z that was well-contained. The fact that E_z went to zero in a smooth curve in the z-direction instead of a discontinuous curve was due to the numerical calculation, and revealed that the error introduced by performing finite sums and integrals was negligible. Figure 3 shows the surface electric fields that resulted from testing the AS-1 solution. Because the AS-1 solution only has a singularity (peak) on the order of 1/s, it was not necessary to resort to a hybrid computation to obtain a reliable check. As can be seen, the E_ϕ field has significant non-zero content and the extent of the corresponding E_z field is much broader than that of the modal solution.

Figure 4 shows the effects of testing the AS-3 solution using the hybrid method. The hybrid method was used because of the higher-order source region terms present in the AS-3 solution. The resultant surface E_ϕ field is very small and displays a rippled character.

The fields that resulted from testing the AS-4 solution using the hybrid technique were essentially identical to those of Fig. 4, although there were slight numerical differences.

DISCUSSION OF E-FIELD TEST

The E-Field Test represents an example of global test. That is, the test is applied once, and the results determine if the field everywhere is correct. If the approximation does accurately predict the surface magnetic fields, then the corresponding electric fields will re-create those obtained by testing the analytic H-transform, within the constraints of numerical accuracy. When results do not duplicate those of the analytic standard, then one can only say that some error is present. Specifying what the error is, or in what region it occurs, is very difficult, perhaps impossible, from observing the results of a single test. However, by comparing the results from tests of different approximate solutions, some insight can be gained.

The results presented here provide a good example of what can be learned from comparing test results of different solutions. Figures 3 and 4 reveal that the E-Field Test is sensitive to the source region behavior of the proposed solution and relatively insensitive to the large-path length behavior. The surface E_ϕ fields of Fig. 3, corresponding to the AS-1 solution, are quite different from those of Fig. 4 resulting from the AS-4 solution, despite the fact that their large-path length behaviors are identical. The approximate solutions of Figs. 3 and 4 differ only in the $(ks)^{-2}$ and $(ks)^{-3}$ source region terms. This point is reinforced by the fact that Fig. 4 represents results from a test of either the AS-3 and AS-4 solutions. As $kR \to \infty$, the source regions of the AS-3 and AS-4 solutions both go to a planar-type singularity. Along theta equal to ninety degrees, however, the AS-3 shows attenuation as $(ks)^{-1/2}$ for large path lengths, while the AS-4 solution shows attentuation as $(ks)^{-1}$. The E-Field Test was unable to discern between the AS-3 and AS-4 solutions, despite the significant differences in large path length behavior.

In summary, the resulting E-field after performance of the test does provide a good qualitative measure of how well the asymptotic solution satisfies the E-field boundary condition. Indeed, application of the test to the transform of the exact modal solution did result in fields that satisfied the boundary condition quite well. In addition, application of the test to approximate asymptotic solutions showed that they did not test as well as the exact solution. In this respect, the E-Field Test does provide a good "global test" of a proposed solution, i.e., if the solution is accurate everywhere, the solution will show good results from the test.

On the other hand, when comparing the relative accuracy of approximate solutions, the source region behavior appears to be more critical for satisfaction of the test than the large path length behavior -- asymptotic solutions that contain higher-order terms ($1/s^2$, $1/s^3$) in the source region H-field do satisfy the E-Field Test better than solutions that contain only terms on the order of $1/s$. The large-path length behavior did not appreciably affect the test results.

Emphasis of the source region is not necessarily detrimental to the E-Field Test, for it is likely that the source region will heavily influence the computation of values of physical parameters of interest, such as the mutual impedance between two slots on a curved surface. Wire antennas provide a good analogy, for the source region behavior dominates calculation of the self- and mutual-impedance, and the current behavior at the end plays less of a role. An E-field boundary condition check of a wire antenna would not be influenced very much by the current far away from the source, but be source region sensitive. It is not surprising, then, that the E-field check reveals little about the local character of the solution, but instead provides more of a global test. The point should be made, however, that a large path length solution for the antenna current would probably be of very limited use, since both the impedance behavior and the radiation pattern derived from this type of asymptotic solution would be grossly in error. The E-Field Test proposed in this paper is able to distinguish between a solution which has a better overall behavior on the entire surface and an alternate solution which is only good in local isolated regions, but has large errors in other regions where the current is significant.

Some numerical difficulties were encountered during execution of the test. The high-order source-region terms made it quite difficult for the FFT to accurately compute the integral involved. Since use of the FFT requires equally spaced sampling over the interval, a sampling sufficient to accurately evaluate the peak resulted in matrix sizes too large for some computers (CDC CYBER 74, for example). For solutions whose source-region behavior went to a planar-type singularity in the limit, however, analytic evaluation of the planar spectrum allowed completion of the test. A solution which contains higher-order source region terms but does not go to planar type behavior in the limit is quite difficult to test by this method unless it has an analytical transform (or unless the investigator has an extraordinarily large computer). However, such a behavior would not be expected to be physically meaningful anyhow.

The AS-2 solution is a good example of this, for it contains the $1/s^2$ source region term, but not the $1/s^3$ term. This solution could be tested by modifying its source-region behavior by the addition of the $1/s^3$ term and performing a hybrid computation. This was not done, since comparison of the AS-3 and AS-4 tests indicates that solutions with the same limiting source region behavior yield virtually identical E-Field Tests.

To sum up, it appears that the E-Field Test can provide a measure of the accuracy of a proposed asymptotic solution. An FFT is employed so that the evaluation of the integrals involved may be efficiently performed, but some care must still be taken in the computation. The test reveals that solutions which have planar-type, source-region behavior in the limit satisfy the E-field boundary condition better than those that only have terms on the order of 1/s. The test is, however, relatively insensitive to the large path-length behavior of a solution.

INTEGRAL E-FIELD TEST

Introduction

The E-Field Test just described was qualitative in nature and highly dependent on the source region accuracy of the proposed solution. The E-Field Test was global in nature and unable to discriminate between solutions which had similar characteristics in the source regions, but varied in their long-path-length behavior. The "Integral E-Field Test" described here is an attempt to achieve a "local" test. This test is also based on satisfaction of the electric field boundary condition at the perfectly conducting surface.

The Integral E-Field Test is quantitative in nature, is straightforward in application, and displays a mixed local/global nature. Under some circumstances, the test lends itself to application as an iterative equation for point-by-point improvement of a proposed solution.

Inspiration for the Integral E-Field Test came from the observation that Lorentz Reciprocity allows the use of test dipoles that can be located at the tester's discretion. A new equation results from each new location of the test dipole (or dipoles), which opens the possibility of achieving a "local" test. It is seen that the Integral E-Field Test displays a mixed local/global character, in general, but that along theta equal to ninety degrees the source region of a proposed solution can be essentially excluded from contributing to the accuracy test, yielding a very strong local or point-test character. Under those circumstances the Integral E-Field Accuracy Test also lends itself to formulation as an iterative equation, allowing point-by-point improvement of a proposed solution.

Formulation of the Integral E-Field Test

The Integral E-Field Test may be formulated using Green's Identities, Generalized Lorentz Reciprocity, or from first principles using Maxwell's Equations and Gauss' Law. The defining equation for the Integral E-Field Test is given by:

$$-\iint_S (\overline{E}^A \times \overline{H}^B - \overline{E}^B \times \overline{H}^A) \cdot d\overline{a} = \iiint_V (\overline{H}^B \cdot \overline{M}^A - \overline{H}^A \cdot \overline{M}^B) dv \tag{17}$$

where fields and sources from two different environments, A and B, are related in a single equation. In order to achieve an accuracy test, Environment A will denote the environment of the approximate solution, that is, an infinitely long, perfectly conducting circular cylinder in free space. An infinitesimal phi-directed magnetic dipole, $\overline{M}^A$, radiates in the presence of the cylinder (Fig. 1a). Environment B is made up entirely of free space. A magnetic dipole source (or sources), $\overline{M}^B$, is placed in Environment B, giving rise to electric and magnetic fields $\overline{E}^B$ and $\overline{H}^B$ (Fig. 5). The fields in Environment A are termed "asymptotic fields" because they are given by the approximate asymptotic solution. The fields in Environment B are termed "test fields" because they are employed to evaluate the accuracy of a proposed solution. The volume V is defined as being enclosed by surface S, which is made up of concentric circular cylinders, S_c and S_∞. Cylinder S_c has radius $R + \delta$, where δ is vanishingly small, and S_∞ has infinite radius. In Environment A, S_c and S_∞ are placed so that their axes coincide with the axis of the perfectly conducting cylinder; while in Environment B, S can be placed anywhere as long as it does not intersect any of the sources, $\overline{M}^B$. Since the sources in both Environments A and B are of finite extent, the fields $\overline{E}^A$, $\overline{H}^A$, $\overline{E}^B$, and $\overline{H}^B$ obey the radiation condition - with the result that the surface integral over S_∞ is zero. Equation 17 can then be rewritten as

$$\int_{S_c}\int \overline{E}^A \times \overline{H}^B \cdot d\overline{a} = \int_{S_c}\int \overline{E}^B \times \overline{H}^A \cdot d\overline{a} - \iiint_V (\overline{H}^B \cdot \overline{M}^A - \overline{H}^A \cdot \overline{M}^B) dV \quad . \tag{18}$$

The accuracy test is accomplished by arguing that the exact solution to a dipole radiating in the presence of, but not on, the perfectly conducting cylinder would satisfy the boundary condition that the tangential electric field is zero everywhere on the perfectly conducting cylinder surface. Taking δ to be vanishingly small, the surface S_c almost coincides with the conducting surface so that the magnitude of the left-hand side of Equation 18 can be made as small as desired, approaching zero in the limit as δ goes to zero. The amount that the right-hand side of Equation 18 differs from zero is termed the "error," ε, and is used as a basis for comparing proposed solutions to the problem embodied in Environment A,

$$\varepsilon = \int_{\substack{S_c \\ \delta\to 0}}\int \overline{E}^B \times \overline{H}^A \cdot d\overline{a} - \iiint_V (\overline{H}^B \cdot \overline{M}^A - \overline{H}^A \cdot \overline{M}^B) dv \quad . \tag{19}$$

If delta-function type sources are used for both Environments A and B (and they are implied by the proposed solutions of Environment A), then the volume integral of Equation 19 reduces to a sampling operation and becomes

$$\varepsilon = \int_{\substack{S_c \\ \delta\to 0}}\int \overline{E}^B \times \overline{H}^A \cdot d\overline{a} - H^B(M^A) + H^A(M^B) \tag{20}$$

where the symbol $H^i(M^j)$ denotes the operation $\overline{H}^i(x_j, y_j, z_j)\cdot\hat{M}^j$ in which the fields of Environment i are sampled at the position of the sources in Environment j, and the dot product is taken with the unit vector parallel to the "j" source. Now, if ε is sufficiently small, Equation 20 can be cast into the form of an iterative equation

$$H^{A(1)}(M^B) = H^B(M^A) - \int_{\substack{S_c \\ \delta\to 0}}\int \overline{E}^B \times \overline{H}^{A(0)} \cdot d\overline{a} = H^{A(0)}(M^B) - \varepsilon \quad . \tag{21}$$

Thus, the field at a point (the location of source M^B) can be updated by modifying the proposed approximate solution by the error term associated with that point.

Observe that the exact solution would satisfy $\varepsilon = 0$ in Equation 20, so that using Equation 21 to iterate would not change the field value.

In review, the formulation of the Integral E-Field Accuracy Test has been accomplished through the use of well-known properties of electromagnetic fields. The resulting equation is straightforward, involves known quantities, and, in general, consists only of sampling operations and numerical integration. In some cases, the test can lend itself to point-by-point improvement of a proposed solution. Local or point-test character is embodied in two ways: 1) By the explicit sampling operation of the surface magnetic field, and 2) by the fact that the near field of the test dipole tends to emphasize the local magnetic fields in the integral of Equation 20. The test, in general, cannot be entirely local, however, for the integral of Equation 20 covers the entire surface of the cylinder. In particular, unless the test electric fields can be constructed to be nearly zero in the region around the asymptotic source, then the large magnetic fields in the asymptotic source region will certainly contribute to the integral and thus affect the test results. Thus, one would expect the Integral E-Field Test to have a mixed local/global character.

It is proper at this point to note that while the development of the Integral E-Field Test has been carried out in the space domain, spectral domain calculations are an alternative when using the test. Specifically, Parseval's Theorem can be used to change the space integral of Equation 20 to an integral in the transform domain. Performing the integration in the transform domain is particularly valuable if a proposed solution is to be tested at a large number of points, for in that case, "moving" the test dipole around can be accomplished by adding the proper phase shift terms to the integrand of Equation 20. It is the authors' experience that use of transform domain integration results in a considerable gain in computational efficiency over spatial integration. To justify use of the spectral domain, though, the asymptotic solution must be tested at enough points to recover the expense of obtaining the spectral domain representation of the proposed solution.

Practical Considerations for the Integral E-Field Accuracy Test

While Equation 19, which defines the test, appears to lend itself to direct evaluation, several practical aspects of the evaluation must be considered. First, an important step in formulating Equation 19 was that the tangential E-field vanish for a source radiating in the presence of the cylinder, while the published solutions are for the case of a dipole radiating on the cylinder surface. To perform the test, the proposed solutions were extended to the case of a dipole radiating a distance Δ away from the cylinder by modifying the path length s used in determining the surface field. The modified path length s' used in Equations 3-7 was computed from $s' = \sqrt{s^2 + \Delta^2}$, where s is the surface path length. In addition to satisfying the assumption made in formulating the test, raising the dipole allows numerical calculation of the integral in Equation 20, since the H^A fields have a finite peak and are not singular (source region H^A fields are singular for a dipole located on the surface).

Location of the test dipole(s) with respect to the surface coinciding with the perfectly conducting cylinder (this surface is denoted S_c^B) greatly influences the test results. A test dipole placed so that it sampled the surface magnetic field of the approximate solution would seem to lend a local character to the test. However, a test dipole so located would introduce singular E^B fields, which makes numerical integration more difficult. Accordingly, the test dipole is, in general, located a Δ distance away from the surface corresponding to the cylinder surface. The E^B fields on the surface of integration are calculated exactly, but the H^A field at M^B is approximated by the H^A field at the point on S_c^B closest to the location of M^B. In other words, if the sampling operation $H^A(M^B)$ required finding $H^A(R + \Delta, \phi_t, Z_t)$, this would be approximated by $H^A(R, \phi_t, Z_t)$. The error introduced by this approximation was checked by applying the test to a problem with a known solution, that of a magnetic dipole radiating over an infinite ground plane. For Δ equal to one sixteenth of a wavelength, the error introduced by this assumption was negligible compared to ε's computed in the cylinder test.

Observe that the user has complete freedom to place one or several test dipoles in Environment B. For example, two test dipoles could be used, located at $(R + \Delta, \phi_{TST}, Z_{TST})$ and $(R + \Delta, \phi_{TST}, -Z_{TST})$. This would have the effect of creating a plane of zero E^B fields that coincide with the H^A source region. Evaluation of the integral of Equation 20 shows that this would decrease the contribution of the source region H_Z^A fields (due to zero E_ϕ^B) and emphasize the source region H_ϕ^A field. As this illustrates, use of multiple-dipole test configurations generally involves a trade-off of some kind, so that the simple single test dipole located Δ away from S_c^B has been chosen as the "standard" test.

Results of the Integral E-Field Test

Representative results of the Integral E-Field Test are presented in Figs. 6-13. These represent application of the test to the AS-1, AS-2 and AS-3 solutions along theta equal to ninety degrees and along theta equal to zero degrees. A single test dipole was used for Figs. 6-8 and 11-13. For tests along $\Theta = \pi/2$, along the cylinder axis, a two-test-dipole configuration was also used with phi-directed test dipoles located at $(R + \Delta, 0, Z_{TST})$ and $(R - \Delta, 0, Z_{TST})$. The test equation under these conditions becomes

$$\varepsilon = \int_{S_c} \int_{\delta \to 0} (E_Z^B H_\phi^A - E_\phi^B H_Z^A) da - H_\phi^B(M^A) + H_\phi^A(M^B) \tag{22}$$

The figures present the H_ϕ field to be tested, error results, and H_ϕ after application of the iterative equation. Several factors combine to increase confidence in the test results along $\Theta = 90°$. First, the magnitude of the error term is relatively small with respect to the magnitude of $H^{A(0)}(M^B)$, so that the criterion for successful iteration is close to being met. Second, for both single- and dual-dipole tests, the asymptotic source-region field contributes very little to the integral that makes up the error term. For the single dipole, this is easily seen from the fact that for $Z_{TST} >> \Delta$ along $\Theta = 90°$, the E^B field evaluated at $(R,0,0)$ will be primarily ρ-directed with a very small E_Z or E_ϕ component. At the same time, the test field E^B directly underneath the test dipole will have a very large E_z component, emphasizing the local H_ϕ field. For two dipoles located at $(R \pm \Delta, 0, Z_{TST})$, the plane of zero tangential electric fields is tangent to the cylinder at $\phi = 0$, thus reducing the asymptotic source region contribution. Local H_ϕ field emphasis does not occur with the two-dipole test, however, since the plane of zero tangential E-fields extends through $Z = Z_{TST}$. Thus, for $Z_{TST} >> \Delta$ along

$\Theta = 90°$, the Integral E-Field Test should have a very strong local character and be almost unaffected by the source-region accuracy of the proposed solution.

Single-dipole tests along angles other than $\Theta = 90°$ will introduce an asymptotic source region contribution to the integral of Equation 20, with the amount of contribution dependent on the path length and Θ-angle. The local E-fields in the vicinity of the test dipole remain large, so that it is expected that the test will display a mixed local/global character. The results of Figs. 11, 12, and 13 reflect this; for while the zeroth-order magnitudes are relatively close, there is a large variance in the associated errors. If the test were truly local, the associated errors should also be relatively close. Examination of Figs. 11 and 13 reveals that along $\Theta = 0°$ the iterative equation is divergent instead of convergent. Close examination of the results of the evaluation of the integral of Equation 20 reveals that the contribution from the vicinity of the source dominated the contribution from the rest of the surface, including the region around the test dipole. The integral of Equation 20, in turn, is the highest contributor to the error, ε. Thus, for $\Theta = 0°$, the Integral E-Field Test retains local character only in the sampling operation. The large source-region contribution drives the error term up so that its magnitude is not small compared to $H^{A(0)}(M^B)$ and the iterative equation diverges.

One may be tempted to employ two dipoles located at ($x_t^{(1)} = R\cos\phi_t$, $y_t^{(1)} = R\sin\phi_t$, Z_t) and at ($x_t^{(2)} = 2R - x_t^{(1)}$, $y_t^{(2)} = y_t^{(1)}$, Z_t) to obtain asymptotic source-region cancellation. This introduces the problem of determining the scattered field at the location of the second dipole. Unfortunately, if surface equivalent currents are used to find the scattered field at the test dipole, then it can be shown that the test is identical to a single-dipole test. That is, the use of surface equivalent currents to find $H^A(M^B)$ always leads to the result that $\varepsilon = 0$, for <u>any</u> proposed solution <u>when</u> M^B is entirely located external to S_c^B. (Observe that for $\Theta = 90°$, the second dipole lies outside of v, and this problem is not encountered.) Use of other approximations to find the "scattered field" external to the cylinder would introduce an unknown error, so use of remotely located dipoles was abandoned for angles other than $\Theta = 90°$.

Thus, the "Integral E-Field Test" displays, in general, a mixed local/global character which changes according to the location of the test dipole. For $\Theta = 90°$ the test has almost entirely a local character, with very little contribution from the asymptotic source region. Under these conditions, the error term is relatively small compared to $H^{A(0)}(M^B)$ and the iterative equation converges. For angles other than $\Theta = 90°$ the asymptotic source region begins to contribute to the integral, resulting in a mixed local/global nature. At the same time, errors in the source region tend to increase the magnitude of the error term, so that it is no longer small compared to $H^{A(0)}(M^B)$. In this case, the iterative equation shows less of a tendency to converge and may diverge.

<u>Summary of Integral E-Field Test</u>

Generalized Lorentz Reciprocity has been employed to formulate an accuracy test that can be used to check any proposed solution. The test is straightforward in application and general in nature, that is, not limited to testing only cylindrical geometries. The test is quantitative in nature, and requires only numerical integration for implementation. Either the spectral domain or the spatial domain can be employed for computation of the integral involved. For testing a large number of points, economics tend to favor computation in the spectral domain, while for testing a small number of points spatial domain integration is indicated.

The Integral E-Field Test has local character in that the error computed is associated with a single point, although the entire solution contributes to the computation of the error term. In cases where the error is small compared to the value of the proposed solution, the test lends itself to use as an iterative equation.

When the test is applied to the problem of magnetic dipole radiation in the presence of an infinite,conducting cylinder, it was found that tests along the cylinder axis lend themselves to iteration. Along the axis, solutions that vary as $(ks)^{-1/2}$ test better than solutions that vary as $(ks)^{-1}$ for large path lengths. Tests at points other than on the axis generate error terms that are large with respect to the approximate field value, so that the iterative equation is not convergent. The source regions of each of the proposed solutions were found to contribute significantly to the large error terms for tests of points off the cylinder axis.

OVERVIEW OF E-FIELD AND INTEGRAL E-FIELD TESTS

Two tests based on satisfaction of the E-field boundary condition have been presented. Both have been applied to proposed solutions to the problem of a magnetic dipole radiating in the presence of an infinitely long, perfectly conducting circular cylinder. The E-Field Test is based on relating the spectral domain of the surface magnetic field to that of the surface electric field and then evaluating the surface E-field. The test is effected by observing how well the resultant E-field satisfies the surface boundary condition, and, as such, is a direct test of boundary condition satisfaction. The E-Field Test has a global character in that it can be applied to a proposed solution only once. For the solutions tested so far, the E-Field Test seems most sensitive to the source region of a proposed solution and relatively insensitive to large path length behavior. One cannot say that only the source region is tested, though, because gross errors in a proposed solution might evidence themselves in the E-Field Test.

The Integral E-Field Test can be formulated from Generalized Lorentz Reciprocity and employs "test" dipoles situated in free space. The surface electric field corresponding to the solution being tested is never computed, so that the Integral E-Field Test is, in effect, an indirect boundary condition check. It has local character in that the test results can be associated with the field at a particular point, and the test can be repeated many times at different points for the same proposed solution. Results of the Integral E-Field Test display, in general, a mixed local/global character. For tests on or near the cylinder axis, the local character is very strong; while for tests in the "deep shadow" region, less local character is retained and the asymptotic source region becomes dominant. Under some circumstances, including tests along the cylinder axis, the Integral E-Field Test lends itself to use as a convergent iterative equation. The Integral E-Field Test formulation is quite general, and application is not limited to cylindrical structures.

In summary, the E-Field and Integral E-Field Tests appear to be complementary. One is global and the other has a mixed global/local nature. One test is highly source region sensitive and relatively insensitive to large-path-length behavior. For the other test, the source region can essentially be excluded from contributing to the result. Finally, the possibility of iterative improvement of a proposed solution is offered, if some relatively stringent conditions can be met.

ITERATIVE ACCURACY TEST FOR BOUNDARY CONDITIONS

Both the E-Field Test and the Integral E-Field Test were applied to asymptotic solutions of surface magnetic fields. The asymptotic solutions tested predicted both the near field and long-path-length behavior of a magnetic dipole radiator. A more traditional and wide-spread use of asymptotic solutions is to predict the far-field behavior of scattering bodies. For analyzing complex scattering bodies, the engineer today can draw on a range of asymptotic techniques, including different "uniform" solutions. In analyzing a complex body, the investigator is likely to combine contributions from straight and curved edges, and possibly diffracted rays from curved surfaces. A great deal of work is being performed today to generate "diffraction coefficients" for the various diffracting mechanics needed to complete the total solution of the far field. This section presents an iterative method based on spectral domain techniques for solving the far field and induced currents of a scattering body. An important feature of this method is that at each step of the iterative method a boundary condition check is incorporated so that the accuracy of the solution can be evaluated. This boundary condition check at the first iteration also provides a means of evaluating the benefit of a proposed solution. In order to illustrate use of this test, it is applied to check a vertex diffraction coefficient for scalar-wave incidence that has been proposed by Albertsen [12]. Although the application to be illustrated is a scalar problem, the extension to vector electromagnetics problems is straightforward.

The Corner Diffraction Coefficient

The corner diffraction term investigated here gives the field diffracted from the vertex when a scalar plane wave is incident on an infinite, acoustically soft quarter-plane, Fig. 14. Such a problem has been addressed by Kraus and Levine [13], Keller [14], and Radlow [15]. Recently, Albertsen used sequential application of the Weiner-Hopf technique and saddle-point integration to extract a corner diffraction coefficient. Albertsen gives the total scattered field from the quarter-plane as

$$u(x,y,0) = \frac{1}{(2\pi)^2} \int_{-\infty+\mu_i}^{\infty+\mu_i} d\mu \int_{-\infty+\lambda_i}^{\infty+\lambda_i} d\lambda \, \frac{M^{++}(\mu,\lambda)M^{++}(\mu_0,\lambda)M^{++}(\mu,\lambda_0)M^{++}(\mu_0,\lambda_0)}{\sqrt{k^2 - \mu^2 - \lambda^2}} \tag{23}$$

$$\times\, e^{-ix\mu}\, e^{-iy\lambda}/[(\lambda + \lambda_0)(\mu + \mu_0)] \quad .$$

The scattered field due to the corner only is given by

$$U_c^s(r,\theta,\phi) = \left(\frac{e^{i\pi/4}}{8\pi k}\right)^2 \frac{e^{ikr}}{r} D_D(\theta_0,\phi_0,\theta,\phi) \quad . \tag{24}$$

The corner diffraction coefficient D_D can be determined from

$$D_D(\theta_0,\phi_0,\theta,\phi) = -4 \, \frac{M(\theta_0,\phi_0,\theta,\phi)}{(\mu + \mu_0)(\lambda + \lambda_0)}$$

where

$$M(\theta_0,\phi_0,\theta,\phi) = M^{++}(\mu,\lambda) \times M^{++}(\mu,\lambda_0) \times M^{++}(\mu_0,\lambda) \times M^{++}(\mu_0,\lambda_0) \tag{26}$$

where

$$\mu = -k \sin\theta \cos\phi \quad , \qquad \mu_0 = -k \sin\theta_0 \cos\phi_0$$

$$\lambda = -k \sin\theta \sin\phi \quad , \qquad \lambda_0 = -k \sin\theta_0 \sin\phi_0 \quad .$$

The M^{++} functions are found from

$$M^{++}(\mu,\lambda) = \sqrt[4]{\frac{k}{2}}\, \sqrt{1 - d_1}\; e^{\tau} \tag{27}$$

where

$$\tau = -\frac{1}{4\pi i}\Big\{ \mathrm{Dilog}(1 - d_1) - \mathrm{Dilog}(1 + d_1)$$

$$+ \mathrm{Dilog}(1 - d_2) - \mathrm{Dilog}(1 + d_2) - i\pi \log(-\zeta_2)$$

$$- (\log(-id_1) - i \tfrac{\pi}{2}) \log(1 + d_1)$$

$$+ (\log(-id_1) + i \tfrac{\pi}{2}) \log(1 - d_1)$$

$$+ (\log(id_2) - i \tfrac{\pi}{2}) \log(1 - d_2)$$

$$- (\log(id_2) + i \tfrac{\pi}{2}) \log(1 + d_2)\Big\} \qquad (28)$$

and

$$d_1 = \zeta_1 K_1, \ d_2 = \zeta_1 K_2$$

$$\left.\begin{matrix}\zeta_1\\ \zeta_2\end{matrix}\right\} = -\frac{1}{k}\left(\lambda \mp i\sqrt{k^2 - \lambda 2}\right)$$

$$\left.\begin{matrix}K_1\\ K_2\end{matrix}\right\} = \frac{1}{k}\left(i\mu \pm \sqrt{k^2 - \mu^2}\right)$$

and the Dilog function is defined by Dilog $(z) = -\int_1^z \frac{\mathrm{Log}\ \xi}{\xi - 1}\, d\xi$. In practice, this corner diffracted field would be added to the edge diffracted field and the geometrical optics field in order to obtain the total field. It is desired, then, that this test assess the contribution of the corner diffraction term to satisfaction of the surface boundary condition (total field equals zero).

Testing Procedure

In order to assess the contribution of corner diffraction to the satisfaction of the boundary condition, a procedure introduced by Ko and Mittra [2] was used. A truncation operator θ is employed, which is defined by $\theta(A) = \int A\ \delta(\vec{r}-\vec{r}_s)dr$, $\vec{r}_s \epsilon S$, where δ is the Dirac delta function and S is the planar scattering surface. The complementary operator $\hat{\theta}$ given by $\hat{\theta}(A) = A - \theta(A)$ is also utilized.

For an acoustically soft scatterer the boundary condition requires that the total field be zero on the scatterer surface. The truncation operators can be used to express the total scattered field in the plane of the scatterer as the sum of a known field [$\theta(-U^i)$ to satisfy the boundary condition which requires that the scattered field exactly equal the negative of the incident field on the surface of the scatterer] and a field which exists only outside the surface and is estimated by a proposed solution ($\hat{\theta}(F)$), so that

$$U^s = \theta(-U^i) + \hat{\theta}(F) \qquad . \qquad (29)$$

A scalar scattering "current," J, is defined by $J = \left(\frac{\partial U(x',y',0^-)}{\partial z} - \frac{\partial U(x',y',0^+)}{\partial z}\right)$, $x',y' \in S$, and the scattered field is given by

$$J * G = U^s, \tag{30}$$

where G is a Green's function operator and * denotes convolution. The Fourier transform of the above equation results in

$$\tilde{J} = \tilde{G}^{-1}[\theta(-U^i) + \hat{\theta}(F)] \tag{31}$$

where ˜ denotes the transform domain. Because $\tilde{G}^{-1}$ is known, and F can be estimated from corner and edge diffraction terms, a surface current can be obtained from $J = \mathcal{F}^{-1}\{\tilde{J}\}$, where $\mathcal{F}^{-1}$ indicates inverse Fourier transformation. The current J is then truncated to the scatterer, $J_t = \theta(J)$, and the associated scattered field determined:

$$U_t = \mathcal{F}^{-1}\{\tilde{G} \cdot \tilde{J}_t\} \quad . \tag{32}$$

Note that U_t extends over a wide range and includes fields external to the scatterer, while J_t has been truncated to the scatterer surface. U_t now provides a basis for checking satisfaction of the boundary condition, for U_t can be compared to $(-U^i)$ on the scatterer surface.

This procedure lends itself well to iteration, for the scattered field can be updated by $U^{(j)} = \hat{\theta}(U_t^{(j-1)}) + \theta(-U^i)$, following which the scattering current associated with this field can be found from

$$J^{(j)} = \mathcal{F}^{-1}\{\tilde{G}^{-1}\tilde{U}^{(j)}\} \quad . \tag{33}$$

Boundary condition satisfaction for this current can again be checked following truncation of the scattering current and determination of the resulting scattered field.

The iteration-boundary condition check can be carried out by the following procedure:

(1) Obtain a first estimate of the transform of the scattered field in the plane of the scatterer from (7) as

$$U^{(0)} = \theta(-U^i) + \hat{\theta}(F) \tag{34}$$

where the F function is given from edge and/or corner diffraction terms.

Observe that $\theta(-U^i)$ is exact as determined by the surface boundary condition, while $\hat{\theta}(F)$ is approximate, since it has been determined from a proposed asymptotic solution.

(2) Determine the associated scattering current from

$$\tilde{J}^{(0)} = \mathcal{F}^{-1}\{\tilde{G}^{-1} \cdot \tilde{U}^{(0)}\} \tag{35}$$

where G is given by exp(ikR)/4πR and R is the distance between the "source" and "observation" points. Note that in this application, R is specialized so that all observation points lie in the plane of the scatterer.

(3) Truncate the current to the scatterer surface

$$J_t^{(0)} = \theta(J^{(0)}) \quad . \tag{36}$$

(4) Determine the scattered field in the plane of the scatterer by

$$U_t^{(0)} = F^{-1}\{\tilde{G} \cdot \tilde{J}_t^{(0)}\} \quad . \tag{37}$$

Satisfaction of the boundary condition on the scatterer surface can be checked at this point.

(5) Update the scattered field so that the boundary condition is satisfied.

$$U^{(1)} = \theta(-U^i) + \hat{\theta}(U_t^{(0)}) \quad . \tag{38}$$

Note that this step is necessary because $\theta(U_t^{(0)})$ has been determined from J_t, and will not, in general, satisfy the boundary condition. This step is the parallel to (12), except that the external field estimate has been updated by the iteration procedure, replacing the initial guess.

(6) Continue the iteration by repeating Steps (2) - (5), using $U^{(1)}$ in place of $U^{(0)}$.

Obviously, the iteration can be repeated as many times as desired, or until convergence of the scattering current and scattered field is obtained. The accuracy test is incorporated into Step 4 where the scattered field is obtained at the surface of the scatterer and can easily be examined to determine boundary condition satisfaction. Note that after the boundary condition check has been carried out, in order to continue the iteration, the scattered field on the surface is discarded and replaced by the known, exact scattered field.

Testing the Corner Diffraction Coefficient

In order to test the contribution of corner diffraction to satisfaction of the boundary condition on the surface by the method previously described, it is necessary to find the field exterior to the scatterer ($\hat{\theta}F$) associated with corner diffraction. This is accomplished in the following manner. The far field as scattered from the vertex of a quarter-plane is determined on a grid of observation points, which include points described by imaginary angles. This effectively gives the transform of the field in the plane of the scatterer, since the two are related by

$$F\{U^s(x',y',0)\}\Big|_{\substack{\alpha=\sin\theta\cos\phi \\ \beta=\sin\theta\sin\phi}} = \frac{Q}{\cos\theta} U^F \quad , \tag{39}$$

where $U^s(x',y',0)$ is the scattered field in the plane of the scatterer, U^F is the scattered far field, Q is a constant, and θ is a standard spherical coordinate observation angle. A Fast-Fourier Transform (FFT) algorithm is used to find the planar scattered field by inverting $F\{U^s(x',y',0)\}$. The complementary operator $\hat{\theta}$ is employed to obtain the scattered field exterior to the quarter plane. This is the "F" function desired. Finally, an FFT is again employed to get $\theta\hat{F}^c$, the transform of the F function due to the corner; this term contributes to the transform of the scattered field used in Step (1).

In order to aid in evaluation of the effect of the corner diffraction coefficient on boundary condition satisfaction, two error terms are defined. A global error term ε is defined as $\varepsilon_j = \sqrt{\sum_{n=1}^{N} |[U_j(x_n,y_n) + U^i(x_n,y_n)]|^2/N}$ where x_n and y_n are the positions of the sample points on the scatterer and U_j denotes the scattered field from Equation 37 at the jth iteration. A local corner error term Δ_j^α was defined similarly, differing from ε_j only in that (x_n, y_n) was confined to the 0.25λ by 0.25λ region at the αth corner. The error terms that resulted from using $\theta(\hat{F}^c) \equiv 0$ (no corner contribution) were compared to those that resulted when a single vertex term was included and to error terms when contributions from all four corners were counted.

The boundary condition test was carried out on a two wavelength by two wavelength plate. The testing procedure was carried through three iterations; Figs. 15 and 16 present the scattered field at each iteration for a normally incident scalar wave. Shown is the scattered field for a constant y value cut through the plate. Figure 15 presents a cut near the edge at y ≃ 0.1 wavelength, while Fig. 16 shows a cut near the plate center at y = 1.0 wavelength. Figure 17 presents a representative scattering "current" very near the edge at y ≃ 0.03λ. Because the constant y cuts cannot represent the behavior of the total scattered field, Tables 1 and 2 present results of the "error terms" ε and Δ. Note that both tables include results when only one corner diffraction coefficient was included. The corner error terms of Table 2 are taken at the corner at which a single vertex diffraction coefficient was included.

TABLE 1 Total RMS Error, ε, for Normal Incidence

	ε_1 1st Iteration	ε_2 2nd Iteration	ε_3 3rd Iteration
No Vertices	.223	.127	.048
One Vertex	.207	.121	.047
Four Vertices	.156	.103	.043

TABLE 2 Local Corner RMS Error, Δ, for Normal Incidence

	Δ_1^1 1st Iteration	Δ_2^1 2nd Iteration	Δ_3^1 3rd Iteration
No Vertices	.373	.205	.069
One Vertex	.360	.192	.067
Four Vertices	.294	.159	.067

Discussion of Results

The figures and tables indicate that use of vertex diffracted fields does improve boundary condition satisfaction. Of greater interest, however, is the performance of the Iterative Accuracy Test for Boundary Conditions procedure. The authors' experience is that both scattering currents and scattered fields are well converged after three iterations as illustrated in Figs. 15-17.

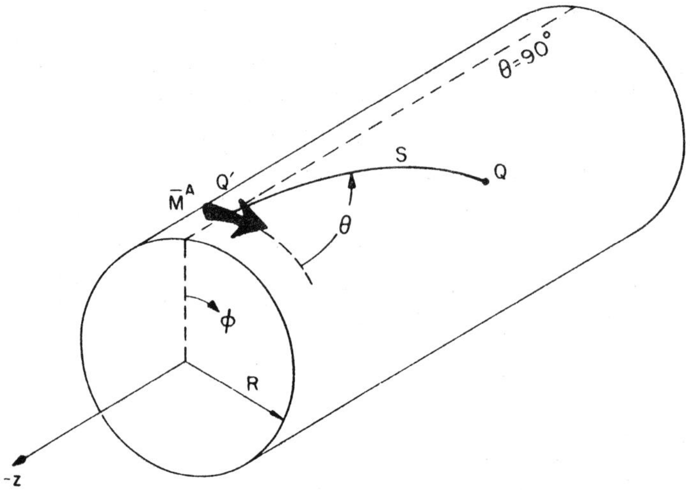

Fig. 1a Geometry of asymptotic solutions: Magnetic dipole in the presence of a perfectly conducting infinite cylinder.

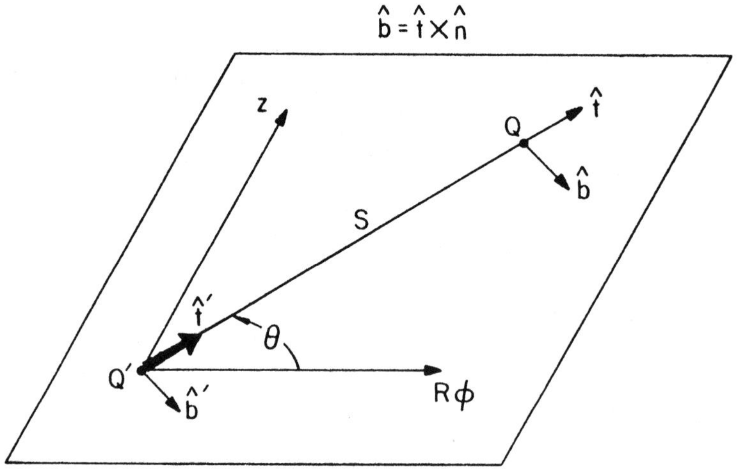

Fig. 1b Geometry of asymptotic solutions: Magnetic dipole in the presence of a perfectly conducting infinite cylinder.

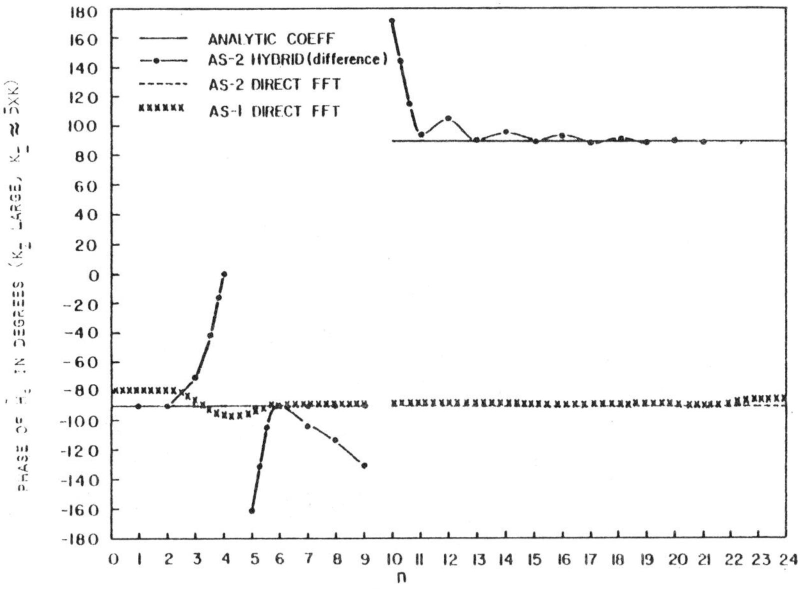

Fig. 2 Effect of hybrid computation on phase of Hϕ transform.

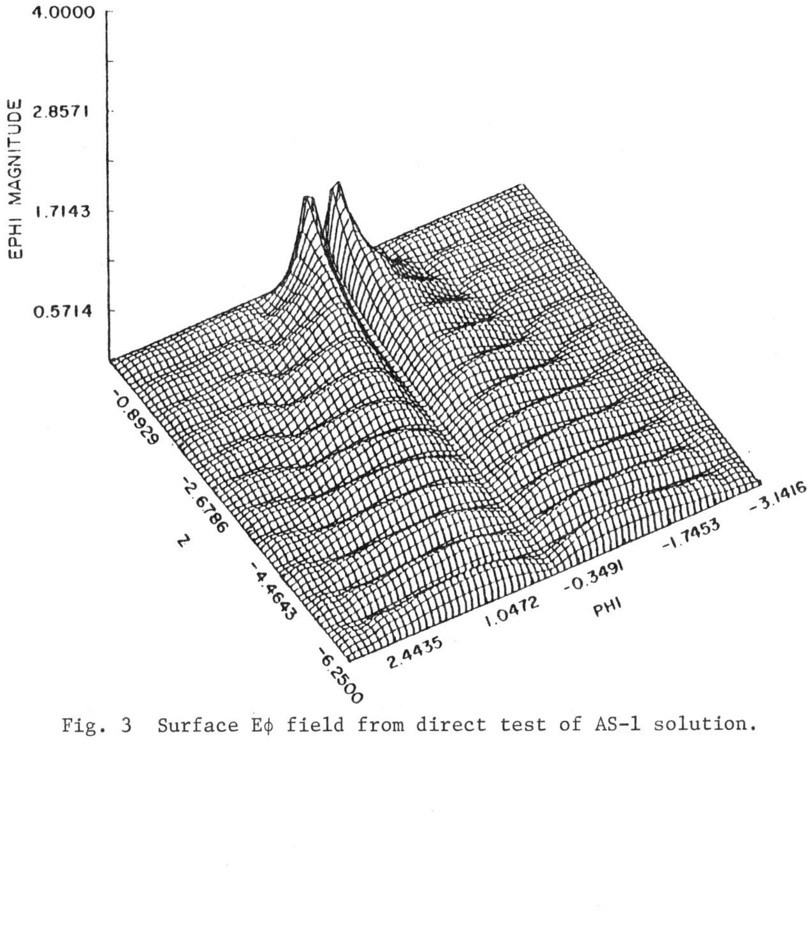

Fig. 3 Surface Eϕ field from direct test of AS-1 solution.

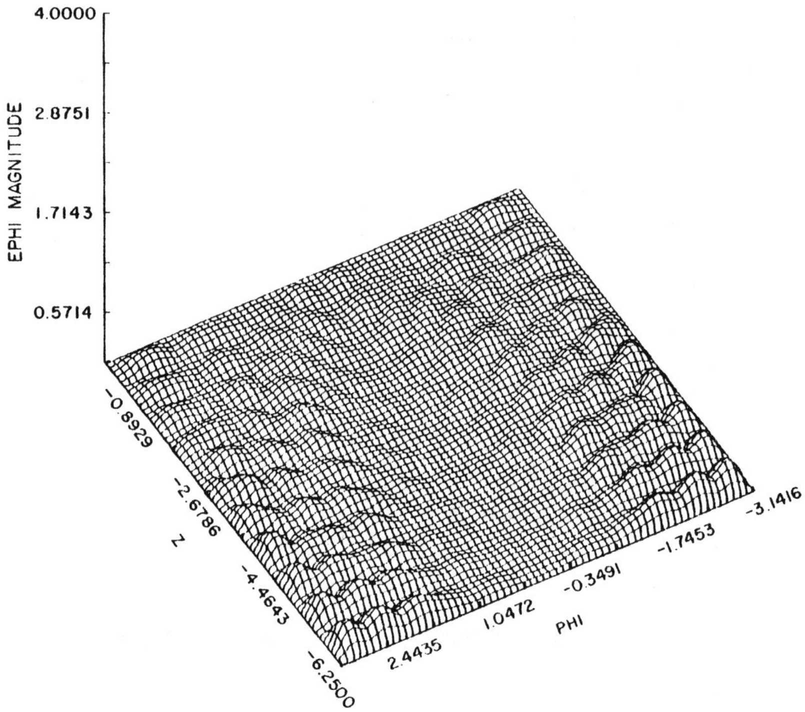

Fig. 4 Surface Eϕ field from hybrid test of AS-3 solution.

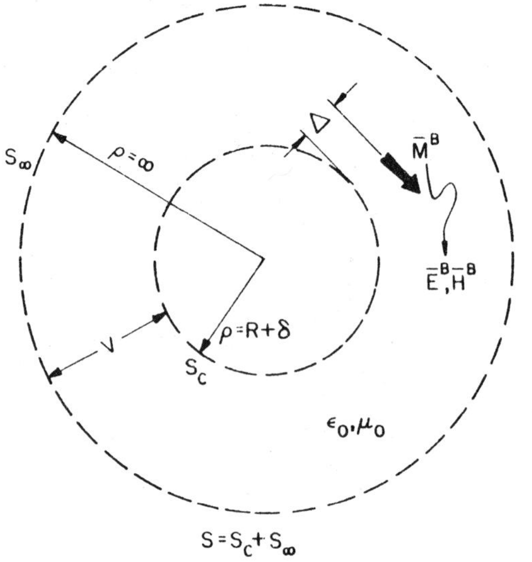

Fig. 5 Free-space test environment: Placement of "reciprocity volume" with respect to test dipole.

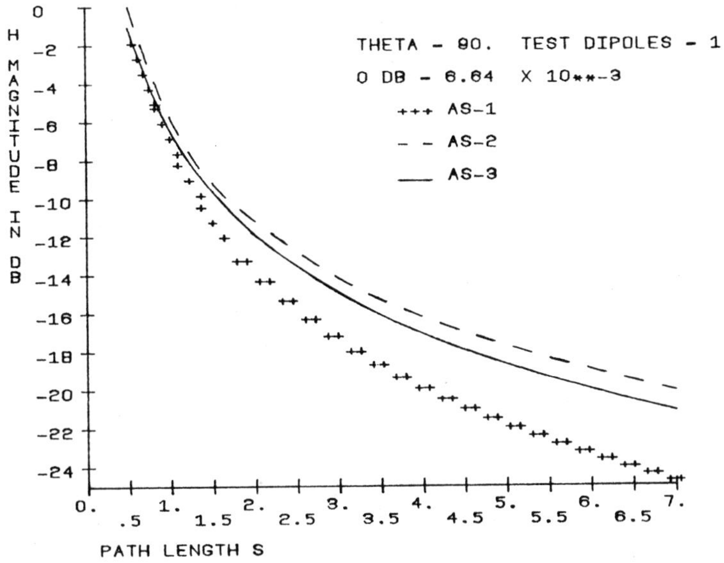

Fig. 6 Magnitude of zeroth-order H_ϕ field along $\Theta = 90°$.

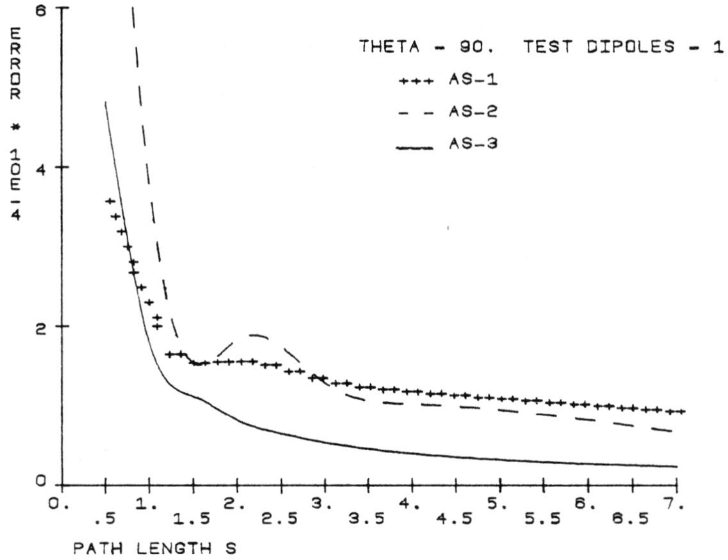

Fig. 7 Results of Integral E-Field Test along $\Theta = 90°$: Single test dipole.

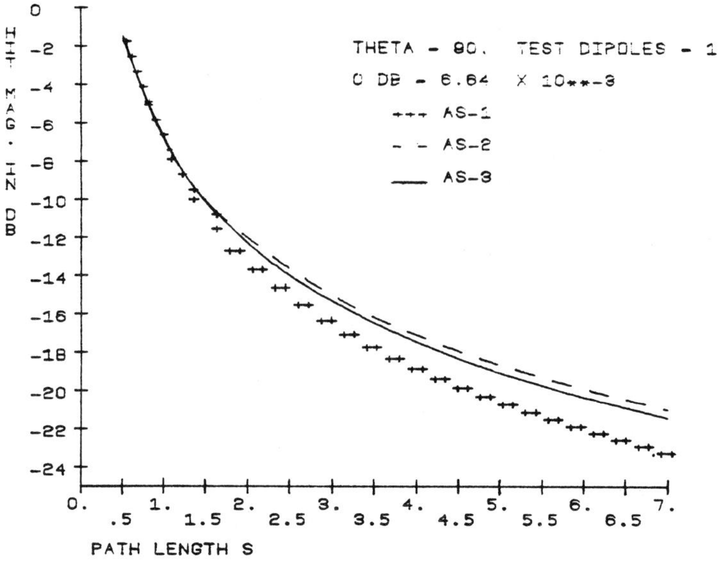

Fig. 8 Iterated magnitudes of H_ϕ field along $\Theta = 90°$: Single test dipole.

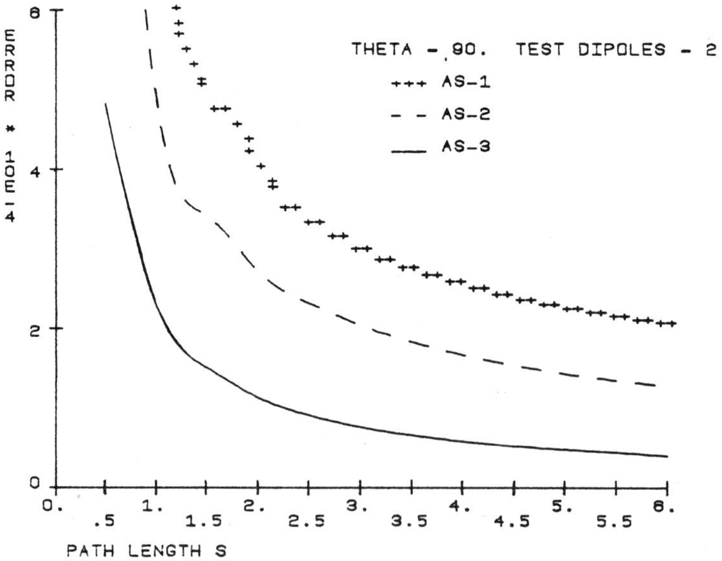

Fig. 9 Results of Integral E-Field Test along $\Theta = 90°$: Test dipole and planar image dipole.

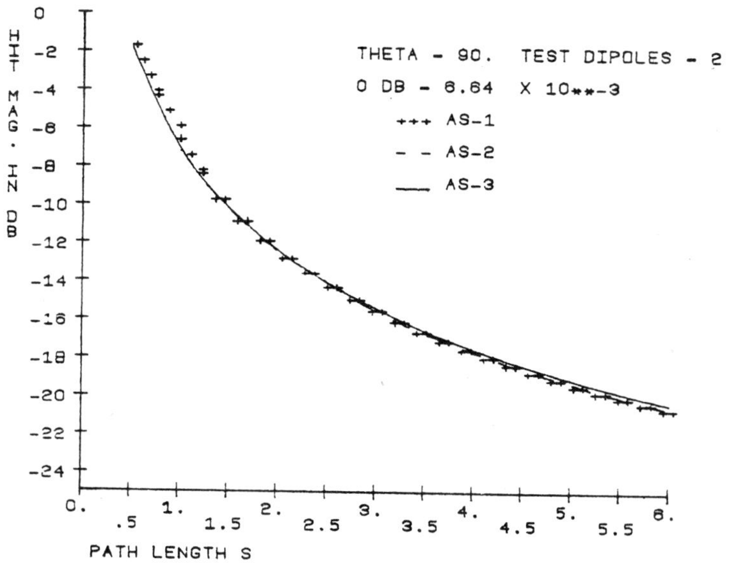

Fig. 10 Iterated magnitudes of H_ϕ field along $\Theta = 90°$: Test dipole and planar image dipole.

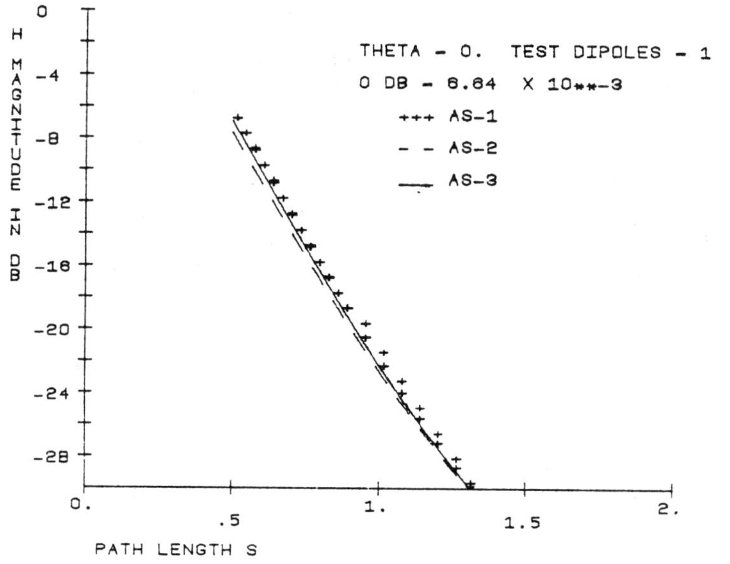

Fig. 11 Magnitude of zeroth-order H_ϕ field along $\Theta = 0°$.

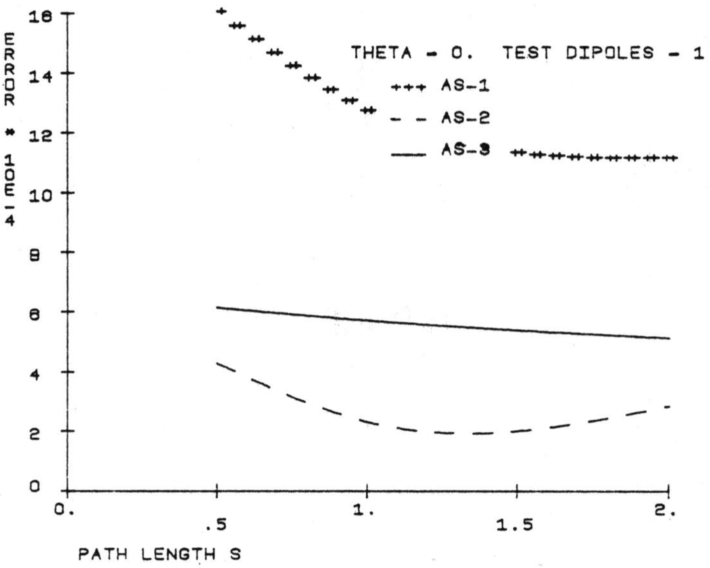

Fig. 12 Results of Integral E-Field Test along Θ = 0°.

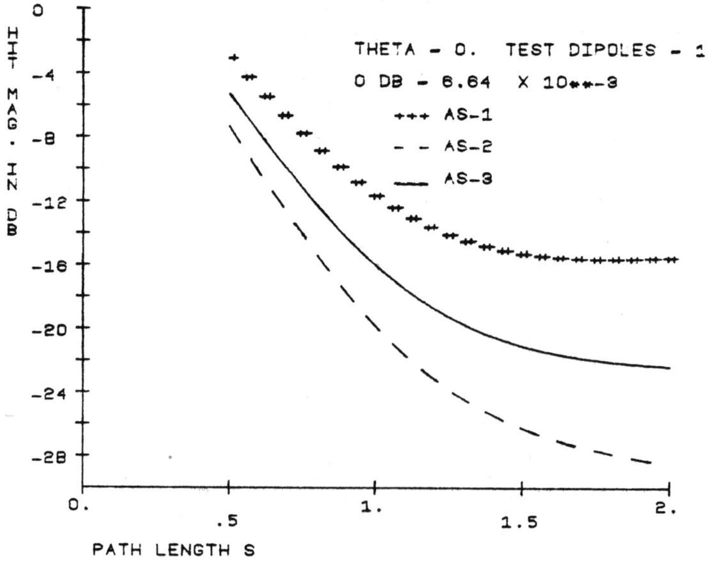

Fig. 13 Iterated magnitudes of H_ϕ field along $\Theta = 0°$: Single test dipole.

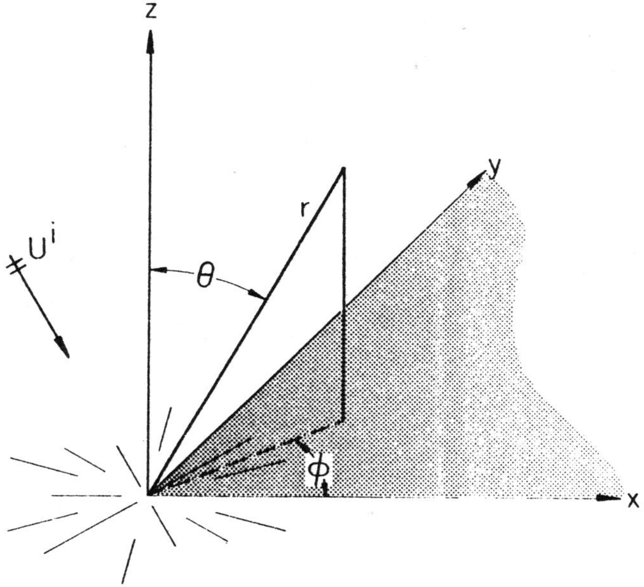

Fig. 14 Scalar wave incident on an acoustically soft quarter-plane.

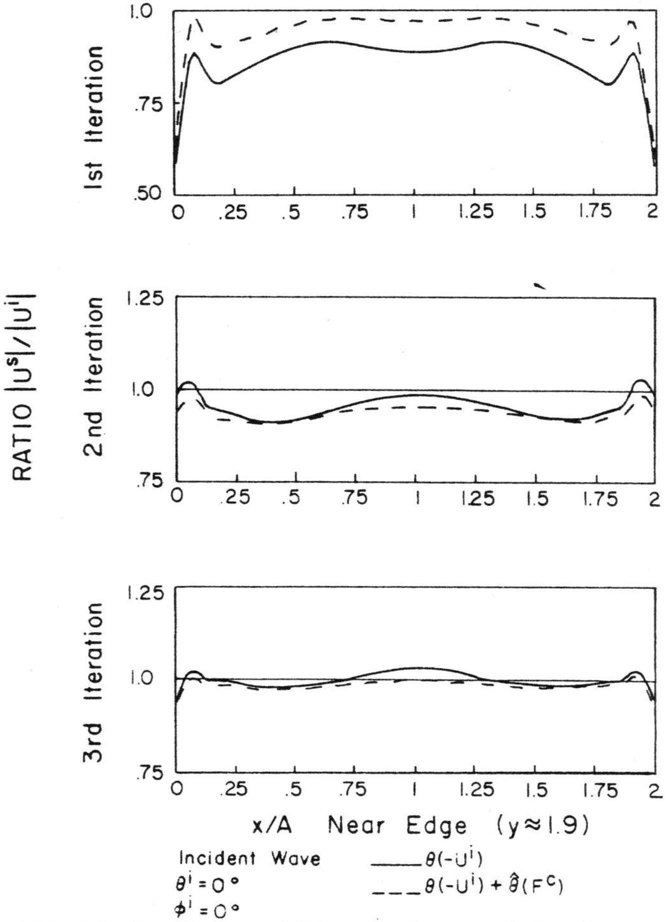

Fig. 15 Boundary condition check near plate edge, normal incidence.

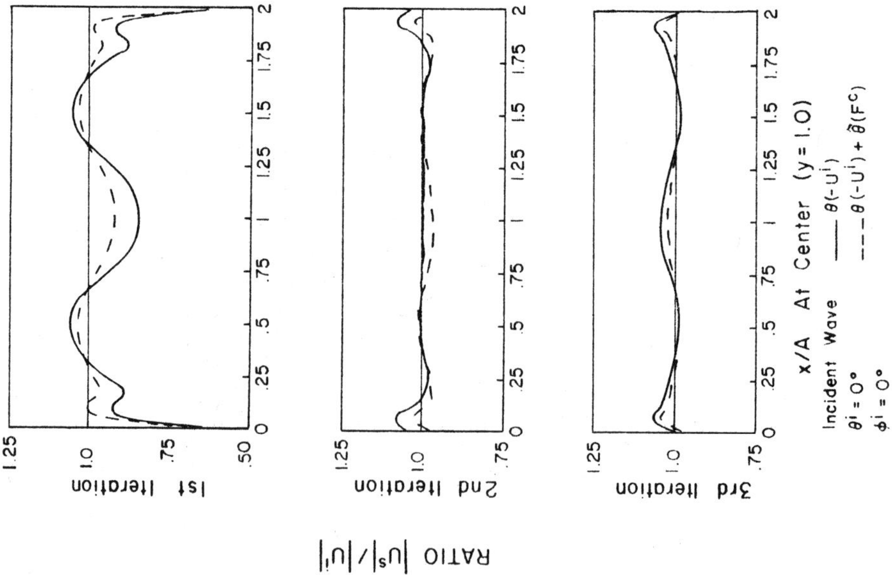

Fig. 16 Boundary condition check at plate center, normal incidence

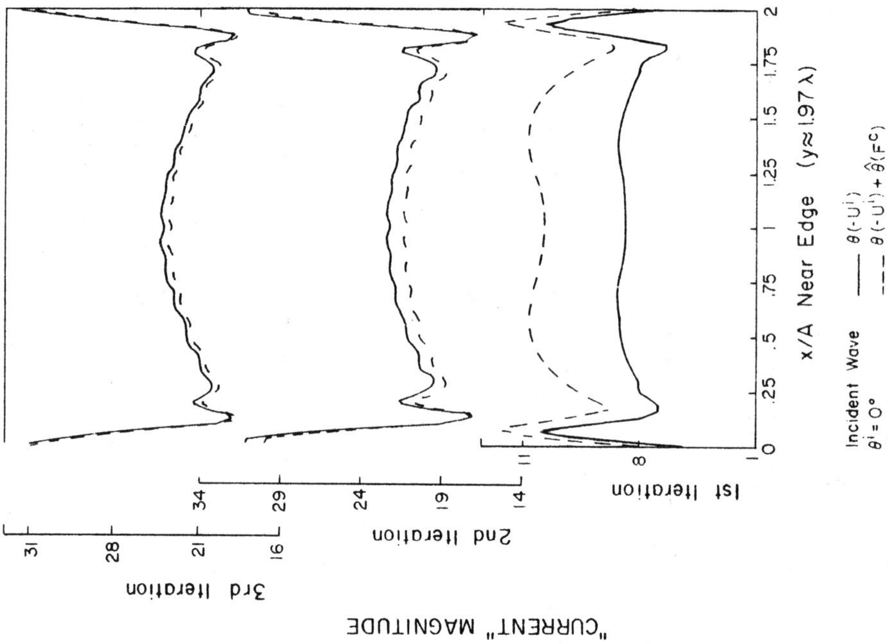

Fig. 17 Scattering currents near plate edge normal incidence.

In addition, this procedure seems to be relatively insensitive to the initial approximation; indeed, use of an identically zero scattered field external to the scatterer is an acceptable initial approximation.

For use in electromagnetics problems, then, the Iterative Accuracy Test for Boundary Conditions can serve a dual role. The contribution of a proposed diffraction coefficient to boundary condition satisfaction can be determined by carrying the test through one iteration for two different cases: First using the far field as given by physical optics plus (possibly) other known diffracted fields (such as edge diffraction) and, second, using the far field of the first case plus that added by the proposed diffraction coefficient. Examination of the surface fields indicates the change in boundary condition satisfaction due to the solution under test. Alternatively, the iterative procedure can be used to solve for the far field and scattering currents without resorting to asymptotic techniques. If the body is such that the external scattered field is appreciably different from zero, then asymptotic solutions can be used to formulate the initial guess, with a possible increase in the rate of convergence.

It should be pointed out here that solution for the scattering currents inherent in this method is an important attribute. Given the scattering currents, the engineer is able to compute both far fields and near fields; and the fields so determined are guaranteed to be regular and free of singularities.

SUMMARY

Three different procedures for testing the effect of asymptotic solutions on boundary condition satisfaction have been presented. Two are formulated entirely using spectral domain techniques, while the third retains the option of performing the test entirely in the spatial domain. Two tests are best suited for testing asymptotic solutions to surface fields, while the third can also be employed to test asymptotic solutions that give the scattered far field. One of the tests is particularly well-suited for use as an iterative equation and seems relatively insensitive to the initial approximation. In contrast, the iterative aspects of the Integral E-Field Test are limited to particular cases where certain conditions are met. Two of the tests are global in nature and test the entire solution, while the "Integral E-Field Test" displays both global test and local test characteristics.

With communications, radar, and missile systems moving toward use of higher and higher frequencies, it seems certain that use of asymptotic solutions by engineers will increase. The tests presented here provide a means by which new proposed solutions can be evaluated.

ACKNOWLEDGMENT

The work reported in this paper was supported in part by the Office of Naval Research under Grant N00014-75-C-0293 and, in part, by the Joint Services Electronics Program under Grant DAAB-0772-C-0259

REFERENCES

[1] R. Mittra, Y. Rahmat-Samii and W. L. Ko, Spectral Theory of Diffraction, Appl. Phys. 10, 1, (1976).

[2] W. L. Ko and R. Mittra, A New Approach Based on a Combination of Integral Equation and Asymptotic Techniques for Solving Electromagnetic Scattering Problems, IEEE Trans. Antennas Propagat. AP-25, 187, (1977).

[3] M. Tew and R. Mittra, Accuracy Tests for Asymptotic Solutions to Radiation from a Cylinder, Coordinated Sciences Laboratory Report UILU-Eng 77-2251, University of Illinois at Urbana-Champaign, (1977).

[4] R. Mittra and M. Tew, Accuracy Test for High-Frequency Asymptotic Solutions, IEEE Trans. Antennas Propagat. AP-27, 62, (1979).

[5] L. L. Bailin, The Radiation Field Produced by a Slot in a Large Circular Cylinder, IRE Trans. AP-3, 128, (1955).

[6] R. F. Harrington (1961) Time-Harmonic Electromagnetic Fields, McGraw-Hill, New York.

[7] Y. Hwang and R. G. Kouyoumjian, The Mutual Coupling Between Slots on an Arbitrary Convex Cylinder, ElectroScience Laboratory Semi-Annual Report 2902-21, The Ohio State University, (1975).

[8] P. H. Pathak, Analysis of a Conformal Receiving Array of Slots in a Perfectly-Conducting Circular Cylinder by the Geometrical Theory of Diffraction, ElectroScience Laboratory Technical Report ESL 3735-2, The Ohio State University, (1975).

[9] Z. W. Chang, L. B. Felsen, and A. Hessel, Surface Ray Methods for Mutual Coupling in Conformal Arrays on Cylinder and Conical Surface, Polytechnic Institute of New York Final Report (1976).

[10] S. W. Lee and S. Safavi-Naini, Asymptotic Solution of Surface Field due to a Magnetic Dipole on a Cylinder, Electromagnetics Laboratory Technical Report No. 76-11, University of Illinois at Urbana-Champaign, (1976).

[11] S. W. Lee and S. Safavi-Naini, Approximate Asymptotic Solution of Surface Field due to a Magnetic Dipole on a Cylinder, IEEE Trans. Antennas Propagat. AP-26, 593, (1978)

[12] N. C. Albertsen, Technical University of Denmark, personal communication.

[13] L. Kraus and L. Levine, Diffraction by an Elliptic Cone, Comm. Pure Appl. Math. 14, 49, (1961).

[14] J. B. Keller, Geometrical Theory of Diffraction, J. Opt. Soc. Amer. 52, 116, (1962).

[15] J. Radlow, Note on the Diffraction at a Corner, Archive for Rational Mechanics 19, 62, (1965).

Part 5
Geometrical Theory of Diffraction

APPLICATION OF RAY THEORY TO DIFFRACTION OF ELASTIC WAVES BY CRACKS

J. D. Achenbach, A. K. Gautesen[†] and H. McMaken
The Technological Institute, Northwestern University,
Evanston, Illinois 60201

ABSTRACT

The field generated when a high-frequency ultrasonic wave is diffracted by a flat crack is analyzed on the basis of elastodynamic ray theory. The diffraction of incident rays by a crack-edge, as described by the geometrical theory of diffraction (GTD) and its uniform asymptotic extension (UAT), is discussed. Particular attention is devoted to a method by which the crack-opening-displacement is computed on the basis of ray theory, and the scattered field is subsequently obtained by the use of a representation integral. Results are presented for slits and penny-shaped cracks. These results are compared with numerically computed "exact" results. Good agreement is already obtained at moderate values of the frequency.

INTRODUCTION

The solution to the direct diffraction problem, that is, the computation of the field generated when an ultrasonic wave is diffracted by a known flaw, is a necessary preliminary to the solution of the inverse problem. In recent years several analytical methods have been developed to investigate scattering of elastic waves by interior cracks as well as by surface-breaking cracks, in both the high- and the low-frequency domains. The appeal of the high-frequency approach is that the probing wavelength is of the same order of magnitude as the length-dimensions of the crack. This gives rise to characteristic interference phenomena.

The analytical work discussed in this paper is concerned with the direct problem. Diffraction of time-harmonic signals by cracks is analyzed on the basis of linearized elasticity theory for a homogeneous, isotropic solid. The propagation of waves in such solids is a classical area of investigation. The equations governing elastodynamic theory, and several pertinent mathematical techniques to obtain solutions, have been discussed in detail in Ref.[1].

From the theoretical point of view a flat crack is a planar surface across which the displacement can be discontinuous. The exact mathematical formulation of the elastodynamic field generated by the presence of a crack is rather complicated if the displacement discontinuities are not known a-priori. That is the case in diffraction problems. Diffraction of an incident wave by a crack is a mixed boundary value problem, whose exact solution satisfies one or more generally singular integral equations for the displacement discontinuities. Only for a semi-infinite crack can an analytical solution conveniently be obtained, as shown by Maue [2].

[†]Permanent address: Department of Mathematics, Southern Methodist University, Dallas, TX 75275

In this paper we investigate diffraction of elastic waves by using elastodynamic ray theory. The approach presented here is valid for $\omega a/c_L >> 1$, and at points where $S/a > 1$. Here ω is the circular frequency, a is a length dimension of the crack, c_L is the velocity of longitudinal waves, and S is the distance from a crack edge.

GEOMETRICAL THEORY OF DIFFRACTION

We start with a brief review of the Geometrical Theory of Diffraction (GTD) for elastodynamic diffraction of an incident wave by a crack. The theory was originally developed by Keller [3] for acoustic and electromagnetic edge diffraction. A qualitative description precedes a more quantitative one.

A wavefront is defined as a surface of constant phase. In a homogeneous, isotropic, linearly elastic solid the rays are straight lines, which are normal to the wavefronts. An unbounded solid can support rays of longitudinal and transverse wave motion. These rays are denoted as L-rays and T-rays, respectively. The free surface of a solid can, in addition, support rays of surface-wave motion, which are denoted as R-rays. The total field at an observation point is given by the sum of the fields associated with all rays passing through that point. The theory is valid for incident waves whose wavelength is short when compared to the dimensions of the crack, but comparisons of the theory with numerical results has shown that quite acceptable results are achieved when these two dimensions are of the same order of magnitude.

The leading order term in the asymptotic expansion of the total field is the Geometrical Elastodynamics (GE) field which consists of the incident wave in the illuminated region together with the specular reflections from the crack surface. This field is discontinuous at shadow boundaries and at the boundaries of zones of reflected rays. It is identically zero in the shadow zone.

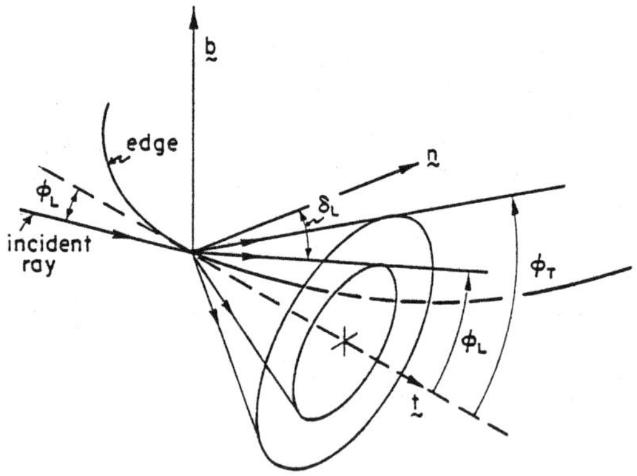

Fig. 1 Incident ray and cones of diffracted rays

Signals in the shadow zone as well as additional signals elsewhere arise from diffracted body-wave rays. When an incident ray intersects the edge of a crack it diffracts two cones of body-wave rays, as well as rays of surface motion which are discussed below (see Fig. 1). The inner (outer) cone consists of infinitely many rays of longitudinal(transverse) motion. The half angles of these cones are related to the angle of incidence by Snell's law of edge diffraction (see Eq.(2.2) below). The fields associated with these diffracted rays are smaller than the incident field by a factor proportional to the square root of the wavelength.

The GTD fields are singular at (1) shadow boundaries and boundaries of zones of reflected rays, and (2) at caustics (which are defined as the envelope of the rays). Near the boundaries of the shadow zone and the zones of specular reflection the GTD fields together with the GE fields can be corrected by the use of a Fresnel integral term to provide a smooth transition across these boundaries, see e.g. Refs.[4] and [5].

The crack edge, where all the diffracted body-wave rays intersect, is a caustic of these rays. The theory fails at this caustic, but the field near the crack edge is generally not of interest. We remark that Achenbach and Gautesen [6] have shown how to modify the theory to predict the singular fields (the stress intensity factors). For other caustic surfaces Ludwig [7] and Kravstov [8] have shown how to construct uniform corrections near such surfaces.

As noted above the incident ray also diffracts rays of surface motion (one on each face of the crack) which propagate at the speed of Rayleigh surface waves. The field associated with these rays is of the same order of magnitude in wavelength as that associated with the incident wave. The angle that the diffracted surface rays make with the crack edge is again related to the incident angle by Snell's law of edge diffraction, see Eq.(2.5) below. It is convenient to consider these rays in pairs - one pair of symmetric and the other of antisymmetric rays with respect to the plane of the crack. In the following discussion no distinction is made between these two pairs of rays, since both exhibit the same phenomena.

When surface rays intersect the edge of the crack, they diffract two cones of body-wave rays, and they are reflected without any mode conversion between the symmetric and antisymmetric modes. The half angles of the diffracted cones of rays are again related to the incident angle by Snell's law of edge diffraction, Eq.(2.10) below. The fields associated with these secondary diffracted body-wave rays are only singular at the caustics, and the remarks made above about the primary diffracted body-wave rays apply here as well. Also the angles of incident and reflection of surface-wave rays are equal, and the reflection coefficient is independent of the wave length.

We now give a more quantitative description of the theory. Further details as well as explicit expression for all diffraction and reflection coefficients can be found in Achenbach et al [9,10].

<u>Primary diffracted body-wave rays</u>.

We use the superscript α = L (longitudinal), TV (transverse-vertically polarized), TH (transverse-horizontally polarized) to denote the nature of the motion on the incident body-wave ray. Likewise, we use the subscript β = L,T to denote the nature of the motion on the diffracted cone of body-wave rays. To the incident body-wave ray we assign two angles of incidence $(\phi_\alpha, \theta_\alpha)$ with ϕ_α being the angle between the incident ray and the unit tangent $\underline{t}$ to the crack edge, and θ_α being the angle between the principal unit normal $\underline{n}$ to the crack edge and the projection of the incident ray onto the plane normal to $\underline{t}$, see Fig. 1. The direction of a diffracted ray is given by

$$\underline{e}_\beta = \cos\phi_\beta \, \underline{t} + \sin\phi_\beta \, (\cos\theta \, \underline{n} + \sin\theta \, \underline{b}) \tag{2.1}$$

where $\underline{b}$ is the unit binormal to the crack edge. The half angle ϕ_β of the diffracted cone of rays is related to the incident angle ϕ_α by Snell's law of edge diffraction:

$$c_\beta^{-1} \cos \phi_\beta = c_\alpha^{-1} \cos\phi_\alpha \ , \tag{2.2}$$

and $\theta \in [0,2\pi]$ is a parameter which defines the individual rays and represents the angle that the projection of the diffracted ray onto the plane normal to the tangent $\underline{t}$ makes with the normal $\underline{n}$.

The displacement fields on the diffracted body-wave rays are

$$\underline{u}_\beta^\alpha = e^{i\omega S_\beta/c_\beta}[S_\beta(1+S_\beta/\rho_\beta^\alpha)]^{-\frac{1}{2}} D_\beta^\alpha(\theta;\phi_L,\theta_L)\ \underline{i}_\beta^\alpha\ U^\alpha \tag{2.3}$$

Here (as well as below) U^α defines the amplitude of the incident wave at the point of diffraction (or reflection), the unit vector $\underline{i}_\beta^\alpha$(or $\underline{i}^\alpha$) relates the directions of displacement of the diffracted fields to those of the incident field, D_β^α are the diffraction coefficients, and S_β are the distances along the diffracted rays from the point of diffraction to the observation point. Also ρ_β^α(or ρ_R) is the distance measured from the crack edge along a ray to the other caustic in Eq.(2.3)

$$\rho_\beta^\alpha = -\ a \sin^2\phi_\beta\ [a(d\phi_\beta/dS) \sin\phi_\beta + \cos\delta_\beta]^{-1} \tag{2.4}$$

where a is the radius of curvature of the edge at the point of diffraction, s is distance measured along the crack edge and δ_β are the angles between the relevant diffracted rays and the normal to the crack edge, see Fig. 1.

We remark that for the fields defined by (2.3) the caustic surfaces are given by $S_\beta = 0$ (the crack edge) and $S_\beta = -\rho_\alpha^\beta$, and that the diffraction coefficients D_β^α are singular for those values of θ for which a diffracted ray coincides with an incident or reflected ray.

Diffracted surface-wave rays.

Here we use the subscript β = RS(RA) to denote the symmetric (antisymmetric) surface-wave rays. The angle ϕ_R that the surface-wave rays make with the tangent $\underline{t}$ to the crack edge is related to the incident angle ϕ_α by Snell's law of edge diffraction

$$c_R^{-1} \cos\phi_R = c_\alpha^{-1} \cos\phi_\alpha \tag{2.5}$$

The displacements on the diffracted surface wave rays are

$$u_\beta^\alpha = e^{i\omega S_R/c_R} (1+S_R/\rho_R)^{-\frac{1}{2}}\ D_\beta^\alpha(\phi_L,\theta_L)\ \underline{i}_\beta^\alpha\ U^\alpha \tag{2.6}$$

where the distance to the caustic is

$$\rho_R = -\ a \sin\phi_R\ (a\ d\phi_R/ds + 1)^{-1} \tag{2.7}$$

The principal difference between Eq.(2.3) and Eq.(2.6) is in the factor $S_\beta^{-\frac{1}{2}}$ which reflects the three-dimensional (spherical) growth and decay in Eq.(2.3) versus two-dimensional (cylindrical) growth and decay in Eq.(2.6).

Reflection of surface-wave rays.

Here α = RS, RA defines the motion on the incident surface-wave rays as well as on the reflected rays, since there is no mode conversion. In addition, the angles of incidence and reflection are the same. The incident field is defined by Eq.(2.6). The fields on the reflected rays are given by

$$\underline{u}_{\alpha}^{\alpha} = e^{i\omega S_R/c_R} (1+S_R/\rho_R)^{-\frac{1}{2}} R_{\alpha}(\phi_R)\, \underline{i}^{\alpha} U^{\alpha} \tag{2.8}$$

where $R_{\alpha}(\phi_R)$ is the reflection coefficient and

$$\rho_R = - a \sin\phi_R \, (a\, d\phi_R/ds + 1)^{-1} \tag{2.9}$$

This last formula is the same as that given by Eq.(2.7), but here $\phi_R(s)$ is the given angle of incidence while in Eq.(2.7) ϕ_R is computed from Eq.(2.5).

Body-wave rays generated by diffraction of surface-wave rays.

Here we use α = RS, RA to denote the nature of the incident surface-wave rays and β = L, T to denote the nature of the motion on the diffracted body-wave rays. The direction of a diffracted ray is again given by Eq.(2.1) with the half angle ϕ_{β} of the diffracted cone of rays related to incident angle ϕ_R by Snell's law of edge diffraction

$$c_{\beta}^{-1} \cos\phi_{\beta} = c_R^{-1} \cos\phi_R \tag{2.10}$$

We remark that if the angle ϕ_{β} is imaginary, then the corresponding cone of diffracted rays is absent. The displacement fields are given by

$$\underline{u}_{\beta}^{\alpha} = e^{i\omega S_{\beta}/c_{\beta}} [S_{\beta}(1+S_{\beta}/\rho_{\beta}^{\alpha})]^{-\frac{1}{2}} D_{\beta}^{\alpha}(\theta;\phi_R)\, \underline{i}_{\beta}^{\alpha} U^{\alpha} \tag{2.11}$$

The distances to the caustics are given by Eq.(2.4) where here ϕ_{β} is determined by Eq.(2.10) instead of Eq.(2.2). We remark that in Eq.(2.11) the diffraction coefficient is continuous in θ, but the caustic surfaces remain.

We briefly summarize how the diffraction and reflection coefficients have been obtained. We consider the canonical problem of incidence of a plane wave of type α = L, TV or TH upon a plane semi-infinite crack. The solution is obtained by using Fourier transforms and the Wiener-Hopf technique, see [9]. The exact solution is expanded asymptotically for large distances from the crack edge. This solution is then compared to that predicted by Eq. (2.3) and (2.6) with $\rho_{\beta}^{\alpha} = \rho_R = \infty$ (since $a = \infty$ and ϕ_{β} is constant). This yields explicit expressions for the diffraction coefficients for incidence of body-wave rays. The remaining coefficients are associated with incidence of surface-wave rays. These coefficients are obtained in the same manner by changing the incident wave in the canonical problem to a Rayleigh surface wave. Details can be found in Ref. [10].

In Figs. 2 and 3 the absolute values of the diffraction coefficients D_L^L and D_T^L are plotted for $\phi_L = \pi/2$. It is noted that D_L^L is singular at $\theta = \theta_L$, i.e., at the shadow boundary.

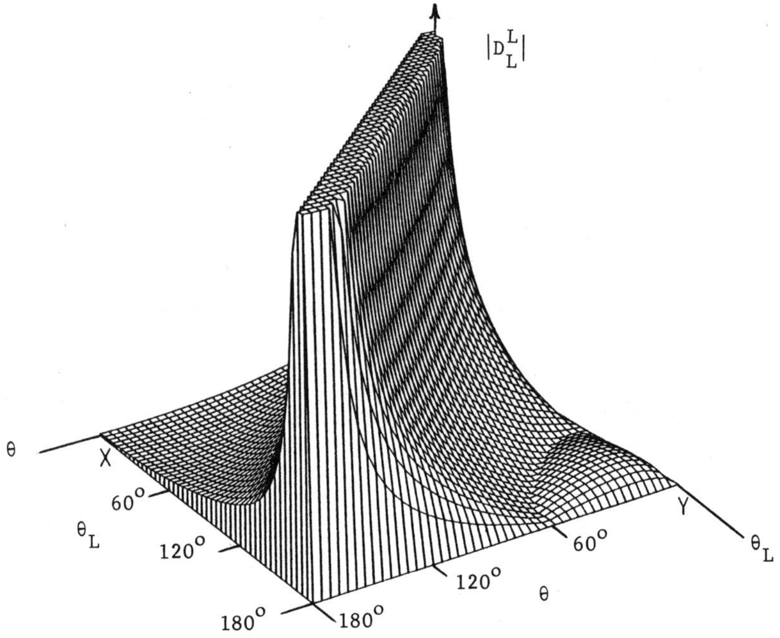

Fig. 2 Absolute value of the diffraction coefficient D^L_L for $\nu = 0.25$

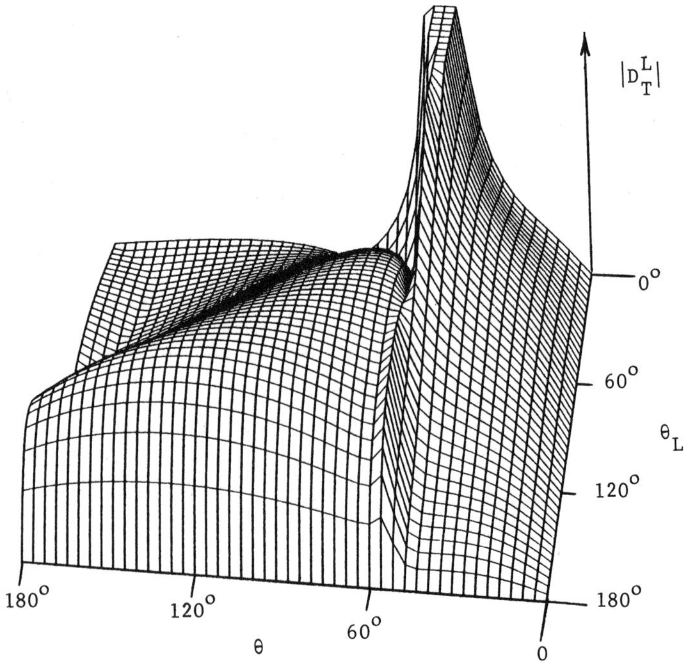

Fig. 3 Absolute value of the diffraction coefficient D^L_T for $\nu = 0.25$

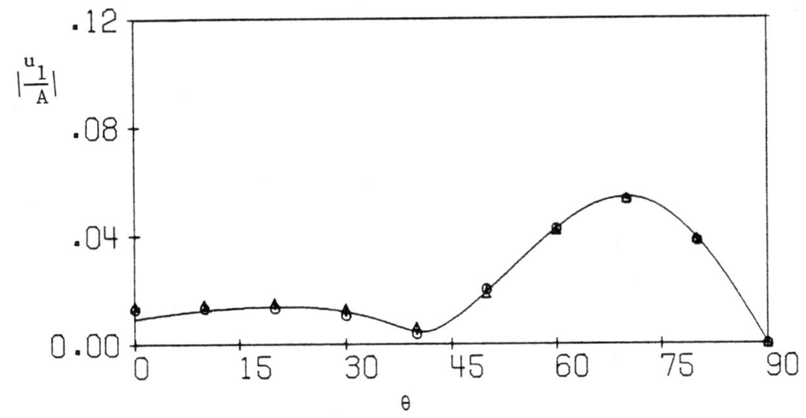

Fig. 4 Scattered longitudinal x_1-displacement for normal incidence of a longitudinal wave; $r/a = 10$, $k_L a = 5$; —— GTD, △ hybrid theory, ○ exact results.

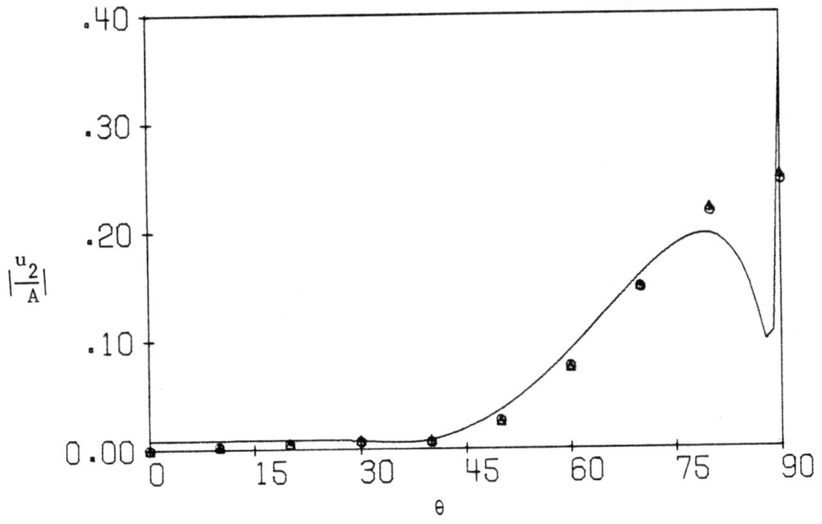

Fig. 5 Scattered longitudinal x_2-displacement for normal incidence of a longitudinal wave; $r/a = 10$, $k_L a = 5$; —— GTD, △ hybrid theory, ○ exact results.

A HYBRID METHOD

In this section a hybrid method is discussed. In this method the crack-opening displacement (COD) is computed on the basis of elastodynamic ray theory, and the diffracted field is subsequently obtained by the use of a representation theorem. The advantage of this approach is that the trouble with ray theory at shadow boundaries and boundaries of zones of specular reflection is eliminated, and caustics only need to be dealt with on the faces of the crack.

Elastodynamic representation theorem

The field generated by scattering of incident waves by an obstacle with surface S can be expressed in terms of a representation integral over S. For a stress-free crack with plane faces A^+ and A^- the representation integral can be simplified. If the total field is written as $\underline{u}^t = \underline{u}^{in} + \underline{u}^{sc}$, where $\underline{u}^{in}$ is the incident field and $\underline{u}^{sc}$ is the scattered field, then at an arbitrary field point $\underline{x}$ the latter can be expressed in the form

$$u_j^{sc}(\underline{x}) = \int_{A^+} \tau_{i3;j}^{G}(\underline{x}-\underline{X})\, \Delta u_i^{sc}(\underline{X})\, dA(\underline{X}) \tag{3.1}$$

Here the crack is in the X_1X_2-plane and the positive X_3-axis is pointing into the material at the A^+ crack-face. Also $\Delta u_i(X)$ is the crack-opening displacement defined by

$$\Delta u_i^{sc}(\underline{X}) = (u_i^{sc})^{A^+} - (u_i^{sc})^{A^-} , \tag{3.2}$$

and

$$\tau_{i3;j}^{G} = \text{tensor of rank three} , \tag{3.3}$$

which represents the stress-components at $X_3 = 0$ due to a unit load in the X_j direction at the point defined by $\underline{X} = \underline{x}$.

Provided that the crack-opening displacement can be adequately approximated, Eq.(3.1) may be expected to give a good approximation to the scattered field. In this section we employ GTD to compute an approximation to the COD.

Crack-opening displacement

Four principal difficulties must be overcome, in the hybrid method presented here. First, GTD predicts unbounded COD's at the crack edge. It is however, seen below by comparison with exact numerical results that this effect is negligible; if necessary, the results of Ref.[6] can be used to correct in part for this singularity. Secondly, caustics remain, although they are reduced by one dimension. In GTD caustic surfaces occur. To compute the COD, only caustic curves which are confined to the crack faces are encountered. Thirdly Eq.(3.1) must be numerically integrated which becomes progressively more difficult with decreasing wave lengths. Finally, the computation of the COD becomes more complicated if "boundary-waves" are included to achieve the desired accuracy.

We now give a brief description of the terms included in the COD, with emphasis on the boundary-wave terms. The COD can be represented by

$$\Delta u_j = \Delta u_j^{GE} + \Delta u_j^{S} + \Delta u_j^{TH} + \Delta u_j^{BL} + \Delta u_j^{BT} \tag{3.4}$$

The first term is the geometrical elastodynamics (GE) contribution to the COD. On the illuminated crack face it consists of the incident wave and the specular reflections. The GE contribution vanishes on the crack face at the shadow side. The second term consists of the diffracted and reflected surface waves which have been described in the previous section. This term is of order one in wavelength with respect to the incident wave. The third term is the contribution to the COD from the diffracted body waves. It includes only horizontally polarized transverse rays, and it is of order the square root in wavelength with respect to the incident wave. We note that the longitudinal diffraction coefficients and those parts of the transverse diffraction coefficients which give rise to vertically polarized waves vanish on the crack faces. The last two terms are the boundary-wave contributions to the COD. In principle they are of order three halves power in wavelength with respect to the incident wave. However, at moderate wave numbers, their amplitude can be large.

Boundary-waves occur because the diffracted body waves do not satisfy the boundary conditions of vanishing traction on the crack faces. The transverse boundary-wave (which is generally known as the "head-wave") and the diffracted longitudinal body-wave combine to satisfy the boundary condition of vanishing tangential tractions on the crack faces. The longitudinal boundary-wave and the diffracted transverse body-wave combine to satisfy the boundary condition of vanishing normal tractions on the crack faces. More details can be found in a paper by Gautesen [11]. From the mathematical point of view, boundary-waves represent branch-point contributions to the inverse Fourier transforms of the displacement fields. The amplitude of the longitudinal boundary-wave is large at moderate wave numbers due to the proximity of the Rayleigh pole to the branch point. The amplitude of the transverse boundary-wave is large at moderate wave numbers because the branch point is close to the extraneous roots of the rationalized Rayleigh function.

The canonical problem for boundary-waves is again the one of incidence of a plane wave upon a semi-infinite crack. To simplify the discussion, we take the incident angle $\phi_\alpha = \pi/2$, where ϕ_α is defined in Fig. 1.

We first discuss the longitudinal boundary-wave. The COD can be expressed as

$$\Delta u_j = e^{ik_T x_1} \int_{-\infty}^{\infty} e^{ik_T \eta x_1} [A_j(\eta) + B_j(\eta)\, \eta^{\frac{1}{2}}] (\eta - s_0)^{-1} \, d\eta \tag{3.5}$$

where $A_j(\eta)$ and $B_j(\eta)$ are analytic in a neighborhood of $\eta = 0$, $s_0 = c_T/c_R - 1$, and $k_T = \omega/c_T$. The integrand in (3.5) without the exponential term is closely related to the Fourier transform of the COD. For incidence of a longitudinal wave, it is given explicitly in Ref. [9] by $\omega\, U_j^{\mp}(\xi)/(\pi c_T)$ with ξ related to η by $\xi = -(\eta+1)c_T$.

As Poisson's ratio ν varies from 0 to 0.5, the parameter s_0 decreases monotonically from 0.160 to 0.047. For the standard asymptotic analysis about the branch point $\eta = 0$ to be valid it is required that $s_0 k_T x_1 >> 1$. However a more careful asymptotic evaluation yields

$$\Delta u_j \sim \Delta u_j^S + \Delta u_j^{BL} + O([k_T x_1]^{-5/2}) \tag{3.6}$$

where

$$\Delta u_j^{BL} = e^{ik_T x_1}\{-2s_0^{-1} B_j(0) \int_{-\infty}^{0} \eta^{\frac{1}{2}}\, e^{ik_T \eta x_1} \, d\eta + 2s_0^{-1} B_j(s_0) \int_{-\infty}^{0} \eta^{3/2} (\eta - s_0)^{-1} \, e^{ik_T \eta x_1} \, d\eta\} \tag{3.7}$$

The first term in Eq.(3.6) which arises from the pole at $\eta = s_o$, in the Rayleigh surface-wave contribution to the COD, while the second term is the modified contribution from the longitudinal boundary-wave.

Performing the integration in Eq.(3.7) yields

$$\Delta u_j^{BL} = e^{ik_T x_1}\{\pi^{\frac{1}{2}} s_o^{-1} e^{-i\pi/4} B_j(0)(k_T x_1)^{-3/2} + s_o^{\frac{1}{2}} B_j(s_o)\, F(s_o k_T x_1)\} \tag{3.8}$$

where

$$F(t) = \pi^{\frac{1}{2}} e^{-i\pi/4} [\, 2it^{-\frac{1}{2}} - t^{-3/2} + 4\, e^{it} \int_{t^{\frac{1}{2}}}^{\infty} e^{-is^2}\, ds] \tag{3.9}$$

The first term in Eq.(3.8) is the standard boundary-wave contribution. The second term in Eq.(3.8) is $O([k_T x_1]^{-5/2})$ for $s_o k_T >> 1$. In the limit as $s_o \to 0$ with $k_T x_1$ fixed Eq.(3.8) reduces, however, to

$$\Delta u_j^{BL} \sim \pi^{\frac{1}{2}} e^{-i\pi/4} [2i(k_T x_1)^{-\frac{1}{2}} B_j(0) - (k_T x_1)^{-3/2} B_j'(0)] \tag{3.10}$$

It is then clear that for s_o sufficiently small the amplitude of the longitudinal boundary-wave is $O([k_T x_1]^{-\frac{1}{2}})$.

The Rayleigh-wave speed is the reciprocal of the positive zero of the Rayleigh function, $R_+(s)$,

$$R_\pm(s) = (2s^2 - c_T^{-2})^2 \pm 4s^2(c_L^{-2} - s^2)^{\frac{1}{2}}(c_T^{-2} - s^2)^{\frac{1}{2}} \tag{3.11}$$

The rationalized Rayleigh function $R_+(s)R_-(s)$ is a sixth-degree polynominal which factors as

$$R_+(s)\, R_-(s) = -\, 16(c_T^{-2} - c_L^{-2})(s^2 - c_R^{-2})(s^2 - c_1^{-2})(s^2 - c_2^{-2}) \tag{3.12}$$

where c_1^{-1} and c_2^{-1} are the extraneous roots with positive real parts that cause difficulty in computing the transverse boundary-wave because their magnitudes are close to the speed of longitudinal waves.

To study the contribution from the transverse boundary-wave we express the COD as

$$\Delta u_j = e^{ik_L x_1} \int_{-\infty}^{\infty} e^{ik_L \eta x_1} \sum_{i=1}^{2} \{A_{ij}(\eta) + B_{ij}(\eta)\, \eta^{\frac{1}{2}}\}(\eta - s_i)^{-1}\, d\eta \tag{3.13}$$

where $A_{ij}(\eta)$ and $B_{ij}(\eta)$ are analytic in a neighborhood of $\eta = 0$, $s_i = c_L/c_i - 1$, and $k_L = \omega/c_L$. For incidence of a longitudinal wave, the integrand without the exponential term is given explicitly in Ref.[9] by $\omega\, U_j^+(\xi)/(\pi c_L)$ with ξ related to η by $\xi = -(\eta+1)c_L^{-1}$. This integral is similar to the one in Eq.(3.5), with an analogous difficulty occuring when s_i is small. The important difference in this case is that the apparent pole at $\eta = s_i$ is absent, since

$$A_{ij}(s_i) + B_{ij}(s_i)\, s_i^{\frac{1}{2}} = 0\ . \tag{3.14}$$

The branch point at $\eta = 0$, then yields the transverse boundary-wave contribution to the COD (usually called the head-wave) as

$$\Delta u_j^{BT} \sim e^{ik_L x_1} \sum_{i=1}^{2} \{\pi^{\frac{1}{2}} s_j^{-1} e^{-i\pi/4} B_{ij}(0)(k_L x_1)^{-3/2} + s_j^{\frac{1}{2}} B_{ij}(s_i) F(s_i k_L x_1)\} \quad (3.15)$$

EXACT ANALYSIS IN 2-D GEOMETRY

To check the accuracy of the approximate methods described in the previous sections we compute exact numerical results for the two-dimensional problem shown in Fig. 6. The angle of incidence of the plane incident wave is θ_α, where α is either L or T. We follow as closely as possible the notation of Mal [12], who treated the case of normal incidence of a longitudinal wave. The integral equation given by Eq.(30) of Ref. [12], which Mal solved numerically, is of the same general form as Eq.(4.18) below, except for the forcing function which depends on the angle of incidence. It should be noted that there are several misprints in the development of Ref.[12] leading to the integral equation stated in Eq.(30), but the equation is correct.

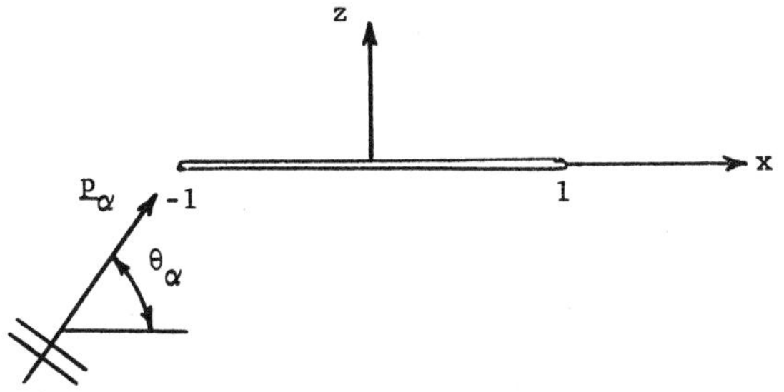

Fig. 6 Plane wave incident on a slit

The incident wave of amplitude A_α is defined by

$$\underline{u}_\alpha^{in} = A_\alpha \underline{d}_\alpha \exp[ik_\alpha \underline{p}_\alpha \cdot \underline{x}] \quad (4.1)$$

where

$$\underline{p}_\alpha = \cos\theta_\alpha \underline{i}_x + \sin\theta_\alpha \underline{i}_z \quad (4.2)$$

while

$$\underline{d}_L = \underline{p}_L, \text{ and } \underline{d}_T = \underline{i}_y \times \underline{p}_T \quad (4.3a,b)$$

for incident longitudinal and incident transverse waves respectively. The tractions generated by the incident wave at the crack faces are

$$\left(\tau_z^{in}, \tau_{zx}^{in}\right) = (p_o, \tau_o)\exp(ik_\alpha x\cos\theta_\alpha), \quad (4.4)$$

where for α = L

$$p_o = i\mu A_L k_L (c_L^2/c_T^2 - 2\cos^2\theta_L) \quad (4.5)$$

$$\tau_o = i\mu A_L k_L \sin 2\theta_L, \quad (4.6)$$

and for α = T

$$p_o = i\mu A_T k_T \sin 2\theta_T \tag{4.7}$$

$$\tau_o = i\mu A_T k_T \cos 2\theta_T \tag{4.8}$$

In Eqs.(4.1)-(4.8) $k_L = \omega/c_L$ and $k_T = \omega/c_T$.

As usual, we express the total field $\underline{u}^t$ as $\underline{u}^t = \underline{u}^{in} + \underline{u}^{sc}$, where the tractions corresponding to the scattered field $\underline{u}^{sc}$ on the crack faces cancel the tractions generated by the incident field. The scattered field is expressed as

$$\underline{u}^{sc} = \underline{\nabla}\varphi + \underline{\nabla}\psi \times \underline{i}_y \tag{4.9}$$

where φ and ψ satisfy the reduced wave equation with wavenumber k_L and k_T, respectively. The boundary value problem for the scattered field is divided into two parts, with the following conditions on $z = 0$, $|x| < 1$:

Symmetric problem: $\tau_z = -\tau_z^{in}$ (4.10a)

$$\tau_{zx} = 0 \tag{4.10b}$$

Antisymmetric problem: $\tau_z = 0$ (4.11a)

$$\tau_{zx} = -\tau_{zx}^{in} \tag{4.11b}$$

We first solve the symmetric problem. Following Mal [12], we take

$$\varphi = -\int_{-\infty}^{\infty} (k^2 - \tfrac{1}{2}k_T^2)\, P(k)\, \nu_L^{-1} \exp[ikx - \nu_L |z|]dk \tag{4.12}$$

$$\psi = i \,\mathrm{sgn}z \int_{-\infty}^{\infty} k\, P(k) \exp[ikx - \nu_T |z|]\, dk \tag{4.13}$$

where $\nu_\alpha = (k^2 - k_\alpha^2)^{\frac{1}{2}}$. Here $P(k)$ is an unknown function which can be written as

$$P(k) = P_e(k) + P_o(k) \tag{4.14}$$

where $P_e(P_o)$ is an even (odd) function of k. Upon application of the boundary conditions and separation of the even and odd functions of x, we then arrive at the following two pairs of equations:

$$\int_0^{\infty} P_e(k) \cos kx\, dk = 0,\ x > 1 \tag{4.15a}$$

$$\int_0^{\infty} F_s(k)\, P_e(k) \cos kx\, dk = -(4\mu)^{-1}\, p_o \cos(k_\alpha\, x \cos\theta_\alpha)\ ,\ 0 < x < 1 \tag{4.15b}$$

and

$$\int_0^{\infty} P_o(k) \sin kx\, dk = 0,\ x > 1 \tag{4.16a}$$

$$\int_0^{\infty} F_s(k)\, P_o(k) \sin kx\, dk = -(4\mu)^{-1}\, p_o \sin(k_\alpha\, x \cos\theta_\alpha),\ 0 < x < 1 \tag{4.16b}$$

where $F_s(k) = k^2\nu_T - (k^2-\frac{1}{2}k_T^2)/\nu_L$. Equations (4.15a) and (4.16a) insure that the displacements are continuous in the plane $z = 0$ for $|x| > 1$.

We first reduce (4.15) to an integral equation of the second kind. Following Mal [12] we set

$$P_e(k) = -p_o[2\mu(k_T^2 - k_L^2)]^{-1}\int_0^1 \xi p(\xi)\, J_o(k\xi)\, d\xi. \tag{4.17}$$

This leads to the integral equation

$$p(\xi) + \int_0^1 \eta F_s(\xi,\eta)p(\eta)\, d\eta = J_o(k_\alpha\, \xi \cos\theta_\alpha),\ 0 \le \xi \le 1 \tag{4.18}$$

where

$$F_s(\xi,\eta) = -\int_0^\infty [k - 2F_s(k)/(k_T^2-k_L^2)]\, J_o(k\xi)J_o(k\eta)\, dk \tag{4.19}$$

It has been shown in Ref.[12] that $F_s(\xi,\eta)$ can also be expressed as an integral over a finite interval. For normal incidence $\cos\theta_\alpha = 0$, and Eq.(4.18) reduces to

$$p_n(\xi) + \int_0^1 \eta F_s(\xi,\eta)\, p_n(\eta)\, d\eta = 1,\ 0 \le \xi \le 1 \tag{4.20}$$

which is the equation solved numerically in Ref. [12].

The solution to Eqs.(4.16a,b) is now related to the solutions to Eqs.(4.18) and (4.20). Integration of Eqs.(4.16a,b) with respect to x yields

$$\int_0^\infty k^{-1}\, P_o(k)\cos kx\, dk = 0,\ x > 1 \tag{4.21a}$$

$$\int_0^\infty F(k)\, k^{-1}\, P_o(k)\cos kx\, dk = -\frac{p_o[\cos(k_\alpha x \cos\theta_\alpha) - C]}{4\mu k_\alpha \cos\theta_\alpha},\ 0 < x < 1 \tag{4.21b}$$

where C is an arbitrary constant of integration. In Eq.(4.21a) the arbitrary constant vanishes since the integral vanishes as x approaches infinity. It is noted that the first term on the right-hand side of Eq.(4.21b) is proportional to the right-hand side of (4.15b), and that the second term is proportional to the right-hand side of (4.15b) for $\theta_\alpha = \pi/2$. Thus we conclude from Eqs. (4.15), (4.17), (4.18) and (4.20), that

$$P_o(k) = -\frac{k\, p_o}{2\mu k_\alpha(k_T^2-k_L^2)\cos\theta_\alpha}\int_0^1 \xi[p(\xi) - Cp_n(\xi)]\, J_o(k\xi)\, d\xi \tag{4.22}$$

Finally, $p(\xi)$ and $p_n(\xi)$ are related to the crack opening displacement defined by $2u_z^s(x,0^+) \equiv U_z(x)$, and C is determined. Substitution of Eqs.(4.17) and (4.22) into Eq.(4.14) and subsequently into Eqs.(4.12) and (4.13) yields from Eq.(4.9) that

$$U_z(x) = \frac{p_o}{\mu(1-k_L^2/k_T^2)} \left\{ \int_{|x|}^{1} \left[\xi\, p(\xi) - \frac{ix}{k_\alpha \cos\theta_\alpha} [p'(\xi)-Cp_n'(\xi)] \right] \frac{d\xi}{(\xi^2-x^2)^{\frac{1}{2}}} \right.$$

$$\left. + \frac{ix}{k_\alpha \cos\theta_\alpha} \frac{1}{(1-x^2)^{\frac{1}{2}}} [p(1) - Cp_n(1)] \right\}, \ |x| < 1 \tag{4.23}$$

where a prime denotes differentiation with respect to ξ. By requiring that $U_z(\pm 1) = 0$, we see that

$$C = p(1)/p_n(1). \tag{4.24}$$

For the anti-symmetric problem, we take

$$\varphi = i \ \mathrm{sgn} z \int_{-\infty}^{\infty} kQ(k) \exp[ikx - \nu_L |z|] \, dk \tag{4.25}$$

$$\psi = \int_{-\infty}^{\infty} \nu_T^{-1} (k^2 - \tfrac{1}{2} k_T^2) \, Q(k) \exp[ikx - \nu_T |z|] \, dk \tag{4.26}$$

Upon setting

$$Q(k) = Q_e(k) + Q_o(k) \tag{4.27}$$

and applying the boundary conditions, two pairs of integral equations similar to Eqs.(4.15) and (4.16) are obtained. Proceding as before, we find an expression for the crack opening displacement (defined by $2u_x(x,0^+) \equiv U_x(x)$) similar to Eq.(4.23). The result is

$$U_x(x) = \frac{\tau_o}{\mu(1-k_L^2/k_T^2)} \int_{|x|}^{1} \left[\xi q(\xi) - \frac{ix}{k_\alpha \cos\theta_\alpha} \left[q'(\xi) - \frac{q(1) q_n'(\xi)}{q_n(1)} \right] \right] \frac{d\xi}{(\xi^2-x^2)^{\frac{1}{2}}} \tag{4.28}$$

Here $q(s)$ and $q_n(\xi)$ satisfy the integral equations

$$q(\xi) + \int_0^1 \eta F_a(\xi,\eta) \, q(\eta) \, d\eta = J_o(k_\alpha \xi \cos\theta_\alpha), \ 0 \le \xi \le 1 \tag{4.29}$$

$$q_n(\xi) + \int_0^1 \eta F_a(\xi,\eta) \, q_n(\eta) = 1, 0 \le \xi \le 1 \tag{4.30}$$

where

$$F_a(\xi,\eta) = - \int_0^{\infty} [k - 2F_a(k)/(k_T^2 - k_L^2)] \, J_o(k\xi) \, J_o(k\eta) \, dk \tag{4.31}$$

and

$$F_a(k) = k^2 \nu_L - (k^2 - \tfrac{1}{2} k_T^2)^2 / \nu_T \tag{4.32}$$

After some manipulation $F_a(\xi,\eta)$ can be expressed as an integral over a finite interval.

The Fredholm integral equations of the second kind given by Eqs.(4.18), (4.20), (4.29) and (4.30) were solved numerically. Since only the forcing function changes for oblique incidence, the same technique was used as described in Ref. [5] for the case of normal incidence. This technique employs infinite series for $F_s(\xi,\eta)$ and $F_a(\xi,\eta)$, whose n-th terms are of order $n^{-2}(k_\beta\xi/4n)^{2n}\exp(2n)$. The integral equations were solved by partitioning the interval [0,1] into N equal sub-intervals. The unknown functions $p(\xi)$, $p_n(\xi)$, $q(\xi)$ and $q_n(\xi)$ were taken as constants over these sub-intervals. Substitution of the step-function approximations into the pertinent integral equations resulted in a system of linear algebraic equations which was solved numerically.

The crack opening displacements were subsequently obtained from Eqs.(4.23) and (4.28) by numerical integration. Finally, the displacement fields were calculated by substitution of the COD's into the representation integral (3.1). In the latter integral an integration by parts was performed on those terms in the crack open displacements which are odd functions of the integration variable. This was followed by an integration by parts with respect to ξ in the integrals (4.23) and (4.28) defining the COD's. In this manner numerical differentiation of the function $p(\xi)$, $p_n(\xi)$, $q(\xi)$ and $q_n(\xi)$ was avoided.

Figures 7 and 8 show the exact crack-opening displacements for normal incidence of longitudinal and transverse waves respectively. Clearly, as $k_La = \omega a/c_L$ increases waveforms develop on the crack faces. The methods developed in the third section of this paper, provide, however, a very good approximation to the COD's.

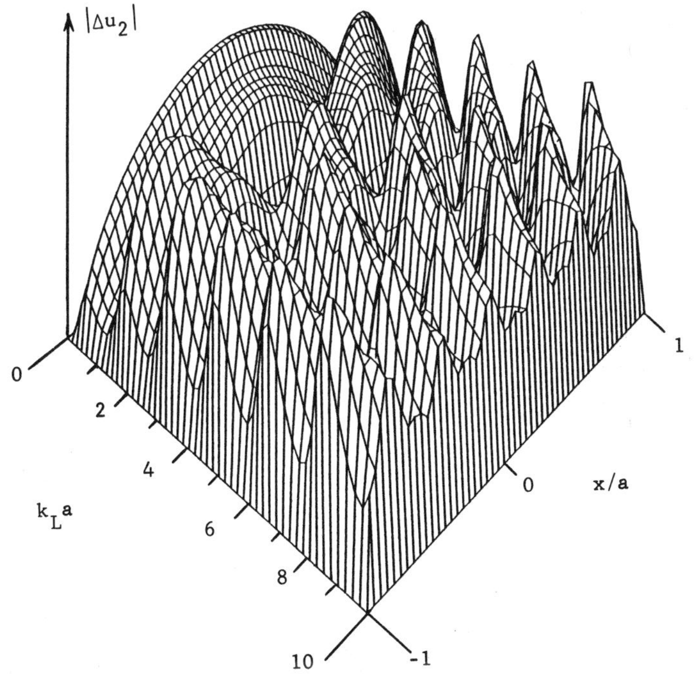

Fig. 7 COD for normal incidence of a longitudinal wave

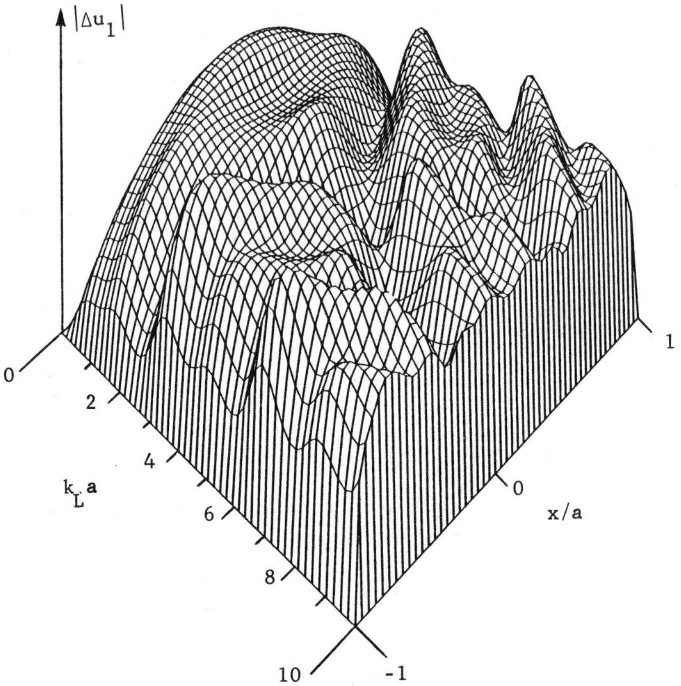

Fig. 8 COD for normal incidence of a transverse wave

RESULTS

For a penny-shaped crack and normal incidence of a longitudinal wave, comparisons between the scattered fields computed by GTD, the hybrid theory, and exact theory, respectively, are shown in Figs. 4 and 5. It is seen that GTD provides a quite adequate approximation for $r/a = 10$ and $k_L a = 5$. For a slit similar comparisons, but versus the dimensionless frequency, are shown in Fig. 9. The agreement is better for the scattered longitudinal wave than for the scattered transverse wave.

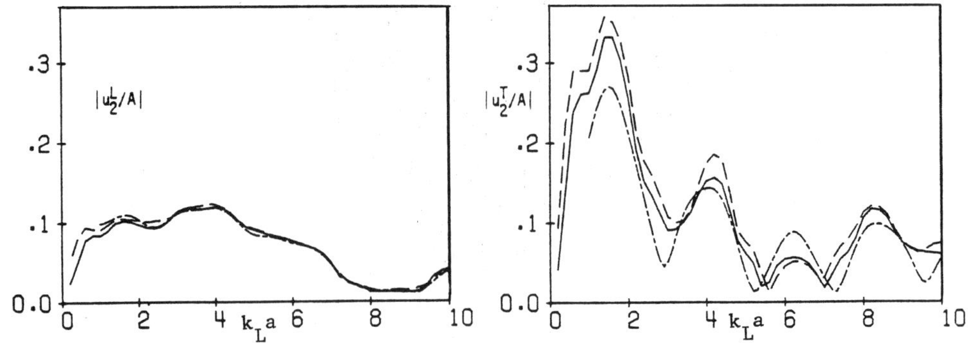

Fig. 9 Displacements in x_2- direction due to scattered wave versus $k_L a$; $r/a = 10$, $\nu = 0.25$, angle of incidence $\theta_L = 60^o$, angle of observation $\theta = 60^o$; —— exact, - - - hybrid method; - — - —— GTD.

ACKNOWLEDGEMENT

This work was carried out in the course of research sponsored by the Center for Advanced NDE operated by the Science Center, Rockwell International for the Advanced Research Project Agency and the Air Force Materials Laboratory under Contact F33615-74-C-5180.

REFERENCES

1. Achenbach, J.D. (1973) Wave Propagation in Elastic Solids, North-Holland Amsterdam - New York.

2. A. W. Maue, Die Beugung elastischer Wellen an der Halbebene, Zeitschrift für angewandte Mathematik und Mechanik 33, 1 (1953).

3. J. B. Keller, A geometrical theory of diffraction, Calculus of Variations and its Applications, McGraw-Hill, New York (1958).

4. R. M. Lewis and J. Boersma, Uniform asymptotic theory of edge diffraction, J. Math. Phys. 10, 2291 (1969).

5. J. D.Achenbach, A. K. Gautesen, and H. McMaken, Application of elastodynamic ray theory to diffraction by cracks, Modern Problems in Elastic Wave Propagation, Wiley-Interscience, New York (1978).

6. J. D. Achenbach and A. K Gautesen, A ray theory for elastodynamic stress-intensity factors, J. Appl. Mech. 45, 123 (1978).

7. D. Ludwig, Uniform asymptotic expansions at a caustic, Comm. Pure Appl. Math. 19, 215 (1966).

8. Y. A. Kravtsov, A modification of the geometrical optics method for a wave penetrating a caustic, Izv. vuz Radiofiz. (USSR) 8, 659 (1965).

9. J. D. Achenbach and A. K. Gautesen, Geometrical theory of diffraction for three-D elastodynamics, J. Acoust. Soc. Am. 61,413(1977).

10. A. K. Gautesen, J. D.Achenbach and H. McMaken, Surface wave rays in elastodynamic diffraction by cracks, J. Acoust. Soc.Am. 63, 1824, (1978).

11. A. K.Gautesen, On matched asymptotic expansions for two-dimensional elastodynamic diffraction by cracks, WAVE MOTION 1, 127 (1979).

12. A. K. Mal, Interaction of elastic waves with a Griffith crack, Int. J. Eng. Sc. 8, 769 (1970).

THE UNIFORM GEOMETRICAL THEORY OF DIFFRACTION AND ITS APPLICATION TO ELECTROMAGNETIC RADIATION AND SCATTERING

R.G. Kouyoumjian, P.H. Pathak, and W.D. Burnside
The Ohio State University ElectroScience Laboratory
Department of Electrical Engineering
Columbus, Ohio 43212

I. INTRODUCTION

When a radiating object is large in terms of a wavelength, the scattering and diffraction are found to be essentially a local phenomenon associated with specific parts of the object, e.g., specular reflection, shadow boundaries, and edges. The high-frequency approach to be described in these notes employs rays in a systematic way to describe this phenomenon. It was originally developed by Keller and his associates at the Courant Institute of Mathematical Sciences [1,2,3] and is referred to as the geometrical theory of diffraction (GTD). The GTD is an extension of geometrical optics in which rays diffracted from edges, vertices and convex surfaces are introduced through a generalization of Fermat's Principle. In its original form, the GTD fails in the transition regions adjacent to shadow and reflection boundaries. About fifteen years ago work began at The Ohio State University [4] to overcome this and other limitations in the GTD. The result today is the new uniform GTD (UTD) which provides expressions for the diffracted field so that the total high-frequency field is continuous at shadow and reflection boundaries. Uniform expressions have been found for electromagnetic fields diffracted from edges and vertices in perfectly-conducting surfaces and for the diffracted fields due to sources either on or off perfectly-conducting convex surfaces. These solutions have greatly enhanced the accuracy and utility of the method, which can be readily demonstrated in the treatment of a number of simple shapes. In many cases the method works surprisingly well on radiating objects as small as a wavelength or so in extent.

II. FORMULATION OF THE THEORY

The treatment of high-frequency radiation to follow is for the most part restricted to perfectly-conducting objects located in isotropic, homogeneous and anisotropic media. The method presented, however, can be extended to objects with impedance surfaces or penetrable surfaces in inhomogeneous and anisotropic media.

A. Geometrical Optics

The geometrical optics field is the leading term in the asymptotic high-frequency solution of Maxwell's equations, and so it not unexpectedly provides the leading term in the GTD (and UTD) high-frequency approximations. Let us begin by examining this field and how it is used in these theories. The geometrical optics field is the sum of the leading terms in the incident and reflected fields. These fields can be expanded in Luneberg-Kline series for large ω of the form

$$\overline{E} \sim \exp(-jk\psi) \sum_{m=0}^{\infty} \frac{\overline{E}_m}{(j\omega)^m} \qquad (2.1)$$

where an $\exp(j\omega t)$ time dependence is assumed and k is the wavenumber of the medium. Substituting the preceding expansion into the vector wave equation for the electric field and integrating the resulting transport equation for m=0 [5], [6], the leading term in (2.1) is found to be

$$\bar{E}(s) \sim \exp[-jk\psi(s)]\bar{E}_0(s) = \bar{E}_0(0)\exp[-jk\psi(0)]\sqrt{\frac{\rho_1\rho_2}{(\rho_1+s)(\rho_2+s)}}\exp(-jks) \quad (2.2)$$

in which s=0 is taken as a reference point on the ray path, and ρ_1, ρ_2 are the principal radii of curvature of the wavefront at s=0 as shown in Fig. 1.

Equation (2.2) is commonly referred to as the geometrical-optics field, because it could have been determined in part from classical geometrical optics. Specifically, the quantity under the square root, the divergence factor, follows from conservation of power in a tube of rays; in addition, we note that the eikonal equation could have been deduced from Fermat's principle, a fundamental postulate of classical geometrical optics. As is well known, classical geometrical optics ignores the polarization and wave nature of the electromagnetic field; however, the leading term in the Luneberg-Kline asymptotic expansion is seen to contain this missing information.

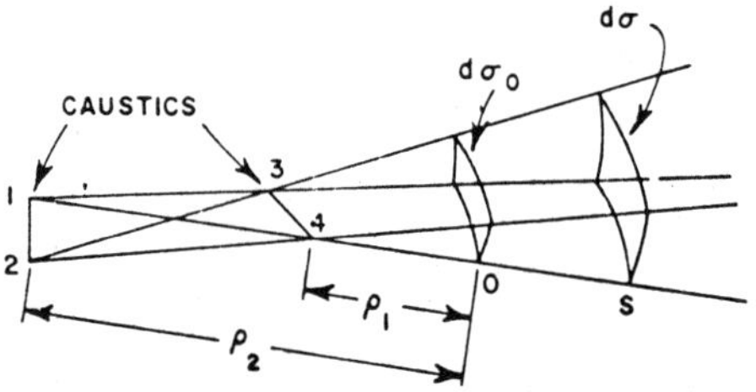

Fig. 1. Astigmatic tube of rays.

It is apparent that when $s=-\rho_1$ or $-\rho_2$ (2.2) becomes infinite so that it is no longer a valid approximation. The intersection of the rays at the lines 1-2 and 3-4 of the astigmatic tube of rays is called a caustic. As we pass through a caustic in the direction of propagation, the sign of $\rho+s$ changes and the correct phase shift of $+\pi/2$ is introduced naturally.

From $\nabla\cdot\bar{E}=0$, one obtains

$$\hat{s} \cdot \bar{E}_0 = 0 \,, \quad (2.3)$$

so that the electric vector of the geometrical optics field is perpendicular to the ray path. Next, employing the Maxwell curl equation $\nabla \times \bar{E} = -j\omega\mu\bar{H}$, it follows from (2.1) that the leading term in the asymptotic approximation for the magnetic field is

$$\bar{H} \sim Y_c \hat{s} \times \bar{E} \quad (2.4)$$

where $Y_c = \sqrt{\varepsilon/\mu}$ is the characteristic admittance of the medium, $\hat{s}$ is a unit vector in the direction of the ray path, and $\bar{E}$ is given by (2.2).

Let a high-frequency electromagnetic wave be incident on a smooth curved perfectly-conducting surface S as depicted in Fig. 2. The geometrical-optics electric field reflected at Q_R on S (see Fig. 2) has the form given by (2.2). Choosing Q_R to be the reference point, it follows from the boundary condition for the total electric field on S that

$$\bar{E}_0^r(0)\exp[-jk\psi^r(0)] = \bar{E}^i(Q_R)\cdot\bar{\bar{R}} = \bar{E}^i(Q_R)\cdot[\hat{e}_\parallel^i\hat{e}_\parallel^r - \hat{e}_\perp\hat{e}_\perp] \qquad (2.5)$$

in which $\bar{E}^i(Q_R)$ is the electric field incident at Q_R and $\bar{\bar{R}}$ is the dyadic reflection coefficient with $\hat{e}_\perp$ the unit vector perpendicular to the plane of incidence and $\hat{e}_\parallel^i$, $\hat{e}_\parallel^r$ the unit vectors parallel to the plane of incidence as shown in Fig. 2.

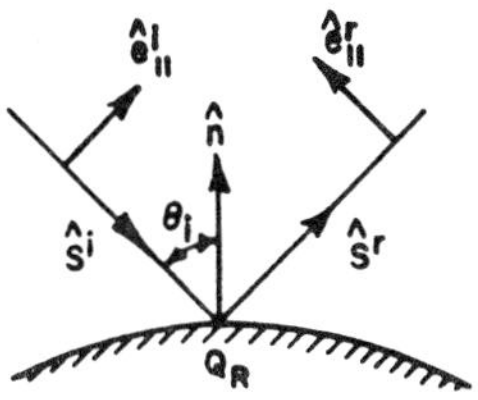

Fig. 2. Reflection at a curved surface.

In matrix notation

$$R = \begin{bmatrix} 1 & 0 \\ 0 & -1 \end{bmatrix} \qquad (2.6)$$

From (2.2) and (2.5)

$$\bar{E}^r(s) = \bar{E}^i(Q_R)\cdot\bar{\bar{R}}\sqrt{\frac{\rho_1^r\rho_2^r}{(\rho_1^r+s)(\rho_2^r+s)}}\exp(-jks) \qquad (2.7)$$

in which ρ_1^r, ρ_2^r are the principal radii of curvature of the reflected wavefront at Q_R. It can be shown that

$$\frac{1}{\rho_1^r} = \frac{1}{2}\left(\frac{1}{\rho_1^i} + \frac{1}{\rho_2^i}\right) + \frac{1}{f_1} \qquad (2.8)$$

$$\frac{1}{\rho_2^r} = \frac{1}{2}\left(\frac{1}{\rho_1^i} + \frac{1}{\rho_2^i}\right) + \frac{1}{f_2} \;. \qquad (2.9)$$

Expressions for f_1 and f_2 are given in [7].

In principle, the geometrical-optics approximation can be improved by finding the higher order terms $\bar{E}_1^r(R)$, $\bar{E}_2^r(R)$, ..., in the reflected field, but in general it is not easy to obtain these from the higher order transport equations. Furthermore, these terms do not correct two serious defects of the geometrical optics:

a) the zero field in the shadow region,
b) the discontinuity in the field at the shadow and reflection boundaries.

At the present time additional postulates are required to remove these limitations. They are given in the next subsection.

B. Postulates

To overcome limitation (a) of the geometrical-optics field pointed out at the end of the last subsection, it is necessary to introduce an additional field, the diffracted field. Keller [1,2,3] has shown how the diffracted field may be included in the high-frequency solution as an extension of geometrical optics. The postulates of Keller's theory, commonly referred to as the geometrical theory of diffraction (GTD), are summarized as follows.

(1) The diffracted field propagates along rays which are determined by a generalization of Fermat's principle to include points on the boundary surface in the ray trajectory.

(2) Diffraction like reflection and transmission is a local phenomenon at high frequencies.

(3) The diffracted wave propagates along its ray so that
(a) power is converved in a tube (or strip of rays),
(b) the phase delay along the ray path equals the product of the wave number of the medium and the distance.

The rays diffracted from an opaque object such as that depicted in Fig. 3 and which pass through the field point P or P' are found from the generalized Fermat's principle. The notion that points on the boundary surface may be included in the ray trajectory is not new. Imposing the condition that a point Q_R on a smooth curved surface be included in the ray path between the source and observation point is a time-honored method for deducing the reflected ray and the law of reflection. It seems reasonable to extend the class of such points as KELLER has done. Diffracted rays are initiated at points on the boundary surface where the incident geometrical-optics field is discontinuous, i.e., at points on the surface where there is a shadow or reflection boundary of the incident field. Examples of such points are edges Q_E, vertices and points Q_1 at which the ray incident from 0 is tangent to the curved surface. Note that a surface ray also may be excited at the edge Q_E. These diffracted rays like the geometrical-optics rays follow paths which make the optical distance between the source and field points an extremum, usually a minimum. Thus the portion of the ray path which traverses a homogeneous medium is a straight line, and if a segment of the ray path lies on a smooth surface, it is a surface extremum or geodesic.

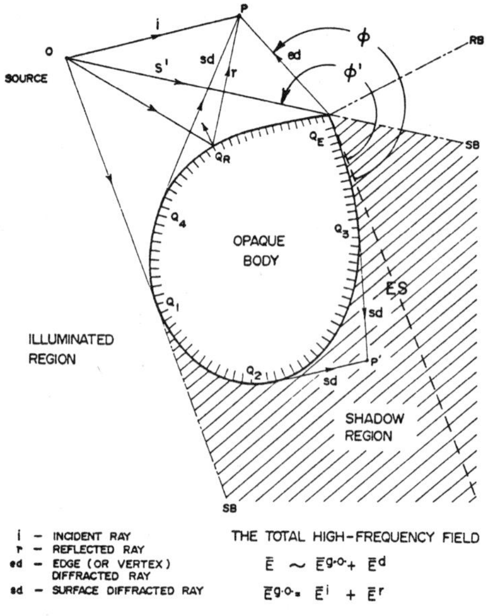

Fig. 3. Rays reflected and diffracted from an opaque object.

It is apparent that the rays provide a natural coordinate system for calculating the field; furthermore, the total high-frequency field at an observation point is synthesized from the fields of all the rays passing through that point.

Postulate 2 is essential for the method to be valid, i.e., the frequency must be sufficiently high that the complex radiation problem breaks down into a number of simple problems associated with specific parts of the object.

The uniform GTD(UTD) unlike the GTD requires that the diffracted field compensate the discontinuity in the geometrical optics field at a shadow or reflection boundary so that the total high-frequency field is everywhere continuous away from the radiating body. Thus the UTD diffracted fields assume their largest values at and near these boundaries, where their strength is comparable with the geometrical field.

III. EDGE DIFFRACTION

A. Curved Wedge

Let us consider the field radiated from a point source at O and observed at P in the presence of a perfectly-conducting curved wedge, as shown in Fig. 3. Applying the generalized Fermat's principle, the distance along the ray path OQ_EP between O and P, which includes the edge point Q_E on its path, is a minimum and the law of edge diffraction

$$\hat{s}' \cdot \hat{e} = \hat{s} \cdot \hat{e} \tag{3.1}$$

results. Here $\hat{e}$ is a unit vector directed along the edge, and $\hat{s}'$ and $\hat{s}$ are unit vectors in the directions of incidence and diffraction, respectively. The above equation also follows from the requirement that the incident and diffracted fields be phase matched along the edge. If the incident ray strikes the edge obliquely, making an angle β_0 with the edge, as shown in Fig. 4a the diffracted rays lie on the surface of a cone whose half angle is equal to β_0. The position of the diffracted ray on this conical surface is given by the angle ϕ, and the direction of the ray incident on the edge, by the angles ϕ' and β_0; these angles are defined in Fig. 4a and b.

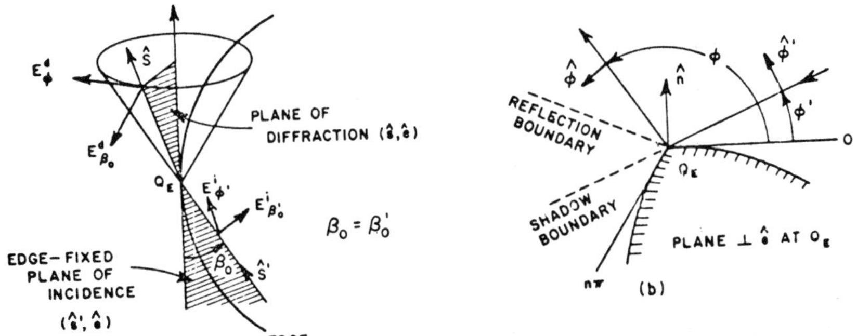

Fig. 4. Diffraction at a curved edge.

The expression for the electric field of the edge-diffracted ray is given by

$$\bar{E}^d(s) = \bar{E}^i(Q_E)\cdot\bar{\bar{D}}(\phi,\phi';\beta_0')\sqrt{\frac{\rho}{s(\rho+s)}}\,e^{-jks} \quad (3.2)$$

The dyadic diffraction coefficient for a perfectly-conducting wedge has been obtained by Kouyoumjian and Pathak; their work is described in [7,8] and will only be summarized here. The dyadic diffraction coefficient is found from the asymptotic solution of canonical problems, which in this case involves the illumination of the wedge by plane, cylindrical, conical and spherical waves. The solution of these canonical problems serves as a basis for deducing the dyadic diffraction coefficient for arbitrary wavefront illumination and for the more general case where there are curved edges and curved surfaces. It is shown in [7,8] that the caustic distance ρ in (3.2) is

$$\frac{1}{\rho} = \frac{1}{\rho_e^i} - \frac{\hat{n}_e\cdot(\hat{s}'-\hat{s})}{a\sin^2\beta_0'} \quad (3.3)$$

in which ρ_e^i is the radius of curvature of the incident wavefront in the plane which contains $\hat{s}'$ and $\hat{e}$ the unit vector tangent to the edge at Q_E; $\hat{n}_e$ is the unit vector normal to the edge at Q_E and directed away from the center of curvature; a is the radius of curvature of the edge at Q_E, $a>0$.

Let us introduce an edge-fixed plane of incidence containing the incident ray and the edge and a plane of diffraction containing the diffracted ray and the edge. The unit vectors $\hat{\phi}'$ and $\hat{\phi}$ are perpendicular to the edge-fixed plane of incidence and the plane of diffraction, respectively. The unit vectors $\hat{\beta}_o'$ and $\hat{\beta}_0$ are parallel to the edge-fixed plane of incidence and the plane of diffraction, respectively, and

$$\hat{\beta}_0' = \hat{s}' \times \hat{\phi}', \quad \hat{\beta}_0 = \hat{s} \times \hat{\phi} .$$

Thus the coordinates of the diffracted ray (s,β_0,ϕ) and incident ray (s',β_0',ϕ') are spherical coordinates except that the incident (radial) unit vector points toward the origin Q_E as shown in Fig. 4.

For each type of edge illumination mentioned previously, it is shown in [8] that the dyadic diffraction coefficient can be represented simply as the sum of two dyads, in the ray-fixed coordinates as

$$\bar{\bar{D}}(\phi,\phi';\beta_0') = -\hat{\beta}_0'\hat{\beta}_0 D_s(\phi,\phi';\beta_0') - \hat{\phi}'\hat{\phi} D_h(\phi,\phi';\beta_0') , \quad (3.4)$$

where D_s is the scalar diffraction coefficient for the acoustically soft (Dirichlet) boundary condition at the surface of the wedge, and D_h is the scalar diffraction coefficient for the acoustically hard (Neumann) boundary condition.

Expressions for the scalar diffraction coefficients which are valid at all points away from the edge (excluding $\phi'=0$ or $n\pi$) are

$$D_{\substack{s\\h}}(\phi,\phi';\beta_0') = \frac{-e^{-j\frac{\pi}{4}}}{2n\sqrt{2\pi k}\sin\beta_0'}$$
$$\times\left[\cot\left(\frac{\pi+(\phi-\phi')}{2n}\right)F[kL^i a^+(\phi-\phi')]\right.$$
$$+\cot\left(\frac{\pi-(\phi-\phi')}{2n}\right)F[kL^i a^-(\phi-\phi')]$$
$$\mp\left\{\cot\left(\frac{\pi+(\phi+\phi')}{2n}\right)F[kL^{ro}a^+(\phi+\phi')]\right.$$
$$\left.\left.+\cot\left(\frac{\pi-(\phi+\phi')}{2n}\right)F[kL^{rn}a^-(\phi+\phi')]\right\}\right], \tag{3.5}$$

where

$$F(X) = 2j\ \sqrt{X}\ e^{jX} \int_{\sqrt{X}}^{\infty} e^{-j\tau^2}\, d\tau \ . \tag{3.6}$$

Let $\phi\pm\phi'=\beta$, then

$$a^{\pm}(\beta) = 2\cos^2\left(\frac{2n\pi N^{\pm}-\beta}{2}\right) \tag{3.7}$$

in which $N^{\pm}$ are the integers which most nearly satisfy the equations

$$2\pi nN^+ - \beta = \pi \ . \tag{3.8}$$

$$2\ nN^- - \beta = -\pi \ . \tag{3.9}$$

Note that N^+ and N^- each have two values. For exterior edge diffraction ($1<n\leq 2$), the value of N^+ or N^- at each boundary is included in Table 1 for convenience; these values can be used throughout their respective transition regions.

Table I

	The cotangent is singular when	value of N at the boundary
$\cot\left(\frac{\pi+(\phi-\phi')}{2n}\right)$	$\phi=\phi'-\pi$, a SB surface $\phi=0$ is shadowed	$N^+=0$
$\cot\left(\frac{\pi-(\phi-\phi')}{2n}\right)$	$\phi=\phi'+\pi$, a SB surface $\phi=n\pi$ is shadowed	$N^-=0$
$\cot\left(\frac{\pi+(\phi+\phi')}{2n}\right)$	$\phi=(2n-1)\pi-\phi'$, a RB reflection from surface $\phi=n\pi$	$N^+=1$
$\cot\left(\frac{\pi-(\phi+\phi')}{2n}\right)$	$\phi=\pi-\phi'$, a RB reflection from surface $\phi=0$	$N^-=0$

At a shadow or reflection boundary one of the cotangent functions in the expression for D given by (3.5) becomes singular; the other three remain bounded. Even though the cotangent becomes singular, its product with the transition function can be shown to be bounded. The location of the boundary at which each cotangent becomes singular is given in Table 1. The distance parameters are given by

$$L^i = \frac{s(\rho_e^i+s)\rho_2^i\rho_1^i \sin^2\beta_o'}{\rho_e^i(\rho_1^i+s)(\rho_2^i+s)} , \tag{3.10}$$

$$L^r = \frac{s(\rho^r+s)\rho_2^i\rho_1^i \sin^2\beta_o'}{\rho^r(\rho_1^r+s)(\rho_2^r+s)} , \tag{3.11}$$

where ρ_1^r and ρ_2^r are the principal radii of curvature of the reflected wavefront at Q_E, and ρ^r is the distance between the caustics of the diffracted ray in the direction of reflection. It may be found from (3.3) with $\hat{s}=\hat{s}'-2(\hat{n}\cdot\hat{s}')\hat{n}$. The additional superscripts o and n on L in (3.5) denote that the radii of curvature (and caustic distance ρ) are calculated at the reflection boundaries $\pi-\phi'$ and $(2n-1)\pi-\phi'$, respectively.

Grazing incidence, where $\phi'=0$ or $n\pi$ must be treated separately. Only the ordinary wedge is considered here. In this case $D_s=0$, and the expression for D_h given by (3.5) is multiplied by a factor of 1/2. The need for the factor of 1/2 may be seen by considering grazing incidence to be the limit of oblique incidence. At grazing incidence the incident and reflected fields merge, so that one half the total field propagating along the face of the wedge toward the edge is the incident field and the other half is the reflected field. Nevertheless, in this case it is clearly more convenient to regard the total field as the "incident" field.

B. Slope Diffraction

It is assumed in (3.2) that the incident field $\overline{E}^i(Q_E)$ has a slow spatial variation (except for the phase along the incident ray). If this is not the case, Kouyoumjian and Hwang [9,10] have shown that a higher-order term must be included in the UTD expression for the edge diffracted field. Employing matrix notation the expression for the edge diffracted field then becomes

$$\begin{bmatrix} E^d_{\beta_o} \\ E^d_{\phi} \end{bmatrix} = \left\{ \begin{bmatrix} -D_s & 0 \\ 0 & -D_h \end{bmatrix} \begin{bmatrix} E^i_{\beta_o'} \\ E^i_{\phi'} \end{bmatrix} + \begin{bmatrix} -d_s & 0 \\ 0 & -d_h \end{bmatrix} \begin{bmatrix} \frac{\partial}{\partial n'} E^i_{\beta_o'} \\ \frac{\partial}{\partial n'} E^i_{\phi'} \end{bmatrix} \right\} \sqrt{\frac{\rho}{s(\rho+s)}}\, e^{-jks} \tag{3.12}$$

where

$$d_{s,h} = \frac{1}{jk\sin\beta_o} \frac{\partial}{\partial \phi'} D_{s,h} . \tag{3.13}$$

The partial derivative with respect to the distance n' is taken in the direction $\hat{\phi}'$ normal to the edge fixed plane of incidence. The first term in (3.12) is just (3.4) combined with (3.2) in matrix form; it makes the total high-frequency field continuous at shadow and reflection boundaries. The second term, referred to as the slope diffraction term, makes the first derivatives of the total high-frequency field continuous (or nearly so) at these boundaries. Thus if this term is omitted, a "kink" may appear in the calculated pattern at a shadow or reflection boundary. A discontinuity of this type is evidence that the slope diffraction term should be included in the UTD solution.

The above formula has been used to calculate the far-zone pattern of a slot directed perpendicular to the edge as shown in Fig. 5; in this case $\rho=s'=d$, the distance of the slot from the edge. It is seen that the pattern calculated from the UTD solution is in excellent agreement with that calculated from the eigenfunction solution, even though the slot is only $\lambda/8$ from edge. For comparison, in Fig. 6 the far-zone pattern is shown for the case where the axis of the slot is parallel to the edge, and again the UTD pattern is seen to be in excellent agreement with the eigenfunction pattern, even though $d=\lambda/8$.

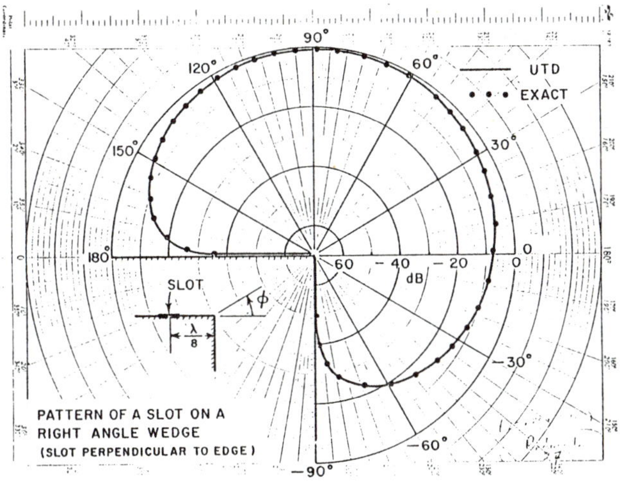

Fig. 5. Diffraction of the field of an infinitesimal slot by a right-angle wedge.

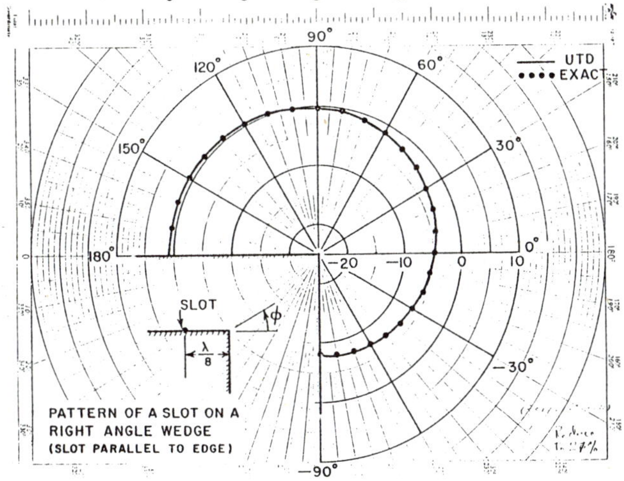

Fig. 6. Diffraction of the field of an infinitesimal slot slot by a right-angle wedge.

C. Vertex Diffraction

The diffraction at a vertex formed by two intersecting edges is shown in Fig. 7; it will be referred to as the corner diffraction problem. It is included in the section on edge diffraction because the corner is treated as a terminated edge here (just as an edge can be regarded as a surface termination).

A diffraction coefficient is needed that is numerically efficient in order for the corner effect to be of any practical use in complicated modeling problems. A solution has been proposed which is based on the asymptotic evaluation of the radiation integral which employs the equivalent edge currents that would exist in the absence of the corner as developed by Burnside and Pathak [11]. The corner diffraction term is then found by appropriately (but at present empirically) modifying the asymptotic result for the radiation integral which is characterized by a saddle point near an end point. This diffraction coefficient is still in the initial stages of its development. However, it has been shown to very successfully predict the corner effect for numerous plate structures. For this reason, it is discussed here as a good engineering approximation to the problem.

The corner diffracted fields associated with one corner and one edge in the near field with spherical wave incidence are given by

$$\begin{Bmatrix} E^c_{\beta_0} \\ E^c_{\phi} \end{Bmatrix} \sim \begin{Bmatrix} IZ_0 \\ MY_0 \end{Bmatrix} \frac{\sqrt{\sin\beta_c \sin\beta_{oc}}}{\cos\beta_{oc}-\cos\beta_c} F[kL_c a(\pi+\beta_{oc}-\beta_c)] \frac{e^{-jks}}{4\pi s} \tag{3.14}$$

where

$$\begin{Bmatrix} I \\ M \end{Bmatrix} = - \begin{Bmatrix} E^i_{\beta'_0}(Q_C) \\ E^i_{\phi'}(Q_C) \end{Bmatrix} \begin{Bmatrix} C_s(Q_E)Y_0 \\ C_h(Q_E)Z_0 \end{Bmatrix} \sqrt{\frac{8\pi}{k}}\, e^{-j\frac{\pi}{4}} \tag{3.15}$$

and

$$C_{s,h}(Q_E) = \frac{e^{-j\frac{\pi}{4}}}{2\sqrt{2\pi k}\sin\beta_0} \left\{ \frac{F[kLa(\beta^-)]}{\cos\frac{\beta^-}{2}} \left| F\left[\frac{La(\beta^-)/\lambda}{kL_c a(\pi+\beta_{oc}-\beta_c)}\right]\right| \mp \frac{F[kLa(\beta^+)]}{\cos\frac{\beta^+}{2}} \left| F\left[\frac{La(\beta^+)/\lambda}{kL_c a(\pi+\beta_{oc}-\beta_c)}\right]\right| \right\} \tag{3.16}$$

The function $F(x)$ was defined earlier, $a(\beta)=2\cos^2(\beta/2)$ where $\beta^{\pm}=\phi\pm\phi'$ and $L=s's''\sin^2\beta/(s'+s'')$ and $L_c=s_c s/(s_c+s)$ for spherical wave incidence. The function $D_{s,h}(Q_E)$ is a modified version of the diffraction coefficient for the half-plane case ($n=2$). The modification factor,

$$\left| F\left[\frac{La(\beta)/\lambda}{kL_c a(\pi+\beta_{oc}-\beta_c)}\right]\right| ,$$

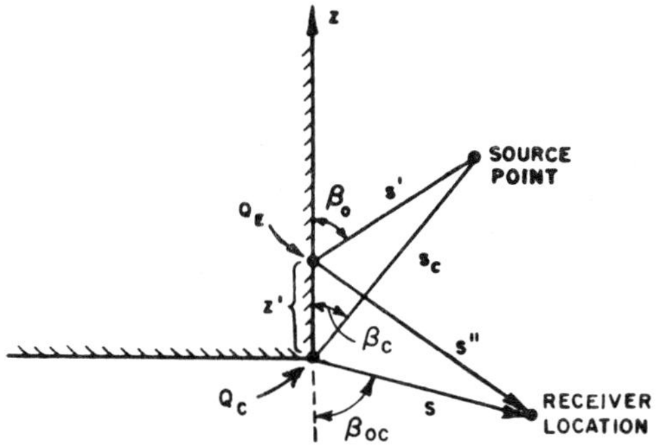

Figure 7. Geometry for corner diffraction problem.

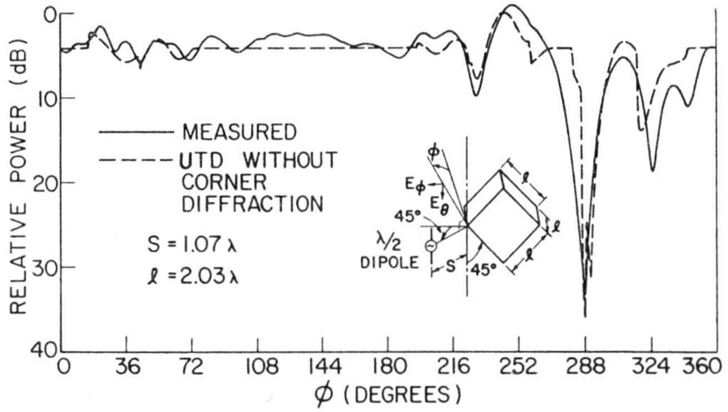

a. Without corner diffraction.

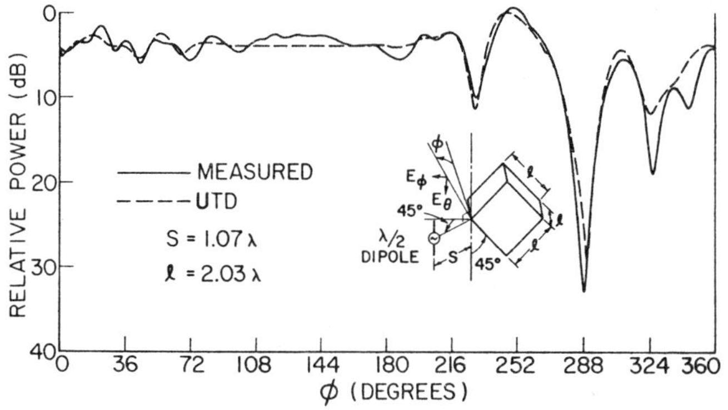

b. With corner diffraction.

Fig. 8. Comparison of measured and calculated E_θ radiation pattern for a dipole near a box in the indicated plane.

is an empirically derived function that insures that the diffraction coefficient will not change sign abruptly when it passes through the shadow boundaries of the edge. There is also a corner diffraction term associated with the other edge forming the corner and is found in a similar manner.

If the vertex involves three intersecting edges, as at the corner of a cube, a third term associated with the third edge must be added. In addition the preceding expressions for the half-plane edge must be generalized to the wedge. The utility of the present result is evident in Fig. 8, where the pattern of an electric dipole radiating in the presence of a cube is presented.

IV. SURFACE DIFFRACTION

When an incident ray system emanating from a source strikes a smooth, perfectly-conducting convex surface as shown in Fig. 9, it produces a system of rays reflected from that surface; at grazing, i.e., at Q_1, the incident ray merges with the reflected ray giving rise to a surface ray which propagates into the shadow region along a geodesic on the convex surface according to the generalized Fermat's principle. The field associated with the surface ray exhibits an exponential decay in the deep shadow region due to the continual leakage of energy along the surface ray resulting from diffracted rays shedding tangentially from the surface ray. One notes that surface rays can be excited by sources which are located either on or off a smooth convex surface; they can also be excited by the illumination of an edge, or other geometrical or electrical discontinuities in an otherwise smooth convex surface.

This section will focus on recently developed UTD solutions for calculating the electromagnetic fields excited by sources which lie either off or on a smooth convex surface. In all of the UTD solutions for convex surfaces presented below, it will be assumed that the principal radii of curvature of the surface R_1 and R_2 are large in terms of a wavelength so that the parameter $m=\left(\frac{k\rho_g}{2}\right)^{1/3}$ is large at all points on the surface ray; in addition it is assumed that m is slowly varying along the ray path. Here ρ_g is the surface radius of curvature along the surface ray. These UTD solutions for the arbitrary convex surface all employ GTD rays, and they have been deduced from the asymptotic solutions to appropriate canonical problems as employed in the GTD procedure.

A. Both Source and Observation Points are off the Smooth Convex Surface

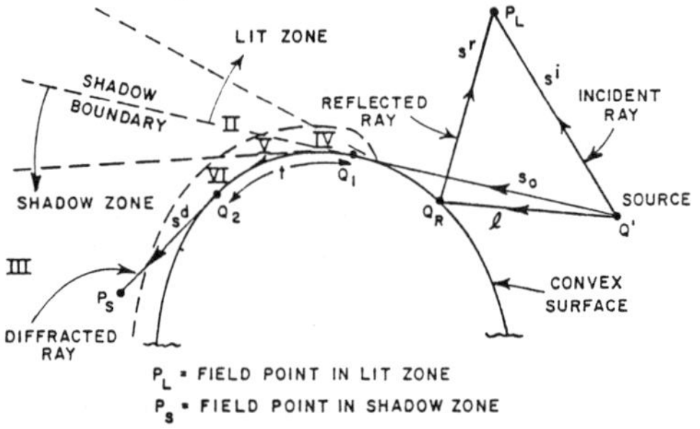

Fig. 9. The rays and regions associated with scattering by a smooth convex surface.

The geometrical configuration of this problem is as hhown in Fig. 9. The details of the construction of this solution together with some experimental verification are discussed in [12]. The final results for the total electric field which exists in the presence of a convex surface illuminated by an incident ray optical field $\bar{E}^i$ are as follows:

$$\bar{E}(P_L) \sim \bar{E}^i(P_L) + \bar{E}^i(Q_R)\cdot[\mathcal{R}_s \hat{e}_\perp \hat{e}_\perp + \mathcal{R}_h \hat{e}^i_\parallel \hat{e}^r_\parallel] \sqrt{\frac{\rho_1^r \rho_2^r}{(\rho_1^r+s^r)(\rho_2^r+s^r)}}\, e^{-jks^r}; \tag{4.1}$$

for P_L in the <u>lit</u> region,

and

$$\bar{E}(P_S) \sim \bar{E}^i(Q_1)\cdot[\mathcal{D}_s \hat{b}_1 \hat{b}_2 + \mathcal{D}_h \hat{n}_1 \hat{n}_2] \sqrt{\frac{\rho_2^d}{s^d(\rho_2^d+s^d)}}\, e^{-jks^d} \tag{4.2}$$

; for P_S in the <u>shadow</u> region.

The quantities within brackets involving $\mathcal{R}_{s,h}$ in equation (4.1) and $\mathcal{D}_{s,h}$ in (4.2) may be viewed as generalized, dyadic coefficients for surface reflection and diffraction, respectively. It is noted that (4.1) and (4.2) are expressed invariantly in terms of the unit vectors fixed in the reflected and surface diffracted ray coordinates. The unit vectors $\hat{e}^i$, $\hat{e}^r$, and $\hat{e}$ in (4.1) have been defined in Section IIA. At Q_1 let $\hat{t}_1$ be the unit vector in the direction of incidence, $\hat{n}_1$ be the unit outward normal vector to the surface, and $\hat{b}_1=\hat{t}_1 \times \hat{n}_1$; at Q_2 let a similar set of unit vectors be defined with $\hat{t}_2$ in the direction of the diffracted ray. In the case of surface rays with zero torsion, $\hat{b}_1=\hat{b}_2$. The distances s^r and s^d along the reflected and surface diffracted ray paths are shown in Fig. 9. It is noted that ρ_1^r and ρ_2^r have been defined in Section IIA, and ρ_2^d is the wavefront radius of curvature of the surface diffracted ray evaluated in the $\hat{b}_2$ direction at Q_2. The $\mathcal{R}_{s,h}$ and $\mathcal{D}_{s,h}$ in (4.1) and (4.2) are given by

$$\mathcal{R}_{s,h} = -\left[\sqrt{\frac{-4}{\xi^L}}\, e^{-j(\xi^L)^3/12} \left\{\frac{e^{-j\frac{\pi}{4}}}{2\sqrt{\pi}\,\xi^L}[1-F(X^L)] + \hat{P}_{s,h}(\xi^L)\right\}\right], \tag{4.3}$$

for the <u>lit</u> region

and

$$\mathcal{D}_{s,h} = -\left[\sqrt{m(Q_1)m(Q_2)}\sqrt{\frac{2}{k}}\left\{\frac{e^{-j\frac{\pi}{4}}}{2\sqrt{\pi}\,\xi^d}[1-F(X^d)] + \hat{P}_{s,h}(\xi^d)\right\}\right]\sqrt{\frac{d\eta(Q_1)}{d\eta(Q_2)}}\, e^{-jkt}, \tag{4.4}$$

for the <u>shadow</u> region

The function F appearing above has been introduced in Section IIIA dealing with edge diffraction (see (3.6)). The Fock type surface reflection function $P_{s,h}$ is related to the $\binom{\text{soft}}{\text{hard}}$ Pekeris function $\binom{p^*}{q^*}$ by [13,14]

$$\hat{P}_{s,h}(\delta) = \begin{Bmatrix} p^*(\delta) \\ q^*(\delta) \end{Bmatrix} e^{-j\frac{\pi}{4}} - \frac{e^{-j\frac{\pi}{4}}}{2\sqrt{\pi}\,\delta}, \quad \text{(Note that } \delta=0 \text{ at SB).} \tag{4.5a, 4.5b}$$

It is shown in [12] that exterior to the SB transition region, the F function approaches unity so that only $P_{\substack{s\\h}}$ in (4.3) and (4.4) dominates allowing (4.1) and (4.2) to reduce uniformly to the GO and Keller's surface diffracted fields, respectively of the GTD solution [3,4,6]. Near and at the SB, the F function dominates; it is entirely responsible for ensuring the continuity of (4.1) and (4.2) at the SB. Finally, this UTD result remains accurate outside the paraxial regions of cylindrical or elongated type surfaces: a different solution is required for handling the paraxial regions. A typical example showing the accuracy of this UTD solution is illustrated in Fig. 10.

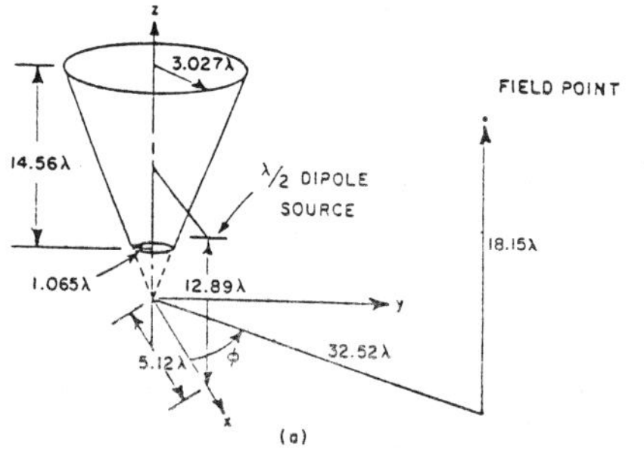

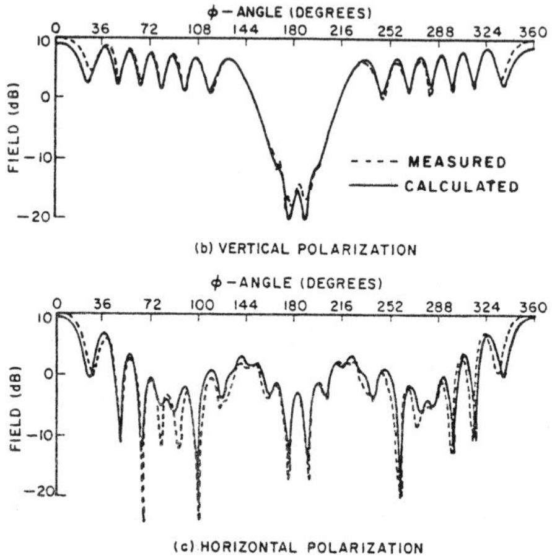

Figure 10. Radiation patterns of an electric dipole near the frustum of a cone.

B. Source Point on but Observation Point Off the Convex Surface

In this configuration as shown in Fig. 11, the source is located on the surface; whereas, the observation point is located at least a few wavelengths from the surface even though it may be in the near zone of that surface. This radiation problem is directly related by reciprocity to the problem of calculating the fields induced on the convex surface by a source which is located off the surface.

The high-frequency electric field is given by $d\bar{E}_{\substack{m\\e}} = \hat{n}dE^n_{\substack{m\\e}} + \hat{b}dE^b_{\substack{m\\e}}$ for points away from the convex surface in both the shadow and lit regions. Expressions for the above n and b directed electric field components have been deduced from a careful study of the cylinder and sphere canonical problems in which higher order terms are retained in the asymptotic solutions; in addition, experimental results for sources on a spheroid were helpful in the generalization to the convex surface. An important step in the present generalization of the canonical solutions to treat the arbitrary convex surface is based on the observation that the effect of surface ray torsion is localized to the source region at least to first order, and its effect on the radiation fields in both the shadow and the shadow boundary transition regions can be described explicitly in terms of a torsion factor T_Q at the source. The torsion factor thus affects only the launching of the surface ray field at Q'; whereas, the diffraction of the surface ray at Q is not affected by surface ray torsion to first order. Expressions for the fields $d\bar{E}_{\substack{m\\e}}$ radiated

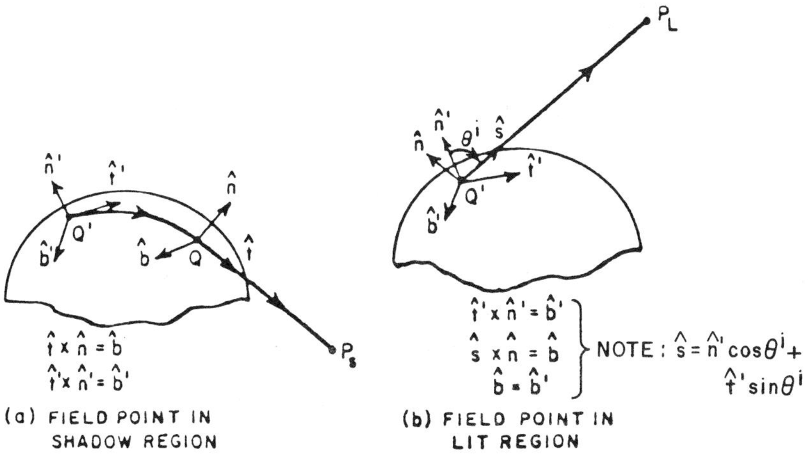

Figure 11. Rays emanating from a source on a convex surface.

by $d\bar{p}_m^e$ are developed in detail in [15]; the final results are presented below. It is noted that the form of the generalized dyadic launching coefficient $\bar{\bar{L}}$ can be immediately deduced from these expressions.

(a) <u>magnetic current moment, $d\bar{p}_m(Q')$ case:</u>

$$dE_m^n(P_L) = C[d\bar{p}_m \cdot \hat{b}')(H^\ell + T_0^2 \mathcal{F} \cos\theta^i) + (d\bar{p}_m \cdot \hat{t}')T_0 \mathcal{F} \cos\theta^i] \frac{e^{-jks}}{s} + O[m_\ell^{-2}] , \tag{4.6}$$

for $P=P_L$ in the <u>lit region</u>

$$dE_m^b(P_L) = C[(d\bar{p}_m \cdot \hat{b}')T_0 \mathcal{F} + (d\bar{p}_m \cdot \hat{t}')(H^\ell \cos\theta^i + \mathcal{F})] \frac{e^{-jks}}{s} + O[m_\ell^{-2}, m_\ell^{-3}]. \tag{4.7}$$

and

$$dE_m^n(P_S) = C(d\bar{p}_m \cdot \hat{b}')H\, e^{-jkt} \left[\frac{\rho_g(Q')}{\rho_g(Q)}\right]^{-1/6} \sqrt{\frac{d\psi_o}{d\eta(Q)}} \sqrt{\frac{\rho_2^d}{s(\rho_2^d + s)}}\, e^{-jks} + O[m^{-2}], \tag{4.8}$$

for $P=P_S$ in the <u>shadow region</u>

$$dE_m^b(P_S) = C[(d\bar{p}_m \cdot \hat{b}')T_0 S + (d\bar{p}_m \cdot \hat{t}')S] e^{-jkt} \left[\frac{\rho_g(Q')}{\rho_g(Q)}\right]^{-1/6} \sqrt{\frac{d\psi_o}{d\eta(Q)}} \tag{4.9}$$

$$\cdot \sqrt{\frac{\rho_2^d}{s(\rho_2^d + s)}}\, e^{-jks} + O[m^{-2}, m^{-3}].$$

(b) electric current moment, $d\overline{p}_e(Q')$ case:

$$dE_e^n(P_L) = CZ_0 dp_e(Q')\sin\theta^i \left[\frac{H^\ell + T_0^2 S^\ell \cos\theta^i}{1+T_0^2\cos^2\theta^i}\right] \frac{e^{-jks}}{s} + O[m_\ell^{-2}] , \quad (4.11)$$

for $P=P_L$ in the lit region

$$dE_e^b(P_L) = CZ_0 dp_e(Q')\sin\theta^i \; T_0 F \frac{e^{-jks}}{s} + O[m_\ell^{-2}] . \quad (4.12)$$

and

$$dE_e^n(P_S) = CZ_0 dp_e(Q') H \, e^{-jkt} \left[\frac{\rho_g(Q')}{\rho_g(Q)}\right]^{-1/6} \sqrt{\frac{d\psi_o}{dn(Q)}} \sqrt{\frac{\rho_2^d}{s(\rho_2^d+s)}} \; e^{-jks} + O[m^{-2}] \quad (4.13)$$

for $P=P_S$ in the shadow region

$$dE_e^b(P_S) = CZ_0 dp_e(Q') T_0 S \, e^{-jkt} \left[\frac{\rho_g(Q')}{\rho_g(Q)}\right]^{-1/6} \sqrt{\frac{d\psi_o}{dn(Q)}} \sqrt{\frac{\rho_2^d}{s(\rho_2^d+s)}} \; e^{-jks} + O[m^{-2}] . \quad (4.14)$$

The previous expressions reduce to the geometrical optics field in the deep lit region and to the GTD field in the deep shadow region. The expressions for the lit and shadow regions join smoothly at the shadow boundary. As expected, they reduce to the asymptotic solutions for the circular cylinder and sphere cases, but the higher order terms in m must be retained to pass smoothly to these two limiting cases. As the radii of curvature of the surface become infinite $T_0=0$, and (4.6), (4.7), (4.11) and (4.12) simplify to the field of magnetic and electric current moment sources on a ground plane.

In addition, this solution has been tested by applying it to calculate the radiation from slots and monopoles on perfectly-conducting circular and elliptic cylinders, cones, and spheroids [15]. In Fig. 12, the patterns of a circumferential slot and a monopole are calculated and measured in the plane tangent to the spheroid at the source location. Note that the E_b component is due to the spheroidal surface; it would vanish if the source were on a ground plane.

C. Both Source and Observation Points on the Convex Surface

In this problem, the observation point P_S of Fig. 11 is moved to the point Q on the perfectly-conducting convex surface, with the source still being positioned at Q' on that surface. This problem is of interest in the calculation of the mutual coupling between a pair of antennas on a convex surface. For convex surfaces, the mutual coupling calculation reduces to finding the electromagnetic surface fields at Q which are excited by a source at Q' on the surface.

The details of the construction of this solution are given in [16]; the final results are presented below for the surface magnetic and electric fields $d\overline{H}_{m,e}(Q)$ and $d\overline{E}_{m,e}(Q)$ at Q, respectively due to sources $\overline{p}_{m,e}(Q')$. It is noted that the $d\overline{H}_{m,e}(Q)$ and $d\overline{E}_{m,e}(Q)$ are expressed invariantly in terms of the unit vectors which are

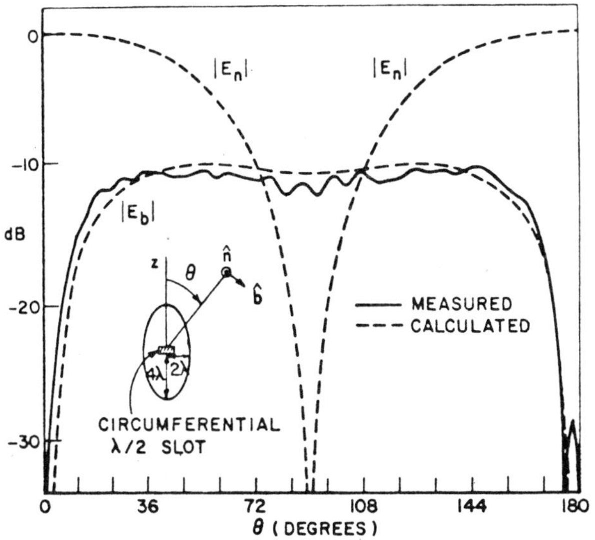

Figure 12a. Radiation patterns of a half-wavelength circumferential slot in a prolate spheroid calculated and measured in the shadow boundary plane.

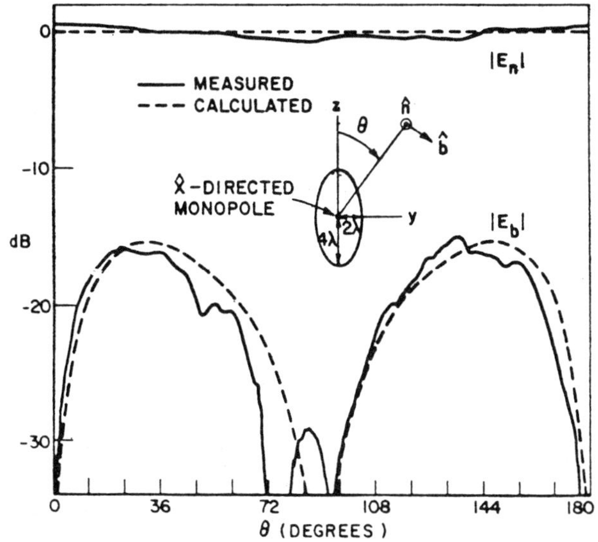

Figure 12b. Radiation patterns of a quarter-wavelength monopole on a prolate spheroid calculated and measured in the shadow boundary plane.

fixed in the surface ray coordinates at Q' and Q as in Fig. 11, and these fields remain uniformly valid along the ray including the immediate vicinity of the source.

(a) $d\bar{p}_m(Q')$ case:

$$d\bar{H}_m(Q) = C d\bar{p}_m(Q') \cdot \Bigg[\hat{b}'\hat{b} \left(\left[1 - \frac{j}{kt}\right] \mathcal{V}(\xi) + D^2\left(\frac{j}{kt}\right)^2 [\Lambda_s \mathcal{U}(\xi) + \Lambda_c \mathcal{V}(\xi)] + \tilde{T}_o^2 \frac{j}{kt} [\mathcal{U}(\xi) - \mathcal{V}(\xi)] \right) + \hat{t}'\hat{t} \left(D^2 \frac{j}{kt} \mathcal{V}(\xi) + \frac{j}{kt} \mathcal{U}(\xi) - 2\left(\frac{j}{kt}\right)^2 [\Lambda_s \mathcal{U}(\xi) + \Lambda_c \mathcal{V}(\xi)] \right) + (\hat{t}'\hat{b} + \hat{b}'\hat{t}) \left(\tilde{T}_o \frac{j}{kt} [\mathcal{U}(\xi) - \mathcal{V}(\xi)] \right) \Bigg] D\, G(kt) \quad . \tag{4.15}$$

$$d\bar{E}_m(Q) = C Z_o d\bar{p}_m(Q') \cdot \Bigg[\hat{b}'\hat{n} \left(\left[1 - \frac{j}{kt}\right] \mathcal{V}(\xi) + \tilde{T}_o^2 \frac{j}{kt} [\mathcal{U}(\xi) - \mathcal{V}(\xi)] \right) + \hat{t}'\hat{n} \left(\tilde{T}_o \frac{j}{kt} [\mathcal{U}(\xi) - \mathcal{V}(\xi)] \right) \Bigg] D\, G(kt) \quad . \tag{4.16}$$

In the above equations, D and G(kt) are defined by

$$D = \sqrt{\frac{t\, d\psi_o}{d\eta(Q)}} \quad ; \qquad G(kt) = 2(Z_o)^{-1} \frac{e^{-jkt}}{t} \quad . \tag{4.17);(4.18}$$

The $\tilde{T}_o$ is defined by

$$\tilde{T}_o = \pm \left| T_o(Q')T_o(Q) \right| \quad , \tag{4.19}$$

where the minus (-) sign in (4.19) is chosen if $T_o(Q')<0$ and/or $T_o(Q)<0$; otherwise the positive (+) sign is chosen. The generalized soft and hard Fock functions $\mathcal{U}(\xi)$ and $\mathcal{V}(\xi)$ are defined as

$$\mathcal{U}(\xi) = \left(\frac{kt}{2m(Q')m(Q)\xi}\right)^{3/2} U(\xi) \quad ; \quad U(\xi) = \xi^{3/2} \frac{e^{j\frac{3\pi}{4}}}{\sqrt{\pi}} \int_{\infty e^{-j2\pi/3}}^{\infty} d\tau \frac{W_2'(\tau)}{W_2(\tau)} e^{-j\xi\tau} \ , \tag{4.20a);(4.20b}$$

$$\mathcal{V}(\xi) = \left(\frac{kt}{2m(Q')m(Q)\xi}\right)^{1/2} V(\xi) \quad ; \quad V(\xi) = \xi^{1/2} \frac{e^{j\frac{\pi}{4}}}{2\sqrt{\pi}} \int_{\infty e^{-j2\pi/3}}^{\infty} d\tau \frac{W_2(\tau)}{W_2'(\tau)} e^{-j\xi\tau} \ . \tag{4.21a);(4.21b}$$

The Fock functions $U(\xi)$ and $V(\xi)$ are tabulated in [13]; a useful summary of the large and small argument approximations for these functions is found in [17]. In order to interpolate smoothly between the canonical cylinder and sphere solutions, the weight factors Λ_C and Λ_S have been introduced heuristically into the solution for the arbitrary convex surface. These weight factors must be such that $\Lambda_S=1$ and $\Lambda_C=0$ for a sphere; whereas, $\Lambda_S=0$ and $\Lambda_C=1$ for a cylinder. An initial choice for Λ_S and Λ_C appears to be:

$$\Lambda_S = \sqrt{\frac{R_2(Q')R_2(Q)}{R_1(Q')R_1(Q)}} \; ; \qquad \Lambda_C = 1 - \Lambda_S \; , \qquad (4.22a);(4.22b)$$

where R_1 and R_2 are the principal surface radii of curvatures.

(b) $d\bar{p}_e(Q')$ case:

$$d\bar{H}_e(Q) = -CZ_0 d\bar{p}_e(Q')\cdot \left[\hat{n}'\hat{b}\left(\left[1 - \frac{j}{kt}\right]\hat{V}(\xi) + \tilde{T}_0^2 \frac{j}{kt}[\hat{U}(\xi)-\hat{V}(\xi)]\right) + \hat{n}'\hat{t}\left(\tilde{T}_0 \frac{j}{kt}[\hat{U}(\xi)-\hat{V}(\xi)]\right)\right] D\, G(kt) \; . \qquad (4.23)$$

$$d\bar{E}_e(Q) = -\, CZ_0^2 d\bar{p}_e(Q')\cdot\hat{n}'\hat{n}\left(\hat{V}(\xi) - \frac{j}{kt}\hat{V}(\xi) + \left(\frac{j}{kt}\right)^2[\Lambda_S\hat{V}(\xi)+\Lambda_C\hat{U}(\xi)] + \tilde{T}_0^2 \frac{j}{kt}[\hat{U}(\xi)-\hat{V}(\xi)]\right) D\, G(kt) \; . \qquad (4.24)$$

It may be easily verified from the property $\hat{V}(\xi)\to 1$ and $\hat{U}(\xi)\to 1$, as $\xi\to 0$, that the above results for $d\bar{H}_{\substack{m\\e}}$ and $d\bar{E}_{\substack{m\\e}}$ reduce in the limit of vanishing surface curvature to the known, exact results for the fields on a planar, perfectly-conducting surface. Some typical results indicating the accuracy of the above solution are presented in Fig. 13 for a cone geometry; it is seen from these calculations that the solutions indeed agree well with both the eigenfunction (modal) series solution and with measurements available in [18]. In these calculations, the present solution is referred to as the OSU solution; whereas, those in [18] and [19] are referred to as the UI and the G-S-PB solutions, respectively.

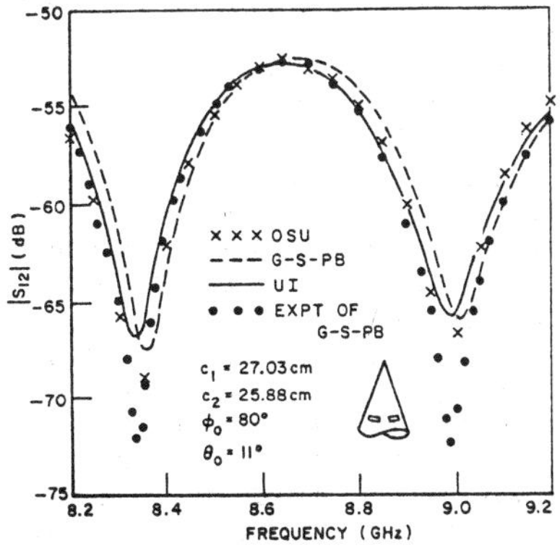

Figure 13. Coupling coefficient S_{12} between two circumferential slots on a cone vs. frequency. The radial separation between the slots is C_1-C_2 and angular separation is ϕ_0. The cone half angle is θ_0.

V. APPLICATIONS

In the previous sections, the UTD solutions for the general wedge and convex curved diffraction solutions were presented. These solutions provide the basic building blocks from which one can simulate much more complex structures. The concept applied here is that all scattered fields can be treated as ray optical. This being the case one can obtain the scattered field from a convex surface in terms of a ray optical field which in turn can diffract from a curved edge, for example. Recall that the UTD solutions for the wedge and curved surface are expressed in terms of general incident ray optical wavefronts. This concept allows one to develop high frequency numerical solutions for very complex electromagnetic problems. However, one must approach this panacea with caution. First, one cannot always assume that the scattered field from one structure which illuminates another is ray optical. Such a case is the double diffraction problem where the source, receiver, and two edges align. Second, one must be aware of the limitations of the UTD and limited number of UTD diffraction solutions. This problem area manifests itself in terms of the UTD model used to simulate the actual structure. Since one does not have a UTD solution for all possible problems, he must attempt to approximate a given structure by a simpler UTD model which can be analyzed efficiently. This leads to two questions:

1) can one accurately analyze the simpler UTD model?
2) does the simpler UTD solution accurately model the actual situation?

From our point of view, the second question is all important and must normally be answered through an experimental verification process. In fact, the experimental verification actually answers both questions.

Based on our fifteen years experience in this area, one can approach the development of UTD solutions in a preferred manner. First, one should start with a UTD model which is known to be valid for a similar problem or with a simpler two-dimensional model in order to examine the significant features of the problem. One, then, proceeds to more complex configurations until he is satisfied that the UTD model resembles the actual situation. This is normally done in terms of critical comparisons with experimental results. One thing to keep in mind is that the UTD is a high frequency solution and will begin to fail as critical dimensions approach a wavelength. Thus, one can make critical tests of the solution at the lower frequency limit which simultaneously verifies the model representation and the UTD approximations.

In order to illustrate the above approach, let us consider the pattern analysis of airborne antennas realizing that most antennas are mounted along the fuselage center-line. With this in mind, a simple two-dimensional UTD model can be developed to study roll plane patterns. Using this model, the effect of the fuselage shape, engines, and wings were examined. As a result of that study it was ascertained that one must treat the fuselage cross-section in terms of an elliptic shape, the engine effects were minimal, and the three-dimensional outline of the wing was very significant. At this point, the UTD model was extended by simulating the fuselage with an elliptic cylinder and wings by finite flat plates. Thus, our UTD solution was extended to three dimensions as shown in Fig. 14. With the development of this model, the UTD solution was compared with scale model aircraft measurements such as shown in Fig. 15. Based on numerous comparisons with various aircraft, antennas, and frequencies, the wing curvature did not play a significant role and the jet engine scattering was minimal for center-line antenna locations. This resulted in the basic roll plane aircraft model shown in Fig. 16a. A similar approach was used to analyze the elevation pattern with the resulting model shown in Fig. 16b. As presented in [19] these two models were used to develop a full volumetric pattern analysis for airborne antennas. Some of the results of that solution are presented in Fig. 15.

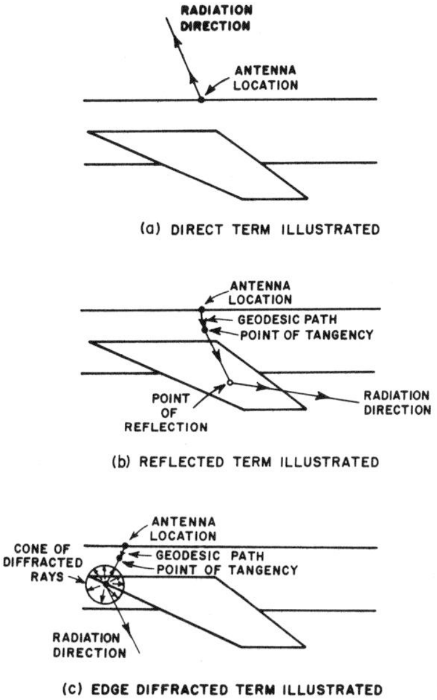

Figure 14. Dominant rays used in radiation pattern computation.

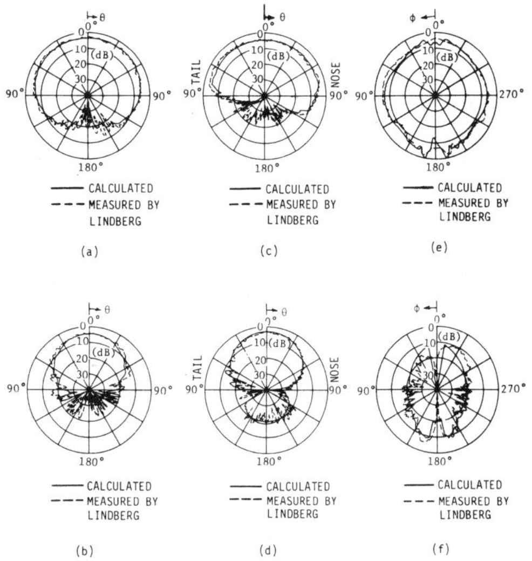

Figure 15. Radiation patterns of Lindberg crossed-slot antenna mounted at Station 470 on KC-135 aircraft. (a) Roll plane pattern (E_θ). (b) Roll plane pattern (E_ϕ). (c) Elevation plane pattern (E_ϕ). (d) Elevation plane pattern (E_θ). (e) Azimuth plane pattern (E_θ). (f) Azimuth plane pattern (E_ϕ).

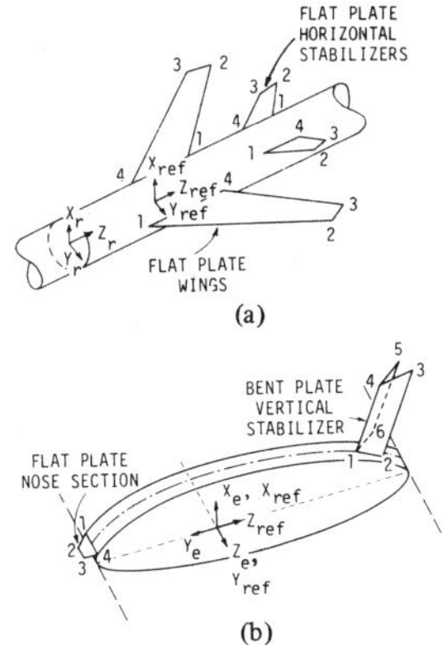

Figure 16. (a) Illustration of roll plane model. (b) Illustration of elevation plane model.

One should realize that he must study the simpler geometries and build to the more complex in that he cannot solve the problem exactly. If one starts with a complex model, he is most likely going to pay a premium for a given result especially if a far simpler model can solve the same problem. The costs of complex solutions are rather obvious:

1) large complex codes
2) inefficient solutions.

Given that one has found a simple UTD model to represent a given situation, how can he develop an efficient solution to such a problem? The efficiency for most three-dimensional problems is dictated by the speed at which one computes the various ray paths. As an example of an efficient ray path, one can refer to Ref. [20] for the ray path from a source on an elliptic cylinder, along a surface geodesic, subsequently diffracted from the curved surface, where it is finally edge diffracted to the receiver.

Another aspect of efficiency improvement has to do with the structure of the program. Considering the constraints of small, medium, and large computers, it is very advantageous to write computer codes which do not hop back-and-forth through the whole program. This is necessary in small computers because one must overlay various sections of the code into a small amount of memory. It is important in medium computers because they usually use paging algorithm which only allocate a certain number of pages per program. For big computers, they employ cache memory which is small but extremely fast. This implies the following for large UTD numerical solutions:

1) define all fixed geometry aspects associated with the problem at the outset and store it in "COMMON",
2) define bounds on various terms such that complete ray paths do not have to be found before one decides to include a term or not,
3) use a single array to store the pattern data which is updated each time a new term is added, and
4) have the code compute each UTD term (such as plate reflected field) for the complete pattern before going to the next term.

Using this structure, significant computations will be performed within smaller portions of the total code before proceeding to the next portion. This will reduce the amount of overlaying, decrease the amount of paging, and utilize cache memory to its fullest potential.

Let us consider a second major code development being done at the Ohio State University ElectroScience Laboratory. This code allows one to treat an antenna in the presence of a set of plates and a finite elliptic cylinder. This is a very general code such that one can model a wide variety of scattering problems. For example, it has been used to study wing mounted airborne antenna problems, a ship mounted antenna configuration, a communication van antenna farm, a radar antenna system mounted on a tank, etc.

The code is divided into three large sections. The first section contains the major scattered fields associated with the individual flat plates and the interactions between the different plates. These include the direct field, the singly reflected fields, doubly reflected fields, the single diffracted fields, the reflected-diffracted fields, and the diffracted-reflected fields. The diffracted fields include the normal diffracted fields as well as slope diffraction, a newly developed heuristic corner diffracted field and slope-corner diffracted field. The double diffracted fields are not included at present, but a warning is provided wherever this field component might be important. This is usually only a small angular section of space. This field may be included later whenever time and effort permit. The second section contains the major scattered fields associated with the elliptic cylinder. This includes the direct field, if not already computed in the plate section, the reflected field, the transition field, the deep shadow fields, the reflected field from the end caps, and the diffracted field from the end cap rims. The diffracted field from the end cap rim is not at present corrected in the pseudo caustic regions. This is where three diffraction points on the rim coalesce into one. This is only important in small angular regions in space and is not deemed appropriate to be included at the present time. An equivalent current method could be used for this small region but it is rather time consuming to use for the benefits derived from it for such a general code. The third section contains the major scattered fields associated with the interactions between the plates and cylinder. This includes, at present, the fields reflected from the plates then reflected or diffracted from the cylinder, the fields reflected from the cylinder then reflected from the plates, and the fields diffracted from the plates then reflected from the cylinder. These terms have been found to be sufficient for engineering purposes when analyzing wing-mounted aircraft antennas as well as many other structures.

The subroutines for each of the scattered field components are all structured in the same basic way. First, the ray path is traced backward from the chosen observation direction to a particular scatterer and subsequently to the source using either the laws of the reflection or diffraction. Each ray path, assuming one is possible, is then checked to see if it is shadowed by any structure along the complete ray path. If it is shadowed the field is not computed and the code proceeds to the next scatterer or observation direction. If the path is not interrupted the scattered field is computed using the appropriate UTD solutions. The fields are then superimposed in the main program. This shadowing process is often speeded up by making various decisions based on bounds associated with the geometry of the structure. This type of knowledge is used wherever possible.

The shadowing of rays is a very important part of the UTD scattering code. It is obvious that this approach should lead to various discontinuities in the resulting pattern. However, the UTD diffraction coefficients are designed to smooth out the discontinuities in the fields such that a continuous field is obtained. When a scattered field is not included in the result, therefore, the lack of its presence is apparent. This can be used to advantage in analyzing complicated problems. Obviously in a complex code the importance of the neglected terms are determined by the size of the so-called gliches or jumps in the pattern trace. If the gliches are small no additional terms are needed for a good engineering solution. If the gliches are large it may be necessary to include more terms in the solution. In any case the user has a gauge with which he can examine the accuracy of the results and is not falsely led into believing a result is correct when in fact there could be an error.

There have been many codes developed at the ElectroScience Laboratory over the past fifteen years using the UTD, two of which have been briefly described here.

REFERENCES

[1] J.B. Keller, "The Geometric Optics Theory of Diffraction," presented at the 1953 McGill Symp. Microwave Optics, A.F. Cambridge Res. Cent., Rep. TR-59-118 (II), pp. 207-210. 1959.

[2] J.B. Keller, "A Geometrical Theory of Diffraction," in Calculus of Variations and Its Applications, L.M. Graves, Ed., New York: McGraw-Hill, 1958, pp. 27-52.

[3] J.B. Keller, "Geometrical Theory of Diffraction," J. Opt. Soc. Amer., Vol. 52, pp. 116-130, 1962.

[4] R.G. Kouyoumjian, "The Geometrical Theory of Diffraction and Its Applications", in Numerical and Asymptotic Techniques in Electromagnetics, R. Mittra, Ed., New York, Springer-Verlag, 1975.

[5] J.B. Keller, R.M. Lewis and B.D. Seckler, "Asymptotic Solution of Some Diffraction Problems," Commun. Pure Appl. Math., Vol. 9, pp. 207-265, 1956.

[6] R.G. Kouyoumjian, "Asymptotic High Frequency Methods," Proc. IEEE, Vol. 53, pp. 864-876, Aug. 1965.

[7] R.G. Kouyoumjian and P.H. Pathak, "A Uniform Geometrical Theory of Diffraction for an Edge in a Perfectly Conducting Surface," Proc. IEEE, Vol. 62, pp. 1448-1461, 1975.

[8] P.H. Pathak and R.G. Kouyoumjian, "The Dyadic Diffraction Coefficient for a Perfectly Conducting Wedge," ElectroScience Lab., Dept. Elec. Eng., Ohio State Univ., Columbus, Ohio, Rep. 2183-4, June 5, 1970.

[9] Y.M. Hwang and R.G. Kouyoumjian, "A Dyadic Diffraction Coefficient for an Electromagnetic Wave Which is Rapidly Varying at an Edge," USNC-URSI 1974 Annual Meeting, Boulder, CO., Oct. 1974.

[10] R.G. Kouyoumjian, Y.M. Hwang and R. Tiberio, "A Uniform Geometrical Theory of Diffraction for an Edge Illuminated by a Field with Rapid Spatial Variation," to appear.

[11] W.D. Burnside and P.H. Pathak, "A Corner Diffraction Coefficient," to appear.

[12] P.H. Pathak, W.D. Burnside and R.J. Marhefka, "A Uniform GTD Analysis of the Diffraction of Electromagnetic Waves by a Smooth Convex Surface," submitted for publication to IEEE Trans. Antennas and Propagation.

[13] N.A. Logan, "General Research in Diffraction Theory," vol. I, LMSD-288087; and Vol. II, LMSD-288-88, Missiles and Space Division, Lockheed Aircraft Corporation, 1959.

[14] J.J. Bowman, T.B.A. Senior and P.L.E. Uslenghi, Eds., Electromagnetic and Acoustic Scattering by Simple Shapes, Amsterdam, The Netherlands: North Holland Publ., 1969.

[15] P.H. Pathak, N. Wang, W.D. Burnside and R.G. Kouyoumjian, "A Uniform GTD Solution for the Radiation from Sources on a Perfectly-Conducting Convex Surface," to appear. (Also paper with above title was presented at the 1979 IEEE APS/URSI Meeting in Seattle, Wash., June 18-22, 1979).

[16] P.H. Pathak and N. Wang, "Surafce Fields of Sources on a Perfectly-Conducting Convex Surface," to appear. (Also see P.H. Pathak and N. Wang, "An Analysis of the Mutual Coupling Between Antennas on a Smooth Convex Surface ," Final Rep. 784583-7, Oct. 1978, The Ohio State Univ. ElectroScience Lab., Dept. Elec. Engr., prepared under Contract N62269-76-C-0554 for Naval Air Development Center, Warminster, PA.)

[17] S.W. Lee, "Mutual Admittance of Slots on a Cone; Solution by Ray Technique," IEEE Trans. on AP-26, No. 6, pp. 768-773, Nov. 1978.

[18] K.E. Golden, G.E. Stewart and D.C. Pridmore-Brown, "Approximation Techniques for the Mutual Admittance of Slot Antennas on Metallic Cones," IEEE Trans. on Antenna and Propaga., Vol. AP-22, pp. 43-48, 1974.

[19] C.L. Yu, W.D. Burnside and M.C. Gilreath, "Volumetric Pattern Analysis of Airborne Antennas," IEEE Trans., AP-26, pp. 636-641, Sept. 1978.

[20] W.D. Burnside, N. Wang and E.L. Pelton, "Near Field Pattern Computations for Airborne Antennas," Rep. 784685-4, June 1978, The Ohio State Univ. ElectroScience Lab., Dept. Elec. Engr.; prepared under Contract N00019-77-C-0299 for Naval Air Systems Command.

Part 6
Resonances and Singularity Expansions

RESONANCE SCATTERING PREDICTIONS AND MEASUREMENTS

Louis R. Dragonette and Lawrence Flax

ABSTRACT

The major features of the acoustic scattering function (i.e. pressure amplitude vs frequency) for metal spheres and cylinders in water are related to the free modes of vibration of the body. This well founded observation was mathematically formalized by separating the normal mode series expression for the scattering function into rigid body terms and resonance terms, and an explicit expression for the resonance widths was obtained. The frequency position and amplitude of the scattering features related to resonances having broad widths can be accurately measured, and can be utilized to determine the geometric and elastic properties of submerged targets. A direct relationship between the individual circumferential waves predicted by creeping wave theory and resonances has been estabished, and this association is especially useful in determining the properties of the leaky Rayleigh surface wave.

The acoustic scattering by simply shaped metal objects in water can be described in terms of the free body resonances of the target (Ref. 1). Scattering curves can be used to predict resonances and resonances can be utilized to predict scattering behavior. The geometry of the problems considered here is described in Fig. 1, which shows an infinitely long cylinder illuminated by a plane incident wave traveling in a direction perpendicular to its axis. As will be demonstrated later, the theoretical solution to the infinite cylindrical problem given an excellent description of the reflection from a finite cylinder in this geometry. The simple shapes discussed here will include both cylindrical and spherical targets, and all problems considered will involve targets submerged in water.

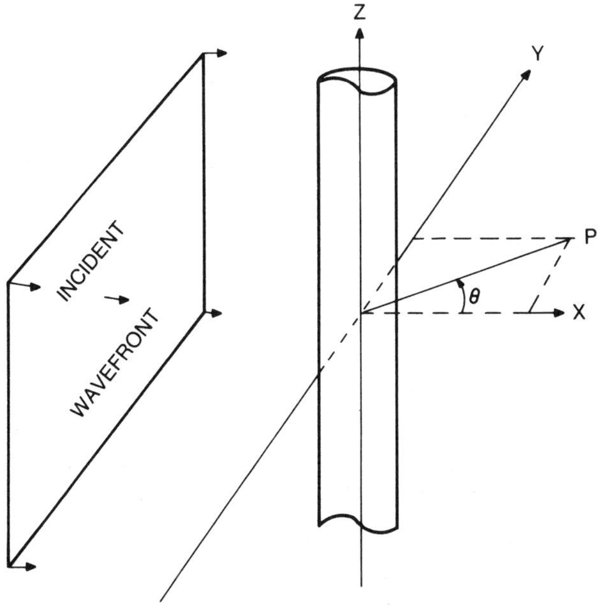

Fig. 1. The geometry of the problem. A plane wave illuminates a cylinder whose central axis is perpendicular to the direction of the incoming wave.

It has been demonstrated (Ref. 1) that the reflection from a solid metal target, whose density and sound speeds are larger than the density and speed of sound in water, can be described in terms of a rigid body reflection term and a resonance term. A thin hollow metal target, on the other hand can at low *ka* be described in terms of a soft background on to which the resonances are superimposed. The familiar rigid and soft scattering solutions for an infinite cylinder are seen in Fig. 2. The quantity plotted in Fig. 2 is the form function $f_\infty(\pi,ka)$, which is the normalized far field backscattered pressure. Equations (1) and (2) define f_∞ for cylinders and spheres respectively.

$$f_\infty(\pi,ka) = (a/2r)^{1/2}(p_r/p_o) \tag{1}$$

$$f_\infty(\pi,ka) = (a/2r)\,(p_r/p_o) \tag{2}$$

In Eqs. 1 and 2, a is the radius, r is the range between target and field point p_o is the amplitude of the incident plane wave and p_r is the amplitude of the backscattered reflection. The k is the wavenumber of the incident sound wave in water ($k = 2\pi/\lambda$).

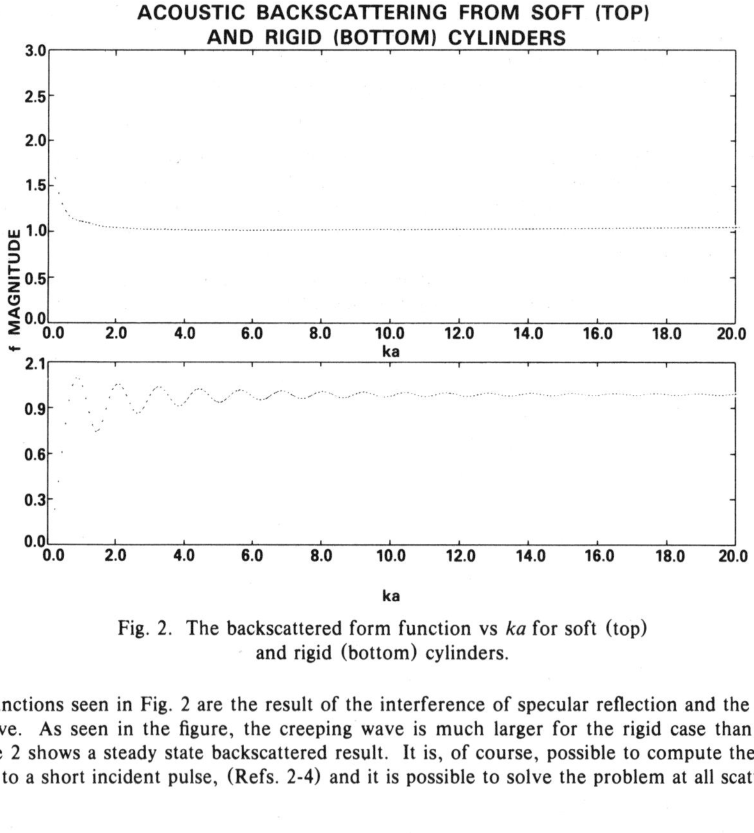

Fig. 2. The backscattered form function vs *ka* for soft (top) and rigid (bottom) cylinders.

The form functions seen in Fig. 2 are the result of the interference of specular reflection and the diffracted or creeping wave. As seen in the figure, the creeping wave is much larger for the rigid case than for the soft case. Figure 2 shows a steady state backscattered result. It is, of course, possible to compute the response of the cylinder to a short incident pulse, (Refs. 2-4) and it is possible to solve the problem at all scattering angles (Refs. 5-8).

The computed response of a rigid cylinder to a two cycle incident pulse, at a center dimensionless frequency $k_o a = 10$, is seen in Fig. 3. The response at scattering angles of 45° and 75° is shown. The scattered echo is made up of the specular reflection and a creeping wave whose relative amplitude and position are functions of the scattering angle. The echoes computed in Fig. 3 are obtained from

$$g_s(ka,\ \theta) = [a/2r]^{1/2} f_\infty(ka,\ \theta) g_i(ka) \tag{3}$$

In Eq. (3) that g_i is the Fourier transform of the incident time pulse and g_s is the transform of the scattered echo. The scattered time pulse $p_r(\tau)$ seen in Fig. 3 was obtained by computing the transform of Eq. (3)

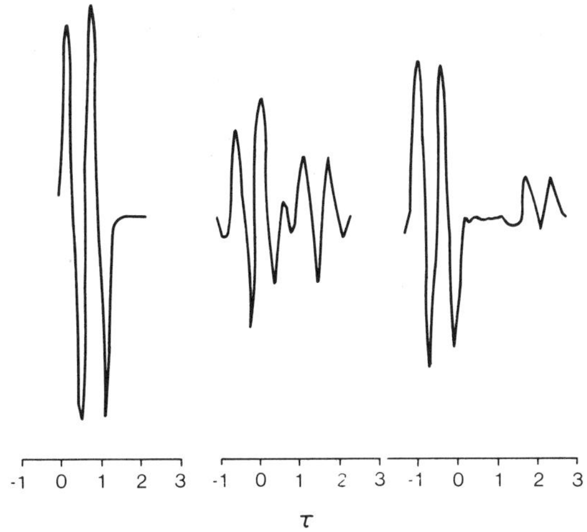

Fig. 3. The computed response of a rigid cylinder to a two cycle incident pulse (far left) at scattering angles of 45° (center) and 75° (right). The center frequency is $k_0 a = 10$.

using fast Fourier transform techniques. The quantities ka and τ are dimensionless frequency and dimensionless time transform pairs.

$$\tau = (ct - a)/r \tag{4}$$

By definition (Eq. 4), τ is equal to 0 when the incident pulse reaches the position of the center of the target. References 2-4 discuss this approach in detail.

A measurement made on a metal cylinder in air with $k_0 a = 10$ is seen in Fig. 4. Theory and experiment are in excellent agreement as expected, since the impedance difference between air and aluminum closely approximate rigid boundary conditions. Cylindrical results are seen in Figs. 2-4 but the descriptions are much the same for solid spheres (Refs. 2-5).

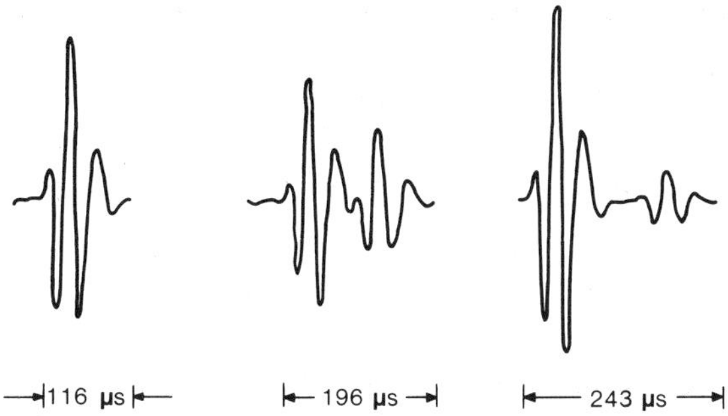

Fig. 4. The measured response of an aluminum cylinder in air at a center frequency $k_0 a = 10$. The incident pulse and the responses at scattering angles of 45° and 75° are given from left to right respectively.

For targets in water, rigid boundary conditions are never satisfied. The form function for a tungsten carbide sphere in water is given in Fig. 5. This is about as close to a rigid reflection as can be found in water. For $ka < 7.4$, the form function curve closely approximates that of a rigid sphere. At $ka \sim 7.4$ a null occurs in f_∞ and this null is related to the (2,1) resonance of a tungsten carbide sphere. In general the form function curves for all solid metal spheres and cylinders whose densities and velocities are greater than that of water will satisfy the rigid cylinder curve closely until the ka value is reached at which the (2,1) resonance is excited. Experimentally it has been determined that elastic effects related to resonances having broad resonance widths can be accurately measured both as regards amplitude and position. The points in Fig. 5 are quantitative measurements made on a tungsten carbide sphere and the accuracy with which the frequency (ka) position of the resonance effects can be measured permits the prediction of material velocities to within 0.5%.

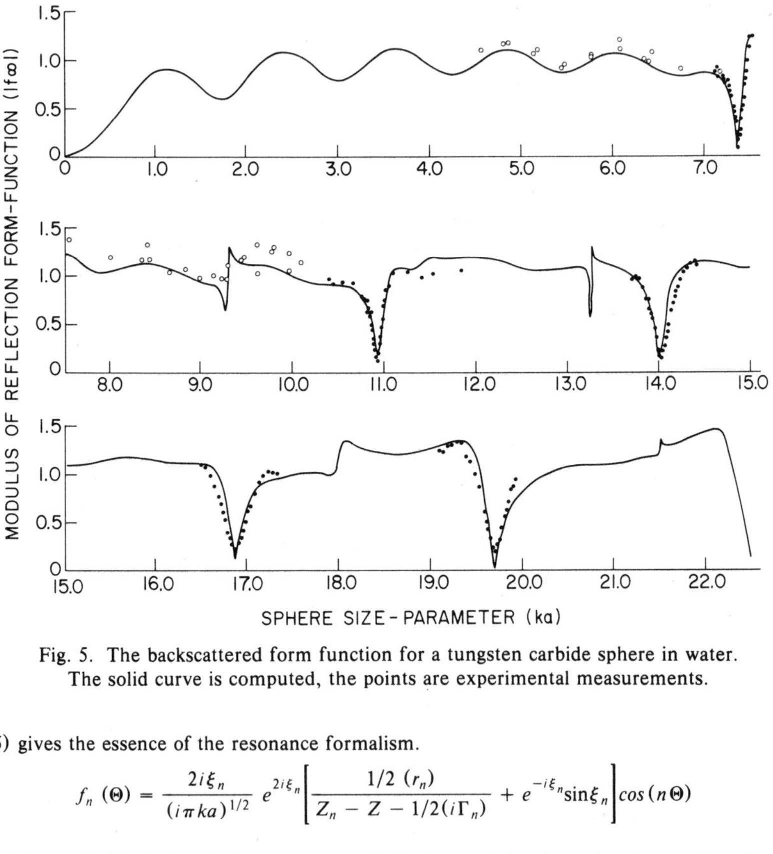

Fig. 5. The backscattered form function for a tungsten carbide sphere in water. The solid curve is computed, the points are experimental measurements.

Equation (5) gives the essence of the resonance formalism.

$$f_n(\Theta) = \frac{2i\xi_n}{(i\pi ka)^{1/2}} e^{2i\xi_n}\left[\frac{1/2\,(r_n)}{Z_n - Z - 1/2(i\Gamma_n)} + e^{-i\xi_n}\sin\xi_n\right]\cos(n\Theta) \tag{5}$$

In Eq. (5) f_n is an individual partial wave, i.e., a single term of the form function series. The ξ_n is the scattering phase shift for a rigid cylinder; the Γ_n is the resonance width, $Z = ka$; and Z_n is the ka value at resonance.

Equation (5) is derived and discussed completely in (Ref. 1). The significant point is that the second term in the bracket belongs solely to the rigid boundary conditions and the first term contains the resonance information. Reference 1 also contains the formulas for computing resonance widths.

The form function for an aluminum cylinder is given in Fig. 6. In this figure the major features of the form function are identified with the free body resonances by the labels (n,l). Here n is the normal mode number and l is the eigenfrequency. The curve for aluminum varies widely from the rigid body solution (Fig. 2) for

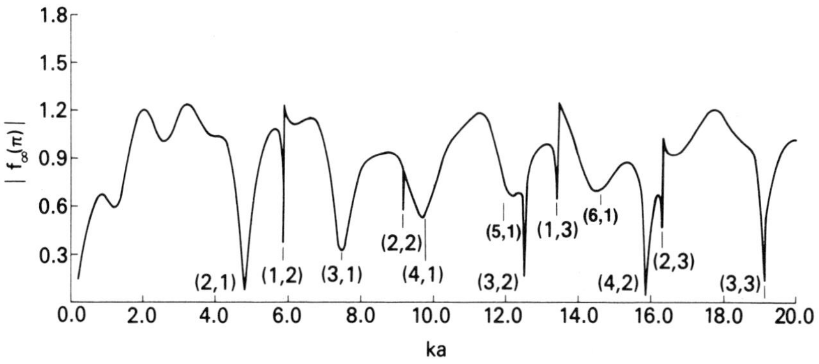

Fig. 6. The computed form function for an aluminum cylinder in water. The effects of free body resonances are identified by the labels (n,l).

$ka > 4.7$; however, as will be demonstrated below in Figs. 7 and 8, the resonance formalism gives an accurate description of the physics of scattering from aluminum even though the aluminum response varies more substantially from the rigid than the response of tungsten carbide (Fig. 5).

Figure 7 compares the $n = 2$ and $n = 3$ terms (i.e. f_2 and f_3) from the normal mode series describing the rigid, the soft, and the elastic aluminum form functions. Note the similarity between the rigid and the elastic curves. These curves are identical except in the ka ranges where the normal mode resonances are excited.

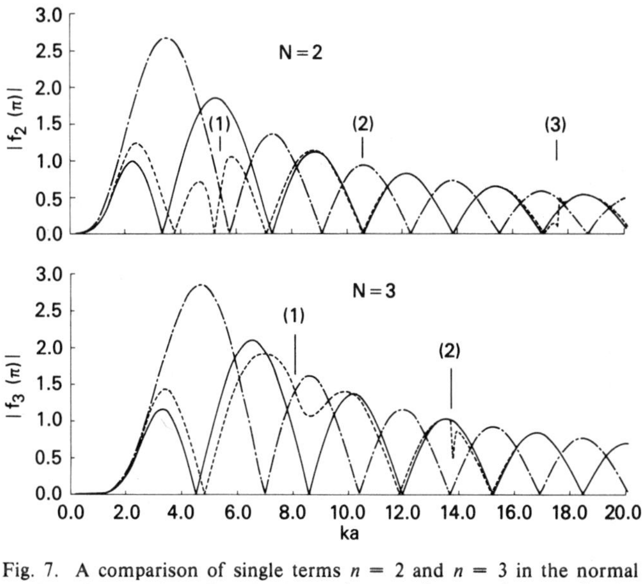

Fig. 7. A comparison of single terms $n = 2$ and $n = 3$ in the normal mode series solutions for aluminum (— — —), rigid (———), and soft (—·—) cylinders.

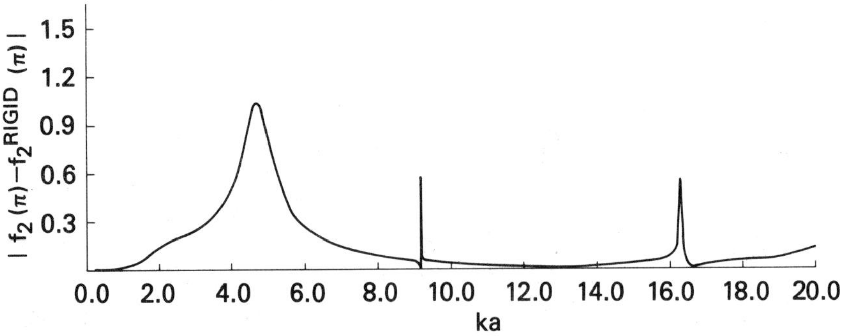

Fig. 8. The $n = 2$ term in the rigid cylinder solution is subtracted from the $n = 2$ term in the aluminum solution isolating the (2,1), (2,2), and (2,3) resonances.

The curve seen in Fig. 8 shows a plot of the $n = 2$ term for the rigid normal mode series solution substracted from the $n = 2$ term for the solution describing aluminum, i.e., $f_2^{\text{elastic}}(\pi,ka)$ - $f_2^{\text{rigid}}(\pi,ka)$. The plot seen in Fig. 8 shows that the amplitude is nearly zero except in regions where the (2,1), (2,2) and (2,3) resonances are excited. Figures 7 and 8 also demonstrate that the (2,1) and (3,1) resonances have broad resonance widths and effect f_∞ over a larger frequency range, then the other modes identified in Figs. 7 and 8.

The relationship between the circumferential waves predicted by creeping wave theory and the resonances identified in Figs. 6-8 adds some insight into the physical mechanisms involved in the excitation of the resonances (Ref. 9). In (Ref. 9), it was established that resonances having the same eigenfrequency, l, belonged to families, and that these familites could each be identified with an individual elastic circumferential wave as enumerated by Doolittle et al. (Ref. 10). All resonances having a particular eigenfrequency lable l are associated with the R_l^{th} circumferential wave. Explained in terms of circumferential waves, resonances occur at frequencies for which the circumference of the cylinder (or sphere) is an integer number of circumferential wavelengths. Thus, the (2,1) mode occurs when the cylinder circumference is exactly two wavelengths of the R_1 circumferential wave, the (3,1) mode when the circumference is 3 wavelenths etc. A similar description applies to the $(n,2)$ resonances, which are related to the R_2 circumferential wave, etc. The n is related to the number of circumferential nodes while the l is related to the number of radial nodes. The $(n,1)$ resonances are of special interest; these are related to the R_1 circumferential wave or leaky Rayleigh surface wave (Ref. 11). This relationship and an examination of Fig. 6 indicate that the Rayleigh circumferential wave contribution to backscattering is significant only for ka values below $ka = 15$. Previous attempts to isolate this wave had been made at much higher ka values (Refs. 12, 13). Experimental isolation of the Rayleigh surface wave was attempted based on the predictions discussed above, and Fig. 9 and 10 show the detection of the wave at k_oa values of 10.4 and 13.5. In both figures the backscattered specular reflection from the solid aluminum cylinder is followed by the echo radiated by the Rayleigh surface wave. At $k_oa = 10.4$ two circumferential transits are observed whereas at $k_oa = 13.5$ the attenuation is already large enough that only the single transit of the wave can be seen. The Rayleigh wave is seen to be 180° out of phase with the specular reflection as observed (Ref. 14) and predicted (Ref. 15) previously. In this ka range the velocity of the Rayleigh wave is a function of frequency (Refs. 16, 17) which leads to a distortion in pulse shape. Reference 18 describes the conditions in which Figs. 9 and 10 are observed.

The Rayleigh surface wave is most strongly observed over the frequency range which includes the (2,1) and (3,1) resonances. It is attenuated by radiation into the water (Ref. 19) and as frequency increases the cylinder circumference becomes larger with respect to a wavelength and the Rayleigh wave ceases to give a major contribution to the backscattered form function. Likewise, the penetration depth of the Rayleigh wave into the cylinder is a function of frequency. As frequency is increased, the cylinder radius changes from a fraction of the Rayleight wavelength to multiples of wavelengths. Thus, it should be possible to screen out the the presence of flaws and also to obtain information on their position, by observing the shifts in the (n,1) resonances caused by the existance of a flaw. Figure 11 shows an overlay of the form function for a solid iron cylinder and a hollow iron cylinder with a central cavity equal to 0.1a, (i.e., $b/a = 0.1$) where b is the radius of the inner air core. The frequency position of the (2,1) resonance effect is seen to shift significantly while

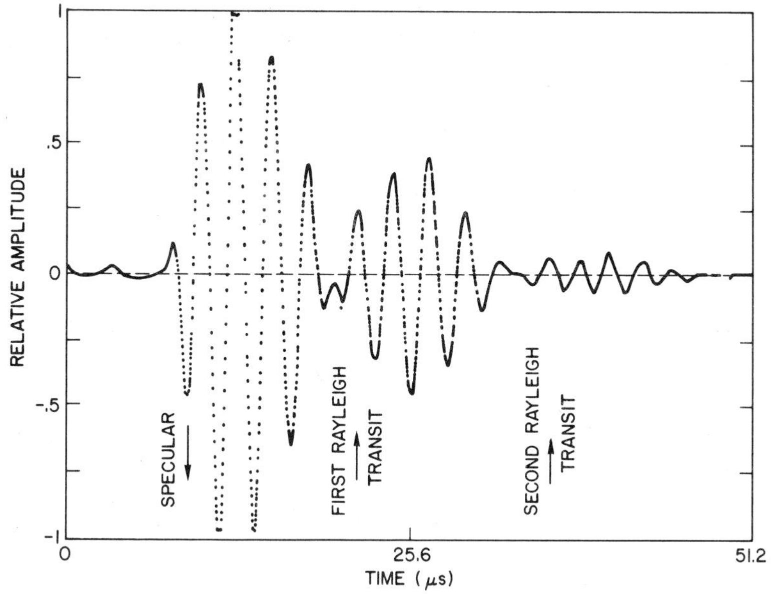

Fig. 9. The experimental observation of backscattering due to Rayleigh surface wave on an aluminum cylinder at $k_o a = 10.4$.

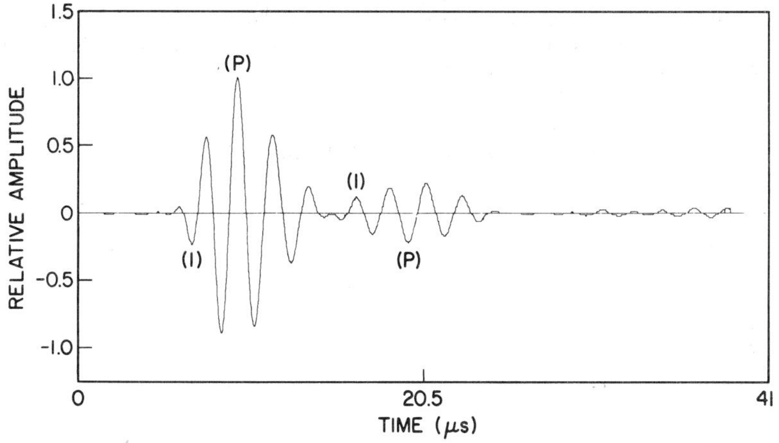

Fig. 10. The experimental observation of backscattering due to the Rayleigh surface wave on an aluminum cylinder at $k_o a = 13.5$.

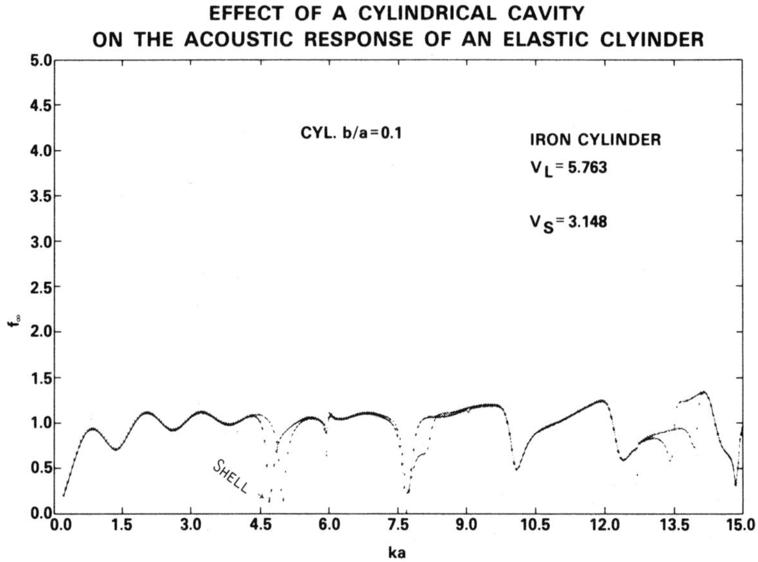

Fig. 11. Computation of the form function for a solid iron cylinder and an iron cylinder with a hollow core of radius $b/a = 0.1$.

the (3,1) resonance effect changes in amplitude but not in frequency position. If the hole were closer to the surface the (3,1) resonance should be shifted. Calculations such as that set forth in Fig. 11 showed measurable shifts for $b/a \gtrsim 0.05$. No off axis computations were attempted.

It was stated earlier that the infinite cylindrical theory describes the results for a finite cylinder in the geometry described in Fig. 1. This is demonstrated by the results seen in Fig. 12, which compares infinite theory for an elastic aluminum cylinder and experimental values obtained from short aluminum cylinders. The experimental method and conditions are delineated in detail in Reference 20. In Fig. 12 the theoretical curves are plotted with a grid $ka = 0.05$ which is the limit of the experimental method utilized in these measurements.

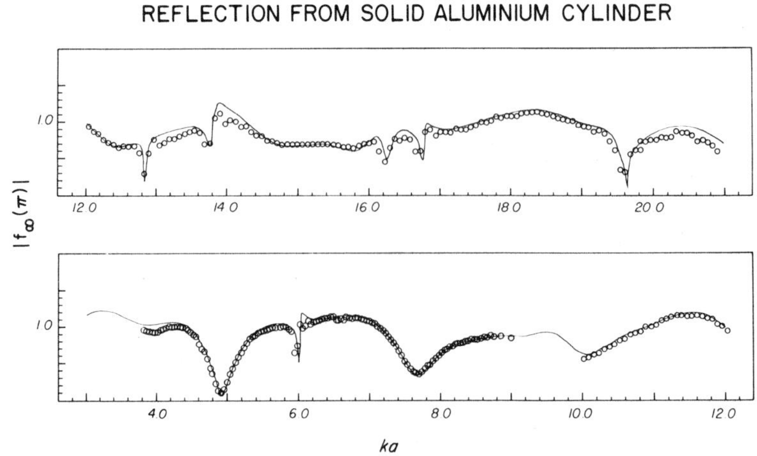

Fig. 12. Comparison of the computed form function for an infinite aluminum cylinder and experimental measurements on short cylinders.

Resonances on thin air filled metal cylindrical shells have also been investigated (Ref. 21). For a thin aluminum shell the geometric background term passes from a soft background at low ka, through a transition region, to a rigid region which extends to $ka \rightarrow \infty$. The extent of the soft and transition regions depends on the shell thickness and material. A particular example is seen in Fig. 13 which gives the form function for an aluminum shell with $b/a = 0.99$ over the ka range $0.2 \leqslant ka \leqslant 50$. In this case the soft background region extends from $0 < ka \lesssim 25$; the transition region from $25 \lesssim ka \lesssim 35$; and the rigid region encompasses $ka > 35$.

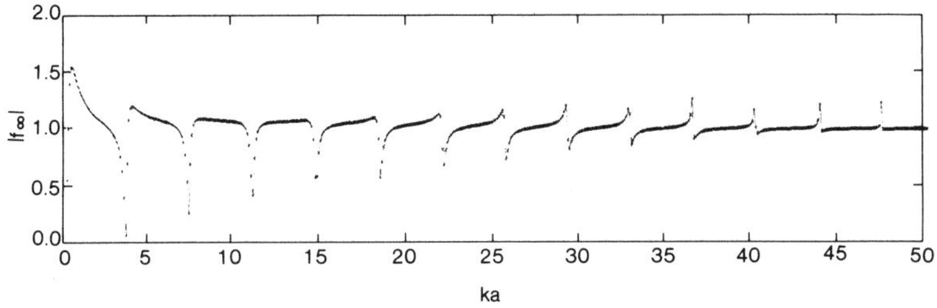

Fig. 13. The form function for a hollow air filled cylindrical shell with $b/a = 0.99$.

Over the soft background region the specular reflection from the shell is 180° out of phase with the incident wave and in the rigid region the specular reflection is in phase with the incident wave. The resonances are described in terms of circumferential waves, and all the resonances observed in Fig. 13 are related to a single circumferential wave. As in the case of the Rayleigh wave on a solid cylinder, the resonances occur when the circumference is an integral number of wavelengths. Thus, the speed of the circumferential wave involved in the form function given in Fig. 13 can be obtained directly from the figure as

$$c^*/c = \Delta ka = 3.7 \qquad (6)$$

Here c^* is the phase velocity of the circumferential wave, c is the speed of sound in water and Δka is the spacing between resonances. A computation of the backscattered echo reflected when a two cycle pulse (center frequency $k_o a = 10$) illuminates the sheel described in Fig. 13 is given in Fig. 14. The incident wave is at the right, and the specular reflection (out of phase with the incident) and the successive transits of the single circumferential are observed.

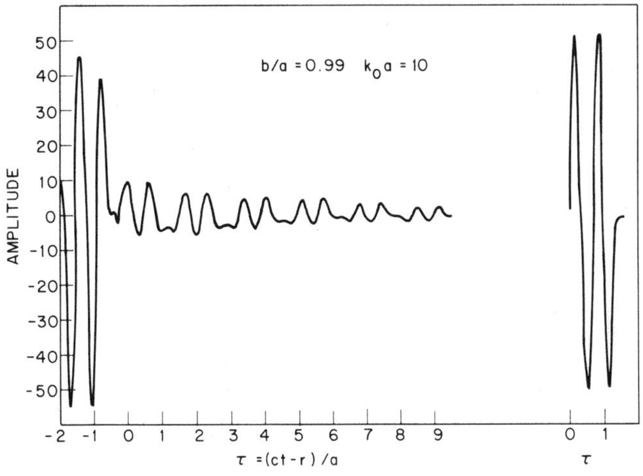

Fig. 14. The computed backscattered response of an aluminum shell with $b/a = 0.99$. The incident pulse is seen at the right $k_o a = 10$).

The resonance theory gives an excellent descriptive and predictive tool in dealing with the scattering by submerged solid and hollow objects with material properties as described. Its usefulness in problems where the scatterer is made of materials whose density or sound speed (shear or compressional) is lower than that of the surrounding fluid is open to question.

REFERENCES

1. L. Flax, L. R. Dragonette, and H. Überall, Theory of elastic resonance excitation by sound scattering, *J. Acoust. Soc. Am.* 63, 723 (1978).

2. C. M. Davis, L. R. Dragonette, and L. Flax, Acoustic scattering from silicone rubber cylinders and spheres, *J. Acoust. Soc. Am.* 63, 1694 (1978).

3. Anthony J. Rudgers, Acoustic pulses scattered by a rigid sphere immersed in a fluid, *J. Acoust. Soc. Am.* 45, 900 (1969).

4. R. Hickling, Analysis of echoes from a solid elastic sphere in water, *J. Acoust. Soc. Am.* 34, 1582 (1962).

5. J. J. Faran, Jr., Sound scattering by solid cylinders and spheres, *J. Acoust. Soc. Am.* 23, 405 (1951).

6. Lawrence Flax and Werner G. Neubauer, Acoustic reflection from layered elastic absorptive cylinders, *J. Acoust. Soc. Am.* 61, 307 (1977).

7. Luise S. Schuetz and Werner G. Neubauer, Acoustic reflection from cylinders—nonabsorbing and absorbing, *J. Acoust. Soc. Am.* 63, 513 (1977).

8. G. C. Gaunaurd, Sonar cross section of a coated hollow cylinder in water, *J. Acoust. Soc. Am.* 61, 360 (1977).

9. H. Uberall, L. R. Dragonette, and L. Flax, Relation between creeping waves and normal modes of vibration of a curved body, *J. Acoust. Soc. Am.* 61, 711 (1977).

10. R. D. Doolittle, H. Uberall, and P. Ugincius, Sound scattering by elastic cylinders, *J. Acoust. Soc. Am.* 43, 1 (1968).

11. G. V. Frisk and H. Uberall, *J. Acoust. Soc. Am.* 59, 46 (1976).

12. R. E. Bunney, R. R. Goodman, and S. W. Marshall, Rayleigh and Lamb waves on cylinders, *J. Acoust. Soc. Am.* 46, 1223 (1969).

13. Werner G. Neubauer and Louis R. Dragonette, Observation of waves radiated from circular cylinders caused by an incident pulse, *J. Acoust. Soc. Am.* 48, 1135 (1970).

14. Werner G. Neubauer, Ultrasonic reflection of a bounded beam at Rayleigh and critical angles for a plane liquid-solid interface, *J. Appl. Phys.* 44, 48 (1973).

15. H. L. Bertoni and T. Tamir, Unified theory of Rayleigh-angle phenomena for acoustic beams at liquid-solid interfaces, *Appl. Phys.* 2, 157 (1973).

16. E. K. Sittig and G. A. Coquin, Visualization of plane-strain vibration modes of a long cylinder capable of producing sound radiation, *J. Acoust. Soc. Am.* 48, 1150 (1970).

17. I. A. Viktorov, *Rayleigh and Lamb Waves*, Plenum, New York, 1967.

18. Louis R. Dragonette, The influence of the Rayleigh surface wave on the backscattering by submerged aluminum cylinders, *J. Acoust. Soc. Am.* 63, (June 1979).

19. K. Dransfeld and E. Salzmann, in *Physical Acoustics, Volume VII*, edited by Warren P. Mason and R. N. Thurston, Academic Press, New York, 1970.

20. H. D. Dardy, J. A. Bucaro, L. S. Schuetz, and L. R. Dragonette, Dynamic wide-bandwidth acoustic form-function determination, *J. Acoust. Soc. Am.* 62, 1373 (1977).

21. Louis R. Dragonette, Evaluation of the relative importance of circumferential or creeping waves in the acoustic scattering from rigid and elastic solid cylinders and from cylindrical shells, Naval Research Laboratory Report 8216 (1978).

RESONANCES IN ACOUSTIC AND ELASTIC-WAVE SCATTERING

G. C. Gaunaurd
Naval Surface Weapons Center, White Oak
Silver Spring, MD 20910

H. Überall
Naval Surface Weapons Center, White Oak
Silver Spring, MD 20910
and
Physics Department, Catholic University
Washington, DC 20064

ABSTRACT

We have studied resonance effects in the scattering of acoustic waves from submerged elastic objects, and of elastic waves from cavities and elastic inclusions. In this contribution, we present some results on the scattering of compressional and shear waves from fluid-filled spherical cavities, on acoustic scattering from air-filled elastic cylindrical shells coated with absorbing material, and on acoustic scattering from air bubbles in water.

INTRODUCTION

The resonance theory of acoustic and elastic-wave scattering, patterned after the earlier theories of nuclear resonance scattering (ref. 1,2), has recently been applied to classical scattering problems, such as the scattering of sound waves from elastic bodies (refs. 3-6) or from air bubbles in water (ref.7), and the scattering of elastic waves (of both compressional and shear type) from fluid-filled cavities in solids (refs.8-13), as well as the reflection of acoustic waves from layers (refs. 14,15). The success of resonance theory consists in its capability of explaining a usually very involved frequency dependence of the scattering cross section in terms of a geometrical, non-resonant background contribution in each partial wave (or normal mode), together with a series of resonance contributions in each partial wave which interfere with the background; in addition, all these normal modes interfere with each other in the total or differential cross section. Besides this practical achievement of the resonance theory in analyzing and labeling these resonances, it also furnishes a theoretical explanation of the physical origin of the resonances, It shows them to be caused by a phase-matching of repeatedly circumnavigating surface waves (for the case of finite or semi-infinite bodies), these waves being analogous to the Regge poles of Nuclear Physics (ref.2), and manifesting themselves by the recurrence of resonances at different frequencies in the successive partial waves. In the case of resonances in elastic-layer transmission and reflection, these are caused in the usual way (ref.16) by the coincidence of the trace velocity of the incident wave with the characteristic waves propagating in the free plate; but for the case of finite-body resonances, these can likewise be formulated in terms of a phase-velocity coincidence argument (ref.17).

In the present review, we would like to outline recent progress in our studies of resonance scattering, mainly on the topic of elastic-wave scattering from fluid-filled spherical cavities imbedded in non-absorptive or in viscous media, but also on some effects of sound scattering from air bubbles, or from air-filled elastic cylindrical shells coated with absorbing material.

THEORY

The theory of resonance scattering has, e.g. for the case of elastic-wave scattering from spherical fluid-filled cavities, been outlined elsewhere (ref. 8), with the following results. If the scalar potential of the incident plane compressional wave is expanded in the form

$$\Phi_{inc} = \sum_{n=0}^{\infty} i^n (2n+1) j_n(Kr) P_n(\cos\Theta) \quad (1)$$

then the scattered compressional wave is given by

$$\Phi_{sc} = \sum_{n=0}^{\infty} i^n A_n (2n+1) h_n^{(1)}(Kr) P_n(\cos\Theta), \quad (2)$$

K being the compressional wave number in the elastic medium. We may introduce the corresponding element of the S-matrix by

$$S_n = 1 + 2A_n \quad (3)$$

which in resonance theory has the form

$$S_n = \exp(2i\phi_n) \prod_{\ell=1}^{\infty} \frac{x - x_{n\ell} + \frac{1}{2} i\Gamma'_{n\ell}}{x - x_{n\ell} + \frac{1}{2} i\Gamma_{n\ell}} . \quad (4)$$

Here, $x = Ka$ where a is the radius of the cavity, $x_{n\ell}$ is the ℓth resonance frequency of the nth partial wave, and $\Gamma_{n\ell}$ are corresponding widths. The S-matrix element of scattering from an evacuated cavity has the form

$$S_n^{(o)} = \exp(2i\phi_n), \quad (5)$$

and appears here as a factor of S_n; accordingly, the scattering amplitude in Eq. (2) is given by

$$A_n = S_n^{(0)} \left[\sum_{\ell=1}^{\infty} \frac{\frac{1}{2} M_{n\ell}}{x - x_{n\ell} + \frac{1}{2} i\Gamma_{n\ell}} + ie^{-i\phi_n} \sin\phi_n \right] \quad (6)$$

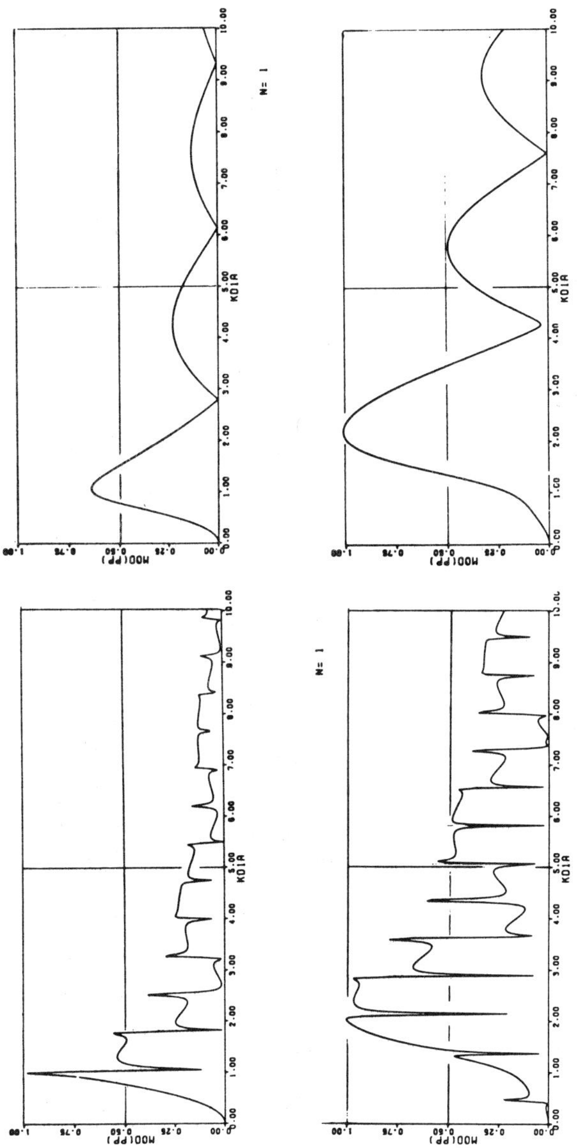

Fig. 1a. Modulus of partial wave P→P scattering amplitudes (left column), and of nonresonant background (right column) for partial waves n = 0 (top) and n = 1 (bottom), for compressional waves scattered from a water-filled spherical cavitiy imbedded in an aluminum matrix, plotted vs. Ka (ref.8).

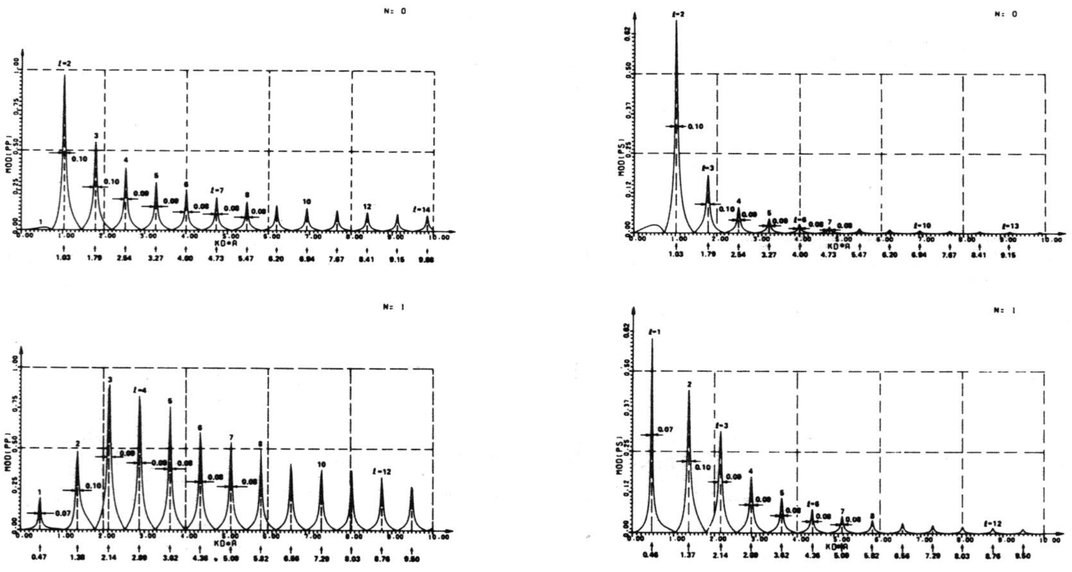

Fig. 1b. Modulus of resonance amplitudes for P-wave scattering from a water-filled spherical cavity in al aluminum matrix; P→ P resonances(left column) and P→ S resonances (right column) are plotted vs. Ka, for partial waves n = 0, top, and n = 1, bottom (ref. 9).

demonstrating that the nth partial wave amplitude consists of a geometrical non-resonant background contribution, given by the second term in parentheses in Eq. (6), with a series of superimposed (and interfering) resonances given by the sum in that same equation. Numerical illustrations of these amplitudes will be given in the following.

VISCOELASTIC SCATTERING FROM A FLUID-FILLED CAVITY

The scattering of elastic compressional (P) waves from a spherical fluid-filled cavity, generating scattered P waves, or scattered shear (S) waves (mode conversion), has been dealth with by us on the basis of resonance theory (refs.8-12), as well as the scattering of shear waves which also may mode-convert into P waves (ref.13).

Figure 1a presents the modulus of the partial-wave scattering amplitude A_n (left column) for P-wave scattering without mode conversion from a water-filled spherical cavity in an aluminum matrix, for n = 0 (top) and n = 1 (bottom), showing resonances interfering with the smooth background. The right column of that figure presents the background alone, which was obtained by considering an evacuated cavity. The curves are plotted vs. x = Ka. When that background is subtracted from the total partial wave amplitude before taking the modulus, the resonances alone are left over; they are shown in Fig. 1b without mode conversion (left column), and with mode conversion, i.e. P→S scattering (right column). This figure demonstrates that in these two cases, the resonance frequencies are the same (since they always originate from the eigenvibrations of the fluid filling the cavity), but that the resonance amplitudes have different heights since the excitation mechanism is different. It is seen that corresponding resonances appear in different partial waves (labeled correspondingly in Fig. 1b), but shift to higher frequencies as n increases. As we shall see later, these "Regge recurrences" (Ref.2) are the manifestations of families of surface waves that circumnavigate the cavity, and that cause the resonances at the integer values n where they constructively interfere by phase matching.

The effect of absorption present in the matrix medium is exhibited in Fig. 2 where the n = 1 partial wave amplitude modulus is plotted for an air-filled cavity in rubber. The resonances are here very narrow. The absorption (described by imaginary parts in the Lamé constants of the rubber) is taken to increase when proceeding from top to bottom of Fig. 2 It is seen that the resonances are little affected by increasing absorption, but that the diffraction minima of the background are successively filled in. This is to be expected since the background depends on the (absorptive) matrix material while the resonances depend on the (non-absorptive) cavity filler.

The elastic-wave backscattering cross section is proportional to the squared modulus of the sum in Eq. (2), so that here, all the modes interfere. We have plotted this quantity vs. Ka up to 10 at $\Theta = 180^o$ in the far field, summing over n from n = 0 up to 15 in Fig. 3. While earlier, the seemingly irregular structure of this plot would have been hard to interpret, our analysis aids in locating the resonances which remain quite distinct in this plot (as labeled). The plot again refers to a water-filled cavity in aluminum. We have demonstrated (ref.12) that an acquisition of the resonance spectrum from the analysis of the scattered echoes will allow an identification of the fluid filler in the cavity (by its density and sound speed), and further analysis will also identify geometrical shapes of cavities and complex compositions of cavity fillers.

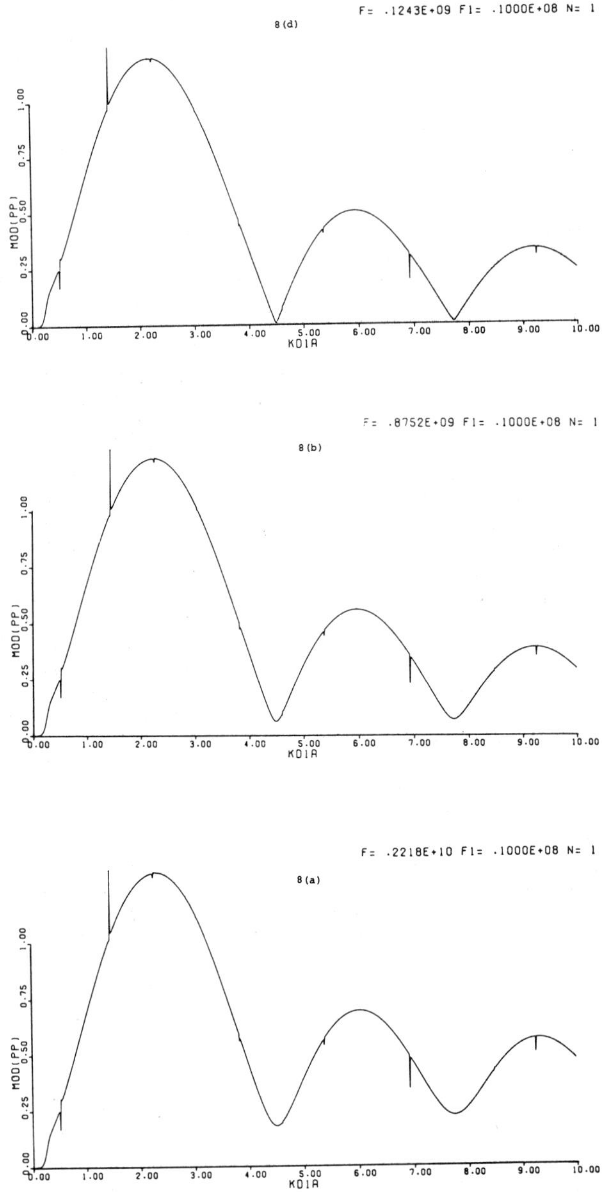

Fig. 2. Modulus of P→P scattering amplitude for a spherical air-filled cavity in rubber, which is taken to be non-absorptive (top), and with values of absorption increasing towards the bottom (ref. 8).

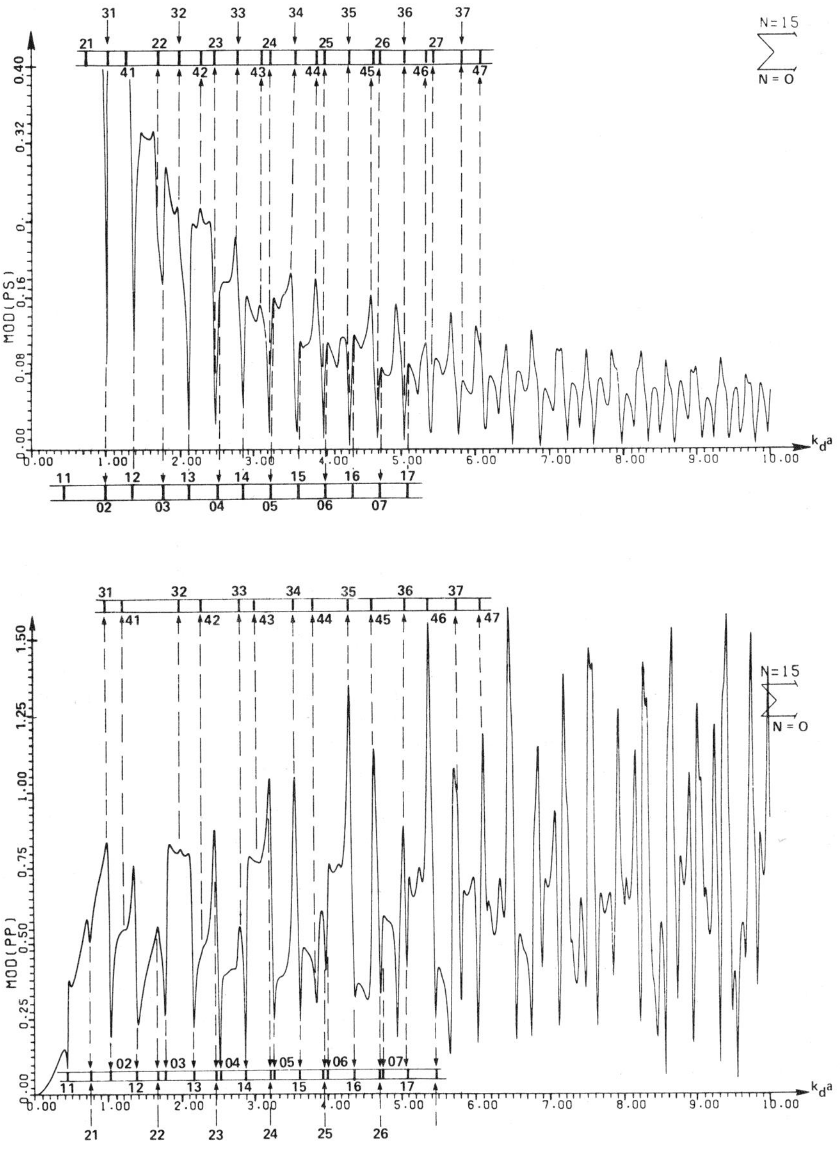

Fig. 3. Modulus of summed (n = 0....15) P⟶S (top) and P⟶P (bottom) scattering amplitude for a spherical water-filled cavity in an aluminum matrix. Resonances are identified and labeled (ref. 9).

It has been shown (refs. 5,9,18) that if in the scattering amplitude coefficient A_n (x) both n and x are considered continuous (complex) variables, then Φ_{sc} of Eq. (2) may be represented in terms of a series of circumferential waves (labeled by ℓ), propagating over the surface of the scattering object with phase velocity ratios

$$c_\ell (x)/c_d = x/\text{Re}\,(n_\ell + \tfrac{1}{2}) \tag{7}$$

where c_d is the compressional-wave speed in the matrix, and n_ℓ are the resonance positions in $|A_n (x)|$ if x is held fixed, and n is considered the variable. Equation (7) describes the dispersion curves of the cirfumferential waves, which are shown in Fig. 4 for ℓ = 1 through 12. The quantities n_ℓ will depend on the frequency x, and every time the latter coincides with a resonance frequency, Re n_ℓ will coincide with an integer n so that (see Eq. 7) a half-integer number n + ½ of circumferential wavelengths will fit over the circumference of the spherical cavity. This leads to perfect phase matching (and hence to a resonant reinforcement of the surface wave during its repeated circumnavigations, i.e. to the excitation of the resonances) since it can be shown that upon passage of the surface waves through the convergence points of the sphere, a quarter-wavelength phase jump takes place (refs. 9,18-20).

As just mentioned, the scattering amplitudes A_n (x) may be considered functions of two continuous variables x (frequency)and n (mode number). In Fig. 5, we present $|A_n(x) - A_n\,(\text{backg.})|$ in the form of three-dimensional plots for P→P (Fig. 5a, top), P→S (Fig. 5a, bottom), S→P (Fig. 5b, top) and S→S (Fig. 5b, bottom) scattering, plotted vs. x increasing towards the lower right, and n increasing towards the upper rear. The same resonances (with different amplitudes) occur in all these plots, manifesting themselves as parallel ridges (see especially Fig. 5b) that are inclined to either axis. (The jagged nature of the ridges is due to the unitarity property of the S-matrix which causes zeroes in addition to the resonance poles). If these surfaces are sliced at constant n, the frequency resonances of Figs. 1,2 are obtained. If the surfaces are sliced at constant frequency x, the same ridges give rise to resonances n_ℓ in the mode number variable which determine the propagation constants and dispersion curves (Fig. 4) of the circumferential waves (refs. 5, 9, 18).

For air-filled spherical cavities in lossless rubber, Fig. 6 presents (a) the total monopole (n = 0) scattering amplitude, (b) the isolated resonances, and (c) the background (ref. 11). Note that the background here contains the "giant monopole resonance" which is due to the presence of the shear modulus in the rubber matrix, as discussed earlier (ref. 21); it is therefore not a resonance in the sense we usually employ here, i.e. being due to oscillations of the cavity filler. However, a small portion of the monopole fundamental resonance (left most spike in Fig. 6b) indeed is due to the compressibility of the air-filler, as are also all the overtones.

The effect of an absorptive (imaginary) part in the shear modulus of the matrix manifests itself in a very interesting manner, as shown in Fig. 7. This figure plots the normalized resonance half-width Γ_0 of the resonance vs. the resonance frequency f_0 (refs. 10,11) resulting in a sloping straight line which fits the data (ref.21) very well, while the neglect of shear absorption would lead to the frequency-independent value c_s/c_d , i.e. the ratio of S and P waves speeds. A measurement of the monopole resonance widths will thus constitute a means of

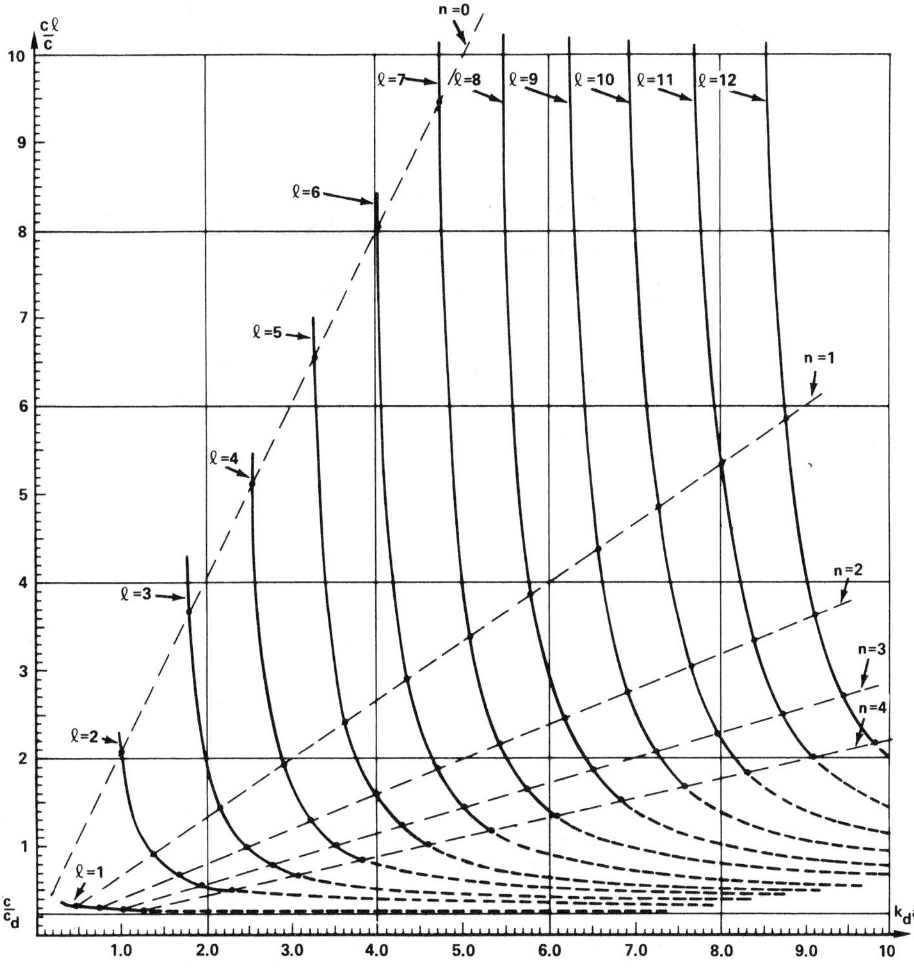

Fig. 4. Dispersion curves of surface waves on a water-filled spherical cavity in aluminum (ref. 9).

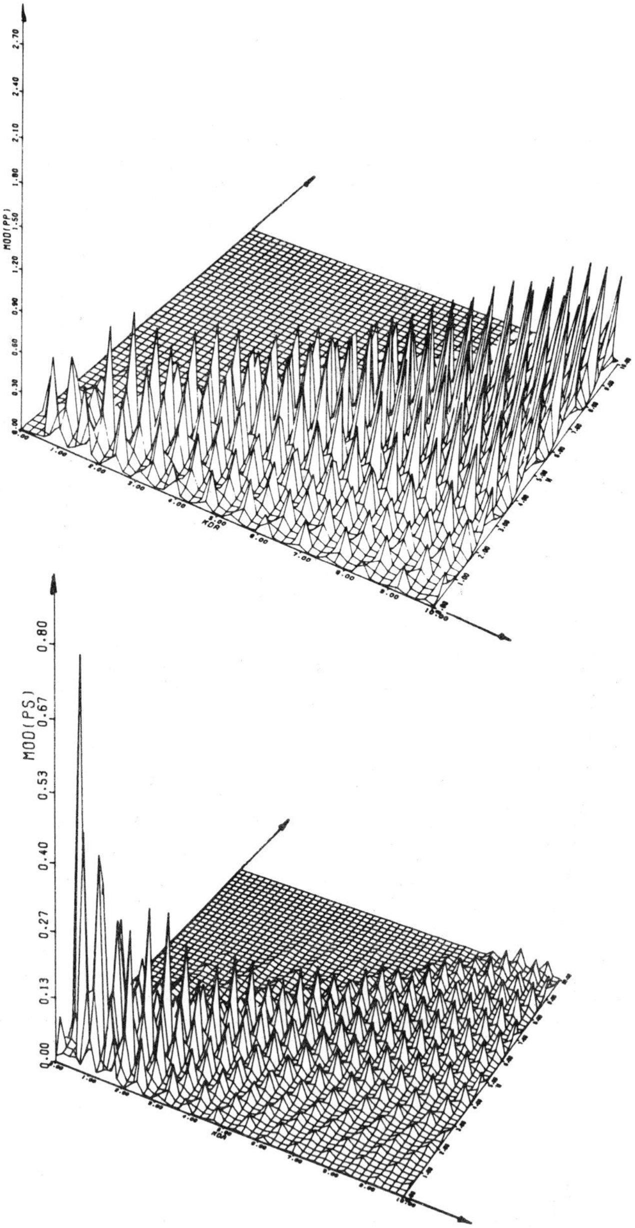

Fig. 5a. Three dimensional plot of $|A_n(x)|$ (after removal of background) vs. variables x and n, for P⟶P, top, and P⟶S scattering, bottom (ref. 9).

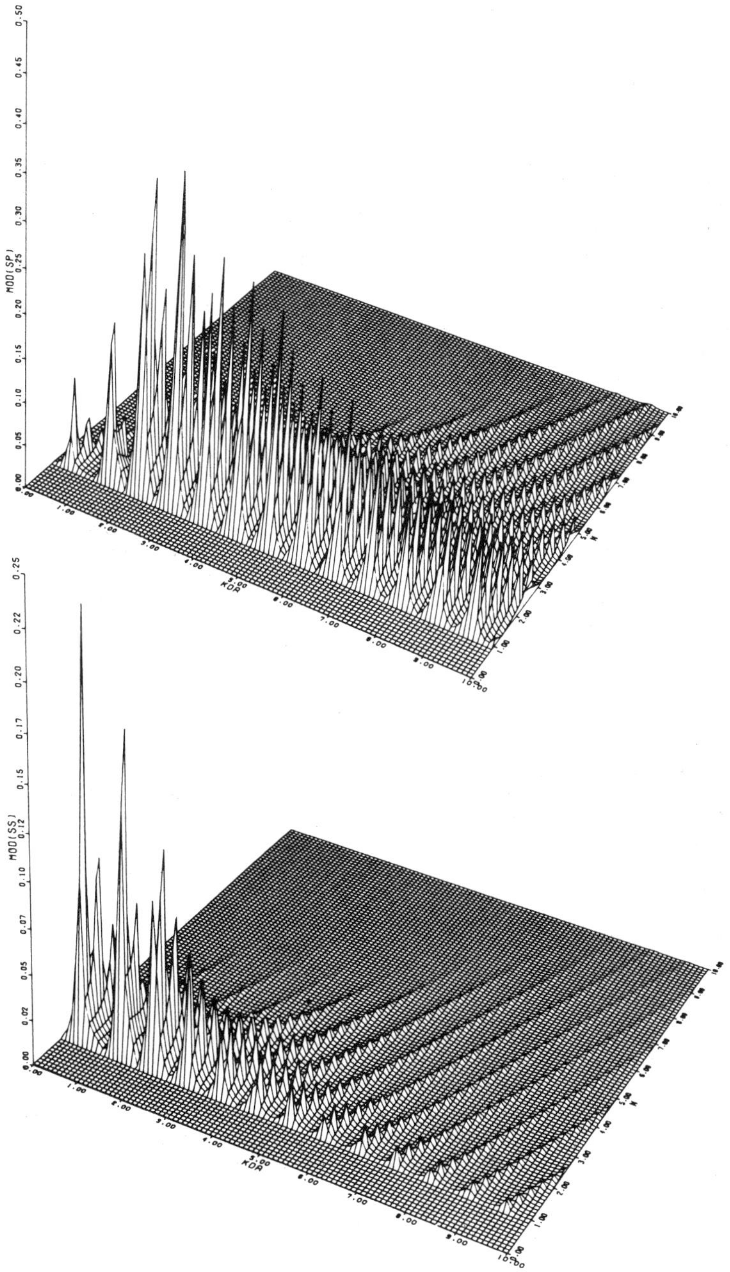

Fig. 5.b Three dimensional plot of $|A_n(x)|$ (after removal of background) vs. variables x and n, for S→P, top and S→S scattering, bottom (ref. 13).

determinig the shear absorption of the matrix material.

ACOUSTIC SCATTERING

In conclusion, we shall discuss some examples of acoustic scattering that were considered by us recently. A very simple example is that of sound scattering from air bubbles in water (ref.7). Figure 8 shows the target strenght (backscattering cross section) of a 1000-micron radius air bubble plotted vs. frequency f, exhibiting the giant monopole resonance (which, in contrast to the case of an air-filled cavity in rubber, stems entirely from the compressibility of the air content of the bubble. Hence it is a genuine resonance, in our sense, and not a part of the background). In addition, one sees the higher resonances which are both overtones of the giant monopole, or are higher-multipolarity (dipole, quadrupole...) resonances and their overtones (ref.11).

Our other example of acoustic scattering is that of sound normally incident upon an air-filled elastic cylindrical shell which has an exterior rubber coating. Fig.9 presents the partial-wave amplitudes $n = 0$ through 3 (from top) for an air-filled aluminum shell of outer radius 26.59cm, a thickness of 7.98cm, coated by layer of rubber 0.08cm thick. The left column shows the total amplitudes, and the right column the resonances. The latter are largely the same as those of a similar aluminum shell without coating, as given by us earlier (ref.4), but extra spikes at the values $ka = 33$, for $n = 1$ through 3 ($k = 2\pi f/c$, c = sound velocity in water, a = outer radius of coating) now appear in Fig. 8 which are due to the coating.

CONCLUSIONS

We have reviewed some recent progress in the scattering of elastic and acoustic waves from obstacles in solid and in fluid media. The analysis is based on the resonance theory of acoustic and viscoelastic wave scattering, with the latter being now extended to also account for incident shear waves. We recently also used the theory in the resonance scattering analysis of sound waves interacting with gas bubbles in water, and with air-filled submerged cylindrical shells covered with layers of viscoelastic materials. In all these situations, it is possible to isolate, display and analyze the scatterer's resonances in either the frequency or in the mode order domains. These resonances have a physical interpretation in terms of surface waves circumnavigating the scatterers, and the parameters that describe the isolated resonances also describe the complete behavior of the creeping waves including the angular decay rate, Regge trajectories, and dispersion curves. On the other hand, the creeping waves traveling around the obstacle provide a physical explanation of the very origin of the resonance, and the general interplay between creeping waves and resonances is a cause-effect relation. This work finds many applications in geophysics, medical ultrasonics and oil prospecting, as well as in acoustic target discrimination problems.

ACKNOWLEDGEMENTS

The support of the Naval Surface Weapons Center's Independent Research Board, and of the Office of Naval Research (Code 421) is acknowledged. We wish to thank Mr. John Barlow who programmed many of the computations shown in the graphs of this paper, including the three-dimensional diagrams.

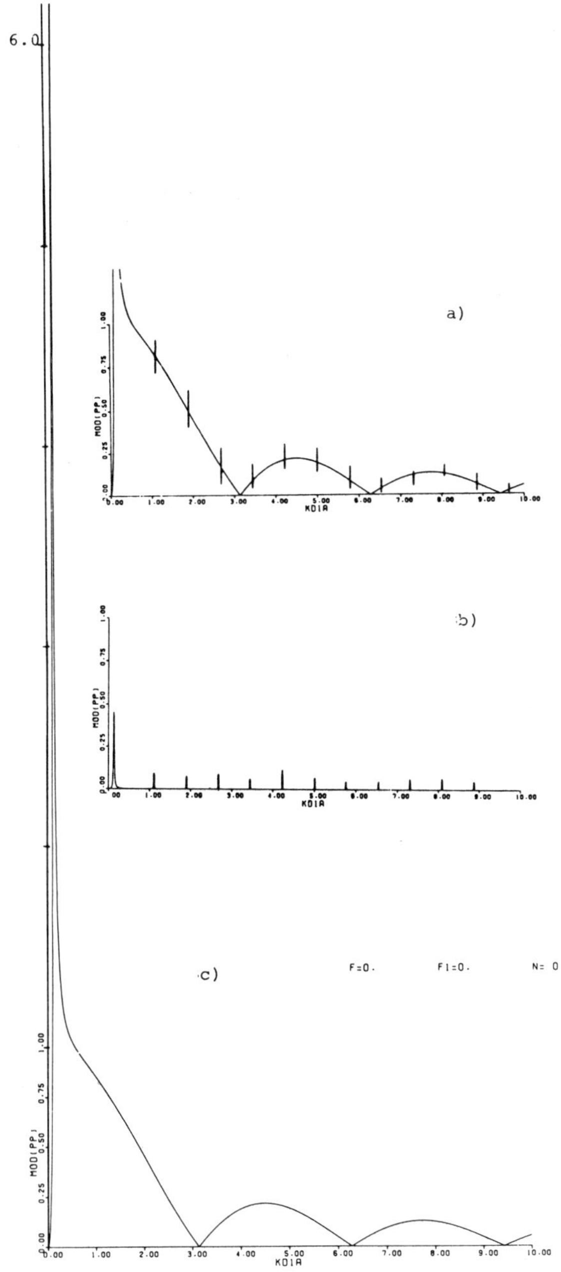

Fig. 6. Monopole scattering amplitude from an air-filled cavity in lossless rubber: (a) total amplitude, (b) isolated resonances of air filler, and (c) background containing the giant monopole resonance of the cavity wall oscillations (ref. 11).

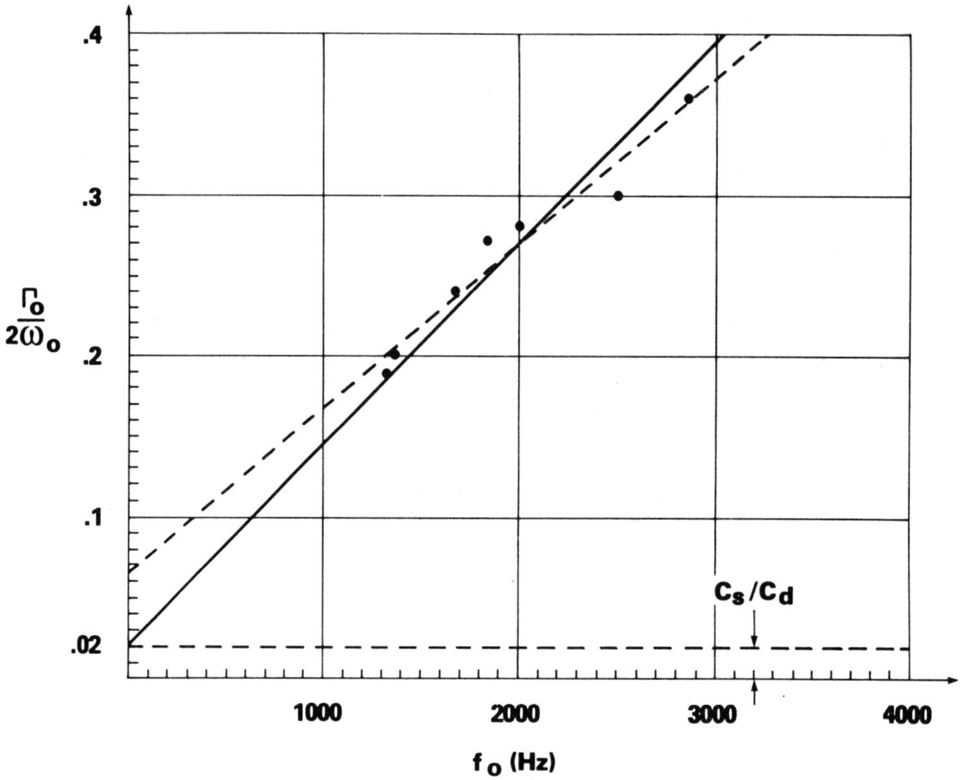

Fig. 7. Plot of normalized monopole resonance half-widths Γ_o versus resonance frequencies (ref.10) for the scattering of viscoelastic P-waves from air-filled spherical cavities in lossy FJ-95 rubber, matching the experimentally measured data points (ref. 21).

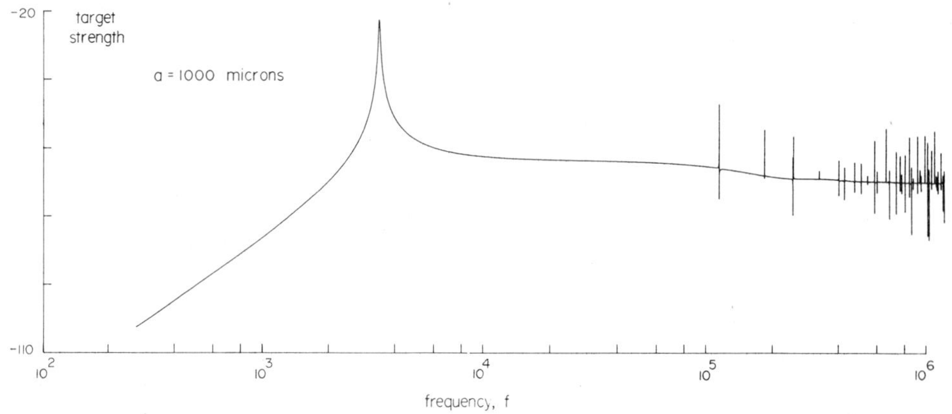

Fig. 8. Target strength of an air bubble (of 1000 micron radius) in water, plotted vs. the frequency (ref. 7).

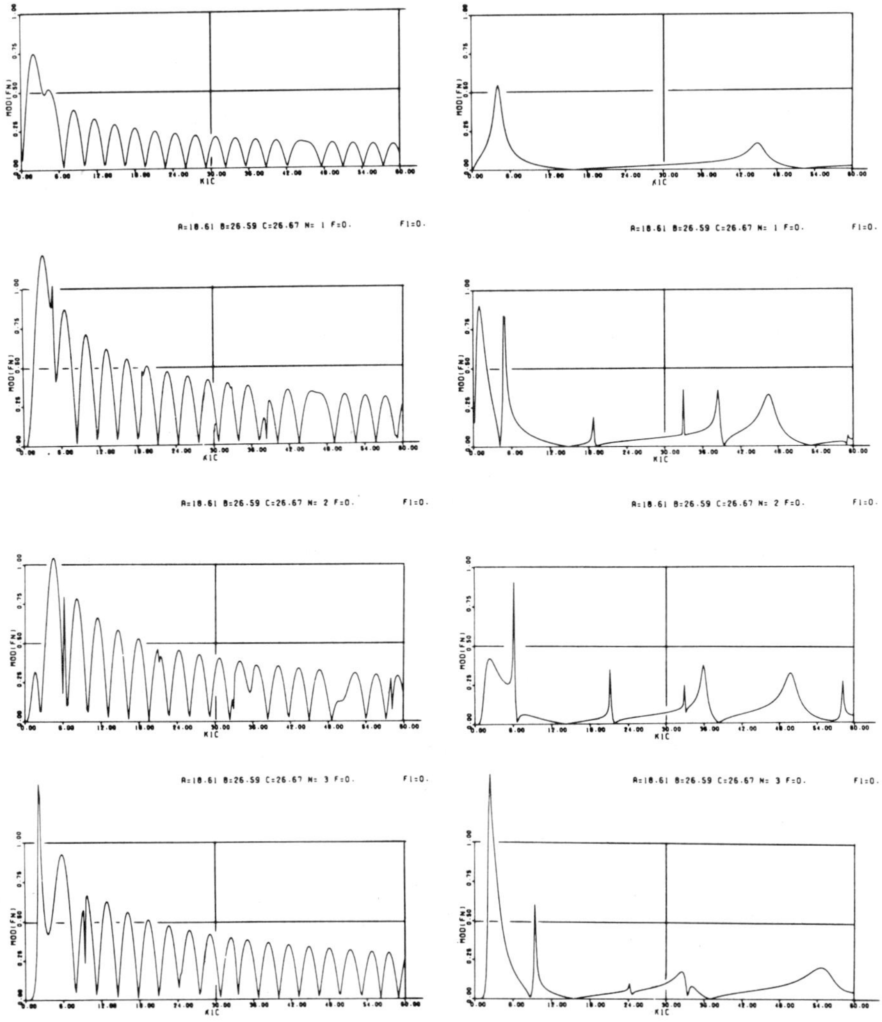

Fig. 9. Plot of scattering amplitudes n = 0 through 3 (from top), for an air-filled cylindrical aluminum shell coated with rubber. Left column: modulus of total amplitudes; right column: resonances of shell and coating, after subtraction of background.

REFERENCES

1. G. Breit and E. P. Wigner, Capture of slow neutrons, Phys. Rev. 49,519(1936).

2. K. W. McVoy, Regge poles and strong absorption in heavy-ion and alpha-nucleus scattering, Phys. Rev. C 3, 1104 (1971).

3. L. Flax, L. R. Dragonette, and H. Überall, Theory of elastic resonance excitation by sound scattering, J. Acoust. Soc. Am. 63, 723 (1978)

4. J. D. Murphy, E. D. Breitenbach, and H. Überall, Resonance Scattering of acoustic waves from cylindrical shells, J. Acoust. Soc. Am. 64, 677 (1978).

5. J. D. Murphy, J. George, A. Nagl and H. Überall, Isolation of the resonant component in acoustic scattering from fluid-loaded elastic spherical shells, J. Acoust. Soc. Am. 65, 368 (1979).

6. G. Gaunaurd and H. Überall, Suppression of resonant modes from the backscattered echoes of acoustically coated air-filled cylindrical shells in water, Proceed. of IV. Symposium on Ship-related acoustic R & D, Monterey, CA 1979 (in press).

7. K. A. Sage, J. George, and H. Überall, Multipole resonances in sound scattering from gas bubbles in a liquid, J. Acoust. Soc. Am. (in press).

8. G. C. Gaunaurd and H. Überall, Theory of resonant scattering from spherical cavities in elastic and viscoelastic media, J. Acoust. Soc. Am. 63, 1699(1978)

9. G. Gaunaurd and H. Überall, Numerical evaluations of modal resonances in the echoes of compressional waves scattered from fluid-filled spherical cavities in solids, J. Applied Phys. 52 (in press).

10. G. Gaunaurd, K. P. Scharnhorst, and H. Überall, A new method to determine shear absorption using the viscoelastodynamic resonance scattering formalism, J. Acoust. Soc. Am. 64, 1211 (1978).

11. G. Gaunaurd, K. P. Scharnhorst, and H. Überall, Giant monopole resonances in the scattering of waves from gas-filled spherical cavities and bubbles, J. Acoust. Soc. Am. 65, 573 (1979).

12. G. Gaunaurd and H. Überall, Deciphering the Scattering code contained in the resonance echoes from fluid-filled cavities in solids, SCIENCE (in press) 1979.

13. D. Brill, G. Gaunaurd, and H. Überall, Resonance theory of elastic shear-wave scattering from spherical fluid obstacles in solids, J. Acoust. Soc. Am. (submitted), 1979.

14. R. Fiorito and H. Überall, Resonance Theory of acoustic reflection and transmission through a fluid layer, J. Acoust. Soc. Am. 65, 9 (1979).

15. R. Fiorito, W. Madigosky, and H. Überall, Resonance theory of acoustic waves interacting with an elastic plate, J. Acoust. Soc. Am. (submitted), 1979.

16. L. Cremer, Über die Analogie zwischen Einfallswinkel und Frequenz problemen, Arch. elektr. Übertragung 1, 28 (1947).

17. H. Überall, L. R. Dragonette, and L. Flax, Relation between creeping waves and normal modes of vibration of a curved body, J. Acoust. Soc. Am. 61, 711 (1977)

18. H. Überall, Jacob George, Ameenah R. Farhan, G. Mezzorani, A. Nagl, and K. A. Sage, Dynamics of acoustic resonance scattering from spherical targets: application to gas bubbles in fluids, J. Acoust. Soc. Am. (in press), 1979.

19. M. C. Junger, Comments on "Phase changes and pulse deformation in acoustics", J. Acoust. Soc. Am. 45, 518 (1969).

20. A. R. Farhan, J. George, and H. Überall, Regge poles in the collective model of the nuclear giant multipole resonances, Nucl. Phys. A305, 189 (1978).

21. E. Meyer, K. Brendel, and K. Tamm, Pulsation oscillations of cavities in rubbers, J. Acoust. Soc. Am. 30, 1116 (1958).

THE SINGULARITY AND EIGENMODE EXPANSION METHODS WITH APPLICATION TO EQUIVALENT CIRCUITS AND RELATED TOPICS

Carl E. Baum and B. K. Singaraju
Air Force Weapons Laboratory, Kirtland AFB, NM 87117

ABSTRACT

Broadband transient electromagnetic problems have been of great interest in the last few years primarily because of nuclear electromagnetic pulse (EMP) interaction and radar identification problems. Singularity expansion (SEM) and Eigenmode Expansion Methods (EEM) are new trends in solving transient electromagnetic coupling problems.

In this paper we present some of the rudimentary concepts of SEM in the form of very simple formulae. Techniques for obtaining the SEM parameters, such as: the natural frequencies, natural modes and coupling coefficients, from the frequency-domain integral-equation formulation, the time-domain integral-equation formulation and from measured data are discussed.

An exciting and very new application of the SEM as applied to an electromagnetic scattering problem is the construction of single port equivalent circuits for the scatterer/antenna using the SEM parameters. Although both open-circuit and short-circuit boundary value problems can be formulated, we discuss the short-circuit boundary value problem in this paper. We define the short-circuit boundary value problem, show some representations for the driving point admittance in the form of pole (taken in conjugate pairs) admittances and modified pole admittances and define the source terms associated with these formulations. We discuss some realizability considerations for the pole and modified pole admittances.

In order to exhibit our technique, we consider a thin linear dipole antenna and derive simple analytical formulae for the short-circuit current and the driving-point admittance. By considering the pole and modified pole admittances, we show the realizability of the driving point pole admittances and thereby show a representation for the driving point pole admittance in terms of inductors, capacitors and resistors.

Equivalent circuits can also be constructed from SEM parameters that are numerically calculated or calculated from frequency domain or time domain measured data. We discuss some of these procedures.

INTRODUCTION

This paper deals with some of the applications of the Singularity Expansion Method (SEM) and the Eigenmode Expansion Method (EEM). Over the last few years, there has been an increasing interest in transient electromagnetic scattering problems. Principal reasons for this new interest has been due to the nuclear electromagnetic pulse (EMP) interaction problems and radar classification, identification problems. These are wideband phenomena in which a significant number of resonant frequencies of the scatterer could be excited by the incident electromagnetic pulse.

The singularity expansion method is a procedure for finding the response of a scatterer to an incident electromagnetic pulse in terms of the resonant frequencies or poles of the scatterer and their associated residues. Since the advent of SEM [1] in 1971, there has been a literal explosion in the numerical and analytical work dealing with the applications of SEM to transient electromagnetic problems. This broad spectrum of applications includes EMP interaction problems, simulator design and radar identification, to name a few. In this paper we will confine ourselves to a discussion of some of the applications of SEM and EEM with the principal emphasis on single port equivalent circuits.

We define a bilateral Laplace transformed quantity $\tilde{f}(s)$ (indicated by a ~ over the quantity) of f(t) as

$$L[f(t)] \equiv \tilde{f}(s) \equiv \int_{-\infty}^{+\infty} f(t)\, e^{-st}\, dt \tag{1}$$

$$f(t) \equiv L^{-1}[\tilde{f}(s)] = \frac{1}{2\pi j}\int_{\Omega_o - j\infty}^{\Omega_o + j\infty} \tilde{f}(s)\, e^{st}\, ds \tag{2}$$

where s is the complex frequency defined as

$$s \equiv \Omega + j\omega \tag{3}$$

The bilateral Laplace transform is assumed to converge in the strip $\Omega_1 \leq \Omega_0 \leq \Omega_2$ Analytic continuation of the function can be made for values outside the strip except at the singularities. For the inversion integral, the Bromwich contour is chosen such that $\Omega_1 \leq \Omega_0 \leq \Omega_2$.

In terms of the bilateral Laplace transformed quantities, Maxwell's equations in free space can be written as

$$\begin{aligned}
\nabla \times \tilde{\vec{E}}(\vec{r},s) &= -s\mu_o\, \tilde{\vec{H}}(\vec{r},s) - \tilde{\vec{J}}_m(\vec{r},s) \\
\nabla \times \tilde{\vec{H}}(\vec{r},s) &= s\varepsilon_o\, \tilde{\vec{E}}(\vec{r},s) + \tilde{\vec{J}}(\vec{r},s) \\
\nabla \cdot \tilde{\vec{B}}(\vec{r},s) &= \nabla \cdot (\mu_o \tilde{\vec{H}}(\vec{r},s)) = \tilde{\rho}_m(\vec{r},s) \\
\nabla \cdot \tilde{\vec{D}}(\vec{r},s) &= \nabla \cdot (\varepsilon_o \tilde{\vec{E}}(\vec{r},s)) = \tilde{\rho}(\vec{r},s)
\end{aligned} \tag{4}$$

We also define

$$\begin{aligned}
\gamma &\equiv \frac{s}{c} \equiv \text{complex propagation constant} \\
Z_o &\equiv \sqrt{\frac{\mu_o}{\varepsilon_o}} \equiv \text{wave impedance of free space} \\
c &\equiv \text{speed of light (in free space)} \\
\mu_o &\equiv \text{permeability of free space} \\
\varepsilon_o &\equiv \text{permittivity of free space}
\end{aligned} \tag{5}$$

We write the general integral equation for an antenna or scatterer as

$$\left\langle \tilde{\overleftrightarrow{\Gamma}}(\vec{r},\vec{r}\,';s) \;\dot{;}\; \tilde{\vec{J}}(\vec{r}\,',s) \right\rangle = \tilde{\vec{I}}(\vec{r},s) \tag{6}$$

where $\tilde{\vec{J}}(\vec{r}\,',s)$ is the current distribution (volume, surface or line as the case may be) on the scatterer, $\tilde{\overleftrightarrow{\Gamma}}(\vec{r},\vec{r}\,';s)$ is the dyadic kernel associated with the integral equation and is related to the Green's function $G_0(\vec{r},\vec{r}\,';s)$ and $\tilde{\vec{I}}(\vec{r},s)$ is the incident (for a scattering problem) or source (for an antenna problem) field. In (6) <,> indicates integration over common coordinates in a volume, surface or line integral as the case may be; the symbol above the comma indicates the type of multiplication specified.

BASIC THEORY OF SEM

In this section, we present some rudimentary concepts pertaining to SEM. For a more detailed discussion, the readers are referred to some of the original reports and review papers [1-6].

Basics of SEM

SEM can be viewed as a generalization of some circuit concepts to an electromagnetic scattering problem, where the response function is expanded in terms of the singularities of the scatterer along with the natural modes and coupling coefficients. The singularities of the scatterer include in general the poles (natural or resonant frequencies), essential singularities, branch cuts and singularity at infinity (entire function) [4]. Although both electric and magnetic field integral equations can be expanded in terms of the SEM parameters, we will confine our present discussion to the electric field or impedance integral equation given by

$$\left\langle \tilde{\overleftrightarrow{Z}}(\vec{r},\vec{r}\,';s) \;\dot{;}\; \tilde{\vec{J}}(\vec{r}\,',s) \right\rangle = \tilde{\vec{E}}^{(inc)}(\vec{r},s) \tag{7}$$

where the dyadic impedance kernel $\tilde{\overleftrightarrow{Z}}(\vec{r},\vec{r}\,';s)$ is defined via the free-space dyadic Green's function [7] as

$$\begin{aligned}\tilde{\overleftrightarrow{Z}}(\vec{r},\vec{r}\,';s) &= s\mu_0\, \tilde{\overleftrightarrow{G}}_0(\vec{r},\vec{r}\,';s) \\ &= s\mu_o(\overleftrightarrow{1} - \frac{1}{\gamma^2}\nabla\nabla)\, \tilde{G}_0(\vec{r},\vec{r}\,';s)\end{aligned} \tag{8}$$

with the scalar free-space Green's function $\tilde{G}_0(\vec{r},\vec{r}\,';s)$ defined as

$$\tilde{G}_0(\vec{r},\vec{r}\,';s) = \frac{e^{-\gamma|\vec{r}-\vec{r}\,'|}}{|\vec{r}-\vec{r}\,'|} \tag{9}$$

and $\overleftrightarrow{1}$ is the identity dyadic.

In SEM or EEM applications, a more convenient form of representation of (7) is for a delta function excitation $\tilde{\vec{I}}(\vec{r},s)$ with the impulse response being represented by

$$\left\langle \tilde{\vec{\vec{\Gamma}}}(\vec{r},\vec{r}\,';s) \; ; \; \tilde{\vec{U}}(\vec{r}\,',s) \right\rangle = \tilde{\vec{I}}(\vec{r},s) \tag{10}$$

It is clear that once the impulse response of the scatterer is known, the response for an arbitrary excitation can be easily determined.

In finding the "standard" solution [8] (non-SEM type), one either solves (7) or (10) with known analytic techniques or casts these equations in the moment method (MoM) formulation [9,10] for a numerical solution. In general, if the incident pulse shape or direction is changed, we will have to again solve the integral equation for each pulse shape or direction of incidence. In the singularity expansion method, we calculate the natural or free characteristics of the scatterer and express the solution as

$$\tilde{\vec{U}}(\vec{r},s) = \sum_{\alpha} \tilde{\eta}_{\alpha}(\vec{1}_1,s_{\alpha}) \; \vec{\nu}_{\alpha}(\vec{r})(s - s_{\alpha})^{-m_{\alpha}} + \tilde{\vec{W}}(\vec{r},s) \tag{11}$$

where s_α is the αth complex natural frequency (a pole of the response) of the scatterer. The natural frequencies are a characteristic of the shape of the scatterer and are not dependent upon the incident field. m_α is the multiplicity of the αth pole, $\vec{\nu}_\alpha(\vec{r})$ is the natural mode of the scatterer corresponding to the natural frequency s_α. The natural mode is a description of the current or charge distribution (or any other electromagnetic response quantity) associated with the scatterer in the natural state and is independent of the forcing function. $\tilde{\eta}_\alpha(\vec{1}_1,s)$ is the αth coupling coefficient corresponding to the natural frequency s_α and natural mode $\vec{\nu}_\alpha$; $\vec{1}_1$ is the direction of the incident wave. The coupling coefficient evaluated at $s = s_\alpha$ represents the strength of the natural oscillations in terms of the object and incident wave parameters; it is not a function of the spatial coordinates. Note that $\tilde{\eta}_\alpha$ is the only quantity that is a function of the incident field. $\tilde{\vec{W}}(\vec{r},s)$ represents singularities other than poles, such as essential singularities and branch points. Nature of $\tilde{\vec{W}}(\vec{r},s)$ for an arbitrary scatterer has not been established in any general way. For finite size perfectly conducting scatterers of sufficiently simple description, it has been shown that $\tilde{\vec{W}}(\vec{r},s)$ is an entire function, i.e., has no singularities in the finite s plane [2]. For the remainder of this paper, we will confine our discussions to poles of multiplicity one, i.e., $m_\alpha = 1$ in (11).

Natural modes and natural frequencies are the solutions of the homogeneous integral equation

$$\left\langle \tilde{\vec{\vec{\Gamma}}}(\vec{r},\vec{r}\,';s_\alpha) \; ; \; \vec{\nu}_\alpha(\vec{r}\,') \right\rangle = \vec{0}, \quad \left\langle \vec{\mu}_\alpha(\vec{r}\,') \; ; \; \tilde{\vec{\vec{\Gamma}}}(\vec{r},\vec{r}\,';s_\alpha) \right\rangle = \vec{0} \tag{12}$$

or in the MoM formulation

$$(\tilde{\Gamma}_{n,m}(s_\alpha)) \cdot (\nu_n)_\alpha = (0_n), \quad (\mu_n)_\alpha \cdot (\tilde{\Gamma}_{n,m}(s_\alpha)) = (0_n) \tag{13}$$

Natural frequencies are those s_α for which $\tilde{\vec{\vec{\Gamma}}}^{-1}(\vec{r},\vec{r}\,';s_\alpha)$ is not defined. This translates in MoM formulation to those s_α for which the determinant $\det(\tilde{\Gamma}_{n,m}(s_\alpha)) = 0$. Over the last few years, a number of techniques [6,11,12] including Newton-Raphson and contour integration have been developed in finding the locations of s_α. Once the natural frequencies are located, natural modes are found from (13) by way of standard matrix solution techniques.

Coupling coefficients are classified according to whether they are obtained by a series expansion of the response (class I) or from the inverse kernel (class II).

Class I coupling coefficients are given by

$$\tilde{\eta}_\alpha(\vec{I}_1,s) = \frac{\left\langle \vec{\mu}_\alpha(\vec{r})\ ;\ \tilde{\vec{I}}_{o_\alpha}(\vec{r}) \right\rangle}{\left\langle \vec{\mu}_\alpha(\vec{r})\ ;\ \left.\frac{\partial \tilde{\vec{\Gamma}}}{\partial s}(\vec{r},\vec{r}\,';s)\right|_{s=s_\alpha}\ ;\ \vec{\nu}_\alpha(\vec{r}\,') \right\rangle} e^{-(s-s_\alpha)t_o} \tag{14}$$

with t_o (the turn-on time) potentially a function of position. Class II coupling coefficients are given by

$$\tilde{\eta}_\alpha(\vec{I}_1,s) = \frac{\left\langle e^{-(s-s_\alpha)t_o}\, \vec{\mu}_\alpha(\vec{r})\ ;\ \tilde{\vec{I}}(\vec{r},s) \right\rangle}{\left\langle \vec{\mu}_\alpha(\vec{r})\ ;\ \left.\frac{\partial}{\partial s}\tilde{\vec{\Gamma}}(\vec{r},\vec{r}\,';s)\right|_{s=s_\alpha}\ ;\ \vec{\nu}_\alpha(\vec{r}\,') \right\rangle} \tag{15}$$

with t_o potentially dependent on both $\vec{r},\vec{r}\,'$. More details on these basic SEM quantities can be found in cited references.

A question which naturally arises at this juncture is, which coupling coefficients should one use? Basic problem with the coupling coefficients is the convergence of the time domain waveforms at early times. No satisfactory explanation exists at the present time that will conclusively give preference to one class of coupling coefficients over the other for a general problem.

Natural Frequencies, Natural Modes and Coupling Coefficients from Time Domain Solutions

We have discussed procedures for obtaining the natural frequencies, natural modes and coupling coefficients from the frequency domain integral equations. We now discuss a procedure for finding the natural frequencies, natural modes and coupling coefficients from the time domain integral equation solutions.

In this procedure, we calculate the time domain response of the scatterer to an incident pulse. The incident pulse rise time is chosen such that a large number of the resonant frequencies of the scatterer are excited. Note that not all resonant frequencies of the scatterer may be excited for a given angle of incidence of the excitation pulse. To compensate for this deficiency, the angle of incidence will have to be varied over a wide range of angles and the response calculated at a large number of points on the scatterer. This later requirement is important if we want to find the natural mode distribution on the body. Having found the time domain response at various points on the scatterer, the technique of Prony [13] is used to find the natural frequencies and coupling coefficients and thereby the natural mode. We will briefly discuss the classical Prony's method and discuss an example of a time domain thin wire integral equation problem.

Classical Prony's technique is simply an approach to fitting some time domain data f(t) with damped exponentials. Let g(t) be the time domain approximation resulting from the model

$$g(t) = \sum_{\alpha=1}^{N} R_n\, e^{s_\alpha t}\, u(t - t_o) \tag{16}$$

We enforce the condition that the model g(t) equal f(t) at 2N equidistant sample points (with t_0 usually taken as 0)

$$g(mT) = f(mT) \qquad m = 1,2,\ldots,2N \tag{17}$$

where T is the sampling interval. We also define a complex variable z_i such that

$$z_i = e^{s_i T} \qquad i = 1,2,\ldots,N \tag{18}$$

It can be shown that z_j, j=1,2,...,N, are the roots of a polynomial whose coefficients a_n are given by the solution to the set of linear equations

$$\sum_{n=1}^{N} f_{n+i} a_n = -f_{n+i} \qquad i=1,2,\ldots,N \tag{19}$$

Having found the polynomial coefficients, z_i are found resulting in

$$s_i = \frac{\ell n\, z_i}{T} \tag{20}$$

R_i are now calculated with the knowledge of z_i and the use of (17), (18), and (19).

The classical Prony's technique works reasonably well in determining the natural frequencies and coupling coefficients in the absence of noise in the data. A host of iterative, least squares and z transform procedures have been proposed, with limited success, to alleviate the problems with the classical Prony's technique in the presence of noise [14,15,16,17]. Readers are referred to extensive references on this subject [18].

The time domain response of the thin wire geometry shown in Fig. 1 is calculated using the time domain integro differential equation (Pocklington type). Using Prony's technique, the natural frequencies, natural modes and coupling coefficients are calculated. Shown in Table 1 are the comparisons of the natural frequency locations calculated by three different techniques [19]. Exhibited in Figs. 2, 3, and 4 are the real and imaginary parts of the natural modes and the coupling coefficients, respectively. Comparing these with the direct frequency domain technique calculations indicates that the accuracy is reasonably good.

TABLE 1 Comparison of the Natural Frequencies Calculated by Three Different Techniques [19]

N	Contour Integration Technique [11]	Newton-Raphson Technique [6]	Time-Domain Integral Equation Technique [19]
1	-.0828±j.9251	-.082±j.926	-.0819±j.919
2	-.1212±j1.9117	-.120±j1.897	-.121±j1.896
3	-.1491±j2.8835	-.147±j2.874	-.149±j2.879
4	-.1713±j3.8741	-.169±j3.854	-.1717±j3.866
5	-.1909±j4.8536	-.188±j4.835	-.1916±j4.852
6	-.2080±j5.8453	-.205±j5.817	-.2093±j5.839
7	-.2240±j6.8286	-.220±j6.800	-.2256±j6.821
8	-.2383±j2.8212	-.234±j7.783	-.2386±j7.794
9	-.2522±j8.8068	-.247±j8.767	-.2575±j8.775
10	-.2648±j9.8001	-.260±j9.752	-.2951±j9.735

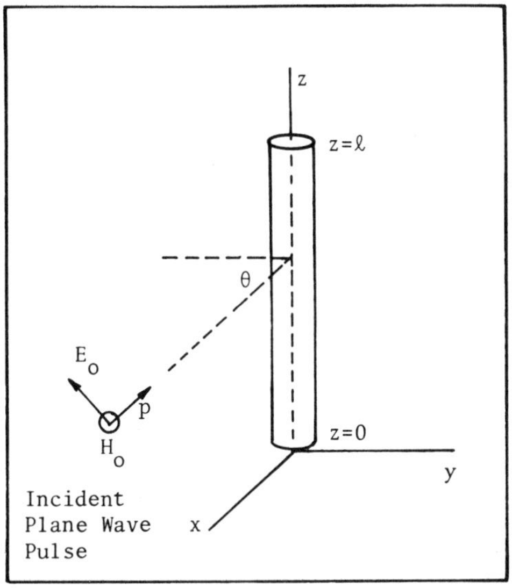

Fig. 1. Thin wire geometry

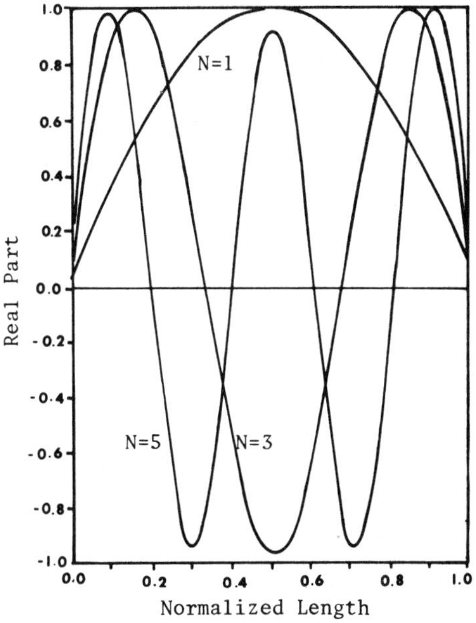

Fig. 2. Real part of the normalized natural mode [19]

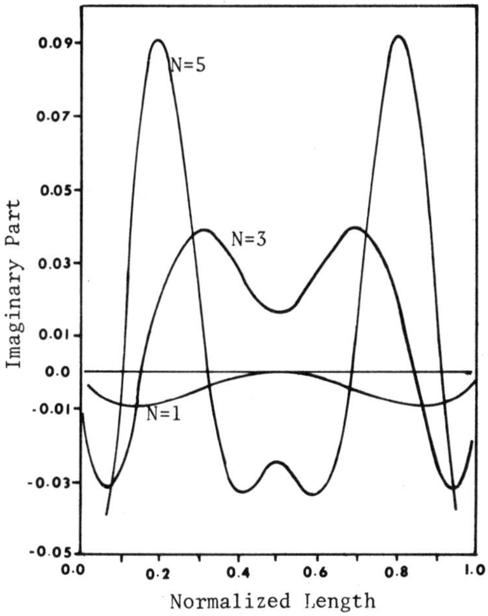

Fig. 3. Imaginary part of the normalized natural model [19]

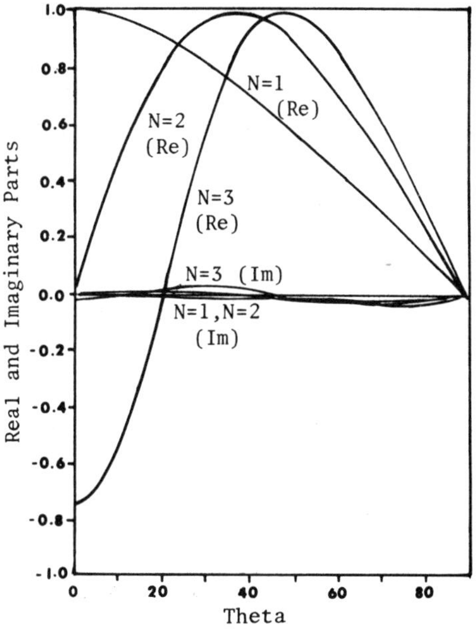

Fig. 4. Real and imaginary parts of normalized coupling coefficients [19]

However, this technique has some drawbacks compared to the direct frequency domain techniques. The principal problem has been that no second layer natural frequencies have yet been found by this technique. It has been observed that the higher order layer natural frequency contribution is important in assuring convergence at early times. Since these natural frequencies are highly damped (compared to the first layer), extraction of these natural frequencies and coupling coefficients from the time domain data is difficult if not impossible. Nevertheless, this is a useful technique in obtaining the SEM parameters from the time domain integral equation approach. This technique can also be used in calculating the natural frequencies, natural modes and coupling coefficients from measured transient data. For scatterers that are too complicated to accurately calculate the response of, this appears to be an important area of future research because the SEM parameters are physical observables. The reader should note that in some cases singularities other than poles are expected; the above considerations for poles can at least be conceptually generalized to other types of singularities in the form of distributed poles [4]. Specific numerical procedures need to be developed and tried.

APPLICATIONS OF SEM IN CONSTRUCTING EQUIVALENT CIRCUITS FOR ANTENNAS AND SCATTERERS

Introduction

The question which naturally arises is, why would one want to construct an equivalent circuit for an electromagnetic scattering/radiating problem? Under certain circumstances, such a representation could be helpful in providing

1. Physical insight
2. Computational convenience
3. Capability of using well-established circuit transformation techniques
4. Combination of electromagnetic analysis with physical circuit elements, transmission lines, etc., which are constructed as part of an antenna or scatterer
5. Use of existing computerized circuit analysis programs
6. Physical construction of equivalent circuits for use in pulsers for special types of EMP simulators
7. Radar target detection and camouflage techniques

The electromagnetic theoretist's urge to construct equivalent circuits from Maxwell's equations is nothing new. James Clerk Maxwell [20] himself alludes to lumped parameters such as inductors and capacitors in his now famous work. Transmission lines have been thought of in terms of distributed circuit parameters consisting of inductors, capacitors and resistors for small sections of transmission lines [21]. The first serious attempt to construct equivalent circuits from Maxwell's equations was made by Gabriel Kron [22]. His procedure was to expand Maxwell's equations in suitable orthogonal curvilinear coordinate systems and identify the capacitance and inductance for each differential element; however, there are several drawbacks in this procedure.

Using the natural frequencies (resonances), Schelkunoff [23] has attempted to construct lumped parameter circuits for the input impedance of antennas at each of the natural frequencies. This was based on function-theoretic techniques and the assumption that impedances are analytic functions in the complex frequency plane. He also conjectured on the existence of certain representations for driving point and transfer impedances. In a recent paper, Bucci and Franceschetti [24] have constructed driving point admittance equivalent circuits for a spheroidal antenna in a dispersive medium and have also calculated the transient response using the natural modes in spherical coordinates. In this technique, however, gyrators had to be used in the equivalent circuits.

The equivalent circuit construction procedure we will discuss is based on Baum's original work [25]. Although equivalent circuits can be constructed for both short circuit and open circuit boundary value problems, we will treat short circuit boundary value problems only. We will discuss two forms of equivalent circuits for the short circuit boundary value problem, discuss realizability conditions, and exhibit some single-port equivalent circuits for a thin linear dipole antenna. We will also discuss a procedure for constructing equivalent circuits from measured data. More details on these procedures can be found elsewhere [26].

Basic Theory

We will now define the short-circuit boundary value problem and discuss some realizability considerations. A more detailed discussion can be found in the references cited [25,26].

The short-circuit boundary value problem. In order to construct a single-port equivalent circuit for the antenna/scatterer, we define a single port on the body. At these terminals we construct a Norton equivalent circuit with $\tilde{Y}_a(s)$ being the antenna driving point admittance, and $\tilde{Y}_T(s)$ the terminal admittance including any terminal loading. The terminal admittance does not enter into our analysis and, hence, may even be nonlinear. The driving point admittance is defined as the ratio of the short-circuit current to the open-circuit voltage at the terminals of the antenna. With the gap geometry defined in Fig. 5, if the current on the body is $\vec{J}(\vec{r},s)$, gap width Δ, gap electric field $\tilde{\vec{E}}_g(\vec{r},s)$ is assumed uniform in the direction $\vec{1}_g$ (arbitrary), the driving point admittance is given by

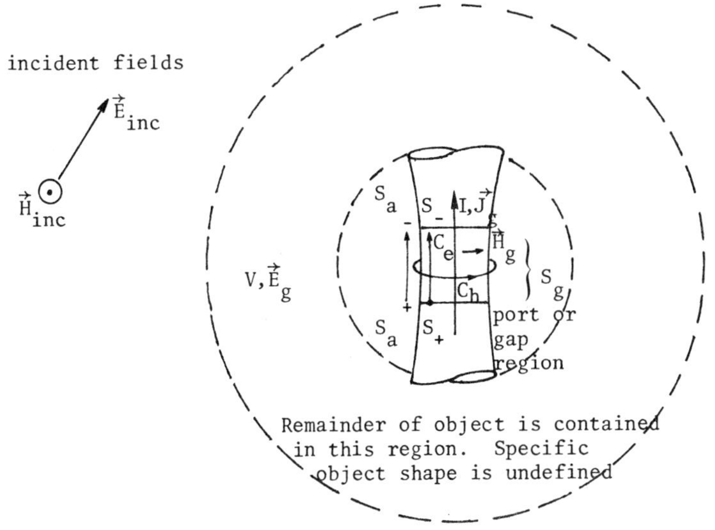

Fig. 5. Antenna or scatterer with single port.

$$\tilde{Y}_a(s) = - \frac{\left\langle \tilde{\vec{J}}(\vec{r},s) \; ; \; \vec{e}_g(\vec{r}) \right\rangle_g}{\tilde{V}(s)} \tag{21}$$

$$\vec{e}_g = - \frac{1}{\Delta} \vec{1}_g \quad , \quad \tilde{\vec{E}}_g(\vec{r},s) \equiv -\tilde{V}(s) \, \vec{1}_g \tag{22}$$

and $\tilde{V}(s)$ is the gap voltage. Equation (21) is valid under even more general conservative forms of $\vec{e}_g(\vec{r})$. In defining this admittance, we assume that the displacement current in the gap is small and that most of the current is due to charge motion. If the gap displacement current is not small, we will have to include the gap capacitance as part of the terminal loading of the antenna/scatterer. In terms of an incident electric field $\tilde{\vec{E}}_{inc}(\vec{r},s)$, we can write the short-circuit current density in the SEM form for a perfectly conducting body as

$$\tilde{\vec{J}}_{sc}(\vec{r},s) = \sum_{\alpha_{sc}} \frac{\tilde{\eta}_{\alpha_{sc}}(s_{\alpha_{sc}}) \, \vec{\nu}_{\alpha_{sc}}(\vec{r})}{(s - s_{\alpha_{sc}})} + \tilde{\vec{W}}_{sc}(\vec{r},s) \tag{23}$$

$$\tilde{\eta}_{\alpha_{sc}}(s_{\alpha_{sc}}) = \frac{\left\langle \vec{\mu}_{\alpha_{sc}}(\vec{r}) \; ; \; \tilde{\vec{E}}_s(\vec{r},s_{\alpha_{sc}}) \right\rangle_{a+g}}{\left\langle \vec{\mu}_{\alpha_{sc}}(\vec{r}) \; ; \; \frac{\partial}{\partial s} \tilde{\vec{\Gamma}}(\vec{r},\vec{r}\,';s) \Big|_{s=s_{\alpha_{sc}}} \; ; \; \vec{\nu}_{\alpha_{sc}}(\vec{r}\,') \right\rangle_{a+g}}$$

where the subscript sc indicates that the SEM parameters are for the short-circuit boundary-value problem. To keep the notation simple, we will drop this subscript with the implied understanding that the problem being treated is the short-circuit boundary-value problem. The subscript g indicates that the integration is over the gap region while the subscript a+g indicates integration over the antenna and the gap.

For both admittance and short-circuit current through the gap, the gap current is defined via

$$\tilde{I}(s) \equiv \left\langle \tilde{\vec{J}}(\vec{r},s) \; ; \; \vec{1}_g(\vec{r}) \right\rangle_g \tag{24}$$

Also (23) can be used for both short-circuit current and admittance provided that the source electric field $\tilde{\vec{E}}_s$ is taken to correspond to the incident electric field (over a+g) and to the gap electric field (zero outside g) for the scattering or antenna problems, respectively.

Using (21), (23), and (24) the antenna driving point admittance can be written as

$$\tilde{Y}_a(s) = \frac{1}{Z_o} \left\{ \sum_{\alpha} \frac{\tilde{a}_{\alpha}}{(s - s_{\alpha})} + \tilde{Y}_e(s) \right\} \tag{25}$$

$$a_\alpha = Z_o \frac{\left\langle \vec{\mu}_\alpha(\vec{r}) \; ; \; \vec{e}_g(\vec{r}) \right\rangle_g \left\langle \vec{\nu}_\alpha(\vec{r}) \; ; \; \vec{e}_g(\vec{r}) \right\rangle_g}{\left\langle \vec{\mu}_\alpha(\vec{r}) \; ; \; \frac{\partial}{\partial s} \tilde{\vec{\Gamma}}(\vec{r},\vec{r}';s)\Big|_{s=s_\alpha} \; ; \; \vec{\nu}_\alpha(\vec{r}') \right\rangle_{a+g}} \tag{26}$$

where Z_o is the free space impedance, $\tilde{Y}_e(s)$ is termed the admittance entire function and $\tilde{a}_\alpha$ the normalized admittance coupling coefficient. If we assume, without any loss of generality, that the gap electric field is in the $\vec{1}_g$ direction, $\tilde{a}_\alpha$ can be rewritten as

$$\tilde{a}_\alpha = \frac{Z_o}{\Delta^2} \frac{\left\langle \vec{\mu}_\alpha(\vec{r}) \; ; \; \vec{1}_g \right\rangle_g \left\langle \vec{\nu}_\alpha(\vec{r}) \; ; \; \vec{1}_g \right\rangle_g}{\left\langle \vec{\mu}_\alpha(\vec{r}) \; ; \; \frac{\partial}{\partial s} \tilde{\vec{\Gamma}}(\vec{r},\vec{r}';s)\Big|_{s=s_\alpha} \; ; \; \vec{\nu}_\alpha(\vec{r}') \right\rangle} \tag{27}$$

Corresponding to each of the poles, we can also define a driving point pole admittance $\tilde{Y}_{a_\alpha}(s)$

$$\tilde{Y}_{a_\alpha}(s) \equiv \frac{1}{Z_o} \frac{\tilde{a}_\alpha}{(s - s_\alpha)} \tag{28}$$

Source terms for each pole admittance can be defined for a given incident field $\tilde{\vec{E}}_{inc}(\vec{r},s)$

$$\tilde{\vec{E}}_{inc}(\vec{r},s) = E_o \sum_p \tilde{f}_p(s) \, \tilde{\vec{\delta}}_p(\vec{r},s) \tag{29}$$

where the subscript p indicates different polarization directions, $\tilde{f}_p(s)$ the generalized incident waveform, $\tilde{\vec{\delta}}_p(\vec{r},s)$ the generalized spatial dependence. The short-circuit current can then be written as

$$\tilde{I}(\vec{r}) = \sum_\alpha \left\{ \sum_p \tilde{f}_p(s) \, \tilde{V}_{\alpha,p} \right\} \tilde{Y}_{a_\alpha}(s) + \left\{ \sum_p \tilde{f}_p(s) \, \tilde{V}_{e,p}(s) \right\} \tilde{Y}_e(s) \tag{30}$$

where $V_{\alpha,p}$, the pole voltage source for a given polarization (or more general spatial dependence) is

$$\tilde{V}_{\alpha,p} = E_o \frac{\left\langle \vec{\mu}_\alpha(\vec{r}) \; ; \; \tilde{\vec{\delta}}_p(\vec{r},s_\alpha) \right\rangle_{a+g}}{\left\langle \vec{\mu}_\alpha(\vec{r}) \; ; \; \vec{e}_g(\vec{r}) \right\rangle_g} \tag{31}$$

We interpret $\sum_p \tilde{f}_p(s) \, \tilde{V}_{\alpha,p}$ as the pole voltage source, $\sum_p \tilde{f}_p(s) \, \tilde{V}_{e,p}(s)$ as the entire function voltage source. For the remainder of this paper, we ignore the entire function contribution and deal with the pole admittances and sources.

Some realizability considerations. We assume that the object into whose ports we are looking is passive. The driving point impedance $\tilde{Z}_{d.p.}(s)$ (and admittance $\tilde{Y}_{d.p.}(s) = \tilde{Z}_{d.p.}^{-1}(s)$) must be positive real (P.R.) for it to be realizable with resistors, inductors, capacitors and possibly ideal transformers. A rational driving point function is P.R. if [27,28]

1. The function is real for real values of s ($\omega = 0$)
2. The function is analytic in the right half plane
3. $j\omega$ axis poles ($\Omega = 0$) are simple and have positive real residues at the poles
4. Real part of the function is not negative on the $j\omega$ ($\Omega = 0$) axis and as $s \to \infty$ in the RHP (right half plane) $[\mathrm{Re}[s] \geq 0]$

Numerous test procedures have been developed to determine if a driving point function is P.R. We will not discuss the general realizability considerations, but confine our discussions to two simple cases.

For a perfectly conducting body that is passive, the driving point quantity, i.e., the sum of all pole terms and the entire function is P.R. This of course implies that an equivalent circuit representation exists, at least in a limiting or approximation sense, and one can synthesize a representation. Since the synthesis procedure is not unique, for a given scatterer/antenna a multiplicity of equivalent circuits can be synthesized for a given driving point representation. Our aim is to construct equivalent circuits for the driving point pole admittances (taken in conjugate pairs). Note that driving point pole admittance may not be P.R. even if the driving point admittance is P.R.

If we consider the pole admittances of the form (28), $\tilde{Y}_{a_\alpha}$ is P.R. iff

$$\frac{\tilde{a}_{\alpha_R}}{|\tilde{a}_{\alpha_I}|} \geq \frac{\omega_\alpha}{|\Omega_\alpha|} \tag{32}$$

where $\omega_\alpha > 0$, $\tilde{a}_{\alpha R}$ and $\tilde{a}_{\alpha I}$ are the real and imaginary parts, respectively, of $\tilde{a}_\alpha$. Geometrically, this implies that for the poles at s_α, the residue must fall in the shaded area shown in Fig. 6. However, if the pole is on the real axis (i.e., $\omega_\alpha = 0$), the residue $\tilde{a}_\alpha$ must be real and non-negative and hence must be on the real axis.

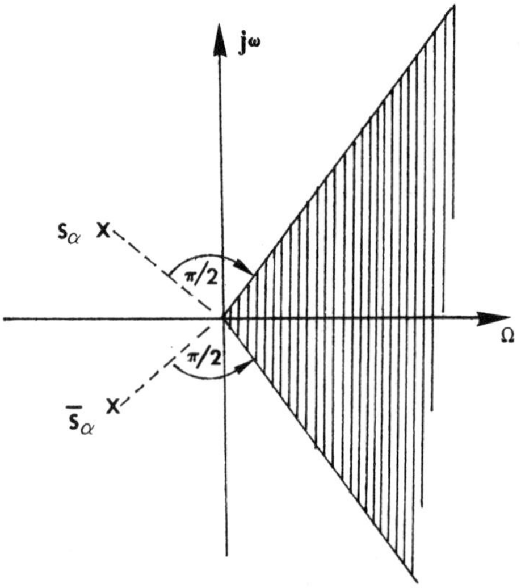

Fig. 6. Geometrical interpretation of the realizability condition (admittance residue $\tilde{a}_\alpha$ and its conjugate $\bar{\tilde{a}}_\alpha$ should lie in the shaded area).

Pole admittances shown in (28) can be modified as

$$\tilde{Y}'_{a_\alpha}(s) = \frac{1}{Z_0}\left\{\frac{\tilde{a}_\alpha}{s - s_\alpha} + \frac{\tilde{a}_\alpha}{s_\alpha}\right\} = \frac{1}{Z_0}\frac{\tilde{a}_\alpha s}{s - s_\alpha} \tag{33}$$

known as modified pole admittance. Note that in this form the entire function is modified by $-(1/Z_0)(\tilde{a}_\alpha/s_\alpha)$. If the driving point admittance approaches a constant value at high frequencies, as is the case with a thin dipole antenna, the entire function contribution is zero and hence no entire function is necessary. A modified pole admittance $\tilde{Y}_{a_\alpha}(s)$ is P.R. iff

$$2|\Omega_\alpha|(\tilde{a}_{\alpha_R}|\Omega_\alpha| - \tilde{a}_{\alpha_I}\omega_\alpha) \geq \tilde{a}_{\alpha_R}(\Omega_\alpha^2 + \omega_\alpha^2) \tag{34}$$

No simple geometrical interpretation exists for (34), however, we can rewrite (33) as

$$\tilde{Y}'_{a_\alpha}(s) = \frac{1}{Z_0}\left\{\tilde{a}_\alpha + \frac{\tilde{a}_\alpha s_\alpha}{s - s_\alpha}\right\} \tag{35}$$

which can be interpreted as a resistance in parallel with a pole admittance. The first term in (35) is realizable if its real part is non-negative, while the second term is realizable if it satisfies the ratio condition (32). Although a representation of the form (35) is useful in obtaining a simple criterion for the P.R. properties of (33), it is not very useful in the synthesis procedure for a thin linear dipole antenna. It should be noted that other choices of an additive constant in (33) are also in principle allowed in the definition of modified pole admittance.

In the case of the pole admittance formulation, as $s \to 0$ the driving point admittance approaches a constant value while in the case of the modified pole admittances it approaches zero. Depending upon the low-frequency behavior of the antenna/scatterer, one chooses the appropriate representation.

In general, if the low and high frequency behavior of the admittance is known, more efficient representations for the driving point admittance

$$\tilde{Y}_a(s) = \sum_\alpha \frac{a_\alpha}{s - s_\alpha} + \tilde{Y}_e(s) \tag{36}$$

can be written as

$$\tilde{Y}_a(s) = \sum_\alpha \frac{a_\alpha}{s - s_\alpha} + \sum_{n=0}^{\infty} a_n s^n \tag{37}$$

with all a_n real. If we consider a truncated set of poles (not all poles considered), $\lim_{s\to\infty} \tilde{Y}_a(s) \to$ constant, then a_1 through a_∞ can be set to zero leading to a resistor of $1/\tilde{a}_0$ ohms in parallel with the pole admittances. It has been shown that a non-zero value of a_0 is required for the truncated pole series to well approximate the admittance, at least at low and resonant frequencies [29].

We now have a systematic way of establishing that the given pole or modified pole admittances are P.R. and thereby proving that they are indeed realizable. This, however, does not say much about the form of the resulting network, nor does it establish a procedure for synthesizing one. Fortunately, network synthesis is a well established electrical engineering discipline and a variety of techniques including Butt-Duffin procedure, Brune synthesis, Miyata synthesis and Darlington synthesis among others are available [28,30,31,32].

Equivalent Circuits for a Thin Linear Dipole Antenna

In general, we can construct equivalent circuits for a scatterer/antenna from the SEM parameters calculated analytically or numerically. It is clear that these SEM parameters can be found from measured data also. We will now consider a thin linear dipole antenna and discuss some of these procedures. More details can be found elsewhere [26].

Analytical calculation of pole admittances. Pocklington's integral equation for a thin wire is given by

$$\left(\frac{d^2}{dz^2} - \frac{s^2}{c^2}\right)\int_0^L \frac{\tilde{I}(z',s)}{4\pi R}\, e^{-sR/c}\, dz' = -s\varepsilon_0\, \tilde{E}_z^{(inc)}\,(z,s) \tag{38}$$

$$R^2 = (z - z')^2 + a^2 \tag{39}$$

We denote the operator as $\tilde{\mathcal{X}}$. The geometry of the thin wire is shown in Fig. 1. Considering the first layer natural frequencies to be those given in Table 2,

TABLE 2 Pole Locations $[(s_\alpha L)/(\pi c)]$ in the Complex Frequency Plane for the Thin Wire of 2a/L = .01, Determined by the Contour Integration Method [11]. (The Numbers in Parenthesis in this Table are Reproduced for Comparison from Tesche [6].)

n	Layer 1	Layer 2	Layer 3
1	$-.0828\pm j.9251$ $(-.082\pm j.926)$	$-2.1687\pm j.349\times10^{-11}$ $(-2.174\pm j0.0)$	$-4.0993\pm j.394\times10^{-7}$
2	$-.1212\pm j1.9117$ $(-.120\pm j1.897)$	$-2.500\pm j1.3329$ $(-2.506\pm j1.347)$	$-4.5142\pm j1.4979$
3	$-.1491\pm j2.8835$ $(-.147\pm j2.874)$	$-2.7342\pm j2.4680$ $(-2.725\pm j2.477)$	$-4.8285\pm j2.7472$
4	$-.1713\pm j3.8741$ $(-.169\pm j3.854)$	$-2.9146\pm j3.5334$ $(-2.890\pm j3.544)$	$-5.0693\pm j3.8894$
5	$-.1909\pm j4.8536$ $(-.188\pm j4.835)$	$-3.0454\pm j4.5757$ $(-3.025\pm j4.581)$	$-5.2851\pm j5.0070$
6	$-.2080\pm j5.8453$ $(-.205\pm j5.817)$	$-3.1640\pm j5.6097$ $(-.3139\pm j5.603)$	$-5.4647\pm j6.0811$
7	$-.2240\pm j6.8286$ $(-.220\pm j6.800)$	$-3.2659\pm j6.6221$	$-5.6277\pm j7.1478$
8	$-.2383\pm j7.8212$ $(-.234\pm j7.783)$	$-3.3562\pm j7.6405$	$-5.772\pm j8.1901$
9	$-.2552\pm j8.8068$ $(-.247\pm j8.767)$	$-3.4376\pm j8.6466$	$-5.9045\pm j9.2351$
10	$-.2648\pm j9.8001$ $(-.260\pm j9.752)$	$-3.5108\pm j9.6555$	

and the corresponding natural modes to be approximated as

$$\tilde{I}_\alpha(z) = \sin(\frac{\alpha\pi}{L} z) = \mu_\alpha(z) \;, \quad \alpha = 1,2,3,\ldots \tag{40}$$

the denominator of the coupling coefficient (residue) can be written as

$$\left\langle \tilde{I}_\alpha, \frac{d\tilde{\mathcal{X}}}{ds}\Big|_{s=s_\alpha}, I_\alpha \right\rangle \approx \frac{\Omega_f s_\alpha}{2\pi c^2}\left[\left\langle \tilde{I}_\alpha(z), \tilde{I}_\alpha(z) \right\rangle + 0(\frac{1}{\Omega_f})\right] \tag{41}$$

where Ω_f is the fatness factor given by $\Omega_f = 2\ln(L/a)$. Neglecting terms of order Ω_f^{-1}, we can write (41) as

$$\left\langle \tilde{I}_\alpha, \frac{d\tilde{\mathcal{X}}}{ds}\Big|_{s=s_\alpha}, \tilde{I}_\alpha \right\rangle \approx \frac{\Omega_f s_\alpha L}{\pi c^2} \tag{42}$$

Hence the driving point admittance for a thin dipole antenna can be written as

$$\tilde{Y}(s') = \frac{1}{\Delta^2}\frac{4}{\Omega_f Z_o}\sum_\alpha \left\langle I_\alpha, 1 \right\rangle_g^2 \frac{1}{s' - s'_\alpha} \tag{43}$$

where the radian frequency s is normalized as $s' = sL/\pi c$, and

$$\left\langle I_\alpha, 1 \right\rangle_g^2 = \begin{cases} (\frac{2L}{\alpha\pi})^2 \sin^2(\frac{\alpha\pi}{2L}\Delta) & \alpha = \text{odd} \\ 0 & \alpha = \text{even} \end{cases} \tag{44}$$

Note that (43) is appropriate for any gap width, however, as a consequence of (44), the sum is over odd α only. If the gap is small, i.e., $\Delta \ll 2L/\alpha\pi$, (44) can be simplified as

$$\left\langle I_\alpha, 1 \right\rangle^2 \approx \begin{cases} \Delta^2 & \alpha = \text{odd} \\ 0 & \alpha = \text{even} \end{cases} \qquad \Delta \ll \frac{2L}{\alpha\pi} \tag{45}$$

resulting in

$$\tilde{Y}(s) = \sum_\alpha \frac{\tilde{a}_\alpha}{s' - s'_\alpha} \qquad \alpha = \text{odd}, \;\; \Delta \ll \frac{2L}{\alpha\pi} \tag{46}$$

with the pole admittance coupling coefficient $\tilde{a}_\alpha$ given by

$$\tilde{a}_\alpha = \frac{4}{\Omega_f Z_o} \tag{47}$$

We note that the admittance coupling coefficients $\tilde{a}_\alpha$ are real and positive and hence pole admittances are realizable pole pair by pole pair. It can be shown that in the small gap approximation (46) results from the short-circuit current on the scatterer if the source term is eliminated. This has been found to be true for a thin linear dipole antenna and a helix [33].

Shown in Figs. 7 through 12 are the real and imaginary parts of the conjugate pole pair admittances and the driving point admittance. Comparing these with the measurements and calculations by other procedures [34] indicates that they compare within 10%, thereby attesting to the accuracy of the approximations made. Using Brune's synthesis procedure, these conjugate pole pair admittances can be shown to have the realization shown in Fig. 13. Comparing these pole pair equivalent circuits with those obtained by Schelkunoff [23], the principal difference is in the inclusion of G_α in our procedure. Note that as $s' \to 0$, the real part of the admittance approaches a small but non-zero value. This of course is incorrect for a thin linear dipole antenna. This is corrected by the inclusion of a constant term as an entire function (not realizable in this case) or by modification of the pole admittances. An alternative method follows the open circuit boundary value problem and corresponding pole impedances [25].

Analytical calculation of modified pole admittances. Starting with the Pocklington's equation, as in the case of the pole admittances, the modified pole admittances for a thin linear dipole antenna are given by

$$\tilde{Y}(s') = \frac{1}{\Delta^2} \frac{4}{\Omega_f Z_o} \sum_\alpha \left\langle I_\alpha, 1 \right\rangle_g^2 \frac{s'}{s'_\alpha (s' - s'_\alpha)} \qquad \alpha = \text{odd} \tag{48}$$

where terms of order Ω_f^{-1} have been neglected. If the gap is small, (48) reduces to

$$\tilde{Y}(s') = \frac{4}{\Omega_f Z_o} \sum_\alpha \frac{s'}{s'_\alpha (s' - s'_\alpha)} \qquad \alpha = \text{odd} \tag{49}$$

Comparing this with (46), the modified pole admittances and pole admittances have the same residue at $s' = s'_\alpha$. In the limit $s' \to 0$, the modified pole admittances become zero thereby satisfying the condition for the driving point admittance for a thin linear dipole antenna. We note that (49) is obtained from (46) by way of

$$\tilde{Y}(s') = \frac{4}{\Omega_f Z_o} \sum_\alpha \left[\frac{1}{s' - s'_\alpha} + \frac{1}{s'_\alpha} \right] \tag{50}$$

where the entire function contribution is modified by $-s_\alpha^{-1}$. If we compare the pole admittances with the modified pole admittances, the modified pole admittances are obtained from the pole admittances by the addition of $(1/s'_\alpha)$. Note that the real part of s_α is negative, which implies that the real part of the modified pole admittance is obtained from the pole admittances by the addition of $-2|\Omega_\alpha|/(\Omega_\alpha^2 + \omega_\alpha^2)$. Although this yields the correct behavior for the low frequencies, this will make the real part of the modified pole admittances negative at high frequencies. This implies that the modified pole admittances are non-P.R. and, hence, non-realizable. An interesting axiom is that if the pole admittances (modified pole admittances) are P.R., corresponding modified pole admittances (pole admittances) will not be P.R. assuming the pole admittance coupling coefficients to be real.

Calculation of source terms. We have shown that driving point admittances in the pole admittance formulation are P.R. and hence realizable. Let us now consider the voltage source terms associated with these pole and modified pole admittances. If we conjugate pole or modified pole pair terms, the short-circuit current associated with the pole pair α can be written as

$$\tilde{I}_\alpha(s) = \tilde{I}_{\alpha_+}(s) + \tilde{I}_{\alpha_-}(s) \tag{51}$$

where α_+, α_- are the conjugate pole pairs and implicit in (51) is the location of the short-circuited gap. In terms of the incident field, (51) can be written as

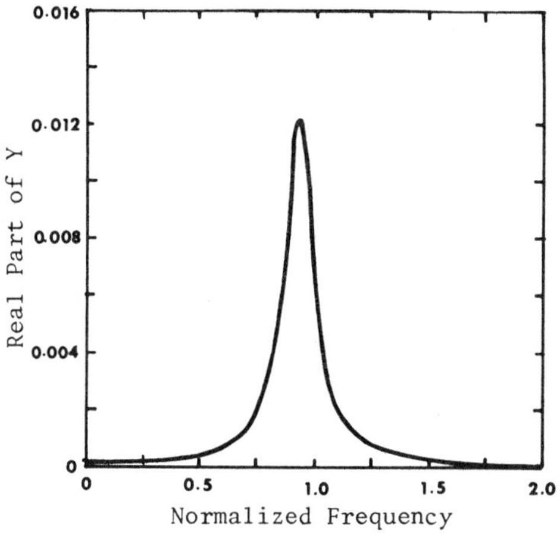

Fig. 7. Real part of the pole admittance for the first conjugate pole pair ($\Omega_f = 10.6$)

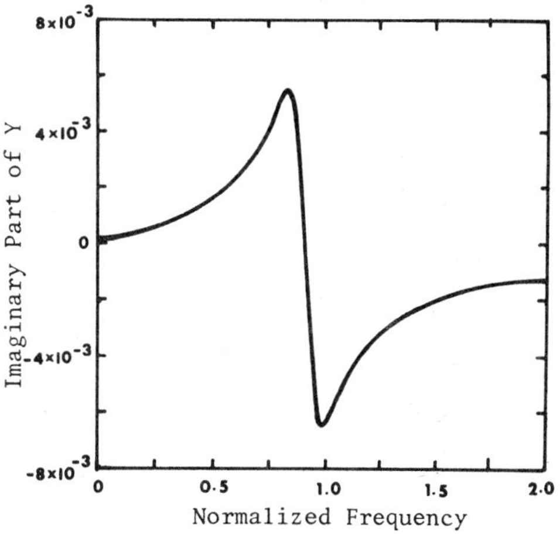

Fig. 8. Imaginary part of the pole admittance for the first conjugate pole pair ($\Omega_f = 10.6$)

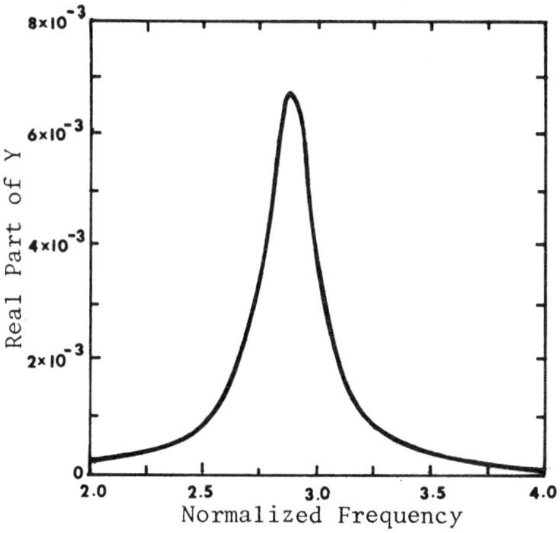

Fig. 9. Real part of the pole admittance for the third conjugate pole pair ($\Omega_f = 10.6$)

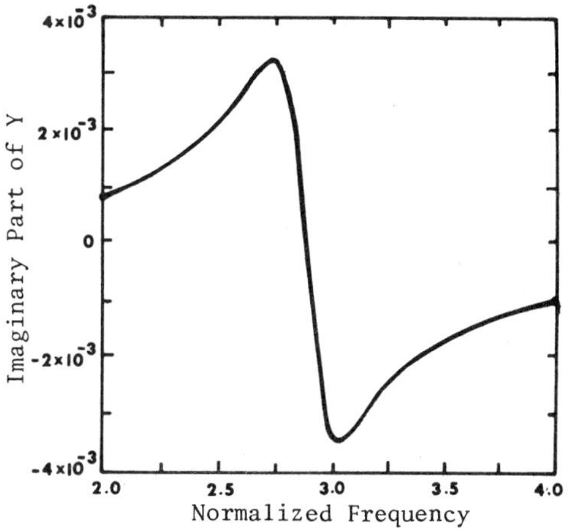

Fig. 10. Imaginary part of the pole admittance for the third conjugate pole pair ($\Omega_f = 10.6$)

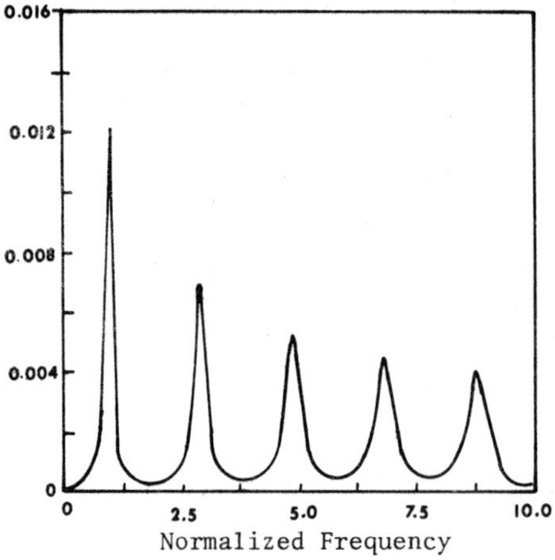

Fig. 11. Real part of the pole admittance for the first nine conjugate pole pairs (Ω_f= 10.6)

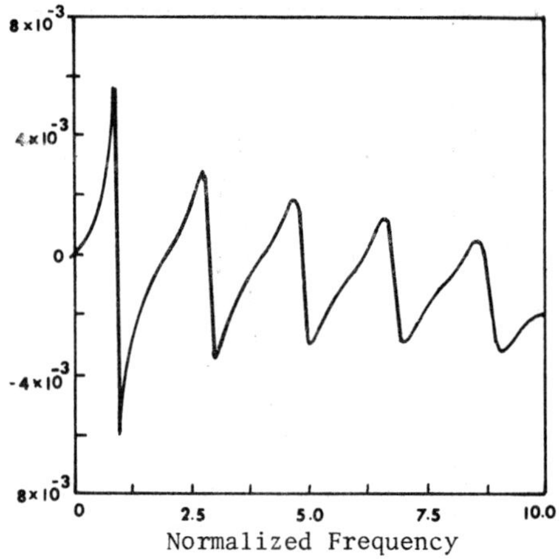

Fig. 12. Imaginary part of the pole admittance for the first nine conjugate pole pairs (Ω_f= 10.6)

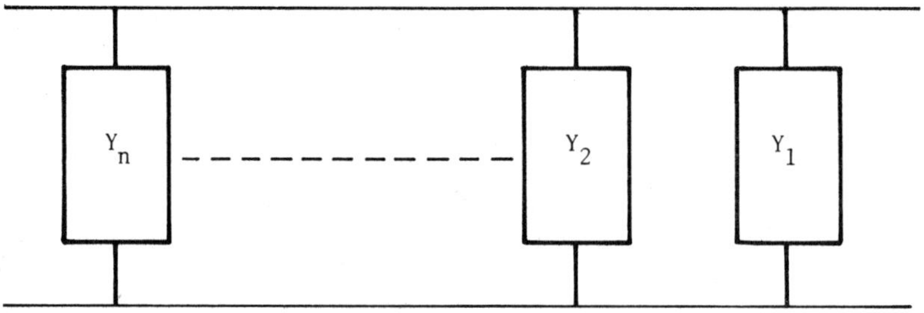

$$L_\alpha = \frac{\Omega_f Z_o}{8}$$

$$C_\alpha = \frac{8}{\Omega_f Z_o} \frac{1}{\omega_\alpha'^2}$$

$$R_\alpha = \Omega_f \frac{|\Omega_\alpha'| Z_o}{8}$$

$$G_\alpha = \frac{8}{\Omega_f Z_o} \frac{|\Omega_\alpha'|}{\omega_\alpha'^2}$$

Y_α = conjugate pole pair admittance (Normalized)

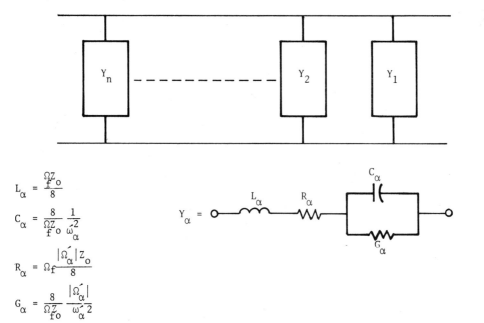

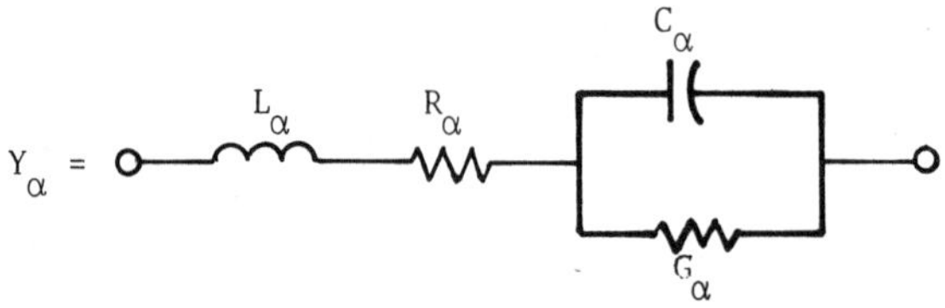

Fig. 13. Network realization for the pole admittance formulation

$$\tilde{I}_{\alpha}(s) = \sum_{p} \tilde{f}_{p}(s)\left[V_{\alpha_+,p}\tilde{Y}_{\alpha_+}(s) + V_{\alpha_-,p}\tilde{Y}_{\alpha_-}(s)\right] \tag{52}$$

The open circuit voltage associated with the conjugate pair is given by the ratio of the short-circuit current to the driving point pair admittance. Hence, the voltage source coefficient associated with the conjugate pair admittance $V_{\alpha,p}(s)$ is given by

$$\tilde{V}_{\alpha,p}(s) = \frac{V_{\alpha_+,p}\tilde{Y}_{\alpha_+}(s) + V_{\alpha_-,p}\tilde{Y}_{\alpha_-}(s)}{\tilde{Y}_{\alpha_+}(s) + \tilde{Y}_{\alpha_-}(s)}$$

$$= \mathrm{Re}\left[V_{\alpha_+,p}\right] + j\,\mathrm{Im}\left[V_{\alpha_+,p}\right]\left[\frac{\tilde{Y}_{\alpha_+}(s) - \tilde{Y}_{\alpha_-}(s)}{\tilde{Y}_{\alpha_+}(s) + \tilde{Y}_{\alpha_-}(s)}\right] \tag{53}$$

where the prefixes Re, Im denote real and imaginary parts, respectively, and the voltage source coefficients are given by

$$V_{\alpha,p} = \frac{E_o\left\langle \vec{\mu}_{\alpha}(\vec{r})\ ;\ \tilde{\vec{\delta}}_{p}(\vec{r},s_{\alpha})\right\rangle_{a+g}}{\left\langle \vec{\mu}_{\alpha}\ ,\ \vec{e}_{g}\right\rangle_{g}} \tag{54}$$

with the incident field representation given by (29) and antisymmetric [35] with respect to the gap as assumed. Note that in (53) if the natural modes are real, and if the $\tilde{\vec{\delta}}_p$ is a plane wave (or sum of the same) with zero phase (or t = 0 defined) at the center of the gap, the voltage source coefficients are real constants, and $V_{\alpha,p}(s)$ is not a function of s, which is a significant advantage for realizability. In our present case, (53) simplifies to a simple voltage source in series with our conjugate pair admittances.

Comments on other procedures for constructing equivalent circuits. We have shown that under certain analytical approximations some very simple formulations can be derived for the driving point admittances and have also shown that these driving point admittances are pole pair by pole pair realizable. One can solve the Pocklington or Hallén's equations for a thin wire and calculate the pole pair admittances more "accurately" using the MoM formulation. In our procedure the residues for the pole admittances were real as were the natural modes. It is possible that in the numerical procedure these quantities may be complex, thereby making the pole pair and modified pole pair admittances unrealizable. If this happens, it simply means that the real part of the pole or modified pole pair admittance is negative over a small portion of the frequencies. There are resistive padding techniques [30] that are available in the synthesis procedure that will make these admittances realizable.

In addition to numerical techniques, one can use measured data, such as the terminal voltage of an antenna to construct equivalent circuits. In this procedure, one calculates the natural frequencies, coupling coefficients, etc. from measured time-domain or frequency-domain data via Prony or some modified Prony technique. Once these SEM parameters are calculated, one can proceed in a direction akin to our procedures discussed earlier and construct equivalent circuits for the driving point admittance. Although not directly related to our technique of constructing the equivalent circuits for conjugate pair admittances Schaubert [36] showed a realization procedure for the driving point admittance from the measured data.

SUMMARY

This paper has tried to illustrate some of the general concepts in the construction of equivalent circuits at an antenna/scatterer port by application to the specific example of a thin linear dipole antenna with centered gap. The analytic approximations for the thin antenna have yielded simple expressions for residues, given the numerically determined natural frequencies. This example has shown at least approximate realizability of an equivalent circuit for such an antenna, thereby showing the applicability of the technique to at least some antennas/scatterers.

Various other types of equivalent circuits are possible, including ones based on eigenimpedances and on open-circuit forms of the scattering/antenna boundary value problem [25,4]. The present example was chosen for simplicity. Note that the general formulas do not necessarily require numerical or analytic solution for any given antenna/scatterer. The required SEM parameters may be also determined experimentally.

Since these techniques are relatively new there is much research to be done to understand all the theoretical possibilities and practical limitations. Both specific examples need to be worked, and more basic insight into the SEM and EEM (eigenmode expansion method) formulas are needed for progress. The reader may expect to see some important advances in the next few years.

REFERENCES

[1] C. E. Baum, "On the Singularity Expansion Method for the Solution of Electromagnetic Interaction Problems," Interaction Note 88, Dec. 1971.

[2] L. Marin, R. W. Latham, "Analytical Properties of the Field Scattered by a Perfectly Conducting Finite Body," Interaction Note 92, Jan.1972.

[3] Baum, C. E. (1976), The Singularity Expansion Method, in Transient Electromagnetic Fields, L. B. Felsen, ed., Springer-Verlag, New York.

[4] Baum, C. E. (1978), Toward an Engineering Theory of Electromagnetic Scattering: The Singularity and Eigenmode Expansion Methods, in Electromagnetic Scattering, P.L.E. Uslenghi, ed., Academic Press, New York.

[5] C. E. Baum, Emerging technology for transient and broadband analysis and synthesis of antennas and scatterers, Proc. IEEE, 64, 1598, 1976.

[6] F. M. Tesche, "On the Singularity Expansion Method as Applied to Electromagnetic Scattering from Thin Wires," Interaction Note 102, April 1972. (Also as F. M. Tesche, On the analysis of scattering and antenna problems using the singularity expansion technique, IEEE Transactions, AP-21, 52, 1973).

[7] Tai, C. T. (1971), Dyadic Green's Functions in Electromagnetic Theory, Intext Educational Publishers, Scranton, Pa.

[8] Jones, D. S. (1964), The Theory of Electromagnetism, Pergamon Press, New York.

[9] Harrington, R. F. (1973), Field Computation by Moment Method, McMillan Co., New York.

[10] Mittra, R. (ed) (1973), Computer Techniques for Electromagnetics, Pergamon Press, New York.

[11] B. K. Singaraju, D. V. Giri and C. E. Baum, "Further Developments in the Application of Contour Integration to the Evaluation of the Zeros of Analytic Functions and Relevant Computer Programs," Mathematics Note 42, March 1976.

[12] D. V. Giri and C. E. Baum, "Application of Cauchy's Residue Theorem in Evaluating the Poles and Zeros of a Complex Meromorphic Function and Apposite Computer Programs," Mathematics Note 55, May 1978.

[13] M. L. VanBlaricum and R. Mittra, "Techniques for Extracting the Complex Resonances of a System Directly from its Transient Response," Interaction Note 301, Dec. 1975.

[14] D. G. Dudley, "Fitting Noisy Data with a Complex Exponential Series," Mathematics Note 51, March 1977.

[15] H. J. Price, "An Improved Prony Algorithm for Exponential Analysis," Mathematics Note 59, Nov. 1978.

[16] G. Scrivner, B. K. Singaraju, Prony Analysis Utilizing Optimization Concepts and Iterative Improvements, IEEE Mini-symposium on Modal Analysis of Experimental Data, Albuquerque, March 1977.

[17] B. K. Singaraju, G. Scrivner, Z transforms, IEEE Mini-symposium on Modal Analysis of Experimental Data, Albuquerque, March 1977.

[18] Proceedings of the IEEE Mini-symposium on Modal Analysis of Experimental Data, Albuquerque, March 1977.

[19] J. T. Cordaro, K. S. Cho, Private communication.

[20] Maxwell, J. C. (1965), Treatise of Electricity and Magnetism, Vols. I, II, Dover, New York.

[21] King, R.W.P. (1956), Transmission Line Theory, McGraw-Hill, New York.

[22] G. Kron, Equivalent circuit of the field equations of Maxwell - I," Proc. of IRE, 32, 289, 1944.

[23] S. A. Schelkunoff, Representation of impedance functions in terms of resonant frequencies, Proc. of IRE, 32, 83, 1944.

[24] O. M. Bucci and G. Franceshetti, Input admittance and transient response of spheroidal antennas in dispersive media, IEEE Transactions, AP-22, 526, 1974.

[25] C. E. Baum, "Single Port Equivalent Circuits for Antennas and Scatterers," Interaction Note 295, March 1976.

[26] B. K. Singaraju, C. E. Baum, "A Procedure for Constructing Single Port Equivalent Circuits from the SEM Solution," Interaction Note, to be published.

[27] O. Brune, Synthesis of a finite two-terminal network whose driving-point impedance is a prescribed function of frequency, Jour. of Mat. Phys., X, 191, 1930.

[28] Guilleman, E. A. (1949), The Mathematics of Circuit Analysis, Wiley, New York.

[29] F. M. Tesche, "Application of the Singularity Expansion Method to the Analysis of Impedance Loaded Linear Antennas," Sensor and Simulation Note 177, May 1973. (Also as F. M. Tesche, Far-field response of a step-excited linear

antenna using SEM, IEEE Transactions, AP-23, 834, 1975.)

[30] Weinberg, L. (1962), Network Analysis and Synthesis, McGraw-Hill, New York.

[31] Chen, W. H. (1964), Linear Network Design and Synthesis, Mc-Graw Hill, New York.

[32] Van Valkenberg, M. E. (1960), Introduction to Modern Network Synthesis, John Wiley, New York.

[33] B. K. Singaraju, R. L. Gardner, "Transient Response of a Helical Antenna," Interaction Note 297, July 1976.

[34] King, R.W.P. (1956), Theory of Linear Antennas, Harvard University Press, Cambridge, Mass.

[35] C. E. Baum, "Interaction of Electromagnetic Fields with an Object which has an Electromagnetic Symmetry Plane," Interaction Note 63, March 1971.

[36] D. H. Schaubert, Application of Prony's method to time-domain reflectometer data and equivalent circuit synthesis, IEEE Transactions, AP-27, 180, 1979.

A CRITIQUE OF THE SINGULARITY EXPANSION AND EIGENMODE EXPANSION METHODS

C. L. Dolph, V. Komkov and R. A. Scott
University of Michigan, Ann Arbor, Michigan 48109

INTRODUCTION

The role of complex singularities in scattering problems is a rich and diverse one [see Dolph and Scott (1)]. Baum (2) has been of the prime movers in the United States of the subjects of the title. He has been helped in his efforts by the work of Marin (3) and Tesche (4). In the Soviet Union their counterparts are Voitovič, Kacenelenbaum and Sivov (5). There appears, however, to be an important difference between the two countries in that in the USSR mathematicians, beginning with A. G. Ramm, carefully examined the formalism of the approaches. Such efforts have culminated in a book with an interdisciplinary spirit entitled: The Generalized Methods of Eigen Vibrations in the Theory of Diffraction (5). The book, which also contains an extensive appendix by Agranovič entitled: Spectral Properties of Diffraction Problems, has been reviewed by the first author (with considerable help from the second author). This review should appear shortly in Mathematical Reviews, and an effort is underway to have the text itself translated into English. It is fortunate that much of the pertinent Soviet work has appeared in English translation in the Journal of Radio Engineering and Electron Physics. The readers attention should be called to the two papers of Ramm [(6), (7)] and the series of papers by Agranovič [(8), (9), (10)]. All of this work was stimulated by the pioneering paper of Kacenelenbaum (11) and its sequel by Voĭtovič, Kacenelenbaum and Sivov (12). There also is an important paper by Agranovič and Golubeva (13). [A word of caution must be injected here. In a relevant paper of Golubeva (14), the word "proposition" is translated as "conjecture," which one must admit does change the flavor. Since each "conjecture" is followed by a full proof, the reader must be alert to the translation problems.]

SINGULARITY EXPANSION METHOD(SEM)

The Singularity Expansion Method and its relationship to the mathematical theory of scattering has recently been analyzed by Dolph and Cho (15) and consequently no exhaustive detail on the technique is required here. Instead, a summary of the results of that work will be given.

The Singularity Expansion Method is a generalization of well-known techniques of linear circuit theory in which the singularities of a transfer function are used to determine the transient response by the Heaviside expansion theorem. In electromagnetic theory, the singularities are found by first applying a two-sided Laplace transform, parameter s, to the Maxwell equations and then constructing an integral equation for the scattered field. Complex singularities $\{s_n\}$ appear as poles of the inverse of this equation and are determined from the non-trivial solutions of the corresponding homogeneous integral equation.

The scattering operator is a unitary operator on a Hilbert space. Its Laplace transform, the scattering matrix, is analytic in the right-hand s-plane including the axis and is meromorphic in the left-half plane. It has poles in the left-half plane at those values of s for which there exist non-trivial outgoing solutions of the reduced Maxwell's equations satisfying the boundary conditions. These discrete values of s are complex eigenvalues of the exterior scattering problem. Shenk and Thoe (16) have established a one-to-one correspondence between the poles of the S-matrix and the poles of the integral equation.

Some of these poles occur on sheets of Riemann surfaces and serious difficulties arise in interpreting their meaning. This fact can be deduced from the study of the integral equation. Only the Green's Kernel can be continued analytically into such "forbidden domains". The resolvent can not.

While it is difficult to relate these concepts in general since most of the work in SEM involves a formalism in which neither spaces nor properties of the integral equation are given, this can be done for the papers of Marin (3). He uses a Hilbert space consisting of tangential currents on the surface of a convex body to discuss the magnetic field integral equation. He deduces a Fredholm integral equation of the second kind from this and uses Carleman's Fredholm theory for the determination of the natural modes. The solution of the corresponding homogeneous integral equation contains both exterior and interior resonances, the latter being purely imaginary.

To relate the scattering matrix to the integral equation one makes use of its representation as a compact Fredholm integral operator. The kernel of this operator is a transmission coefficient arising from the asymptotic form of the scattered field. Since its determination involves the solution of the scattering problem, rather than use the above representation, it is simpler to use the one-to-one correspondence between the kernel of this representation and the solution of the vector wave equation. It can be given in terms of a vector integral equation which is more general than the magnetic field integral equation. Fredholm theory is now used. Recall that if A is a compact operator (such as the above mentioned operator defining the integral equation) then the first part of the Fredholm alternative states that $A\phi = f$ has a unique solution ϕ if $A\phi = 0$ has only the trivial solution. Here the first part of the Fredholm alternative yields a unique solution to the general equation for the right-hand s-plane including the imaginary axis. The analytic Fredholm theorem for compact operators which is given, for example, in Reed and Simon (17), then implies the analytic and meromorphic properties of the scattering matrix discussed above.

For the magnetic field integral equation, one must use the second part of the Fredholm alternative, namely: If $A\phi = 0$ has non-trivial solutions then:

(i) $A\phi = 0$, $A^*\psi = 0$, where A^* is the adjoint of A, have the same finite number of solutions.

(ii) For $A\phi = f$ to have a solution ϕ_f, f must be orthogonal to all of the solutions of $A^*\psi = 0$.

Since the solution to $A\phi = 0$ can be added to ϕ_f, then clearly the process is not unique and no one-to-one map between the scattering matrix and the integral equation exists for the same half-plane. The non-trivial solutions of the corresponding homogeneous integral equation occurring for complex $\{s_n\}$ do, however, correspond to poles of the scattering matrix but the interior resonances corresponding to purely imaginary $\{s_n\}$ are not poles of the scattering matrix nor do they appear in the solution for the scattered field. They are consequently method dependent. The exterior resonance corresponding to complex $\{s_n\}$ are method independent and intrinsic to the scattering body.

Since a one-to-one correspondence fails to exist, it is conceivable that there are complex poles of the scattering matrix which are not given by the solutions of the homogeneous magnetic integral equation corresponding to complex $\{s_n\}$. No examples, however, in which this occurs are known and thus the problem remains open.

Since it is generally believed that the scattering matrix contains all observable information about the scattering process, the above observations substantiate the claim that the complex singularities are intrinsic to the scatterer whereas those which are purely imaginary are not.

Further difficulty arises in that most problems treated by the SEM formalism invole numerical techniques applied to matrix equations obtained from finite approximations to the integral equations. Hence, Hilbert space solutions are not really appropriate. Fortunately, the results can be given in terms of solutions in the Banach space of continuous functions. In this space the Fredholm theorem as given by Steinberg (18) can be used to discuss the magnetic field integral equation and to establish the one-to-one map between the scattering matrix and the associated set of vector integral equations. This map between the scattering matrix and the integral equation has been carried out for convex, perfectly conducting bodies only [see Lax and Phillips (19), Brakkage and Werner (20), and Dolph and Cho (15)]. Certain other aspects of the SEM formalism have also been investigated by Dolph and Cho (15) and serious doubts were raised regarding those parts of it where integral equations of the first kind are used. Also, there appears to be a confusion in the SEM literature between the eigenvalues of the integral equations and those of the vector wave equations and a belief persists that what is true for the finite matrix equations holds in the limit.

Though said elsewhere, it seems worth repeating that areas in which the formalism needs further investigation include:

I. The construction of variational principles useful for providing estimates for the location of the poles in the SEM and possibly of use in establishing their existence for off-axis poles not covered by the known Lax-Phillips results (21) for the scalar case and their generalization by Beale (22) for the electromagnetic case.

II. An investigation of integral equations of the first kind as used is SEM and a possible justification of their use through regularization methods, such as those of Tihonov (23).

III. The extension to the electromagnetic case of theorems sufficient to guarantee that all poles are simple. This should include generalizations of the theorems of Steinberg (18) and Howland (24).

IV. An investigation of the entire functions which arise, through the Mittag-Leffler theorem, in the formalism. This should include the determination of conditions under which they do not occur and their explicit form when they do occur.

V. The creation of a systematic theory of the asymptotic contribution of branch lines. As can be seen from the discussion in Dolph-Scott (1), such a theory could have implications in many areas.

At this point a few remarks on the T-matrix would seem to be in order. The T-matrix formalism for scattering has proved to be an efficient way of obtaining numerical results in a number of complicated problems [see Bolomey and Wirgin (25)]. It was used by Waterman (26) in 1969 and subsequent publications include papers by Peterson and Ström[(27), (28), (29)], Ström [(30), (31), (32)] and Varatharajulu and Pao (33). However there is a relationship between the T and S matrices, namely, T = S - 1, as stated in Wu and Ohmura (34), and physicists in the past have used S in preference to T.

EIGENMODE EXPANSION METHOD

Baum (2) has advanced the idea of synthesizing transient responses by means of the eigenmodes of integral equations of the first kind describing the system response. Considerable caution must be exercised in this approach however, since as Ramm [(6), (7)] has pointed out, incorrect results can sometimes arise due to the non-self-adjoint property of the operators that are treated. At this point it appears certain that the theory of non-self-adjoint operators can be used to contribute substantially to the understanding and limitation of the formalism of EEM (and SEM). While the idea of using non-self-adjoint operator theory in scattering problems is not new---it was already suggested by Dolph (35) in 1960--it does not appear to have been used in connection with SEM and EEM in the English literature. This is rather surprising since this theory has been imployed in connection with scalar diffraction problems in papers translated from the Russian and reprinted in the journal Radio Engineering and Electron Physics beginning after the formalism presented in the paper by Voĭtovič, Kacenelenbaum and Sovov (12).

While this latter paper contained a formalism similar to EEM for both the scalar and electromagnetic case, including dielectric problems, its subsequent interpretation, at least in translated papers known to the writers, in terms of non-self-adjoint operator theory have been limited to the scalar problem. In the simplest case one attempts to construct a formal solution to the system

$$(\Delta + k^2)u = f \tag{1}$$

$$u = 0 \text{ , on a smooth convex scatterer } \Gamma,$$

satisfying the radiation condition as a series $u = u_0 + \Sigma A_n \phi_n(\underline{x})$. Here u_0 is the incident field and $\{\phi_n(\underline{x})\}$ are the eigenfunctions of a compact integral operator A, which for in three dimensions, is given explicitly by

$$A\phi_n = \int_\Gamma G_0(\underline{x}-\underline{y})\phi_n(\underline{y})d\sigma = \lambda_n(k)\phi_n(\underline{x}) , \tag{2}$$

where the free space Green's function G_0 is

$$G_0 = \frac{1}{4\pi}\frac{e^{ik|\underline{x}-\underline{y}|}}{|\underline{x}-\underline{y}|} . \tag{3}$$

It was further assumed that the coefficients in the above expansion could be determined by the Fourier coefficient formula

$$A_n = \frac{\int_\Gamma (u-u_0)\phi_n d\sigma}{\int_\Gamma \phi_n^2 d\sigma} \tag{4}$$

Ramm (7) interpreted and clarified these results by using a Hilbert space $L_2(\Gamma)$ with the usual Hermitian inner product

$$(f,g) = \int_\Gamma f(\underline{x})\overline{g}(\underline{x})d\sigma \tag{5}$$

Since G_0 is real symmetric but complex valued, this implies that

$$(A\phi,\psi) = (\phi,\overline{A}\psi) \tag{6}$$

so that the operator A is non-self-adjoint.

Since the compact operator A is non-self-adjoint, it may have root vectors instead of simple eigenvectors. That is for a given λ, there may exist an integer $p > 1$ such that $(A-\lambda I)^p\phi = 0$ for some ϕ, while $(A-\lambda I)^q\phi \neq 0$ for all $q < p$. (In the matrix case this happens when non-simple elementary divisors occur and requires the use of the Jordan normal form rather than the diagonal form in the canonical representation of the matrix.) However, Ramm was able to show that while the system of root vectors was always complete in $L_2(\Gamma)$, the simple form of the coefficients given above would only occur if the surface Γ were such that A defined over it was a normal operator, that is $AA^* = A^*A$ (this condition is necessary and sufficient for A to be a diagonal operator). The normality of the operator can be tested as follows: For example, consider the operator $A: H \to H$, where

$$Af = \frac{1}{4\pi}\int_\Gamma \frac{\exp(ik|s-t|)}{|s-t|} f(t)dt \tag{7}$$

A is normal provided the integral

$$\int_\Gamma \frac{\sin(k|x-t| \; |y-t|)}{|x-t| \; |y-t|} d\sigma \;, \quad x,y\epsilon\Gamma \tag{8}$$

vanishes, a condition that Ramm was able to show was true if Γ is a sphere. In complicated problems it may have to be tested by direct computation.

If the operator A is not normal, considerable complexity arises in the theory and computational schemes. The problem is that in general one has to work in an infinite dimensional analog of the Jordan normal form, since A fails to be diagonalizable. The lemma of Schur [see Gohberg and Krein (36)] states that any completely continous operator A mapping a Hilbert H space onto itself can be represented in a triangular form. Specifically, there exists an orthagonal basis ω_j of H such that

$$A\omega_j = \sum_{i\leq j} a_{ji}\omega_i \tag{9}$$

where $a_{ji} = (A\omega_j,\omega_j) = \lambda_j$, (3) denoting inner product and λ_j being an eigenvalue of A.

There are many areas which need detailed mathematical investigation. For example Voĭtovič, Kacenelenbaum and Sivov consider several variational principles which produce stationary solutions. These are reminisicent of those of Schwinger and McFarlane, the latter occuring in the problem of anomalous propagation through the atmosphere. Attempts to make mathematical sense of these occur in Dolph (37), Dolph and Ritt (38) and Dolph, McLaughlin and Marx (39). The max-min characterization of the last paper unfortunately depends upon the dimension of the approximating space and so it badly needs reformulation. The only other pertinent work it seems to be that of Morawetz (40) where as discussed in Dolph-Scott (1) a variational principle is given implicitly.

In connection with variational principles, the following facts should be mentioned. The variational approach given in (5) has been briefly reviewed by Dolph in (41). Voĭtovič, Kacenelenbaum and Sivov introduced the ε-approach by considering an approximate scattering problem (the ε-problem) and the corresponding systems of equations

$$\begin{aligned} \Delta u + k^2 \varepsilon u &= 0 \quad \text{in } V^+ \\ \Delta u + K^2 u &= 0 \quad \text{in } V^- \\ u^+ - u^- \big|_{S_\varepsilon} &= 0 \\ u\big|_S &= 0 \\ \frac{\partial u^+}{\partial n} - \frac{\partial u^-}{\partial n} &= 0\big|_{S_\varepsilon}, \end{aligned} \tag{10}$$

leading to a miminization of the functional

$$L(u) = \int_V (\nabla u)^2 dV - k^2 \int_{V^-} u^2 dV - k^2 \varepsilon \int_{V^+} u^2 dV \tag{11}$$

A vigorous justification of such techniques still remains an open problem. Similar variational principles were given by J. Schwinger and H. Levine (42). [Also see Kato (43), and Dolph (37)]. As in the case discussed by Dolph and Ritt (38), or Dolph (37), it can be conjectured that the real and imaginary parts of the unknown function near the stationary point lie on a saddle-like surface, but the orientation of such a saddle is unknown.

REFERENCES

1. C. L. Dolph and R. A. Scott (1978) Electromagnetic Scattering (Ed. P.L.E. Uslenghi), Academic Press, New York.

2. C. E. Baum, The emerging technology for transient and broad band analysis and synthesis of antennas and scatterers, Proc. I.E.E.E. 64, 1588 (1976).

3. L. Marin, Natural mode representation of transient scattered fields, I.E.E.E. Trans. Ant. and Prop. 21, 809 (1973).

4. F. M. Tesche, On the analysis of scattering and antennas problems using the singularity expansion technique, I.E.E.E. Trans. Ant. and Prop. 21, 57 (1973).

5. N. N. Voĭtovič, B. Z. Kacenelenbaum and A. N. Sivov, The Generalized Methods of Eigen Vibrations in the Theory of Diffraction, Nauka, Moscow, 1976.

6. A. G. Ramm, Exterior problems of diffraction, Radio Eng. and Elect. Phys. 17, 1064 (1972).

7. A. G. Ramm, Eigenfunction expansion of a discrete spectrum in diffraction problems, Radio Eng. and Elect. Phys. 18, 364 (1973).

8. M. S. Agranovič, Non-self-adjoint operators in the problem of diffraction by a dialectric body and similar problems, Radio Eng. and Elect. Phys. 19, 34 (1974).

9. M. S. Agranovič, Non-self-adjoint integral operators with kernels of the Green function type and diffraction problems related to them, Radio Eng. and Elect. Phys. 20, 26 (1975).

10. M. S. Agranovič, Non-self-adjoint operators in certain scalar problems of diffraction at a smooth closed surface, Radio Eng. and Elect. Phys. 20, 61 (1975).

11. B. Z. Kacenelenbaum, Expansion of forced oscillations of non-closed systems in eigenfunctions of a discrete spectrum, Radio Eng. and Elect. Phys. 14, 19 (1969).

12. N. N. Voĭtovič, B. Z. Kacenelenbaum and A. N. Sivov, Surface current method for constructing systems of eigenfunctions of a discrete spectrum in diffraction problems, Radio Eng. and Elect. Phys. 15, 577 (1970).

13. M. S. Agranovič and Z. N. Golubeva, Some problems for Maxwell systems with a spectral parameter in the boundary condition, Sov. Math. Dok. 12, 1614 (1976).

14. Z. N. Golubeva, Some scalar diffraction problems and their related non-self-conjugate operators, Radio Eng. and Elect. Phys. 21, 219 (1976).

15. C. L. Dolph and S. K. Cho, On the relationship between the singularity expansion method and the mathematical theory of scattering, submitted to : I.E.E.E. Trans. Ant. and Prop.

16. N. Shenk and D. Thoe, Resonant states and poles of the scattering matrix for perturbations of $-\Delta$, J. Math. Anal. and App. 37, 467 (1972).

17. M. Reed and B. Simon (1972), Methods of Mathematical Physics, Academic Press, New York.

18. S. Steinberg, On the poles of the scattering operator for the Schrödinger equation, Arch. Rat. Mech. Analy. 38, 278 (1970).

19. P. D. Lax and R. S. Phillips (1967), Scattering Theory, Academic Press, New York.

20. H. Brakkage and P. Werner, Über das Dirichletsche Aussenraumproblem für die Helmholtzche Schwingungsgleichung, Arch. Math. 16, 325 (1965).

21. P. D. Lax and R. S. Phillips, Decaying modes for the wave equation in the exterior of an obstacle, Comm. Pure Appl. Math. 22, 737 (1969).

22. J. T. Beale, Purely imaginary scattering frequencies for exterior domains, Duke Math. J. 41, 607 (1974).

23. A. Tihonov, Solutions to incorrectly formulated problems and the regularization method, Sov. Math. 4, 1034 (1964).

24. J. S. Howland, Simple poles of operator valued functions, Math. Anal. and Appl. 36, 12 (1971).

25. J. C. Bolomey and A. Wirgin, Numerical comparison of the Green's function and the Waternam and Rayleigh theories of scattering from a cylinder with arbitrary cross-section, Proc. I.E.E.E. 121, 794 (1974).

26. P. C. Waternam, New formulation of acoustic scattering, J. Acoust. Soc. Amer. 45, 1417 (1969).

27. B. Peterson and S. Ström, T-matrix for electromagnetic scattering from an arbitrary number of scatterers and representations of E(3), Phys. Rev. D8, 3661 (1973).

28. B. Peterson and S. Ström, T-matrix formulation of electromagnetic scattering from multilayered scatters, Phys. Rev. D, 10, 2620 (1974).

29. B. Peterson and S. Ström, Matrix formulation of acoustic scattering from multilayered scatterers, J. Acoust. Soc. Amer. 57, 2 (1975).

30. S. Ström, T matrix for electromagnetic scattering from an arbitrary number of scatterers with continuously varying electromagnetic properties, Phys. Rev. D, 10, 2685 (1974).

31. S. Ström, On the integral equations for electromagnetic scattering, Amer. J. Phys. 43, 1060 (1975).

32. S. Ström, Quantum-mechanical scattering from an assembly of nonoverlapping potentials, Phys. Rev. D 13, 3485 (1976).

33. V. Varatharajulu and Y.-H. Pao, Scattering matrix for elastic waves. I. theory, J. Acoust. Soc. Amer. 60, 556 (1976).

34. T.-Y. Wu and T. Ohmara (1962), Quantum theory of scattering, Prentice Hall, New Jersey.

35. C. L. Dolph, Recent developments in some non-self-adjoint problems in mathematical physics, Bull. Amer. Math. Soc. 67, 1 (1961).

36. I. C. Gohberg and M. G. Krein (1969) Introduction to the theory of linear nonselfadjoint operators, A.M.S. Translations 18, Rhode Island.

37. C. L. Dolph, A saddle point characterization of the Schwinger stationary points in exterior scattering problems, J. Soc. Indust. Appl. Math. 5, 89 (1957).

38. C. L. Dolph and R. K. Ritt, The Schwinger variational principles for one-dimensional quantum scattering, Math. Zeit. 65, 309 (1956).

39. C. L. Dolph, J. E. McLaughlin and I. Marx, Symmetric linear transformations and complex quadratic forms, Comm. Pure Appl. Math. VII, 621 (1954).

40. C. S. Morawetz, On the modes of decay for the wave equation in the exterior of a reflecting body, Proc. Roy. Irish Acad. A, 72, 113 (1972).

41. C. L. Dolph, A review of the Voĭtovič, Kacenelenbaum and Sivov book (reference [5]), To appear in Mathematical Reviews-August 1979 issue.

42. J. Schwinger and H. Levine (1972), On the theory of diffraction by an aperture in an infinite plane screen, The theory of electromagnetic waves, Interscience, New York.

43. T. Kato, Note on Schwinger's variational principle, Progress of theoretical physics, (Japan) 6, 295 (1951).

ESTIMATION OF THE COMPLEX NATURAL RESONANCES OF A CLASS OF SUBSURFACE TARGETS

L.C. Chan, D.L. Moffatt, L. Peters, Jr.
The Ohio State University ElectroScience Laboratory
Department of Electrical Engineering
Columbus, Ohio 43212

I. ABSTRACT

This chapter summarizes a study in the characterization of subsurface targets by their complex natural resonances. The methodology in estimating the complex natural resonances of these targets is emphasized. A method for extracting the resonances from the backscattered waveforms is derived for completeness. This method is known as Prony's method (10-18,24) and, when applied to measured data, it is extremely sensitive to the values of its parameters. An approach to solve certain of these problems will be presented. Prony's method is applied to the measured backscattered waveforms and an analysis of the resulting resonances is presented.

The concept of using complex natural resonances for target characterization is developed from the fact that all finite-size objects have resonances that depend on their physical characteristics such as size, shape and composition as well as the medium surrounding the object. These resonances, however, are independent of the excitation (1,2,22,23). As a useful but inexact analogy, in circuit theory, the form of the transient response of a lumped linear circuit may be determined from the knowledge of the resonances and the corresponding residues of the response function in the complex frequency plane. The actual transient response of the circuit is then simply a summation of all the residues multiplied by the inverse transforms of the resonances. In 1965, Kennaugh and Moffatt (3) generalized the impulse response concept to include the distributed parameter scattering problems and suggested that a lumped circuit representation, at low frequencies or long time, was possible. Later, similar and more formal representations have been designated as the Singularity Expansion Method (SEM) (2). While some questions remain, the hypothesis is generally supported by the fact that the transient backscattered waveforms from the subsurface targets received by a video pulse radar (4) appear to be dominated by a few exponentially damped sinusoids. Based on this concept, a subsurface target can be characterized by a set of complex natural resonances which is independent of the location and orientation of the video pulse radar system. These resonances, however, are dependent on ground condition. Such a characterization is attractive for it catalogs a target by a small set of complex numbers.

II. SUBSURFACE RADAR TARGET MEASUREMENTS

Five targets of similar size were buried at the same depth of 5 cm (2 inches, measured from the ground surface to the nearest target surface). Fig. 1 shows the geometry of the targets.

The average dc conductivity within 30 cm (12 inches) of the ground surface measured at the target sites ranged from 30 mS/m for wet ground to about 20 mS/m for dry ground and 10 mS/m for icy ground. The relative dielectric constant measured at approximately 100 MHz ranged from 25 for wet ground to 16 for dry ground and 9 for icy ground.

Backscattered waveforms were obtained using a subsurface video pulse radar system, which basically consisted of a 150 ps video pulser, a pair of crossed-dipole

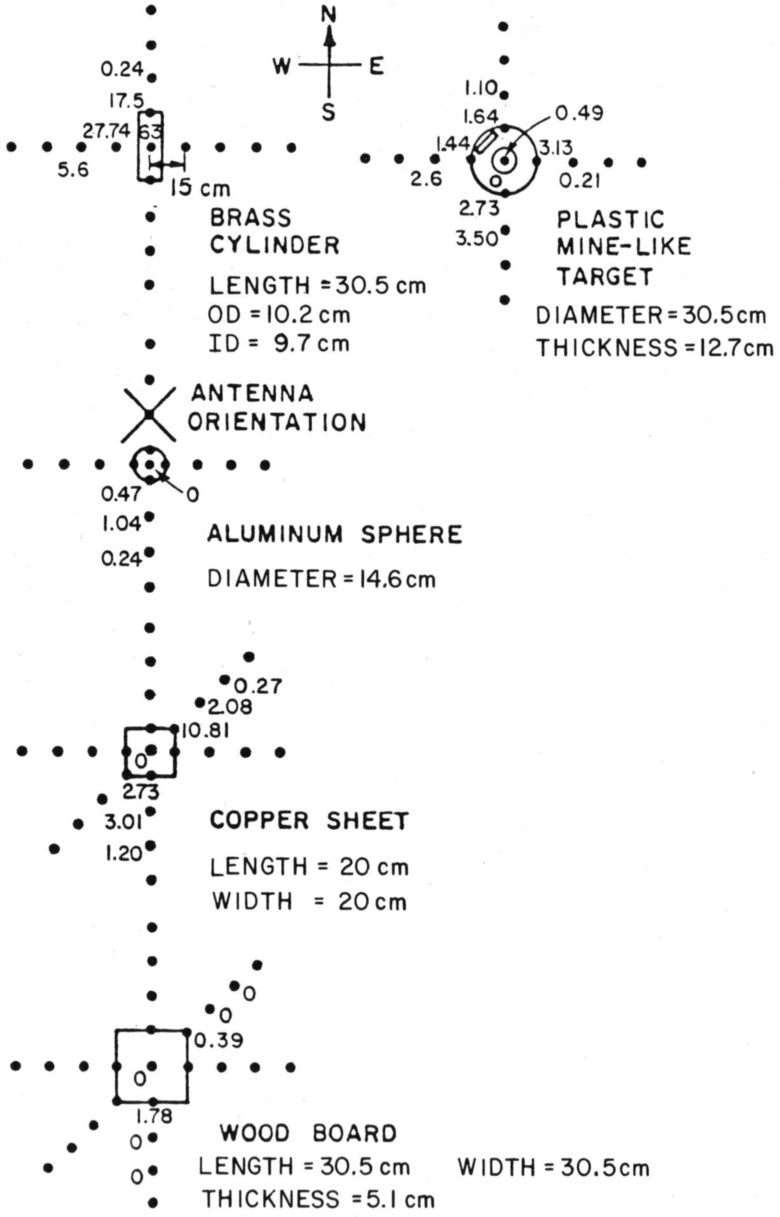

Fig. 1. The subsurface targets, measurement locations, antenna orientation and S/C estimates.

antennas and a sampling oscilloscope. Details of this system and the application of the measured waveforms for target identification purposes have been reported elsewhere (4,19) and thus will not be repeated here. Measurements were made at different ground locations with respect to the various targets. Locations of the antenna center for these measurements are shown as dots in Fig. 1. At each location a backscattered waveform was obtained using a standard antenna orientation (see Fig. 1). Measurements were obtained with the antenna center vertically above the center and edges of the targets. Beyond the target edges, measurements were made at the regular interval of 15 cm (6 inches). During the data-taking period the ground condition changed from wet to dry and to icy. Data were obtained for each ground condition to gauge the effects of the changing ground condition on the complex natural resonances of the subsurface targets.

All waveforms collected by the video pulse radar used in this study consist of 256 samples in a time window of 50 ns. The (hardware) basic sampling period T_B is 0.2 ns, giving a sampling frequency of 5.12 GHz.

A set of measured waveforms and their Fast Fourier Transforms (FFT) (8) for the mine-like target at various antenna locations in a wet ground is shown in Figs. 2 and 3.

The following important generalizations with regard to these waveforms are made:

1. All time-domain waveforms exhibit transient behavior in the late-time region where only the natural response of the target exists. This transient behavior is of prime importance, and, as will be shown in Section III, dictates the characterization method for these targets.

2. The strong peaks in the FFT's of the waveforms indicate the possible existence of resonance behavior in the backscattered waveforms. These peaks may be a good approximate measure of the imaginary parts of the complex resonances of the targets in situ. Note that while the time-domain waveforms from different antenna locations over the mine-like target change noticeably, the locations of the strong peaks of their FFT's stay relatively unchanged (see the vertical dotted lines in Fig. 3), indicating the complex natural resonances of the target are excitation invariant. This was anticipated and is the most attractive feature of the target characterization method.

3. As a reasonable quantitative parameter for comparison of signal and clutter* levels, the following definition of signal-to-clutter ratio (S/C) was used in this study.

$$\frac{S}{C} \triangleq \frac{E_T}{\bar{E}_{NT}} = \frac{E_M - \bar{E}_{NT}}{\bar{E}_{NT}} \qquad (1)$$

where

E_T is the energy of the target signal,

E_M is the energy of the measured waveform, and

$\bar{E}_{NT}$ is the statistical mean energy of the ensemble of clutter or no-target waveforms used to estimate $\bar{E}_{NT}$. A no-target waveform is a received

*Here the term clutter is used to represent the total unwanted extraneous signal present in the waveform.

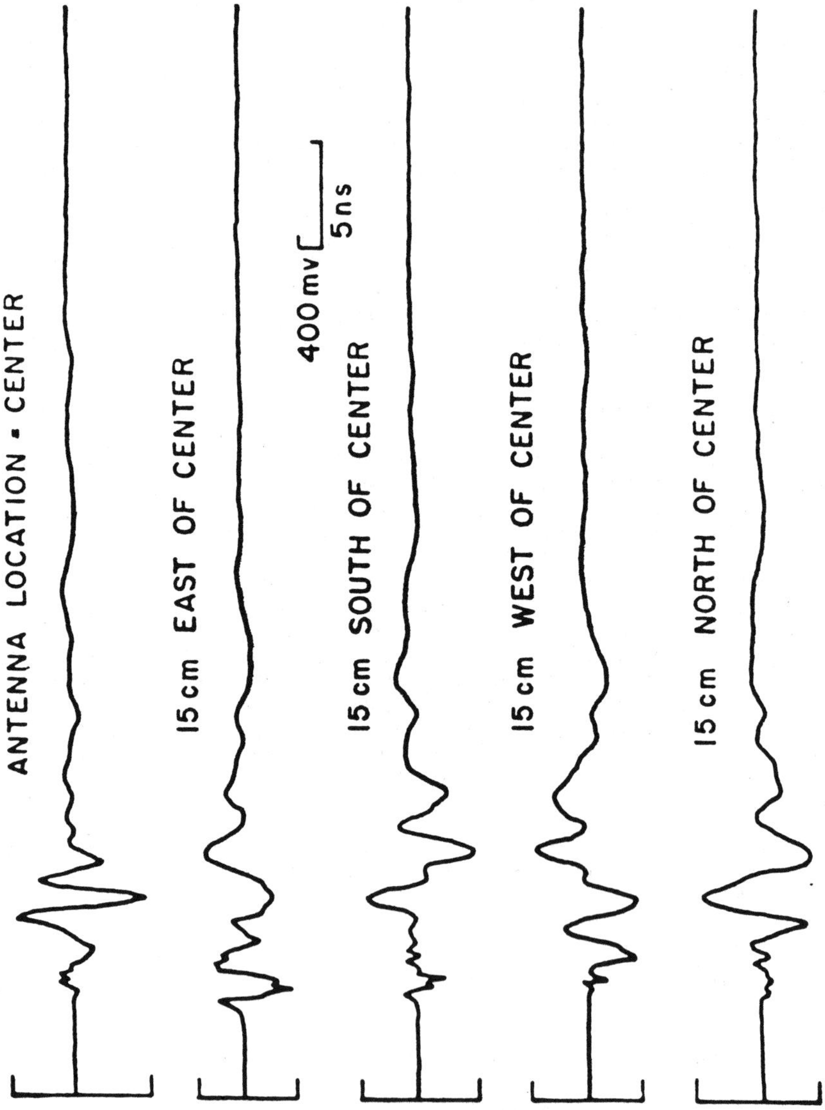

Fig. 2. Processed waveforms from the mine-like target at different antenna locations in wet ground.

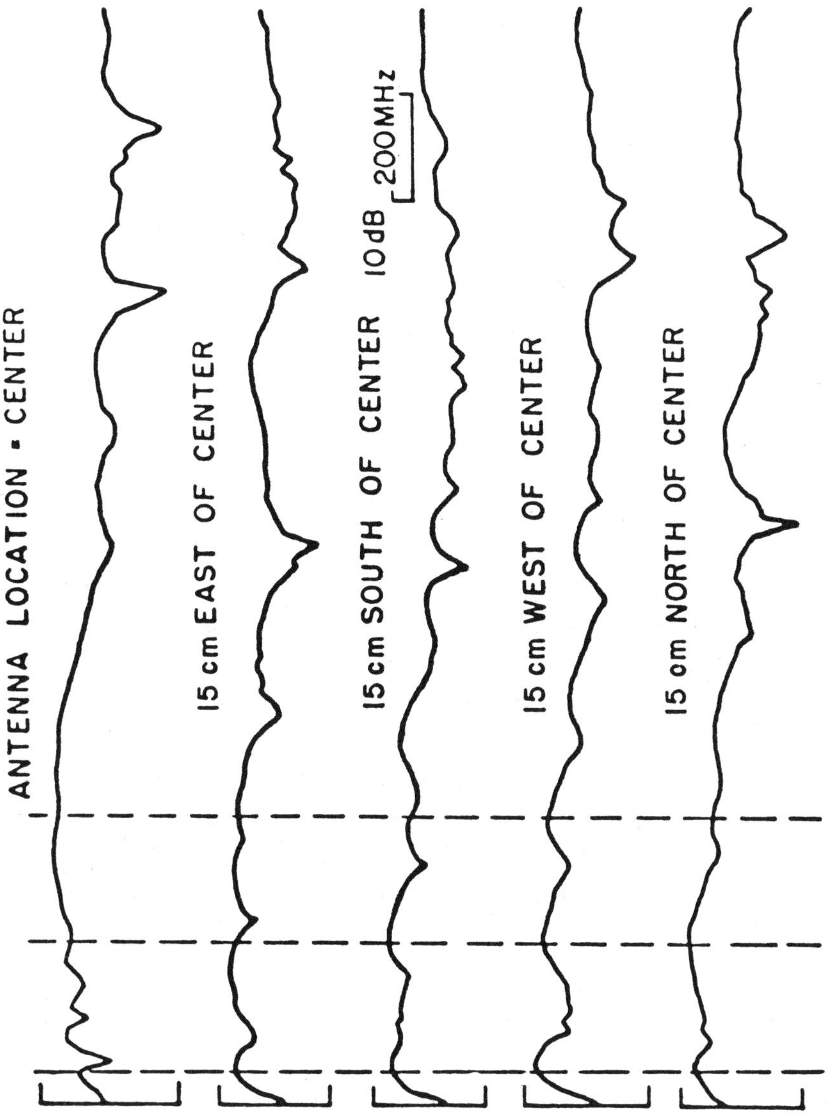

Fig. 3. FFT of the processed waveforms from the mine-like target.

waveform with no target present within the radar range. In this study, we considered only single-target situations, hence a collection of 51 no-target measurements taken at various locations in the vicinity of the target sites was used as the ensemble of clutter waveforms.

The energy of a waveform was defined and estimated as follows.

$$E \triangleq \frac{\sum_{t=t_s}^{t_e} r^2(t)}{(t_e-t_s)/t_B}\ ; \quad t = iT_B \tag{2}$$

where $r(t)$ is the waveform under consideration and t_s, t_e are the start and end time, respectively, of the interval of interest. In this study t_s was taken to be the time at which the absolute maximum of the waveform occurred and t_e was taken to be the time at which the balun reflection occurred. In the video pulse radar system, a balun was used to connect the unbalanced impulse generator to the balanced crossed-dipole antenna. The impedance mismatch at this connection was the source of the balun reflection. The balun reflection was considered as clutter and was thus not included in the interval for signal processing and analysis. Reasons for the above choices of t_s and t_e are given in Section III-C.

The mean clutter level was evaluated as

$$\overline{E}_{NT} = \frac{\sum_{i=1}^{51} \sum_{t=t_{si}}^{t_{ei}} \left[r_{NTi}^2(t)/(t_{ei}-t_{si})/T_B\right]}{51} \tag{3}$$

where $r_{NTi}(t)$ is the ith no-target waveform in the ensemble and t_{si}, t_{ei} are the start and end time, respectively of the ith no-target waveform.

Using the above definition, the signal-to-clutter ratio at various antenna locations over the various targets were evaluated and are given in Fig. 1. It was found that the signal-to-clutter ratio for the mine-like target ranged from 0.21 to 3.50 depending on the antenna location. The brass cylinder and the wood board targets had the highest and lowest signal-to-clutter ratios, respectively.

The three items mentioned above: the transient behavior, the complex natural resonances and the signal level of the backscattered waveforms are of prime importance in subsurface target characterization. Their dependence on the changing ground condition complicates the characterization process. A comparison between the brass cylinder waveform in dry and icy ground given in Fig. 4 clearly shows the effects of changing ground condition. Although both waveforms exhibit similar transient behavior, the time intervals between the zero crossings are different, indicating a shift in the locations of the target resonances. The amplitudes of the two waveforms are also different.

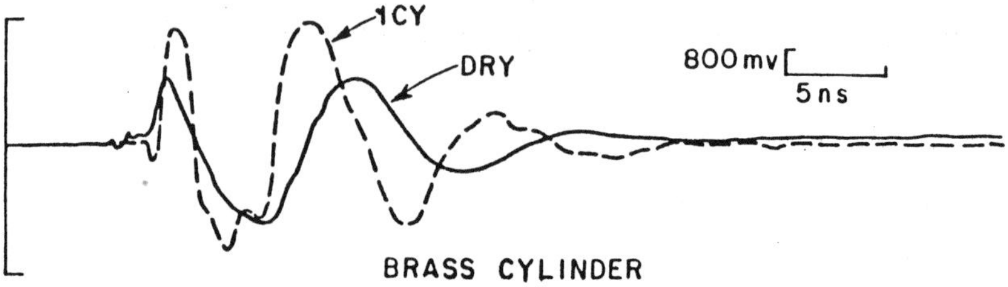

Fig. 4. Processed waveforms from the brass cylinder in dry and icy ground.

III. ESTIMATION OF THE COMPLEX NATURAL RESONANCES OF THE SUBSURFACE TARGETS VIA PRONY'S METHOD

The backscattered waveforms from the subsurface targets received by the pulse radar system are good approximations to the impulse responses of the targets. Furthermore, they appear to be dominated by a few exponentially damped sinusoids, and thus can be represented as

$$r(t) = \sum_{n=1}^{N} a_i e^{s_i t} \tag{4}$$

where r(t) is the received transient waveform, s_i's are the complex resonant frequencies or pole locations in the complex frequency plane. These have, by common usage in this representation, become designated as complex resonances or more simply as resonances. These various terms will be used for s_i in this chapter. a_i's are the corresponding residues and N is the number of complex resonances within the frequency band of the radar system.

In order to exploit Equation (4), it is necessary to first determine the values of the complex natural resonances of the targets. The method used here extracts the resonances of a target directly from its transient response. This method is known as Prony's method, which was first derived by Prony in 1795 (10), and was later suggested by Van Blaricum, et al., for extracting the pole singularities of transient waveforms in 1975 (11).

A. Derivation of Prony's Method

In discrete form, Equation (4) can be written as

$$r(kT_B) = \sum_{i=1}^{N} a_i\ e^{s_i KT_B},\ K = 0,1,2... \tag{5}$$

where K is the sampling index and T_B is the basic hardware sampling period of our measurement system (see Section II). For an exact solution of the 2N unknowns a_i's and s_i's, we can set up 2N (nonlinear) equations by using 2N sample values of $r(KT_B)$. Prony's method uses 2N uniform samples

$$r_n = r(nT) = \sum_{i=1}^{N} a_i\ e^{s_i nT}\ ;\quad n = 0,1,2...,M = 2N-1 \tag{6}$$

where T, the Prony interval, is the interval between the samples used along the waveform. Since no waveform interpolation was exercised, T was equal to integer multiples of T_B. N, the number of target resonances excited by the interrogating frequencies within the system bandwidth, is in general an unknown. In this study the value of N was found by a trial-and-error process outlined in Section III-C. Equation (6) is a set of 2N nonlinear equations in the z_i's where $z_i = e^{s_i T}$. Let $z_1, z_2 ... z_N$ be the roots of the algebraic equation

$$\alpha_0 + \alpha_1 z^1 + \alpha_2 z^2 + \dots \alpha_N z^N = 0, \tag{7}$$

so that the left hand side of Equation (7) is equal to the product

$$(z-z_1)(z-z_2)\ \dots\ (z-z_N) = 0, \tag{8}$$

that is,

$$\sum_{m=0}^{N} \alpha_m z^m = \prod_{i=1}^{N} (z-z_i) = 0. \tag{9}$$

Thus, if we can evaluate α_m, then z_i can be obtained by a simple factorization of an Nth degree polynomial. To solve for α_m, we obtain from Equation (6)

$$\sum_{m=0}^{N} \alpha_m r_{K+m} = \sum_{m=0}^{N} \alpha_m \left(\sum_{i=1}^{N} a_i z_i^{K+m} \right) ; \; K = 0,1,2...M-N$$

Interchanging the order of the summation yields

$$\sum_{m=0}^{N} \alpha_m r_{K+m} = \sum_{i=1}^{N} a_i z_i^k \sum_{m=0}^{N} \alpha_m z_i^m .$$

From Equation (9), we see that the summation inside the parenthesis of the above equation is zero, thus, we arrive at the desired linear, homogenoeus difference equation

$$\sum_{m=0}^{N} \alpha_m r_{K+m} = 0; \quad K = 0,1,2...M-N . \tag{10}$$

Thus, the sample values of r(t) satisfy an Nth order linear homogenoeus difference equation. This difference equation is commonly referred to as the Prony difference equation.

The Prony difference equation is linear and homogeneous, and can be used to solve for the N+1 coefficients, i.e., α_m's. In the classical Prony's method, these coefficients are obtained by setting α_N=1 and solving the resulting matrix equation by matrix inversion. That is,

$$\overset{A}{\begin{bmatrix} r_0 & r_1 & r_2 & \cdots & r_{N-1} \\ r_1 & r_2 & r_3 & \cdots & r_N \\ \cdot & & & & \\ \cdot & & & & \\ \cdot & & & & \\ r_{M-N} & r_{M-N+1} & r_{M-N+2} & \cdots & r_{M-1} \end{bmatrix}} \overset{B}{\begin{bmatrix} \alpha_0 \\ \alpha_1 \\ \cdot \\ \cdot \\ \cdot \\ \alpha_{N-1} \end{bmatrix}} = \overset{C}{\begin{bmatrix} r_N \\ r_{N+1} \\ \cdot \\ \cdot \\ \cdot \\ r_M \end{bmatrix}} . \tag{11}$$

Note that for M=2N-1, A is a square symmetric circulant matrix and is readily invertable. Standard computer routines such as GELG (43) can be used to do the matrix inversion. Once the α_m's are determined, the next step is to solve for the N values of z_i. These z_i's are obtained by finding the roots of Equation (9). The N roots are complex numbers and because r(t) is real, these complex roots appear in complex conjugate pairs. The polynomial root finding process can be eaisly performed by using standard routines such as Muller (44,45).

It is now trivial to obtain the poles s_i. The poles are simply given by

$$s_i = \frac{1}{T} \ln(z_i). \tag{12}$$

The final step in Prony's method is to determine the value of the residues a_i's. To do this, we simply solve the matrix equation embodied in Equation (6). In matrix form this set of equations is written as

$$\overset{D}{\begin{bmatrix} 1 & 1 & \cdots & 1 \\ z_1 & z_2 & \cdots & z_N \\ z_1^2 & z_2^2 & \cdots & z_N^2 \\ \vdots & \vdots & & \vdots \\ z_1^{N-1} & z_2^{N-1} & \cdots & z_N^{N-1} \end{bmatrix}} \overset{E}{\begin{bmatrix} a_1 \\ a_2 \\ \vdots \\ a_N \end{bmatrix}} = \overset{F}{\begin{bmatrix} r_0 \\ r_1 \\ \vdots \\ r_{N-1} \end{bmatrix}} \tag{13}$$

where now the only unknowns are the elements of the residue matrix E.

The above derivation of Prony's method is valid only when all natural resonances present are simple poles. For multiple-order poles, a slight modification is necessary in solving for the residues (4,12). However, in this study we did not find it necessary to postulate multiple-order poles.

In summary, Prony's method solves for the complex natural resonances and the corresponding residues associated with the back-scattered time-domain waveforms from a system of nonlinear equations (Equation (6)) by breaking it down into three simple steps:

1. Solve for the values of α_m's of the linear Equation (11) by matrix inversion.
2. Solve for the poles by factoring the polynomial of Equation (9).
3. Solve for the residues from the linear Equation (13) by matrix inversion.

The derivation of Prony's method is simple enough. However, its application to the measured backscattered waveforms is a much more complicated process. The following section outlines some of the difficulties.

B. Clutter Effects in Prony's Method

Prony's method has been found to be extremely sensitive to clutter. Its ability to extract the complex natural resonances of a waveform accurately is severely inhibited by the presence of clutter (13-15). Since Prony's method is a process of curve fitting, in the presence of clutter, it will give a set of poles which fit the clutterly transient response but will not necessarily represent the complex natural resonances of the target. Various signal-processing techniques have been applied to reduce the effect of clutter in Prony's method (13-15), with the most commonly used being the least-square error technique. With it, Equation (11) in the previous section is solved in the least-square sense. In this case, M samples are used in lieu of 2N samples where $M>2N$. Thus, the matrix A becomes rectangular, and Equation (11) is solved by the pseudo-inverse technique

$$A^T AB = A^T C \tag{14}$$

or

$$\Phi B = D \tag{15}$$

where

$$\Phi = A^T A$$

and

$$D = A^T C .$$

Since Φ is the signal covariance matrix of size N by N, it is real, symmetric and positive definite, and is thus readily invertible to yield the value of α_m's.

A second technique applied to reduce clutter in Prony's method was brought about by the observation that, in solving for the N+1 α_m's in the N homogeneous equations (11), we can, instead of setting $\alpha_N=1$, require that

$$\sum_{m=0}^{N} \alpha_m^2 = 1 \tag{16}$$

Such an approach leads to the eigenvalue method (15,16).

Instead of setting the leading coefficients $\alpha_N=1$, we can of course set any of the N+1 coefficients to 1 in solving for the α_m's of Equation (10). Such a constraint leads to the interpolation version of Prony's method (4,17).

The classical Prony's method, the eigenvalue method and the interpolation version of Prony's method were all considered in this study.*

Numerous other signal-processing techniques have been applied (13,14), thus far however, no completely satisfactory result has been reported using measured data. In the following section a systematic procedure that is given good results for the present data is outlined. This procedure was used in extracting the complex natural resonances of the backscattered waveforms discussed in Section II and yielded our best result to date. As we will see, the procedure does indeed provide satisfactory target characterization.

C. Applying Prony's Method to the Measured Backscattered Waveforms

In applying Prony's method, one approach is to pre-determine the following parameters:

1. N, the number of poles present in the waveform. Van Blaricum (12-14) suggested a method which relies on the fact that the (N+1)th eigenvalue of the signal covariance matrix of size N+1 by N+1 should equal σ^2, the variance of the additive gaussian stationary and uncorrelated noise. Such a method does not seem practical for our measured data in which the clutter is nonstationary (transient).

2. T, the Prony interval. Obviously, undersampling (T too large) will almost surely bring aliased results. It was also found that oversampling produces extraneous high-frequency poles.

3. t_s, t_e the start and end time of the fitting interval. t_s must lie in the time region where the forced response portion of the backscattered waveform has ended and t_e must lie in the region where clutter effects are not dominant.

4. M, the number of sample points used in the fitting process. M determines the amount of overspecification on the system of Equations (11).

*A complete discussion of these signal-processing problems can be found in (21).

In the presence of clutter, it was found that the accuracy of the extracted resonances was found to be extremely sensitive to the values of the above five parameters (4,15,16). For the method to yield an accurate solution, we have to find the "right" set of values for these parameters. The approach developed in this study is to vary these parameters (over a reasonable range) and assume that the "right" values of the parameters corresponding to the "desired" resonances are those that allow the closest approximation to the measured waveform in the time domain. That is, a calculated waveform is developed from the resonances and residues found, and this waveform is compared to the original waveform point by point over the fitting interval $[t_s, t_e]$ and the total squared error found. The solution which affords the smallest total squared error is considered to be the "desired" solution. The ranges of these parameters in this study were fixed as follows:

1. The number of "significant" peaks in the Fast Fourier Transform of the waveforms is usually a good measure of N and since the number of "significant" peaks is between 2 and 7 in all waveforms considered, the range of N was chosen to be from 4 to 14 (we assumed that one peak corresponded to at most two poles).

2. Shannon's sampling theorem constrains the maximum value of T, while the bandwidth of the radar system (<500 MHz)* constrains the minimum value of T. The values of T were chosen to be $3T_B$, $5T_B$ ... $10T_B$ corresponding to minimum and maximum Nyquist frequencies of 256 MHz and 768 MHz, respectively.

3. The values of t_s, were chosen to be t_{max}, $(t_{max}+2T_B)$, $(t_{max}+4T_B)$...$(t_{max}+8T_B)$ where t_{max} was the time at which the absolute maximum of the waveform occurred. It was found that such a choice ensured a decaying nature in almost all the waveforms considered. t_e was chosen to be the time at which the balun reflection occurred (see Section II).

4. Since a vastly oversized M would result in an unstable solution from Prony's method (12,15,18), the values of M were chosen to be 2N, 3N, ... 6N.

Each set of values of the five parameters (N,T,t_s,t_e,M) will give a set of complex resonances when Prony's method is applied to a waveform. This set of complex resonances maximizes the fit between the approximated waveform $r_A(t)$ and the measured waveform $r_M(t)$ in the interval $[t_s, t_s+(M-1)T]$ with the sampling interval of T. For all complex resonances resulting from all possible sets of (N,T,t_s,t_e,M) in the chosen range, the "desired" set of complex resonances is chosen to be the one which minimizes the total normalized point-by-point squared error ε over the error-calcalating interval. The error ε is defined as

$$\varepsilon = \left\{\sum_t [r_M(t)-r_A(t)]^2\right\} \Big/ \left\{\sum_t [r_M^2(t)+r_A^2(t)]\right\} \; ; \; t=iT_B. \qquad (17)$$

In this study, $r_A(t)$ was generated via the method of linear prediction (25), where

$$r_A(t+m_0T) = \sum_{\substack{m=0 \\ m\neq m_0}}^{N} \frac{-\alpha_m}{\alpha_{m_0}} \, r_M(t+mT) \qquad (18)$$

*See Reference (4).

In Equation (18), $r_A(t)$ and $r_M(t)$ are the approximated and the measured waveform, respectively, the α_m's are the difference equation coefficients obtained from the Prony's method, and m_o is an index chosen for suppression of clutter effects. In this study, m_o was chosen to be the coefficient of maximum magnitude. With Equation (18), the error ε can be expressed as

$$\varepsilon = \frac{\displaystyle\sum_{t=t_s+m_oT}^{t_e-NT+m_oT} \frac{1}{\alpha_{m_o}^2} \left[\sum_{m=0}^{N} \alpha_m r_M(t+mT) \right]^2}{\displaystyle\sum_{t=t_s+m_oT}^{t_e-NT+m_oT} \left[r_A^2(t)+r_M^2(t) \right]} \quad ; \; t = iT_B \tag{19}$$

From Equations (10) and (19), we note that the error ε is zero when the measured waveform is free of clutter and is perfectly characterized by Equation (4). ε should be small when $r_M(t)$ is closely approximated by Equation (4).

IV. THE EXTRACTED RESONANCES OF THE SUBSURFACE TARGETS

The locations of extracted resonances of the mine-like target at different antenna locations in icy ground are plotted in Fig. 5 (here only poles in the upper left half s plane are shown). From Fig. 5, we make the following observations:

1. The extracted resonances tend to form "clusters". Some possible clusters are shown in Fig. 5. The formation of these clusters are based on the obviousness of clustering of the resonances and the known fact that the accuracy in determining the real parts of the extracted resonances is normally poor. A cluster can contain at most one pole extracted from a waveform. Poles with residues which are three orders down in magnitude compared to the maximum residue are discarded. Poles which are remote from the clustered groups are excluded. Beyond an obvious weighing dictated by the actual pole locations no real significance should be attached to the shape of the closed contour surrounding each cluster.

2. Only a small number of clusters or resonances are present.

3. The variation in the real parts of the resonances within a cluster is generally greater than the variation in their imaginary parts. There is at least one more major factor, besides the ever-present clutter, that causes such variations, namely, the target-antenna interactions. At the shallow depth of 5 cm, for most antenna locations considered, the targets are in the near field of the antenna for the entire bandwidth (<500 MHz) of our radar system.

4. An additional factor contributes to the variations in the extracted resonances from the mine-like target, namely, its complex structure. This target possesses the most complex structure of all targets considered.

5. The phenomenon of certain resonance(s) being weakly excited in certain radar aspects is evident. Some weakly exicted resonances are not extracted.

In this study, a subsurface target was characterized by the set of average extracted resonances. Averaging was performed over all the extracted resonances in each cluster. For the mine-like target, the average extracted resonances are shown as solid dots in Fig. 5. Parameters such as the variation from the average of each pole within the cluster is not meaningful because of its causes

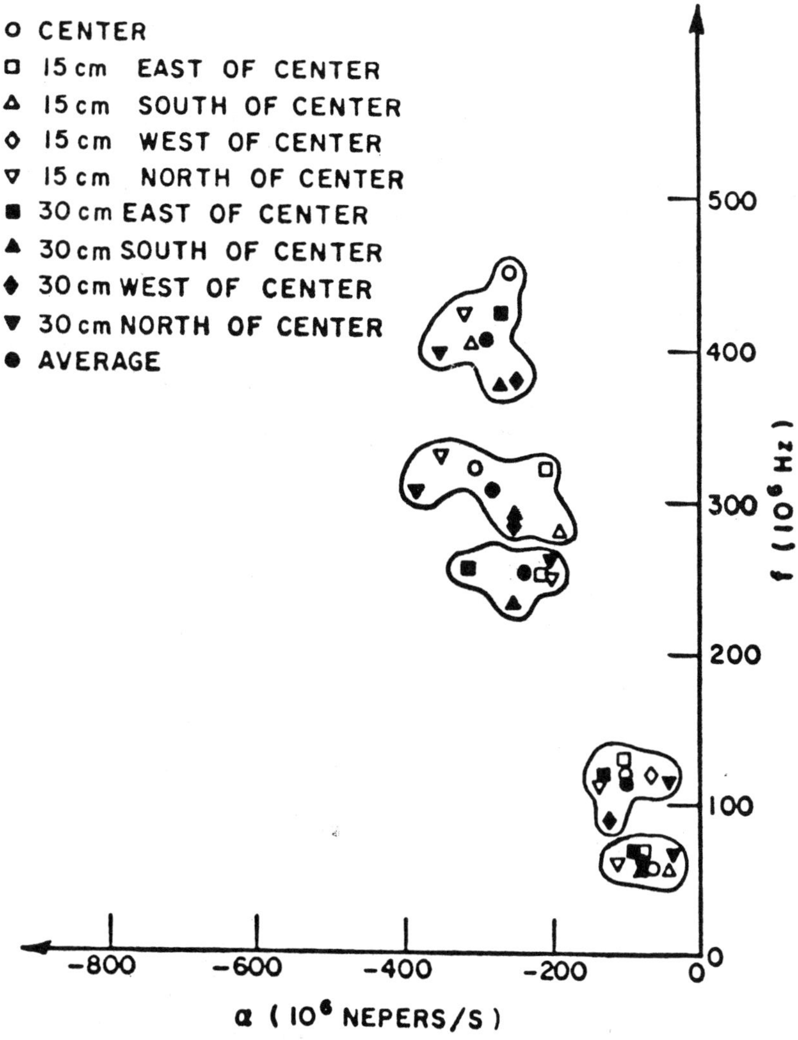

Fig. 5. Location of the extracted resonances of the mine-like target at different antenna locations in icy ground.

which include, besides the effects of clutter, the possible variations in the pole excitation at the various antenna locations. Slight pole variation due to the variations in the antenna locations is possible for the finite exponential sum representation of the target's transient response is only an approximation and that the targets considered are located in the close vicinity of the radar system.

The extracted resonances shown in Fig. 5 were obtained using classical Prony's method (i.e., $\alpha_N=1$). Classical Prony's method was found to extract poles with tighter clusterings among the results given by other methods under the constraint $\alpha_m=1, m=0,1...N$. The eigenvalue method provided results similar to those given by the Classical Prony's method. Thus, no clear-cut choice of method was discernible. Accordingly, the extracted resonances shown in this chapter were the results of either of these two methods.

In order to see the effects of the changing ground condition on the locations of the extracted resonances, the average resonances of the mine-like target in different ground conditions are tabulated in Table 1 and are plotted in Fig. 6. From Fig. 6, we see that there were five (pairs) extracted resonances. The imaginary parts of the extracted resonances were relatively insensitive to changes in ground condition. This seems to imply the resonances of the mine-like target are internal resonances.

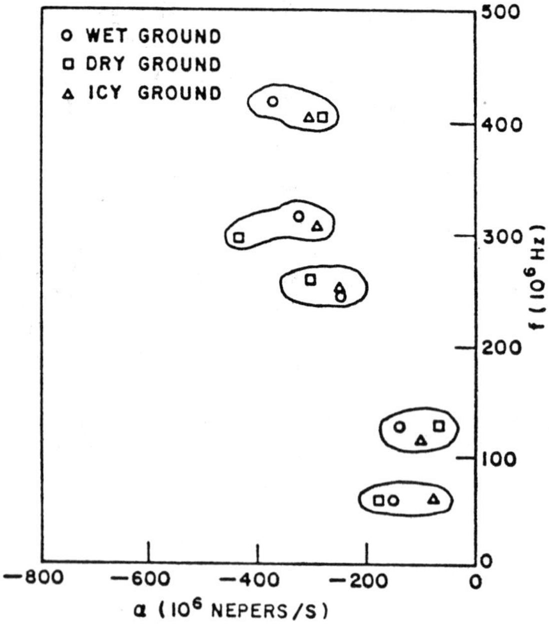

Fig. 6. Locations of the average extracted resonances of the mine-like target in different ground conditions.

The implications of the fact that there were five (pairs) resonances extracted from the mine-like target waveforms is significant. It means that this target can now be characterized by a finite-order system.

TABLE 1
AVERAGE EXTRACTED RESONANCES OF THE MINE-LIKE TARGET IN DIFFERENT GROUND CONDITIONS

ICY GROUND		DRY GROUND		WET GROUND	
POLE REAL PART*	POLE IMAG PART*	POLE REAL PART	POLE IMAG PART	POLE REAL PART	POLE IMAG PART
-.7493116E8	±.6347621E8	-.1755790E9	±.5769154E8	-.1502978E8	±.6074772E8
-.9981995E8	±.1146405E9	-.6843569E8	±.1287960E9	-.1339401E9	±.1286094E8
-.2416503E9	±.2535799E9	-.3048629E9	±.2609797E9	-.2434968E9	±.2418420E9
- 2809195E9	±.3074791E9	-.4321735E9	±.2957980E9	-.3207980E9	±.3180883E9
-.2885261E9	±.4076659E9	-.2797264E9	±.4037985E9	-.3661906E9	±.4218155E9

*Real and Imaginary parts of the extracted resonances shown in Tables 1, 2 and 3 are in Nepers/s and Hz, respectively.

Not all the extracted resonances are related to the scattering mechanisms of the mine-like target. In fact, the lowest resonance was found to be the antenna resonance of the system. The antenna resonance was extracted from almost every waveform of all targets considered.

In contrast to the case of the plastic mine-like target, the brass cylinder was found to possess external resonances. Table 2 lists the average extracted resonances of the brass cylinder in two different ground conditions. Locations of these resonances are also plotted in Fig. 7. From Fig. 7 we see the following effects of the changing ground condition on the extracted resonances of the brass cylinder. First, the antenna resonance is insensitive to changes in the ground condition. This may be attributed to the fact that the arms of the crossed-dipole antenna were not in electrical contact with the ground surface. Second, the imaginary parts of the three higher-frequency resonances increased significantly when the ground changed from dry to icy. This is to be expected, because the resonances of the brass cylinder are external resonances. The increase in the imaginary parts of these external resonances indicates a decrease in the value of the dielectric constant of the ground. Third, the real part of the three higher-frequency resonances generally decreased as the ground changed from dry to icy, indicating that icy ground in this case was more lossy. The increase in loss seems to be the reason for the absence of the real cylinder pole in icy ground.

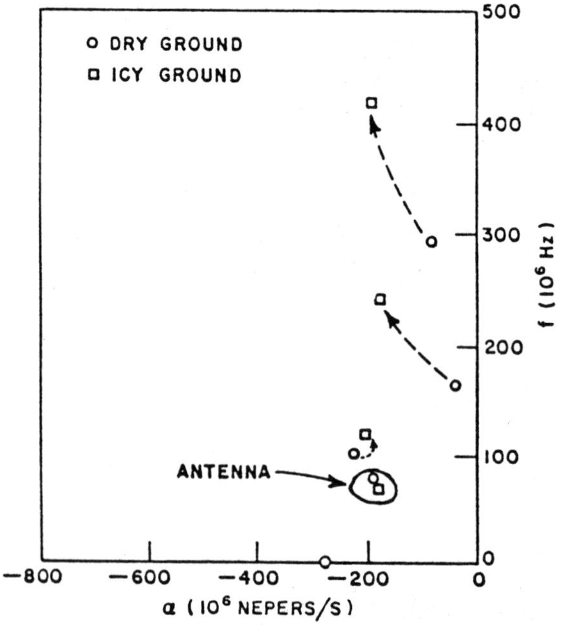

Fig. 7. Average extracted resonances of the brass cylinder in different ground conditions.

TABLE 2
AVERAGE EXTRACTED RESONANCES OF THE BRASS CYLINDER
IN DIFFERENT GROUND CONDITIONS

ICY GROUND		DRY GROUND		WET GROUND	
POLE REAL PART	POLE IMAG PART	POLE REAL PART	POLE IMAG PART	POLE REAL PART	POLE IMAG PART
		-.2769107E9	.0000000		
-.1754246E9	±.7320464E8	-.1904656E9	±.7836537E8	-.1514337E8	±.6581380E8
-.1993071E9	±.1208164E9	-.2233100E9	±.9971379E8	-.2173419E9	±.9414942E8
-.1763301E9	±.2462725E9	-.0480810E9	±.1681942E9	-.2332553E9	±.2106191E9
-.1882724E9	±.4224952E9	-.0701791E9	±.2964066E9	-.6256987E8	±.3249909E9

TABLE 3
AVERAGE EXTRACTED RESONANCES OF THE ALUMINUM SPHERE,
COPPER SHEET AND THE WOOD BOARD

ALUMINUM SPHERE DRY GROUND		COPPER SHEET DRY GROUND		WOOD BOARD WET GROUND	
POLE REAL PART	POLE IMAG PART	POLE REAL PART	POLE IMAG PART	POLE REAL PART	POLE IMAG PART
-.1571540E9	±.6774348E8	-.1542518E9	±.6444038E8	-.1017222E9	±.6758739E8
-.9679531E8	±.1115148E9	-.1801754E9	±.9524389E8	-.5267251E8	±.1368924E9
-.2448631E9	±.3035603E9	-.2092095E9	±.1634348E9	-.1346969E9	±.1986688E9
-.5818507E8	±.3973762E9	-.2065430E9	±.2889621E9	-.2283739E9	±.2903170E9
				-.1330163E9	±.4125691E9

The average extracted resonances of the aluminum sphere, copper sheet and wood board are tabulated in Table 3. From Table 3, we see that the antenna resonance is present in the waveforms of all targets. Note that the extracted resonances of the five targets considered lie in the same general region of the complex frequency plane and are only marginally separated. Such is expected to some extent because all targets considered have (again marginally) similar sizes.

The locations of the target resonances are related to the scattering mechanisms of the target. For subsurface targets, these relationships are complicated by the presence of the air-ground interface, the ground condition and the characteristics of the transient antenna system. For shallow targets the near-field effects and the target-antenna interactions further complicate the picture. In this chapter, we do not intend to explore these relationships. However, it is interesting that, despite the complex electromagnetic scattering situation, the backscattered waveforms of the subsurface targets can still be characterized by a finite exponential series. In fact, the complex exponents in the finite series (i.e., the complex resonances) can be used as discriminants for target identification purposes (4,19,20).

V. CONCLUSIONS

The concept of characterizing subsurface targets by their complex natural resonances has been demonstrated using real radar measurements. Prony's method in conjunction with the parametric-optimization approach outlined in this chapter extracts the complex natural resonances from the measured transient responses of the subsurface targets with reasonable accuracy. It was found that the subsurface targets (metallic and non-metallic) considered in this study can be characterized by a finite number of complex natural resonances that reside inside the bandwidth of the video pulse radar system. The locations of the complex resonances are independent of the radar location, however, they are dependent on ground condition.

REFERENCES

1. D. A. Hill, "Electromagnetic Scattering Concepts Applied to the Detection of Targets Near the Ground," Report 2971-1, September 1970, The Ohio State University ElectroScience Laboratory, Department of Electrical Engineering; prepared under Contract F19628-70-C-0125 for Air Force Cambridge Research Laboratories.

2. C. E. Baum, "On the Singularity Expansion Method for the Solution of Electromagnetic Interaction Problems," Interaction Note 88, December 11, 1971.

3. E. M. Kennaugh and D. L. Moffatt, "Transient and Impulse Response Approximations," Proc. IEEE, Vol. 53, August 1965, pp. 893-901.

4. L. C. Chan, "Subsurface Electromagnetic Target Characterization and Identification," Ph.D. Dissertation, 1979, The Ohio State University, Department of Electrical Engineering.

5. J. D. Young and R. Caldecott, "A Portable Detector for Plastic Pipe and Other Underground Objects," Report 404X-1, September 1973, The Ohio State University ElectroScience Laboratory, Department of Electrical Engineering; prepared for the Columbia Gas Systems Service Corporation, Columbus, Ohio.

6. L. C. Chan, "A Digital Processor for Transient Subsurface Radar Target Identification," Report 479X-3, December 1975, The Ohio State University ElectroScience Laboratory, Department of Electrical Engineering; prepared for Columbia Gas System Service Corporation, Columbus, Ohio.

7. C. W. Davis, III,"The Design and Analysis of a Transient Electromagnetic Antenna for Subsurface Radar Applications," M.Sc. Thesis, The Ohio State University ElectroScience Laboratory, Department of Electrical Engineering, June 1975.

8. E. O. Brigham, The Fast Fourier Transform, Prentice-Hall, Inc., 1974.

9. G. R. Cooper and C. D. McGillam, Method of Signals and System Analysis, Holt, Rinehard and Winston, Inc., 1967.

10. R. Prony, "Essai expérimental et analytique sur les lois de la dilatabilite de fluides elastiques et sur celles del la force expansive de la vapeur de l'alkool, a differentes témperatues," J. L'Ecole Polytech, (Paris), Vol. 1, No. 2, 1795, pp. 24-76.

11. M. L. Van Blaricim and R. Mittra, "A Technique for Extracting the Poles and Residues of a System Directly from Its Transient Response," IEEE Trans. on Antennas and Propagation, Vol. AP-23, No. 6, November 1975, pp. 777-781.

12. M. C. Van Blaricum, "Techniques for Extracting the Complex Resonances of a System Directly from Its Transient Response," Ph.D. Dissertation, 1976, University of Illinois at Urbana-Champaign, Department of Electrical Engineering.

13. M. L. Van Blaricum, "An Analysis of Existing Prony's Method Techniques," Mini-Symposium of Modal Analysis of Experimental Data, Albuquerque, New Mexico, March 1977.

14. M. L. Van Blaricum and R. Mittra, "Problems and Solutions Associated with Prony's Method for Processing Transient Data," IEEE Trans. on Antennas and Propagation, Vol. AP-26, No. 1, January 1978, pp. 174-182.

15. D. L. Moffatt, L. C. Chan and G. A. Hawisher, "Characterization of Subsurface Electromagnetic Soundings," Report 4490-1, September 1977, The Ohio State University ElectroScience Laboratory, Department of Electrical Engineering; prepared under Grant No. ENG76-04344 for National Science Foundation.

16. L. C. Chan and D. L. Moffatt, "Characterization of Subsurface Electromagnetic Soundings," Report 4490-2, December 1978, The Ohio State University ElectroScience Laboratory, Department of Electrical Engineering; prepared under Grant No. ENG76-04344 for National Science Foundation.

17. Uwe Dibbern, "Speech Analysis by Wiener Filtering with Interpolation," 1974, Phillips Forschungslaboratorium Hamburg GmbH, 2 Hamburg, Germany 54.

18. R. N. McDonough, "Matched Exponents for the Representation of Signals," D.Eng., Dissertation, 1963, The Johns Hopkins University, Department of Electrical Engineering.

19. L. C. Chan, D. L. Moffatt and L. Peters, Jr., "A Characterization of Subsurface Radar Targets," Special Issue on Exploration Geophysics, Proc. of IEEE, July, 1979.

20. L. C. Chan and L. Peters, Jr., "Electromagnetic Mine Detection and Identification," Report 4722-1, December 1978, The Ohio State University ElectroScience Laboratory, Department of Electrical Engineering; prepared under Contract DAAK70-77-C-0114 for U.S. Army Mobility Equipment Research and Development Command, Fort Belvoir, Virginia.

21. "Joint Services Electronics Program," Report 710816-1, December 1978, The Ohio State University ElectroScience Laboratory, Department of Electrical Engineering; prepared under Contract N00014-78-C-0049 for Dept. of the Navy, Office of Naval Research, Arlington, Virginia.

22. R. K. Mains and D. L. Moffatt, "Complex Natural Resonances of an Object in Detection and Discrimination," Report 3424-1, June 1974, The Ohio State University ElectroScience Laboratory, Department of Electrical Engineering; prepared under Contract F19628-72-C-0203 for Hanscom Air Force Base, Massachusetts.

23. C. W. Chuang and D. L. Moffatt, "Complex Natural Resonances of Radar Targets Via Prony's Method," Report 3424-3, April 1975, The Ohio State University ElectroScience Laboratory, Department of Electrical Engineering; prepared under Contract F19628-72-C-0203 for Hanscom Air Force Base, Massachusetts.

24. J. N. Brittingham, E. K. Miller, J. L. Willows, "The Derivation of Simple Poles in a Transfer Function from Real-Frequency Information," April 6, 1976, Lawrence Livermore Laboratory, University of California/Livermore; prepared under Contract W-7405-Eng-48 for U. S. Energy Research and Development Administration.

25. J. D. Markel and A. H. Gray, Jr., Linear Prediction of Speech, Springer-Verlag, 1976.

ACKNOWLEDGMENTS

The work reported in this paper was supported in part by The Ohio State University Research Foundation Contracts DAAK70-77-C-0114 with U. S. Army Mobility Equipment Research and Development Command, Ft. Belvoir, Virginia, and Joint Services Electronics Program N00014-78-C-0049 with Department of the Navy, Office of Naval Research, Arlington, Virginia.

Part 7

Finite Element Method

RECENT DEVELOPMENTS OF THE UNIMOMENT METHOD

Kenneth K. Mei and John F. Hunka
Department of Electrical Engineering and Computer Science
University of California, Berkeley, California 94720

Shu-Kong Chang
EMtec Engineering, Inc., Box 679, Berkeley, California 94701

INTRODUCTION

The unimoment method (1) was developed with the objective of solving electromagnetic scattering problems consisting of penetrable inhomogeneous targets. It has since been applied to scattering by dielectric cylinders (2), inhomogeneous loading of biconical antennas (3), scattering by bodies of revolution (4), scattering and penetration of advanced composite material (5), scattering by buried obstacles (6) and multiple scattering (7). In this paper, we shall present the basic techniques of the unimoment method, the formulas associated with various applications, and some recent obtained results particularly for buried obstacles and two body scattering. Finally, we shall speculate on the possibility of composing a complex scatterer from the results of isolated simple scatterers via the multiple scattering technique.

BASIC TECHNIQUES

The motivation of the unimoment method is to solve scattering problems consisting of penetrable inhomogeneous targets. Its principle idea is to approach the solution via differential equations directly rather than the integral equations. The most appealing advantage of the differential equations is that the equations are basically unchanged whether the target is homogeneous or inhomogeneous, yet inhomogeniety puts great strain on the integral equation method. The elliptical partial differential equation resulting from the associated Maxwell's equations can be solved numerically with conventional finite difference or finite element methods. However, the concomitant of the finite methods are the problems of numerical stability and that of radiation conditions. The mesh of the finite methods must be terminated at a finite surface, while the radiation conditions are to be enforced in the far field zone. Numerous ways of enforcing the radiation conditions have been suggested, e.g., McDonald, et al. (8) and Sylvester, et al. (9) used integral equations and Mei, et al. (10) used harmonic expansions, in which the finite difference or finite element equations are directly coupled to the integral equation or the harmonic expansion. Those approaches, however, add complications to numerical solutions in that the resulting matrix is partially full (from integral equations) and partially sparse (from finite difference). The unimoment method provides a simple numerical strategy, which separates the sparse matrix from the full matrix, thus making the problem more manageable.

The unimoment method is most conveniently viewed as a modification of the classical method of the separation of variables, wherein the modal solutions of the differential equation inside a separable surface is replaced by a sequence of computer generated solutions. A typical example of the essence of unimoment method is that of a two-dimensional scattering by an arbitrary dielectric cylinder. If the medium inside the circle C of Fig. 1, were uniform one would use the separation of variables to expand the interior and fields by cylindrical harmonics,

$$\phi(\bar{r}) = \sum_{n=0}^{\infty} a_n^{e,o} J_n(k\rho) \begin{cases} \cos n\phi \\ \sin n\phi \end{cases} \qquad \rho \leq a \tag{1}$$

$$\phi(\bar{r}) = \phi^{inc}(\bar{r}) + \sum_{n=0}^{\infty} b_n^{e,o} H_n^{(2)}(k\rho) \begin{cases} \cos n\phi \\ \sin n\phi \end{cases} \qquad \rho \geq a \tag{2}$$

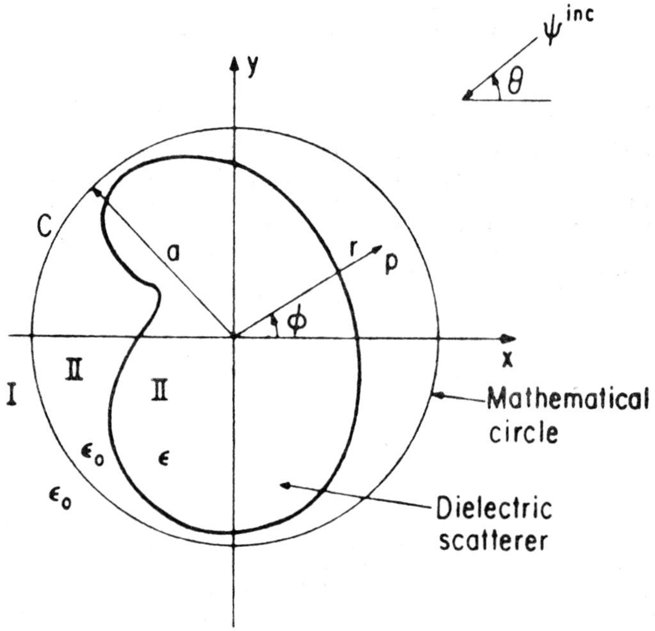

Fig. 1. Scattering geometry of a dielectric cylinder.

When $\psi^{inc}(\bar{r})$ is expanded in Fourier series, and the continuity of $\phi(\bar{r})$ and $\frac{\partial\phi(\bar{r})}{\partial r}$ are enforced on C, the coefficients $a_n^{e,o}$ and $b_n^{e,o}$ can be determined. If the medium inside C is not uniform, the expansion in (1) will not be available and the method of separation of variable fails. The unimoment method of solving the same problem proceeds essentially in parallel with the method of separation of variable, except the interior solution is expanded by a set of numerically generated solutions $\psi_n(\bar{r})$. That is,

$$\phi(\bar{r}) = \sum_{n=0}^{N} A_n \psi_n(\bar{r}) \tag{3}$$

where $\psi_n(\bar{r})$ can be generated by finite difference equations (3), finite element equations (2), Monte Carlo methods (11) or partially in harmonics (4). To generate $\psi_n(\bar{r})$, one needs to solve inhomogeneous wave equations within C, and for the convenience of doing so, we assign a set of linearly independent functions on C and solve the fields inside C numerically for each case. From the numerical solutions we may obtain the normal derivatives $\frac{\partial\psi_n(\bar{r})}{\partial n}$ on C. Of course, one rarely needs to solve the interior problem N times for N number of $\frac{\partial\psi_n(\bar{r})}{\partial n}$ on C. In fact, the inversion may be obtained for part of the interior nodes. In this case, our interest is to obtain $\frac{\partial\psi_n(\bar{r})}{\partial n}$, $\psi_n(\bar{r})$ on the boundary, therefore, the inversion may result in a matrix operator, such that when operated on the $\psi_n(\bar{r})$ on the boundary nodes produces $\psi_n(\bar{r})$ on the nodes next to the boundary. We may then use finite difference to obtain $\frac{\partial\psi_n}{\partial n}$ on the boundary. The numerical details of the finite element or finite difference and the sparse matrix inversion techniques are given in (2), (3), and (4).

The differences between the unimoment method and the method of moment are not limited to numerical. Indeed, the problems often have to be formulated entirely differently for the two methods. Since the numerical procedures involved with the finite element method are well known, we shall skip that and rather discuss some basic formulas used in the unimoment method and the motivation of their developments.

VARIATIONAL FORMULAS FOR AXIALLY SYMMETRIC MAXWELL'S EQUATIONS

The simplest three dimensional problem is one consisting of a body of revolution. With this specialization, each azimuthal mode can be treated separately and thus a three-dimensional problem can be decomposed to several two-dimensional problems. In order to apply the finite element method to such a problem it is necessary to formulate axially symmetric Maxwell's equations in variational form. There are numerous ways Maxwell's equations can be put in variational form, and most of them are quite general. We wish to seek one which is special for axially symmetrical scatterers and valid for inhomogeneous media, because the equations which take advantage of the symmetry are, in general, more economical to compute.

In most classical analysis, solutions to Maxwell's equations are obtained via potentials. It is also advantageous numerically to use potentials, which can often be represented by scalar functions. For axially symmetric problems, the most natural functions that can be used as potentials are the azimuthal components of the electric and magnetic field. The rest of the field components can be derived for each azimuthal mode of $e^{jm\phi}$ variation as:

$$\left.\begin{aligned}
E_z &= j\,\frac{1}{f_m}\left[mR\,\frac{\partial E_\phi}{\partial Z} - \eta_o\mu_r R\,\frac{\partial (RH_\phi)}{\partial R}\right]\\
E_\rho &= j\,\frac{1}{f_m}\left[m\,\frac{\partial (RE_\phi)}{\partial R} + \eta_o\mu_r R^2\,\frac{\partial H_\phi}{\partial Z}\right]\\
H_z &= j\,\frac{1}{f_m}\left[mR\,\frac{\partial H_\phi}{\partial Z} + \frac{1}{\eta_o}\,\varepsilon_r R\,\frac{\partial (RE_\phi)}{\partial R}\right]\\
H_\rho &= j\,\frac{1}{f_m}\left[m\,\frac{\partial (RH_\phi)}{\partial R} - \frac{1}{\eta_o}\,\varepsilon_r R^2\,\frac{\partial E_\phi}{\partial Z}\right]
\end{aligned}\right\} \qquad (4)$$

where $f_m(R,Z) = \mu_r \varepsilon_r R^2 - m^2$

$$R = k_o\rho; \quad Z = k_o z, \quad \eta_o = \sqrt{\mu_o/\varepsilon_o}$$

We have used cylindrical coordinates for the interior fields, because the integration involved in the finite element would be simpler than those using spherical coordinates.

The following coupled differential equations are obtained by substituting the field components of (4) into Maxwell's equations,

$$\frac{\partial}{\partial Z}\left[\frac{m}{f_m}\frac{\partial(RE_\phi)}{\partial R}\right] + \frac{\partial}{\partial Z}\left[\frac{\eta_o\mu_r R}{f_m}\frac{\partial(RH_\phi)}{\partial Z}\right] - \frac{\partial}{\partial R}\left[\frac{m}{f_m}\frac{\partial(RE_\phi)}{\partial Z}\right] + \frac{\partial}{\partial R}\left[\frac{\eta_o\mu_r R}{f_m}\frac{\partial(RH_\phi)}{\partial R}\right] + \eta_o\mu_r H_\phi = 0 \tag{5}$$

$$-\frac{\partial}{\partial Z}\frac{\eta_o m}{f_m}\frac{\partial(RH_\phi)}{\partial R} + \frac{\partial}{\partial Z}\left[\frac{\varepsilon_r R}{f_m}\frac{\partial(RE_\phi)}{\partial Z}\right] + \frac{\partial}{\partial R}\left[\frac{\eta_o m}{f_m}\frac{\partial(RH_\phi)}{\partial Z}\right] + \frac{\partial}{\partial R}\left[\frac{\varepsilon_r R}{f_m}\frac{\partial(RE_\phi)}{\partial R}\right] + \varepsilon_r E_\phi = 0 \tag{6}$$

If we define the differential operator, ∇_a, as

$$\nabla_a = \hat{\rho}\frac{\partial}{\partial R} + \hat{z}\frac{\partial}{\partial Z}$$

the differential equations (5) and (6) may be arranged to the following form,

$$\nabla_a \cdot \frac{1}{f_m}\left[\eta_o\mu_r R\nabla_a(RH_\phi) + \hat{z}\times m\nabla_a(RE_\phi)\right] + \eta_o\mu_r H_\phi = 0 \tag{7}$$

$$\nabla_a \cdot \frac{1}{f_m}\left[\varepsilon_r R\nabla_a(RE_\phi) - \hat{z}\times\eta_o m\nabla_a(RH_\phi)\right] + \varepsilon_r E_\phi = 0 \tag{8}$$

Equations (5), (6) and (7), (8) are exactly the same and they may be used interchangeably.

For the interior problems of the unimoment method the differential equations are solved by imposing the Dirichlet boundary conditions. That is, the functions $E_\phi(r=a,\theta)$ and $H_\phi(r=a,\theta)$ are assigned over the spherical surfaces. The uniqueness of the solution is followed immediately by the usual way of proving the uniqueness theorem as that presented in the textbooks by Stratton (12) and by Harrington (13).

In order to facilitate the Ritz finite element method for the interior solutions, the variational formulations corresponding to the differential equations will be established. The variational principles dictate that the solution of (5) and (6) is the stationary functional of the Lagrangian integral (14, p. 275).

$$L = \iint_S F(R,Z,f_1,f_2,f_3,f_4,f_5,f_6)\, dr\, dz \tag{9}$$

where the function F should take the following quadratic form so that the finite element method will result in a system of linear equations. That is,

$$F = \sum_{i=1}^{6} \sum_{j=1}^{6} g_{ij} f_i f_j \tag{10}$$

where

$$f_1 = RE_\phi \tag{11}$$

$$f_2 = \eta_o RH_\phi \tag{12}$$

$$f_3 = \frac{\partial (RE_\phi)}{\partial R} \tag{13}$$

$$f_4 = \eta_o \frac{\partial (RH_\phi)}{\partial R} \tag{14}$$

$$f_5 = \frac{\partial (RE_\phi)}{\partial Z} \tag{15}$$

$$f_6 = \eta_0 \frac{\partial (RH_\phi)}{\partial Z} \tag{16}$$

and $g_{ij}(R,Z)$ are coefficient functions.

For the Dirichlet boundary value problems the requirement that L be stationary corresponds to the Lagrange-Euler equations (15, p. 277)

$$-\frac{\partial F}{\partial f_1} + \frac{\partial}{\partial R}\left(\frac{\partial F}{\partial f_3}\right) + \frac{\partial}{\partial Z}\left(\frac{\partial F}{\partial f_5}\right) = 0 \tag{17}$$

and

$$-\frac{\partial F}{\partial f_2} + \frac{\partial}{\partial R}\left(\frac{\partial F}{\partial f_4}\right) + \frac{\partial}{\partial Z}\left(\frac{\partial F}{\partial f_6}\right) = 0 \tag{18}$$

The coefficient functions $g_{ij}(r,Z)$ are obtained by substituting (10) into (17) and (18), and then comparing the resulting equation resepctively with (5) and (6). The resulting function F for the Lagrangian integral (9) is

$$F = \frac{R}{2f_m}\left[\varepsilon_r\left(\frac{\partial (RE_\phi)}{\partial R}\right)^2 + \varepsilon_r\left(\frac{\partial (RE\ )}{\partial Z}\right)^2 + \eta_o^2\mu_r\left(\frac{\partial (RH_\phi)}{\partial R}\right)^2 + \eta_o^2\mu_r\left(\frac{\partial (RH_\phi)}{\partial Z}\right)^2 + 2m\eta_o \frac{\partial (RE_\phi)}{\partial R}\frac{\partial (RH_\phi)}{\partial Z} - 2m\eta_o \frac{\partial (RE_\phi)}{\partial Z}\frac{\partial (RH_\phi)}{\partial R}\right] - \frac{R}{2}\left[\varepsilon_r E_\phi^2 + \eta_o^2\mu_r H^2\right] \tag{19}$$

Using the differential operator, ∇_a, the functional F takes the following form

$$F = \frac{\varepsilon_r R}{2f_m}\nabla_a(RE_\phi)\cdot\nabla_a(RE_\phi) + \frac{\mu_r R\eta_o^2}{2f_m}\nabla_a(RH_\phi)\cdot\nabla_a(RH_\phi) + 2m\eta_o\,\hat{\phi}\cdot\nabla_a(RE_\phi)\times\nabla_a(RH_\phi) - \frac{R}{2}\left(\varepsilon_r E_\phi^2 + \eta_o^2\mu_r H_\phi^2\right) \tag{20}$$

The range S of the variational integral (9) is on the meridional plane and within the artificial sphere $(r < a)$. Specifically, S is the interior of a circular disc of radius $r = a$. Only one half of the disc on one side of the symmetrical axis is required to integrate in all computations because of the symmetrical properties. In this case, the boundary conditions at the straight side of the half-disc are required. Since

$$e^{jm\pi} = \begin{cases} 1 & \text{if m is even} \\ -1 & \text{if m is odd.} \end{cases}$$

we find that, when m is odd, $E_\phi(\phi=\pi+\phi_o)$ is equal to $-E_\phi(\phi_o)$. This implies that E_ϕ is an odd-symmetric function with respect to z-axis. Consider the actual E-field normal to the meridional plane. The reference direction of E_ϕ is reversed on the other side of the z-axis. Hence the field normal to the meridional plane is an even-symmetric function. This means that its normal derivative is zero. That is,

$$\frac{\partial E_\phi}{\partial n} = -\frac{\partial E_\phi}{\partial \rho} = 0 \quad \text{at z-axis.} \tag{21}$$

A similar argument can be applied to H_ϕ when m is odd

$$\frac{\partial H_\phi}{\partial n} = -\frac{\partial H_\phi}{\partial \rho} = 0 \quad \text{at z-axis.} \tag{22}$$

Similarly, for even values of m (23)

$$E_\phi = 0 \quad \text{on the z-axis}$$

$$H_\phi = 0 \quad \text{on the z-axis.} \tag{24}$$

The conditions of (21) to (24) are simply homogeneous Neumann and Dirichlet conditions. The functional (20) together with the boundary conditions provide the necessary formulation to generate the interior fields via the finite element method.

The sequence of interior solutions obtained from the finite element method, when coupled with the conventional spherical harmonics expansions of the exterior fields should result in the scattering solutions in free space. A typical result of such a calculation is shown in Fig. 2. The agreement between computation and measurement is quite remarkable, indicating the soundness of the unimoment method.

LOSSY GROUND

It is quite evident in the unimoment method that we can use a variety of exterior expansions to solve scattering by a target in the presence of another object. The repertoire of known exterior expansions for that purpose include circular cylinders, spheres, infinite cone, etc. These exterior expansions are essentially classical solutions of scattering by separable targets using multipole fields.

The exterior expansions involving lossy ground is an important topic which deserves special attention because it can be applied to many problems in

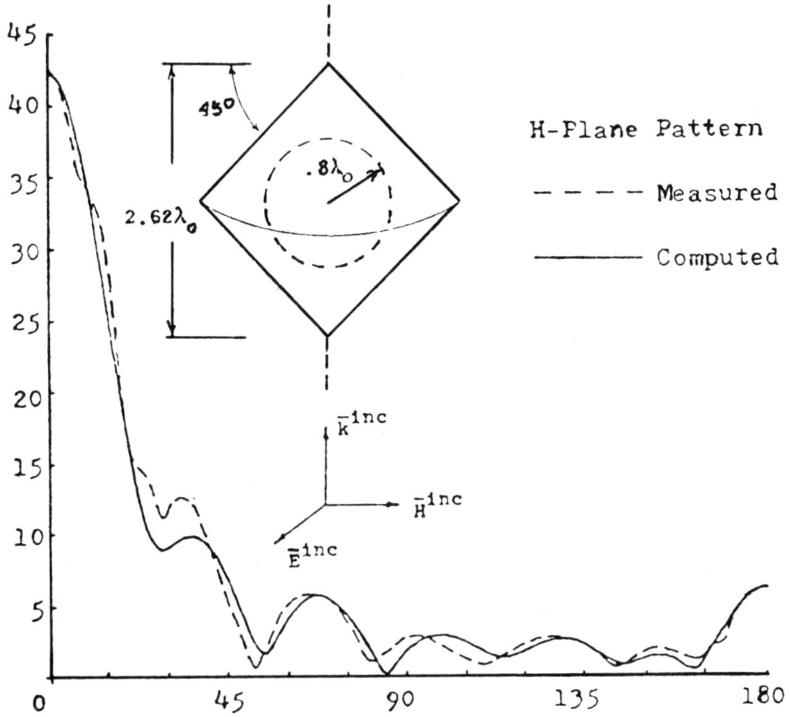

Fig. 2. Scattered far field of a hollow dielectric body $\varepsilon_r = 2.61$

geophysical research as well as in electromagnetics. The mathematical problem of finding the exterior expansions including a lossy, semi-infinite ground is to find the fields due to spherical multipole sources, which satisfy the boundary conditions of the air-earth interface and the radiation conditions. These lead to generalized Sommerfeld integrals, which we shall describe in this section.

The geometry of the air-ground interface problem is shown in Fig. 3. The objective is to find the electromagnetic fields using a spherical multipole as a primary source at the origin which is a distance d below the earth surface. The electromagnetic fields are to be derived from potentials A_m (TE) and A_e (TM) for each azimuthal mode. The electric and magnetic fields in terms of these vector potentials are given by

$$\overline{E} = \nabla \times A_m \vec{a} + \frac{1}{j\omega\varepsilon} \nabla \times \nabla \times A_e \vec{a} \tag{25}$$

$$\overline{H} = \nabla \times A_e \vec{a} - \frac{1}{j\omega\mu} \nabla \times \nabla \times A_m \vec{a} \tag{26}$$

where A_e and A_m are solutions of scalar wave equations,

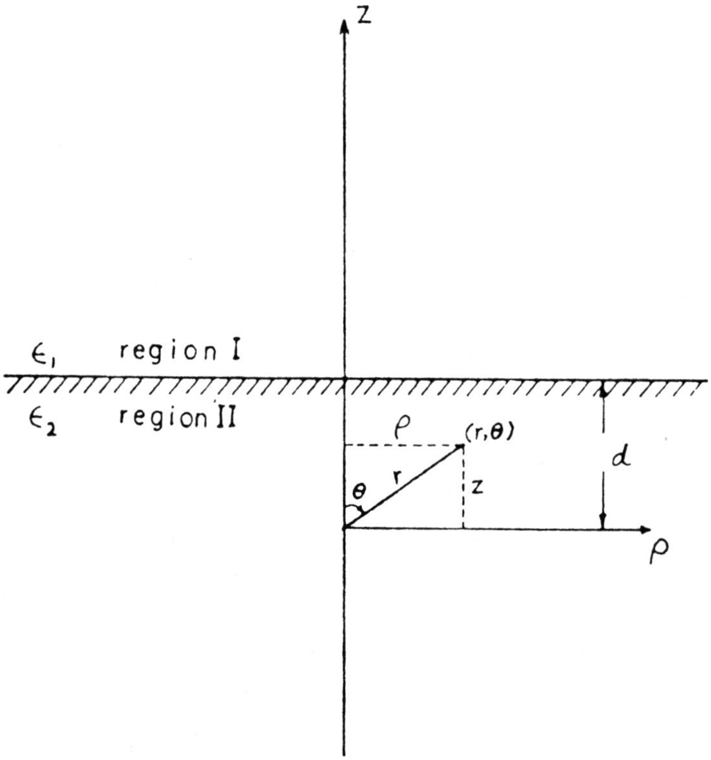

Fig. 3. Coordinates in the meridional plane ($\phi = 0$) and $\phi = \pi$)

$$(\nabla^2 + k^2) \begin{Bmatrix} A_e \\ A_m \end{Bmatrix} = 0 \tag{27}$$

if $\vec{a}$ is a constant vector. Normally $\vec{a}$ would be a unit vector in any of the x, y and z directions. The special case in spherical coordinate requires $\vec{a}$ to be the radial vector $\vec{a} = \vec{r}$ in order to preserve eq. (27) for the potentials.

In the unimoment method, we have used the sphere as a termination surface for the finite mesh. It is necessary that the modal sources be confined within a finite sphere. Thus, the spherical multipoles are chosen as primary sources. On the other hand, to satisfy the planar boundary conditions of the air-ground interface, the cylindrical modal potentials are desirable. This combination of restrictions induces us to use spherical functions for A_m and A_e, and the unit vector $\hat{z}$ for $\bar{a}$.

In order to find the multipole fields which satisfy the boundary conditions of the air-ground interface we must find the Fourier Bessel representation of the multipole potentials, i.e.,

$$h_n^{(2)}(kr)P_n^m(\cos\theta) = (\mathrm{sign}\, z)^{m+n} \int_0^\infty f_{mn}(\lambda) J_m(\lambda\rho) e^{-\mu|z|} d \tag{28}$$

where $\mu = \sqrt{\lambda^2 - k^2}$, $\mathrm{Real}(\mu) \geq 0$.
It has been shown by Chang and Mei (15) that $f_{mn}(\lambda)$ can be conveniently obtained by recurrence formulas,

$$f_{m+1,m+1}(\lambda) = (2m+1) \frac{\lambda}{k} f_{mm}(\lambda) \tag{29}$$

with

$$f_{0,0}(\lambda) = j \frac{\lambda}{k\mu} \tag{30}$$

and

$$f_{m,n+1}(\lambda) = \frac{(2n+1)}{(n-m+1)} \left[\frac{\mu}{k} f_{m,n}(\lambda) + \frac{(n+m)}{2n+1} f_{m,n-1}(\lambda)\right] \tag{31}$$

with

$$f_{m,m-1}(\lambda) = 0 \tag{32}$$

Using the above given recurrence formulas, it is a few simple algebraic steps away from the exterior expansions for a lossy ground, or that for a lossy half space. They are the generalization of the well known Sommerfeld's integrals for spherical multipoles, hence are appropriately termed generalized Sommerfeld's integrals. For buried obstacles the potentials in region II (containing the source) may be separated into a prime component and a secondary component. For the TE potentials

$$A_m^{II} = A_m^{II\ prime} + A_m^{II\ sec.} \tag{33}$$

where

$$A_m^{II\ prime} \vec{a} = \hat{z}\, h_n^{(2)}(k_2 r)\, P_n^m(\cos\theta)\, e^{\pm jm\phi} \tag{34}$$

$$A_m^{II\ sec} \vec{a} = \hat{z}\, e^{\pm jm} \int_0^\infty \frac{\mu_2-\mu_1}{\mu_2+\mu_1} f_{mn}(\lambda)\, J_m(\lambda\rho)\, e^{-\mu_2(2d-z)} \tag{35}$$

And, in region I, the potential A_m^I is,

$$A_m^I \vec{a} = \hat{z}\, e^{\pm jm\phi} \int_0^\infty \frac{2\mu_1}{\mu_2+\mu_1} f_{mn}(\lambda)\, e^{-\mu_2 d-\mu_1(z-d)} \tag{36}$$

Similarly integrals are found for the TM potentials,

$$A_e^{II\ sec} \vec{a} = \hat{z}\, e^{\pm jm\phi} \int_0^\infty \frac{\varepsilon_1\mu_2-\varepsilon_2\mu_1}{\varepsilon_1\mu_2+\varepsilon_2\mu_1} f_{mn}(\lambda) J_m(\lambda\rho) e^{-\mu_2(2d-z)}\, d\lambda \tag{37}$$

$$A_e^I \vec{a} = \hat{z}\, e^{\pm jm\phi} \int_0^\infty \frac{2\varepsilon_1\mu_2}{\varepsilon_1\mu_2+\varepsilon_2\mu_1} f_{mn}(\lambda) J_m(\lambda\rho) e^{-\mu_2 d-\mu_1(z-d)}\, d\lambda \tag{38}$$

It is important to point out that the combination of spherical functions and unit vector $\vec{a} = \hat{z}$ in (25) and (26) does not result in a complete set of electromagnetic fields. Consequently, it is necessary to add a set of horizontal rotating potentials. They are,

$$A_m^{II\ prime}\ \vec{a} = (\hat{x}\pm j\hat{y})\ h_{m-1}^{(2)}(k_2 r)\ P_{m-1}^{m-1}(\cos\theta)\ e^{\pm j(m-1)\mu} \tag{39}$$

$$\vec{A}_m^{II\ sec} = (\hat{\rho}\pm j\hat{\phi})\ A_m^s + A_{mz}^s\ \hat{z} \tag{40}$$

with

$$A_m^s = e^{\pm jm\phi}\int_0^\infty \frac{\varepsilon_1\mu_2-\varepsilon_2\mu_1}{\varepsilon_2\mu_1+\varepsilon_1\mu_2}\ f_{m-1,m-1}(\lambda)\ J_{m-1}(\lambda\rho)e^{-\mu_2(2d-z)}\ d\lambda \tag{41}$$

$$A_{mz}^s = e^{\pm jm\phi}\int \frac{(\varepsilon_1-\varepsilon_2)\ 2\mu_2\lambda}{(\mu_1+\mu_2)(\varepsilon_2\mu_1+\varepsilon_1\mu_2)} f_{m-1,m-1}(\lambda)e^{-\mu_2(2d-z)}\ d\lambda \tag{42}$$

$$\vec{A}_m^I = (\hat{\rho}\pm j\hat{\phi})\ A_m^t + A_{mz}^t\ \hat{z}$$

with

$$A_m^t = e^{\pm jm\phi}\int_0^\infty \frac{2\varepsilon_2\mu_2}{\varepsilon_2\mu_1+\varepsilon_1\mu_2}\ f_{m-1,m-1}(\lambda)\ J_{m-1}(\lambda\rho)e^{-\mu_2 d-\mu_1(z-d)}\ d\lambda \tag{43}$$

$$A_{mz}^t = e^{\pm jm\phi}\int_0^\infty \frac{(\varepsilon_1-\varepsilon_2)\ 2\mu_2\lambda}{(\mu_1+\mu_2)(\varepsilon_2\mu_1+\varepsilon_1\mu_2)} f_{m-1,m-1}(\lambda)J_m(\lambda\rho)e^{-\mu_2 d-\mu_1(z-d)}\ d\lambda \tag{44}$$

$$A_e^{II\ prime}\ \vec{a} = (\hat{x}\pm jy)\ h_{m-1}^{(2)}\ (k_2 r)\ P_{m-1}^{m-1}(\cos\theta)\ e^{\pm j(m-1)\phi} \tag{45}$$

$$\vec{A}_e^{II\ sec} = (\hat{\rho}\pm j\hat{\phi})\ A_e^s + \hat{z}\ A_{ez}^s \tag{46}$$

with

$$A_e^s = e^{\pm jm\phi}\int_0^\infty \frac{\mu_2-\mu_1}{\mu_2+\mu_1}\ f_{m-1,m-1}(\lambda)\ J_{m-1}(\lambda\rho)\ e^{-\mu_2(2d-z)}\ d\lambda \tag{47}$$

$$A_{ez}^s = e^{\pm jm\phi}\int_0^\infty \frac{(\varepsilon_1-\varepsilon_2)2\mu_2\lambda}{(\mu_1+\mu_2)(\varepsilon_2\mu_1+\varepsilon_1\mu_2)} f_{m-1,m-1}(\lambda)J_m(\lambda\rho)e^{-\mu_2(2d-z)}\ d\lambda \tag{48}$$

$$\vec{A}_e^I = (\hat{\rho}\pm j\hat{\phi})\ A_e^t + \hat{z}\ A_{ez}^t$$

with

$$A_e^t = e^{\pm jm\phi}\int_0^\infty \frac{2\mu_2}{\mu_2+\mu_1}\ f_{m-1,m-1}(\lambda)J_{m-1}(\lambda\rho)e^{-\mu_2 d-\mu_1(z-d)}\ d\lambda \tag{49}$$

$$A_{ez}^t = e^{\pm jm\phi}\int_0^\infty \frac{(\varepsilon_1-\varepsilon_2)\ 2\mu_2}{(\mu_1+\mu_2)(\varepsilon_2\mu_1+\varepsilon_1\mu_2)}\ f_{m-1,m-1}(\lambda)J_m(\lambda\rho)e^{-\mu_2 d-\mu_1(z-d)}\ d\lambda \tag{50}$$

Equations (28)-(50) give the complete account of the exterior potentials of Maxwell's equations associated with lossy half space. They have been used with the unimoment method to find the scattering from buried targets. Some representative results of the buried scatterer configuration of Fig. 4 are shown in Fig. 5.

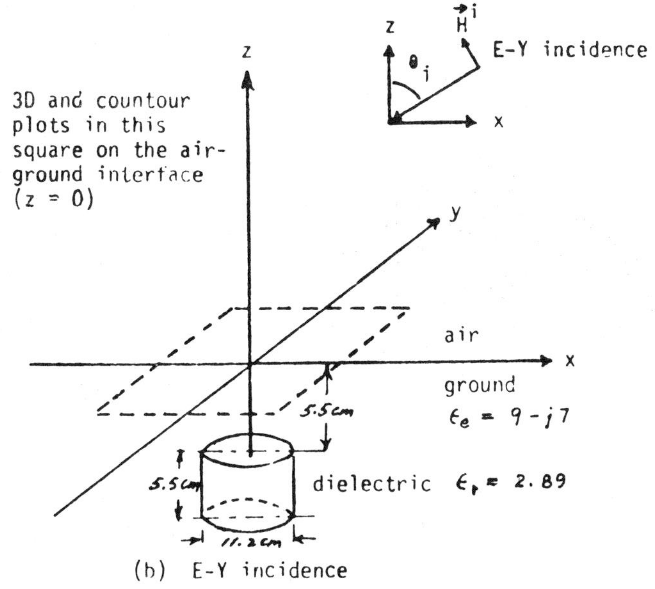

Figure 4. The square on the ground plane in which the 3D and contour plots of the scattered electric fields will be shown.

MULTIPLE SCATTERING

The electromagnetic scatterings of two or more targets near each other is an interesting problem by itself. They can be applied to discriminate single targets from a cluster of targets. Computationally multiple scattering techniques can also be used to solve scattering from complex targets from the result of scattering from simple targets such as the bodies of revolution of the previous sections. Multiple scattering may be solved as a single self-contained problem or as an iterative problem. We have chosen the latter in consideration of lesser demand on memory and time.

Consider two targets as shown in Fig. 6. Each of them is a body of revolution but the ensemble is not axially symmetric. It is apparent that to solve this ensemble of targets as a single body would require a true three dimensional solution where the azimuthal modes cannot be decoupled. An iterative technique of solving this two body scattering shall only require the solutions of the two individual targets which have been described in the previous sections.

Our iteration first solves the scattering from target one as a single scatterer. The scattered field is then added to the incident fields of target 2 and the scattered fields of target 2 are found. The scattered fields of target 2 are then used as incident fields of target 1 to find the secondary scattered fields of target 1. The process continues until a convergence criterion is satisfied. The iteration steps are listed in Table I for clarity. The total scattered fields of each target is the sum of all the iterated results. It is noted that in the unimoment method described in previous sections, the finite element equations are essentially inverted, thus each iteration amounts to a matrix operation on a vector, which is computationally quite economical.

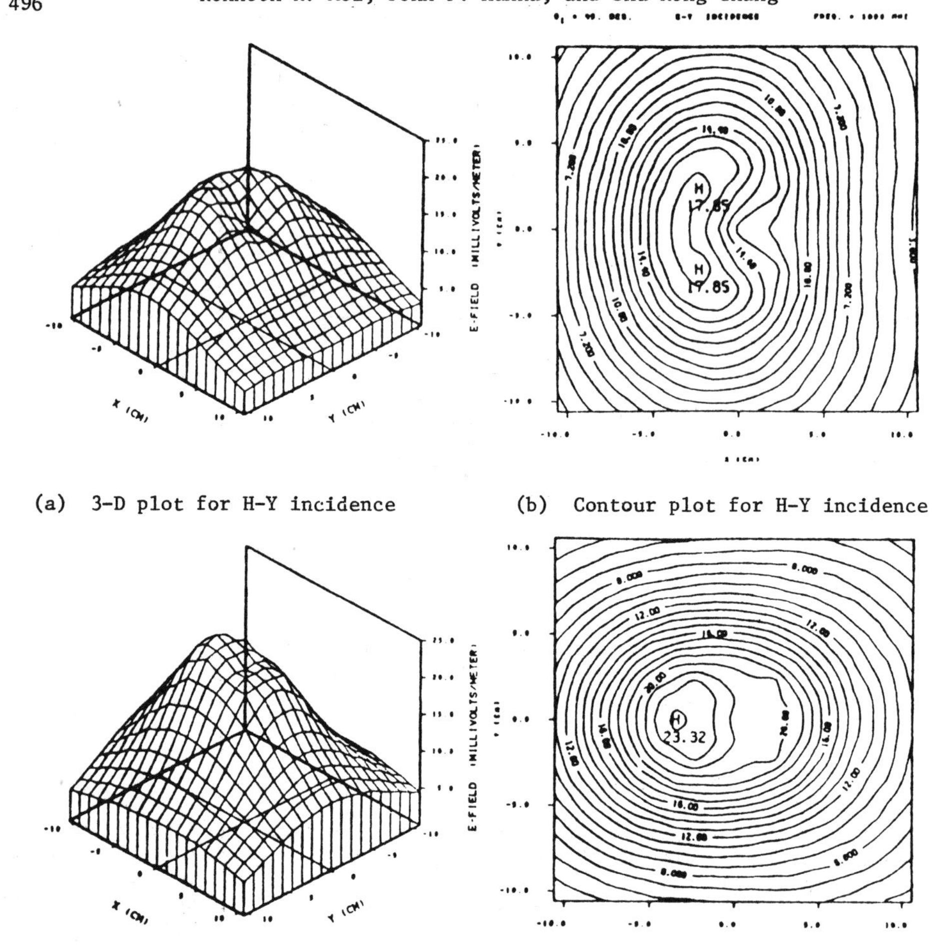

(a) 3-D plot for H-Y incidence (b) Contour plot for H-Y incidence

(a) 3-D plot for E-Y incidence (b) Contour plot for E-Y incidence

Figure 5. 3-D and contour plots of scattered E-field amplitude on the earth surface ($\theta_i = 45°$, frequency = 1000 MHz)

The success of the iterative solutions depends largely on the classical addition theorems of spherical harmonics, which transforms the scattered fields of one target from origin 0' to 0 of the other target, and vice versa. Numerically computing such transformation would be too expensive.

Referring to Fig. 6, the coordinates transformations involves a rotation from the target axis to the common z z' axis, a translation from 0' to 0 and a rotation from the z z' axis to the target axis of the target at origin 0.

The addition theorem of rotation is given in terms of Euler angles α, β, ν as shown in Fig. 7. We will use only a right handed rotation. Referring to Fig. 7 the fixed system is (XYZ) while the final rotated system is $(\tilde{X}\tilde{Y}\tilde{Z})$. The first rotation is $(0 \leq \alpha \leq 2\pi)$ about the Z axis. The second rotation is

TABLE 1

	SCATTERER 1		SCATTERER 2	
ITERATION	INCIDENT FIELD	SCATTERED FIELD	INCIDENT FIELD	SCATTERED FIELD
1	$\bar{E}_i$	$\bar{E}_{S1}^{(1)}$	—	—
2	—	—	$\bar{E}_i' + \bar{E}_{S1}^{(1)'}$	$\bar{E}_{S1}^{(2)'}$
3	$E_{S1}^{(2)}$	$\bar{E}_{S2}^{(1)}$	—	—
4	—	—	$\bar{E}_{S2}^{(1)'}$	$\bar{E}_{S2}^{(2)'}$
5	$\bar{E}_{S2}^{(2)}$	$\bar{E}_{S3}^{(1)}$	—	—
6	—	—	$\bar{E}_{S3}^{(1)'}$	$\bar{E}_{S3}^{(2)'}$

$\beta(0 \le \beta \le \pi)$ about the $\underset{\sim}{Y'}$ (also Y") axis. The third rotation is $\nu(0 \le \gamma \le 2\pi)$ about the $\underset{\sim}{Z}$ (also Z") axis. Using these definitions, it is easy to show that

$$\begin{bmatrix} \tilde{x} \\ \tilde{y} \\ \tilde{z} \end{bmatrix} = \begin{bmatrix} \cos\nu & \sin\nu & 0 \\ \sin\nu & \cos\nu & 0 \\ 0 & 0 & 1 \end{bmatrix} \begin{bmatrix} \cos\beta & 0 & -\sin\beta \\ 0 & 1 & 0 \\ \sin\beta & 0 & \cos\beta \end{bmatrix} \begin{bmatrix} \cos\alpha & \sin\alpha & 0 \\ -\sin\alpha & \cos\alpha & 0 \\ 0 & 0 & 1 \end{bmatrix} \begin{bmatrix} x \\ y \\ z \end{bmatrix}$$

From Stein (16) and Edmonds (17), the rotational addition theorem for the spherical harmonics is written as

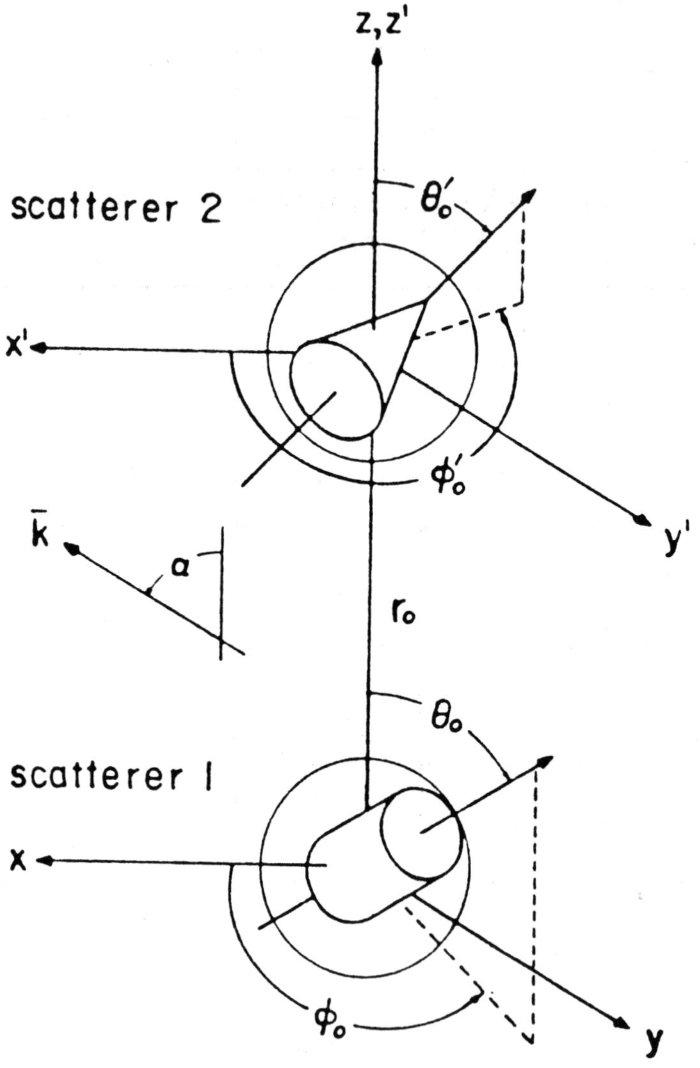

Fig. 6. Coordinate systems for describing the general three dimensional two-body-of-revolution scattering problem.

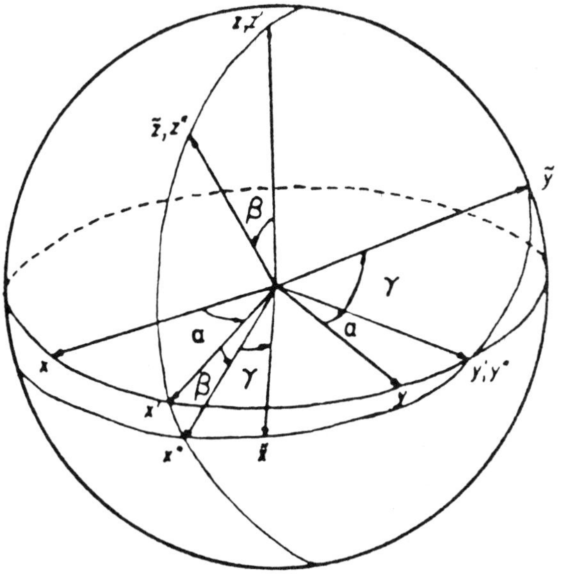

Figure 7. Rotation of coordinate axes through successive Euler angle rotations

$$Y_{nm}(\theta,\phi) = \sum_{\mu=n}^{n} Y_{n\mu}(\tilde{\theta},\tilde{\phi})\ D_{nm}^{(n)}\ (\alpha\beta\nu) \tag{52}$$

where

$$Y_{nm}(\theta,\phi) = (-1)^m \left[\frac{2n+1}{4\pi}\ \frac{(n-m)!}{(n+m)!}\right]^{1/2} P_n^m(\cos\,\theta)\ e^{jm\phi} \tag{53}$$

$$D_{\mu m}^{(n)}(\alpha\beta\nu) = e^{j\mu\nu} d_{\mu m}^{(n)}(\beta)\ e^{jm\alpha} \tag{54}$$

$$d_{\mu m}^{n}(\beta) = \left[\frac{(n+\mu)!\ (n-\mu)!}{(n+m)!\ (n-m)!}\right]^{1/2} \sum_{\sigma=\sigma_{min}}^{\sigma_{max}} \binom{n+m}{n-\mu-\sigma}\binom{n-m}{\sigma} (-1)^{n-\mu-\sigma}$$

$$\cdot(\cos\ \beta/2)^{2\sigma+\mu+m}\ (\sin\ \beta/2)^{2n-2\sigma-\mu-m} \tag{55}$$

$$\begin{aligned} \sigma_{min} &= \max\ \{0,\ -(m+\mu)\} \\ \sigma_{max} &= \min\ \{n-m,\ n-\mu\} \end{aligned} \tag{56}$$

The translated addition theorem for the scalar spherical wave functions is,

$$Z_n(kr)P_n^m(\cos\theta)\, e^{jm\phi} = j^{-n} \sum_{\nu=0}^{\infty} \sum_{\mu=-\mu}^{\nu} \sum_{p=p_{max}}^{p_{min}} \Big\{ j^{\nu+p}(2\nu+1)(-1)^{\mu}$$

$$a(m,\mu|p,n,\nu)\; Z_p(kr_>)\; P_p^{m+\mu}(\cos\theta_>)\; e^{j(m+\mu)\phi_>}$$

$$j_\nu(kr_<)\; P_\nu^{-\mu}(\cos\theta_<)\; e^{j-\mu\phi_<} \Big\} \tag{57}$$

where

$$\begin{bmatrix} r_> = r' & r_< = r_o \\ \theta_> = \theta' & \theta_< = \theta_o \\ \phi_> = \phi' & \phi_< = \phi_o \end{bmatrix} \quad r' \geq r_o \tag{58}$$

$$\begin{bmatrix} r_> = r_o & r_< = r' \\ \theta_> = \theta_o & \theta_< = \theta' \\ \phi_> = \phi_o & \phi_< = \phi' \end{bmatrix} \quad r' \leq r_o \tag{59}$$

$$\left.\begin{aligned} P_{min} &= |n-\nu| \\ P_{max} &= n+\nu \end{aligned}\right\} \tag{60}$$

The coefficient $a(m,\mu|p,n,\nu)$ is defined by the expansion,

$$P_n^m(\cos\theta)\; P_r^{\mu}(\cos\theta) = \sum_{p=p_{min}}^{P_{max}} a(m,\mu|p,n,\nu)\; P_p^{m+\mu}(\cos\theta) \tag{61}$$

where

$$a(m,\mu|p,n,\nu) = (-1)^{m+\mu}\;(2p+1)\left\{\frac{(n+m)!\;(\nu+\mu)!\;(p-m-\mu)!}{(n-m)!\;(\nu-\mu)!\;(p+m+\mu)!}\right\}^{1/2}$$

$$\begin{pmatrix} n & \nu & p \\ 0 & 0 & 0 \end{pmatrix}\begin{pmatrix} n & \nu & p \\ m & \mu & -m-\mu \end{pmatrix} \tag{62}$$

where

$\begin{pmatrix} j_1 & j_2 & j_3 \\ m_1 & m_2 & m_3 \end{pmatrix}$ is the Wigner 3-j symbol,

$$\begin{pmatrix} j_1 & j_2 & j_3 \\ m_1 & m_2 & m_3 \end{pmatrix} = (-1)^{j_1-j_2-m_3} \, \delta(m_1+m_2-m_3) \left\{ \frac{(j_1+j_2+j_3)! \, (j_1-j_2+j_3)! \, (-j_1+j_2+j_3)!}{(j_1+j_2+j_3+1)!} \right\}^{1/2}$$

$$[(j_1+m_1)! \, (j_1-m_1)! \, (j_2+m_2)! \, (j_2-m_2)! \, (j_3+m_3)! \, (j_3-m_3)!]^{1/2}$$

$$\sum_{Z_{min}}^{Z_{max}} \frac{(-1)^z}{z! \, (j_1+j_2-j_3-z)! \, (j_1-m_1-z)! \, (j_2+m_2-z)! \, (j_3-j_2+m_1+z)! \, (j_3-j_1-m_2+z)!} \tag{63}$$

where

$$\begin{aligned} Z_{min} &= \text{Max}\{0, \, j_2-j_3-m, \, j_1-j_3+m_2\} \\ Z_{max} &= \text{Min}\{j_1+j_2-j_3, \, j_1-m_1, \, j_2+m_2\} \end{aligned} \tag{64}$$

With the above addition theorems for scalar spherical harmonics, it is now possible to get the addition theorem for vector spherical harmonics. The details may be found in reference (18). Indeed, those are all one needs to compute multiple scattering once the single scattering results are obtained from the unimoment method.

Figure 8 shows the results of scattering by two conducting finite cylinders placed at right angles to each other. The cylinders are identical at 1λ in diameter and height. The agreement between measurement and calculation is remarkable. Figure 9 shows the result of scattering by the same two finite cylinders in contact with each other, and again the agreement between calculation and measurement is most gratifying.

SCATTERER COMPOSITION

The promising results of the iterative multiple scattering technique for contacting targets prompts us to speculate on the future application of the multiple scattering to scatter composition, where a complex target is composed from several bodies of revolution. It is noted that the result of unimoment solution of scattering by an isolated body of revolution is a matrix operator, which can be easily stored in disks of modern minicomputers. The iterative part is very undemanding on memory. It is possible to carry out the multiple scattering by a mini-computer now available. A repertoire of unimoment solutions for a number of canonical scatterers should enable a mini-computer to generate solutions to a variety of scattering problems at a very low cost.

CONCLUSION

The unimoment method started with a modest objective to solve electromagnetic scattering by inhomogeneous bodies of revolution and has now been developed to solve scattering by buried objects, which shall have many applications in geophysical research in addition to electromagnetic scattering. Although it is possible, in principle, to extend the unimoment method to an arbitrary three-dimensional problem (non-axially symmetric), the cost of solving a set of three-dimensional finite element or finite difference equations would be excessive. In this paper, we have demonstrated that an iterative multiple scattering technique can be applied to solve a class of non-symmetric scattering problems, Indeed, it

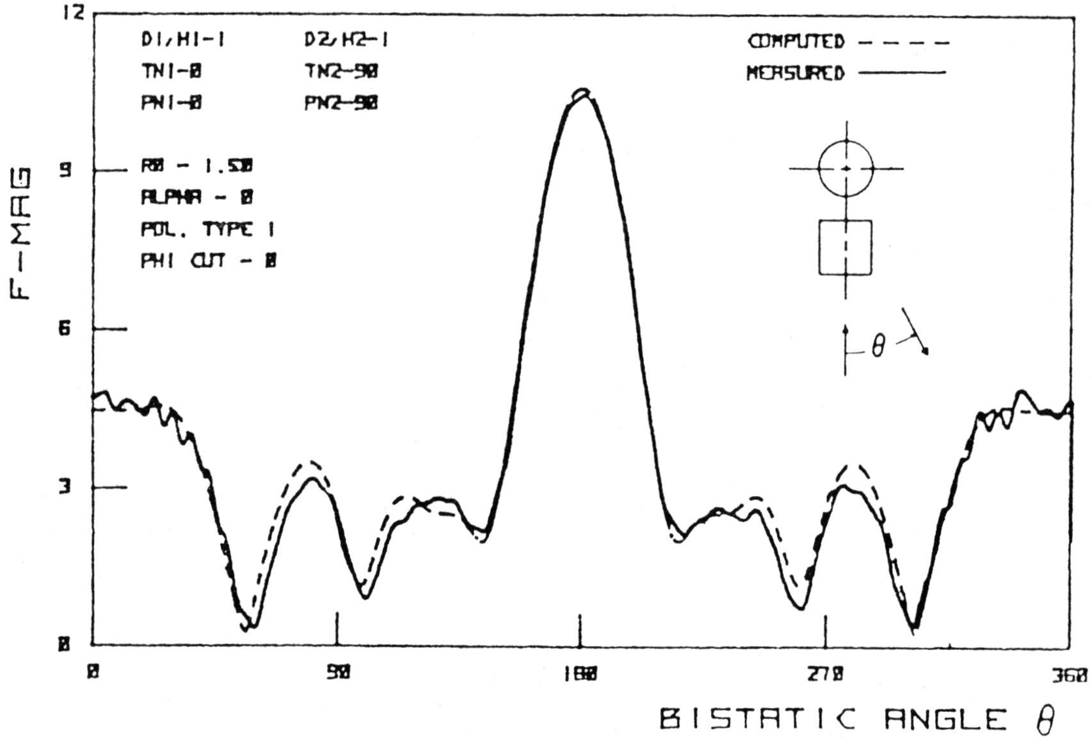

Fig. 8. Computed and measured far field scattering by two perfectly conducting finite cylinders

is now possible to synthesize scatterers via multiple scattering by a mini-computer if a repertoire of unimoment solutions is available for that purpose.

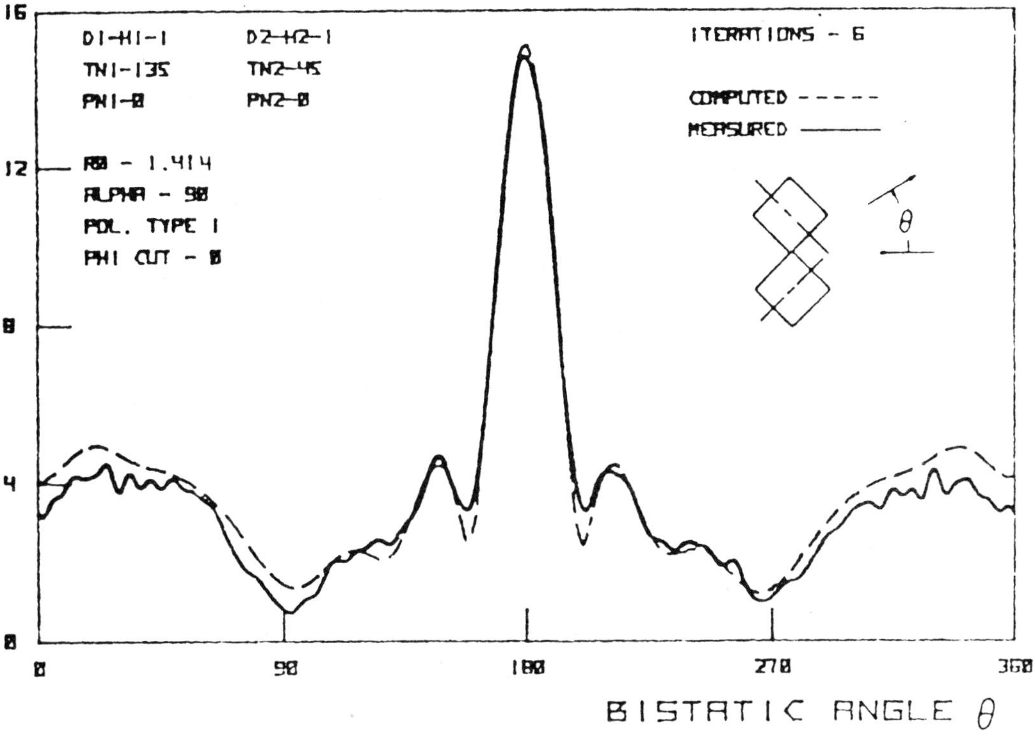

Fig. 9. Computed and measured far field scattering by two perfectly conducting finite cylinders. Touching case with no addition theorem violation.

ACKNOWLEDGEMENT

This work is supported by the U.S. Army Research Office under Grant No. DAAG 29-77-G-0021 and the National Science Foundation under Grant No. ENG 76-22296-A01.

REFERENCES

1. Mei, K.K., "Unimoment method of solving antenna and scattering problems," IEEE Trans. on Antennas and Propagation, AP-24, No. 6, pp. 760-766, Nov. 1974.

2. Chang, S.K. and K.K. Mei, "Application of the unimoment method to electromagnetic scattering of dielectric cylinders," IEEE Trans. on Antennas and Propagation, AP-24, No. 1, pp. 35-42, Jan. 1976.

3. Stovall, R. and K.K. Mei, "Application of a unimoment technique to a biconical antenna with inhomogeneous dielectric loading," IEEE Trans. on Antennas and Propagation, AP-23, No. 3, pp. 335-342, May 1975.

4. Morgan, M.A. and K.K. Mei, "Finite-element computation of scattering by inhomogeneous penetrable bodies of revolution," IEEE Trans. on Antennas and Propagation, AP-27, No. 2, pp. 202-214, March 1979.

5. Kao, H. and K.K. Mei, "Scattering by advanced composite bodies of revolution," Nuclear Electromagnetic Conference, University of New Mexico, Conference Record, p. 35, June 1978.

6. Chang, S.K., "On Electromagnetic Wave Scattering by Buried Obstacles", Ph.D. Dissertation, University of California, Berkeley, California, June 1977.

7. Hunka, J.F. and K.K. Mei, "Numerical computation of electromagnetic scattering by two bodies of revolution," URSI/USNC Meeting, Boulder, CO, p. 111, Nov. 1978.

8. McDonald, B.H. and A. Wexler, "Finite-element solution of unbounded field problems," IEEE Trans. Microwave Theory and Tech., MTT-20, pp. 841-847, Dec. 1972.

9. Silvester, P. and M.S. Hsieh, "Finite-element solution of two-dimensional exterior field problems," Proc. Inst. Elect. Engr., 118, pp. 1743-1747, Dec. 1971.

10. Mei, K.K., R. Stovall and D. Tremain, "Difference coupling method of solving radiation problems," URSI Fall Meeting, Digest, p. 81, Sept. 1971.

11. Coffey, E.L. and W.L. Stutzman, "Hybrid computer solutions of electromagnetic scattering problems," IEEE AP-S International Symposium, University of Massachusetts, Amherst, Digest, pp. 82-85, Oct. 1976.

12. Stratton, J.A., Electromagnetic Theory, McGraw-Hill, New York, 1941.

13. Harrington, R.F. Time Harmonic Electromagnetic Fields, McGraw-Hill, New York, 1961.

14. Morse, P.M. and H. Feshbach, Methods of Theoretical Physics, Vol. I, McGraw-Hill, New York, 1953.

15. Chang, S.K. and K.K. Mei, "Generalized Sommerfeld Integrals and Field Expansions in Two-Medium Half Spaces", submitted to IEEE Trans. on Antennas and Propagation.

16. Stein, S. "Addition theorems for spherical wave functions," Quarterly of Applied Mathematics, 19, No. 1, pp. 15-24, April 1961.

17. Edmonds, A.R., Angular Momentum in Quantum Mechanics, 2nd ed. Princeton University Press, Princeton, NJ, 1960.

18. Hunka, J.F. "Electromagnetic scattering by two bodies of revolution," Ph.D. Dissertation, University of California, Berkeley, 1979.

BOUNDARY LAYER AND FINITE ELEMENT TECHNIQUES APPLIED TO WAVE PROBLEMS

Chiang C. Mei
Parsons Laboratory, Department of Civil Engineering
Massachusetts Institute of Technology
Cambridge, Mass. 02139, U.S.A.

ABSTRACT

Two techniques for wave problems are discussed in this paper. The first technique involves boundary layers and aims at an approximate analytical solution when there is a small parameter arising from the existence of vastly contrasting scales. Examples described are (1) grazing incidence of short elastic waves by a slender cavity, and (2) thermo-elastic waves interacting with a free surface. In both examples, boundary layers not only simplify the mathematics but sharpen the physical picture. The second techniques is a numerical one using finite elements in a way particularly suited for exterior problems. It is very advantageous when a finite part of the medium is inhomogeneous and when the frequency is not necessarily high or low. Two theoretical properties of this method are discussed and applications to a variety of scattering and radiation problems are outlined.

PART I. PROBLEMS BY BOUNDARY LAYER TECHNIQUES

INTRODUCTION

Although the idea of boundary layers began with Prandtl's work for viscous fluid flows at high Reynolds numbers, the notion of inner and outer approximations can and have now been applied profitably to a wide range of physical problems. In fact, traces of the underlying idea can already be found in the early literature of acoustic waves, although the formalism of inner and outer approximations and of systematic improvement to successively higher orders are products of modern development. For example, Rayleigh (1870) already recognized the advantage of using different approximations for the near and far fields of a small body (or hole) in scattering problems. He argued in essence that in the neighborhood of the scatterer Helmholtz equation can be approximated with a relative error of $O(ka)^2$ by Laplace equation which can easily and formally be solved to satisfy the condition on the scatterer. Thus in the near field one has a static problem. In the far field where the wavelength is the controlling length scale, the full Helmholtz equation must be used, but the body or hole now appears so small that the formal solution can be written as an integral of sources or an expansion of multipoles which satisfy the radiation condition. Unknown coefficients in each part are found by requiring the two parts to match smoothly in the intermediate zone. A later example of this approach may be found in Lamb (1932, p. 534 ff) for sound waves incident on a wire mesh. In recent works the formalism of matched asymptotics has been applied beyond the leading order (See for example, many papers by Datta and associates (1977), Viswanathan and Sharma (1977) for elastic waves. For acoustic waves the reviews by Tuck (1977) and by Lesser and Crighton (1975) are germane.) Further examples may be found for resonant scattering of SH waves by a thin structure above the found (Mei and Foda, 1979) or of long ocean waves in a harbor by Mei and Ünlüata (1976). In addition to being an efficient method to get an approximate solution, the merit of boundary-layer or matched asymptotic approximation is to enable a thorough study of the physical implications analytically.

A very recent *unified slender-body theory* by Newman (1979) uses matched asymptotics and is capable of treating radiation of sound or water waves of arbitrary frequency. Its potential application to other fields is great, and the reader may refer to his own account.

In this part of the paper we shall describe two problems of high frequency waves: (1) scattering of elastic waves at grazing incidence by a slender cavity, and (2) propagation and scattering of thermo-elastic waves. In the first problem a small paramter arises from the geometry of the scattering, i.e., the slenderness ratio, while in the second problem the small parameter is the ratio between the short wave period and the long diffusion time, a situation rather close to the boundary layer of Stokes in an oscillating viscous fluid.

1. SHORT P WAVES INCIDENT ALONG THE AXIS OF A SLENDER CAVITY (Mei, 1979a)

Let a slender two-dimensional cavity be situated along the x-axis between $x = 0$ and $x = L$. The maximum width of the cavity is $O(\delta L)$ where $\delta << 1$. What is the effect of an incident plane P wave from $x \sim -\infty$? A roughly similar problem is important in hydrodynamics where one wishes to know the response of a ship heading into sea waves; present theories for this problem are, however, complicated and not in complete agreement.

The boundary value problem may be stated as follows. Using the usual potentials ϕ and ψ for plane strain and normalizing all distances by the cavity length L and time t by ω^{-1} we have the field equations

$$(\nabla + \alpha^2)\phi = 0 \qquad (\nabla + \alpha^2 + \beta^2)\psi = 0 \qquad (1.1.a,b)$$

where

$$\alpha^2 = \omega^2 L^2 \rho/(\lambda + 2\mu) \qquad \alpha^2 + \beta^2 = \omega^2 L\rho/\mu \qquad (1.2.a,b)$$

The incident wave is given by

$$\phi = e^{i\alpha x} \qquad x = 0 \qquad (1.3.a,b)^*$$

The time factor exp(-it) is omitted: Assuming symmetry about the x-axis, the boundary condition is

$$\tau_{ij} n_j = 0 \qquad \text{on} \qquad y = \pm\delta h(x) \qquad (1.4)^*$$

where $h(0) = h(1)$ and $h = O(1)$. It will now be assumed that

$$\alpha, \beta = O(1/\delta) >> 1 \qquad (1.5)$$

Consider the half-space $y > 0$ only. We divide the space into the far field

---

*When i appears as a subscript, it means 1, 2 or 3 in reference to the cartesian coordinates, otherwise it represents $\sqrt{-1}$.

$x,y = 0(1)$, the intermediate far field ($x = 0(1)$, $y = 0(\delta^{1/2})$) and the near field ($x = 0(1)$, $y = 0(\delta)$). The reason for the intermediate outer field is that for short waves past a long body along its axis, it is intuitively obvious that the P waves will be essentially propagating forward, hence can be expressed as

$$\phi = A(x,y)e^{i\alpha x} \tag{1.6}$$

But for large α, Eq. (1.1.a) implies that there is a region where the following approximation is valid

$$2i\alpha A_x + A_{yy} \cong 0 \tag{1.7}$$

Since $2iaA/A_{xx} = 0(\delta^{-1}) >> 1$. The domain implied by (1.7) is ($x = 0(1)$, $y = 0(\delta^{1/2})$), i.e., the intermediate far field and this approximation is known in other contexts as the *parabolic approximation* (Tappert, 1977).

1.1 The Intermediate Far Field

Formalizing the parabolic approximation we introduce

$$X = x \qquad Y = y/\delta^{1/2} \tag{1.8}$$

and (1.6) and (1.7) may be written in terms of (1.8) with $A = A(X,Y)$. Equation (1.7) preserves its form if α is replaced by a where $a \equiv \alpha\delta$. Because of the parabolic nature we impose the initial condition

$$A = 0 \qquad \text{along} \qquad X = 0 \tag{1.9}$$

Symmetry about the x-axis demands that $\partial\phi/\partial y = 0$ on $y = 0$, implying,

$$\frac{\partial A}{\partial Y} = 0 \qquad Y = 0 \qquad X \notin [0,1] \tag{1.10}$$

Furthermore, we require,

$$A \to 1 \qquad Y \to \infty. \tag{1.11}$$

Formally the solution may be expressed in terms of $\partial A/\partial Y$ on $Y = 0$, $0 < X < 1$

$$A = 1 + \frac{1}{2}\int_0^X \frac{d\xi}{ia}\left.\frac{\partial A}{\partial Y}\right|_0 \left[\frac{2a}{\pi(X-\xi)}\right]^{1/2} \exp\left[\frac{iaY^2}{2(X-\xi)} - \frac{i\pi}{4}\right] \tag{1.12}$$

If $X > 1$ we simply replace the upper limit of the preceeding integral by 1. The inner approximation of this solution is

$$A = A(X,0) + Y\left.\frac{\partial A}{\partial Y}\right|_0 + \dots \tag{1.13.a}$$

where

$$A(X,0) = 1 + \frac{1}{2}\int_0^X \frac{d\xi}{ia}\left.\frac{\partial A}{\partial Y}\right|_0 \left[\frac{2a}{\pi(X-\xi)}\right]^{1/2} e^{-i\pi/4} \tag{1.13.b}$$

These will be used for matching with the P-wave in the near field.

SV waves are expected to emanate from the cavity by mode conversion. In view of (1.1.b) we assume that

$$\psi = B(x,y)e^{i(\alpha x+\beta y)} \tag{1.14}$$

i.e., the wave rays are inclined along the vector (α,β). It is convenient to introduce an orthogonal coordinate system (ξ,η) with ξ parallel to the wave rays, thus

$$x = \xi \cos \theta - \eta \sin \theta, \quad y = \xi \sin \theta + \eta \cos \theta \tag{1.15}$$

where $\tan \theta = \beta/\alpha$. Eq. (1.14) may then be written as

$$\psi = B(\xi,\eta)\exp[i(\alpha^2 + \beta^2)^{1/2}\xi] \tag{1.16}$$

Substituting (1.16) into (1.1.b) we easily find that

$$B_\xi = O(\delta B) \quad \text{so that} \quad B = B(\eta) \tag{1.17}$$

to the leading order. The value $B(\eta)$ will be found by matching with the SV wave in the near field. It is interesting that the amplitude B is invariant along a ray.

1.2 The Neighborhood of the Cavity

Let us define the region $0 < x < 1$ and $y \leqq O(\delta)$ to be the near field, introduce the near field variables

$$\bar{X} = X = x \qquad \bar{Y} = y/\delta = Y/\delta^{1/2} \tag{1.18}$$

and denote the near field potentials by $\bar{\phi}$ and $\bar{\psi}$. The governing equations may be written as

$$\bar{\phi}_{\bar{Y}\bar{Y}} + \delta^2\bar{\phi}_{\bar{X}\bar{X}} + a^2\bar{\phi} = 0 \quad , \quad a = \alpha\delta = O(1)$$

$$\bar{\psi}_{YY} + \delta\psi_{\bar{X}\bar{X}} + (a^2 + b^2)\bar{\psi} = 0 \quad , \quad b = \beta\delta = O(1) \tag{1.19}$$

Assume

$$\bar{\phi} = \bar{A}(\bar{X},\bar{Y})e^{i\alpha x} \quad , \quad \bar{\psi} = \bar{B}(\bar{X},\bar{Y})e^{i(\alpha x+\beta y)} \tag{1.20}$$

To the leading order we have

$$\bar{A}_{\bar{Y}\bar{Y}} = O(\delta\bar{A}) \ , \quad \bar{B}_{\bar{Y}\bar{Y}} + 2ib\bar{B}_{\bar{Y}} = O(\delta\bar{B}) \tag{1.21}$$

so that the inner approximation is

$$\bar{A} = F + D(\bar{Y} - h) \qquad \bar{B} = E\, e^{-ibh} \tag{1.22}$$

where D, E, F are functions of X only. To apply the boundary condition on the cavity surface we note first that the components of the unit normal are

$$\begin{pmatrix} n_1 \\ n_2 \end{pmatrix} = \begin{pmatrix} -\delta h_{\bar{X}} \\ 1 \end{pmatrix} [1 + \delta^2(h_{\bar{X}})^2]^{-1/2} \tag{1.23}$$

To the leading order Equations (1.4) become

$$\tau_{12} = O(\delta) \qquad \tau_{22} = O(\delta) \quad \text{on } \bar{Y} = h \tag{1.24}$$

Relating the stress components to $\bar{\phi}$ and $\bar{\phi}$, the boundary conditions imply

$$2iaD + (a^2 - b^2)E = 0 \ , \quad -\lambda af + 2\mu bE = 0 \tag{1.25}$$

Thus E and F may be solved in terms of D.

The solutions (1.22) with the relations (1.25) are the Goodier-Bishop wave system in which P and SV waves co-exist near the plane boundary of an infinite half space. The algebraic growth of the P wave amplitude renders the solution of little value in the strict half space problem, but is quite acceptable here as an inner solution. The outer limit of the inner solution is

$$A = F + D\bar{Y}\ , \quad B = E\ e^{-ibh}\ , \quad \bar{Y} >> h \qquad (1.26)$$

Matching (1.13) with (1.26) we get two equations for A and D. Simple algebra leads to

$$D = \left.\frac{\partial A}{\partial \bar{Y}}\right|_0 = \delta^{1/2} \left.\frac{\partial A}{\partial Y}\right|_0 \qquad (1.27.a)$$

$$D + \frac{e^{-3i\pi/4}}{(\pi/\Delta)^{1/2}} \int_0^X d\xi \frac{D}{(X-\xi)^{1/2}} = -\lambda \frac{a^2-b^2}{4\mu b} \qquad 0 < X < 1 \qquad (1.27.b)$$

where

$$\Delta^{-1/2} = \frac{\lambda(a^2-b^2)}{4\mu b} \left(\frac{1}{2a\delta}\right)^{1/2} \qquad (1.27.c)$$

Equation (1.27.b) is an Abel integral equation for D(X), with the exact solution

$$D = -\frac{\lambda(a^2-b^2)}{4\mu b} w[(1-i)(X/2\Delta)^{1/2}]\ , \quad \text{where } w(z) = e^{-z^2}\text{erfc}(-iz) \qquad (1.28.a)$$

For $X/\Delta = O(X/\delta) >> 1$,

$$D \cong \delta^{1/2}(2a/\pi X)^{1/2} \qquad (1.29)$$

which is monotonically decreasing in the direction of waves. For $X/\Delta << 1$

$$D = \frac{-\lambda(a^2-b^2)}{4\mu b} [1 + (1+i)\left(\frac{X}{\pi\Delta}\right)^{1/2} + \ldots] \qquad (1.30)$$

so that D approaches a finite constant at the leading edge. From (1.25), we find

$$E = -\frac{2iaD}{a^2-b^2} \qquad F = -\frac{4\mu bD}{\lambda(a^2-b^2)} \qquad (1.31)$$

Thus P waves in the near and the intermediate far field are known. The SV wave in the near field is also known through (1.22). To match with (1.17.b) we require $B(\eta)|_{\bar{Y}=0} = \bar{B}(\bar{X},\bar{Y})|_{\bar{Y}>>1}$. Since $x = X = -\eta/\sin\theta$ when $Y = 0$ we conclude that

$$B = \frac{2i\lambda a}{4\mu b} e^{-ibh} w[(1-i)(-\eta/\Delta \sin\theta)^{1/2}]\ , \quad \text{for } 0 < \eta < \sin\theta \qquad (1.32)$$

Thus the SV wave in the intermediate far field is also found.

Further analysis is, however, needed to smooth the transition around the first ray (from x = 0) and the last ray (from x = 1) of SV wave field. Thus around these two rays two boundary layers must be fitted. For each boundary layer this is done by first rotating the coordinates system so that the ray is an axis. In terms of the new coordinates the approximate equation in each transition boundary layer is again of the form Eq. (1.7). This is not surprising because these transition zones are analogous to the shadow boundaries of a knife edge (Malyazhinets, 1959;

Budal and Keller, 1960). The trailing transition zone about the ray from x = 1 has the conventional Fresnel behavior; the amplitude of the SV approaches a constant, in an undulatory way, to the left, and zero, monotonically, to the right. But the leading transition zone about the ray from x = 0 has a different structure. The amplitude of the SV wave approaches zero to the left but $X^{1/2}$ to the right (cf.(1.29)). The solution here can be written in terms of Bessel functions of order 1/4 and gives rise to diffraction fringes towards the incident P waves, hence in the shadow of the SV waves. For details reference is made to Mei (1979a).

Near the cavity tips these approximations break down, but one may in principle find the local behavior by solving the exact equations for an infinite wedge. We remark that Rayleigh waves are not excited along the cavity surface because they do not depend on x as $e^{i\alpha x}$.

For many practical purposes the knowledge of the near field is sufficient. If the scattering amplitude $F(\theta)$ of the P wave in the far field $(x,y > 0(1))$ is of interest where $F(\theta)$ is defined by

$$\phi_\infty \cong \frac{F(\theta)}{\sqrt{r}} e^{ikr} \qquad r^2 = x^2 + y^2 \tag{1.33}$$

Then, as pointed out by Tuck (1979) $F(\theta)$ may be found by matching with the intermediate far field:

$$\phi_\infty(y<<x) = \phi(X>>1,\ Y>>1) \tag{1.34}$$

Using (1.12) the right hand side of (1.34) is

$$\phi = A\, e^{i\alpha x} = e^{i\alpha x} + \left[\frac{1}{2}\int_0^1 \frac{d\xi}{ia}\left.\frac{\partial A}{\partial Y}\right|_0 e^{-ia(Y/X)^2\xi}\right]\sqrt{\frac{2a}{\pi X}}\, e^{i\alpha x + iaY^2/2X}\, e^{-\pi/4}$$

$$= e^{i\alpha x} + \left[\frac{1}{2}\int_0^1 \frac{d\xi}{ia}\left.\frac{\partial A}{\partial Y}\right|_0 e^{-i\alpha\tan^2\theta\xi}\right]\sqrt{\frac{2a}{\pi x}}\, e^{i\alpha x + i\alpha y^2/2x}\, e^{-i\pi/4} \tag{1.35}$$

The left hand side of (1.34) is from (1.33)

$$\phi_\infty^2 = \frac{F(\theta)}{\sqrt{x}} e^{i\alpha(x+y^2/2x)} \tag{1.36}$$

The scattered wave amplitude is therefore

$$F(\theta) = \left[\frac{1}{2}\int_0^1 d\xi \left.\frac{\partial A}{\partial Y}\right|_0 e^{-i\alpha\tan^2\xi}\right]\sqrt{\frac{2}{\pi\alpha\delta}}\, e^{-3i\pi/4} \tag{1.37}$$

The physical picture is therefore brought to light with the help of boundary layers.

Similar analysis for a three-dimensional but axially symmetric cavity may be worthwhile. If the cavity is replaced by an inclusion of different elastic constants, the analysis given here can be modified. One would then anticipate possible resonances in the inclusion for certain values of the elastic properties and the wavelength. A simpler problem of shallow water waves incident on a long and submerged island (or a canyon or an iceberg) has been recently treated by Mei and Tuck (1979). The matching of inner and outer region leads to an Abel integral equation with variable coefficients, which must then be solved numerically.

2. BOUNDARY LAYERS IN DYNAMIC THERMO-ELASTICITY AND PORO-ELASTICITY

The equations of Biot (1956) for poro-elasticity or mechanics of fluid-filled porous elastic solid can be used as the basis of theoretical soil and rock mechanics when the deformation is *infinitesimal*. The main ingredients are Darcy's law for the fluid motion and Hooke's law for the solid skeleton. The motion of the two media are, however, coupled and the governing equations are complicated indeed, even for isotropic and homogeneous cases. As a consequence, very few analytical solutions are available except for the cases of uni-directional propagation exp(ikx-ωt) where the governing equations can be reduced to an ordinary differential equation in the transverse direction. Still the algebraic work of handling this high order ordinary differential equation and its boundary value problem is immense and the extraction of physical information is difficult.

Recently Mei and Foda (1979b) use the fact that for frequency ranges typical of sea waves or seismic waves, the low permeability of earth material limits the relative motion between fluid and solid only within a thin boundary near the free surface of the porous solid. A boundary layer analysis has been developed which makes it possible to solve many problems of practical interest. Now the governing equations of thermal elastic waves may be shown to be a special case of the poro-elastic waves. From a casual survey of the current literature, it appears that the dominant approach in thermo-elastic waves is to search for the formally exact solution and then to seek various approximations of the exact solution (Nowacki (1978)). Enormous patience is required to carry out the analysis of relatively simple two-dimensional problems. In problems of technological interest, one is likely to need solutions for scattering by finite inclusions or cavities of general geometry. It is then advantageous to focus attention on certain specific ranges of parameters and seek to approximate the governing equations from the outset, in order to facilitate the solution.

We illustrate the essential ideas for thermo-elastic waves in a medium with a free surface. It will be demonstrated that for a sufficiently high frequency, thermal diffusion is confined within a thin boundary layer near the free surface and that in the interior of the medium the temperature is essentially governed by static conduction. Simple ways of solving scattering problems will then be outlined.

The governing equations for thermo-elasticity are as follows (Nowacki (1978))

$$\mu\nabla^2\vec{u} + (\lambda + \mu)\nabla\nabla\cdot\vec{u} = \gamma\nabla T + \rho\vec{u}_{tt} \tag{2.1}$$

$$\nabla^2 T - \frac{1}{\kappa} T_t = \eta\nabla\cdot\vec{u}_t \tag{2.2}$$

$$\sigma_{ij} = 2\mu\varepsilon_{ij} + \lambda\varepsilon_{kk}\delta_{ij} - \gamma T\delta_{ij} \tag{2.3}$$

where $\vec{u}$ = solid displacement, T = temperature rise above a static datum Θ_o, i.e., $T = \Theta - \Theta_0$, κ = thermal diffusivity, $\gamma = \alpha(\lambda + 2\mu/3)$ where α = linear thermal expansion coefficient, $\eta = \gamma\Theta_0/\lambda_0$ where λ_0 = heat conductivity and $\gamma = \lambda_0/c_\varepsilon$, c_ε = specific heat for constant strain.

We define an outer field whose length scale is L which can be the wavelength or the typical dimension of a scatterer. For thermal effects to couple nontrivially with elastic effects one must scale the variables as follows

$$\vec{x} = L\vec{x}^* , \quad t = t^*/\omega, \tag{2.3.a}$$

$$T = T_o T^* , \quad \vec{u} = (\gamma T_o L/\mu)\vec{u}^* \tag{2.3.b}$$

where $T_o = \max(\Theta - \Theta_o)$ or,

$$T = (\mu U/\gamma L)T^* \ , \quad \vec{u} = U\vec{u}^* \tag{2.3.c}$$

Equation (2.3.b) is appropriate if T_o is given (thermally driven waves) while Eq. (2.3.c) is appropriate if U is given (mechanically driven waves). Either way the governing equations (2.1) and (2.2) may be written as

$$\nabla^2\vec{u} + \frac{\lambda+\mu}{\mu}\nabla\nabla\cdot\vec{u} = \nabla T + \frac{\rho\omega^2 L^2}{\mu}\vec{u}_{tt} \tag{2.4}$$

$$\frac{\kappa}{\omega L^2}\nabla^2 T = T_t = \beta\nabla\cdot\vec{u}_t \tag{2.5}$$

where the symbols * have been omitted for brevity and

$$\beta = \eta\gamma\kappa/\mu = \gamma^2\Theta_o/\mu C_\varepsilon \tag{2.6}$$

We assume that the base temperature Θ_o and the frequency are so high or the diffusivity κ so low that

$$\beta = 0(1) \ , \quad \frac{\kappa}{\omega L^2} = 0(\delta/L)^2 << 1 \tag{2.7}$$

Then to the leading order, (2.5) may be approximated by

$$-T = \beta\nabla\cdot\vec{u} = \beta\varepsilon \tag{2.8.a}$$

or in physical variables

$$-T^o = \eta\kappa\nabla\cdot\vec{u}^o \tag{2.8.b}$$

where the superscript "o" stands for "outer". Thus the volume dilatation ε is directly coupled with temperature perturbation. Taking the divergence of (2.4) and using (2.8) we get, in normalized variables,

$$[1 + (\lambda+2\mu)/\mu\beta]\nabla^2 T = (\rho\omega^2 L^2/\mu\beta)T_{tt} \tag{2.9.a}$$

In dimensional form, we have, to the leading order

$$\nabla^2 T^o = [\rho/(\lambda+2\mu+\beta\mu)]T^o_{tt} \tag{2.9.b}$$

The volume dilatation $\varepsilon = \nabla\cdot\vec{u}$ of course satisfies (2.9) also. This means that T or ε behaves as a compressional wave with the wave speed C_p

$$C_p = [(\lambda+2\mu+\lambda\mu)/\rho]^{1/2} \tag{2.10}$$

which is greater than that of pure elastic wave $(C_p)_E = [(\lambda+2\mu)/\rho]^{1/2}$. Taking the curl of (2.4) and using (2.8.a) we get, in normalized variables,

$$\nabla^2\vec{\Omega} = (\rho\omega^2 L^2/\mu)\vec{\Omega}_{tt} \tag{2.11.a}$$

In physical variables we have

$$\nabla^2\vec{\Omega}^o = (\rho/\mu)\vec{\Omega}^o_{tt} \tag{2.11.b}$$

implying that $\vec{\Omega}^o$ propagates at the same sheave wave velocity as the pure elastic case

$$C_s = (\rho/\mu)^{1/2} \tag{2.12}$$

We may now define an effective λ_e by

$$\lambda_e = \lambda + \beta\mu \tag{2.13}$$

and a corresponding effective Poisson ratio

$$\nu_e = \lambda_e/2(\lambda_e+\mu) = (\lambda+\beta\mu)/2(\lambda+2\mu+\beta\mu) \tag{2.14}$$

For $\beta \to 0$, $\nu_e \to \nu$, but for $\beta \to \infty$, $\nu_e \to \frac{1}{2}$.

The important feature is that in the outer field, once $\vec{u}$ is found, T is determined by (2.8). Now this rigid relationship makes it impossible in general for the traction and the temperature to meet the boundary conditions on the free surface. Since in obtaining (2.8) higher order derivatives are omitted, we expect a correction of the boundary layer type near the free surface. Let us suppose that inside the boundary layer of thickness $O(\delta)$ the total solution is given by

$$\vec{u} = \vec{u}^o + \vec{u}^b \qquad T = T^o + T^b \tag{2.15}$$

All boundary-layer terms with superscript b are assumed to vary with y with the length scale $O(\Delta)$ and to vanish outside the boundary layer. To be specific, we let $y = 0$ to be the free surface. First of all, we observe that inertia $\rho\vec{u}^b_{tt}$ is unimportant in the boundary layer since

$$\rho\vec{u}^b_{tt}/\mu \frac{\partial^2\vec{u}^b}{\partial y^2} \sim \rho\omega^2\delta^2/\mu^2 \sim (\delta/L)^2 << 1$$

Equation (2.1) then contains only static terms.

$$\mu\nabla^2\vec{u}^b + (\lambda+\mu)\nabla\nabla\cdot\vec{u}^b = \gamma\nabla T^b + O(\delta/L)^2$$

Taking the curl we get

$$\nabla^2\nabla\times\vec{u}^b \cong O(\delta/L)^2$$

Since $\nabla^2 \sim \partial^2/\partial y^2$ we have $\nabla\times\vec{u}_b \cong 0$ throughout the boundary layer. Considering each component

$$\frac{\partial u_i^b}{\partial y} - \frac{\partial u_2^b}{\partial x_i} = 0 \qquad i = 1,3$$

we conclude that

$$u_i^b/u_2^b = O(\delta/L) \tag{2.16}$$

Thus the tangential displacement is much less than the transverse displacement.

From the transverse component of (2.1) the dominant terms are

$$\mu \frac{\partial^2 u_2^b}{\partial y^2} + (\lambda+\mu) \frac{\partial^2 u_2^b}{\partial y^2} = \gamma \frac{\partial T^b}{\partial y} \tag{2.17}$$

after neglecting inertia. The tangential components (i = 1,2) of Eq. (2.1) are dominated by

$$\mu \frac{\partial^2 u_i^b}{\partial y^2} + (\lambda+\mu) \frac{\partial^2 u_i^b}{\partial x_i \partial y} = \gamma \frac{\partial T^b}{\partial x_i} \tag{2.18}$$

Lastly, Eq. (2.2) is dominated by

$$\mu \frac{\partial^2 T^b}{\partial y^2} - \frac{1}{\kappa} \frac{\partial T^b}{\partial t} = \eta \frac{\partial^2 u_2^b}{\partial y \partial t} \tag{2.19}$$

Integrating (2.17) we get

$$(\lambda+2\mu) \frac{\partial u_2^b}{\partial y} = \gamma T^b \tag{2.20}$$

Substituting this into (2.19) we get

$$\frac{\partial^2 T^b}{\partial y^2} = (\frac{1}{\kappa} + \frac{\eta\gamma}{\lambda+2\mu}) \frac{\partial T^b}{\partial t} \tag{2.21}$$

For harmonic time dependence $T^b \sim e^{-i\omega t}$ the solution is easily written as

$$T^b = A(x_i) \exp[\frac{1-i}{\sqrt{2}} \frac{y}{\delta} - i\omega t] \tag{2.22}$$

Clearly when $y >> -\delta$, T^b vanishes exponentially, hence

$$\delta = \{\omega[\frac{1}{\kappa} + \eta\gamma/(\lambda+2\mu)]\}^{-1/2} \tag{2.23}$$

is the measure of boundary layer thickness. $A(x_i)$ is so far unknown.

In terms of A, u_i^b can be found from (2.20)

$$u_2^b = \frac{\gamma}{\lambda+2\mu} A(\frac{1-i}{\sqrt{2}} \frac{1}{\delta})^{-1} \exp(\frac{1-i}{\sqrt{2}} \frac{y}{\delta} - i\omega t) \tag{2.24}$$

and from (2.18)

$$u_i^b = \frac{\gamma}{\lambda+2\mu} \frac{\partial A}{\partial x_i} (\frac{1-i}{\sqrt{2}} \frac{1}{\delta})^{-2} \exp(\frac{1-i}{\sqrt{2}} \frac{y}{\delta} - i\omega t) \tag{2.25}$$

From (2.3) the boundary layer corrections of the stress field are, to the leading order,

$$\sigma_{ii}^b \cong \lambda \frac{\partial u_2^b}{\partial y} - \partial T^b = - \frac{2\mu}{\lambda+2\mu} \partial T^b$$

$$\sigma_{22}^b = (\lambda+\mu) \frac{\partial u_2^b}{\partial y} - \partial T^b = - \frac{\mu}{\lambda+2\mu} \partial T^b$$

$$\sigma_{i2}^b = 2\mu(\frac{\partial u_i^b}{\partial y} + \frac{\partial u_2^b}{\partial x_i}) = \frac{4\mu\gamma}{\lambda+2\mu}(\frac{1-i}{\sqrt{2}} \frac{1}{\delta})^{-1} \frac{\partial A}{\partial x_i} \exp(\frac{1-i}{\sqrt{2}} \frac{y}{\delta} - i\omega t) \tag{2.26.a,b,c}$$

The preceeding analysis completes the primary step of the approximation, and the above results can be useful to propagation as well as scattering problems. As a possible application one may sketch the additional steps needed for the stress concentration near a circular cylinderical cavity of radius a due to a plane incident wave. This problem was studied by Ignaczak and Nowacki (see Nowacki (1978)) using the classical method of partial wave expansions to the full thermo-elastic equations. Their details are very involved. Let the boundary condition for the total traction on the cavity surface be

$$\sigma_{rr} = \sigma_{r\theta} = 0, \quad T = 0 \quad \text{on} \quad r = a \tag{2.27}$$

or indeed any prescribed functions of the angle θ. As long as $\delta/a << 1$ the boundary layer results apply directly with a-r replacing y and aθ replacing x. From (2.25) and (2.24) we get on r = a,

$$\sigma^o_{rr} = -\sigma^b_{rr} = \frac{\mu\gamma}{\lambda+2\mu} A\ e^{-i\omega t}$$

$$\sigma^o_{r\theta} = -\sigma^b_{r\theta} = -\frac{4\mu\gamma}{\lambda+2\mu}\left(\frac{1-i}{\sqrt{2}}\frac{1}{\delta}\right)^{-1}\frac{1}{a}\frac{\partial A}{\partial\theta}\ e^{-i\omega t}$$

$$T^o = -T^b = -A \tag{2.28}$$

where A = A(aθ). Equations (2.28) serve as the boundary conditions of the outer problem. The displacement u^o of the outer problem can be expressed in terms of the compressional wave potential ϕ and the shear wave potential ψ by partial wave expansions. Invoking the stress boundary conditions (2.28.a,b) the expansion coefficients are then solved in terms of A. Now the outer temperature field T^o is given by (3.9.b). Invoking the temperature boundary condition (2.28.b) A can be determined. Because of the circular geometry and the simplicity of (2.27) the solution should be explicit.

While the details physics are left for the readers interested in thermo-elasticity, we remark that the corresponding problem in poro-elasticity has been worked out in this way by Mei and Foda (1979), for the case of $ka << 1$ but $\delta/a << 1$ where Kirsch's static solution is used for the outer solution near the cavity.

The remarkable point of this method is that the analysis is really no more complicated than that of classical elastodynamics. This reduction of labor suggests that one can now tackle the really difficult problems of mixed boundary value problems. Such work is underway for poro-elastic waves.

PART II. FINITE ELEMENT TECHNIQUE - A HYBRID ELEMENT METHOD

3. DESCRIPTION OF THE HYBRID-ELEMENT METHOD FOR SHALLOW WATER WAVES

In many scattering problems the neighborhood of the scatterer is complicated due to irregularities of boundary shape and of inhomogeneity in the medium. In general the governing equation contains variable coefficients. This renders the boundary value problem difficult to solve by even the integral equation method. The reason is that it is now difficult to find a Green's function which satisfies the governing equation and some boundary conditions (e.g., the radiation condition) or the fundamental solution which satisfies only the governing equation. This situation, of course, calls for a discrete method.

For arbitrary distribution of inhomogeneity, the flexibility of finite elements offers the greatest advantage. However, for exterior problems it is still uneconomical to discretize a very large domain. Therefore it is wise to mix analytic and finite element representations in regions where each is the most natural. This idea originates from the interior problems of a finite solid with a crack where the neighborhood of the crack tip is singular and is best represented analytically. This kind of method has been called by Tong et al. (1973) the *hyrbid element method* (HEM) and the analytical region, the *super-element*. Consider for example the two-dimensional scattering of long waves in shallow water with varying depth h(x,y). Let the region of variable depth and geometric irregularity be Ω, which is bounded by C. Within Ω the governing equation is

$$\nabla\cdot(h\nabla\phi) + \frac{\omega^2}{g}\phi = 0 \qquad \vec{x} \in \Omega \tag{3.1}$$

with the boundary condition on a rigid scatterer:

$$h\frac{\partial\phi}{\partial n} = 0 \qquad \vec{x} \in B \tag{3.2}$$

Let the region outside C be $\bar{\Omega}$ where $h = h_o$ = constant. The governing equation is

$$\nabla^2\bar{\phi} + k^2\bar{\phi} = 0 \qquad k^2 = \omega^2/gh_o \qquad h_o = \text{constant} \tag{3.3}$$

If the incident wave is denoted by ϕ_I

$$\phi_I = A\,e^{ikr\cos\theta} = A\sum_n \varepsilon_n i^n J_n(kr)\cos n\theta \tag{3.4}$$

The scattered wave $\phi_s = \phi - \phi_I$ must be outgoing at infinite, i.e.,

$$\sqrt{kr}\left(\frac{\partial}{\partial r} - ik\right)\phi_s \to 0 \qquad kr \to \infty \tag{3.5}$$

Along the border C it is necessary that

$$\phi = \bar{\phi}\ , \quad \frac{\partial\phi}{\partial n} = \frac{\partial\bar{\phi}}{\partial n} \tag{3.6}$$

where $\vec{n}$ is defined to be a unit normal outward from Ω.

To proceed further along the hybrid idea, some freedom exists in the choice of analytical representations of the super-element and the manner of enforcing the matching conditions (3.6). We only describe the option of Chen and Mei (1974) who represented the scattered waves in the super-element by partial wave expansions:

$$\bar{\phi}_s = \bar{\phi} - \bar{\phi}_I = \sum_n (\alpha_n\cos n\theta + \beta_n\sin n\theta)H_n^{(1)}(kr) \tag{3.7}$$

which satisfies (3.3) and (3.5) identically and introduced a localized variational principle for the whole problem with the functional

$$J(\phi,\bar{\phi}) = \iint_\Omega \frac{1}{2}\left[h(\nabla\phi)^2 - \frac{\omega^2}{g}\phi^2\right] + \int_C \frac{1}{2}h\,\bar{\phi}_s\frac{\partial\bar{\phi}_s}{\partial n} - \int_C h\phi\frac{\partial\bar{\phi}_s}{\partial n} - \int_C h\phi\frac{\partial\bar{\phi}_I}{\partial n} + \int_C h\phi_I\frac{\partial\bar{\phi}_I}{\partial n} \tag{3.8}$$

which involves integrals over Ω and along C; the symbols for area element dS and the line element $d\ell$ being omitted for brevity. The first variation of this functional is

$$\delta J = -\iint_\Omega\left(\nabla\cdot h\nabla\phi + \frac{\omega^2}{g}\phi\right)\delta\phi - \int_B h\frac{\partial\phi}{\partial n}\delta\phi + \int_C h\left(\frac{\partial\phi}{\partial n} - \frac{\partial\bar{\phi}}{\partial n}\right)\delta\phi + \int_C h(\bar{\phi} - \phi)\frac{\partial\delta\bar{\phi}}{\partial n} + \frac{1}{2}\int_C h\left(\bar{\phi}_s\frac{\partial\delta\bar{\phi}_s}{\partial n} - \delta\bar{\phi}_s\frac{\partial\bar{\phi}_s}{\partial n}\right) \tag{3.9}$$

The last integral is identically zero by using Green's theorem on $\bar{\phi}_s$ and $\sigma\bar{\phi}_s$ over the region $\bar{\Omega}$. It is now seen that the stationarity of J implies and is implied by (3.1) as the Euler-Lagrange equation, (3.2) as a natural boundary condition, and most importantly (3.6) as two natural boundary conditions. Upon using the finite element representation for ϕ, for example

$$\phi = \sum \phi_i N_i(x,y) \tag{3.10}$$

where ϕ_i is the nodal potential and N_i the shape functions, and define the unknown vector as

$$\{\phi\}^T = \{\{\phi_i\}^T, \{\alpha_n, \beta_r\}^T\}^T \tag{3.11}$$

Equation (3.9) leads to an inhomogeneous matrix equation

$$[K]\{\phi\} = \{F\} \tag{3.12}$$

Now the stiffness matrix K is banded and symmetric. If the curve C is chosen to be a circle, the orthogonality of {cos nθ , sin nθ} further simplifies some of the matrix elements considerably. The matrix question is readily solved.

The method was first developed for the resonant scattering by harbors and islands, and can be easily modified for cases where there is a sharp corner (e.g., the tip of a thin breakwater). Thus one introduces a small circle Ω_D centered at the sharp tip. The circle must be small enough so that h = constant and the wall boundaries are straight; thus the governing equation can be solved as a Fourier series expansions with the proper singularity represented analytically. Let us denote the circular arc bounding Ω_D by D. Equations (3.8) is modified by the addition of

$$-\iint_{\Omega_D} \frac{1}{2}\left[h(\nabla\phi_D)^2 - \frac{\omega^2}{g}\phi_D^2\right] - \int_D h\phi \frac{\partial\phi_D}{\partial n} \tag{3.13}$$

where Ω is defined to exclude Ω_D.

A large scale application of this method has been made by Houston (1978) to calculate the effects of tsunami on the Hawaiian Islands. The transient incident wave for Alaskan and Chilean earthquakes are first Fourier analyzed into 15 discrete harmonics. Each harmonic is treated as a scattering problem. Finite elements of varying sizes are used in accordance with the ocean depth. The calculated responses agreed remarkably well with the wave records at four different stations.

4. TWO THEORETICAL PROPERTIES OF THE HYBRID ELEMENT METHOD

An important problem in developing any numerical method is to devise means to check the accuracy. In the past many investigators have used the global identities such as reciprocity, global energy conservation, etc., as gauges of accuracy. Whether this is really helpful depends on the numerical scheme itself. Aranha (1978) has now shown that the method of §4.a satisfies all global identities identically (see Aranha, Mei and Yue, 1979). On the other hand, the method of integral equations using Green's function is a powerful alternative for scattering problems in homogeneous media. Nevertheless, it is well known since Lamb (1932) that such methods can lead to the so-called *irregular frequencies* which leads to the ill-conditioning of the matrix equation approximating the integral equation. This is best illustrated by the simple example of radiation from an axially symmetric pulsating circular cylinder. The boundary condition on the cylinder is

$$\frac{\partial\phi}{\partial r} = U \qquad r = a \tag{4.1}$$

The exact solution satisfying (3.3), (3.5) and (4.1) is easily obtained:

$$\phi = (U/k)H_o^{(1)}(kr)/H_o^{(1)'}(ka) \tag{4.2}$$

Now if one represents ϕ by distributing sources along the cylinder B,

$$\phi(\vec{x}) = \oint_B \sigma(\vec{x}')G(\vec{x}|\vec{x}')ds' \qquad \vec{x} \notin B \tag{4.3}$$

where

$$G = \frac{1}{4} H_0^{(1)}(k|\vec{x}-\vec{x}'|) \tag{4.4}$$

Then (4.1) leads to the integral equation for σ

$$-U = \underset{r\to a}{\ell in} \oint \frac{\partial G}{\partial r'} \, ds' \tag{4.5}$$

which can be solved for the source strength

$$\sigma = -\frac{2i}{ka} \frac{U}{J_0(ka)} \frac{1}{H_0^{(1)'}(ka)} \tag{4.6}$$

When (4.6) is substituted in (4.3), (4.2) is recovered. However, while ϕ is perfectly well-behaved for all k, the source strength is singular at the zeroes of $J_0(ka)$ which are the eigen values of the fictious interior problem defined by (3.3) within the circle r = a and the Dirichlet condition $\phi = 0$ on r = a. These eigen values are also the eigenvalues of the homogeneous version of the integral equation (4.5). Thus the solution of Eq. (4.5) is non-unique for certain frequencies, which means that the approximating matrix is ill-conditioned. Although there are practical ways to remove this trouble, it is nevertheless a nuisance since for arbitrary geometry one does not know *a priori* what these irregular frequencies are.

Now in the paper of Aranha et al. it is also proved that the present hybrid element method satisfies all reciprocity relations identically and gives a unique solution for all frequencies, therefore the matrix is never ill-conditioned. The proof of both properties is facilitated by replacing the variational principle by the weak formulation. We only outline the arguments here. Denote $H^1(\Omega)$ as the Sobolev space which is made up of functions square integrable in Ω, and $C^\infty(\bar{\Omega})$ as the space of functions expressible in the form of Eq. (3.7). The weak formulation of the problem is: Find $\phi \in H^1(\Omega)$ and $\bar{\phi} \in C^\infty(\bar{\Omega})$ such that

$$\forall \quad \psi \in H^1(\Omega): \quad \iint_\Omega h\nabla\phi\cdot\nabla\psi - \frac{\omega^2}{g}\iint_\Omega \phi\psi - \int_C \psi \frac{\partial\bar{\phi}}{\partial n} = 0 \tag{4.7}$$

and

$$\forall \quad \bar{\psi} \in C^\infty(\bar{\Omega}): \quad \int_C (\phi - \bar{\phi})\bar{\psi} = 0 \tag{4.8}$$

It is easy to verify that (4.7) and (4.8) are equivalent to (3.9) if one takes $\psi = \delta\phi$ and $\bar{\psi} = \partial\delta\bar{\phi}/\partial n$ which are admissible.

Now if we take $\psi = \phi^*$ in (4.7) and $\bar{\psi} = \partial\bar{\phi}^*/\partial n$ in (4.8) where * represents the complex conjugate, we can easily prove that

$$Im \int_C \bar{\phi} \frac{\partial\bar{\phi}^*}{\partial n} = 0 \tag{4.10}$$

Since $\bar{\phi}$ is proportional to fluid pressure and $\partial\bar{\phi}/\partial n$ is the fluid velocity normal to C, Eq. (4.10) states that net energy flux across C is zero, therefore it is the law of energy conservation which is one of many identities satisfied by the exact solution. The remarkable point is that ϕ in Ω and $\bar{\phi}$ in $\bar{\Omega}$ are only approximate subject to discretization error; energy is nevertheless conserved no matter how coarse the discretization. Thus satisfaction of this law provides no indication of the discretization error! Aranha et al. further showed that all other reciprocity relations are also satisfied by the HEM solution.

If we further let there be no incident wave $\bar{\phi}_I = 0$, take C to be a circle and use (3.7) in (4.10) we get

$$Im \int_C \bar{\phi} \frac{\partial \bar{\phi}^*}{\partial n} = 4h_o \sum_n (\frac{1}{\varepsilon_n} |\alpha_n|^2 + |\beta_n|^2) = 0 \qquad (4.11)$$

Thus $\alpha_n = \beta_n = 0$ for all n which means $\bar{\phi} \equiv 0$ in $\bar{\Omega}$. Now (4.8) implies

$$\phi = 0 \qquad \text{on C} \qquad (4.12)$$

The resulting weak problem now corresponds to the strong problem with an elliptic equation in Ω and zero Cauchy conditions on C. It is not suprising that $\phi \equiv 0$ in Ω. Details of the proof are given in Aranha et al.

5. ELASTIC WAVES IN AN INFINITE SPACE

We discuss below the variational formulation for a radiation problem in an infinite elastic space which is homogeneous except in a finite region. No computational results are yet available.

Let the elastic space be divided into Ω and $\bar{\Omega}$ by the surface C. Ω is a finite domain enclosing all inhomogeneities and scatterers. Within Ω the governing equation is

$$\frac{\partial \tau_{ij}}{\partial x_j} + \rho\omega^2 u_i = 0 \qquad \text{in } \Omega \qquad (5.1)$$

with

$$\tau_{ij} = C_{ijpq}\varepsilon_{pq} \qquad \tau_{pq} = \frac{1}{2}(\frac{\partial u_p}{\partial x_q} + \frac{\partial u_q}{\partial x_p}) \qquad (5.2)$$

The coefficient C_{ijjp} are functions of x_i. On the interior boundary B of Ω the surface traction is prescribed:

$$\tau_{ij}n_j = T_i \qquad (5.3)$$

We denote the stress and displacements in $\bar{\Omega}$ by $\bar{\tau}_{ij}$ and $\bar{u}_i$ which may be represented by partial wave expansions. The governing equations are similar to (5.1) and (5.2) but it is assumed that C_{ijpg} is an isotropic constant tensor. In addition, $\bar{\tau}_{ij}$ and $\bar{u}_i$ are outgoing waves at infinity. On the bordering surface C, we require that

$$u_i = \bar{u}_i \qquad \tau_{ij}n_j = \bar{\tau}_{ij}n_j \qquad (5.4)$$

where n_j points outward from Ω.

The localized functional is

$$J = \iiint_\Omega (W - \frac{1}{2}\rho\omega^2 u_j u_j) - \iint_B T_i u_i + \iint_C (\frac{1}{2}\bar{u}_i - u_i)\bar{\tau}_{ij}n_j \qquad (5.5)$$

where

$$W = \frac{1}{2} C_{ijpq}\, \varepsilon_{ij}\, \varepsilon_{pq} \qquad (5.6)$$

is the strain energy. Using the fact that

$$\frac{\partial W}{\partial \varepsilon_{ij}} \delta\varepsilon_{ij} = \frac{\partial}{\partial x_j}(\tau_{ij}\, \delta u_i) - \frac{\partial \tau_{ij}}{\partial x_j}\delta u_i \qquad (5.7)$$

it is easy to show that the first variation of J is

$$\delta J = -\iiint_\Omega \left(\frac{\partial \tau_{ij}}{\partial x_j} + \rho\omega^2 u_j\right)\delta u_j + \iint_B (\tau_{ij}n_j - T_i)\delta u_i$$

$$+ \iint_C (\tau_{ij} - \bar{\tau}_{ij})n_j \delta u_i + \iint_C (\bar{u}_i - u_i)\delta\tau_{ij}n_j$$

$$+ \frac{1}{2}\iint_C (\delta\bar{u}_i\bar{\tau}_{ij}n_j - \bar{u}_i\delta\bar{\tau}_{ij} - n_j) \tag{5.8}$$

The last integral involves only the analytical representations $\bar{u}_i$ and $\bar{\tau}_{ij}$. Applying Green's formula to $(\bar{\tau}_{ij},\bar{u}_i)$ and $(\delta\bar{\tau}_{ij},\delta\bar{u}_j)$ over the entire super-element bounded within by C and without by an infinitely large sphere C_∞, it is easy to show that

$$\left\{\iint_C + \iint_{C_\infty}\right\}[\delta\bar{u}_i\,\bar{\tau}_{ij} - \bar{u}_i\delta\bar{\tau}_{ij}]n_j = 0 \tag{5.9}$$

Because the radiation condition is satisfied by both $(\bar{\tau}_{ij},\bar{u}_{ij})$ and $(\delta\bar{\tau}_{ij},\delta\bar{u}_j)$, the integral over C_∞ vanishes (Mei (1978)), hence the last integral in (5.8) vanishes. It is now clear that $\delta J = 0$ implies, and is implied by (5.1) as the Euler-Lagrange equation, (5.3) and (5.4) as the natural boundary conditions. Thus the stationary of J is equivalent to the original boundary value problem.

Having deduced the variational principle, the numerical work is similar to §3 and is almost as straightforward as an interior elastodynamic problem. Although requiring further proof, the global identities such as thos discussed in Mei (1978), and the uniqueness of solution are likely to held here just as in §4.

In half space problems, partial wave expansion is not possible, but one may in principle use the much more complicated integral representation via Green's function. The numerical work will, of course, be very involved. Similar work in water waves have been succeeded by Jami et al. (1978). There are of course problems where the super-element is still more difficult and analytical representation is impractical. An example is a footing on an elastic layer of finite depth in which the determination of *partial waves* which consist of propagating and evanescent modes is itself a hard task. Waas (1972) devised an effective approach which divides the super-elements into thin horizontal layers. In each layer the vertical variation is approximated by a simple polynomial, but the horizontal variation is governed by a wave equation which can be solved to satisfy the radiation condition. Thus this technique treats the super-element semi-discretely, but is particularly suited for a horizontal layer with vertical variation in elastic properties. Similar ideas should be profitable in long water waves scattered by a continental shelf or waves in stratified fluids.

6. EXTENSIONS TO OTHER WAVE RADIATION AND SCATTERING PROBLEMS

The hybrid element method of §3 has also been developed for two-dimensional surface water waves in a vertical plane by Bai (see Bai and Yeung (1974)) and for three-dimensional surface water waves by Yue, Chen and Mei (1978). More recently Mei, Foda and Tong (1979) made further extensions to the radiation of acoustic waves in water due to forced vibrations of a submerged elastic structure. The aim of the problem is to analyze the response of offshore structures to earthquake excitation. If the sea bottom surrounding the structure is not horizontal only in a finite neighborhood then we introduce a vertical cylinder C so that outside C (i.e., the super-element $\bar{\Omega}$) the fluid velocity potential $\bar{\phi}$ can be given an analytical representation which satisfies the Helmholtz equation, has zero pressure on the sea surface, zero vertical velocity on the sea bottom and behaves as out-going waves at infinity.

The fluid velocity potential ϕ within C, i.e., inside Ω and the stresses δ_{ij} and displacements u_i in the structure must be such that the following functional is extrenum

$$J = \iiint_{\Omega_s} (W - \frac{1}{2} \rho_s \omega^2 u_j u_j) + \frac{\rho_w}{2} \iiint_{\Omega_w} [(\nabla\phi)^2 - k^2\phi]$$

$$- \iint_S i\omega\rho_w \phi u_j n_j + \rho_w \iint_C (\frac{1}{2}\bar{\phi} - \phi) \frac{\partial\bar{\phi}}{\partial n} \qquad (6.1)$$

where Ω_s is the solid volume, Ω_w the water volume in Ω, S the wetted surface of the structure. Finite element solutions have been demonstrated for a plate dam, and a vertical tower. More complicated geometries can be, however, treated in a straightforward way.

For these structural-fluid-interaction problems, many global reciprocity theorems can also be derived, and they are again satisfied by the present hybrid element method as shown by Mei (1979b) in the manner of Aranha.

ACKNOWLEDGMENTS

This paper is based on recent research by the author and several of his colleagues. It is a pleasure to acknowledge the support at various stages by the Fluid Mechanics and Earthquake Engineering Programs of the National Science Foundation, the Office of Naval Research, the National Oceanic and Atmospheric Administration and the MIT Sloan Fund.

REFERENCES

Aranha, J.A. (1978) Theoretical analysis of the hybrid element method in water wave problems, Ph.D. Thesis, Part II, Department of Civil Engineering, Mass. Inst. of Technology.

Aranha, J.A., C.C. Mei and D.K.P. Yue (1979) Some properties of a hybrid element method for water waves, Int. J. Numerical Methods in Engineering (in press).

Bai, J.K. and R.W. Yeung (1974) Numerical solutions of free-surface flow problems, Proc. 10th Symp. Naval Hydrody., Cambridge, Mass. 609-647.

Biot, M.A. (1956) Theory of propagation of elastic waves in a fluid-saturated porous solid, Part I, J. Acoust. Soc. Amer. 28, 168-191.

Budhal, R.N. and J.B. Keller (1960) Boundary layer problems in diffraction theory, Comm. Pure Appl. Math. 13, 165-184.

Chen, H.S. and C.C. Mei (1974) Oscillations and wave forces in a man-made harbor in the open sea, Proc. 10th Symp. Naval Hydrody., Cambridge, Mass. 573-596.

Datta, S.K. (1977) Diffraction of plane elastic waves by ellipsoidal inclusions, J. Acoust. Soc. Amer. 61, 1432-1437.

Houston, J.R. (1978) Interaction of tsunamis with the Hawaiian Islands calculated by a finite-element numerical model, J. Phys. Oceanog. 8, 93-102.

Jami, A., M. Lenoir, D. Martin and M. Polyzakis (1978) Quelques apllication d'une nouvelle formulation variationelle pour le couplage entre une method d'elements finis et une representation integrale, Rapport de Recherche 108, Ecole Nat. Sup. de Techn. Avancee.

Lamb, H. (1932) Hydrodynamics, Dover, New York.

Lesser, M.B. and D.B. Crighton (1975) Physical acoustics and the method of matched asymptotic expansions, Phys. Acoust. II, 69-149, ed. by Mason.

Malyazhinets, G.D. (1959) Development in our concepts of diffraction phenomena, Sov. Phys. Uspekhi 69, 749-758.

Mei, C.C. (1978) Extentions of some identities in elastodynamics with rigid inclusions, J. Acoust. Soc. Amer. 64, 1514-1522.

Mei, C.C. (1979a) Grazing incidence of short elastic waves on a slender cavity, Wave Motion (in press).

Mei, C.C. (1979b) Reciprocity relations for offshore structures vibrating at acoustic frequencies, to be presented at BOSS Conf. on Big Offshore Structures, London.

Mei, C.C. and M.A. Foda (1979a) An analytical theory of resonant scattering of SH waves by thin overground structures, Earthq. Eng. Struc. Dyn. 3, 3-19.

Mei, C.C. and M.A. Foda (1979b) Wave induce responses in a fluid-filled poro-elastic solid with a free surface--a boundary layer theory (submitted for publication).

Mei, C.C., M.A. Foda and P. Tong (1979) Exact and hybrid element solutions for the vibration of thin elastic structure seat on the sea floor, Applied Ocean Research 1 (in press).

Mei, C.C. and E.O. Tuck (1979) Forward scattering by long thin bodies, Univ. of Adelaide Appl. Math Report T7901 (submitted for publication).

Mei, C.C. and Ü. Ünlüata (1976) Resonant scattering by a harbor with two coupled basins, J. Eng. Math. 10, 333-353.

Newman, J.N. (1979) The theory of ship motion, Advances in Applied Mechanics 18, 221-283, ed. by C.S. Yih.

Nowacki, W. (1978) Dynamic Problems of Thermoelasticity, Noordhoff, Amsterdam.

Rayleigh, Lord (1870) On the theory of resonance, Phil. Trans. 161, 77-118 (Coll. Works p. 33).

Rayleigh, Lord (1897) On the incidence of aerial and electric waves upon small obstacles in the form of ellipsoids or elliptic cylinders, and on the passage of electric waves through a circular aperture in a conducting screen, Phil. Mag. 44, 38-53 (Coll. Works p. 305).

Tappert, F. (1977) The parabolic approximation, Wave Propagation and Underwater Acoustics, 224-287, ed. by J.B. Keller and J.S. Papadakis, Springer-Verlag, Berlin.

Tong, P., T.H.H. Pian and S.J. Lasry (1973) A hybrid element approach to crack problems in plane elasticity, Int. J. Numerical Methods in Engineering 7, 297-308.

Tuck, E.O. (1974) Matching problems involving flow through small holes, Advances in Applied Mechanics 15, ed. by C.S. Yih.

Tuck, E.O. (1978) Models for predicting tsunami propagation, Proc. Tsunami Workshop, U.S. Nat. Sci. Foun. (to appear).

Viswanathan, K. and J.P. Sharma (1977) On the matched asymptotic solution to the diffraction of a plane elastic wave by a semi-infinite rigid boundary of finite width, IUTAM Symp. Mod. Probl. in Elastic Wave Propag., ed. by J.D. Ackenbach and J. Miklowitz.

Waas, G. (1972) Analysis methods for footing vibrations through layered media, Ph.D. Thesis, Department of Civil Engineering, Univ. of California, Berkeley.

Yue, D.K.P., H.S. Chen and C.C. Mei (1977) A hybrid element method for diffraction of water waves by three-dimensional bodies, Int. J. Numerical Methods in Engineering 12, 245-266.

Part 8
Multiple Scattering Theories

MULTIPLE SCATTERING THEORY AND THE SELF-CONSISTENT IMBEDDING APPROXIMATION

J. Korringa
Dept. of Physics, The Ohio State Univ., Columbus, OH 43210

The self-consistent imbedding approximation (SCI) is an important tool in the theory of heterogeneous materials. Although much has been written about its justification, it is and remains an intuitive approach, which is attractive because it often gives reasonable results with very little effort. The following elementary derivation illustrates this point.

Let a heterogeneous medium be defined in microscopic detail by a local, position-dependent property C(x). To focus attention I take C(x) to be the local Hooke tensor of elasticity, relating the stress $\sigma(x)$ in the point x to the strain $\varepsilon(x)$ in that same point, $\sigma(x) = C(x)\varepsilon(x)$. The medium is supposed to be homogeneous on a macroscopic scale. This means that the average behavior under static quasi-homogeneous deformation is the same as for a medium with constant $C = C^*$, the so-called effective medium. The same holds for the propagation of elastic waves of a wavelength long compared to the length characterizing the heterogeneity. C^* will be defined by

$$\langle\sigma(x)\rangle \equiv \langle C(x)\varepsilon(x)\rangle = C^*\langle\varepsilon(x)\rangle. \tag{1}$$

Here $\varepsilon(x)$ is the strain of a sample body due to external influences of such a nature that they would produce a uniform strain if the body were homogeneous. $\langle\ \rangle$ means average over an ensemble of identically shaped bodies cut at random from the medium.

The SCI gives an estimate of C^* for a cellular model of the medium. This model is defined as follows. Let the components of the tensor C be stepfunctions, i.e. be piecewise constant with values C_α in simply connected regions (cells) α of volume v_α which together fill all space. In formula:

$$C(x) = \sum_\alpha C_\alpha\theta_\alpha(x), \quad \theta_\alpha(x) = 1, \quad x\epsilon v_\alpha, \quad \theta_\alpha(x) = 0, \quad x\not\epsilon v_\alpha. \tag{2}$$

Equation (1) takes the form

$$\langle\sum_\alpha v_\alpha(C_\alpha - C^*)\bar{\varepsilon}_\alpha\rangle = 0, \tag{3}$$

where $\bar{\varepsilon}_\alpha$ is the volume average of the strain in cell α. $\bar{\varepsilon}_\alpha$ for a given cell depends on C_α, on the shape, size and orientation of the cell and on its surroundings. I admit only a finite set of distinct cells. The ensemble average in Eq.(3) for a very large sample can then be expressed as an average over the orientations and over the different surroundings of the cells:

$$\sum_\alpha v_\alpha(C_\alpha - C^*)\langle\bar{\varepsilon}_\alpha\rangle_{env} = 0. \tag{4}$$

Finally I eliminate the external forces which produce the strain by dividing by the volume average of strain. Defining the "strain enhancement factor" Λ_α by

$$\langle\bar{\epsilon}_\alpha\rangle_{env} = \Lambda_\alpha\langle\epsilon\rangle_V \tag{5}$$

this gives

$$\sum_\alpha v_\alpha(C_\alpha - C^*)\Lambda_\alpha = 0, \tag{6}$$

while from the definition, Eq.(5), one has

$$\sum_\alpha v_\alpha\Lambda_\alpha = 1. \tag{7}$$

The evaluation of Λ_α involves all the details of texture of the medium and can in general not be carried out. The SCI approximation is obtained by approximating Λ_α by the value found from imbedding the cell α in the effective medium to be computed. It is well-known that this implies an extreme form of randomness. For example, incorrect estimates can be expected for a uniform medium with well-spaced inclusions of a different material. If $\Lambda'_\alpha(C'^*)$ is the value of Λ_α found from imbedding in a medium C'^*, then C'^* follows from the implicit equation

$$\sum_\alpha v_\alpha(C_\alpha - C'^*)\Lambda'_\alpha(C'^*) = 0. \tag{8}$$

Besides using extreme randomness, this approximation ignores the effects of the immediate surroundings of a given cell. The portent of this can be gleaned from the fact that the solution of Eq.(8) is independent of the size distribution of the cells: two cells of the same shape and volume give the same contribution as one cell of that shape and twice the volume. Still another disadvantage is, that it requires a cellular model to be used even when such cells are not there. In the above example of a uniform medium with inclusions, one has to attribute shapes not only to the inclusions, but also to the matrix. Any rational choice of these last shapes will have to take into account how the inclusions are distributed in space, and that includes correlations, which we tried to avoid! One usually, resolves this dilemma by assuming "spheres" for these shapes, regardless of the shapes or distribution of the inclusions. In applications of the SCI to cracked rocks (1) we found this choice to be inadequate. Using flat oblate spheroids to describe air-filled cracks, spheres for the solid material, the solution of Eq.(8) invalidates Eq.(7), i.e. one finds that $\Sigma v_\alpha\Lambda'_\alpha \neq 1$. Only when all shapes are spheres does Eq.(8) lead to Eq.(7). This happens to be the case in Bruggeman's original work (2) and most subsequent applications of the SCI. It is difficult to say whether this inadequacy associated with non-spherical shapes is due to the artificial definition of cells or to the effects of variations in the local surroundings, or both. We have not found a simple remedy.

These objections not withstanding, the SCI is often used for its simplicity. In particular, when all cells are ellipsoids, Eq.(8) reduces to an implicit set of algebraic equations which is readily solved with numerical methods. This comes from the fact, of which a special case was already known to Newton, that the strain inside an imbedded ellipsoid with constant C_α is uniform, (3),(4),(5). From this it follows that

$$\Lambda'_\alpha(C^*) = \langle[1 + \Gamma_\alpha(C^* - C_\alpha)]^{-1}\rangle_{av}, \tag{9}$$

where the averaging is over orientations of the ellipsoid and where the tensor Γ_α is the integral of the Green's function of the effective medium, to be discussed later, over the volume of the ellipsoid, cf.(5),

$$\Gamma_\alpha(C^*) = - \int_{V_\alpha} \mathcal{G}(C^*, x-x')dx'. \tag{10}$$

Correlations can, to some extent, be introduced into the SCI by considering composite cells, for example, spheres with one component, surrounded by a concentric spherical shell with the other component. There are many things which have not yet been tried, for which one has to abandon the simplicity due to ellipsoidal shapes. The real problem is that the SCI method is a one-shot approximation, i.e. that it is not the first step in a converging sequence of approximations.

In order to obtain a different perspective, I will now develop the multiple scattering theory as applied to classical properties of heterogeneous materials, as was first done in (6). I will limit my remarks to the static theory and, as before, to elasticity. It is important to generalize to the case that external forces produce a position-dependent mean strain and to define the effective medium in terms of a Hooke's operator, i.e. of a non-local stiffness. In analogy with Eq.(1) one has, for a macrohomogeneous medium,

$$\langle C(x)\varepsilon(x)\rangle = \int C^{eff}(x-x')\langle\varepsilon(x')\rangle dx'. \tag{11}$$

From this one obtains, for $\langle\varepsilon(x')\rangle = \text{constant}$

$$C^* = \int C^{eff}(x-x')dx'. \tag{12}$$

The range of the function $C^{eff}(x-x')$ will be of the order of the grain size, so that C^* suffices for describing macroscopic deformations. A multiple scattering approach that bypasses the introduction of $C^{eff}(x-x')$ (7) does not give the same value as Eq.(12), as will be shown below.

The microscopic equilibrium equation is

$$\frac{\partial}{\partial x_i} C_{ijk\ell}(x) \frac{\partial}{\partial x_k} u_\ell(x) = f_j(x). \tag{13}$$

Here $u_\ell(x)$ is the elastic displacement vector and $f_j(x)$ is an applied force density. For abbreviation, Eq.(13) is written as

$$\nabla C \nabla u = f. \tag{13 bis}$$

To obtain a Lippman-Schwinger type equation, let C^o be a constant elastic tensor, as befits a homogeneous medium, and let $u^o(x)$ be the displacement field that the force density f(x) would produce in a medium with stiffness C^o:

$$\nabla C^o \nabla u^o = f. \tag{14}$$

Subtracting Eq.(14) from Eq.(13 bis) gives

$$\nabla C^o \nabla(u-u^o) = -\nabla(C-C^o)\nabla u. \tag{15}$$

f has thus been eliminated in favor of the displacement field $u^o(x)$.

The infinite space Green's function $g^o(x-x')$ of the medium C^o is defined by

$$\nabla C^o \nabla g^o(x-x') = -I\delta(x-x'), \tag{16}$$

where I is the unit tensor and $g^o(x-x') \to 0$ when $|x-x'| \to \infty$. From Eqs.(15) and (16) one has

$$u(x) = u^o(x) + \int dx' \, g^o(x-x') \nabla (C(x')-C^o) \nabla' u(x'). \tag{17}$$

A partial integration and differentiation of the equation with respect to x gives

$$\varepsilon(x) = \varepsilon^o(x) + \int dx' \, \mathcal{G}^o(x-x')(C(x')-C^o)\varepsilon(x'), \tag{18}$$

where $\mathcal{G}^o$ is the symmetrized second derivation of g^o:

$$\mathcal{G}^o_{ijk\ell} = \partial_{(j} \partial_{((\ell} g^o_{i)k))}. \tag{19}$$

For abbreviation I write

$$\varepsilon = \varepsilon^o + \mathcal{G}^o(C-C^o)\varepsilon. \tag{18 bis}$$

This is the desired integral equation. From its solution $\varepsilon(x)$ one obtains $C^{eff}(x-x')$ by means of Eq.(11), which I write in shorthand as

$$\langle C\,\varepsilon\rangle = C^{eff}\langle\varepsilon\rangle. \tag{11 bis}$$

The brackets mean "ensemble average" which, for an infinite macrohomogeneous medium, is defined by taking the force density f(x) to be fixed in space, i.e. not fixed in the medium, and considering the ensemble of all possible positions related by translations, of the medium with respect to it. $\langle\varepsilon(x)\rangle$ is thus not the space average of strain; this would be zero if, e.g. f(x) was sinusoidal.

Through introduction of the "T"-matrix it will now be shown that C^{eff}, defined by Eq.(11), is independent of ε^o, and therefore also independent of the force density f which served as probe. Let $T(x,x')$ be the operator which, when applied to $\varepsilon^o(x)$, yields the microscopic stress difference $(C(x)-C^o)\varepsilon(x)$:

$$(C-C^o)\varepsilon = T\varepsilon^o. \tag{20}$$

Substituting in the second term of Eq.(18) and multiplying the equation at left with $C-C^o$ gives

$$T\varepsilon^o = (C-C^o)\varepsilon^o + (C-C^o)\mathcal{G}T\varepsilon^o \tag{21}$$

which is satisfied for all $\varepsilon^o(x)$ if the operator T satisfies the equation

$$T = C-C^o + (C-C^o)\mathcal{G}^o T. \tag{22}$$

The explicit form of the first term is $(C(x)-C^o)\delta(x-x')$. As shown first in (6), $C^{eff}(x-x')$ is expressed in terms of the ensemble average $\langle T\rangle$ (which, for a macrohomogeneous medium is also a function of (x-x')) by

$$C^{eff} = C^o\delta(x-x') + \langle T\rangle(1+\mathcal{G}^o\langle T\rangle)^{-1}, \tag{23}$$

for which

$$C^* = C^o + \int dx' \langle T\rangle(1+\mathcal{G}^o\langle T\rangle)^{-1}. \tag{24}$$

(To derive Eq.(23), take the ensemble average of Eq.(20) and substitute in the left-hand side Eq.(11 bis) and then, for $\langle\varepsilon\rangle$, the right-hand side of the ensemble average of Eq.(18 bis) in which first Eq.(20) has been substituted.)

Although Eq.(23) contains C^o explicitly and through $\langle T\rangle$ and $\mathcal{G}^o$, the resulting C^{eff} can not depend on it, as C^o is entirely arbitrary. From Eq.(24) one sees that C^* is that value of C^o for which the space integral of the second term in Eq.(23) vanishes. One can of course not make the entire function $\langle T\rangle$ of $(x-x')$ to vanish. Equation (23) suggests that this can be achieved by taking for C^o a non-local stiffness operator $C^o(x-x')$. The entire formalism leading to Eq.(23) can indeed be generalized to a non-local C^o. It requires the introduction of a Green's function of a non-local medium, which is not usually considered. With this generalization, C^{eff} can be defined as the operator C^o for which $\langle T\rangle = 0$.

In order to make contact with the SCI approximation, the above formalism must be adjusted to a cellular model. With Eq.(2) and a corresponding splitting of T:

$$T(x,x') = \sum_{\alpha} T_{\alpha}(x,x'), \tag{25}$$

in which I take $T_\alpha(x,x') = T(x,x')\theta_\alpha(x)$, i.e. require that in Eq.(25), $T_\alpha(x,x')$ is zero outside v_α, equal to $T(x,x')$ for x inside v_α. Equation (22) gives, because of $\theta_\alpha\theta_\beta = \delta_{\alpha\beta}\theta_\alpha$,

$$T_{\alpha} = \theta_{\alpha}(C_{\alpha}-C^o) + \theta_{\alpha}(C_{\alpha}-C^o)\mathcal{G}^o\sum_{\beta} T_{\beta}, \tag{26}$$

or

$$(1-\theta_{\alpha}(C_{\alpha}-C^o)\mathcal{G}^o)T_{\alpha} = \theta_{\alpha}(C_{\alpha}-C^o) + \theta_{\alpha}(C_{\alpha}-C^o)\mathcal{G}^o\sum_{\beta\neq\alpha} T_{\beta}. \tag{27}$$

Let t_α be the solution of

$$t_{\alpha} = \theta_{\alpha}(C_{\alpha}-C^o) + \theta_{\alpha}(C_{\alpha}-C^o)\mathcal{G}^o t_{\alpha}, \tag{28}$$

i.e.

$$t_{\alpha} = [1-\theta_{\alpha}(C_{\alpha}-C^o)\mathcal{G}^o]^{-1}\theta_{\alpha}(C_{\alpha}-C^o). \tag{29}$$

Applying this same inverse operator to Eq.(27) one therefore has

$$T_{\alpha} = t_{\alpha} + t_{\alpha}\mathcal{G}^o\sum_{\beta\neq\alpha} T_{\beta}, \tag{30}$$

which has the form of the central equation in multiple scattering theory. Comparison between Eqs.(26) and (28) shows that t_α is the T-operator for a medium with $C(x) = C^o + (C_\alpha - C^o)\theta_\alpha(x)$. From t_α one obtains, for that medium, the strain inside v_α with use of Eq.(20), from which the strain outside v_α is found with use of Eq.(18). From this point of view $t_\alpha(x,x')$ is the full "scattering operator" for the cell α imbedded in the medium C^o. In order to obtain the strain enhancement factor, $\Lambda'_\alpha(C^o)$, for cell α, as given in Eqs.(9) and (10), one must take ϵ_0 constant in Eq.(20), with $T \rightarrow t_\alpha$. The right-hand side then reduces to $(\int t_\alpha(x,x')dx')\epsilon_0$ and according to Eq.(28) this integral obeys the equation

$$\int t_{\alpha}(x,x')dx' = \theta_{\alpha}(x)(C_{\alpha}-C^o)(1 + \int\mathcal{G}^o(x-x'')dx''\int t_{\alpha}(x'',x')dx'). \tag{31}$$

By definition, $\Lambda'_\alpha(C^o)$ is obtained from the solution as

$$\Lambda'_\alpha(C^o) = \langle (C_\alpha - C^o)^{-1} \int_{v_\alpha} dx \int t_\alpha(x,x')dx' \rangle_{av}. \tag{32}$$

When v_α is an ellipsoid, the solution of Eq.(31) is a constant. Therefore Eqs.(31) and (32) reduce to Eqs.(9) and (10).

The solution of Eq.(30), summed over all cells and then averaged over the ensemble of all distributions would give the exact C^{eff}, as in Eq.(23). In practice, Eq.(30) is the starting point of approximations. For example, in the case of inclusions in a homogeneous matrix mentioned before, one can take $C^o = C_M$, the elastic tensor of the matrix material. Then α needs to refer only to the inclusions, because $t = 0$ in any region with $C = C_M$, and T_α is easily expressed as a cluster expansion, see Ref. (8),

$$T_\alpha = t_\alpha + t_\alpha \mathcal{G}^o \sum_{\beta \neq \alpha} t_\beta + t_\alpha \mathcal{G}^o \sum_{\beta \neq \alpha} t_\beta \mathcal{G}^o \sum_{\gamma \neq \beta} t_\gamma + \ldots \tag{33}$$

For other systems, such as polycrystals, this does obviously not work. It is not my intention to discuss specific approximation schemes, but rather to use the multiple scattering formalism to shed light on the SCI approximation. This is obtained with the "CPA", i.e. the coherent potential approximation, which, like the SCI, neglects the effects of correlation between the cells (9).

The CPA is based on the approximation $\langle T_\alpha \rangle \approx t_\alpha$, i.e. it neglects the average value of the second term in Eq.(30). However, this term and therefore the merits of this approximation depend on the choice of C^o. The assumption underlying the CPA is that the best choice of C^o is such that $\Sigma t_\alpha \approx \langle T \rangle = 0$, or, loosely speaking, that the imbedding medium is selected in such a manner that there is no scattering in the average over all cells. From Eq.(23) one sees that this means that C^o is equal to the value of C^{eff} to which this approximation gives rise. This condition looks similar to that underlying the SCI. The difference is that $\Sigma t_\alpha = 0$ can only be achieved with a non-local C^o, just as the exact condition $\langle T \rangle = 0$ requires a non-local C^o. To my knowledge computation of t_α for non-local C^o has never been undertaken. A more restrictive approximation introduces the "best" local C^o. There are two possibilities, (a) one applies self-consistency in the sense $C^o \approx \int C^{eff}(x-x')dx' \equiv C^*$ and (b) one applies the condition of no further scattering in the sense $\int \langle T \rangle (x-x')dx' \approx \sum \int t_\alpha(x,x')dx' = 0$. It is seen from Eq.(24) that these criteria are different. Case (b) is the assumption underlying the SCI, i.e. the imbedding medium obtained is the SCI-approximate C^*. However, as Eq.(24) shows, this differs from the CPA-approximate C^* by the integral of the second term, which is not zero in this case and, because it involves the non-local quantity $\langle T \rangle$ (in the approximation Σt_α), can not be calculated in the framework of the SCI. Case (a) is more difficult to apply and requires calculation of the full scattering function $t_\alpha(x,x')$ for all cells.

In summary, it has been shown that the SCI makes an approximation in addition to those explicitly specified in that it assumes that, when $\int \langle T \rangle (x-x')dx' = 0$, then $\int \langle T \rangle (1 + \mathcal{G}^o \langle T \rangle)^{-1} dx'$ is negligible. It remains to be investigated what the consequences of this assumption are for a particular system. If a non-local version of the SCI were developed, this reservation would disappear, because in that case SCI and CPA are equivalent. The introduction of a non-local imbedding medium has the advantage of involving the sizes as well as the shapes of the cells. It loses, however, all the simplicity connected with knowing the explicit form of $\mathcal{G}^o$ and with the constancy of $\int t_\alpha(x,x')dx'$ for ellipsoids. In most applications we lack detailed knowledge of the texture. The above refinements may therefore not be very meaningful and the SCI, imperfect as it is, may just as well be used, albeit in a very discriminating way.

REFERENCES

1. J. Korringa, R. J. J. Brown, D. D. Thompson, and R. J. Runge, Self-consistent imbedding and the ellipsoidal model for porous rocks, J. Geophys. Res., in the press.

2. D. A. G. Bruggeman, Berechnung verschiedener physikalischer Konstanten von heterogenen Substanzen, Ann. d. Physik 24, 636-680 (1935); Thesis, Utrecht (Netherlands), 1930.

3. J. D. Eshelby, The determination of the elastic field of an ellipsoidal inclusion, and related problems, Proc. Roy. Soc. Lond. A241, 376-396 (1957).

4. I. A. Kunin and E. G. Sosnina, Ellipsoidal inhomogeneity in an elastic medium, Soviet Physics-Doklady, Mathematical Physics 16, 534-536 (1972).

5. J. Korringa, I-H. Lin, and R. L. Mills, General theorem about homogeneous ellipsoidal inclusions, Am. J. Phys. 46, 517-521 (1978).

6. R. Zeller and P. H. Dederichs, Elastic constants of polycrystals, Phys. Stat. Sol. B 55, 831-842 (1973).

7. J. Korringa, Theory of elastic constants of heterogeneous media, J. Math. Phys. 14, 509-513 (1973).

8. M. Hori, Statistical theory of effective properties of random heterogeneous materials VII. Comparison of different approaches, J. Math. Phys. 18, 487-501 (1977).

9. P. Soven, Coherent-potential model for substitutional disordered alloys, Phys. Rev. 156, 809-814 (1967).

ELECTROMAGNETIC WAVE SCATTERING BY SMALL BODIES OF AN ARBITRARY SHAPE

A. G. Ramm
Department of Mathematics, University of Michigan

SECTION 1. INTRODUCTION

The theory of wave scattering by small bodies was initiated by Rayleigh (1871). Thompson (1893) was the first to understand the role of magnetic dipole radiation. Since then, many papers have been published on the subject because of its importance in applications. From a theoretical point of view there are two directions of investigation: (i) to prove that the scattering amplitude can be expanded in powers of ka, where $k = 2\pi/\lambda$ and a is a characteristic dimension of a small body (ii) to find the coefficients of the expansion efficiently. Stevenson (1953), Senior and Kleinman can be mentioned among contributors to the first topic. To my knowledge there were no results concerning the second topic for bodies of an arbitrary shape. Such results are of interest in geophysics, radiophysics, optics, colloidal chemistry and solid state theory.

In this paper we study scalar and vector wave scattering by small bodies of an arbitrary shape with the emphasis on practical applicability of the formulas obtained and the mathematical rigor of the theory. For scalar wave scattering by a single body, the main results can be described as follows: (1) Analytical formulas for the scattering amplitude for a small body of an arbitrary shape are obtained; dependence of the scattering amplitude on the boundary conditions is studied (2) An analytical formula for the scattering matrix for electromagnetic wave scattering by a small body of an arbitrary shape is given. Applications of these results are outlined (calculation of the properties of a rarefied medium; inverse radio measurement problem; formulas for the polarization tensors and capacitance) (3) The multiparticle scattering problem is analyzed and interaction of the scattered waves is taken into account. For self-consistent fields in a medium consisting of many particles ($\sim 10^{23}$), integral-differential equations are found. The equations depend on the boundary conditions on the particle surfaces. These equations offer a possibility of solving the inverse problem of finding the medium properties from the scattering data. For about 5 to 10 bodies the fundamental integral equations for the theory can be solved numerically to study the interaction between the bodies.

In section 2 the results on scalar wave scattering are described. In section 3 electromagnetic scattering is studied and the solution of the inverse problem of radio measurements is outlined. In section 4 the many body problem is examined.

SECTION 2. SCALAR SCATTERING BY A SINGLE BODY

Consider the problem

$$(\nabla^2 + k^2)v = 0 \text{ in } \Omega; \quad \left(\frac{\partial v}{\partial N} - hv\right)\Big|_\Gamma = -\left(\frac{\partial u_o}{\partial N} - hu_o\right)\Big|_\Gamma \tag{1}$$

$$v \sim \frac{\exp(ik|x|)}{|x|} f(n,k) \text{ as } |x| \to \infty, \quad \frac{x}{|x|} = n \tag{2}$$

where $\Omega = R^3 \setminus D$, D is a bounded domain with a smooth boundary Γ, N is the outer normal to Γ, u_o is the initial field which is usually taken in the form $u_o = \exp\{ik(\nu,x)\}$. We look for a solution of the problem (1)-(2) in the form

$$v = \int_\Gamma \frac{\exp(ikr_{xt})\sigma(t)}{4\pi r_{xt}} dt, \quad r_{xt} = |x-t| \tag{3}$$

and for the scattering amplitude f we have the formula

$$f = \frac{1}{4\pi} \int_\Gamma \exp\{-ik(n,t)\}\sigma(t,k)dt = \frac{1}{4\pi} \int_\Gamma \sigma_o(t)dt + O(ka), \tag{4}$$

where

$$\sigma(t,k) = \sigma_o(t) + ik\sigma_1(t) + \frac{(ik)^2}{2} \sigma_2(t) + \ldots \tag{5}$$

Putting (3) in the boundary conditions (1) we get the integral equation for σ.

$$\sigma = A(k)\sigma - hT(k)\sigma - 2hu_o + 2\frac{\partial u_o}{\partial N}, \tag{6}$$

where

$$A(k)\sigma = \int_\Gamma \frac{\partial}{\partial N_s} \frac{\exp(ikr_{st})}{2\pi r_{st}} \sigma(t)dt, \quad T(k)\sigma = \int_\Gamma \frac{\exp(ikr_{st})}{2\pi r_{st}} \sigma(t)dt. \tag{7}$$

Expanding σ, A(k) and T(k) in powers k and equating the corresponding terms in (6) we obtain for h = 0, i.e. Neumann boundary condition, the following equations:

$$\sigma_o = A_o\sigma_o \tag{8}$$

$$\sigma_1 = A_o\sigma_1 + A_1\sigma_o + 2\frac{\partial u_{o1}}{\partial N} \tag{9}$$

$$\sigma_2 = A_o\sigma_2 + 2A_1\sigma_1 + A_2\sigma_o + 2\frac{\partial u_{o2}}{\partial N}, \tag{10}$$

where

$$A(k) = A_o + ikA_1 + \frac{(ik)^2}{2} A_2 + \ldots, \quad u_o = u_{oo} + iku_{o1} + \frac{(ik)^2}{2} u_{o2} \tag{11}$$

Expanding f in formula (4) we obtain, up to terms of the second order

$$f = \frac{1}{4\pi}\int_\Gamma \sigma_o dt + ik\{\frac{1}{4\pi}\int_\Gamma \sigma_1 dt + (n, \int_\Gamma \sigma_o(t)tdt)\} + \frac{(ik)^2}{2}\{\frac{1}{4\pi}\int_\Gamma \sigma_2 dt$$

$$+ \frac{2}{4\pi}(n, \int_\Gamma \sigma_1 tdt) + \frac{1}{4\pi}\int_\Gamma \sigma_o(n,t)^2 dt\} \qquad (12)$$

From (8) it follows that $\sigma_o = 0$ and from (9) it follows that $\int_\Gamma \sigma_1 dt = 0$. Some calculations lead to the following final results:

$$f = \frac{ikV}{4\pi}\beta_{pq}n_p \frac{\partial u_o}{\partial x_q}\bigg|_{x=0} + \frac{V}{8\pi}\Delta u_o\bigg|_{x=0} \qquad (13)$$

Usually $u_o = \exp\{ik(\nu,x)\}$ and in this case formula (13) can be written

$$f = -\frac{k^2V}{4\pi}(\beta_{pq}\nu_q n_p + \frac{1}{2}) \qquad (13')$$

In the above V is the volume of the body D and β_{pq} is the magnetic polarization tensor of D. Note that $f \sim k^2a^3$ and the scattering is anisotropic and is defined by the tensor β_{pq}.

For $h = \infty$ (Dirichet boundary condition) the integral equation (6) takes the form:

$$T(k)\sigma = -2u_o. \qquad (14)$$

Hence

$$\int_\Gamma \frac{\sigma_o dt}{4\pi r_{st}} = -u_o\big|_\Gamma .$$

Since $ka \ll 1$ the field $u_o\big|_\Gamma = u_o(x,k)\big|_{x=0}$, where the origin is supposed to be inside the body D. From which it follows, that $\int_\Gamma \sigma_o dt = -Cu_o$,

$$f = -\frac{Cu_o}{4\pi}, \qquad (15)$$

where C is the capacitance of a conductor with a shape D. Hence for the Dirichet boundary condition, $f \sim a$, where a is a characteristic length of D, and the scattering is isotropic.

For $h \neq 0$, following the same line of arguments it is possible to obtain the following approximate formula for the scattering amplitude:

$$f \approx -\frac{hS}{4\pi(1+hSC^{-1})} u_{oo} , \qquad (16)$$

where S = meas(Γ), and C is the capacitance of D. If h is very small ($h \sim k^2a^3$) the formula for f should be changed and the terms analogous to (13) should be taken into account.

SECTION 3. ELECTROMAGNETIC WAVE SCATTERING BY A SINGLE BODY

If a homogeneous body D with parameters ε, μ, σ is placed into a homogeneous medium with parameters ε_0, μ_0, σ_0, then the following formula for the scattering matrix was established by the author

$$S = \frac{k^2V}{4\pi}\begin{bmatrix} \mu_o\beta_{11}+\alpha_{22}\cos\theta-\alpha_{32}\sin\theta, & \alpha_{21}\cos\theta-\alpha_{31}\sin\theta-\mu_o\beta_{12} \\ \alpha_{12}-\mu_o\beta_{21}\cos\theta+\mu_o\beta_{31}\sin\theta, & \alpha_{11}+\mu_o\beta_{22}\cos\theta-\mu_o\beta_{32}\sin\theta \end{bmatrix}, \qquad (17)$$

where S is defined by the formula

$$\begin{pmatrix} f_2 \\ f_1 \end{pmatrix} = S\begin{pmatrix} E_2 \\ E_1 \end{pmatrix} = \begin{pmatrix} S_2 S_3 \\ S_4 S_1 \end{pmatrix}\begin{pmatrix} E_2 \\ E_1 \end{pmatrix} , \qquad (18)$$

θ is the angle of scattering, E_1, E_2 are the components of the initial field, f_1, f_2 are the components of the scattered field in the far field multiplied by $|x|\exp(-ik|x|)$, the plane YOZ is the plane of scattering, $\alpha_{ij} = \alpha_{ij}(\gamma)$, $\gamma = (\varepsilon-\varepsilon_0)/(\varepsilon+\varepsilon_0)$ is the polarization tensor, and $\beta_{ij} = \alpha_{ij}(-1)$ is the magnetic tensor.

If one knows S one can find all values of interest to physicists for electromagnetic wave propagation in a rarefied medium consisting of small bodies. The tensor of refraction coefficients can be calculated by the formula $n_{ij} = \delta_{ij} + 2\pi N k^{-2} S_{ij}(o)$, where N is the number of bodies per unit volume. The tensor $\alpha_{ij}(\gamma)$ can be calculated analytically be means of the formula

$$|\alpha_{ij}(\gamma) - \alpha_{ij}^{(n)}(\gamma)| \le Aq^n, \quad 0 < q < 1, \qquad (19)$$

where A, q are some constants depending only on the geometry of the surface, and

$$\alpha_{ij}^{(n)} = \frac{2}{V}\sum_{m=0}^{n}\frac{(-1)^m}{(2\pi)^m}\frac{\gamma^{n+2}-\gamma^{m+1}}{\gamma-1}b_{ij}^{(m)} , \quad n \ge 1 \qquad (20)$$

In (20)

$$b_{ij}^{(o)} = V\delta_{ij}, \quad b_{ij}^{(1)} = \int_\Gamma\int_\Gamma \frac{N_i(s)N_j(t)dsdt}{r_{st}} , \qquad (21)$$

$$b_{ij}^{(m)} = \int_\Gamma\int_\Gamma dsdt\, N_i(s)N_j(t) \int_\Gamma {}_{m-1} \int_\Gamma \frac{1}{r_{st}} \psi(t_1,t)\ldots\psi(t_{m-1},t_{m-2}).$$

$$dt_1\ldots dt_{m-1} \;;\quad \psi(t,s) \equiv \frac{\partial}{\partial N_t}\frac{1}{r_{st}} \tag{22}$$

In particular

$$\alpha_{ij}^{(1)}(\gamma) = 2(\gamma + \gamma^2)\delta_{ij} - \frac{\gamma^2 b_{ij}^{(1)}}{\pi V}\,,\quad \beta_{ij}^{(1)} = -\frac{b_{ij}^{(1)}}{\pi V}\,. \tag{23}$$

For particles with $\mu = \mu_o$ and ε not very large, so that the depth δ of the skin layer is considerably larger than a, one can neglect magnetic dipole radiation and in the formula (17) for the scattering matrix one can omit terms with multipliers β_{ij}.

The vectors of electric and magnetic polarizations can be found by the following formulas, respectively,

$$P_i = \alpha_{ij}(\gamma)V\varepsilon_o E_j\,,\quad \gamma = \frac{\varepsilon - \varepsilon_o}{\varepsilon + \varepsilon_o}\,, \tag{24}$$

where E_j is the initial field, and

$$M_i = \alpha_{ij}(\bar{\gamma})V\mu_o H_j + \beta_{ij}V\mu_o H_j\,,\quad \bar{\gamma} = \frac{\varepsilon - \varepsilon_o}{\varepsilon + \varepsilon_o}\,, \tag{25}$$

where H_j is the initial field and the second term on the right hand side of equality (25) should be omitted if the skin-layer depth $\delta >> a$.

The scattering amplitudes can be found from the formulae

$$f_E = \frac{k^2}{4\pi\varepsilon_o}[n[P,n]] + \frac{k^2}{4\pi}\sqrt{\frac{\mu_o}{\varepsilon_o}}[M,n]\,, \tag{26}$$

$$f_M = \sqrt{\frac{\varepsilon_o}{\mu_o}}\,[n,f_E]\,, \tag{27}$$

where [A,B] stands for the vector product A x B and P, M can be calculated by formulae (24), (25), (19), (20), (22). If $\delta >> a$ one can neglect the second term on the right-hand side of (26). If is possible to give a simple solution to the following inverse problem which can be called the inverse problem of radiomeasurements. Suppose an initial electromagnetic field is scattered by a small probe. Assume that the scattered field E', H' can be measured in the far field. The problem is to calculate the initial field at the point where the small probe detects E', H'. This problem is of interest in determining the electromagnetic field distribution in antenna apertures. Let us assume for simplicity that for the probe $\delta >> a$, so that

$$E' = \frac{\exp(ikr)}{r}\,\frac{k^2}{4\pi\varepsilon_o}[n[P,n]]\,. \tag{28}$$

From (28) we can find $P - n_1(P,n_1) = E'(n_1)b$ where $b = \frac{\exp(ikr)}{r}\frac{k^2}{4\pi\varepsilon_0}$. A measurement in the n_2 direction, where $(n_1,n_2) = 0$ results in $P - n_2(P,n_2) = E'(n_2)b$. Hence $(n_1,P) = b(E'(n_2),n_1)$. Thus $P = b\{E'(n_1) + n_1(E'(n_2),n_1)\}$. But (*)$P_i = \alpha_{ij}(\gamma)V\varepsilon_0 E_j$. Since V and ε_0 are known and $\alpha_{ij}(\gamma)$ can be calculated by formulae (19), (20) and the matrix α_{ij} is positive definite(because $\frac{1}{2}\alpha_{ij}V\varepsilon_0 E_j E_i$ is the energy) it follows that system (*) is uniquely solvable. Its solution is the desired vector E.

Let us give a formula for the capacitance of a conductor D of arbitrary shape, which proved to be very useful in practice

$$C^{(n)} = 4\pi\varepsilon S^2\left\{\frac{(-1)^n}{(2\pi)^n}\iint_{\Gamma\Gamma}\frac{dsdt}{r_{st}}\underbrace{\int_\Gamma\cdots\int_\Gamma}_{n\text{-times}}\psi(t,t_1)\ldots\psi(t_{n-1},t_n)dt_1\ldots dt_n\right\}^{-1}, \tag{29}$$

$$C^{(0)} = \frac{4\pi\varepsilon_0 S^2}{J} \le C, \quad J \equiv \int_\Gamma\int_\Gamma\frac{dsdt}{r_{st}} \tag{30}$$

It can be proved that

$$|C - C^{(n)}| \le Aq^n, \quad 0 \le q < 1, \tag{30'}$$

where A, q are constants which depend only on the geometry of Γ.

Remark 1. The theory is also applicable for small layered bodies.
Remark 2. Two sided variational estimates for α_{ij} and C were given in [18].

SECTION 4. MANY BODY WAVE SCATTERING

First we describe a method for solving the scattering problem for r bodies, $r \sim 5$-10 and, then we derive an integral differential equation for the self-consistent field in a medium consisting of many ($r \sim 10^{23}$) small bodies. We look for a solution of the scalar wave scattering problem in the form

$$u = u_0 + \sum_{j=1}^{r}\int_{\Gamma_j}\frac{\exp(ikr_{st})}{4\pi r_{xt}}\sigma_j(t)dt. \tag{31}$$

Applying the boundary condition,

$$u|_{\Gamma_j} = 0, \quad 1 \le j \le r \tag{32}$$

we obtain the system of r integral equations for the r unknown functions σ_j. In general this system can be solved numerically. When $d \gg \lambda$, where $d = \min_j d_{ij}, i\ne j$, and d_{ij} is the distance between i=th and j-th body, the system of integral equations has dominant diagonal terms and it can be easily solved by an iterative process, the zero approximation being the initial field u_0. If $ka \gg 1$, $d \gg a$, but not necessarily $d \gg \lambda$, the average field in the medium of small particles can be found from the integral equation

$$u(x,k) = u_o(x) - \int \frac{\exp(ikr_{xy})}{4\pi r_{xy}} q(y)u(y,k)dy . \tag{33}$$

Here $q(y)$ is the average value of $h_j S_j(1 + h_j S_j C_j^{-1})^{-1}$ over the volume dy in the neighborhood of y for bodies with impedance boundary conditions. For $h_j = \infty$ (Dirichlet condition) and identical bodies $q(y) = N(y)C$, where $N(y)$ is the number of the bodies per unit volume and C is the capacitance of a body. For Neumann boundary condition the corresponding equation is the integral-differential equation

$$u(x,k) = u_o(x,k) + \int \frac{\exp(ikr_{xy})}{4\pi r_{xy}} \{B_{pq}(y) \frac{\partial u(y,k)}{\partial y_q} \frac{x_p - y_p}{r_{xy}} +$$

$$\frac{1}{2} \Delta u(y,k)b(y)\}dy. \tag{34}$$

where

$$b(y) = N(y)V , B_{pq}(y) = ikV\beta_{pq}N(y) , \tag{35}$$

V is the volume of a body, and β_{pq} is its magnetic polarization tensor. The solution to equations (33), (34) can be considered as the self-consistent (effective) field acting in the medium.

Equations (33), (34) allows one to solve the inverse problems of the determination of the medium properties from the scattering data. For example, from (33) it follows that the scattering amplitude has the form

$$f = -\frac{1}{4\pi} \int \exp\{-ik(n,y)\}q(y)u(y,k)dy . \tag{36}$$

For a rarefied medium it is reasonable to substitute u by u_o (Born approximation) and to obtain

$$f \approx -\frac{1}{4\pi} \int_{R^3} \exp\{-ik(n,y)\}q(y)u_o(y,k)dy. \tag{37}$$

If $u_o = \exp\{ik(\nu,x)\}$ formula (37) is valid for $k \gg 1$ with the error $O(k^{-1})$ if

$$q(x) \in C_1(R^3, 1 + |x|^{3+\varepsilon}), \ \varepsilon > 0.$$

Hence if f is known for $0 < k < \infty$, $(n-\nu) S_1$, where S_1 is the unit sphere in R^3, the Fourier transform of $q(y)$ is known and q can be uniquely determined. If $q(y)$

is finite i.e. $q(y)$ is equal to zero outside some bounded domain, then f is entire in k and knowing f in any interval $[k_o, k_1]$, $0 < k_o < k_1$, for all $(n-\nu)\in S_1$ one can find f for all $0 < k < \infty$ by analytical continuation and then one can determine $q(y)$.

Let us consider the r-body problem for a few bodies (small r). Assume that the Dirichlet boundary condition holds. Let us look for a solution of the form

$$u(x) = u_o + \sum_{j=1}^{r} \int_{\Gamma_j} \frac{\exp(ik|x-t|)}{4\pi|x-t|} \sigma_j dt \tag{38}$$

The scattering amplitude is equal to

$$f(n,k) = \frac{1}{4\pi} \sum_{j=1}^{r} \exp\{-ik(n,t_j)\} \cdot \int_{\Gamma_j} \exp\{-k(n,t-t_j)\}\sigma_j(t)dt \tag{39}$$

where t_j is some point inside the j-th body. Since $ka \ll 1$ this formula can be rewritten

$$f(n,k) = \frac{1}{4\pi} \sum_{j=1}^{r} \exp\{-ik(n,t_j)\} \cdot Q_j, \tag{40}$$

where

$$Q_j = \int_{\Gamma_j} \sigma_{jo} dt + O(ka) , \quad \sigma_{jo} = \sigma_j \big|_{k=0} . \tag{41}$$

This is the same line of arguments as in §2. Using the boundary condition one gets

$$\sum_{j\neq m, j=1}^{r} \int_{\Gamma_j} \frac{\exp(ik|x_m-t|)}{4\pi|x_m-t|} \sigma_j(t)dt + \int_{\Gamma_m} \frac{\exp(ik|x_m-t|)}{4\pi|x_m-t|} \sigma_m dt = -u_o(x_m). \tag{42}$$

With the accuracy of $O(ka)$ this can be written

$$\int_{\Gamma_m} \frac{\sigma_m(t)dt}{4\pi|x_m-t|} + \sum_{j=1, j\neq m}^{r} \frac{\exp(ikd_{mj})}{4\pi d_{mj}} Q_j = -u_{om}, \quad 1 \leq m \leq r \tag{43}$$

where $d_{mj} = |x_m - t_j|$. If C_m is the capacitance of the m-th body we can rewrite (43)

$$Q_m = -C_m u_{om} - \sum_{j=1, j\neq m}^{r} C_m \frac{\exp(ikd_{mj})}{4\pi d_{mj}} Q_j, \quad 1 \leq m \leq r. \tag{44}$$

This is a linear system from which Q_j, $1 \leq m \leq r$ can be determined. If $d_{mj} C_m^{-1} \gg 1$ this system can be easily solved by an iterative process. If $\{Q_j\}$ are known, then the scattering amplitude can be found from (40).
More details about the described theory the reader can find in the References.

REFERENCES

1. A. G. Ramm, "Iterative solution of integral equation in potential theory", Sov. Phys. Dokl., 186, (1969), 62-65. Math. Rev., 41 #9462.

2. A. G. Ramm, "Approximate formulas for tensor polarizability and capacitance of bodies of arbitrary shape and its applications", Sov. Phys. Dokl., 195, (1970), 1303-1306. Math. Rev. 55 #1947.

3. A. G. Ramm, "Calculation of the initial field from scattering amplitude", Radiotech. i Electron., 16, (1971), 554-556.

4. A. G. Ramm, "Approximate formulas for polarizability tensor and capacitances for bodies of an arbitrary shape", Radiofisika, 14, (1971), 613-620. Math. Rev. 47 #1386.

5. A. G. Ramm, "Electromagnetic wave scattering by small bodies of an arbitrary shape", Proc. Fifth all Union Sympos. on Wave Diffraction, Trudy Math. Inst. Steklova, Leningrad, (1971), 176-186.

6. A. G. Ramm, "Calculation of the magnetization of thin films", Microelectronica, 6, (1971), 65-68. (with Frolov).

7. A. G. Ramm, "Calculation of scattering amplitude of electromagnetic waves by small bodies of an arbitrary form II.", Radiofisika, 14, (1971), 1458-1460.

8. A. G. Ramm, "Electromagnetic wave scattering by small bodies of an arbitrary shape and relative topics", Proc. Intern. Sympos. URSI, Moscow, (1971), 536-540.

9. A. G. Ramm, "Calculation of the capacitance of a parallelepiped", Electricity, 5, (1972), 90-91. (with Golubkova, Usoskin).

10. A. G. Ramm, "On the skin-effect theory", Journ. of Techn. Phys., 42, (1972), 1316-1317.

11. A. G. Ramm, "Calculation of the capacitance of a conductor placed in anisotropic inhomogeneous dialectric", Radiofisika, 15, (1972), 1268-1270. Math. Rev. 47 #2284.

12. A. G. Ramm, "Remark to integral equation theory", Diff. eq., 8, (1972), 1517-1520. Engl. transl. pp.1177-1180; Math. Rev. 47 #2284.

13. A. G. Ramm, "Iterative process to solve the third boundary problem", Diff. eq., 9, (1973), 2075-2079. Math. Rev. 48 #6861.

14. A. G. Ramm, "Light scattering matrix for small particles of an arbitrary shape", Opt. and Spectroscopy, 37, (1974), 125-129.

15. A. G. Ramm, "Scalar scattering by the set of small bodies of an arbitrary shape", Radiofisika, 17, (1974), 1062-1068.

16. A. G. Ramm, "New methods of calculation of the static and quasistatic electromagnetic waves", Proc. Fifth Int. Symp. "Radio-electronics-74", Sofia, 3, (1974), 1-8 (report 12).

17. A. G. Ramm, "Plane wave scattering by the system of small bodies of an arbitrary form", Abstracts of the Eight Int. Acoust. Congr., London, (1974), p. 573.

18. A. G. Ramm, "Estimates of some functionals in quasistatic electrodynamics", Ukrain. Phys. Journ. 5, (1975), 534-543. Math. Rev. 56 #14165.

19. A. G. Ramm, "Investigation of some classes of integral equations and their applications", In collection "Abel inversion and its generalization", Edited by N. Preobrazhensky, Siberian Dep. of Acad. Sci. USSR, Novosibirsk, (1978), 120-179.

MULTIPLE SCATTERING FROM GRATINGS OF COMPLIANT TUBES IN FLUID AND IN A VISCOELASTIC LAYER IMMERSED IN FLUID

Ronald P. Radlinski
Naval Underwater Systems Center, New London, CT 06320

ABSTRACT

Arrays of compliant (squashed) tubes have been used as tuned reflecting baffles to reduce noise or to provide directivity for a transducer. To gain an understanding of the dynamic characteristics of the compliant tube arrays, a two dimensional series formulation has been developed for the scattering of a plane wave from an infinite planar grating of elliptic cylindrical shells in a fluid medium. With the model, one can study variations of insertion loss resulting from tubing of different orientations, aspect ratios, and material constants or from different grating spacings of the array. Analytical predictions and experimental measurements are compared for several cases. The use of several gratings in a baffle is shown to increase the bandwidth of reflectivity. For multiple gratings, the modeling has been restricted to plane waves normally incident to arrays of rectangular tubing. The method of partial domains is used to determine the scattering coefficients. Also shown is the insertion loss that results from encapsulating the compliant elements in a rubber layer. The effects of the complex valued, frequency-dependent, material properties of the elastomer on the array resonances are discussed.

INTRODUCTION

A compliant tube is often described as a long non-circular cylindrical shell which is sealed at the ends. The shells usually have a static compliance from twenty to more than a hundred times that of water. The low sound velocity achieved by packing the compliant devices in a volumetric array was demonstrated by Toulis (1) by constructing an acoustic lens comprised of compliant tubes in water. The reflective properties of compliant tubes were also demonstrated by Toulis who constructed sparsely packed, planar, metal tube arrays as reflecting baffles. Since the thickness of a planar array of tubing is usually less than an acoustic wavelength, the static compliance alone is not sufficient to reflect appreciable energy. Near frequencies corresponding to modal resonances of the shell with net volume displacement, the dynamic compliance of the shell also exhibits a resonance behavior. For an elliptic cylindrical shell, the symmetric flexural modes in a plane perpendicular to the length of the shell have a net volume displacement and such a compliant tube is designed to have the first of these flexural modes in the frequency band where reflectivity is desired. Since, for a right circular cylindrical shell, the higher frequency membrane modes are the first modes with net volume displacement, a planar grating of these shells would have a much greater cross-sectional area than a grating of elliptical shells. Recently, rectangular tubes have been tested as compliant devices. The fabrication of rectangular plate tubing may be considerably simpler than for elliptic cylindrical shells. The following sections describe some analytical methods, other than the T-matrix formulation, that have been used to study the multiple scattering of waves from compliant tube gratings. In these cases, the geometries of the compliant elements are not as general as allowed by the T-matrix methods.

MATHEMATICAL FORMULATION FOR A SINGLE GRATING OF COMPLIANT ELEMENTS IN A FLUID MEDIUM

For a two-dimensional multiple scattering analysis of plane waves incident on a planar grating of elliptically shaped, compliant tubes in fluid, the Burke-Twersky grating models (2), given in terms of the scattering amplitude of a single cylinder, can be directly adapted when the major axes of the elliptical elements are aligned perpendicular to the grating. However, the following formulation for arbitrary orientation of the grating elements was obtained by explicit use of an addition theorem.

Incident and Scattered Pressures

The following presentation summarizes the formulation discussed in reference 3. The equation for the time independent part of an incident plane wave is given by the expression

$$P_i(X, Y) = e^{ik(X\cos\beta_i + Y\sin\beta_i)} \tag{1}$$

where k is the wavenumber, the Y axis is the plane of the grating and β_i is the angle of the incident wave with respect to the gratings positive X axis as shown in Fig. 1. In terms of the coordinates x, y of the sth grating element, the incident wave is expressed as

$$P_i = e^{iskd\sin\beta_i} e^{ik(x\cos\theta_i + y\sin\theta_i)} \tag{2}$$

where d is the grating spacing, and θ_i is the polar angular coordinate of the incident wave with respect to the x axis of the ellipse. The incident wave is then expressed in terms of even and odd elliptical functions as

$$P_i = \sqrt{8\pi} \sum_{n=0}^{\infty} i^n \Big[Se_n(h, \cos\theta_i) Se_n(h, \cos\phi) Je_n(h, \cosh\mu)/Me_n(h) + So_n(h, \cos\theta_i) So_n(h, \cos\phi) Jo_n(h, \cosh\mu)/Mo_n(h) \Big] e^{iksd\sin\beta_i}, \tag{3}$$

where ϕ, μ are respectively the elliptical angular and radial coordinates, Se_n, So_n are the even, odd periodic Mathieu functions, Me_n, Mo_n are normalization functions, Je_n, Jo_n are the radial Mathieu functions of the first kind, h = kf/2 (f is the interfocal distance), and s is an integer element number.

The scattered pressure from the array can be expressed in terms of Mathieu functions as

$$P_r = \sqrt{8\pi} \sum_{s=-\infty}^{\infty} e^{iskd\sin\beta_i} \sum_{n=0}^{\infty} \left[\frac{Ae_n Se_n(h, \cos\theta_i) Se_n(h, \cos\phi_s) He_n(h, \cosh\mu_s)}{Me_n(h)} + \frac{Ao_n So_n(h, \cos\theta_i) So_n(h, \cos\phi_s) Ho_n(h, \cosh\mu_s)}{Mo_n(h)} \right] \tag{4}$$

where He_n, Ho_n are the even, odd radial Mathieu functions of the third kind and Ae_n, Ao_n are the even, odd undetermined scattering coefficients. To satisfy the boundary conditions at a particular elliptical cylinder in the array, the wave functions associated with all the other cylinders must be expressed in terms of the coordinates of that cylinder. This transformation is accomplished by rewriting the elliptical wave functions in terms of circular wave functions (4), using the Graf addition theorem (5) for cylindrical harmonics to express all the wave functions in terms of the coordinates of the reference cylinder. The addition theorem for cylindrical wave functions of the third kind is illustrated in Fig. 2. Finally, the cylindrical functions are transformed back to the elliptical wave functions of that cylinder.

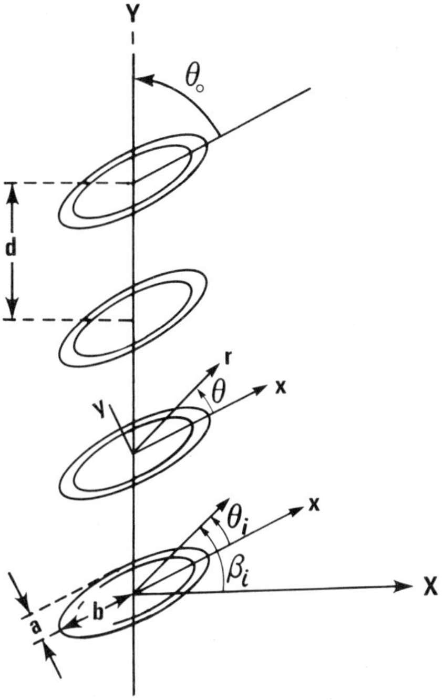

Fig. 1. Geometry of an infinite planar grating of elliptic, cylindrical shells.

The resulting scattering field from a particular configuration, including the multiscattering from all the cylinders in the infinite planar array is expressed in terms of even and odd elliptical functions of the elliptical coordinates of the reference cylinder as

$$\begin{aligned} P_r = {} & \sqrt{8\pi}\sum_{n=0}^{\infty} i^n \left[Ae_n Se_n(\theta_i) Se_n(\phi) He_{\ell n}/Me_n + Ao_n So_n(\theta_i) So_n(\phi) Ho_n/Mo_n\right] \\ & + \sqrt{8\pi}\sum_{\ell=0}^{\infty} i^{\ell} \left\{ Se_\ell(\phi) Je_\ell(\mu)/Me_\ell \left[\sum_{n=0}^{\infty} 2\pi Ae_n Se_n(\theta_i) H_{\ell n}^{(1)}/Me_n \right.\right. \\ & \left. + \sum_{n=0}^{\infty} 2\pi Ao_n So_n(\theta_i) H_{\ell n}^{(3)}/Mo_n \right] + So_\ell(\phi) Jo_\ell(\mu)/Mo_\ell \left[\sum_{n=0}^{\infty} 2\pi Ae_n Se_n(\theta_i) H_{\ell n}^{(2)}/Me_n \right. \\ & \left.\left. + \sum_{n=0}^{\infty} 2\pi Ao_n So'_n(\theta_i) H_{\ell n}^{(4)}/Mo_n \right]\right\} \end{aligned} \tag{5}$$

where the generalized Schlömilch series for an array of elliptic cylinders is given by

$$H^{(1)}_{\ell n} = \frac{1}{2} \sum_0^\infty De_p(h, n) \sum_0^\infty De_m(h,\ell) \left[\cos(p+m)\theta_0 \, \mathcal{H}_{p+m} + \cos(p-m)\theta_0 \, \mathcal{H}_{p-m}\right],$$

$$H^{(4)}_{\ell n} = \frac{1}{2} \sum_0^\infty Do_p(h, n) \sum_0^\infty Do_m(h,\ell) \left[\cos(p+m)\theta_0 \, \mathcal{H}_{p+m} - \cos(p-m)\theta_0 \, \mathcal{H}_{p-m}\right],$$

$$H^{(3)}_{\ell n} = \frac{1}{2} \sum_0^\infty Do_p(h, n) \sum_0^\infty De_m(h,\ell) \left[\sin(p+m)\theta_0 \, \mathcal{H}_{p+m} + \sin(p-m)\theta_0 \, \mathcal{H}_{p-m}\right],$$

and

$$H^{(2)}_{n} = \frac{1}{2} \sum_0 De_p(h, n) \sum_0 Do_m(h,\ell) \left[\sin(p+m)\theta_0 \, \mathcal{H}_{p+m} - \sin(p-m)\theta_0 \, \mathcal{H}_{p-m}\right].$$

where

$$\mathcal{H}_{p\pm m}(kd, \beta_i) = 2 \sum_{s=1}^\infty H_{p\pm m}(skd)\cos\left[(skd\sin\beta_i) - (p\pm m)\pi/2\right]$$

is the circular Schlömilch series, and θ_0 is the angular rotation of the reference cylinder with respect to the vertical axis of the array. $\theta_0 = 0$ corresponds to the major axis of the tubes aligned along the vertical axis of the array, and $H_{p\pm m}(skd)$ is the cylindrical Hankel function. The terms De_n, Do_n, are the even odd coefficients for the angular and radial elliptical radiation functions. The unknown coefficients Ae_n, Ao_n are found by satisfying the boundary conditions at the surface of the ellipse.

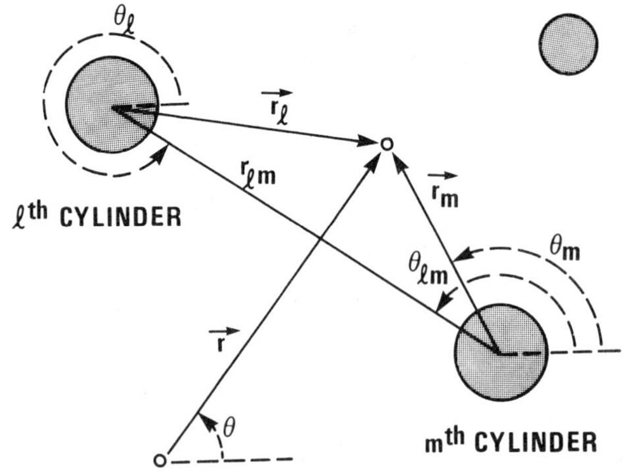

$$H_q(kr_\ell)\, e^{iq\theta_\ell} = \sum_{p=-\infty}^{\infty} J_p(kr_m)\, H_{p-q}(kr_{\ell m})\, e^{ip\theta_m}\, e^{-i(p-q)\theta_{\ell m}}$$

Fig. 2. The additional theorem for cylindrical wave functions of the third kind.

Elastic Model of an Elliptical Shell

The two-dimensional, steady-state force equations for an oval shell of thickness h_s and local radius of curvature of the midsurface of the shell r' in a fluid medium (6) can be expressed as

$$\frac{T}{r'} + \frac{\partial N}{\partial s'} = \rho_s h_s \frac{\partial^2 V}{\partial_t^2} + (P_i + P_r) \tag{6a}$$

and

$$\frac{\partial T}{\partial s'} - \frac{N}{r'} = \rho_s h_s \frac{\partial^2 W}{\partial_t^2} \quad , \tag{6b}$$

where s' is the differential arc length of the midsurface of the shell and T and N are, respectively, the tension and the normal shearing-stress resultants which for a shell are related to the midsurface deflection (V, W) as

$$T = \frac{E_s h_s}{1-\sigma^2}\left(\frac{\partial W}{\partial s'} - \frac{V}{r'}\right) \tag{7a}$$

and

$$N = \frac{\partial G}{\partial s'} = \frac{-E_s h_s^3}{12(1-\sigma^2)} \frac{\partial^2}{\partial s'^2}\left(\frac{\partial V}{\partial s'} + \frac{W}{r'}\right) \quad , \tag{7b}$$

Factor G is the bending-moment resultant about an axis at the midsurface normal to the shell's cross section. The force equations in Eq. 6 relate the normal and tangential component of the deflection of the shell's midsurface (V, W) to all internal reaction stresses and externally applied stresses. The symbols E_s, ρ_s, σ represent the Young's modulus, density, and Poisson ratio of the shell material.

If the analysis is restricted to frequencies smaller than the resonance frequency of the first membrane mode, then only the lower order flexural modes of the cross section of the shell are important. In this case, a single equation can be derived using the inextentional condition generated by setting T = 0 in eq. (7a).

The resulting force equation, including both even and odd shell modes, is written in terms of the flexural eigenfunctions for the normal displacement as given by

$$\rho_s h_s \sum_t \left[(\omega^2 - \omega_{e_t}^2) Ve_t \left(\psi e_t - \frac{1}{r'}\int ds' \int \psi e_t \frac{ds'}{r'}\right)\right.$$
$$\left. + (\omega^2 - \omega_{e_t}^2) Vo_t \left(\psi_{o_t} - \frac{1}{r'}\int ds' \int \psi_{o_t} \frac{ds'}{r'}\right)\right] = P_i + P_r . \tag{8}$$

where

$$V = \sum_t (Ve_t \psi e_t + Vo_t \psi_{o_t})$$

and

$$\psi e_n = \sum_{p=0}^{\infty} Ne_p(\omega, n, a/b) \cos(p\phi)$$

$$\psi o_n = \sum_{p=0}^{\infty} No_p(\omega, n, a/b) \sin(p\phi) ,$$

where Ne_n, No_m are the even, odd expression coefficients of the shells eignfunctions.

respectively, and Ve_t, Vo_t are the modal coefficients of the normal displacement. In addition, the second boundary condition to be satisfied is the continuity of the normal deflection at the shell fluid interface as expressed by

$$\omega^2 \rho b(1 - e^2\cos\phi)^{\frac{1}{2}} \sum_t (Ve_t \psi e_t + Vo_t \psi o_t) = \frac{\partial P_t}{\partial \mu} + \frac{\partial P_r}{\partial \mu} . \tag{9}$$

where e is the eccentricity of the midsurface of the shell and ρ is the fluid density.

As detailed in reference 3, from the orthogonality properties of the Mathieu functions and the model eigenfunctions of the normal deflection, one can eliminate the modal coefficients and subsequently solve for the scattering coefficients. To calculate the transmitted contribution from the entire infinite array, the Sommerfield integral representation of the Hankel function and the Poisson summation formula are used to write the scattered pressure in terms of a sum of propagating and exponentially decaying plane waves (7) as expressed by

$$P_r = 2/(kd) \sum_{p=0}^{\infty} \sum_{s=-\infty}^{\infty} e^{ikX\cos\beta_s - ip\beta_s} (\cos\beta_s)^{-1}$$
$$\times \left\{ \cos\left[kY \sin\beta_s + p(\theta_0 - \pi/2)\right] \sum_{n=0}^{\infty} 2\pi Ae_n Se_n(\theta_i) De_p(h, n)/Me_n \right. \tag{10}$$
$$\left. + \sin\left[kY \sin\beta_s + p(\theta_0 - \pi/2)\right] \sum_{n=0}^{\infty} 2\pi Ao_n So_n(\theta_i) Do_p(h, n)/Mo_n \right\},$$

where $\sin\beta_s = \sin\beta_i + s\lambda/d$. Here, s, such that $|\sin\beta_s|<1$ designates the propagating plane wave modes scattered from the grating. The modes such that $|\sin\beta_s| > 1$ are the evanescent modes which decay exponentially from the grating and thus only contribute in the near field. The values $|\sin\beta_s| = 1$ corresponds to the grazing modes.

If the center to center spacing between array elements is less than a wavelength, only a single plane wave will propagate from the array. If we consider only cases where the incident wave is normal to the grating and restrict the analysis to low frequencies keeping three even and one odd scattering coefficients and no evanescent modes, then the total transmitted pressure is

$$P_t = P_i + P_r \simeq \left\{ 1 + \frac{2}{kd} \left[Ae_o - 2(Ae_2 \cos 2\theta_i \cos 2\theta_0 - Ae_1 \cos\theta_i \sin\theta_0 + Ao_1 \sin\theta_i \cos\theta_0) \right] \right\} e^{ikX} \tag{11}$$

COMPARISON OF THEORY AND EXPERIMENT

The results of the analytic model were compared to measured transmissivity of large planar gratings of plastic tubes. The elliptically shaped tubes were extruded from Lexan polycarbonate thermoplastic ($E_s = 2.38 \times 10^9 N/m^2$, $\rho_s = 1.18 \times 10^3 kg/m^3$, $\sigma = 0.37$). The aspect ratio of a tube was about 2 to 1, with a major axis dimension of about one cm. The thickness of the tube was about 0.13 cm. The first in-air flexural resonance (f_1) of these compliant tubes was measured to be at 22 kHz and the static compressibility was twenty times greater than water.

For the first series of measurements, the plastic tubes were inserted into a 3m by 3m grating frame with their major axes parallel to the direction of the normally incident plane wave. This size panel, which is 10λ by 10λ at 5 kHz, was choosen to minimize any diffraction effects. The insertion loss measurement location of 5 cm behind the panel was beyond the effects of the evanescent modes. The insertion loss (insertion loss = 20 $\log_{10}$|incident pressure/transmitted pressure|) is shown in Fig. 3 for a grating spacing of 0.635 cm (λ_1/d = 10.8). A large insertion loss implies that the transmitted pressure is small.

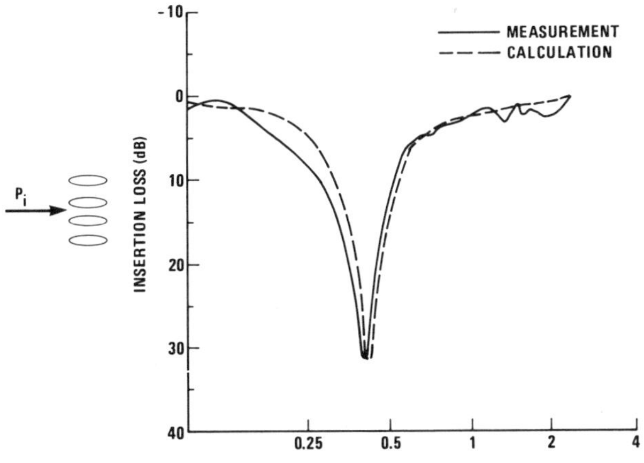

Fig. 3. Calculated and measured insertion loss for a densely packed grating of compliant tubes (θ_0 = π/2, β_i = 0, d = 0.64 cm).

For this array of densely packed compliant tubes, the calculated and measured insertion loss values are in good agreement at all frequencies. The center frequency of the array resonance is a function of both the resonance of a single compliant tube and the radiation loading resulting from the spacing and orientation of the array elements. Because of multiple interactions, the array resonant frequency at f/f_1 = 0.41 is much lower than that of a single tube in water which occurs at approximately f/f_1 = 0.77. The calculated insertion loss represents the contribution of the first three scattering coefficients of the propagating transmitted plane wave. Mathieu functions were evaluated from a computer algorithm developed by Aerospace Research Laboratories (8) and the elliptical Schlömilch series were evaluated to at least order h^2. A quality factor of 25 associated with the mechanical hysteresis in the walls of a single tube was incorporated into the calculations and was found to be the limiting factor for the maximum calculated insertion loss.

The insertion loss for a more sparsely packed array of plastic tubes having their major axes parallel to the direction of the incident plane wave is shown in Fig. 4. With a grating spacing of 2.54 cm, the array resonance approaches that of a single tube in water. Increasing the grating spacing also results in a decrease in the maximum insertion loss and bandwidth of the array resonance.

A measurement was made where the minor axes of the tubes were oriented parallel to the direction of the incident plane wave. For ease of fabrication, a 0.76 m by one m panel was constructed. The measured and calculated insertion losses are shown in Fig. 5 for a 1.27 cm grating spacing. Increase diffraction effects are evident in the measurements for frequencies below the array resonance. In this case, the array resonance is slightly higher than for a single compliant tube in water. The difference between the measured and calculated insertion loss at the array resonance can be attributed to diffraction.

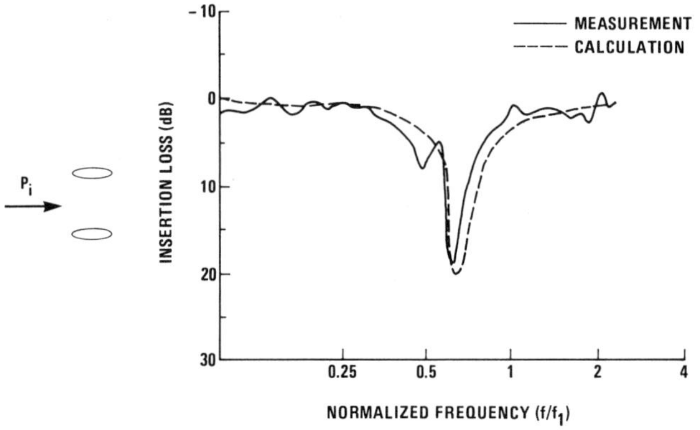

Fig. 4. Calculated and measured insertion loss for a sparsely packed grating of compliant tubes ($\theta_0 = \pi/2$, $\beta_i = 0$, d = 2.54 cm).

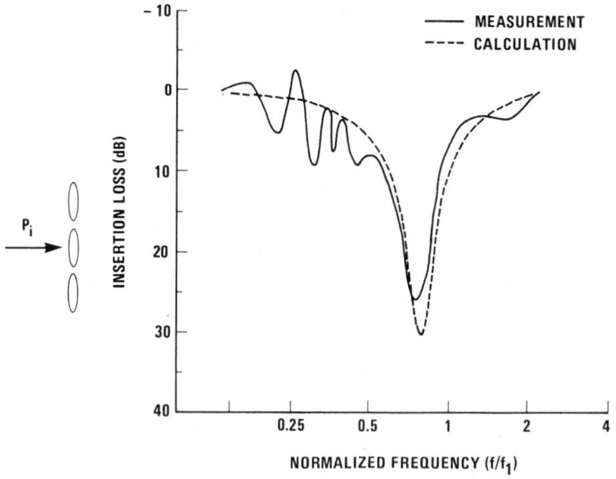

Fig. 5. Calculated and measured insertion loss for a densely packed grating of compliant tubes ($\theta_0 = 0$, $\beta_i = 0$, d = 1.27 cm).

SCATTERING FROM MULTIPLE GRATINGS OF COMPLIANT TUBES IN A FLUID

Experimental Results

In this section, we consider the broadband insertion loss that can be achieved through the use of multiple gratings in a fluid. The wide separations of the center frequencies of the array resonances for various single layer orientations suggests that combining the panels into double layer gratings will result in an insertion loss with a wider bandwidth. The measurements for combinations of the single layers previously considered are shown in Fig. 6. To a first approximation, the insertion loss for double layered gratings can be found by the superposition of the insertion losses for each individual layer. The combinations shown in Fig. 6 produce acoustic filtering over a bandwidth of almost two octaves. If one uses gratings of compliant tubes with resonances at other frequencies, the bandwidth can be extended even further.

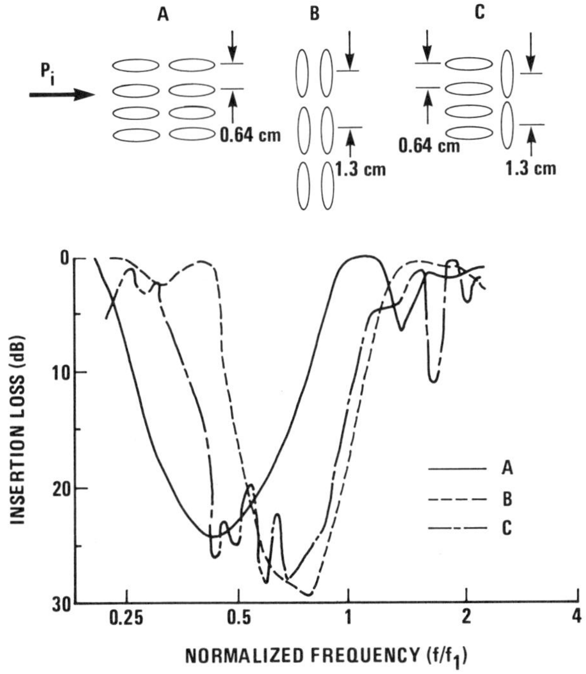

Fig. 6. Normal incident insertion loss measurements for three orientations of double-layered gratings of compliant tubes.

Mathematical Model of Multiple Gratings of Rectangular Tubes in a Fluid Layer

In many applications, one wishes to minimize the thickness of the baffle but retain a wide reflectivity bandwidth. In such situations, multiple layers of tubing are used in the flat configuration shown in Fig. 7. In this figure, a plane wave is shown normally incident to a double grating of rectangular inclusions in a fluid layer. Both gratings have the same spacing d, but multiple gratings of different spacings can also be considered. The waveguide solution is found by matching the boundary conditions between the planes of symmetry shown in Fig. 7. The vertical lines between the planes of symmetry delineate boundaries separating regions where the solutions vary.

The analysis is an extension of the work by Vovk, et al (9) to include multiple planar gratings and to immerse these configurations in a finite thickness fluid layer between two fluid half spaces. Each of the fluid can have a different velocity and density. In an individual element, the two plates may be of different thickness, density, or Young's modulus. In particular, bars with simply supported end conditions are considered and the connections between the plates are assumed rigid and immovable.

The differential equations to be satisfied are the wave equation for the pressure in the fluid and the thin plate equation which is assumed to govern the motion of the bars. The wave equation for the pressure in the fluid is given by

$$\nabla^2 P_i + k_i^2 P_i = 0 \tag{12}$$

where k_i is the wave number of the ith fluid and P_i is the pressure. The thin plate equation for compliant plate motion of the inclusions is given in the form

$$D_j \frac{d^4 W_j(y)}{dy^4} - \mu_j \omega^2 W_j(y) = P_i \tag{13}$$

where D_j is the plate stiffness of the jth plate, μ_j is the mass per unit area of the plate, ω is the circular frequency, and W_j is the displacement of the plate.

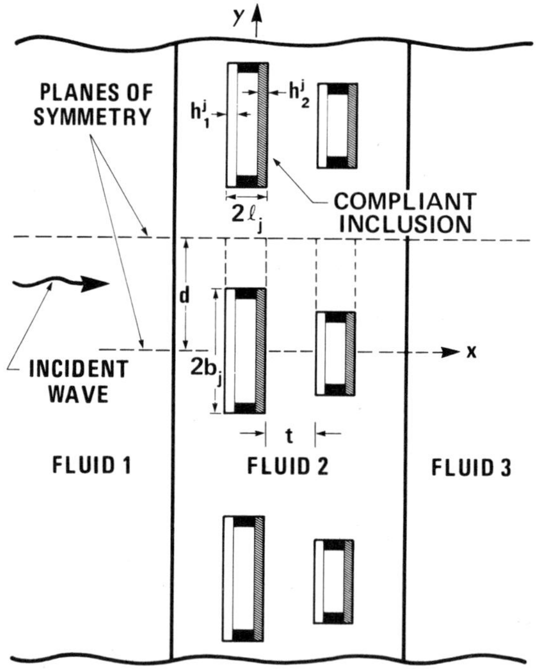

Fig. 7. Double grating of rectangular compliant elements in a fluid layer.

The boundary conditions imposed at the fluid interfaces, symmetry planes, and tubes surfaces are given as follows. At the fluid interfaces, the two boundary conditions to be satisfied are the continuity of pressure

$$P_i = P_j \tag{14}$$

and continuity of normal velocity

$$\frac{1}{\rho_f^i} \frac{\partial P_i}{\partial x} = \frac{1}{\rho_f^j} \frac{\partial P_j}{\partial x} \tag{15}$$

of adjacent i^{th} and j^{th} regions ρ_f is the density of the fluid. The condition on the velocity normal to the symmetry planes is

$$\frac{\partial P}{\partial y} = 0. \tag{16}$$

At the tube surfaces, the three boundary conditions to be met are for the nondeformable supports as given by

$$\frac{\partial P}{\partial y} = 0 \qquad |y| = b_j \quad -\ell_j+x_j<x<x_j+\ell_j \tag{17}$$

the continuity of normal displacement at the flexible plates,

$$\frac{1}{\omega^2\rho_f}\frac{\partial P}{\partial x} = W_j(y) \qquad |y| \le b_j, \tag{18}$$

and the simply supported condition

$$W_j(y) = \frac{d^2W_j(y)}{dy^2} = 0 \qquad |y| = b_j \tag{19}$$

at $x = \pm \ell_j + x_j$ where x_j is the x coordinate at the center of the jth grating.

The solution for the total pressure in the half space of fluid 1 is the sum of the incident plane wave and an infinite series of reflected waves as indicated by the expression

$$P_1 = e^{ik_1x} + \sum_{n=0}^{\infty} R_n e^{-ik^0_n(x - x_0)}(\cos \alpha_n y) \tag{20}$$

where

$$k^0_n = \sqrt{k_1^2 - \alpha_n^2} \quad ; \quad \alpha_n = \frac{n\pi}{d},$$

k_0 is the wavenumber in fluid I, x_0 is the x coordinate of the boundary between fluid 1 and fluid 2, and R_n is a reflection coefficient to be determined. When $k_1^2>\alpha_n^2$ this wave propagates away from the grating. If, however, $k_1^2<\alpha_n^2$, the evanescent wave propagates parallel to the y axis and decays exponentially in the x direction away from the grating. In fluid layer 2 with inclusions, the solution in each of the intervening regions consists of infinite series of waves of the form

$$P_2^{(j)} = \sum_{m=0}^{\infty} [A_m^{(j)}\cos k_m^{(j)}(x-x_j^0) + B_m^{(j)}\sin k_m^{(j)}(x-x_j^0)] \cos\alpha_m^j(y-c_j) \tag{21}$$

where

$$k_m^{(j)} = \sqrt{k_2^2 - (\alpha_m^j)^2} \quad \alpha_m^j = \frac{m\pi}{d-c_j} .$$

Here c_j is a constant determined by region j of the fluid layer, k_2 is the wavenumber in fluid 2, and $A_m^{(j)}$ and $B_m^{(j)}$ are unknowns in region j of the fluid layer, x_j^0 is the x coordinate at the center of the jth region. The transmitted pressure waves in fluid 3 are of particular importance and given by the expression

$$P_3 = \sum_{n=0}^{\infty} T_n e^{ik'_n(x - x_m)} (\cos \alpha_n y) . \tag{22}$$

where

$$k'_n = \sqrt{(k_3)^2 - \alpha_n^2} ,$$

k_3 is the wavenumber in fluid 3, and x_m is the x coordinate of the boundary between fluid 2 and fluid 3 and T_n is the transmission coefficient to be determined.

For the compliant elements, a normal mode solution for the simply supported plates is given by the displacement distribution

$$W_j(y) = \sum_{s=0}^{\infty} W_s^{(j)} \cos\alpha_s^j y \qquad \alpha_s^j = \frac{(2s+1)\pi}{2b_j} \tag{23}$$

where $W_s^{(j)}$ are unknown amplitudes determined from the boundary conditions. If the above expression for the displacement is substituted into the homogeneous plate equation, the natural frequencies of the bars in vacuum are

$$f_s^{(j)} = \frac{h_j}{2\pi}\left(\frac{\pi(2s+1)}{2b_j}\right)^2 \sqrt{\frac{E_j}{12\rho_j(1-\sigma_j^2)}} \tag{24}$$

where ρ_j, E_j, σ_j, $2b_j$, h_j are the density, Young's modulus, Poisson ratio, length, and thickness of the jth plate.

The unknown pressure distributions and displacement amplitudes of the plate motion are found by satisfying the boundary conditions between the various regions. The set of equations arising from the matching conditions is easily transformed on the basis of completeness and orthogonality into an infinite system of linear algebraic equations (10).

Model Calculations

Using the analytical techniques described above, one can investigate the interaction between gratings, transmission resonances of a fluid layer with inclusions, and the effects of changing material constants within the gratings. The calculations presented here are for the maximum grating spacing being less than a wavelength in fluid layer 3. The transmission coefficient is then completely determined by the value of the unknown coefficient T_0. Calculations of T_0 from truncated series determined by satisfying the boundary conditions were shown to converge by retaining four to five values of each unknown in the linear system of equations.

The effects on insertion loss from varying the velocity of the fluid layer are shown in Fig. 8. The insertion loss is plotted versus frequency normalized to the frequency of the first bending resonance of the smaller bars in vacuum (f_1). The distance between the center lines of the two gratings is $0.0625\lambda_1$. A structural damping factor of 0.1 was assumed for the bars. The elastic plates are considered to be identical and constructed of Lexan plastic ($h_1^{(1)} = h_2^{(1)} = 0.0185\lambda_1$). The in-vacuum first structural resonance of the bars was calculated to be at 15 kHz. For a fluid layer velocity of about 1500 m/sec, shown by the solid curve, four distinct array resonances are evident. The first is a split resonance in that the fluid loading on the two plates causes them to resonate at slightly different frequencies. Higher order resonances are also evident in this example. When the velocity of the layer is decreased to about 150 m/sec, as shown by the dashed curve, transmission resonances appear at the higher frequencies. The transmission resonances are associated with layer thickness. When the effective velocity is such that the layer approaches a half wavelength in thickness, increased transmission results. Thus, the lower fluid velocity in the layer can negate the effectiveness of the baffle in reflecting energy. In most uses of compliant elements, the experimental values of insertion loss are reduced by rigid body motion and diffraction, both of which are not considered in this analysis.

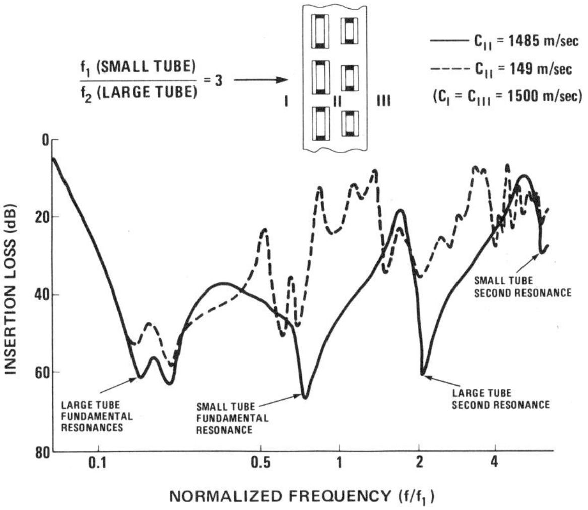

Fig. 8. Variation of insertion loss with sound velocity in fluid layer II.

THE EFFECTS OF EMBEDDING A GRATING OF THE RESONANT DEVICES IN A VISCOELASTIC LAYER SURROUNDED BY FLUID

Mathematical Formulation

If the compliant tube elements are encapsulated in a rubber layer as shown in Fig. 9, the stiffness and damping of the elastomer will alter the performance of the arrays. To study these effects, the waveguide solution considered in the previous section was reformulated to include inhomogeneous plane dilatational and shear waves in the elastic layer. The expressions for the reflected and transmitted pressure waves in the fluid half spaces are identical to those given in section II.

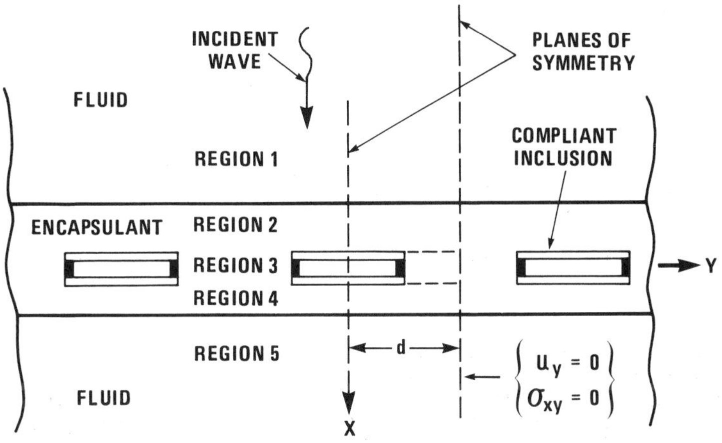

Fig. 9. An encapsulated rectangular compliant element grating.

In the assumed linear viscoelastic medium, the two wave equations to be satisfied are for the scalar ϕ and vector potentials $\vec{A}$ given by

$$\nabla^2\phi + k_p^2\,\phi = 0$$
$$\nabla^2\vec{A} + k_s^2\,\vec{A} = 0, \qquad (25)$$

where

$$k_p = \omega/c_p = \omega/\left[(\lambda+2\mu)/\rho\right]^{1/2}$$

$$k_s = \omega/c_s = \omega/\left[\mu/\rho\right]^{1/2},$$

λ and μ are the frequency dependent, complex valued, material properties that reduce to Lame' constants in the limiting elastic case. The above formulation is equivalent to the Kelvin-Voigt model with harmonic time dependence assumed. In the viscoelastic layer of regions two, three and four of Fig. 9, the scalar and vector displacement potentials for the dilatation ϕ_j and shear waves A_z^j in the jth elastomeric region are written as infinite series of inhomogeneous waves travelling in both directions given by

$$\phi_j = \sum_{m=0}^{\infty}\left[A_m^{(j)} \cos k_{m_p}^{(j)}(x-x_j^0) + B_m^{(j)} \sin k_{m_p}^{(j)}(x-x_j^0)\right] \cos\alpha_m^j(y-c_j)$$

$$A_z^j = \sum_{m=0}^{\infty}\left[C_m^{(j)} \cos k_{m_s}^{(j)}(x-x_j^0) + D_m^{(j)} \sin k_{m_s}^{(j)}(x-x_j^0)\right] \sin\alpha_m^j(y-c_j) \qquad (26)$$

where

$$k_{m_p}^{\;2} = k_p^{\;2} - (\alpha_m^j)^2; \quad k_{m_s}^{\;2} = k_s^{\;2} - (\alpha_m^j)^2; \quad \alpha_m^j = \frac{m\pi}{d-c_j},$$

and c_j is a constant determined by the region. $A_m^{(j)}$, $B_m^{(j)}$, $C_m^{(j)}$, $D_m^{(j)}$ are again scattering coefficients determined by satisfying the boundary conditions at the region interfaces.

The normal and tangential displacements are calculated from the scalar and vector potentials as

$$u_x = \frac{\partial\phi}{\partial x} + \frac{\partial A_z}{\partial y}$$

$$u_y = \frac{\partial\phi}{\partial y} + \frac{\partial A_z}{\partial x} \qquad (27)$$

and the stress as $\sigma_{ij} = \lambda\Delta\delta_{ij} + \mu\left(\dfrac{\partial u_i}{\partial x_j} + \dfrac{\partial u_j}{\partial y}\right) \qquad i = x, y; \quad j = x, y$

where δ is the Kronecker delta and

$$\Delta = \frac{\partial u_x}{\partial x} + \frac{\partial u_y}{\partial y}. \qquad (28)$$

The displacement potentials in the elastic layer are chosen to satisfy the boundary conditions of zero transverse velocity u_y and zero shear stress σ_{xy} at the planes of symmetry and at the plate supports. At the fluid-elastomer interface, the boundary conditions to be satisfied are (1) the shear stress σ_{xy} is zero, (2) the normal deflection u_x is continuous, and (3) the normal stress σ_{xx} in the elastomer must be equal to the negative of the total pressure in the fluid. At the boundaries of region 3, the normal and tangential displacements and stresses are assumed continuous. At the plate boundaries, the normal displacements of the viscoelastic layer are equal to the displacement at the tube walls and for the equation of motion of the plate, the pressure is replaced by the normal stress. Further details of the use of the above conditions to derive an infinite system of linear algebraic equations can be found in reference 11.

Measured and Calculated Results from Encapsulation of Plastic Compliant Tubing

The configuration of plastic tubes considered in Fig. 5 were encapsulated in two different rubbers. The values of the frequency and temperature dependent complex wavenumbers were determined from measurements of the complex moduli of the viscoelastic material. The complex shear modulus was assumed to be one-third that of measured values of the complex Young's modulus. The bulk moduli and the density of the encapsulants were similar to that of the fluid. For the materials selected in this study, the loss tangent of the shear waves is much greater than that for the dilatational waves. For rubbers in general, the complex shear modulus μ varies much more rapidly with frequency and temperature than λ. For a comparison with experimental data of insertion loss from the plastic tubes, fixed end conditions and identical velocity displacements were assumed for the two plates of the tube. To a first approximation, the frequencies and compliant mode shapes of an elliptical tube can be approximated by a plate with fixed end conditions. Uncoupled rigid body displacement of the compliant elements was also included in the analysis.

In Fig. 10, the measured and calculated insertion loss is shown for the plastic compliant tubes encapsulated in a rubber with a low shear modulus of about 10^6N/M^2 and with a 0.2 loss factor. The loss factor, shown as delta, is the ratio of the imaginary part of the complex shear modulus to the real part. A comparison is made for the measured insertion loss of the compliant elements with no encapsulant as shown by the dotted curve, and for the corresponding measured curve of encapsulated array as given by the solid line. The calculated insertion loss for the encapsulated panel is shown by the dashed line. The maximum measured insertion loss of 20 dB for the encapsulated panel is about 6 dB less than that for the unencapsulated tubes. The net effect of this encapsulant is to add a small damping factor. Good agreement is found between measurements and calculation of the insertion loss for the encapsulated layer. From these curves, we can conclude that placing the compliant devices in a rubber material possessing both a low shear modulus and low shear loss factor results in an insertion loss which at the tube array resonance is similar to that for an unencapsulated tube layer of the same spacing.

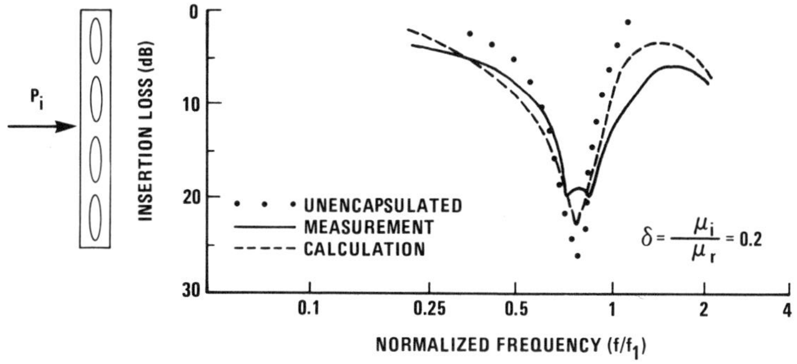

Fig. 10. Measured and calculated insertion loss of a compliant tube grating encapsulated in a low shear modulus material (μ_r=10^6N/m^2, θ_0=0, β_i=0, d= 1.21 cm).

A more detailed investigation of the effects of damping is shown in Fig. 11 by holding the shear modulus constant at $10^6 N/M^2$ and varying the shear loss factor between one and ten. Certain materials such as gels have a realizable loss factor of about ten, but most rubbers have values less than two for the temperatures and frequencies of interest. Note that with increased loss factors, not only does the effective Q of the array resonance decrease, but also, the center array resonance shifts to a lower frequency. If one is interested in minimizing the transmitted pressure near the center array frequency, highly damped material have a detrimental effect.

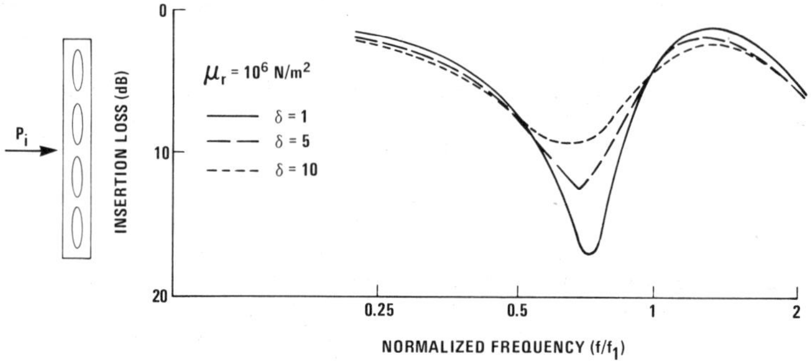

Fig. 11. Calculated insertion loss of a compliant tube grating with respect to loss factor variations (θ_0=0, β_i=0, d=1.27 cm).

We now consider in Fig. 12 the effects of encapsulating the compliant elements in a material whose shear modulus is two orders of magnitude higher than the previous low shear modulus material. The value of the shear modulus for this material is $10^8 N/M^2$ and the loss tangent of 0.4 is of the same order of magnitude as that for the low shear modulus material. The measured insertion loss for the encapsulated panel as given by the solid curve significantly changes from that of a tube layer without an encapsulant as shown by the dotted curve. The net effect of the encapsulant stiffness is to decrease the compliance of the array and provide a better impedance match to water. Thus more of the energy is transmitted through the encapsulated layer. The stiffness of the elastomer also increases the center frequency of the array resonance.

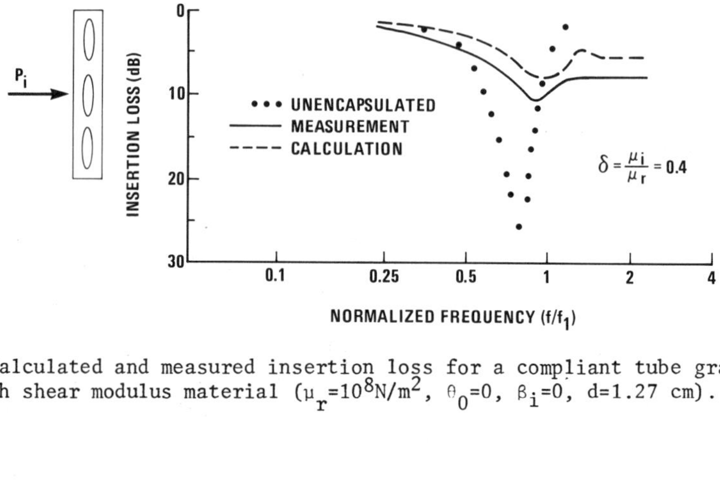

Fig. 12. Calculated and measured insertion loss for a compliant tube grating in a high shear modulus material (μ_r=10^8N/m², θ_0=0, β_i=0, d=1.27 cm).

CONCLUSIONS

In this paper, methods have been described which are used to study the two dimensional multiple scattering of acoustic waves from compliant tube arrays in either fluid or in viscoelastic layers in fluid. In each case, either a shell theory or a plate theory was used to describe the elastic response of the compliant elements. The T-matrix formulation offers an alternative method of investigating the scattering from noncircular cylindrical shells. With this approach, the shells could also be described by two dimensional elasticity theory. Multiple scattering from an array of shells in fluid or elastic material would be of interest. For the rectangular compliant elements, some approximations to the sharp edges would have to be made to conform to the smoothness conditions required for use of the T-matrix.

REFERENCES

1. W. J. Toulis, "Acoustic Refraction and Scattering with Compliant Elements", Journal of the Acoustical Society of America, 29, 1021-1033 (1957).

2. J. E. Burke and V. Twersky, "On Scattering of Waves by the Infinite Grating of Elliptic Cylinders", IEEE Transactions on Antennas and Propagation, 14, 465-480 (1966).

3. G. A. Brigham, J. J. Libuha, and R. P. Radlinski, "Analysis of Scattering from Large Planar Gratings of Compliant Cylindrical Shells", Journal of the Acoustical Society of America, 62, 48-59 (1977).

4. P. M. Morse and H. Feshbach, (1950) Methods of Theoretical Physics, McGraw-Hill, New York, Chaps. 5 and 11.

5. M. Abramowitz and I. A. Stegun, eds., (1970) Handbook of Mathematical Functions, Dover Publications, New York, Chap. 9.

6. A. E. H. Love, (1944) A Treatise on the Mathematical Theory of Elasticity, Dover Publications, New York, Chap. 13.

7. V. Twersky, "Elementary Function Representations of Schlömilch Series", Archive for Rational Mechanics and Analysis, 8, 323-332 (1961).

8. D. S. Clemm, "Algorithm 352: Characteristic Values and Associated Solutions of Mathieu's Differential Equation", Communications of the ACM, 12 (1969)

9. I. V. Vovk, V. T. Grinchenko, and L. A. Kononuchenko, "Diffraction of a Sound Wave by a Plane Grating Formed by Hollow Elastic Bars", Soviet Physics-Acoustics, 32, 113-115 (1976)

10. R. P. Radlinski, "Transmission Loss Through Double Gratings of Simply-Supported Bars in a Fluid Layer", NUSC Tech. Report (in press).

11. R. P. Radlinski, "Scattering of Plane Waves from a Compliant Tube Grating in a Linear Viscoelastic Layer", NUSC Tech. Report 5433 (1977).

STOCHASTIC VARIATIONAL FORMULATIONS OF ELECTROMAGNETIC WAVE SCATTERING*

R. H. Andreo** and J. A. Krill
Applied Physics Laboratory, Johns Hopkins University
Johns Hopkins Road, Laurel, Maryland 20810

ABSTRACT

Vector variational expressions for the scattering of electromagnetic waves from random systems are reviewed. These formulations are based on deterministic variational principles of the general form $T = (1/4\pi)N_1N_2/D$ for the scattering amplitude T. The nonstochastic nature of the incident field allows the statistical moments of T and of the differential scattering cross section $d\sigma/d\Omega = |T|^2$ to be expressed as the stochastic variational principles $(4\pi)^n\langle T^n\rangle = \langle N_1{}^n\rangle\langle N_2{}^n\rangle/\langle D^n\rangle$ and $(4\pi)^{2n}\langle|T|^{2n}\rangle = \langle|N_1|^{2n}\rangle\langle|N_2|^{2n}\rangle/\langle|D|^{2n}\rangle$ for arbitrary scatterer statistics. These new variational principles are readily seen to be inherently more tractable than the direct averages of the deterministic expressions, e.g., $4\pi\langle T\rangle = \langle N_1N_2/D\rangle$, and promise to allow practical application of variational techniques to random scattering problems.

INTRODUCTION - THE STOCHASTIC VARIATIONAL PRINCIPLE

The modeling of electromagnetic wave scattering is of interest in many diverse areas such as sensor systems, light scattering from polymers, aerosols, and rain, and the study of amorphous surfaces. Of special interest are interactive scattering effects such as interference and multiple scattering. In order to investigate such effects numerous approximation methods have been developed.[1] In particular, the variational method[2] is recognized as having certain advantages which distinguish it from other approximation methods. The primary one is its variational invariance, which implies the cancellation of first order errors introduced by the trial approximation of the field at the scatterer. A variational expression for the far-field scattering amplitude T is a ratio of integrals with the form $T = (1/4\pi)N_1N_2/D$. The averaging of this ratio for the case of stochastic scatterers has generally proven intractable,[3,4] so that application of the variational method to random scattering problems has been extremely limited.[4,5]

In 1977, Hart and Farrell[3] demonstrated that this apparent intractability could be overcome. They considered the scalar Neumann scattering problem and proved, for arbitrary scatterer statistics, that the average of the ratio of integrals for T equals the ratio of their individual averages (where the exact fields are to be used in evaluating N_1, N_2, and D), that is, $\langle T\rangle = (1/4\pi)\langle N_1N_2/D\rangle = (1/4\pi)\langle N_1\rangle\langle N_2\rangle/\langle D\rangle$,

*This work was supported by the Department of the Navy, Naval Sea Systems Command under Contract N00024-78-C-5384.
**Speaker

where $\langle\ \rangle$ denotes the average. Moreover, they proved that both forms are valid variational expressions for $\langle T\rangle$. Evaluation of the mean of each individual integral in the ratio is inherently more tractable than the direct average of the ratio.

This new stochastic variational approach was subsequently applied[4,5] to a classical random rough surface[6] consisting of parallel conducting hemicylinders randomly distributed on a conducting plane, in order to test its tractability and to examine its accuracy. Gray, Hart, and Farrell[4] and Krill and Farrell[5] performed variational, first-order perturbational (Born), and exact calculations in the Rayleigh frequency limit, and demonstrated that the variational result is more accurate than the Born approximation and accounts for multiple scattering as well as interference.

In order to account explicitly for the vector nature of the electromagnetic field, i.e., to investigate polarization effects, we have recently developed vector stochastic variational principles for the scattering of a plane electromagnetic wave by perfect conductors as well as by conducting dielectrics. In this paper we will outline the derivation of vector stochastic variational expressions for electromagnetic wave scattering from conducting dielectrics; a more complete description of this work appears in Ref. 7.

ELECTROMAGNETIC WAVE SCATTERING FROM CONDUCTING DIELECTRICS

We consider here the scattering of an incident plane wave [with electric field $\vec{E}^i(\vec{x}) = A\hat{e}_i \exp(i\vec{k}_i\cdot\vec{x})$] by a localized inhomogeneous, anisotropic conducting dielectric. The time harmonic electric field $\vec{E}$ (with frequency ω) satisfies the vector wave equation

$$\vec{\nabla}\times\vec{\nabla}\times\vec{E}(\vec{x}) - k^2\vec{E}(\vec{x}) = \overline{\overline{U}}(\vec{x})\cdot\vec{E}(\vec{x}), \tag{1}$$

where the dyadic operator $\overline{\overline{U}}(\vec{x}) = k^2[\overline{\overline{\varepsilon}}(\vec{x}) + (4\pi i/\omega)\overline{\overline{\sigma}}(\vec{x}) - \overline{\overline{I}}]$ depends on the tensor conductivity $\overline{\overline{\sigma}}(\vec{x})$ and permittivity $\overline{\overline{\varepsilon}}(\vec{x})$, and where $\overline{\overline{I}}$ is the unit dyadic. For the free-space Green dyadic $\overline{\overline{G}}_0(\vec{x},\vec{x}') = (\overline{\overline{I}} + \vec{\nabla}\vec{\nabla}/k^2)\exp(ik|\vec{x}-\vec{x}'|)/4\pi|\vec{x}-\vec{x}'|$, Green's Theorem[8] allows (1) to be rewritten as the familiar integral equation

$$\vec{E}(\vec{x}) = \vec{E}^i(\vec{x}) + \int d^3x'\ \overline{\overline{G}}_0(\vec{x},\vec{x}')\cdot\overline{\overline{U}}(\vec{x}')\cdot\vec{E}(\vec{x}'), \tag{2}$$

the asymptotic form of which yields the vector scattering amplitude

$$\vec{T}(\vec{k}_s,\vec{k}_i) = (1/4\pi A)\overline{\overline{I}}_{\hat{n}}\cdot\int d^3x'\ \exp(-i\vec{k}_s\cdot\vec{x}')\overline{\overline{U}}(\vec{x}')\cdot\vec{E}(\vec{x}'), \tag{3}$$

where the appearance of the projection operator $\overline{\overline{I}}_{\hat{n}} = \overline{\overline{I}} - \hat{n}\hat{n}$ guarantees the asymptotic transversality of the scattered field propagating in the direction $\hat{n} = \vec{k}_s/k$.

This scattering amplitude can be recast into an invariant form by expressing the amplitude A of the incident field in terms of the adjoint field $\tilde{\vec{E}}$ and an adjoint operator $\tilde{\overline{\overline{U}}}$, to yield the *deterministic* variational result

$$T = (1/4\pi)(N_1N_2/D). \tag{4}$$

In (4) $T = \hat{e}_s\cdot\vec{T}$ is the (experimentally measurable) polarization component of the scattering amplitude along $\hat{e}_s$, and the functionals N_1, N_2, and D are

$$N_1 = \int d^3x'[\hat{e}_s\cdot\overline{\overline{I}}_{\hat{n}}\exp(-i\vec{k}_s\cdot\vec{x}')]\cdot\overline{\overline{U}}(\vec{x}')\cdot\vec{E}(\vec{x}'), \tag{5a}$$

$$N_2 = \int d^3x\ \tilde{\vec{E}}(\vec{x})\cdot\tilde{\overline{\overline{U}}}^{Tr}(\vec{x})\cdot[\hat{e}_i\exp(i\vec{k}_i\cdot\vec{x})], \tag{5b}$$

and

$$D = \int d^3x\ \vec{\tilde{E}}(\vec{x})\cdot\tilde{\bar{\bar{U}}}^{Tr}(\vec{x})\cdot\vec{E}(\vec{x}) - \int d^3x \int d^3x'\ \vec{\tilde{E}}(\vec{x})\cdot\tilde{\bar{\bar{U}}}^{Tr}(\vec{x})\cdot\bar{\bar{G}}_0(\vec{x},\vec{x}')\cdot\bar{\bar{U}}(\vec{x}')\cdot\vec{E}(\vec{x}'), \tag{5c}$$

where the superscript Tr denotes matrix transposition. The requirement that the (first) variation of (4) with respect to $\vec{E}$ and $\vec{\tilde{E}}$ vanish, i.e.,

$$\delta T/T = \delta N_1/N_1 + \delta N_2/N_2 - \delta D/D = 0, \tag{6}$$

yields integral equations for the fields $\vec{E}$ and $\vec{\tilde{E}}$ inside the scatterer and, furthermore, suggests the natural choice $\tilde{\bar{\bar{U}}} = \bar{\bar{U}}^{Tr}$. Equations (4) and (5) then show that $\hat{e}_s\cdot\vec{T}(\vec{k}_s,\vec{k}_i;\bar{\bar{U}}) = \hat{e}_i\cdot\vec{T}(-\vec{k}_i,-\vec{k}_s;\bar{\bar{U}}^{Tr})$, which is a reciprocity relation.

For random systems the quantity of interest is generally a statistical moment, the simplest of which is the ensemble average $\langle T\rangle$. The direct average of (4) is

$$\langle T\rangle = (1/4\pi)\langle N_1N_2/D\rangle, \tag{7}$$

which is generally impossible to evaluate except in special simple cases. However, as in Ref. 3, it can be shown that a more tractable variational expression for $\langle T\rangle$ may be obtained with the form

$$\langle T\rangle = (1/4\pi)\langle N_1\rangle\langle N_2\rangle/\langle D\rangle. \tag{8}$$

The proof rests on the observation that for the exact fields $\vec{E}$ and $\vec{\tilde{E}}$ the quantity D/N_2 is simply the nonstochastic amplitude A of the incident plane wave, so that $\langle D\rangle = A\langle N_2\rangle$, and therefore $\langle D/N_2\rangle = A = \langle D\rangle/\langle N_2\rangle$. Substitution of this relation into (7) leads directly to (8). Similarly, D/N_1 is proportional to the amplitude of the incoming plane wave in the integral equation for $\vec{\tilde{E}}$ [obtained from (6)] and is thus also nonstochastic, thereby implying $D/N_1 = \langle D\rangle/\langle N_1\rangle$. Substitution of these expressions into the first-order variation of (8) gives, by virtue of (6), the stationary condition $\delta\langle T\rangle = 0$. Thus, (8) is established as a valid *stochastic* variational expression, in the sense that the expression is exact, and that errors cancel to first order when approximate fields are substituted in place of the exact ones.

In a similar fashion, one can show that

$$\langle T^n\rangle = (1/4\pi)^n\langle N_1{}^n\rangle\langle N_2{}^n\rangle/\langle D^n\rangle \tag{9a}$$

and

$$\langle |T|^{2n}\rangle = (1/4\pi)^{2n}\langle |N_1|^{2n}\rangle\langle |N_2|^{2n}\rangle/\langle |D|^{2n}\rangle \tag{9b}$$

are stochastic variational principles for the arbitrary n-th statistical moments of the scattering amplitude T and of the differential scattering cross section $d\sigma/d\Omega = |T|^2$. Equation (9b) may be applied to obtain a variational expression for the Fourier transform of the probability density function $f(|T|^2)$ of the differential cross section through the relation[9]

$$\mathcal{F}\{f(|T|^2)\} = \sum_{n=0}^{\infty} [(i\nu)^n/n!]\langle |T|^{2n}\rangle, \tag{10}$$

where $\mathcal{F}$ denotes the Fourier transform over parameter ν. A similar expression may be obtained for $\mathcal{F}\{f(T)\}$.

DISCUSSION

We have reviewed the development of vector stochastic variational expressions for the scattering of a plane electromagnetic wave by an inhomogeneous, anisotropic conducting dielectric. These expressions are based on the corresponding deterministic variational formulations and allow the variational evaluation of the statistical moments and probability density functions of T and $|T|^2$. Analogous variational principles with the forms (9) and (10) may be derived for permeable materials (by considering the magnetic field) and for perfect conductors. We observe that a variational expression of the form $\langle T \rangle = (1/4\pi)\langle N_1 \rangle \langle N_2 \rangle / \langle D \rangle$ is inherently more tractable than $\langle T \rangle = (1/4\pi)\langle N_1 N_2 / D \rangle$, as was demonstrated for scalar wave scattering by Gray, Hart, and Farrell[4] and by Krill and Farrell.[5] The potential tractability of the new vector stochastic variational approach shows promise of allowing broader application of variational methods to electromagnetic wave scattering from random systems.

REFERENCES

(1) Ishimaru, A. (1978) Wave Propagation and Scattering in Random Media, Academic Press, New York.
(2) Cairo, L. and T. Kahan (1965) Variational Techniques in Electromagnetism, Gordon and Breach, New York.
(3) Hart, R. W. and R. A. Farrell, "A Variational Principle for Scattering from Rough Surfaces," IEEE Trans. Ant. Prop. AP-25, 708 (1977).
(4) Gray, E. P., R. W. Hart, and R. A. Farrell, "An Application of a Variational Principle for Scattering by Random Rough Surfaces," Radio Science 13, 333 (1978).
(5) Krill, J. A. and R. A. Farrell, "Comparisons between Variational, Perturbational, and Exact Solutions for Scattering from a Random Rough Surface Model," J. Opt. Soc. Am. 68, 768 (1978).
(6) Twersky, V., "On Scattering and Reflection of Electromagnetic waves by Rough Surfaces, IRE Trans. Ant. Prop. AP-5, 81 (1957).
(7) Krill, J. A. and R. H. Andreo, "Vector Stochastic Variational Principles for Electromagnetic Wave Scattering," to be submitted to IEEE Trans. Ant. Prop. (1979).
(8) Tai, C-T. (1971) Dyadic Green's Functions in Electromagnetic Theory, Intext, San Francisco.
(9) Papoulis, A. (1965) Probability, Random Variables, and Stochastic Processes, McGraw-Hill, New York.

Part 9
Experimental Methods and Inverse Scattering Analyses

COMPARISON BETWEEN EXPERIMENTAL AND COMPUTATIONAL RESULTS FOR ELASTIC WAVE SCATTERING

R. K. Elsley, J. M. Richardson, R. B. Thompson, B. R. Tittmann
Rockwell International Science Center
Thousand Oaks, California 91360

ABSTRACT

One of the major applications of elastic wave scattering theory is in the characterization of flaws in structural materials, as needed to implement lifetime predictions based upon fracture mechanics. This paper reviews experimental work performed as a part of the development of nondestructive techniques for this purpose. The basic philosophy for the use of scattering information in flaw characterization is first discussed. A discussion is then given of the experimental techniques used in gathering the data, including the limitations which they impose on the available information. Comparisons of the results to various exact and approximate models are next presented. Finally, the results of the application of several algorithms to invert the data and predict flaw parameters are included. Here, despite the occasional ill-posedness of the inversion problem, many useful estimates of flaw size have been obtained.

I. INTRODUCTION

The level of interest in many of the practical aspects of elastic wave scattering has recently grown rapidly because of its importance in the area of nondestructive evaluation of structural materials (1, 2). Here the elastic waves take the form of ultrasonic signals which are used to determine quantitatively the locations and sizes of flaws.

From the time of emergence of this technology in the early 1940's, the primary emphasis was on flaw detection and location. For these purposes, it was not necessary to consider seriously the elastic nature of the wave motion, and simple scalar models were able to satisfactorily explain most of the phenomena of interest. However, advances in the materials science community placed a new demand on ultrasonic nondestructive testing. As shown in Fig. 1, the discipline of fracture mechanics provides the keystone of a failure prediction methodology which makes it possible to predict the remaining lifetime, before catastrophic failure, of a part. In order to apply this technique, three inputs are needed: stress levels and other parameters defining the environment in which the part must serve, material parameters such as fracture toughness defining its resistance to crack growth, and geometrical properties describing the flaw. The former two can be estimated from design calculations and tabulations of materials test results. However, no good techniques are available for field use to determine flaw sizes and orientations nondestructively.

As a response to this need, a number of research efforts have been initiated to provide the scientific foundation necessary to make such flaw size evaluations. Ultrasonics is an important form of interrogating energy, due to its ability to penetrate deeply into the interior of solid bodies and recover detailed flaw information. Therefore, much of this effort has been placed on improving various aspects of ultrasonic theory and technology.

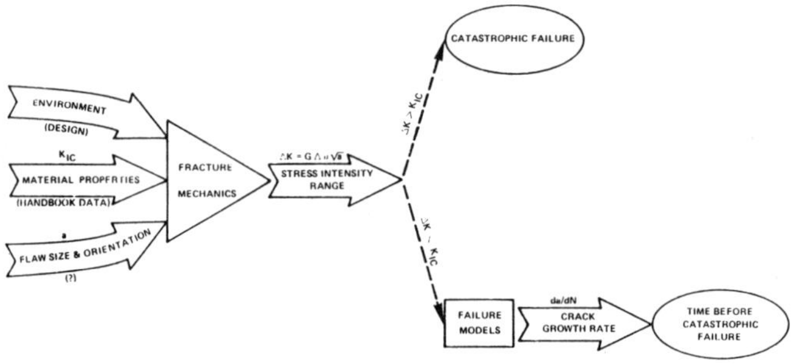

Fig. 1 Role of flaw size measurement in life prediction

In the investigation of the theory, it was found that nearly all practical interpretation of NDE data was being made on the basis of scalar theories developed previously for the case of fluid media. The elastic wave scattering from only a few simple geometries, such as spherical (3) and cylindrical (4, 5) shaped objects had been calculated. However, these results clearly indicated that substantial differences existed between the fluid and elastic cases, and that these had to be taken into account if quantitative information was to be deduced from the ultrasonic signals.

One of the major responses to this need was the Interdisciplinary Program for Quantitative Flaw Definition (6), sponsored jointly by the Defense Advanced Projects Agency and the Air Force Materials Laboratory. Some of the results of this program, produced by the collaborative efforts of a group of university and industrial scientists, are described in this paper.

Figure 2 describes the scientific methodology which has guided this effort. Initial effort was placed on obtaining solutions to the direct scattering problem, in which the scattering properties of flaws of known shape were predicted. Because of the complex shapes of many naturally occurring flaws, considerable effort was placed on the development of approximate theories which could be applied to such shapes. Exact results obtainable for simpler shapes, and experimental results on a variety of shapes, were both used to guide the theoretical development and test its results.

It was also recognized that, even if these problems were completely solved, the results would not be directly applicable to the practical problem of identifying an unknown flaw from a set of ultrasonic signals. Consequently, later emphasis was shifted to the ultimate objective of using the theoretical results as a basis for developing inversion solutions.

The majority of the papers in these proceedings deal with the development of some of the theoretical scattering solutions. This paper will present experimental results which have been obtained to test those theories and to evaluate inversion procedures.

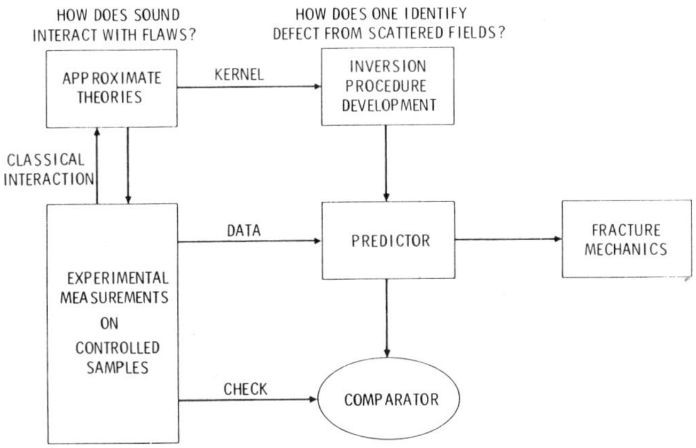

Fig. 2 Philosophy for development of quantitative ultrasonic measurement capability

II. EXPERIMENTAL TECHNIQUE

The relatively slow rate of progress in elastic wave scattering theory had been, in part, caused by the inability to fabricate samples with finite, three dimensional, scattering objects of controlled shape. This has recently been overcome by the use of diffusion bonding techniques (7, 8, 9). A titanium-6Al-4V sample is constructed from two mating sections which are joined by applications of heat and pressure. Prior to bonding, the voids or inclusions are placed at desired locations in the bonding plane by appropriate machining. Under the correct bonding conditions, the original bond plane disappears and all that remains is the intended scattering object in an essentially homogeneous sample.

Figure 3 shows a schematic view of one such sample, which has had its exterior surface machined into a spherical shape. This "trailer hitch" geometry allows the scatterer to be viewed from all angles, so that many possible scattering directions can be studied in a single sample. The spherical exterior also ensures that the metal path traveled is the same in all measurements, so that the angular dependence of scattering can be directly determined at a given frequency with no correction factors needed.

Elastic waves are excited and detected by piezoelectric transducers as shown in the photograph in Fig. 4. These are coupled to the sample by end caps having one planar face on which the transducer is permanently bonded and a second concave surface which mates with the spherical exterior of the sample. For longitudinal waves, a fluid bond can be used. However, for transverse waves the bond must either be extremely viscous or solid.

The flaws have been placed at the center of the samples, and have included a wide variety of shapes. Their dimensions are tabulated at the bottom of Fig. 3.

The data was primarily obtained using a broadband measurement system (10). The transducer was excited by a sharp voltage impulse and an elastic wave of a few cycles was injected into the material. Frequency content typically ranged from 1-10 MHz.

After detection by the same (pulse-echo) or a second (pitch-catch) transducer, the signal was digitized by a fast (100 MHz) A-D converter and transferred to a mini-

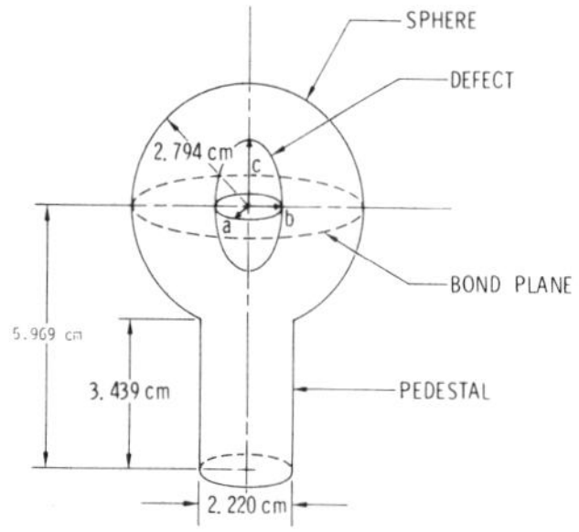

DESCRIPTION	a (μm)	b (μm)	c (μm)	STAMP NO.
PROLATE SPHEROID	200	200	800	40
PROLATE SPHEROID	400	400	800	41
SPHERE	200	200	200	35
SPHERE	400	400	400	35
SPHERE	600	600	600	37
OBLATE SPHEROID	400	400	200	39
OBLATE SPHEROID	400	400	100	38
CIRCULAR DISC	600	600	100	62
ELLIPTICAL DISC	2500	600	250	61
SIMULATED CRACK	600	600	-	67

Fig. 3 Sample configuration and list of samples

Fig. 4 Photograph of measurement apparatus

computer. Therein, a number of operations were performed to improve signal-to-noise and extract information. Included were signal averaging to reduce the random receiver noise, Fourier analysis, and deconvolution to eliminate effects of transducer response from the measurements.

It is important to note some of the limitations of the measurements which limit the fidelity of the data and the range of frequencies and/or flaw sizes for which it can be obtained. First of these is the attenuation of the ultrasonic waves caused by scattering of the energy at grain boundaries in the metal. Not only does this limit the strength of available signals, but it also produces a coherent background noise which can obscure the signals from small flaws. This effectively places an upper limit of about 10 MHz for samples made using a Ti-6Al-4V alloy of the dimensions shown in Fig. 3.

A second limitation is the existence of near field effects. Since the transducer may be viewed to first order as a piston radiator, there will be a series of near field peaks and nulls in its radiation pattern before the beam enters the Fraunhofer diffraction regime in which it begins to spread spherically. The locations of these, in principle, can be calculated. However, in practice such computations are in error due to uncertainties in the driving profile of the transducer. Consequently, they are usually avoided by restricting measurement frequencies below the value for which they approach the flaw position. For a 0.5 inch diameter transducer used with the "trailer hitch" samples, the first near field null occurs at about 8 MHz.

At low frequencies, data collection is limited by the ringing of the transducer after it is excited by the electrical impulse, and by wave vibrations in the end caps. These effects restrict measurements below about 0.5 MHz.

On real parts, these effects are accompanied by other geometrical constraints which may limit the set of angles from which a flaw can be viewed. Because of all these conditions, it is important ultimately to develop techniques for identifying flaws which can successfully operate on limited sets of data.

Other details of the experimental technique, as well as data not presented herein, can be found in References 8 - 12.

III. EXPERIMENTAL RESULTS

The experimental technique was first checked by making a detailed comparison to exact calculations (3) for the scattering from spherical voids and inclusions. Figures 5 - 9 present some of these results which demonstrate the experimental accuracy (10). In Fig. 5, the angular dependence of the longitudinal wave scattering between a pair of transducers is presented. Here, the spherical scatterer is a tungsten carbide inclusion having an 800 μm diameter and measurement is at a frequency of 4.0 MHz. This is an absolute comparison which shows that not only the angular variation but also the absolute value of the scattered signals can be accurately measured (13). Figure 6 shows the backscattered signal as a function of frequency for a spherical void of the same size. Here the agreement between theory and experiment is excellent for frequencies ranging up to about 6 MHz. At that point the experimental results fall below the theory because of the previously mentioned near field effects, since a null in the radiation pattern of the transducer develops at the location of the void.

Figure 7 presents the time domain equivalent of those results for a 1200 μm diameter void in titanium. Here the time harmonic theory has been used, together with a Fourier analysis of the transducer response, to synthesize the time domain behavior. Both the specular reflection from the front surface of the void and the reradiated signal from the creeping wave traveling around its periphery can be seen.

Transverse, as well as longitudinal, waves can be measured. However, the measurements are somewhat more difficult because of the need to establish a solid, or highly viscous, bond between the transducer and the part. Figure 8 illustrates this capability by comparing theory and experiment for the angular variation of 5 MHz shear wave scattering from an 800 μm spherical cavity in titanium. Figure 9 presents measurements of mode converted signals on the same sample, demonstrating the reciprocity between longitudinal to transverse and transverse to longitudinal scattering.

These results establish the ability to measure quantitatively a variety of features of scattered elastic wave signals. The measurements can then be used with high confidence as a tool in guiding and checking more complex or approximate theoretical models.

Before presenting such results, a definition of some of the models of interest is presented in Fig. 10. Here it is noted that, in general, a scattering model can be viewed as a double power series in the two parameters that determine the strength and character of the scattering (14). These are the ratio of flaw size to wavelength, ka, and the fractional change in elastic constants and density of the flaw from those of the host medium, $\Delta C/C$. These are multidimensional parameters in general, but will be treated as scalars for the purposes of simplicity in this discussion. The models illustrated are results of iterative solutions to the volume integral formulation of the scattering problem, as developed by Krumhansl's group at Cornell. In this framework, the Born approximation (15) is equivalent to a simplified power series, in which only the lowest order term in ($\Delta C/C$) is retained. It asymptotically approaches the correct result as the properties of the flaw approach those of the host medium. The quasi-static approximation (16)

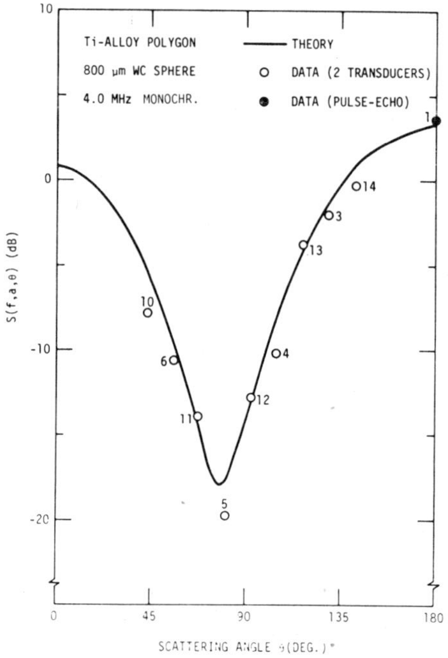

Fig. 5 Absolute comparison of exact theory and experiment for angular variation of longitudinal wave scattering from a WC sphere. 0° is defined to be the forward scattering direction.

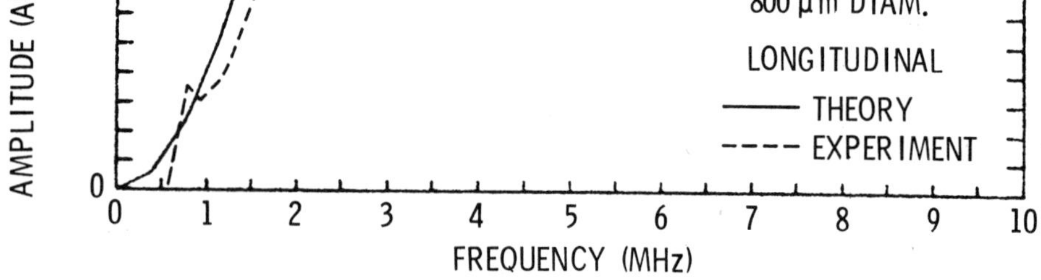

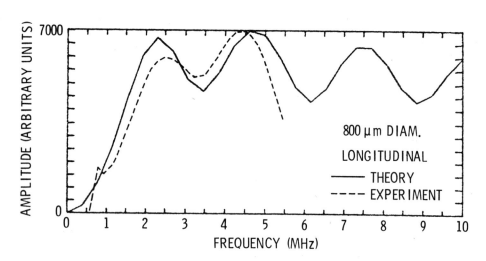

Fig. 6 Comparison of exact theory and experiment for frequency dependence of scattering from a spherical cavity

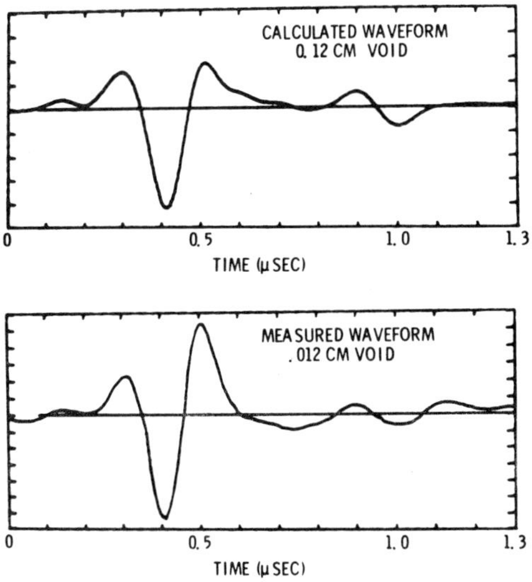

Fig. 7 Comparison of theory and experiment for time dependent scattering from a spherical cavity

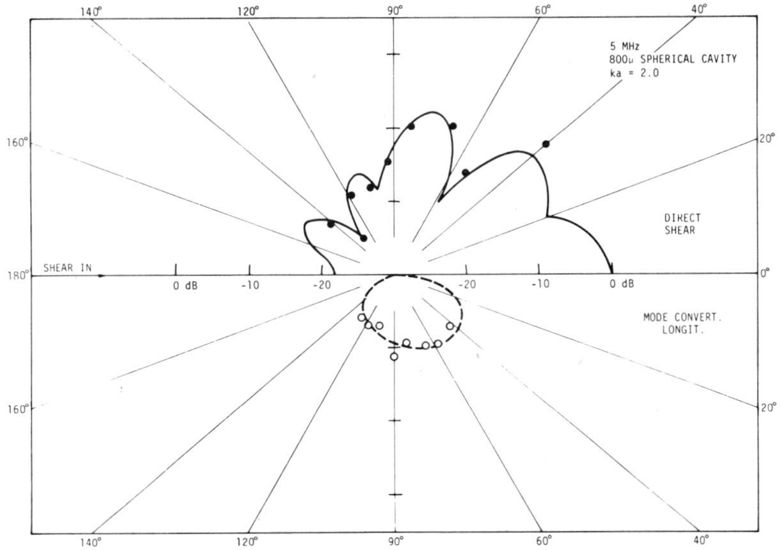

Fig. 8 Comparison of exact theory and experiment for angular variation of shear wave scattering from a spherical cavity

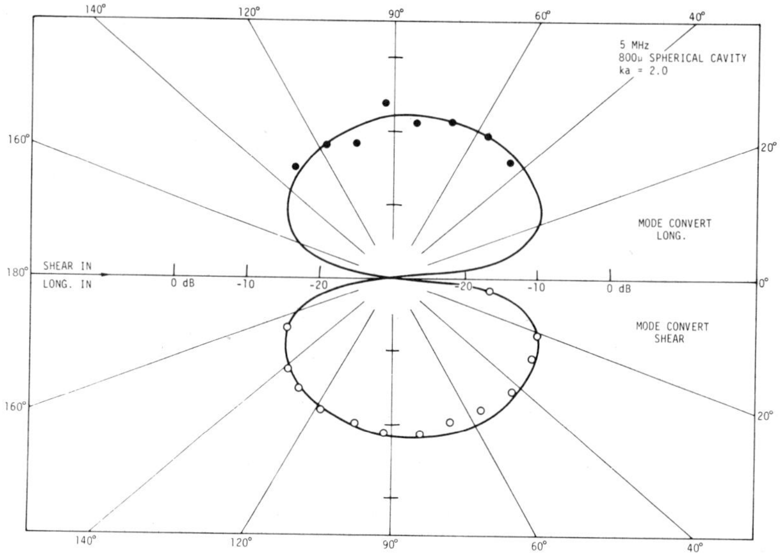

Fig. 9 Demonstration of reciprocity for a spherical cavity

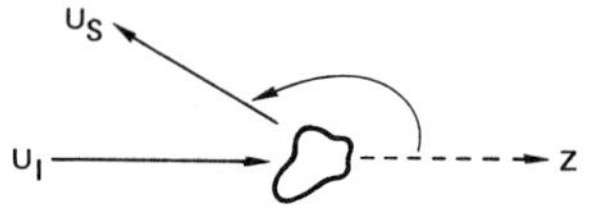

$$U_S = \frac{A}{r} e^{ik_L r} + \frac{B}{r} e^{ik_T r}$$

IN GENERAL:

$$|A| = \sum_{m=2,\, n=1}^{\infty} a_{m,n} (ka)^m \left(\frac{\Delta c}{c}\right)^n$$

$a_{m,n}$ is function of { ANGLE OF INCIDENCE, ANGLE OF SCATTERING, SHAPE

BORN APPROXIMATION:

$$|A| = \frac{\Delta c}{c} \underbrace{\sum_{m=2}^{\infty} a_{m,1} (ka)^m}_{\text{SPATIAL FOURIER TRANSFORM OF OBJECT SHAPE}}$$

QUASI-STATIC APPROXIMATION:

$$|A| = (ka)^2 \underbrace{\sum_{n=1}^{\infty} a_{2,n} \left(\frac{\Delta c}{c}\right)^n}_{\text{LONG WAVE COEFFICIENT } A_2}$$

EXTENDED QUASI-STATIC APPROXIMATION:

$$|A| = \frac{\Delta c}{c} \sum_{m=2}^{\infty} a_{m,1} (ka)^m + (ka)^2 \sum_{n=2}^{\infty} a_{2,n} \left(\frac{\Delta c}{c}\right)^n$$

+ KEY ADDITIONAL HIGHER ORDER TERMS

Fig. 10 Summary of approximate theoretical models

represents the limit in which only the lowest order terms in (ka) are retained, which in this case are quadratic. Hence, accurate results are obtained in the long wavelength limit. Other techniques have also been recently developed which are essentially equivalent to an evaluation of the full, double power series to obtain essentially exact results. Included, is the T-matrix approach (17), which is the subject of several other papers in this proceedings.

In Figs. 11 and 12, experimental results (10) for the scattering from an 800 μm spherical cavity, as measured at 2.25 MHz and 5 MHz, are used to illustrate the strength and weaknesses of the Born approximation (15). These plots show that when ka = 0.9, and for angles close to the backscattered direction, good agreement is observed between exact theory, experiment, and the Born approximation. This occurs despite the fact that the cavity is a strong perturbation in elastic constants, with ($\Delta C/C$) = -1. Thus the experimental results have indicated a practical applicability of the Born approximation in a region which could not be expected on purely theoretical grounds.

At higher frequencies, however, the agreement is not nearly as good, as shown in Fig. 12. This illustrates the limits of applicability of the Born approximation as applied to a cavity.

Based on these results, and experience gained in similar quantum mechanical scattering problems, it has been postulated that the Born approximation can usefully predict angular variations of scattering from voids when averaged over frequency to suppress some of the detailed errors in frequency dependent effects. This implies that it can successfully serve as an easily applied tool to predict experimental trends, although being deficient with respect to highly precise predictions. Figure 13 is an experimental confirmation of this hypothesis for the case of backscattering from an oblate spheroidal cavity (10). Here the angle of measurement has been varied from the spheroid axis, 0° to the spheroid edge, 90°. The theory has been averaged over a range of frequencies equal to the bandwidth of the transducer. Excellent agreement between the two is observed.

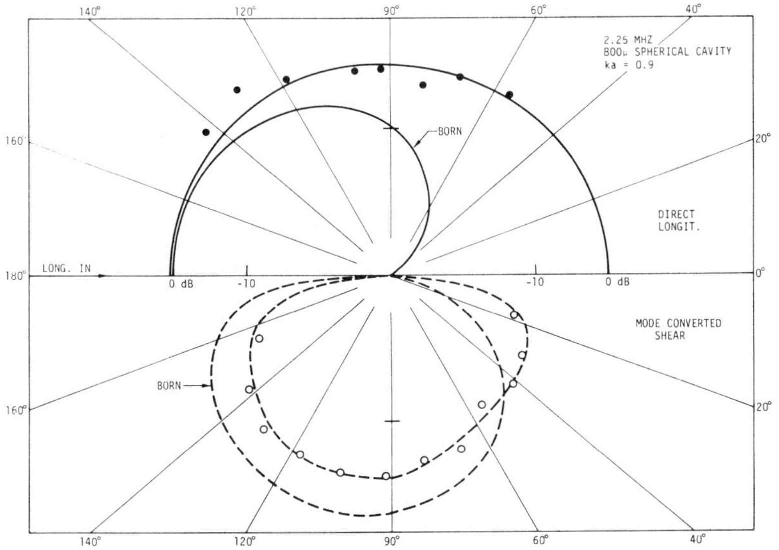

Fig. 11 Comparison of exact theory, Born approximation, and experiment for scattering from a spherical cavity at ka = 0.9

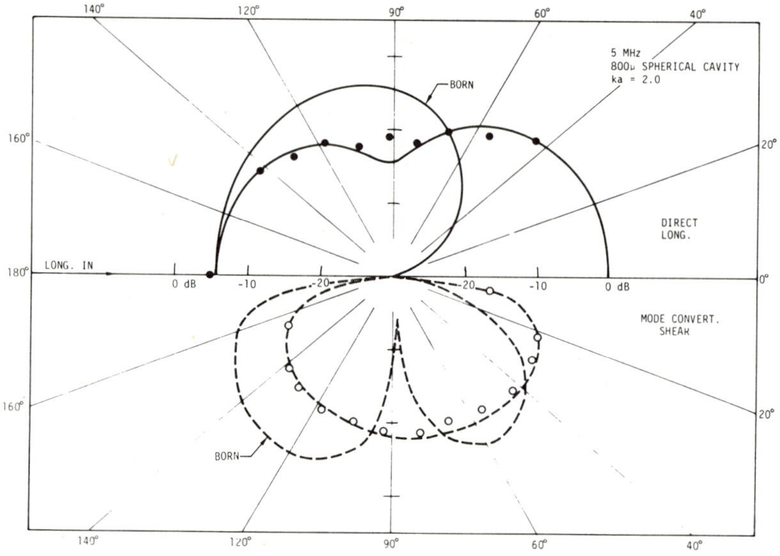

Fig. 12 Comparison of exact theory, Born approximation, and experiment for scattering from a spherical cavity at ka = 2.0

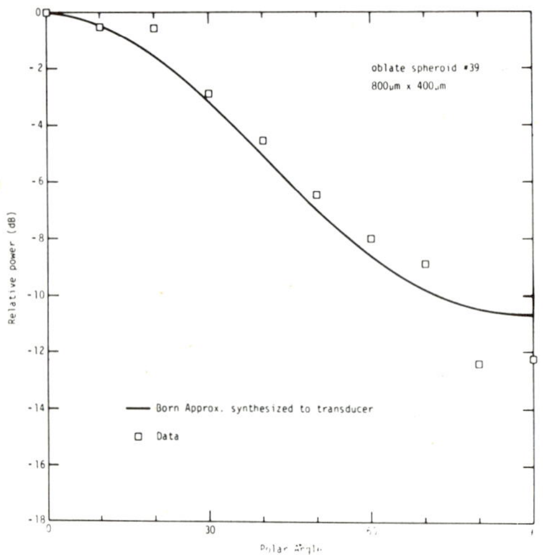

Fig. 13 Comparison of frequency averaged Born approximation and experiment for pulse-echo scattering for oblate spheroidal cavity

A second approximation of interest is the quasi-static approximation (16), which asymptotically approaches the exact results at long wavelengths. The scattering can be characterized by a parameter A_2, which is the coefficient of the leading, quadratic term in a power series expansion of scattered amplitude versus frequency. Figure 14 illustrates the results of such a measurement for the cases of a spherical and an oblate spheroidal cavity (18). Theory and experiment have been fitted with one scalar multiplier for the spherical cavity. No further adjustable parameters were used for the oblate spheroidal case. The excellent agreement obtained confirms the accuracy of the quasi-static model in the long wavelength limit.

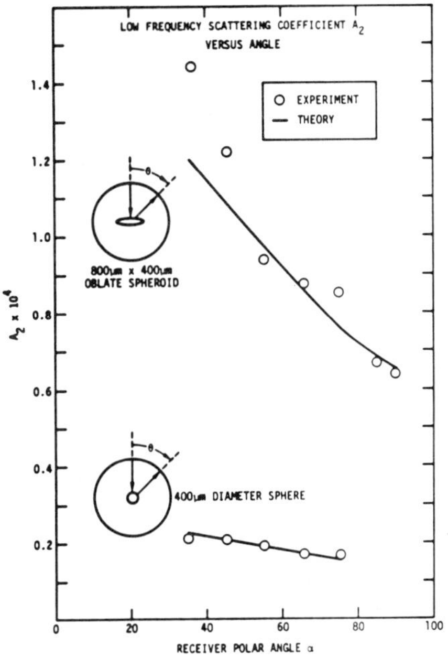

Fig. 14 Comparison of low frequency scattering coefficient A_2 and experiment for spherical and oblate spheroidal void

As noted elsewhere in these proceeding (17), the T-matrix approach represents a more computationally intensive model which is accurate over a broader range (‿C/C) and (ka) than these simpler models. Experimental verifications of these are presented in Figs. 15 and 16, which show the frequency dependence of the ultrasonic backscattered signal for two different oblate spheroidal cavities at three distinct angles of illumination (19). The rather complex frequency spectra are accurately confirmed by the experiment with the exception of shift in the position of a minor peak in the 60° data for the 200 µm by 400 µm spheroid.

IV. INVERSION

As noted in the Introduction, the practical problem which has motivated this research is the need to measure nondestructively geometrical characteristics of flaws, such as size, shape, or orientation, which are needed for the prediction of lifetimes in structures. Once adequate models have been developed for the forward scattering problem, emphasis shifts to the inverse problem, whereby the parameters of an unknown flaw are identified from measured fields. Because of the presence

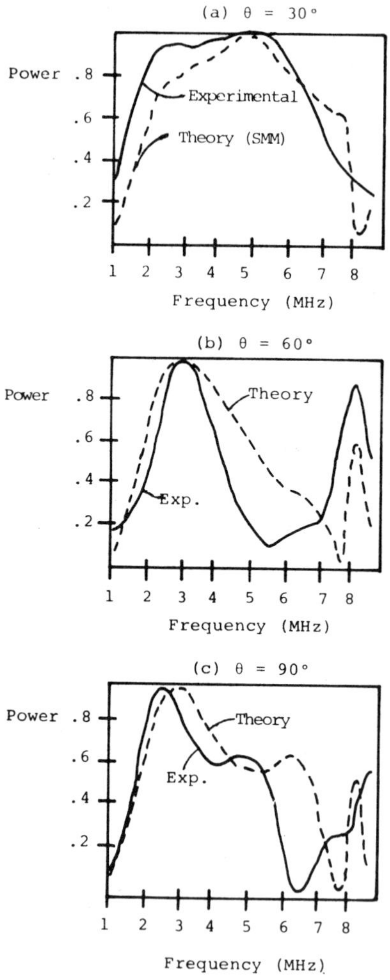

Fig. 15 Comparison of T-maxtrix calculation and experiment for scattering from 100 μm x 400 μm oblate spheroidal cavity

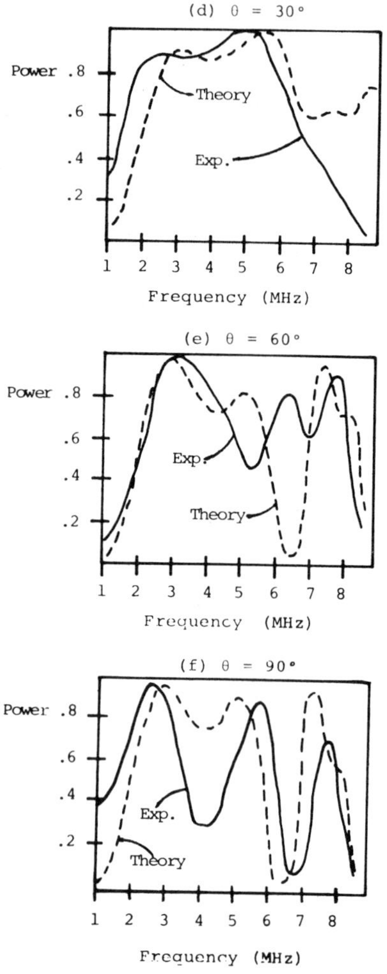

Fig. 16 Comparison of T-matrix calculation and experiment for scattering from 200 μm x 400 μm oblate spheroidal cavity

of experimental error and limitation on the amount of information that can be practically determined, the inversion problem is occasionally ill posed. Nevertheless, a wide variety of practical approaches have been developed for estimating the desired flaw parameters. These are ordered in Fig. 17 in accordance with the ratio of flaw size to wavelength required to apply the technique. For reference, the frequency dependence of the backscattering from a spherical cavity is reproduced at the bottom of the figure. At short wavelengths, it is useful to try to reconstruct a detailed picture of the flaw. Implicit in any such approach is a physical model of the elastic wave-flaw interaction. Imaging systems are usually designed with the implicit assumption that the flaw is a diffuse reflector. That is, any incident ray is assumed to be converted into a divergent cone of rays, emanating from the point of illumination. For cavities with rough surfaces, this model can accurately describe the scattering process and high quality images can be formed. However, for inclusions or flaws whose surfaces are smooth with respect to a wavelength, specular reflections, mode converted signals, reradiation from creeping rays, and internal reverberations can all produce degradations of the reconstruction. One example is the formation of ghost images.

TYPE	REGIME	ADVANTAGES
IMAGING	$ka > 6.3$ ($\lambda < a$)	HIGH INFORMATION CONTENT EASILY INTERPRETED DISPLAY RESOLVES MULTIPLE FLAWS
MODEL BASED RECONSTRUCTION	$ka \geqslant 3$	PHYSICAL PRINCIPLES USED TO IMPROVE RESOLUTION AND TREAT MODE CONVERSION
MODEL BASED ADAPTIVE LEARNING NETWORKS	$0.4 < ka < 3$	MULTIPLE SCATTERING TAKEN INTO ACCOUNT GAIN MORE INFORMATION IN DIFFICULT REGIME
LONG WAVELENGTH SCATTERING	$ka < 0.5$	FRACTURE RELATED PARAMETERS DEDUCED FROM A FEW MEASUREMENTS MAY BE USEFUL IN AUTOMATION

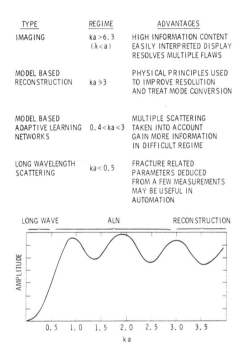

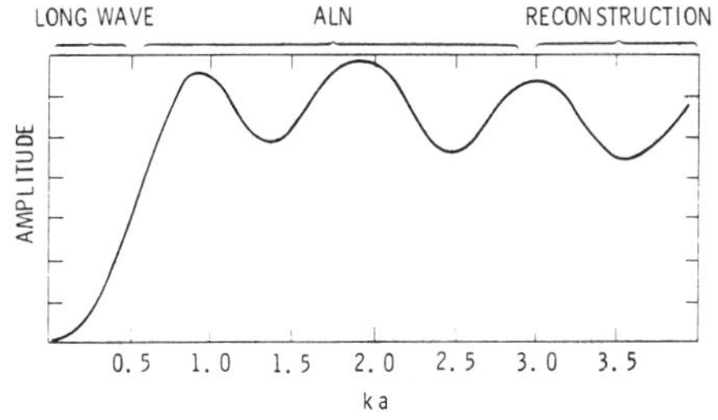

Fig. 17 Inversion techniques

To overcome these problems, it is necessary to base the reconstruction on a more precise model of the elastic wave-flaw interactions. The basic strategy, as illustrated in Fig. 18 for the case of the Born approximation (20), is to use the model to specify a relationship between measurement and the Fourier transform of the object shape function. Inverse Fourier transforms can then be used to reconstruct the object. One such approach, based on the extended quasi-static model, is described by Rose (21) in a subsequent paper in this proceeding. A second, by Bleistein and Cohen (22), is illustrated in Fig. 19. The model used in this case

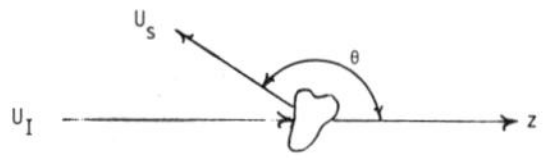

$$U_S = \frac{\vec{A}}{r} e^{iK_L r} + \frac{\vec{B}}{r} e^{iK_T r}$$

$$\vec{A} = \frac{K_L^2}{4\pi}\left[\frac{\delta\rho}{\rho}\cos\theta - \frac{\delta\lambda + 2\delta\mu\cos^2\theta}{\lambda + 2\mu}\right] F\left[\frac{f\sin\theta}{V_L}, 0, \frac{f(\cos\theta - 1)}{V_L}\right]$$

$$\vec{B} = \frac{K_T^2}{4\pi}\left[\frac{2\alpha\delta\mu\cos\theta\sin\theta}{\mu\rho} - \frac{\delta\rho\sin\theta}{\rho}\right] F\left[\frac{F\sin\theta}{V_T}, 0, f\left(\frac{\cos\theta}{V_T} - \frac{1}{V_L}\right)\right]$$

where

$$F(u,v,w) \underline{\Delta} \int_{-\infty}^{\infty} dx \int_{-\infty}^{\infty} dy \int_{-\infty}^{\infty} dz\, f(x,y,z) e^{-i2\pi(ux+vy+wz)}$$

$$f = \begin{cases} 1 & \text{INSIDE OBJECT} \\ 0 & \text{OUTSIDE OBJECT} \end{cases}$$

Fig. 18 Use of Born approximation to relate measurements to Fourier transform of flaw shape function

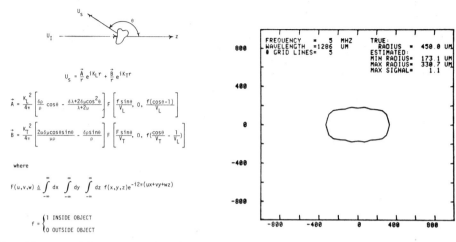

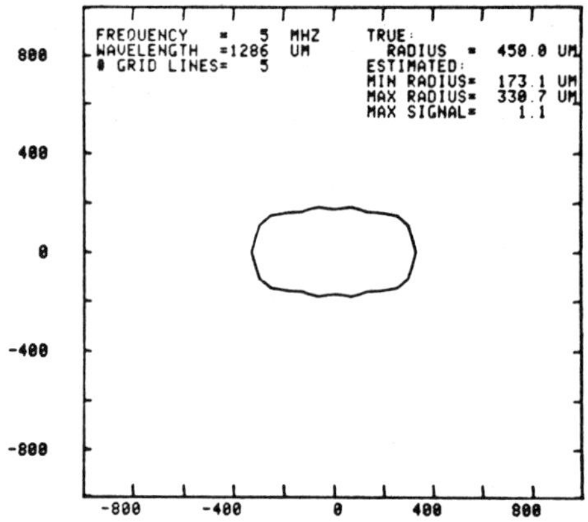

Fig. 19 Reconstruction of oblate spheroid from synthetic data using POFFIS algorithm

was the physical optics approximation, which has been shown (23, 24) to asymptotically approach the exact result, for the case of elastic wave backscattering, in the high frequency limit. Here the shape of an oblate spheroidal cavity has been reconstructed by applying the POFFIS (physical optics far field inverse scattering) algorithm to simulated experimental data (25).

A review of reconstruction procedures has recently been prepared by Lee (24).

At longer wavelength, there is not sufficient information to develop a detailed reconstruction of a flaw shape. Furthermore, because of nonlinearities caused by such effects as strong multiple scattering, analytic solution to the inverse problem are difficult to develop. Consequently, adaptive learning procedures (26) have been utilized as tools to find empirical solutions. Use has been made of the forward scattering models in the training procedure as illustrated in Fig. 20. The output from the training is an empirically derived expression whereby specific flaw characteristics can be predicted from measured features. This takes the form of a polynomial in the measured parameters, whose coefficients are determined by the training process. Table I compares the actual and predicted values of the size, shape, and orientation of oblate spheroidal cavities when the Born approximation was used as the theoretical model for the training process. It will be noted that a good deal of the observed error is systematic, which is believed to be caused by inaccuracies in this model when applied to such a strong scatter. Work is now in progress to retrain the procedures using the more accurate T-matrix calculations, to evaluate the extent of the improvement.

At still lower frequencies, when the wavelength is large with respect to the flaw size, the forward scattering problem can be treated using the quasi-static approximation (16). An analysis of this solution shows that there are, in general, 22 independent parameters which control the flaw-elastic wave interaction. This suggests that, contrary to initial intuition, considerable flaw information can be deduced from an appropriately selected set of long wavelength measurements. This result is a direct consequence of the tensor nature of the fields involved in the scattering process, and is richer in information than the scattering behavior observed in situations governed by a scalar wave equation.

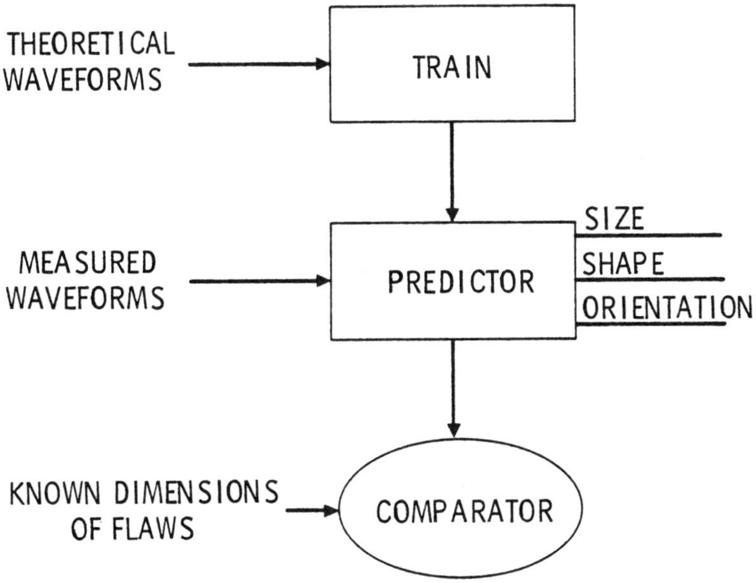

Fig. 20. Procedure for development and evaluation of adaptive learning techniques

The existence of this source of long wavelength information has led to a new set of inversion procedures. As shown in Fig. 21, the frequency dependence of the ultrasonic scattering is first measured and the coefficient, A_2, of the quadratic term in a frequency expansion is deduced. As noted above, A_2 is in general a function of 22 independent parameters of the flaw. Hence, measurements of A_2 at a sufficient number of angles can in principle be used to deduce these parameters.

It is obvious that this is not sufficient information to reconstruct a detailed shape of the flaw. However, recent work has shown that, if independent a priori information is available limiting the flaw to membership in a certain class, the long wavelength information can be used quite effectively. Budiansky and Rice (27) have shown that, for an elliptical crack, three measurements are sufficient to determine orientation and the reduced stress intensity factor, essentially independently of the ellipticity of the flaw. This result is of great significance, since it represents a direct relationship between an elastic wave measurement and a parameter used in the prediction of part lifetime.

An experimental demonstration of the inversion of long wavelength data is illustrated in Table II (28). In this case, the flaw was an oblate spheroidal cavity, and this fact was assumed in the analysis. Thus the number of independent parameters was reduced from 22 to 4. Based on pulse echo and pitch-catch measurements of long wavelength scattering, estimates of three parameters (major axis, minor axis, orientation) of the spheroid were obtained using a least squares estimation procedure. The additional angle was not considered because measurements were made in a symmetry plane. The accuracy of the answer, and small estimated standard derivations, illustrate the power of the approach. The fact that the pulse-echo results are superior, demonstrates the higher leverage that this configuration has on the desired information.

TABLE I Comparison of Actual (True) and Predicted (Meas.) Size and Orientation for Oblate Spheroids Based On Adaptive Learning Results

	Size (in Microns)				Orientation (in Degrees)			
	ALN1		ALN2		ALN3		ALN4	
Defect No.	True A	Meas. $\hat{A}$	True B	Meas. $\hat{B}$	True α	Meas. $\hat{\alpha}$	True β	Meas. $\hat{\beta}$
1	200	154	400	493	0	37	-	197
2	200	163	400	528	30	2	225	215
3	100	100	400	506	80	68	160	165
4	100	114	400	521	0	0	-	-
5	200	202	400	533	80	22	160	174
6	100	126	400	599	30	60	180	196
7	200	114	400	569	30	21	180	178
8	100	130	400	645	30	42	225	218
Average Absolute Error			30		149		23	8
Percent Average Absolute Error			15.0		37.3		25.6	2.2

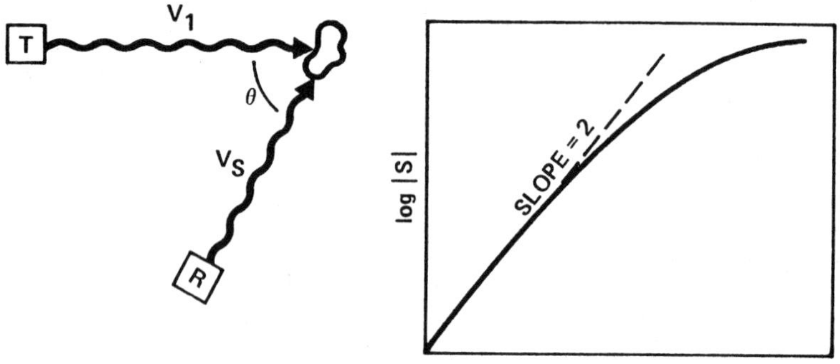

$$|S(\omega,\theta)| = |V_S/V_I| = A_2(\theta)\,\omega^2 + \theta(\omega^4)$$

Fig. 21 Long wavelength measurement technique

TABLE II Estimated Flaw Radii and Orientation

Oblate Spheroid - Pulse Echo Measurements

	True	Estimated	σ
a (μm)	400	395	14
c (μm)	200	200	26
γ_x	0	$<10^{-4}$	.24

Oblate Spheroid - Pitch-Catch Measurements:

	True	Estimated	σ
a (μm)	400	430	31
c (μm)	200	160	99
γ_x	0	$<10^{-4}$	.28

V. CONCLUSIONS

Elastic wave scattering has become an important tool in the quantitative measurement of flaw sizes, and hence, in the prediction of part lifetimes, for structural materials. Direct comparison of theory and experiment has helped guide and establish the accuracy of several scattering algorithms. Present areas of greatest challenge include extending these to treat flaws and parts of complex shapes and in the development of inversion procedures. An important feature of the latter is the optimum use of the available a priori information about the flaw.

ACKNOWLEDGMENT

This work was sponsored by the Center for Advanced NDE, operated by the Rockwell International Science Center, for the Advanced Research Projects Agency and the Air Force Materials Laboratory under Contract F33615-74-C-5180.

REFERENCES

(1) R. B. Thompson and A. G. Evans, "Goals and Objectives of Quantitative Ultrasonics," IEEE Transactions on Sonics and Ultrasonics, SU-23, p. 292, (1976).

(2) D. O. Thompson and R. B. Thompson, "Quantitative Ultrasonics," Discussion of the Royal Society (London) (in press).

(3) R. Truell, C. Elbaum, and B. B. Chick, Ultrasonic Methods in Solid State Physics (Academic, New York, 1969) Chap.3.

(4) R. M. White, "Elastic Wave Scattering at a Cylindrical Discontinuity in a Solid," J. Acoust. Soc. Amer. 30, p. 771 (1958).

(5) Y. H. Pao and C. C. Mow, The Diffraction of Elastic Waves and Dynamic Stress Concentrations (Crane Russack and Co., New York, 1973).

(6) Proceedings of The ARPA/AFML Review of Progress In Quantitative NDE, held July 17-21, 1978 at LaJolla, California, to be published as Air Force Materials Laboratory Technical Report. Also, previous proceedings: AFML-TR-78-55, AFML-TR-77-44, AFML-TR-55-212, AFML-TR-74-235, AFML-TR-73-69.

(7) B. R. Tittmann, H. Nadler, and N. E. Paton, "A Technique for Studies of Ductile Fracture in Metals Containing Voids or Inclusions," Metall. Trans. 7A, 320 (1976).

(8) B. R. Tittmann, E. Richard Cohen, and John M. Richardson, "Scattering of Longitudinal Waves Incident or a Spherical Cavity in a Solid," J. Acoust. Soc. Am. 63, p. 68 (1978).

(9) B. R. Tittmann, E. R. Cohen, H. Nadler, and N. E. Paton "Elastic Wave Scattering from Various Obstacles Imbedded in Solid Media in the Regime .. Obstacle Dimension," 1973 Ultrasonics Symposium Proceedings, (IEEE, New York, 1973) p. 220.

(10) B. R. Tittmann, R. K. Elsley, H. Nadler, and E. R. Cohen, "Experimental Measurements and Interpretation of Ultrasonic Scattering by Flaws," Interdisciplinary Program for Quantitative Flaw Definition, Special Report Third Year Effort (Rockwell International Science Center, Thousand Oaks, California, 1977) p. 82.

(11) B. R. Tittmann, "Mode Conversion and Angular Dependence for Scattering from Voids in Solids," 1975 Ultrasonics Symposium Proceedings, (IEEE, New York, 1975) p. 111.

(12) B. R. Tittmann, "Ultrasonic Scattering Studies for Failure Prediction," 1976 Ultrasonics Symposium Proceedings (IEEE, New York, 1975) p. 74.

(13) B. R. Tittmann, D. O. Thompson, and R. B. Thompson, "Standards for Quantitative Nondestructive Evaluation," in Nondestructive Testing Standards - A Review, ASTM STP 624, p. 295. Also in Research Supplement of Materials Evaluation 35, 75 (1977).

(14) J. E. Gubernatis, private communication.

(15) J. E. Gubernatis, D. Domany, J. A. Krumhansl, M. Huberman "The Born Approximation in the Theory of the Scattering of Elastic Waves by Flaws," J. Appl. Physics 48, 2812 (1977).

(16) J. E. Gubernatis, "Long Wave Approximations for the Scattering of Elastic Waves from Flaws with Applications to Ellipsoidal Voids and Inclusions," J. Appl. Physics (in press).

(17) V. V. Varadan, "Elastic Wave Scattering," this proceedings.

(18) R. K. Elsley, J. M. Richardson, R. B. Thompson, "Determination of Fracture Mechanics Parameters From Elastic Wave Scattering Measurements at Low Frequencies," Interdisciplinary Program for Quantitative Flaw Definition, Special Report Fourth Year Effort," (Rockwell International Science Center, Thousand Oaks, California, 1978) p. 95.

(19) R. Shankar, A. N. Mucciardi, M. F. Whalen, and M. D. Johnson, "Inversion of Ultrasonic Scattering Data to Measure Defect Size, Orientation, and Acoustic Properties," in Proceedings of the ARPA/AFML Review of Progress in Quantitative NDE, AFML-TR-78-55, p. 50.

(20) R. B. Thompson, "Introduction to Defect Characterization by Quantitative Ultrasonics," Proceedings fo the ARPA/AFML Review of Progress in Quantitative NDE," AFML-TR-78-55, p. 15.

(21) J. H. Rose, "Inversion of Ultrasonic Data," this proceedings.

(22) N. Bleistein and J. K. Cohen, "Application of a New Inverse Method to Nondestructive Evaluation," Denv. Research Report, MS-R-7716 (1977).

(23) N. Bleistein and J. K. Cohen, "Application of Inverse Method to Non-Destructive Evaluation of Flaws," Interdisciplinary Program for Quantitative Flaw Definition, Semi-Annual Report (Rockwell International Science Center, Thousand Oaks, California, 1979) p. 112.

(24) D. A. Lee, "Mathematical Principles of Data Inversion," (in preparation).

(25) J. K. Cohen, N. Bleistein, and R. K. Elsley, "Nondestructive Detection of Voids by a High Frequency Inversion Technique," Interdisciplinary Program for Quantitative Flaw Definition, Special Report Fourth Year Effort (Rockwell International Science Center, Thousand Oaks, California, 1978) p. 81.

(26) M. F. Whalen and A. N. Mucciardi, "Inversion of Physically Recorded Ultrasonic Waveforms Using Adaptive Learning Network. Models Trained on Theoretical Data," Interdisciplinary Programs for Quantitative Flaw Definition, Special Report Fourth Year Effort (Rockwell International Science Center, Thousand Oaks, California, 1978) p. 55.

(27) B. Budiansky and J. R. Rice, "On the Estimation of a Crack Fracture Parameter by Long Wavelength Scattering," Trans. ASME, J. Appl. Mech., 45, 453 (1978).

(28) J. M. Richardson, "The Inverse Problem in Elastic Wave Scattering at Long Wavelengths," 1978 Ultrasonic Symposium Proceedings (IEEE, New York, 1978) p. 759.

INVERSION OF CRACK SCATTERING DATA IN THE HIGH-FREQUENCY DOMAIN

J. D. Achenbach, A. Norris and K. Viswanathan*
Northwestern University, Evanston, Illinois 60201

ABSTRACT

Certain approximate methods for the solution of the direct scattering problem in the high-frequency domain, such as physical elastodynamics and geometrical diffraction theory, yield expressions for the scattered fields which suggest the application of Fourier-type integrals to solve inverse problems. In this paper we explore the application of two kinds of inversion integrals to far-field high-frequency longitudinal scattering data from flat cracks of arbitrary shape. The inversion problem becomes relatively simple if some a-priori information, such as the orientation of the plane of the crack or of a plane of symmetry of the crack is available. The general three-dimensional problem will, however, also be considered. Results are checked for an elliptical crack for which "exact" results are available. Some computational technicalities are discussed, and numerical results are presented.

INTRODUCTION

An important method in quantitative non-destructive evaluation of materials (QNDE) is based on scattering of ultrasonic waves by flaws. The presence of a flaw is relatively easy to detect. The determination of the size, shape and orientation of a flaw from the scattered field does, however, pose a challenging inverse scattering problem.

It is known that at high frequencies the far-field generated by a volume scatterer in an acoustic medium is proportional to the Fourier transform of the characteristic function associated with the scatterer. The characteristic function is defined so that it has unit value for every point inside the scatterer and vanishes elsewhere. The Fourier transform parameter which enters in this relation is a function of the wave-number and the angle of observation. A number of studies have recently been devoted to examine the extent to which the far-field data can be used to numerically invert this Fourier transform relation and recover the size, shape and the location of the scatterer. In these studies the possible limitations on the bandwidths of the observed scattered signals has been taken into account, as well as the restricted range of the aperture covered by the angles of observation. For details, the reader is referred to the recent work of Bleistein and Cohen [1].

In the present paper it is assumed that the flaws that are to be detected are of the most dangerous variety, namely, flat cracks. We discuss the inversion of far-field crack-scattering data in the high frequency range by a method which does not involve a three dimensional Fourier inversion but only a single integration in the wave-number domain. The method was introduced in Ref.[2], and it is further discussed in the present paper.

*Permanent address: Defense Science Laboratory, Delhi - 110054, India.

Exact and approximate analytical solutions to the direct scattering problem for flat cracks, which provide indispensable insight into the inverse problem can be found in Ref.[3] and [4].

The inversion of the crack scattering data is achieved by constructing first the plane of the crack and then sets of lines tangential to the crack edge. The results are checked against a known solution for scattering by an elliptical crack. The method is illustrated by numerical examples. The numerical results include an iterative technique for improving the accuracy of the inversion. A comparison with experimental results can be found in Ref.[2].

The analysis is for time-harmonic motions with the time factor exp(-iωt) implied throughout. For a homogeneous, isotropic and linearly elastic solid the displacement gradients $u_{i,j}(\underline{x})$ and the stresses $\tau_{ij}(\underline{x})$ are related by

$$\tau_{ij} = \lambda U_{k,k}\, \delta_{ij} + \mu\, (u_{i,j} + u_{j,i}) \tag{1.1}$$

where λ and μ are the Lame's constants. The equation governing elastic motion is (see Ref.[5])

$$\mu\, u_{i,jj} + (\lambda + \mu)\, u_{j,ji} + \rho\, \omega^2 u_i = -F_i \tag{1.2}$$

where ρ is the mass density, and F_i are the components of the body force distribution (per unit volume). The relation

$$t_\ell = \tau_{k\ell}\, n_k \tag{1.3}$$

gives the components of traction across a surface with a unit normal n_k.

2. SCATTERING OF ELASTIC WAVES AT HIGH FREQUENCIES

As a preliminary to an expression for the far-field generated by scattering of high frequency elastic waves by a crack, we introduce some well-known results from elastodynamic theory.

Basic singular solution

The basic singular solution for Eq.(1.2) is that due to a point load at $\tilde{x} = \tilde{X}$ defined by $\tilde{F} = \tilde{f}\, \delta(\tilde{x} - \tilde{X})$, where δ( ) denotes the Dirac-delta function. The corresponding elastic displacement u_i^G and stresses τ_{ij}^G are given by

$$u_i^G = u_{i;m}^G\, f_m\, , \quad \tau_{ij}^G = \tau_{ij;m}^G\, f_m \tag{2.1}$$

where $u_{i;m}^G$ is well known, see e.g. Ref.[2].

Representation Integral

The field generated by scattering of incident waves by an obstacle with surface S can be expressed in terms of a representation integral over S, see e.g. [2] and [4]. For a stress-free crack with plane faces A^+ and A^- the representation integral can be further simplified. If the total field is written as $\tilde{u}^t = \tilde{u}^{in} + \tilde{u}^{sc}$ where $\tilde{u}^{in}$ is the incident field and $\tilde{u}^{sc}$ is the scattered field, then the latter is given by

$$u_m^{sc}(\tilde{x}) = - \int_{A^+} \tau_{ij;m}^G(\tilde{x}-\tilde{X})\, \Delta u_i^{sc}(\tilde{X})\, n_j\, dA(\tilde{X}) \tag{2.2}$$

where

$$\Delta u_i^{sc} = (u_i^{sc})^{A^+} - (u_i^{sc})^{A^-} \tag{2.3}$$

In Eq.(2.2) $\underset{\sim}{x}$ represents any point outside of the crack, and $\underset{\sim}{n}$ is the outward normal (pointing from the A_+ to the A_- face).

The far-field longitudinal solution

Let us assume that the origin 0 of the coordinate system is close to the crack while the source S of the incident field and the observation point Q are far away, see Fig. 1).

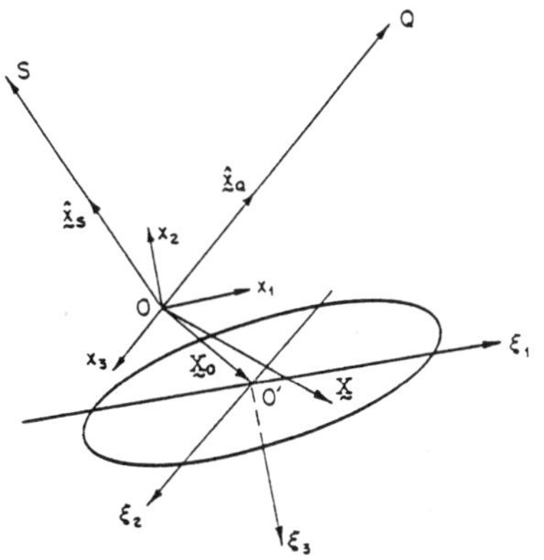

Fig. 1. Flat crack, with source point S and point of observation Q.

Thus if $\underset{\sim}{x}_S$, $\underset{\sim}{x}_Q$, and $\underset{\sim}{X}$ denote the position vectors of S, Q and any point on the crack, and $x_S = |\underset{\sim}{x}_S|$, $x_Q = |\underset{\sim}{x}_Q|$ and $X = |\underset{\sim}{X}|$, then x_S, $x_Q \gg X$. Defining the unit vector $\hat{\underset{\sim}{x}}_Q = \underset{\sim}{x}_Q/x_Q$, we can write

$$|\underset{\sim}{x}_Q - \underset{\sim}{X}| \approx x_Q - (\hat{\underset{\sim}{x}}_Q \cdot \underset{\sim}{X}) \ , \quad x_Q \gg X \tag{2.4}$$

The expression for $\tau^G_{ij;m}$ obtained from Eq.(2.1) then simplifies. Considering only the longitudinal wave we obtain

$$\tau^{G;L}_{ij;m}(\underset{\sim}{x}_Q-\underset{\sim}{X}) \approx ik_L b^{G;L}_{ij;m}(\hat{\underset{\sim}{x}}_Q)\, G_L(x_Q)\, \exp(-ik_L\hat{\underset{\sim}{x}}_Q\cdot\underset{\sim}{X}) \tag{2.5}$$

where

$$b^{G;L}_{ij;m} = (\lambda + 2\mu)^{-1}\, (2\mu\hat{x}_i\hat{x}_j + \lambda\delta_{ij})\, \hat{x}_m \tag{2.6}$$

$$G_L(x) = \frac{1}{4\pi x}\exp(ik_L x) \tag{2.7}$$

$$k_L = \omega/c_L \ ; \qquad c_L^2 = (\lambda + 2\mu)/\rho \tag{2.8a,b}$$

Substituting from Eq.(2.5) into Eq.(2.2) we obtain

$$\left[u_m^{sc}(\underset{\sim}{x}_Q)\right]^L = -ik_L\, b_{ij;m}^{G;L}(\hat{\underset{\sim}{x}}_Q)\; G_L(x_Q)\; n_j\; I_i^L(\hat{\underset{\sim}{x}}_Q) \tag{2.9}$$

where

$$I_i^L(\hat{\underset{\sim}{x}}_Q) = \int_{A^+} e^{-ik_L \hat{\underset{\sim}{x}}_Q \cdot \underset{\sim}{X}}\; \Delta u_i^{sc}(\underset{\sim}{X})\; dA(\underset{\sim}{X}) \tag{2.10}$$

Physical elastodynamics

The displacements u_i on the crack faces are yet unknown. They can, for example, be solved as part of a canonical problem following a ray-theoretic approach as in [4]. For our analysis we retain only the leading contributions arising from the incident field $\underset{\sim}{u}^{in}$ and the specularly reflected body waves, $\underset{\sim}{u}^{re}$, from the illuminated face A^-. This is the physical elastodynamic approximation, equivalent to the well-known physical optics approximation. Thus it can be shown that

$$\Delta\underset{\sim}{u}^{sc} \approx -\,(\underset{\sim}{u}^{in} + \underset{\sim}{u}^{re})^{A^-} \tag{2.11}$$

For the incident field we assume a longitudinal wave from S given by

$$\underset{\sim}{u}^{in} = -\,A\,\hat{\underset{\sim}{x}}_S\; G_L(x_S)\; \exp(-ik_L \hat{\underset{\sim}{x}}_S \cdot \underset{\sim}{X}), \qquad x_S >> X \tag{2.12}$$

The reflected field $\underset{\sim}{u}^{re}$ from A^- can be found from the standard results on reflection of plane waves (see [5]). Thus we obtain

$$\Delta\underset{\sim}{u}^{sc} \approx A\,\alpha(\hat{\underset{\sim}{x}}_S)\; G_L(x_S)\; \exp(-ik_L \hat{\underset{\sim}{x}}_S \cdot \underset{\sim}{X}) \tag{2.13}$$

where

$$\underset{\sim}{\alpha}(\hat{\underset{\sim}{x}}_S) = \underset{\sim}{p}_L^L + R_L^L\, \underset{\sim}{p}^{Lr} + R_T^L\, \underset{\sim}{d}^{Tr} \tag{2.14}$$

Here, $\underset{\sim}{p}_L^L = -\,\hat{\underset{\sim}{x}}_S$ is the propagation vector of the incident wave, while $\underset{\sim}{p}^{Lr}$ and $\underset{\sim}{p}^{Tr}$ are the propagation vectors of the reflected longitudinal and transverse waves, respectively. The corresponding reflection coefficients are defined by R_L^L and R_T^L. The vector $\underset{\sim}{d}^{Tr}$ defines the direction of the displacement of the reflected transverse wave. Expressions for these quantities are given in Ref. [2].

Scattered longitudinal far-field

Substituting Eq.(2.13) into Eq.(2.9) we obtain the scattered longitudinal far-field as

$$\frac{\left[u_m^{sc}(\underset{\sim}{x}_Q)\right]^L}{G_L(x_S)G_L(x_Q)} = -\,A\,\alpha_i(\hat{\underset{\sim}{x}}_S)\; b_{ij;m}^{G;L}(\hat{\underset{\sim}{x}}_Q)\; n_j\; I(k_L) \tag{2.15}$$

where

$$I(k_L) = ik_L \int_{A^+} e^{-ik_L \underset{\sim}{q}\cdot \underset{\sim}{X}}\; dA(\underset{\sim}{X}) \tag{2.16}$$

and

$$\underset{\sim}{q} = (\hat{\underset{\sim}{x}}_S + \hat{\underset{\sim}{x}}_Q) \tag{2.17}$$

is a vector in the bisector-direction of $\overline{OS}$ and $\overline{OQ}$. The far-field dependence on the crack occurs only through the function $I(k_L)$, which depends on the wave-number k_L and the bisector-vector $\underset{\sim}{q}$.

Transformation of co-ordinates

In Fig. 1 we have defined the coordinate system $\underset{\sim}{\xi}$, with origin $0'$ in the plane of the crack, and with the ξ_3-axis normal to that plane. Let $\overline{00'} = \underset{\sim}{X}_o$. Then Eq.(2.16) can be rewritten as

$$I(k_L) = ik_L e^{-ik_L\chi} \iint_{A+} e^{-ik_L \underset{\sim}{q}\cdot\underset{\sim}{\xi}} d\xi_1\, d\xi_2 \tag{2.18}$$

where

$$\chi = \underset{\sim}{q}\cdot\underset{\sim}{X}_o \tag{2.19}$$

Let us consider still another system $\bar{\underset{\sim}{\xi}}$ defined by (see Fig. 2)

$$\bar{\xi}_1 = q_1\,\xi_1 + q_2\,\xi_2 \;;\; \bar{\xi}_2 = -q_2\,\xi_1 + q_1\,\xi_2 \;;\; \bar{\xi}_3 = \xi_3 \tag{2.20}$$

Note that the $\bar{\xi}_1$-axis is parallel to the projection of $\underset{\sim}{q}$ on the crack-plane. The q_i used here are defined in the $\underset{\sim}{\xi}$-system. Then Eq.(2.18) can be reduced to the simple form, see Ref.[2],

$$I(k_L) = -\frac{e^{-ik_L\chi}}{q_1^2+q_2^2} \int_C e^{-ik_L\bar{\xi}_1} d\bar{\xi}_2 \tag{2.21}$$

where C is the edge of the crack. Thus the scattered field is expressed in terms of a distribution of equivalent sources along the edge of the crack. This is analogous to the concept of edge currents in electromagnetic scattering (see for instance [6]).

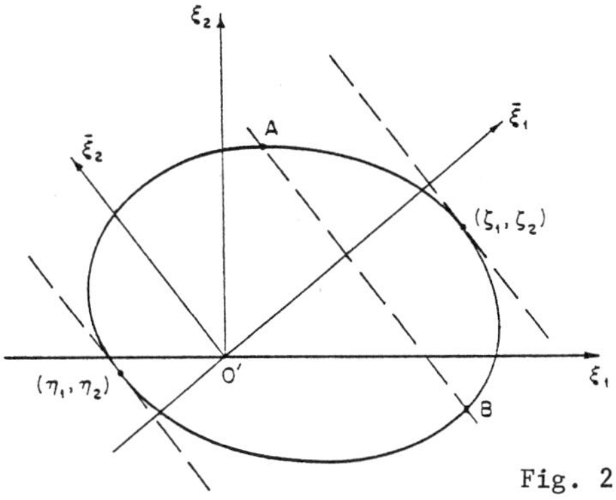

Fig. 2 Coordinates in the plane of the crack

Local coordinates

Another useful approximation results by introducing local coordinates near the points $E(\zeta_1,\zeta_2,0)$ and $F(\eta_1,\eta_2,0)$ on C where the tangents are parallel to the lines $\bar{\xi}_1$ = constraint, (see Fig. 2). For instance, near E $(\zeta_1,\zeta_2,0)$, the points on C can be represented by

$$\bar{\xi}_1 = \zeta_1 + \frac{1}{2!} s^2 \beta_0 + \frac{1}{3!} s^3 \beta_1 \tag{2.22}$$

$$\bar{\xi}_2 = \zeta_2 + s - \frac{1}{3!} s^3 \beta_0^2 \tag{2.23}$$

where s is the arc-length measured from E $(\zeta_1,\zeta_2,0)$ while β_0 and β_1 are the curvature of C and the value of $d\beta_0/ds$ at this point. We next substitute (2.22) and (2.23) into Eq.(2.21) and apply the stationary-phase approximation. The contribution from $E(\zeta_1,\zeta_2,0)$ then becomes

$$I(k_L) \approx - \frac{(2\pi)^{\frac{1}{2}} e^{-i\pi/4}}{(q_1^2+q_2^2)\beta_0^{\frac{1}{2}}} k_L^{-\frac{1}{2}} \exp\{-ik_L(\zeta_1 + \chi)\} \tag{2.24}$$

A similar contribution arises also from $F(\eta_1,\eta_2,0)$.

3. INVERSION INTEGRALS

The different forms of $I(k_L)$ in Eqs.(2.16),(2.21) and (2.24) suggest simple Fourier-type inversion integrals to recover the size, shape and orientation of a crack from the far-field data. We examine two such inversion integrals:

$$1.\quad f_1^*(\underset{\sim}{\lambda}) = \mathcal{I}_1\{f(k_L)\} = \int_{-\infty}^{\infty} e^{ik_L \mathbf{q}\cdot\underset{\sim}{\lambda}} f(k_L)\, dk_L \tag{3.1}$$

$$2.\quad f_2^*(\underset{\sim}{\lambda}) = \mathcal{I}_2\{f(k_L)\} = \int_{-\infty}^{\infty} k_L^{\frac{1}{2}} e^{ik_L \mathbf{q}\cdot\underset{\sim}{\lambda}} f(k_L)\, dk_L \tag{3.2}$$

where $\underset{\sim}{\lambda}$ defines any test-point in the medium.

The inverse operator $\mathcal{I}_1$

Application of the operator $\mathcal{I}_1$ of Eq.(3.1) to the expression for $I(k_L)$ in Eq. (2.16) gives

$$I_1^*(\underset{\sim}{\lambda}) = \int_{A^+} \delta'[\mathbf{q}\cdot(\underset{\sim}{\lambda} - \underset{\sim}{X})]\, dA(\underset{\sim}{X}) \tag{3.3}$$

where $\delta'(\ )$ denotes the derivative with respect to the argument of the δ-function. In Eq.(3.3) we have used the property that

$$\int_{-\infty}^{\infty} e^{ik_L\phi} dk_L = 2\pi\, \delta(\phi) \tag{3.4}$$

The argument of the δ-function in Eq.(3.3) will vanish only when $(\underset{\sim}{\lambda}-\underset{\sim}{X})$ is at right angles to $\mathbf{q}$ for $\underset{\sim}{X}$ on the crack. Thus $I_1^*(\underset{\sim}{\lambda}) = 0$ everywhere except when $\underset{\sim}{\lambda}$ lies within a layer between two planes normal to $\mathbf{q}$. The crack touches the end-planes of the layer. Thus from different observation points Q_n we can obtain a number of such layers or equivalently pairs of parallel end-planes which envelope the crack-edge. The intersection of two such envelopes from, say, two locations of the source point will lead to the crack itself.

The exact nature of $I_1^*(\underset{\sim}{\lambda})$ for any $\underset{\sim}{\lambda}$ within a given layer, and in particular the behavior at the end-planes of the layer will next be examined, from the point of view of identifying such planes from actual data. For this purpose we now apply the operator $\mathcal{J}_1$ to $I(k_L)$ given by Eq.(2.21). This gives

$$\begin{aligned} I_1^*(\underset{\sim}{\lambda}) &= -\frac{1}{q_1^2+q_2^2}\int_C \delta[q.\lambda - \chi - \bar{\xi}_1]\, d\bar{\xi}_2 \\ &= -\frac{1}{(q_1^2+q_2^2)}\int_C \delta[\underset{\sim}{q}.(\underset{\sim}{\lambda}-\underset{\sim}{X}_o) - \bar{\xi}_1]\frac{d\bar{\xi}_2}{d\bar{\xi}_1}\, d\bar{\xi}_1 \end{aligned} \tag{3.5}$$

where we have used the expression for χ given by Eq.(2.19).

Evaluating the integral in Eq.(3.5) by the sifting property of the δ-function we obtain

$$I_1^*(\underset{\sim}{\lambda}) = \begin{cases} \dfrac{-1}{(q_1^2+q_2^2)}\left[\left(\dfrac{d\bar{\xi}_2}{d\bar{\xi}_1}\right)^A + \left(\dfrac{d\bar{\xi}_2}{d\bar{\xi}_1}\right)^B\right]_{\bar{\xi}_1=\varkappa} & ;(\eta_1 < \varkappa < \zeta_1) \\ 0\ ,\ (\varkappa < \eta_1\ ;\ \varkappa > \zeta_1) \end{cases} \tag{3.6}$$

where

$$\varkappa = \underset{\sim}{q}.(\underset{\sim}{\lambda} - \underset{\sim}{X}_o) \tag{3.7}$$

Here the range of variation of $\bar{\xi}_1$, on C is given by $\eta_1 < \bar{\xi}_1 < \zeta_1$, where η_1 and ζ_1 were defined earlier in the context of the local co-ordinate system. In Eq.(3.6) A and B are the points at which the plane $\bar{\xi}_1 = \varkappa$ intersects the crack edge C when $\eta_1 < \varkappa < \zeta_1$. At the extreme positions $\varkappa = \eta_1$ and $\varkappa = \zeta_1$ the gradient terms in Eq.(3.6) become infinite, usually, in the inverse square-root sense, as will later be illustrated by the example of an elliptical crack.

In practice, we can choose $\underset{\sim}{\lambda}$ along the $\underset{\sim}{q}$-direction from 0 and determine the finite layer normal to $\underset{\sim}{q}$ which contains the crack. The singular behavior of the gradient terms will be the principal test for identification of the end-planes of the layers. We further note that although η_1, ζ_1 and $\varkappa$ in Eq.(3.6) and Eq.(3.7) intrinsically depend on the choice of the origin $0'$, the location of the layer is unique for a given $\underset{\sim}{q}$ in the initial co-ordinate system at 0, with reference to which $\underset{\sim}{\lambda}$ is defined. This is easily verified as shown below by taking two arbitrary positions for $0'$ and comparing the values of $\underset{\sim}{\lambda}$ along $\underset{\sim}{q}$ corresponding to the end-planes of the layers for the two choices of $0'$. The same value of $\underset{\sim}{\lambda}$ is reached. To prove this, consider two such origins $0_1'$ and $0_2'$ with $\overline{00_1'} = \underset{\sim}{X}_o^{(1)}$ and $\overline{00_2'} = \underset{\sim}{X}_o^{(2)}$. At each of these points we take the coordinate axes parallel to the $\underset{\sim}{\xi}$-system and we denote these by $\bar{\underset{\sim}{\xi}}^{(k)}$, k = 1 and k = 2, with $\bar{\xi}_3^{(k)} = 0$ on the crack plane. We retain the notation $\underset{\sim}{\xi}$ to denote a coordinate system with reference to a reference origin at some point $0'$.

Let the point $E(\zeta_1,\zeta_2,0)$ in the new systems have the coordinates $(\zeta_1^{(k)},\zeta_2^{(k)},0)$, k=1,2. If the point along $\underset{\sim}{q}$ at which the singular behavior of I_1^* in Eq.(3.6) occurs would be defined by $\underset{\sim}{\lambda}^{(k)}$, k = 1,2 for the two locations of $0'$, then we have to show that $\underset{\sim}{\lambda}^{(1)} = \underset{\sim}{\lambda}^{(2)}$. First we obtain from Eq.(3.6) and Eq.(3.7) that

$$\underset{\sim}{q}\cdot(\lambda^{(k)} - X_o^{(k)}) = \zeta_1^{(k)} \quad , \; k = 1,2 \tag{3.8}$$

Taking the difference of the two relations for k = 1 and k = 2 we obtain

$$\underset{\sim}{q}\cdot(\underset{\sim}{\lambda}^{(2)}-\underset{\sim}{\lambda}^{(1)}) = \underset{\sim}{q}\cdot\overline{0_1'0_2'} + (\zeta_1^{(2)} - \zeta_1^{(1)}) \tag{3.9}$$

Let ℓ and d denote the projections of $\overline{0_1'0_2'}$ along the direction of the $\underset{\sim}{q}$-vector and along the projection of $\underset{\sim}{q}$ on the crack plane, respectively. Then we have

$$\underset{\sim}{q}\cdot\overline{0_1'0_2'} = |\underset{\sim}{q}|\,\ell = (q_1^2+q_2^2)^{\frac{1}{2}}\, d \tag{3.10}$$

Moreover, since E is a fixed point on the crack edge,

$$(\zeta_1^{(2)} - \zeta_1^{(1)}) = -\left\{(\bar{\xi}_1)_{0_2'} - (\bar{\xi}_1)_{0_1'}\right\} \tag{3.11}$$

where the $\bar{\underset{\sim}{\xi}}$-system is defined at the reference origin 0′. We also note that the difference between the $\bar{\xi}_1$ coordinates of any two points is the product of the projection of the line connecting them along the direction of the $\bar{\xi}_1$-axis and the scaling factor of the $(\bar{\xi}_1,\bar{\xi}_2)$-system in the crack-plane, see Eq.(2.20). For the two points $0_2'$ and $0_1'$ the above projection, by definition, is d, while the scaling factor for Eq.(2.20) a and b is $(q_1^2+q_2^2)^{\frac{1}{2}}$. Thus Eq.(3.11) yields

$$(\zeta_1^{(2)} - \zeta_1^{(1)}) = -\,(q_1^2 + q_2^2)^{\frac{1}{2}}\, d \tag{3.12}$$

Employing Eq.(3.10) and Eq.(3.12) in Eq.(3.9) we obtain that $\underset{\sim}{\lambda}^{(2)} = \underset{\sim}{\lambda}^{(1)}$ and hence the proof is completed.

The elliptical crack

The inverse square root singularity of $I_1^*(\underset{\sim}{\lambda})$ in Eq.(3.6) will now be verified for an analytic example in which the crack is of elliptical shape, defined by

$$\frac{\xi_1^2}{a^2} + \frac{\xi_2^2}{b^2} = 1 \quad , \; \xi_3 = 0 \tag{3.13}$$

Then Eq.(2.21) can be evaluated analytically to give

$$I(k_L) = -\,2\pi i\,\frac{ab}{\rho}\,J_1(k_L\rho)\,e^{-ik_L\chi} \tag{3.14}$$

where $J_1(\;)$ is the cylindrical Bessel function and

$$\rho = (a^2\,q_1^2 + b^2\,q_2^2)^{\frac{1}{2}} \tag{3.15}$$

The inverse operator $\mathcal{I}_1$ of Eq.(3.1) when applied to Eq.(3.14) gives

$$I_1^*(\underset{\sim}{\lambda}) = \begin{cases} \dfrac{2\pi ab}{\rho^2}\,\dfrac{\varkappa}{(\rho^2-\varkappa^2)^{\frac{1}{2}}} \;; & |\varkappa| < \rho \\ 0 \;; & |\varkappa| > \rho \end{cases} \tag{3.16}$$

which clearly exhibits the inverse square root singularity at $\varkappa = \pm \rho$, where $\varkappa$ is defined in Eq.(3.7). Conversely it is not difficult to verify that the planes

$$\varkappa = \pm \rho \tag{3.17}$$

touch the ellipse at the following points in the ξ-system:

$$\left(\pm \frac{q_1 a^2}{\rho}, \pm \frac{q_2 b^2}{\rho}, 0\right) \tag{3.18}$$

The inverse operator $\mathcal{J}_2$

It is often desirable to identify the end planes of the layer by a singularity of the Dirac delta function type. To this end we apply the inverse operator $\mathcal{J}_2$ defined by Eq.(3.2) to the stationary phase approximation of $I(k_L)$ given by Eq.(2.24). This gives

$$I_2^*(\tilde{\lambda}) \sim - \frac{(2\pi)^{3/2} e^{-i\pi/4}}{(q_1^2+q_2^2)(\beta_o)_E^{\frac{1}{2}}} \delta(\varkappa - \zeta_1) - \frac{(2\pi)^{3/2} e^{i\pi/4}}{(q_1^2+q_2^2)(\beta_o)_F^{\frac{1}{2}}} \delta(\varkappa - \eta_1) \tag{3.19}$$

These expressions have the desired δ-function behavior across the planes $\varkappa = \zeta_1$ and $\varkappa = \eta_1$

4. NUMERICAL EXAMPLES

The method of inversion of far-field longitudinal crack scattering data is now applied to two analytical examples. In the first example we assume that a plane of symmetry of the crack is known, and that both the source and the receiver are located in that plane. Within the context of that example we also discuss an iterative technique to improve the accuracy of the inversion method. The second example, which is three-dimensional in nature, is concerned with the computation of points on the edge of an elliptical crack, on the basis of Eq.(3.17).

Source and receiver in a plane of symmetry of the crack

The position of the origin 0 of the coordinate system x_i must be close to the crack. The diffracted field is taken in the form given by Eq.(2.24). Note that for this case the flash points defined by $\xi_1 = \zeta_1$ and $\xi_1 = \eta_1$ are located in the plane of symmetry. Adding the contributions from the two flash points we write

$$I(k_L) = - \frac{1}{q_1^2+q_2^2} \left(\frac{2\pi}{k_L}\right)^{\frac{1}{2}} J \tag{4.1}$$

where

$$J = \sum_{j=1}^{2} D_j \, e^{-ik_L(\hat{\tilde{x}}_Q + \hat{\tilde{x}}_S)\cdot \tilde{X}_j} \tag{4.2}$$

$$D_1 = \frac{1}{(\beta_o^{\frac{1}{2}})_1} e^{-i\pi/4}, \quad D_2 = \frac{1}{(\beta_o^{\frac{1}{2}})_2} e^{i\pi/4} \tag{4.3}$$

The position vectors of the crack tips $\underset{\sim}{X}_j (j=1,2)$ are defined in Fig. 3.

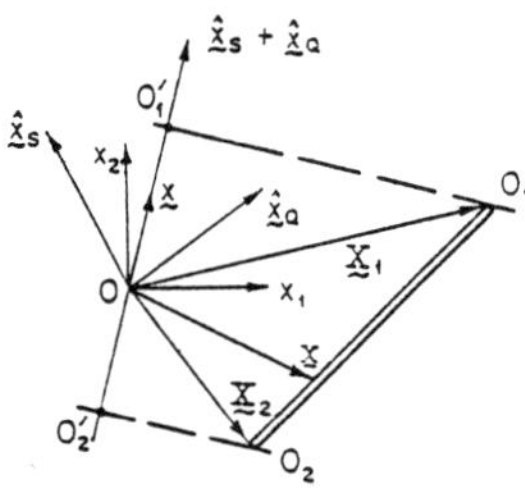

Fig. 3 Source and receiver in plane of symmetry of the crack

Actual crack-scattering data must first be deconvolved into the form given by Eqs.(2.15) with $I(k_L)$ defined by Eq.(4.1). The two exponentials in Eq.(4.2) can, however be obtained by combining information from the arrival time of the first diffracted signal with information obtained from the periodicity of the amplitude spectrum. The arrival time of the first diffracted signal provides the total length $SO_1 + O_1Q$. Since SO and OQ are known, we have

$$\overline{SO_1} + \overline{O_1Q} - \overline{SO}-\overline{OQ} \approx -(\hat{\underset{\sim}{x}}_S + \hat{\underset{\sim}{x}}_Q)\cdot\underset{\sim}{X}_1 \tag{4.4}$$

To investigate the amplitude spectrum, the absolute magnitude of J is computed from (4.2) as

$$|J| = \{|D_1|^2+|D_2|^2+2|D_1||D_2| \sin[k_L(\hat{\underset{\sim}{x}}_Q+\hat{\underset{\sim}{x}}_S)\cdot(\underset{\sim}{X}_1-\underset{\sim}{X}_2)]\}^{\frac{1}{2}} \tag{4.5}$$

Equation (4.5) implies that the amplitude of the longitudinal field is modulated with respect to $k_L = \omega/c_L$ with period

$$P = 2\pi/[(\hat{\underset{\sim}{x}}_Q+\hat{\underset{\sim}{x}}_S)\cdot(\underset{\sim}{X}_1-\underset{\sim}{X}_2)] \tag{4.6}$$

Experimental results for the amplitude spectrum, see e.g. Ref.[7] do indeed show this periodicity. Thus, if $[u_m^{sc}(\underset{\sim}{x}_Q)]^L$ on the left-hand side of Eq.(2.15) represents experimental data, we can define P_Q as the period of $|[u_m^{sc}(\underset{\sim}{x}_Q)]^L|$. By virtue of Eq.(4.6) we then obtain

$$(\hat{\underset{\sim}{x}}_Q+\hat{\underset{\sim}{x}}_S)\cdot(\underset{\sim}{X}_1-\underset{\sim}{X}_2) = 2\pi/P_Q \tag{4.7}$$

Equations (4.4) and (4.7) provide the information to express the diffracted field in the form given by Eq.(4.2).

It follows from (4.2) and (3.19) that the observation at one location Q leads to one strip normal to the bisector of $\overline{OQ}$ and $\overline{OS}$. Thus end-planes of this strip are touched by Σ, where Σ is the trace of the crack in the plane of symmetry. A second point of observation (which can be the point S for backscatter) defines a second strip, whose intersection with the first one yields four points, which provide two possibilities for Σ. The final determination of Σ is obtained by using data from a third point of observation.

Iterative procedure

The exponential terms obtained from experimental data involve actual ray-path lengths, and thus they will differ from the ones derived from Eqs.(4.4) and (4.7) which employ approximations such as in Eq.(2.4). This means that the strips (leading to the crack location) will be shifted by small amounts. However, the points of intersection of these strips can shift considerably from the true crack-tip positions. In order to rectify this error we introduce an iterative scheme.

Let the error terms ϵ_S^j and ϵ_Q^j be defined by

$$\epsilon_S^j = |\underset{\sim}{X}_j - \underset{\sim}{x}_S| - (x_S - \hat{\underset{\sim}{x}}_S \cdot \underset{\sim}{X}_j) \tag{4.8}$$

$$\epsilon_Q^j = |\underset{\sim}{X}_j - \underset{\sim}{x}_Q| - (x_Q - \hat{\underset{\sim}{x}}_Q \cdot \underset{\sim}{X}_j) \tag{4.9}$$

where $\underset{\sim}{X}_j$, $j = 1,2$ are the true crack tip positions as defined in Fig. 3. Also let the positions of these tips calculated by the use of (4.4) and (4.7) be at $\underset{\sim}{X}_j^{(0)}$, $j = 1,2$. Then the following identities hold good:

$$(\hat{\underset{\sim}{x}}_S + \hat{\underset{\sim}{x}}_Q) \cdot (\underset{\sim}{X}_j - \underset{\sim}{X}_j^{(0)}) = \epsilon_S^j + \epsilon_Q^j \quad (j = 1,2) \tag{4.10}$$

$$\hat{\underset{\sim}{x}}_S \cdot (\underset{\sim}{X}_j - \underset{\sim}{X}_j^{(0)}) = \epsilon_S^j \quad (j = 1,2) \tag{4.11}$$

which arise from the longitudinal far-field at Q and the far-field back-scatter at S respectively. The above equations are then solved iteratively as follows. The first approximation is given by $\underset{\sim}{X}_j = \underset{\sim}{X}_j^{(0)}$ where we assume $\epsilon_S^j = 0$, $\epsilon_Q^j = 0$. Next we evaluate ϵ_S^j and ϵ_Q^j at $\underset{\sim}{X}_j^{(0)}$ from Eqs.(4.8) and (4.9). Substituting these in Eqs.(4.10) and (4.11) and solving the resulting equations for $\underset{\sim}{X}_j$ leads to the next higher order solution for the crack-tip positions. This procedure is repeated till the necessary convergence is achieved. Usually four to five iterations are found adequate. This is illustrated by the example of Fig. 4 for which we have assumed $\overline{OQ} = \overline{OS} = 25$, $\overline{OC} = 2$ (where C is the center of the crack), $\theta_2 = \pi/3$, $\theta_3 = \pi$ and $\theta_4 = \pi/2$. The crack half-length is taken as the unit of length. Figure 5 shows the edge-error for the crack edge at O_1 as a function of the angle θ_1. The edge-error is the absolute distance between the true and calculated locations. Both the iterated and non-iterated errors are shown for comparison. Other examples of the iteration procedure are given in Ref.[2].

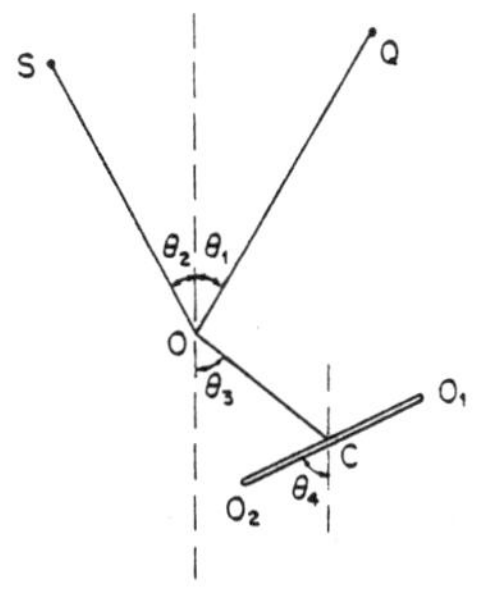

Fig. 4 Geometry for numerical calculations

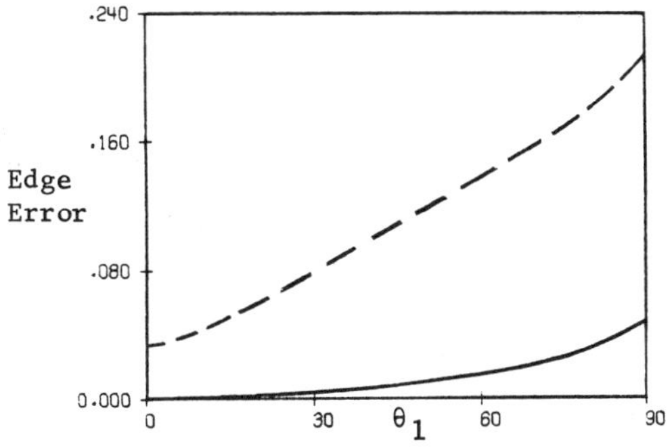

Fig. 5 Edge error at O_1; — — — no iteration; ——— 1 iteration

In band-limited observations the range of k_L that can be covered will be finite, say, $k_- \leq |k_L| \leq k_+$. The corresponding effects on the sharpness of the δ-function behavior which govern our inverse procedure have been discussed in Ref.[2].

The 3-D Crack

As a 3-D example we consider an elliptical crack of semi-major and semi-minor axes a and b respectively. Relative to a coordinate system at the center of the crack, we have

$$x_1^2/a^2 + x_2^2/b^2 = 1 \; , \quad x_3 = 0 \tag{4.12}$$

The diffracted far-field will be assumed in the form discussed earlier, with $I(k_L)$ given by Eq.(3.14) in terms of a Bessel function. For any given pair of source and observation points, the inverse operator $\mathcal{J}_1$ will then lead to a pair of end-planes touching the crack edge as given by Eq.(3.17). The plane of the crack and segments of the crack-edge will now be constructed by using the far-field diffraction data from two source locations S_k, k = 1,2 and a set of observation points Q_n , n = 1, ---20. For our numerical example the spherical coordinates of S_k (k = 1,2) are taken as (25, $\pi/6$, $\pi/2$) and (25, $\pi/6$, $2\pi/3$). The points Q_n are taken at $\underset{\sim}{x}^{(n)}$ where

$$x_1^{(n)} = 10 + (n-5)\sin(\pi/3)\cos(3\pi/4) \tag{4.13}$$

$$x_2^{(n)} = 10 + (n-5)\sin(\pi/3)\sin(3\pi/4) \tag{4.14}$$

$$x_3^{(n)} = (n-5)\cos(\pi/3) \; , \qquad n = 1, ---20 \tag{4.15}$$

A tentative origin 0 in the neighborhood of the crack is taken at (0.2, 0.3, 0.15). The inversion integral then leads to a number of layers for given source and observation points as described earlier. For a source at S_k and an observation point at Q_n the pair of layer-end-planes $\Omega(k,n;p)$, p = 1,2 obtained from Eq.(3.17), are defined by

$$q_1(k,n)\, x_1 + q_2(k,n)\, x_2 = \pm \{a^2 q_1^2(k,n) + b^2 q_2^2(k,n)\}^{\frac{1}{2}} \tag{4.16}$$

where $\pm$ signs correspond to p = 1,2, respectively. The bisector vector $\underset{\sim}{q}(k,n)$ is associated with $\overline{OS_k}$ and $\overline{OQ_n}$.

For large n, each set of planes $\Omega(k,n;p)$ defined by Eq.(4.16) for a given k, forms a prismatic surface, which will touch the crack-edge C at a set of points. These points span a polygon, which approximates a segment C(k;p) of the edge C. The intersection of the two prismatic surfaces for the same p but k = 1 and k = 2, respectively, will lead to points on C common to C(1;p) and C(2;p). We can use these points to generate the crack-plane. The above points are obtained as follows. The prismatic surface formed by the first-set of planes $\Omega(1,n;p)$ will be intersected by the various individual planes of the set $\Omega(2,n;p)$ along a set of polygons $\Gamma(2,n;p)$, which we initially determine. The points where the polygons of this set intersect in 3-D space constitute points for an approximate determination of the crack-plane. These points are easily found by testing where any polygon $\Gamma(2,n;p)$ with a given n is intersected by the remaining planes of $\Omega(2,n;p)$. The plane of the crack can thus be determined, and once this has been achieved the intersection of this plane with the set of planes $\Omega(k,n;p)$ gives the desired tangent lines enveloping the crack edge. For a = 1 and b = 0.5, Table 1 gives the vertices of the polygon formed by the tangent-lines corresponding to $\Omega(1,n;1)$ which envelopes the arc C(1;p) with p = 1. These points evidently lie very close to the crack-plane $x_3 = 0$. The last column of Table 1 shows that these points lie very close to the elliptical boundary as well. A similar set of points corresponding to the arc C(1;p) with p = 2 is also generated by the above calculations, since they define an opposite quadrant of the crack-edge referred to the origin at the center of the crack.

The extent of the crack-edge recovered will depend on the relative locations of S_k and Q_n with respect to the crack-plane. A proper choice can usually be made once the crack-plane is determined from an initial configuration.

x_1	x_2	x_3	$x_1^2/a^2+x_2^2/b^2-1$
.85076	.26265	.00041	-.26033E-03
.82938	.27927	.00036	-.16001E-03
.80476	.29679	.00031	-.40334E-04
.77657	.31505	.00026	.10016E-03
.74455	.33388	.00020	.26192E-03
.70851	.35301	.00014	.44407E-03
.66841	.37211	.00007	.64425E-03
.62437	.39084	.00000	.85866E-03
.57670	.40881	-.00007	.10824E-02
.52589	.42566	-.00014	.13099E-02
.47260	.44107	-.00022	.15359E-02
.41760	.45480	-.00029	.17556E-02
.36175	.46667	-.00036	.19654E-02
.30585	.47661	-.00043	.21628E-02
.25070	.48464	-.00049	.23466E-02
.19695	.49085	-.00055	.25162E-02
.14513	.49538	-.00061	.26719E-02
.09565	.49841	-.00066	.28143E-02
.04877	.50014	-.00071	.29440E-02

Table 1: Computed coordinates of vertices of polygon enveloping the crack edge, for a = 1 and b = 0.5. Coordinates relative to origin at center of elliptical crack.

ACKNOWLEDGEMENT

This work was carried out in the course of research sponsored by the Center for Advanced NDE operated by the Science Center, Rockwell International for the Advanced Research Project Agency and the Air Force Materials Laboratory under Contact F33615-74-C-5180.

REFERENCES

1. N. Bleistein and J. K. Cohen, Inverse methods for reflector mapping and sound speed profiling, in Ocean Acoustics, (J. A. de Santo, editor), Topics in Current Physics, Springer-Verlag, 225-242 (1979).

2. J. D. Achenbach, K. Viswanathan and A. Norris, An inversion integral for crack-scattering data, WAVE MOTION (in press).

3. J. D. Achenbach, A. K. Gautesen and H. McMaken, Diffraction of elastic waves by cracks - analytical results, in Elastic Waves and Non-Destructive Testing of Materials (Y. H. Pao, editor), AMD-29, American Society of Mechanical Engineers, 33-52 (1978).

4. J. D. Achenbach, A. K. Gautesen and H. McMaken, Diffraction of point-source signals by a circular crack, Bull. Seism. Soc. Am. 68, 889 (1978).

5. Achenbach, J. D. (1973) Wave Propagation in Elastic Solids, North-Holland, Amsterdam - New York.

6. E. F. Knott and T.B.A. Senior, Equivalent currents for a ring discontinuity, IEEE Trans. Ant. Prop. AP-21, 693-695 (1973).

7. J. D. Achenbach, L. Adler, D. Kent Lewis and H. McMaken, Diffraction of ultrasonic waves by penny-shaped cracks in metals: theory and experiment, J. Acoust.Soc. Am. (in press).

INVERSION OF ULTRASONIC SCATTERING DATA*

J. H. Rose
Physics Department, University of Michigan, Ann Arbor, MI 48109

R.K. Elsley and B. Tittman
Rockwell International Science Center, Thousand Oaks, CA 91360

V. V. Varadan and V. K. Varadan
Boyd Laboratory, Ohio State University, Columbus, OH 43210

ABSTRACT

The determination of flaw characteristics from ultrasonic scattering amplitudes is discussed in terms of the Born and the extended quasi-static approximation. The resulting inversion algorithms have been tested for both spheres and 2-1 oblate and prolate spheroids. Accurate results are obtained for the size, shape and orientation of the flaw.

INTRODUCTION

Recent developments in ultrasonic scattering theory have been strongly motivated by the non-destructive evaluation needs of the structural materials community. Their primary question is; given a set of ultrasonic measurements (e.g. scattering amplitudes) from some industrial component, when will it break? An intermediate step in answering this question is: given the scattering data, what are the characteristics of the flaws in the piece? Here one would like to know if one has a volume flaw such as a void or inclusion, or if one has a crack. Also, one would like to know the size, shape and orientation of the flaw: and, if it is an inclusion, what it is made of. Answering these questions is what we will refer to as the ultrasonic inversion problem.

The current status of the ultrasonic inversion problem depends upon the ratio of the characteristic size of the flaw (a_o) to the wavelength λ ($k = 2\pi/\lambda$). When the size of flaw is much larger than the wavelength, $ka_o >> 1$, then imaging techniques can be used, and a good deal of progress has been made. In the opposite limit, $ka_o << 1$, there has been some recent progress, both in terms of describing what information can be extracted in principle and in terms of practical algorithms for simply shaped flaws (1). Between these two limits we have the intermediate regime, where the wavelength is on the order of the size of the object. This paper focuses on the intermediate regime and studies the geometric features of single voids.

We will review the theoretical development of an inversion algorithm for the intermediate wavelength case (2). Further we will summarize the progress of our group effort to empirically verify this algorithm. The need for detailed empirical verification stems from the theoretical justification of the algorithm, which is based on perturbative solutions of the wave equation and is valid only if the scattering is sufficiently weak! However, many of the flaws of interest in NDE are anything but weak (e.g. a void). It is not clear how to extend our current inversion algorithm formally to the strong scattering case. However, for voids of simple shape, the algorithm yields good results as we will report.

It is in this empirical verification scheme that recent developments in elastic wave scattering theory (such as the T-matrix method (3)) have a key role to play. For in order to establish the limits of validity of this algorithm and other empirical inversion algorithms, it is desirable to have the scattering amplitudes for a wide range of differently shaped flaws. Particularly interesting for this purpose would be flaws with sharp edges such as cones and pill boxes. Up to the present time we are limited to investigating spherical, and oblate and prolate spheroidal flaws.

The structure of the paper is as follows. In the second section we review the derivation of the algorithm. In section III we indicate how the theory was simplified for the case of ellipsoidally shaped flaws. In the fourth section we report the results of testing the algorithm with experimentally generated data. In the fifth section we report the results of testing the inversion algorithm using scattering amplitudes generated by the T-matrix method for 2-1 oblate and prolate spheroidal voids. Finally in section six we provide a discussion of our results and conclude.

GENERAL THEORY

The algorithm to be discussed below is a procedure for approximately determining the Fourier transform of the characteristic function, $\gamma(\vec{r})$, of the flaw. Here $\gamma(\vec{r})$ is 1 for $\vec{r}$ inside the flaw, and $\gamma(\vec{r})=0$ for $\vec{r}$ outside the flaw. We restrict our review of the theory to the simplest experimental situation. That is we assume a pulse-echo geometry as shown in Fig. 1. Here a longitudinally (or shear) polarized plane wave is incident on the flaw, and the directly backscattered longitudinal (or shear) amplitude is determined. The pulse echo scattering amplitudes can be written for an arbitrarily shaped flaw as

$$A(\vec{k}) = a(\vec{k},\{\mu\})S(2\vec{k})k^2 \tag{2-1}$$

Here $S(2\vec{k})$, the shape factor, is the Fourier transform of the characteristic function of the flaw. The wavevector of the incident wave is denoted by $\vec{k}$ and $a(\vec{k},\{\mu\})$ is a function to be calculated which yields the correct scattering amplitudes A for an arbitrary $\vec{k}$. Here $\{\mu\}$ denotes the material parameters of the host material.

The virtue of writing the scattering amplitudes in the form of Eq. 2-1 is that several approximate theries (4,5) yield very simple forms for the factor $a(\vec{k},\{\mu\})$. In particular we will use the form of $a(\vec{k},\{\mu\})$ which can be derived from the extended quasi-static approximation. In that approximation one takes account of the long wavelength elastic deformation of the flaw correctly, and hence obtains the angular features of the scattering correctly in this limit. For the extended quasi-static approximation $a(\vec{k},\{\mu\})$ is assumed to be independent of $|\vec{k}|$ and given by its long wavelength limit which depends only on the direction of $\vec{k}$, $\hat{k}$, and on $\{\mu\}$. We denote this approximate form of $a(\vec{k},\{\mu\})$ as $a_{QSA}(\hat{k},\{\mu\})$.

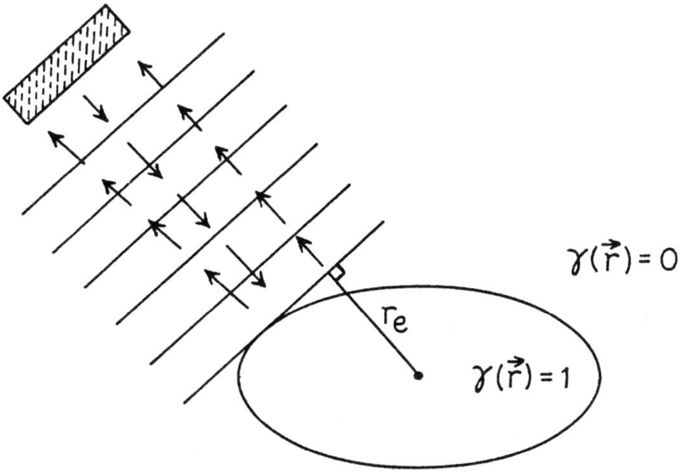

Fig. 1. The geometry of a pulse echo experiment. The distance from the center of the flaw to the tangent plane is the effective radius, r_e, discussed in the text.

Using this approximation we rewrite equation 2.1 as

$$S(2\vec{k}) \approx A(\vec{k})/(k^2 a_{QSA}(\hat{k},\{\mu\})) \tag{2-2}$$

Experimentally, a_{QSA} can be obtained for an arbitrarily shaped object by measurements of the long wavelength scattering amplitudes. In that limit $S(2\vec{k})$ goes to a constant, and a_{QSA} can be determined from the angularly dependent coefficients of A

$$a_{QSA}(\hat{k},\{\mu\}) = \lim_{k\to 0} A(\vec{k})/k^2 \tag{2-3}$$

Once a_{QSA} is obtained we can determine $S(2\vec{k})$ from Eq. 2-2 via an experimental measurement of the backscattered amplitudes. Taking the Fourier transform of $S(2\vec{k})$ then allows us to determine the characteristic function of the flaw, and hence its size, shape and orientation. The major approximation in using a_{QSA} is that we assume that it depends only on $\hat{k}$ and not on $|\vec{k}|$. The characteristic function is given explicitly in terms of the shape function as (6)

$$\gamma(\vec{r}) = \text{const.} \int d^3\vec{k}\; e^{2i\vec{k}\cdot\vec{r}}\; \mathrm{Re}(A(\vec{k}))/(k^2 a_{QSA}(\hat{k},\{\mu\})) \tag{2-4}$$

SIMPLIFIED THEORY FOR ELLIPSOIDALLY SHAPED FLAWS

In the last section we described an approximate procedure for determining the size and shape and orientation of an arbitrary three dimensional flaw. In order to use this inversion technique one requires pulse-echo measurements from all incident directions $\hat{k}$. The characteristic function is then obtained (Eq. 2-4) as an inverse Fourier transform which involves integrating over both $|\vec{k}|$ and $\hat{k}$. For the class of ellipsoidally shaped flaws, one can obtain all relevant information about the flaw by inverting each pulse-echo record independently as discussed below. This avoids the angular integration over $\hat{k}$ in the inverse Fourier transform, and significantly simplifies the application of the algorithm.

In order to illustrate how this simplification comes about, let us consider the weak scattering limit. Then the theory of the last section is rigorously valid and Eq. 2-2 becomes

$$S(2\vec{k}) = \text{const.}\ A_{L\to L}(\vec{k})/k^2 \tag{3-1}$$

We have used the fact that $a(\hat{k},\{\mu\})$ is a constant in the weak scattering limit as a function of $\hat{k}$. For an ellipsoid we know that $S(2\vec{k})$ is given by the following equations

$$S(2\vec{k}) = \frac{\sin(2kr_e) - 2kr_e\cos(2kr_e)}{(2kr_e)^3} \tag{3-2}$$

and

$$r_e = (a_x^2\cos^2\theta\sin^2\phi + a_y^2\cos^2\theta\cos^2\phi + a_z^2\sin^2\theta)^{\frac{1}{2}} \tag{3-3}$$

Here the axes of the ellipsoid are $a = (a_x,a_y,a_z)$, and θ and ϕ define the direction of $\hat{k}$ in spherical coordinates. The angular dependence of the shape factor comes in strictly through the function which we have called $r_e(\theta,\phi)$. In a pulse-echo measurement, the incident direction $\hat{k}$ is kept fixed, and r_e is a constant for that set of data. We note for a fixed incident direction, Eq. 3-2 has the same form as a Fourier transform of a sphere with an effective radius r_e. For each incident direction $\hat{k}$, we obtain r_e in the following way. First we obtain $S(2|\vec{k}|)$ from Eq. 3-1. We then extend $S(2|\vec{k}|)$ to be spherically symmetric in $\vec{k}$-space. Thus we obtain the three dimensional Fourier transform of a sphere of radius $r_e(\theta,\phi)$. This Fourier transform is then inverted to yield the effective radius for that direction. The resulting effective radius (Eq. 3-3) has a simple geometric interpretation as shown in Fig. 1. When a wavefront strikes the surface, it is first tangent at some one point (which is an accumulation point for phase). The radius r_e is the distance from the center of the flaw to the plane of the wavefront. An important consequence of Eq. 3-3 is that pulse-echo measurements along the axis of an ellipsoid yield the axis length directly. For example, a measurement along the a_x axis yields an effective radius equal to a_x. Hence one can obtain the length of the ellipsoid axes directly from three measurements if one knows the orientation of the ellipsoid.

So far we have been discussing the weak scattering limit for the sake of illustration. The appropriate extension to the strong scattering case is straightforward. Equation 2-2 is

$$S(2\vec{k}) \approx \text{const. } A(\vec{k})/(k^2 a_{QSA}(\hat{k},\{\mu\})) \quad (3\text{-}4)$$

For a given incident direction $a(\hat{k},\{\mu\})$ is just a constant since it doesn't depend on $|\vec{k}|$ in the quasi-static approximation. With this approximation we recover Eq. 3-1 and can proceed in an approximate way with the entire procedure which was given above. Of course for a strongly scattering flaw, our analysis is only approximate and must be checked empirically. In the next sections we provide some empirical tests of the strong scattering limit.

INVERSION OF EXPERIMENTAL DATA

We summarize the initial results of testing the algorithm, in its simplified form for ellipsoids, with experimental data. More extensive results and a comprehensive treatment of both experiment and data analysis will be given in Ref. 7. We report results for a spherical void with a radius of 400 microns, and an oblate spheroid with a semi-major axis of 400μ and a semi-minor axis of 200μ. These were machined flaws in the center of large spheres of Ti-6Al-4V. Details of the construction of the flaws and their use as calibration samples are given in Refs. 8 and 9.

The simplified algorithm allows us to treat each pulse-echo measurement separately, and it yields the distance from the center of the flaw to the tangent plane of the incoming wavefront. For the sphere we obtained a single pulse-echo record which suffices to determine the size of the flaw due to its spherical symmetry. However, for the spheroid we only examined the pulse echo record for a measurement along the axis of symmetry.

Before presenting the results, we want to discuss two crucial details of the data analysis scheme. First, for sufficiently small wavevector, k, the phase of the scattering amplitudes must be constant and zero. This reflects the fact that the real part of the scattering amplitude rises as k^2 for small k while the imaginary rises much more slowly. This constraint on the phase allows one to establish the phase of the experimental data, which otherwise would not be entirely determined (7). The second point concerns the effects of limited bandwidths. A lack of low frequency data would leave the phase of the data undetermined as just indicated. A lack of high frequency data causes the characteristic function to be blurred, and this introduces some uncertainty in determining the size of the flaw. In large part the effects of blurring due to the limited bandwidth can be overcome by an appropriate calibration procedure. For the simplified form of the algorithm, the analysis is carried out in terms of equivalent spheres. The effects of a limited high frequency bandwidth on the characteristic function of a sphere can be determined in the following way. We consider the Fourier transform of a sphere in k-space. We then bandlimit it with a rectangular window extending from k=0 to k_{max}. Then we transform it back to r-space. The resulting curves can then be compared to the experimentally determined characteristic functions and thus serve as a calibration for the effects of blurring.

Inverting the pulse echo data for the sphere ($0<ka<4$) where a is the radius. We find a radius of approximately 400μ with an uncertainty of about 40μ. This should be compared to the exact value of 400μ. The inversion of the spheroid data yields an estimate of the semi-minor axis of 220μ with an uncertainty of about 20μ. The exact value is 200μ. We consider these results to be quite encouraging. It is clear however, that considerably more scattering data for other orientations of the spheroids, for other materials and for other types of volume flaws (e.g. inclusions) will be necessary before the algorithm can be considered fully tested. To partially examine these questions we turn to the theoretically generated data of the next section.

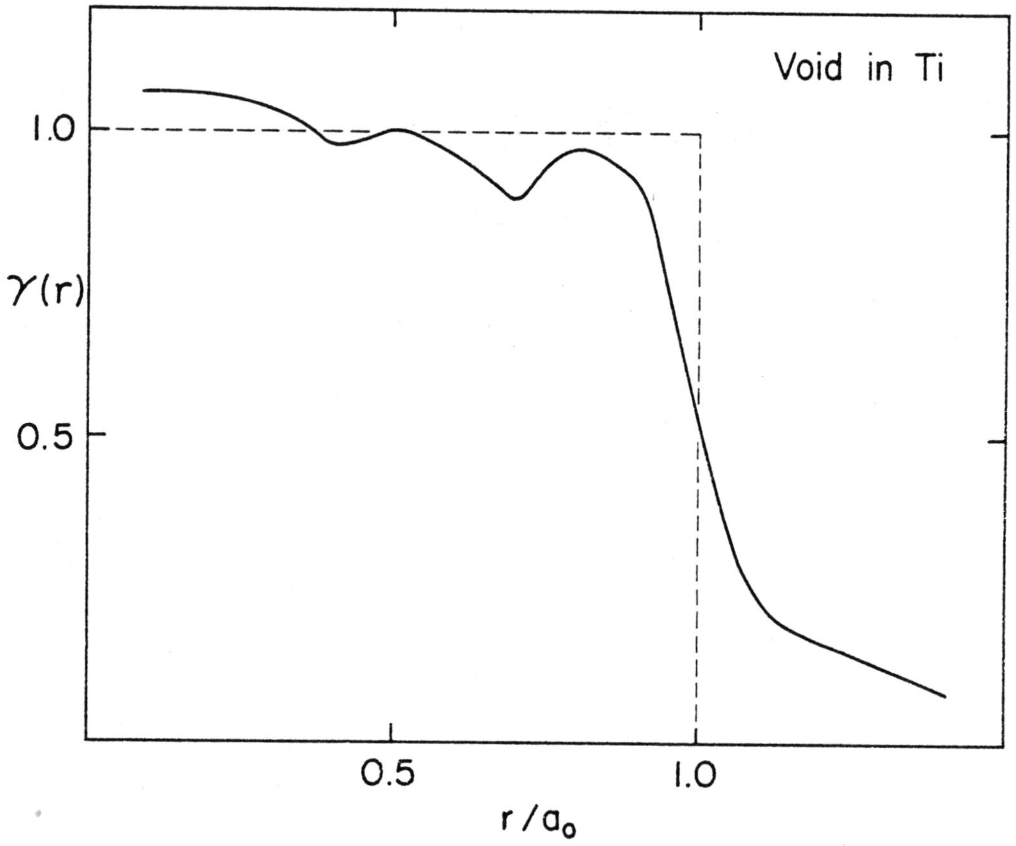

Fig. 2. The calculated characteristic function for a spherical void of radius a_0 in titanium. The result was obtained by inverting theoretical scattering amplitudes with a bandwidth $0<ka_0<10$.

INVERSION OF THEORETICAL DATA

The inversion algorithm was tested for three different flaws in titanium using data generated from theory. The first flaw was a spherical void with 0<ka<10. The second flaw was a 2-1 oblate spheroid with 0<ka<4 (where a denotes the semi-major axis). The third flaw was a 2-1 prolate spheroid with 0<kb<4 (here we define b as the semi-minor axis). The sphere data was generated using the exact theory of Ying and Truell and isotropic elastic constants for titanium. The T-matrix method was used to obtain the scattering amplitudes for the (400μ by 200μ) prolate and oblate spheroidal flaws.

The spherical flaw is considered first. Figure 2 shows the characteristic function obtained from the inversion procedure. Using the 50% point to define the boundary, we find that the radius is determined to within about 5%. We note that the inversion procedure was tested for sensitivity to noise for this spherical flaw and found to be quite insensitive (7).

The preliminary analysis of the spheroidal data is confined to an approximate determination of the semi-major and semi-minor axes using the simplified theory of section III. The simplified theory has the feature that a pulse-echo waveform along one of the axes can be used to determine the radius of an equivalent sphere with the radius of that axis. In figure 3 we show the characteristic function derived from the pulse-echo waveform measured along the semi-minor axis. Figure 4 is the equivalent result for the semi-major axis. Using these results we obtain estimates of 420μ and 210μ for these axes compared to the exact results of 400μ and 200μ.

Similar results for the oblate spheroid are 360μ and 210μ compared to exact values of 400μ and 200μ.

In section IV we calculated the characteristic function for the semi-minor axis of an oblate spheroid from experiment. In this section we computed the same result using scattering amplitudes obtained from the T-matrix method. We now compare both results (with a bandwidth 0<kb<2). The results are shown in Fig. 5, and the agreement is essentially exact.

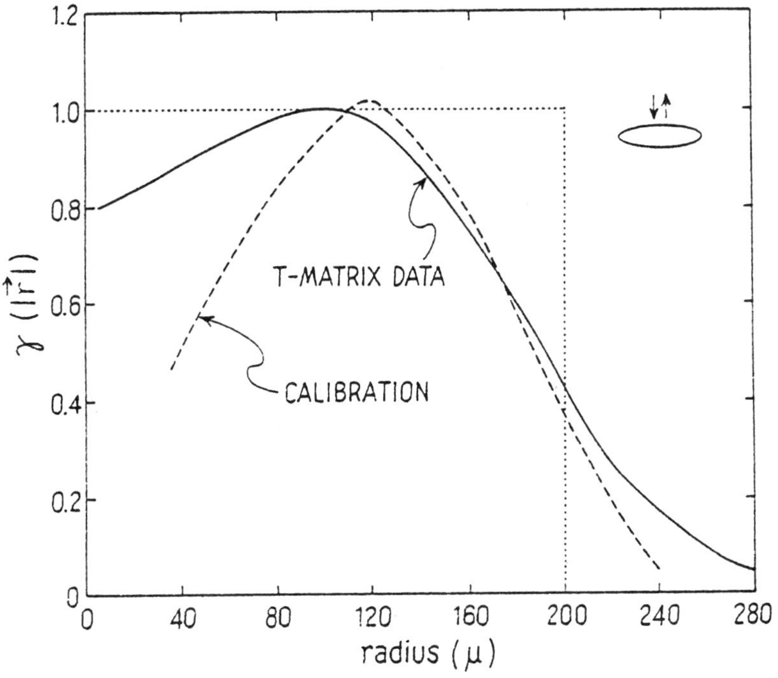

Fig. 3 Calculated characteristic function for the semi-minor axis of the prolate spheroid using theory data with a bandwidth of 0<kb<4.

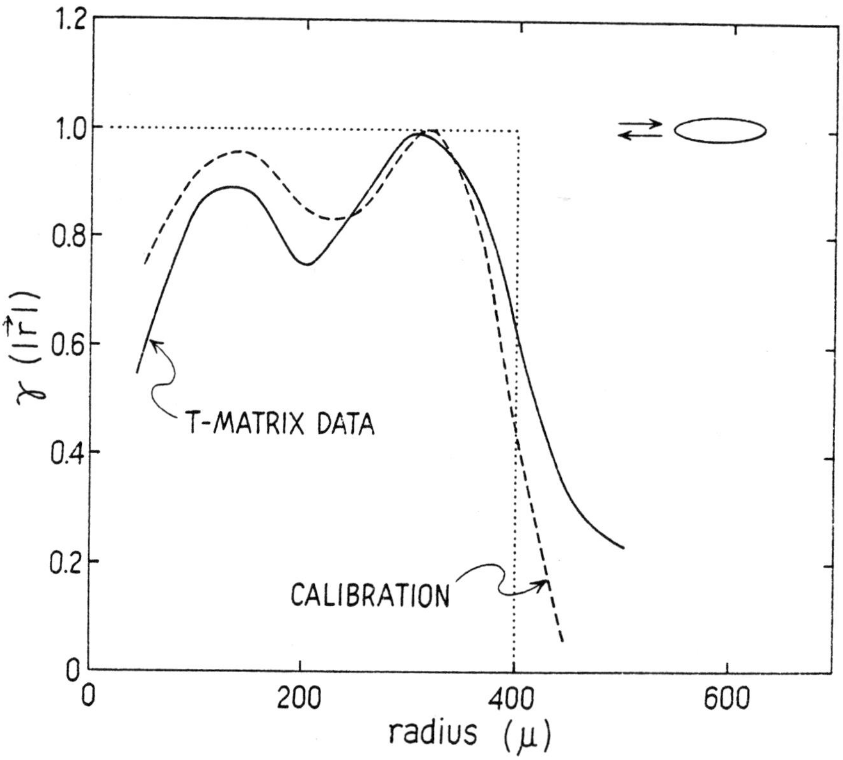

Fig. 4 Calculated characteristic function for the semi-major axis of the prolate spheroid using the theory data with a bandwidth of $0<ka<4$.

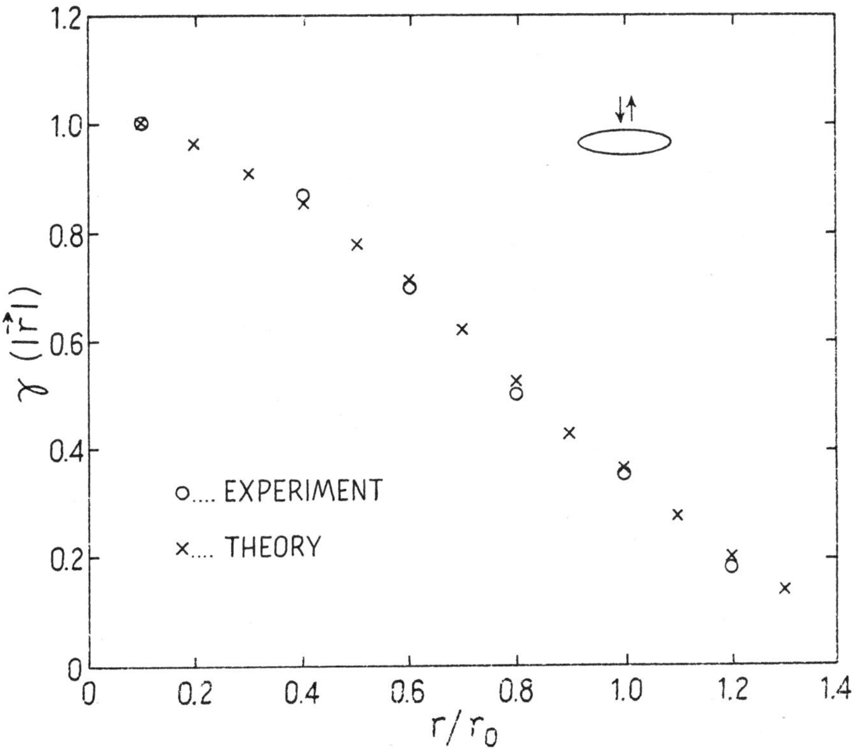

Fig. 5 The calculated characteristic functions for the semi-minor axis of the oblate. Theory and experiment are compared.

CONCLUSION

We have presented an inversion algorithm for the intermediate scattering regime when the size of the flaw is comparable to the wavelength of the ultrasound. Tests of the algorithm were performed for the case of spherical and spheroidal voids in titanium. Good results were obtained for the size and shape of the flaws. These results suggest that this algorithm may be of practical use for the non-destructive testing community in determining the characteristics of volume type flaws in various solids.

REFERENCES

*This research was supported in part by the Center for Advanced NDE operated by the Science Center, Rockwell International.

1. J. M. Richardson, IEEE Ultrasonics Symposium, Cherry Hill, New Jersey 759 (1978).
2. J. H. Rose and J. A. Krumhansl, J. Appl. Phys. 50, 2951 (1979).
3a. V. V. Varadan and Y.-H. Pao, J. Acoust. Soc. Am. 60, 556 (1976).
b. V. V. Varadan, J. Acoust. Soc. Am. 63, 1014 (1978).
c. V. V. Varadan and V. K. Varadan, J. Acoust. Soc. Am., in press.
4. J. E. Gubernatis, E. Domany, J. A. Krumhansl, and M. Huberman, J. Appl. Phys. 48, 2812 (1977).
5. J. E. Gubernatis, Los Alamos Scientific Laboratories Report LA-UR-771339 (1977), unpublished.
6. The characteristic function, $\gamma(\vec{r})$, is real which we insure by inverting only the real part of the scattering amplitude. This procedure for preserving the reality of γ is discussed in reference 7.
7. J. H. Rose, V. V. Varadan, V. K. Varadan, R. K. Elsley, and B. R. Tittman, to be published.
8. B. R. Tittman, E. R. Cohen, J. M. Richardson, J. Acoust. Soc. Am. 63, 68 (1978).
9. B. R. Tittman, R. K. Elsley, H. Nadler, and E. R. Cohen, to be published.

INVERSE SCATTERING AS A TARGET IDENTIFICATION PROBLEM

A. A. Ksienski and Heng-Cheng Lin
The Ohio State University ElectroScience Laboratory
Department of Electrical Engineering
Columbus, Ohio 43212

I. INTRODUCTION

The inverse scattering problem for an arbitrary target cannot be solved exactly in practice. The reason, quite apart from mathematical complexities, is that the scattering function must be known for an infinite range of certain parameters and consequently would require an infinite number of measurements (1). Practical approaches must compromise and restrict the measurements to an available range of the variables which are frequency and observation angle. The results are thus approximate, and require the interpretation of an observer. The observer, assuming a human, uses a priori information in terms of expected target shape and previous experience and declares the identify of the target. The goal of the observer is to correctly identify the target, i.e., to minimize the probability of a wrong decision. A similar process of identification can be done automatically much faster and perhaps with better results, more predictable results, and with considerably more flexibility regarding input data. A common example is the radar image of an airplane that is apparent only to the trained observer. It is possible to train an automatic classifier by feeding all the possible, or probable, radar images relating to the expected types of airplanes and thus duplicate the performance of the human observer with his experience. It is not necessary, however, to be restricted to visual displays or images for an automatic classifier to provide reliable target identification. Any set of features uniquely representing a set of targets will do. Indeed the problem may be treated in a fashion similar to the communication problem of message transmission. If a message is properly encoded and then transmitted through a noisy channel an optimum receiver can be designed to receive and decode the message with minimum probability of error. The target identification problem is more restricted in the encoding aspect since we have a limited access to the target via its radar (or acoustic) illumination and the resulting return. The optimum choice of the illumination parameters corresponds to the optimum communication code. But beyond this point, the signal reception and decision processor may be optimized using the well established techniques of decision theory which are capable of optimum use of any a priori knowledge regarding the targets and their environment and will produce the most reliable decisions consistent with the above conditions. It is also possible to include in the decision process certain cost functions such as is done in target detection, where desired false alarm probability versus detection probability are set in advance. Thus it may be considered more important not to err regarding some target identities as compared to others. Decision theoretic techniques have, indeed, been applied to pattern recognition problems (2), including the target identification (3,4).

The most difficult part of the target identification problem is feature selection since there are no analytical techniques that can be used to obtain an optimum set. The effectiveness of a feature set can only be evaluated after the fact and optimality often can only be tested by an exhaustive search. On the other hand the choice of appropriate features for target identification can be readily made if the physics of the problem is well understood. It is important to note that most pattern recognition techniques are best suited for statistical problems where the patterns are random and are specified in terms of their statistics. In the target identification case the targets are usually specified precisely and their responses to an electromagnetic or acoustic excitation is predictable within measurement errors. The problem is therefore deterministic and the random aspects are due to noise, measurement errors and unknown observation angles. Thus the object is to choose features that would provide reliable identification in the presence of corruptive influences, which is again similar to the communication problem or more precisely a hypothesis testing problem. The various hypotheses relate to the presence of the various possible targets. Given a set of relevant features the procedure then reduces to the choice of the most likely hypothesis based on the parametric values of the features. In the following sections we will discuss the choice of features, the technique used to identify targets based on these features, and the identification results obtained for both simple and complex objects.

II. FEATURE SELECTION

The features selected should be directly related to the physical properties of each target that distinguish it from other targets. Thus if targets differ in volume or size then the features should represent volume or size respectively. In order to assure successful identification one must derive an unequivocal relationship between a target response parameter and the volume or size of the target. Many attempts were made in the past to relate any set of measured responses to a set of targets and determine empirically a relationship based strictly on the data available, called a learning set. Such approaches usually fail since the relationship derived is likely to be accidental. This approach is appropriate for patterns specified statistically since the statistics can be learned from representative samples. This approach is inappropriate for a deterministic problem since there is usually much less room for error in that the targets are often quite similar and statistically would most likely appear as a single class. An example of this is the case for target identification from radar returns. Attempts to use signatures at high frequencies such as x band did not succeed since no inherent relationships were established for feature selection. On the other hand when low frequency scattering returns were used which were shown to be directly related to object sizes and shapes the results were very gratifying (3,4,5,6). In the above case of radar the frequency range of operation was chosen in view of a proven physical relationship. Other frequency ranges may be shown to be relevant to certain target characteristics and accordingly adopted. Once the frequency range is determined the specific frequencies chosen are subject to an optimization search. Of course the number of frequencies used is an appropriate parameter. It is, however, a very costly parameter and at least in the radar case its use must be minimized. Of course, the dimensionality of the feature vector should always be minimized since it simplifies both measurement as well as computational complexity. Examining the features available in the frequency domain, amplitude, phase and polarization are available at each frequency used. More succinctly expressed the complex scattering matrix providing six independent components is available at each frequency. The six components are the real and imaginary components at each polarization and the cross polarized components (4). The "cost" of each feature as well as its effectiveness vary. The amplitude feature is the easiest to obtain and corresponds to the conventional radar cross section, but it is limited in its effectiveness. The phase information is very useful when combined with the amplitude in producing a highly reliable identification, as will be shown below. It is, however, much harder to extract the phase information in addition to requiring a coherent radar, as compared to the amplitude for which an incoherent radar is adequate. Polarization is a very desirable feature since it tends to compensate for observation angle dependence and target orientation. Most targets exhibit polarization dependence due to their shape. For example, a long relatively thin object, such as a missile or airplane fuselage will exhibit a much stronger response to a signal polarized parallel to its long dimension as compared to one polarized perpendicular to it. Having both polarizations assures a relatively strong return independently of the orientation of the objects with respect to the radar.

An alternate set of features to the frequency domain may be obtained by time sampling of the return signal and utilizing the samples as features. Theoretically the time and frequency domain representations of the return signal are equivalent, but the effectiveness of a subset of time samples vs frequency domain features is quite different especially when the subset is small.

Another set of features, quite attractive in principle, are the poles, or resonances, of the object (7) which could represent the object independently of its observation angle. Also the dominant poles are quite small in number, assuring a small dimensionality for the feature vector. So far, however, it has been difficult to extract the required poles with sufficient accuracy to provide reliable target identification.

As may be inferred from the above discussion various sets of features have been tried and the frequency domain features located in the low frequency range have been selected as the most effective ones. Comparative studies were made of the effectiveness of various subsets of features and the results will be presented in section IV.

III. CLASSIFICATION

The classification, or identification, process utilizes the selected features and bases its decision on the numerical values of these features as observed by the receiver. As mentioned in the introduction the process of feature selection is somewhat analogous to the encoding process in communication. It inherently determines the locations of the signals or target returns in feature space or signal space. In the communication encoding process the attempt is made to locate each symbol as far as possible from other symbols so as to minimize the chance that noise and errors will push one symbol into the region of another, thus causing a decision error. The location in feature space of the different targets is represented by continua of points corresponding to the values of the scattering parameters for say the polar observation angles θ and ϕ. They thus form a two dimensional surface in hyperspace of a dimensionality equal to the number of features used (8). It is obvious that it is not possible to locate these distributions corresponding to different targets at arbitrary distances or locations in feature, or signal, space. Indeed they may not even be separable, that is, two different objects may produce the same scattering returns for one or more observation angles. This is particularly true when the targets are similar in shape, such as in airplane identification. If the targets are separable in feature space it is possible to attain arbitrarily low identification errors when the noise and measurement errors are sufficiently low. Targets that are relatively simple in shape and different such as a cube and a sphere, or a cone and a cylinder, can be shown to be separable, indeed linearly separable (see section IV). There are various transformations (2) mostly non linear that can place the various targets at large separations in a transformed space, thus apparently capable of suppressing classification errors due to noise. Such transformations are most often misleading in that they also distribute the noise and measurement errors over large regions of the transformed space and provide no advantage for the target identification problem. If properly treated, however, some transformations can be obtained that do improve performance (9). The difficulties due to the inherent similarities between objects cannot be overcome by data manipulations. In fact the selection of an optimum classification algorithm is far less important than the choice of the appropriate features. This applies particularly to deterministic problems as contrasted to statistical ones. As the data representing the various targets approach each other (for basically similar targets) it becomes more difficult to effect a global separation and, if possible, one should resort to local discrimination. For example, in the case of aircraft in flight it is possible to determine with fair accuracy the aircraft orientation with respect to the radar. This information permits the identification of the aircraft based on local rather than global differences between the scattering characteristics of the alternate aircraft considered. The orientation information reduces substantially the misclassification probabilities and may eliminate class overlap which puts a non zero lower bound on misclassification probability. The above is an example of the importance of utilizing any a priori information available for the classification process.

The classification algorithms used to produce the results shown in section IV were the linear discriminant and the Nearest Neighbor Rule. The linear discriminate (2) searches for a hyperplane in the multidimensional feature space that would separate the classes and then provide a linear decision surface permitting the identification of an observed target return by its location with respect to the hyperplane. An optimum choice of such a hyperplane is one that minimizes the average probability of misclassification. The linear discriminant is straightforward to apply and permits easy computation of the resulting misclassification probabilities. The Nearest Neighbor Rule (2) as its name implies, classifies an unknown target return to a class to which its nearest neighbor belongs. The Nearest Neighbor Rule is a nonparametric classifier in that is does not require knowledge of the signal or noise statistics and is by its very nature a local test. The Nearest Neighbor Rule is thus ideally suited for discrimination of objects of similar overall shapes and only local variations. The next section presents the classification performance for both simple and complex shapes.

IV. PERFORMANCE RESULTS

The classification performance for simple shapes has been previously reported (3,4,9) and only a limited sample will be shown here, mostly for convenience. As to complex shapes, such as airplanes, the performance for four airplanes of similar size has been published (3,4). The results shown here involve a much more extensive group of aircraft in both size and shape varying from a Mig 19 to a B1. The effect of feature selection as well as dimensionality will be demonstrated.

The method of selection of training sets and test sets is as follows: The "training sets" consist of the scattering data obtained either by measurement or computation. In the case of simple shaped objects such as a sphere, cube, cone, etc., the data were obtained by measurement. In the case of airplanes the complex scattering matrix was obtained by computation (10). The training sets provide the information that is needed to choose the appropriate hyperplane in the case of the linear discriminant classifier, and the class representation proximity which will determine class belonging of an unknown target in the Nearest Neighbor classifier. The "test sets" representing the "unknown" are obtained by adding to the training set gaussian noise whose variance is proportional to the scattering cross section of the target. The constant of proportionality is varied and the misclassification probability is plotted as a function of the noise level injected. Thus for σ=0.1 the noise standard deviation is 10% of the average level.*

*The only exceptions to the normalization of noise are the first two figures where the noise standard deviation is given in terms of centimeters which compares directly to the square root of the scattering cross section.

The noise model representing the noise and errors of a radar system should have both additive and multiplicative components. The additive components correspond to corruptive influences such as thermal noise and the multiplicative components reflect such effects as measurement errors and clutter. Because of the mathematical complexity introduced by multiplicative noise, the noise model chosen here is additive only but represents the multiplicative components, which are signal level dependent, by normalizing the noise variance to the signal power level. This normalization of the noise further removes such parameters as range to the target, antenna gains, transmitter power, etc. from affecting the relative strengths of noise vs signal. The statistical nature of the noise model is gaussian due to the assumption that several relatively independent sources of noise and error contribute to the total corruptive influence; invoking the central limit theorem their sum tends to a gaussian distribution.

It should be borne in mind that the precise nature of the noise is less important than its effect on the classification performance. Thus no attempt is made to combat the noise effect by devising schemes taylored to its statistics. The noise serves strictly the purpose of determining the degradation of performance with increasing noise levels, and the assessment of such performance as the selected feature sets are varied.

The classification performance of "simple objects" is represented by, a cube 2.1 cm on the side, a hemispherical boss of diameter 2.385 cm, a 60^0 cone with a base of 4.3 cm, a prolate spheroid 4.77 cm by 2.38 cm., a square plate 2.1 cm on the side and .006 cm thick, a wire .015 cm in diameter and 3 cm long, and 3 spheres, 2 metal ones 2 cm and 3 cm in diameter and a dielectric one having a diameter of 2 cms and a relative dielectric constant $\epsilon_r = 2.208$. All of the objects except for the last one were metal (highly conducting). The features used were amplitude returns at 10 harmonically related frequencies from 1.08 GHz through 10.8 GHz, thus the wavelengths varied from about 2.7 centimeters to 27 centimeters. The sizes of the objects in terms of wavelengths were from a tenth of a wavelength to one wavelength. This frequency range was found to provide the relevant information of a target's size and gross shape characteristics (3).

The scattering measurements were taken at 10^0-15^0 increments in θ and ϕ polar coordinates. The grid density was chosen in each case to assure that any significant changes in scattering parameters were represented. The performance shown in Fig. 1 displays the probability of misclassifying one of the objects, a hemispherical boss from each one of the others. It is representative of other pairwise classifications of the objects listed. As can be seen the probability of misclassification is quite low indicating a good reliability of identification when only alternative targets are possible, namely pairwise classification. When any one out of $K(>2)$ classes are potentially present it is possible to combine the K classes into two sets, one containing only one class and the other consisting of the remaining $(K-1)$ classes. This procedure reduces a multiclass problem to one of pairwise classification, but may cause difficulties. The classes grouped together may not be sufficiently similar to each other, or may not as a group be distinct from the remaining object. It can be seen from Fig. 2 that the misclassification probabilities indeed depend on the groupings and that some do not produce very satisfactory results. The difficulties encountered in the multiclass case do not reflect the close proximity or inseparability in feature space of the objects considered but rather the particular method used for classification. As discussed above, the linear discriminant approach, although straightforward and simple to implement and evaluate, is not very effective unless the different object classes can be easily separated by hyperplanes.

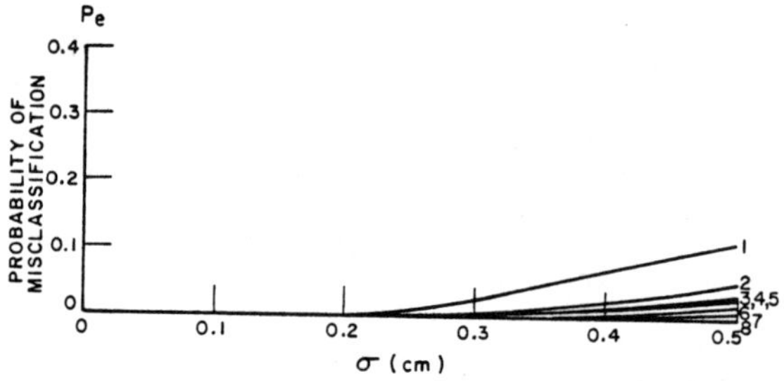

1. half sphere vs. spheroid
2. half sphere vs cube
3. half sphere vs plate
4. half sphere vs 2 cm dielectric sphere
5. half sphere vs 60^0 cone
6. half sphere vs thick wire
7. half sphere vs 3 cm sphere
8. half sphere vs 2 cm sphere

Fig. 1. Average probability of misclassification for pairwise classification by linear discriminant using amplitudes at 10 frequencies.

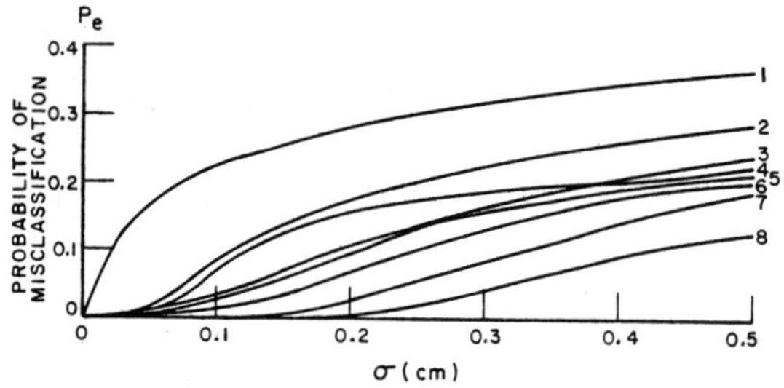

1. 3 cm sphere vs all others
2. plate vs all others
3. cube vs all others
4. 2 cm dielectric sphere vs all others
5. thick wire vs all others
6. half sphere vs all others
7. 60^0 cone vs all others
8. 2 cm sphere vs all others

Spheroid was found not to be linearly separable from the rest combined as one class

Fig. 2. Average probability of misclassification by a linear discriminant where 8 classes are combined together into 1 class and separated from remaining class, using amplitudes at 10 frequencies.

In the present case of multiple classes, such linear separation may not be possible even though single classes do not overlap. Here the more effective approach is the nearest neighbor rule whose effectiveness is demonstrated in Fig. 3. As can be seen, all groups are separable including the spheroid vs all other objects (curve 9). Note that the value of σ is not in centimeters, but has been normalized to the average of the object's signal return which is given as $\overline{A}$ = 3.48 cm. Thus the error level corresponding to σ = 0.5 cm in Fig. 2 corresponds to σ = 0.144 in Fig. 3 indicating negligible error probabilities even in the case of "spheroid vs all others".

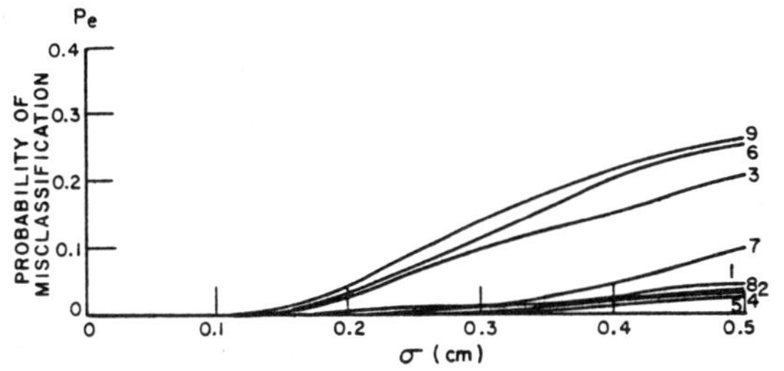

1. 3 cm sphere vs all others
2. plate vs all others
3. cube vs all others
4. 2 cm dielectric sphere vs all others
5. thick wire vs all others
6. half sphere vs all others
7. 60^0 cone vs all others
8. 2 cm sphere vs all others
9. spheroid vs all others

Fig. 3. Average probability of misclassification by N.N. classifier where 8 classes are combined together into 1 class and discriminated from remaining class, using amplitudes at 10 frequencies, $\overline{A}$ = 3.48.

From the above results it is apparent that relatively simply shaped objects can be reliably identified using the 10 frequency amplitude features selected. Although the linear discriminant is not capable of handling the multiclass problem the Nearest Neighbor classifier does provide highly reliable identification even in the presence of considerable amounts of noise and/or errors. The success of the Nearest Neighbor classifier is due to the fact that it considers the various alternative classes in the immediate neighborhood of the unknown point. Thus, as long as each class occupies a disjoint part of feature space reliable identification can be achieved with the precise error probability dependent on signal to noise ratio. Note that no a priori information of observation angle was required for the above shapes. This global separability or disjointness in feature space may not hold for basically similar but complex objects such as airplanes to be considered next. The global inseparability is particularly true when the dimensionality of the feature vector is lowered. Since the frequency parameter, as mentioned above, is the most difficult to come by, the attempt was made to use first a single frequency and utilize not only amplitude but phase and polarization as well. The objective was to demonstrate the classification performance with varying features. The most readily available is the amplitude of the return signal, the next is the polarization diversity. One can present the performance for a fixed linear polarization but its utility is marginal since the relative orientation of the target with respect to the target is unknown and changing with time, resulting in variable incident and reflected polarizations. It was thus decided to assume that both horizontal and vertical polarizations are available and that both are used. Thus the first level of feature complexity to be considered is single frequency amplitude returns at two orthogonal polarizations. The next level involves the introduction of phase information which is used to obtain the quadrature components of the signal. At this stage the feature vector is four dimensional. With an additional frequency the dimensionality can rise to eight. The set of airplanes used for testing the classification performance were: Mig 19, Mig 21, Mig 25, F-104, F-4, F-14, SR-71 and B-1. The scattering data were obtained by computation (10), and the range of frequencies used was from 2 MHz to 24 MHz in increments of 2 MHz. The target sizes ranged from a 13 m fuselage to a 43 m one and wing spans varied from 6 m to 23 m. Thus the target sizes ranged from the order of a wavelength to a small fraction of wavelength, which was consistent with the optimum choice of frequency range. The complex scattering matrix was computed for an observation angle grid of 5° increments in the polar coordinates θ and ϕ, and covered the complete observation sphere. Some information about the aspect angle of the aircraft was assumed to be available from the doppler and range data. These combined with known airplane flight dynamics can provide very close estimates of observation angle. The estimate was assumed to be good only to within $\pm 5^o$ in θ and ϕ. Thus an angular region of 10° in θ and ϕ centered on selected points of the observation sphere were tested such as nose on, side view, bottom view, etc. The scattering matrices for grids of points at $\pm 5^o$ in θ and ϕ for all targets considered were then used as the "learning set" or class representations in feature space. The test sets were obtained by adding to the above data random noise of varying amounts and computing the resulting misclassification probabilities. The level of the noise was normalized to the average scattering return as discussed above. The error probabilities using a Nearest Neighbor classifier were computed by Monte Carlo simulation, since the analytical calculations would involve multidimensional integrations over very complicated boundaries. In the Monte Carlo approach used, 600 random numbers were generated in each case and added to the noise free signals to form test vectors. These were then classified by the Nearest Neighbor rule and the number of misclassification were divided by the number of total trials to provide an estimate of misclassification probability.

In the figures below the following symbols are used to represent the various airplanes: C_1 - F104, C_2 - Mig 19, C_3 - F4, C_4 - Mig 21, C_5 - Mig 25, C_6 - SR71, C_7 - B1, C_8 - F14. As can be seen from Fig. 4 the performance varies substantially with the individual airplane. The curves indicate that when the test sample is a noisy return from an F4 (C_3) it is much more likely to be misclassified than one returned from a B1 (C_7). This may be expected since the B1 is substantially larger than most of the group and will be located in a different portion of feature space, thus being less susceptible to noise and errors. This is particularly true for the lower frequencies where the scattering amplitude is proportional to the target's size (Rayleigh Range). The overall performance is not too impressive, but only single frequency amplitudes are used, namely two features (for the two polarizations). The frequency f_2 is 4 MHz, the subscript indicates the multiple of the fundamental which is 2 MHz. The response at this frequency was found to produce the best performance. Frequency selection is very important as can be seen by comparing to the performance displayed in Fig. 5 at f_7 = 14 MHz. Note that C_1 approaches almost unity error for high levels of noise while C_2 keeps declining with increased noise. This phenomenon

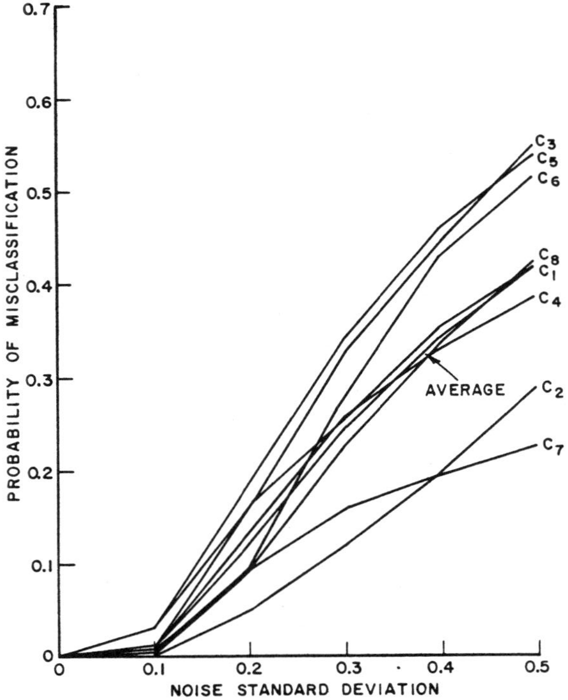

Fig. 4. Probability of misclassification for individual aircraft, using amplitude returns at frequency f_3 and $(0^0,0^0)$ aspect angle (nose on).

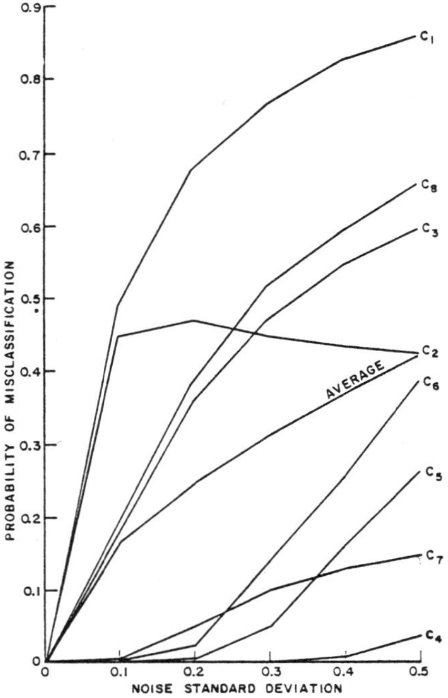

Fig. 5. Probability of misclassification for individual aircraft, using amplitude returns at frequency f_7 and $(0^0,0^0)$ aspect angle (nose on).

is known as bias, which may occur when one class if surrounded by one or more other classes in feature space. Thus as noise is added to the surrounded class its elements are "captured" by the surrounding classes. The reverse happens for a surrounding class and its error probability will drop beyond a certain noise level (11). A comparison of the average error probability for the best four frequencies is shown in Fig. 6. For further comparison another observation angle performance is shown in Fig. 7. Note that performance differences are very substantial even among the best frequencies. Here again f_2 is the optimum frequency.

A very dramatic improvement in performance can be attained by introducing phase information, i.e., utilizing the two quadrature components of the complex return. This is demonstrated in Fig. 8 which presents the misclassification probabilities for the individual airplanes at the optimum frequency f_{12}. It can be seen that even for $\sigma = .3$ where previous errors were extremely high the present errors are negligibly small and even for $\sigma = .5$ the errors are relatively small. The optimum frequency was found to be at the opposite end of the frequency range from the previous cases where amplitudes only were used. The relative performance of the various frequencies for amplitudes alone as well as for complex features is shown in Fig. 9. The average probability of misclassification is seen to be drastically lower for complex returns than amplitudes alone for all except the lowest two frequencies where the performance is approximately the same. The reason for lack of improvement at the low frequencies is due to the fact that in the low Rayleigh range of frequencies the target's scattering phase is very small and has little if any informational content. Utilizing both quadrature components may indeed degrade somewhat the performance due to the introduction of additional noise components in the higher dimensional feature vector. As the phase becomes more significant at increasing frequencies the error is progressively reduced. Note that this improvement in performance with frequency does not occur for the amplitude features. In fact the performance progressively deteriorates as the frequency rises above 4 MHz. The reason for this behavior can be seen from the variation of the scattering amplitudes with frequency for the various airplanes depicted in Fig. 10. As can be seen there is a substantial spread amplitudes among most of the various classes at frequencies below 8 MHz. As the frequency rises above 6 MHz the largest targets B1 (C_7) and SR71 (C_6) go through their first resonance and their return amplitudes start decreasing. From then on, i.e., 8 MHz and up the spread of amplitudes fluctuates and is reflected in the error fluctuation in curve A of Fig. 9.

A comparative evaluation of the various features is summarized in Fig. 11. It is clear that the number of features is not the most important parameter since curve 2 represents the same number of features as curve 3 yet it is far superior in performance. Thus the phase plays a very important role in the identification process (4). The curve No. 4 representing the utilization of two frequency complex returns shows extremely low error probabilities and indicates that the introduction of additional frequencies beyond two would not be necessary, if phase information is available.

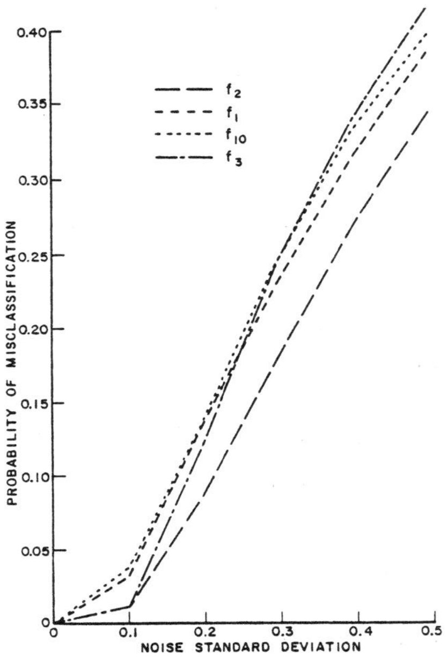

Fig. 6. The average performance at different frequencies, using single frequency amplitude returns. The observation angle is $\theta=0^o$, $\phi=0^o$.

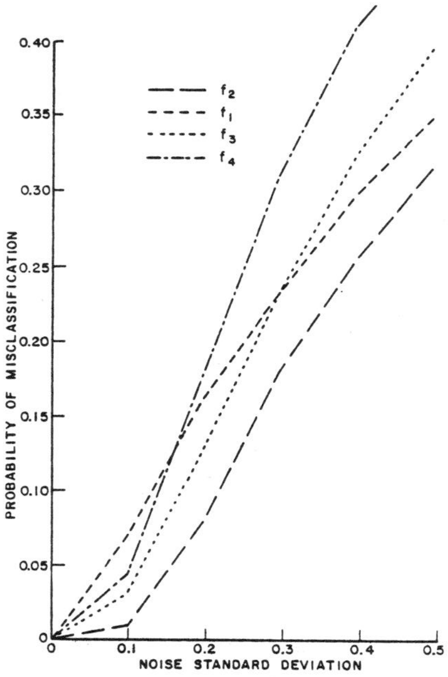

Fig. 7. The average performance at different frequencies, using single frequency amplitude returns. The observation angle is $\theta=90^o$, $\phi=0^o$.

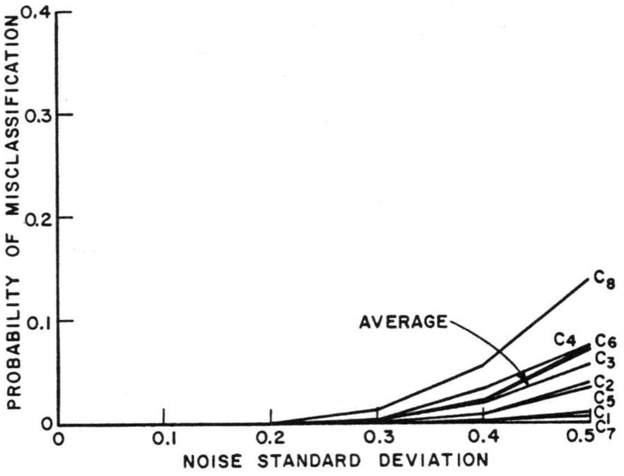

Fig. 8. Probability of misclassification for individual aircraft, using complex returns at frequency f_{12} and $(0^o,0^o)$ aspect angle (nose on).

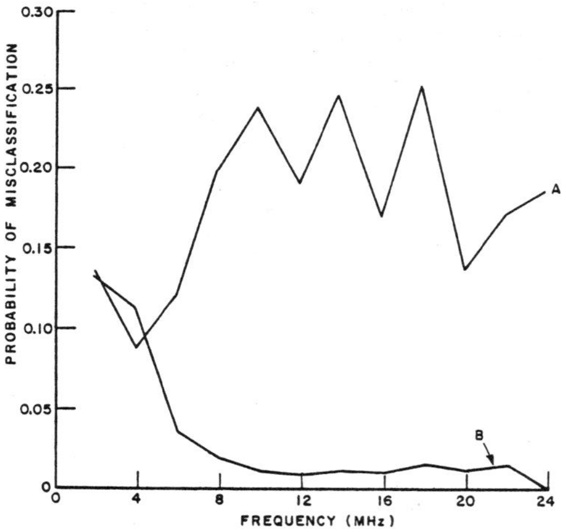

Fig. 9. Comparison of average probability of misclassification, using 3 different sets of features for the eight aircraft at $\theta=0^o$, $\phi=0^o$. The noise added to each signal is 20% of noise free signal.
A = single frequency amplitude returns.
B = single frequency complex returns.

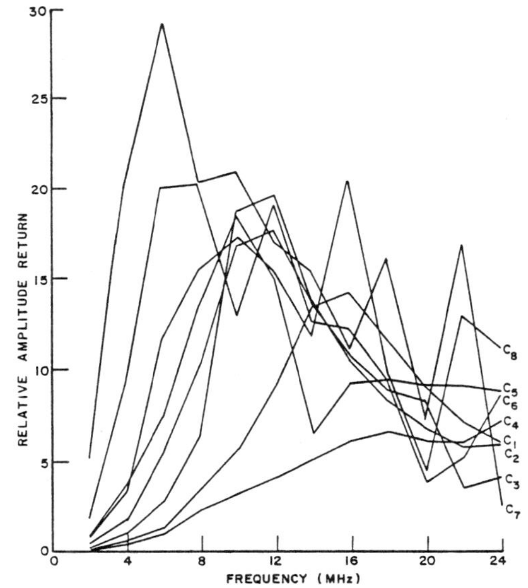

Fig. 10. Amplitude returns of horizontally polarized wave for eight aircraft at $\theta=0^0$, $\phi=0^0$.

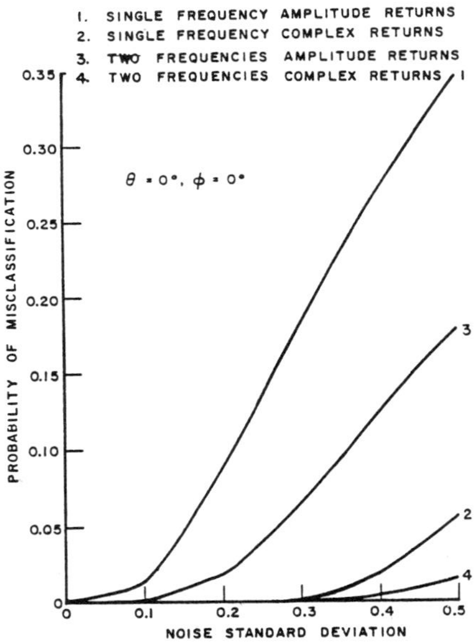

Fig. 11. Performance comparison for different combinations of features at $\theta=0^0$, $\phi=0^0$ employing optimum set of frequency returns.

V. CONCLUSIONS AND COMMENTS

The above discussion and results indicate that reliable target identification is possible utilizing a limited amount of data. The quality of the results, or level of decision reliability regarding the target's identity are dependent on the system's sophistication, such as coherence, phase estimates, frequency agility, and of course signal to noise ratio. Of major importance is also the extent of a priori information available regarding the target such as airplane orientation.

One of the most time consuming tasks in designing a target classifier is the selection of optimum features. Even when the physical characteristics upon which identification is based are well understood, there are still many alternate feature sets to choose from. As was shown the difference in performance may be very substantial between optimum and suboptimum choices of features. The selection process often involves an exhaustive search which may be costly in computer time. The classification process itself, however, when the features have been selected, is quite rapid and easily implementable in real time.

One problem which should be mentioned involves the case of the unexpected target. The assumption is usally made in automatic target identification is that a list of all possible classes, or targets, is available and that the classifier will pick the one most likely to be present, based on received data. The list, however, may not be exhaustive since it may become excessively long. Thus the relatively unlikely targets would not be listed and if one did appear it would be wrongly classified as a member of the list. The problem of the unexpected target may be particularly serious in a situation where the a priori information on possible target shapes is limited because of innovations or secrecy. A satisfactory solution to the problem has been recently obtained (12). The technique provides the alternative of an unlisted target and institutes a search for its physical characteristics if one is detected. The technique can be easily incorporated in conventional classifiers and does not degrade perceptibly the overall performance.

ACKNOWLEDGMENT

The work reported in this paper was supported in part by Grant AFOSR-74-2611 between Air Force Office of Scientific Research and The Ohio State University Research Foundation.

REFERENCES

1. R. M. Lewis, "Physical Optics Inverse Diffraction," IEEE Trans. Antennas Propagat. vol. AP-17, pp. 308-313, May 1969.

2. Richard O. Duda and Peter E. Hart, "Pattern Classification and Scene Analysis," Stanford Research Institute, Menlo Park, California, A Wiley-Interscience Publication.

3. Aharon A. Ksienski, Yau-Tang Lin and Lee James White, "Low-Frequency Approach to Target Identification," Proceedings of the IEEE, Vol. 63, No. 12, December 1975.

4. Y. T. Lin and A. A. Ksienski, "Identification of Complex Geometrical Shapes by Means of Low-frequency Radar Returns," The Radio and Electronic Engineer, Vol. 46, No. 10, pp. 472-486, October 1976.

5. J. D. Young, "Radar Imaging from Ramp Response Signatures," IEEE Transactions on Antennas and Propagation, Vol. AP-24, No. 3, May 1976.

6. David L. Moffatt and Richard K. Mains, "Detection and Discrimination of Radar Targets," IEEE Transactions on Antennas and Propagation, Vol. AP-23, No. 3, May 1975.

7. C. W. Chuang and D. L. Moffatt, "Natural Resonances of Radar Targets Via Prony's Method and Target Discrimination," IEEE Transactions on Aerospace and Electronic Systems, Vol. AES-12, No. 5, September 1976.

8. L. J. White and A. A. Ksienski, "Aircraft Identification Using a Bilinear Surface Representation of Radar Data," Pattern Recognition, Pergamon Press, June 1974, Vol. 6, pp. 35-45.

9. A. G. Repjar, A. A. Ksienski and L. J. White, "Object Identification from Multi-Frequency Radar Returns," The Radio and Electronic Engineer, Vol. 45, No. 3, March 1975.

10. Y. T. Lin and J. H. Richmond, "EM Modeling of Aircraft at Low Frequencies," IEEE Transactions on Antennas and Propagation, Vol. AP-23, No. 1, January 1975.

11. Heng-Cheng Lin and A. A. Ksienski, "Optimum Frequencies for Aircraft Classification," Report 783815-6, January 1979, The Ohio State University Electro-Science Laboratory, Department of Electrical Engineering; prepared under Grant AFOSR-74-2611 for Air Force Office of Scientific Research.

12. Heng-Cheng Lin, "Identification of Catalogued and Uncatalogued Classes," Report 783815-7, December 1978, The Ohio State University ElectroScience Laboratory, Department of Electrical Engineering; prepared under Grant AFOSR-74-2611 for Air Force Office of Scientific Research.

APPROXIMATE IMAGE RECONSTRUCTION FROM TRANSIENT SIGNATURE

J. D. Young
The Ohio State University ElectroScience Laboratory
Department of Electrical Engineering
Columbus, Ohio 43212

ABSTRACT

A technique for generating a target image utilizing electromagnetic transient response signature data is explained and demonstrated. Geometry-related features of the time domain ramp response signature are discussed, and several measured waveforms are presented. The imaging algorithm is discussed. The technique combines features of physical optics inverse diffraction and feature-space target identification approaches. Recent advances including a linear direct search iteration process for optimizing the estimated image quality are discussed. Finally, example images using just 2 look angles separated by 30°, a severely underspecified but practical case, are presented. Possible generalization of this approach is briefly discussed.

I. INTRODUCTION

This paper describes recent progress in the investigation of techniques for generating the image of a radar target from measured transient signature information. Specifically, a spatial reference frame is assumed, and this frame is fixed with respect to the interrogating radar system(s). A single target is assumed to be quasi stationary within this system, centered at the origin. Backscattered signature information is obtained at a small set of directions within the reference frame. Based on these data, a three-dimensional approximate surface is to be reconstructed, from which images for any designated look angle may be produced.

A large body of research concerning Physical Optics Inverse Diffraction is pertinent to this study. In a series of reports Norbert Bojarski (1,2,3,4,5) has formulated a relationship between the monostatic scattered far field cross section of perfect conductors and the geometry of the conductors. The results are based on the physical optics approximation.

If the scatterer is expressed in three dimensions by its characteristic function $\gamma(\overline{x})$ where,

$$\gamma(\overline{x}) = \begin{cases} 1 & \text{inside the body} \\ 0 & \text{outside the body} \end{cases} \tag{1}$$

and if the scattered far field cross section $\sigma(\overline{p})$ is used to define $\Gamma(\overline{p})$ where,

$$\Gamma(\overline{p}) = \sqrt{4\pi}\ \frac{\rho(\overline{p}) + \rho^{*}(-\overline{p})}{|p|^{2}} \tag{2}$$

with $(\overline{p})$ = the field cross section

$\overline{p} = \frac{2\omega}{c}\,\overline{a} = 2\overline{k}$ $\left(\frac{\text{cycles}}{\text{meter}}\right)$

$\overline{k}$ = wavenumber propagation vector in the direction of aspect;

then Bojarski has shown that

$$\gamma(\overline{x}) = \frac{1}{(2\pi)^3} \int_{-\infty}^{\infty} \Gamma(\overline{p})\, e^{i\overline{p}\cdot\overline{x}}\, d^3\overline{x} \quad (3)$$

which is just a 3-dimensional inverse Fourier transform of $\Gamma(\overline{p})$.

Several other authors have made important studies and extensions of this work (6,7,8,9,10). This paper treats two important questions, both of which deserve more study.

1. What is the validity and accuracy of the physical optics assumption on which this technique is based?

2. The three-dimensional Fourier Transform relationship implies that scattering data at all look angles and for all frequencies is necessary for rigorous reconstruction of the target. What can be done when the range of possible look angles is severely restricted (to within 30° of nose-on, for example)?

Section II describes a set of measurements of the complex scattering cross-section over a 60:1 bandwidth for several cone targets. From these data, the Ramp Response Signature is generated. This signature, first described by Kennaugh (11) can be shown to be the Fourier transform of the $\Gamma(\overline{p})$ discussed by Bojarski for a given look angle. Comparison between the measured ramp responses and the actual target profile functions (cross sectional area vs distance) is made.

Section III discusses the imaging of these cone targets. The philosophy of the process goes beyond the inverse diffraction, making use of some a priori geometrical shape information and some target identification concepts. Information having a strong correlation with the target shape is extracted from only certain portions of the time domain signatures. A set of moldable shapes is postulated, which embody assumptions on the target shape. The moldable "limiting surfaces" are related to the signature shape information, and are also controlled by several process parameters. There is a limiting surface for each look angle. The image consists of the surface defined by the intersection of the limiting surfaces. An automated iterative process adjusts the process parameters of the limiting surfaces until the agreement between the images and all input signature information is optimized. The process is demonstrated using just two look angles (0° or nose-on and 30°) for a cone target.

Some general conclusions from this work are contained in Section IV.

II. MEASURED RAMP RESPONSE SIGNATURES

A set of cone target shapes shown in Fig. 1 have been measured and imaged. The Ramp Response Signature for aspect angles from 0° (nose-on) to 30° was generated.

Figure 2 summarizes the geometry and the definition of the canonic time-domain scattering signatures, including the ramp response (12). Given an incident plane wave with its space-time variation respresented by $\overline{E}_i(x,y,z,t)$, a backscattered time domain signal $\overline{E}_s(r,t)$ is observed at a far-field observation point P, at a distance r from the target. The time-domain signature waveform is the retarded-time scattered field value, multiplied by a far-field normalization factor 2r/c. Of course, a complex frequency-domain spectrum is associated with each time domain signature. If the incident signal is an impulse, then the Impulse Response $F_R(t)$ and its impulse response spectrum $F(j\omega)$ are obtained. For the normalization chosen, the frequency-domain backscattered cross-section

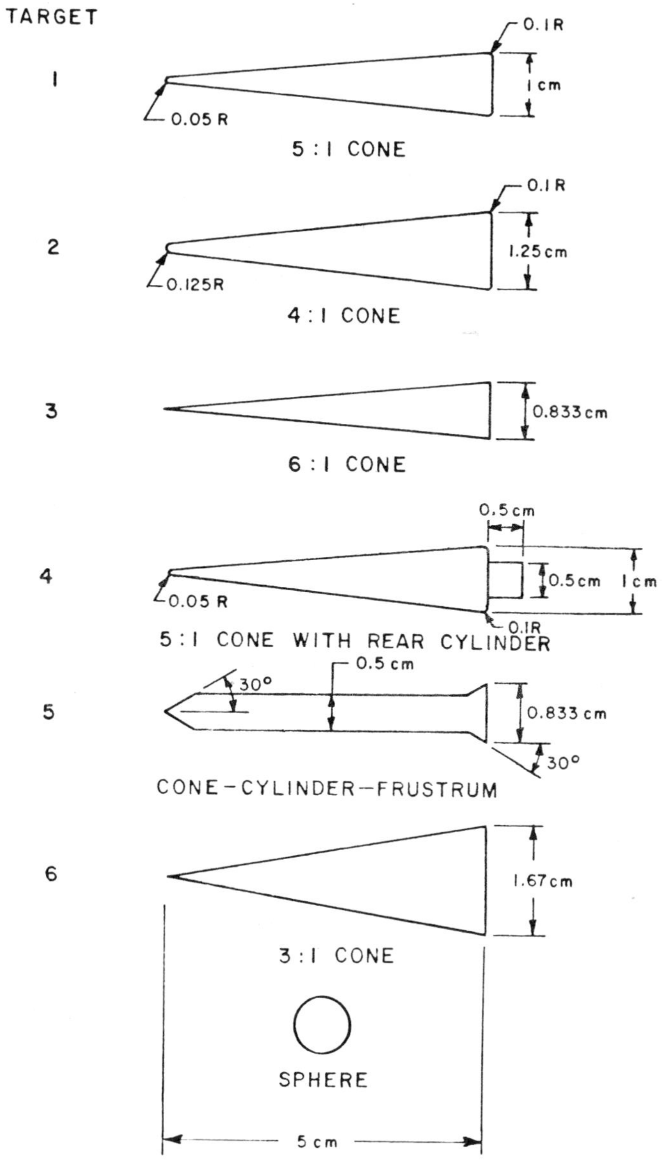

Fig. 1. Outlines of cone targets studied.

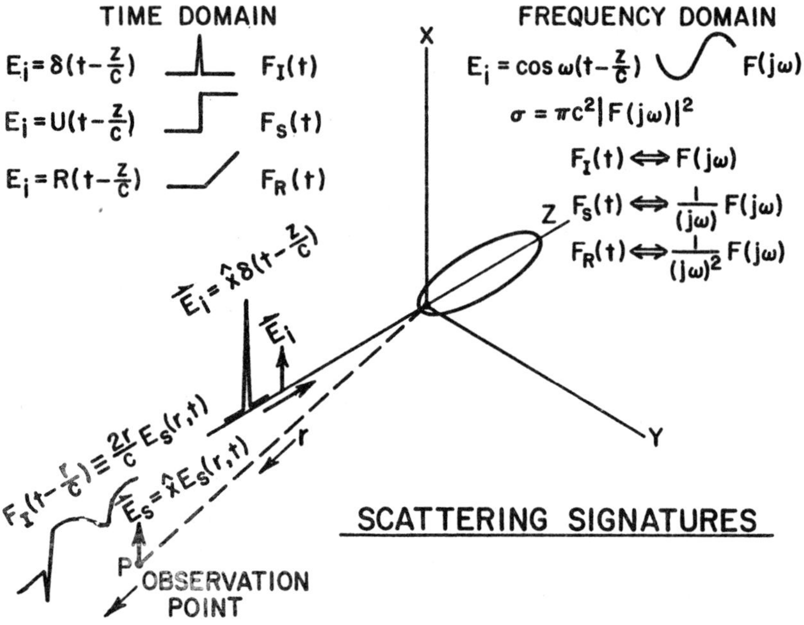

Fig. 2. Geometry for transient signature definitions.

$$\sigma(j\omega) = \pi c^2 |F(j\omega)|^2.$$

The step response is the first integral of the impulse response. The ramp response is the second integral of the impulse response.

There are two important features of the ramp response which make it attractive for identifying the target shape. First, Rayleigh showed that

$$F(j\omega) = \alpha_0 + \alpha_1(j\omega) + K(j\omega)^2,$$

and $\alpha_0 = \alpha_1 = 0$, $K \propto$ target volume. Thus $\int_{-\infty}^{\infty} F_R(t)dt = K \propto$ target volume. Second, physical optics analysis using the Kirchoff current approximation predicts that

$$F_R(t) \propto A(t),$$

where the profile function A(t) is defined as the cross-sectional area intercepted by a transverse plane, moving at a velocity of one-half the speed of light in the same direction as the incident field.

Transient response data can be obtained by direct time-domain measurement, or by complex cross section measurements over a broad bandwidth (at least 10:1) in the frequency-domain and calculation of the inverse Fourier transform. The latter technique was used in this effort (13). Two frequency-domain systems, both covering the range of about 1 to 12 GHz, were used in the measurements.

A very sensitive 10-frequency (1.08, 2.18, ... 10.8 GHz) system was used to measure the response of a set of models 5 cm long. Then a swept-frequency system covering 1 to 8 GHz was used to measure a set of models 40 cm long. Using appropriate calibration factors, a composite signature data set with a 58:1 frequency bandwidth was produced.

A. Ten-frequency Radar System

A block diagram of the experimental ten-harmonic radar backscatter system is shown in Fig. 3. In this system a single L-band microwave source is used to harmonically generate a set of frequencies. The network analyzer receiver local oscillator is used to modulate the higher harmonics (balanced modulators), and also to modulate (lower sideband) the L-band signal going into the reference port of the network analyzer receiver head. The result is that the receiver head of the network analyzer downconverts each of the harmonically related backscattered signals to the same IF frequency. The output of the network analyzer is, in fact, the vector sum of the set of signals being received. The processing

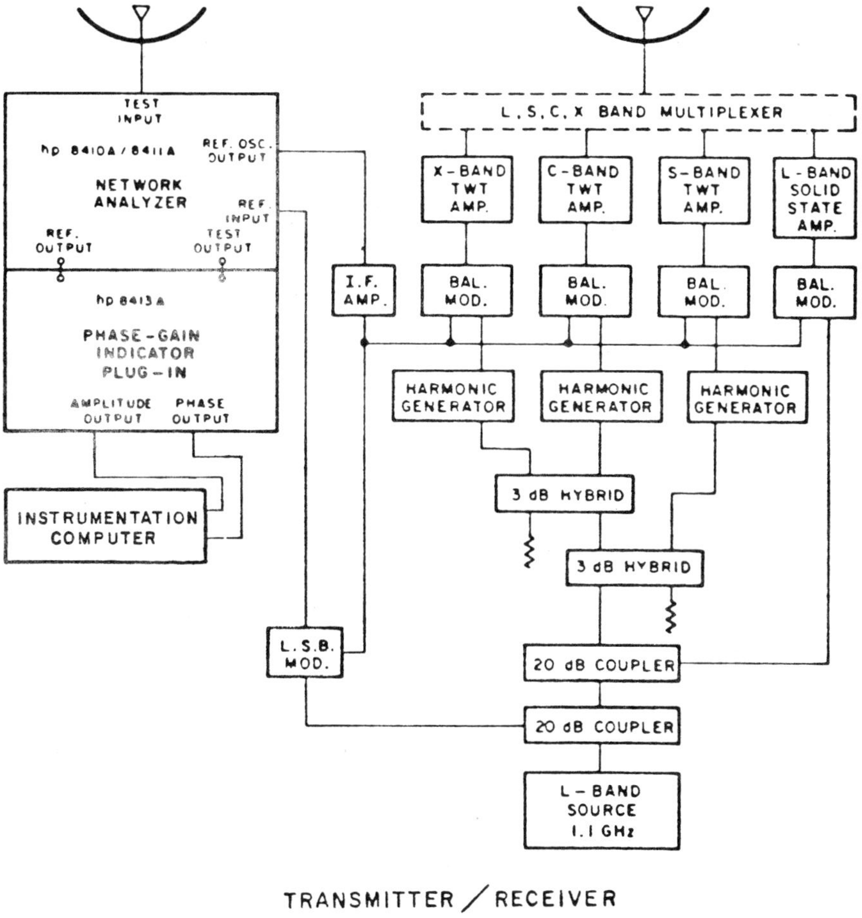

Fig. 3. 10-harmonic scattering cross-section measurement system block diagram.

of the data from this point reduces the contribution of the background and antenna coupling in the received signal and recovers the individual amplitude and phase of the individual harmonics. The operational procedure is to move the target slowly toward the radar antennas and to take amplitude and phase data samples under computer control. Arrays of amplitude and phase versus position data for the targets of interest as well as the empty support pedestal and a calibration and test sphere are stored in the computer. The raw data are Fourier transformed to yield the amplitude and phase of the individual harmonics. Any antenna coupling or scattering from stationary objects is rejected at this point as a DC term. The target measurements then have the empty pedestal (background) data subtracted and the final result is calibrated by using the calibration sphere data. In the end, we have the amplitude of the targets expressed as absolute radar cross section (in cm.) and the phase of the backscattered return relative to the center of the calibration sphere. The test sphere data serves as an estimate of the overall accuracy of the system.

B. The Swept-Frequency Radar System

A block diagram of the swept frequency radar system is shown in Fig. 4. This system is intended for larger target models than the ten frequency system, with data being obtained in finer increments in frequency. In such a system, the target is stationary, and the signal to noise improvement introduced by the moving target data processing system is not available. On the other hand, the targets have radar cross sections approximately 64 times larger due to their 8 times larger dimensions. The basic concept of the swept frequency system and the steps in the measurement of the data are shown below.

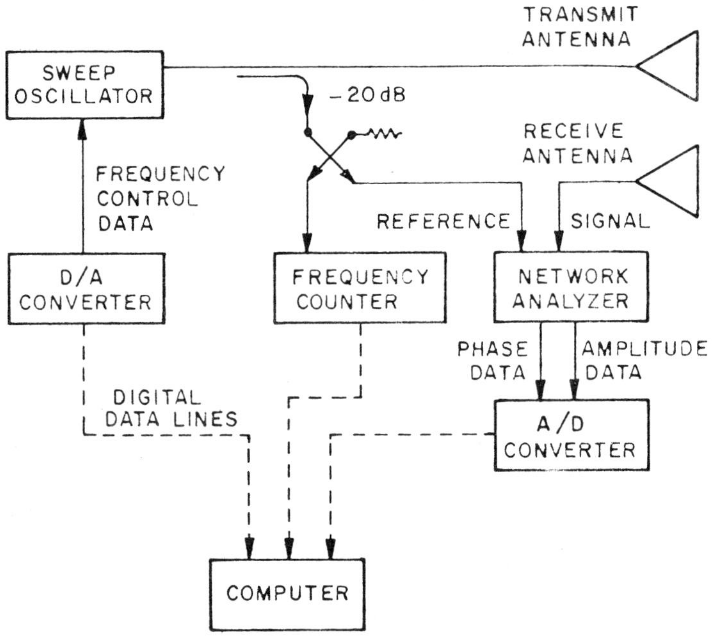

Fig. 4. Swept-frequency scattering cross-section measurement system block diagram.

(1) Measure and store in computer files amplitude and phase readings as the frequency is stepped over a band of frequency. Readings are taken on the targets, on reference and check spheres, and on the background.

(2) Subtract (phasor subtraction) the background data from the various target and sphere data to remove the errors due to various antenna coupling and unwanted scatter effects.

(3) Use computed values of the reference sphere to calibrate the system in terms of absolute radar backscatter cross section (amplitude and phase).

(4) Use data on the check sphere to confirm the overall integrity of the system and the data.

The major problem in the above set of concepts is due to constraints on the frequency repeatability of the swept frequency system. If the background is to be subtracted from the data on the various targets, then the frequency for each step in the frequency sweep must be repeatable to a high degree. The constraint is that the amplitude and phase of the background must be unchanged when the target is introduced. This can be assured if the set of data samples is taken at nearly identical frequencies for the background and for the set of targets. Therefore, microwave frequency counter was interfaced to the instrumentation computer, and the computer controls the frequency of the sweeper as well as the position of the object as the measurement is performed.

C. Typical Target Data

Measured scattering data are compared to exact calculations for a 3.5" diameter metal test sphere in Fig. 5 (14). The data are normalized and calibrated to represent square root of echo area (in cm) and phase with respect to the center of the sphere. Based on these data and other tests, the accuracy of the measured data are estimated to be $\pm 10\%$ in amplitude and $\pm 10^{\circ}$ in phase.

Measured data on a 3:1 sharp cone at nose-on look angle are presented in Fig. 6. Also shown are data measured by Keys and Primich (15) on a target of very slightly different shape, and calculations using the geometrical theory of diffraction (16) and equivalent current analysis techniques (17).

Ramp response waveforms for the six targets of Fig. 1, nose-on incidence, are shown in Figs. 7 and 8. In each case, the data have been normalized as in Fig. 5. The resulting plots of area vs distance are compared to the actual profile function of the objects in each case.

Fair agreement between the ramp responses and profile functions is seen in all cases. It is noted that some reverberation probably due to diffraction at the back face of the cones, is evident in all cases. Also note that these 58 frequency band ramp responses do not contain sufficient resolution to distinguish sharp from rounded tips and edges on the models. However, the presence of a small cylinder, completely in the shadow region, is evident in the response of target #4 compared to the corresponding response of the cone alone - Target #1.

Of primary importance for this imaging study is the fact that, if the reverberations are eliminated by a range-gating process, the time domain ramp response gives a good approximation of the geometrical shape of the target, including the shadow region, with some loss of definition. Thus, out of all parameters which might be extracted from broadband signature data, it constitutes a simple piece of information, highly correlated to the target shape.

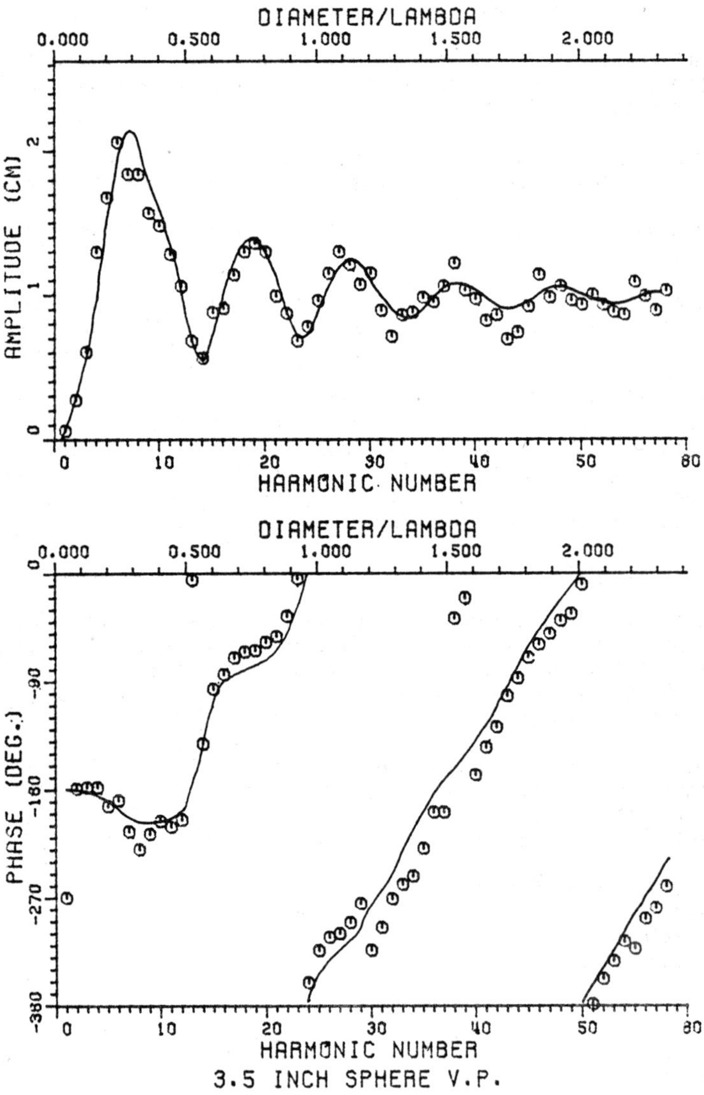

Fig. 5. Complex scattering cross-section of a 35" sphere.

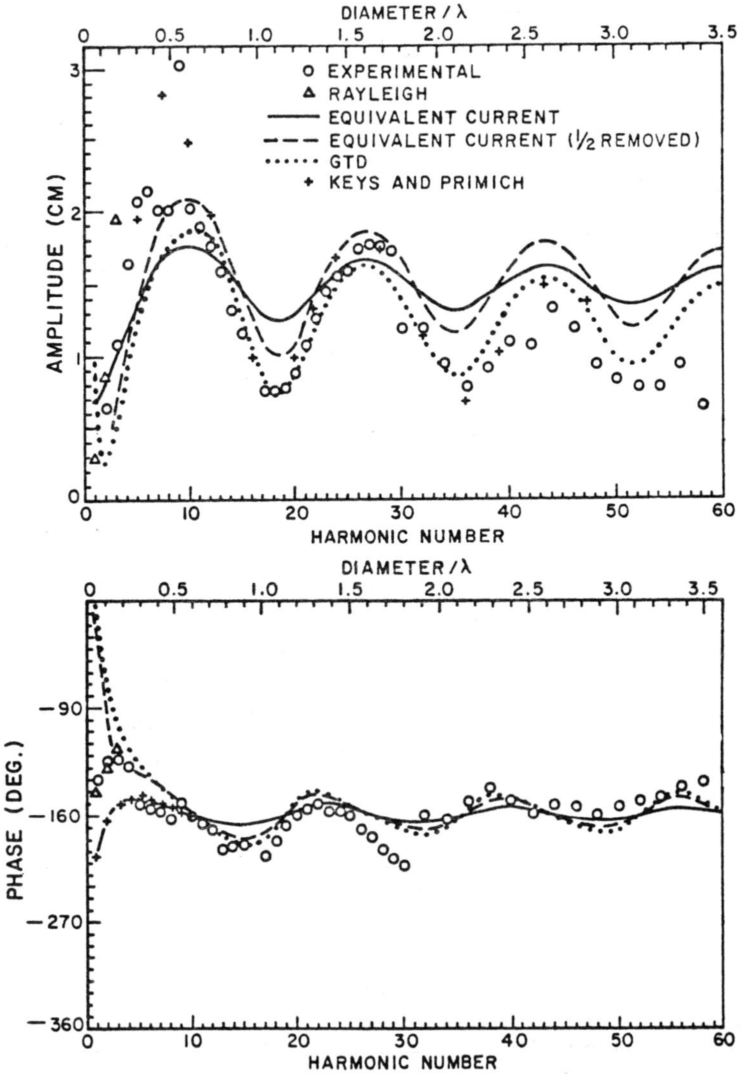

Fig. 6. Complex scattering cross section for a 3:1 cone, nose-on look angle.

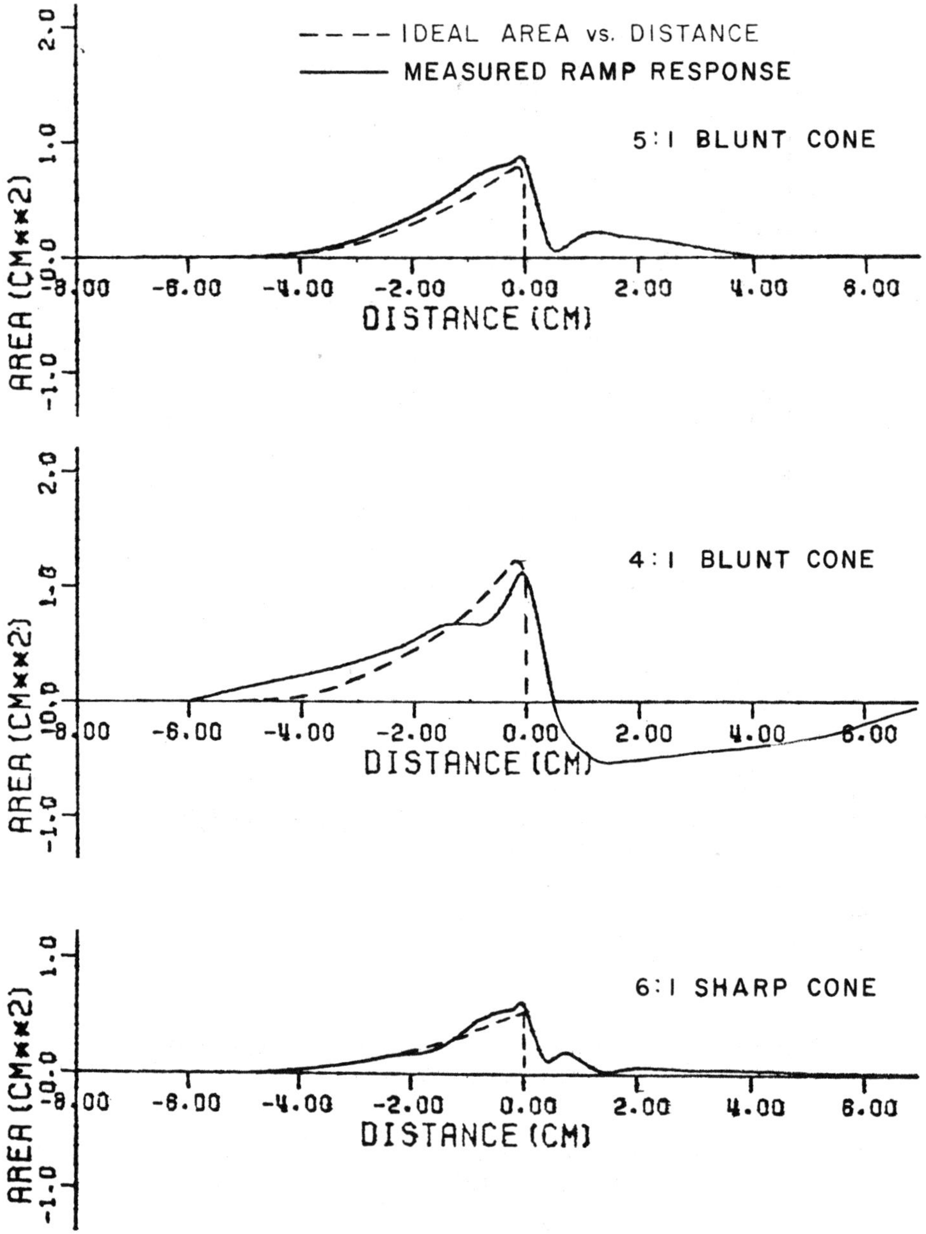

Fig. 7. Measured ramp response waveforms for cones #1,2,3.

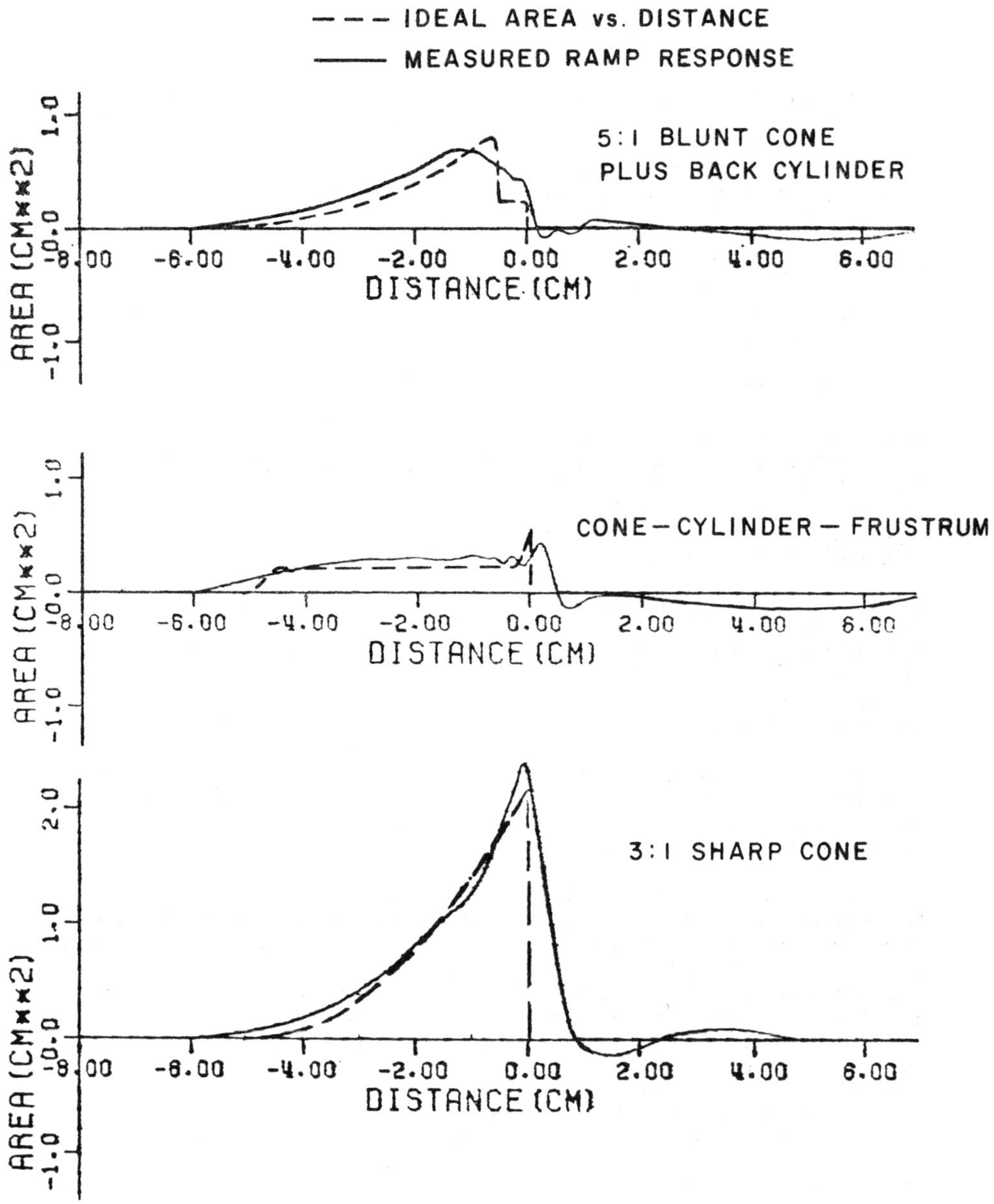

Fig. 8. Measured ramp response waveforms for cones #4,5,6.

III. TARGET IMAGING

The creation of a three-dimensional image from measured scattering signature data is to be discussed here. Specifically, we need to specify the boundaries of the three dimensional characteristic function $\gamma(x,y,z)$ for a target, which in turn permits the creation of an image at any look angle. For the purposes of this discussion, isometric views of the object will be presented, since these seem to best portray the targets in a single picture.

Direct utilization of the Fourier Transform relationship of Bojarski is one approach which might be used to find $\gamma(x,y,z)$. The restricted range of look angles can be viewed as windowing in the spatial frequency domain, a situation which can easily be analyzed using standard Fourier transform theory.

If the reverberations of the ramp response are gated out, and the remainder is used, then its demonstrated characteristics have an important impact on the inverse transform imaging process. Normally, as seen in Equation (1), the input scattering data requires measurement at a pair of opposing look angles, which are combined to form the characteristic scattering function $\Gamma(p)$. However, our results show that the single look angle, after gating out reverberations, is the approximate one-dimensional transform of $\gamma(\overline{p})$. Thus the need for the antipodal look angle is eliminated.

Gross (18) has studied the problem of narrow aspect angle windowing in the physical optics inverse diffraction process. For example, Fig. 9 shows a grey scale plot of a two-dimensional triangle, using ideal scattering data which has been windowed to $\pm 60^{\circ}$ and $\pm 30^{\circ}$ from nose-on. For the 30° window, the "back" face is well defined, but severe degradation of the front point almost completely obscures its trangular shape.

The limiting surface imaging technique is an attempt to improve image quality in cases like the one above, by embodying a priori information into the process. This is done by building the image object from a set of moldable limiting surfaces, one for each look angle (19).

The limiting surface for each look angle has a profile function specified by the measured ramp response. However, its boundaries consist of a shape determined by a combination of a priori and process parameters. For example, the limiting surfaces used here have elliptical cross-sections perpendicular to the look angle whose size is determined by the profile function at the given distance, and whose orientation and alignment are determined by a set of process parameters. The final image is the intersection (the region common to all limiting surfaces) of the limiting surfaces. It has been shown that a wide variety of simple shapes may be accurately formed from these surfaces. In some cases, the nature of the image might be known more specifically, such as an aircraft. Thus a set of moldable wing, fuselage, and empannage shapes might be used instead.

The process parameters used with the elliptical limiting surfaces are shown in Fig. 10 (20). Two of the parameters, θ and ϕ specify the look angle of a particular profile function. The remaining parameters are in general, unknown and require a priori knowledge of the target geometrty or iterative manipulation to achieve the best image. A complete set of these parameters must be specified for each look angle. A straight line along which the centers of ellipses are to be aligned must be specified. This center line is described by two angles, α and β. α is the angle between the center line and the look angle axis. β gives the rotation of the center line about the look angle axis with the y axis as the β=0 reference. γ describes the tilt of the ellipses about the center

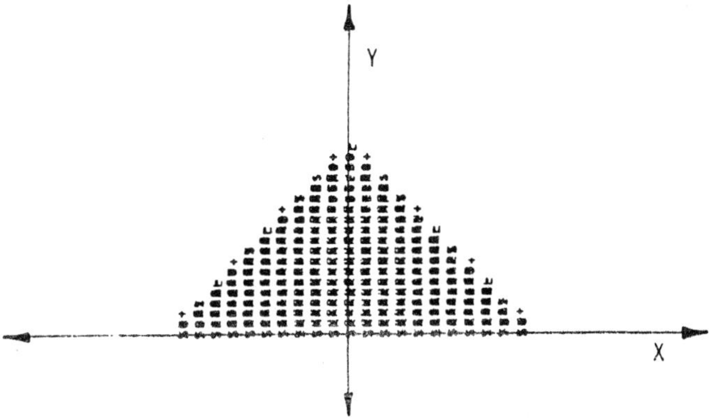

(a)

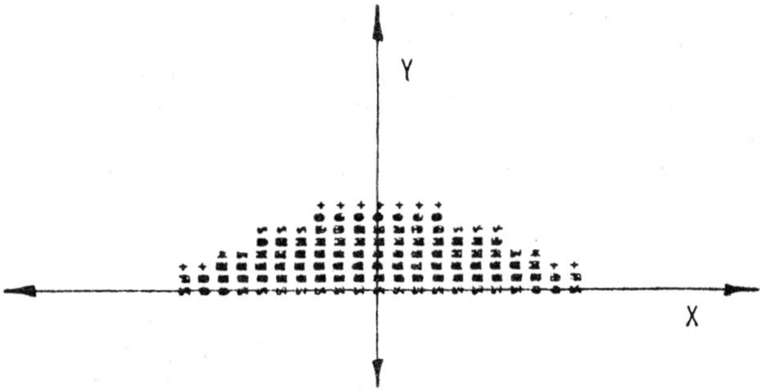

(b)

Fig. 9. Image of triangle using ideal input, $\pm 30^{o}$ and $\pm 60^{o}$ windowing.

line and is also referenced to the y axis. In addition the aspect ratio, R, for the ellipses must be specified.

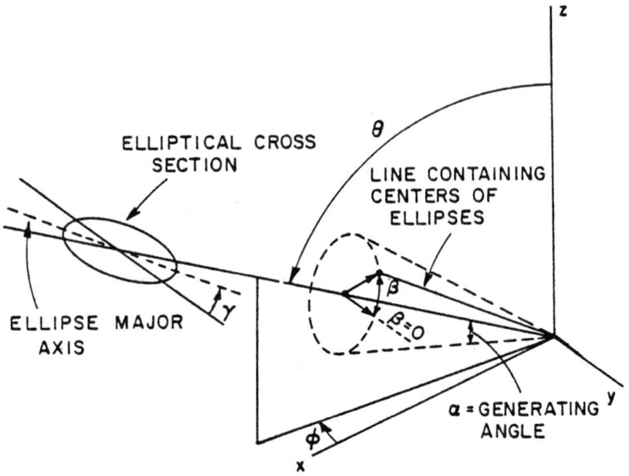

Fig. 10. Specification of angles used in imaging algorithm.

Functionally the imaging algorithm can be described as follows. For a given point in the x-y plane the program iterates through values of z. For each z value the area of an ellipse which passes through that point, using the parameters specified, is calculated for each look angle. If the areas of all the ellipses are found to be less than the values specified by the corresponding input profile functions, then the point is assumed to lie within the volume of the target. The points with minimum and maximum z values which are found to be interior to the target then become a surface contour in a constant x plane. A complete contour is swept out by repeating this process while iterating over y and holding x fixed. Finally a complete set of contours describing the target surface are drawn by iterating over x. Portions of contours which should be visually obscured by preceding contours are omitted to heighten the image's 3-dimensional effect.

The correspondence between an elliptical cross section and profile function is depicted in Fig. 11. The look angle is in the x-y plane with $\phi=45^{0}$. An ellipse perpendicular to the look angle and passing through the point p is shown. The ellipse has area a which is also the value of the profile function at the corresponding point along the look angle axis. Therefore the point p lies precisely on the target's surface. A surface contour containing p is also indicated. When several look angles are used, the final result is the same as if a separate image surface were calculated for each look angle and their intersection taken as the actual image surface.

B. A Criterion for Determining Image Accuracy

One method of determining image quality would be a visual comparison of an image to it's corresponding target. However, a target being imaged is in general unknown. A more objective measure is the volume error between the input and output profile functions. Volume error is defined as:

$$\varepsilon_{vol.} = \frac{\int_{-\infty}^{\infty} |A_{in}(x) - A_{out}(x)|\,dx}{\int_{-\infty}^{\infty} A_{in}(x)\,dx} \tag{4}$$

where $A_{in}(x)$ is the ramp response or input profile functiton and $A_{out}(x)$ is the output profile function for the image along a given look angle. Volume error therefore is expressed as the actual error volume between the input profile function information and the resulting image profile function, divided by the volume corresponding to the ramp response. Volume error is calculated for each look angle and the average is used as a basis for comparing image quality.

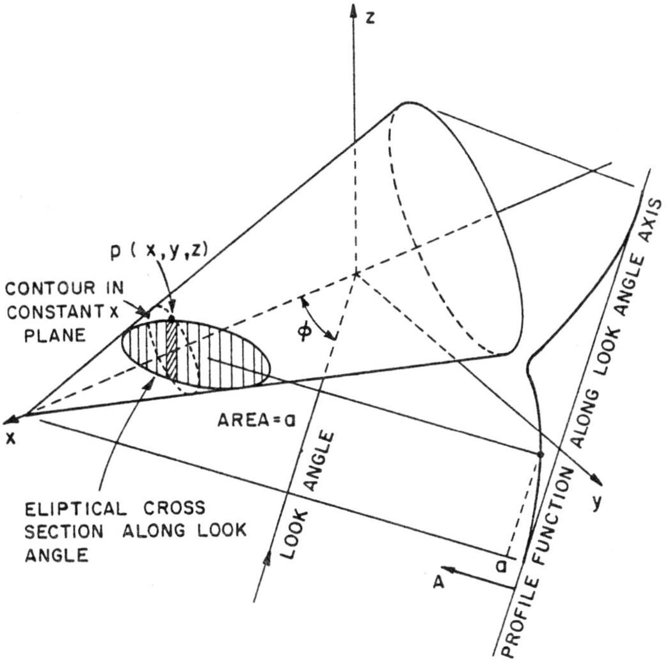

Fig. 11. Illustration of imaging algorithm.

An extensive study was done on the behavior of the images and the image volume error as the process parameters were varied. For example, Fig. 12 shows a plot of volume error vs α_2, one of the process parameters, where a 4:1 cone is being imaged using data at 0^0 (nose-on) and 30^0 look angles. The parameter α_2 controls the major axis of the limiting surface corresponding to the 30^0 look angle. It is seen the volume error is minimized when α_2 is at 30^0, its correct value.

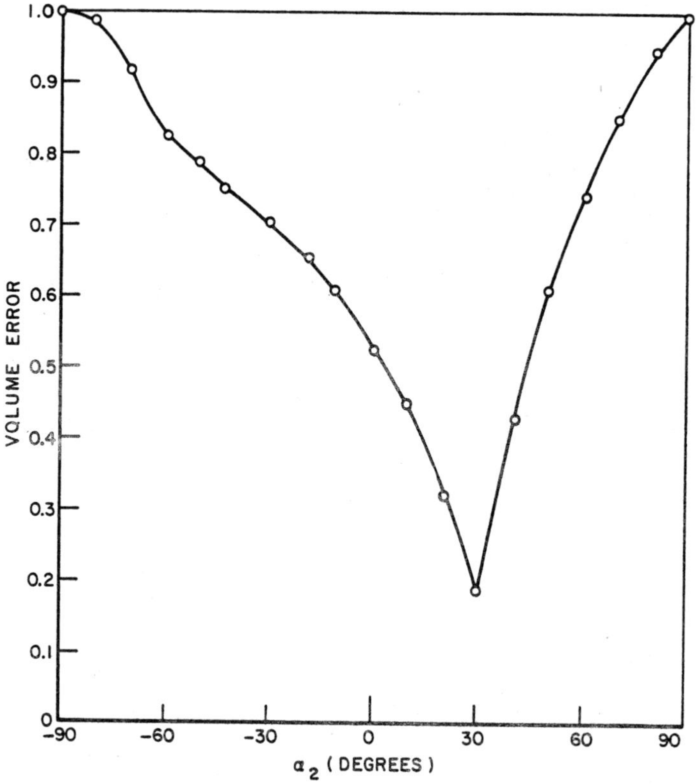

Fig. 12. Volume error vs α_2.

Figure 13 shows what happens to the limiting surfaces in two cases for the volume error calculation above. When $_2 = 30^0$, it is seen that the two limiting surfaces are lined up, and their intersection (the surface lying within both simultaneously) is a relatively accurate image. However, when $_2 = 45^0$, the two surfaces are not aligned. The image which would be produced in this case is seen in Fig. 14. The front tip has been "chopped off" by the misalignment of the two limiting surfaces. The profile functions in Fig. 14 show how this effect also produces a volume error.

The study of the volume error vs all process control parameter variations revealed that the process if relatively "well behaved", i.e., incorrect values of each parameter give measurable increases in the volume error, and volume error increases proportional to process parameter error. Because of this characteristic, an automated iteration procedure was developed to obtain the image, and was evaluated using several specific test cases.

The procedure which was used is a linear direct search method. A linear method is one which uses a set of direction vectors, in this case the parameter coordinate axes, throughout the search. A direct search method is any minimization technique not requiring explicit evaluation of any derivatives of the function

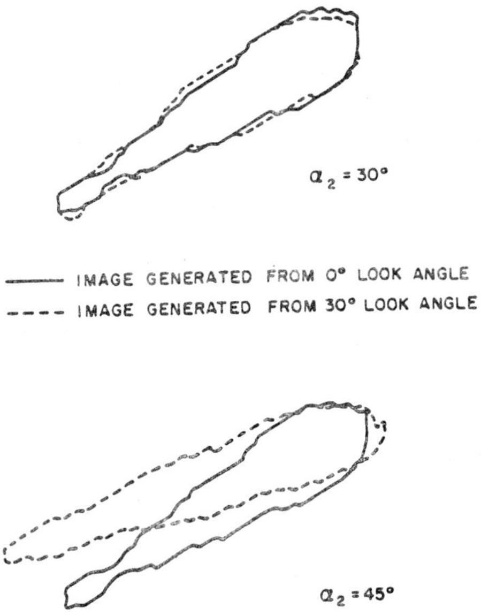

Fig. 13. Example of imaging dependence on α.

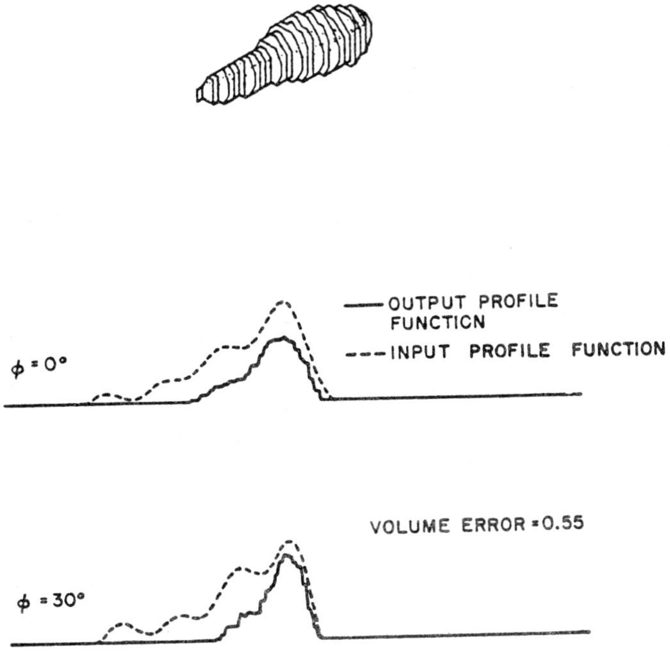

Fig. 14. Image and profile functions of target 2 with limiting surfaces misaligned as in Fig. 13.

being minimized. The procedure begins by determining, at some initial point, which direction along each parameter one would have to travel in order for volume error to decrease. This is done by making small perturbations from the initial point along each parameter direction. Each parameter is then incremented by some fixed amount in the appropriate direction. This is equivalent to moving only roughly in the direction of the gradient. The elliptical cross section aspect ratio is actually incremented logarithmically because of the nature of this parameter. This procedure is repeated until the vicinity of the minimum has been located. The process converges to the minimum by decreasing the increment sizes each time the minimum is passed as explained above.

The possible initial limiting surface parameter combinations can be represented by several cases. These cases demonstrate the performance and restrictions of the iterative algorithm. The ideal 0° and 30° profile functions for target 6 will be used in each example unless otherwise indicated. The first case is when the limiting surface parameters are known for one look angle and unknown for the other. The algorithm will iterate to very nearly the correct values of the unknown parameters. For example, with the parameters for the 0° look angle correct and the initial 30° look angle parameters set at $\alpha_2 = 60^o$, $\beta_2 = 140^o$, $\gamma_2 = 50^o$, and $R_2 = 0.5$, the resulting parameter values are $\alpha_2 = 29.5$, $\beta_2 = 179.7$, $\gamma_2 = -0.25$ and $R_2 = 1.19$. Another example required 20 iterations with the initial parameter increments set at 4.9^o, 4.9^o, 20.1^o and 0.0792 for α_2, β_2, γ_2, and R_2, respectively. Volume error decreased from 0.77 initially to 0.186. A third example used the measured 0°, 29-harmonic ramp response and 30°, 10-harmonic response for target 6. This time 18 iterations were required and the resulting parameter values were $\alpha_2 = 29.38$, $\beta_2 = 179.51$, $\gamma_2 = -9.67$, and $R_2 = 1.083$. The final image and image profile functions are shown in Fig. 15. Here the volume error decreases from 0.729 to 0.164.

Images of the other cone targets, made using just 2 look angles, are shown in Figs. 16 and 17.

IV. SUMMARY AND CONCLUSIONS

Measurement of transient scattering signatures, and their characteristics have been described using cones as an example. The creation of three-dimensional images based on these data have been discussed. Specifically a limiting surface imaging process, implying a priori information and a set of process parameters which are iterated to obtain a shape which best agrees with this input data, has been described and demonstrated.

It is seen that the time-domain transient scattering signature is a useful target identification tool, summarizing all complex scattering cross-section data in a single waveform which is strongly correlated to the target shape. Thus, this waveform should be considered as a source of features for target identification processes, as well as input information for target imaging. The time domain ramp response is an approximate profile function of the target, both in the "lit" and "shadowed" region, and was thus chosen for the imaging process input. If the reverberations are gated out, the measured ramp responses are seen to have characteristics which correspond to the physical optics approximation for scattering by a target.

The limiting surface imaging process is seen to be an efficient way of incorporating a priori information in order to get improved accuracy with severely limited scattering signature data. Depending on the applicaiton, rather generalized or very specific moldable shapes might be used in the process.

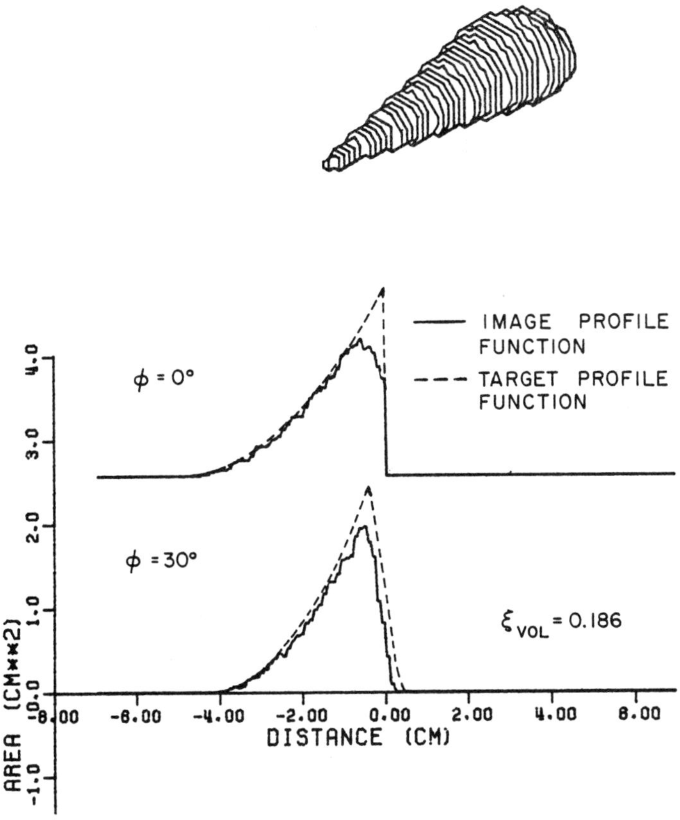

Fig. 15. Final image and profile functions, 3:1 cone.

An objective indicator of image quality was chosen to be the comparison of the image profile functions with those of this input data. Generalizing this concept, it would be possible , where necessary, to use other scattering analysis techniques such as Geometrical Theory of Diffraction or Moment Methods to calculate "virtual" scattering signatures for comparison to input data. In all cases, the time-domain signature format is important to the process, because it shows where the image shape needs to be iteratively changed, and by how much.

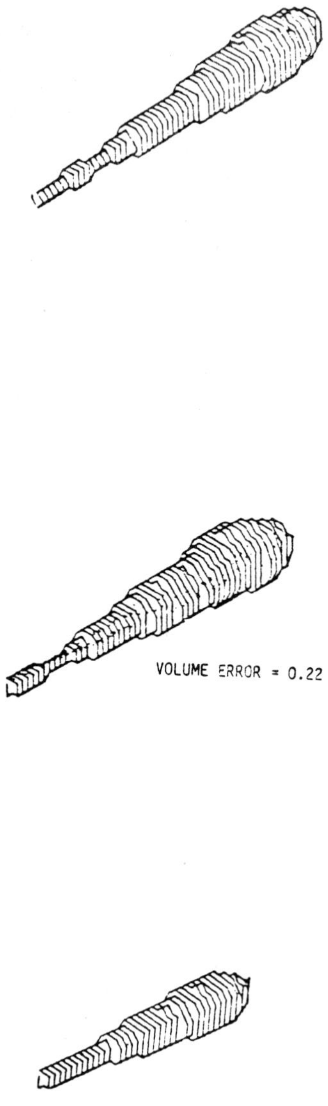

Fig. 16. Images of targets #1,2,3.

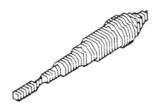

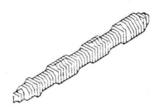

Fig. 17. Images of targets #4, 5.

REFERENCES

1. N. N. Bojarski, "Inverse Scattering," Final Report to Contract N00019-72-C-0462, Department of Navy, Naval Air Systems Command, April 1975.

2. N. N. Bojarski, "Signal Processing Studies and Analysis, Vol. IV - Three Dimensional Electromagnetic Short Pulse Inverse Scattering," Contract No. AF30(602)-3961, Syracuse University Research Corporation, November 1968.

3. N. N. Bojarski, "Signal Processing Studies and Analysis, Vol. V - Electromagnetic Short Pulse Inverse Scattering for Discontinuities in an Area Distribution (Scattering Centers)," Contract No. AF30(602)-3961, Syracuse University Research Corporation, November 1968.

4. N. N. Bojarski, "Signal Processing Studies and Analysis, Vol. VII - Three-Dimensional Electromagnetic Short Pulse Inverse Scattering with Equatorial Derivative Scattering Data," Contract No. AF30(602)-3961, Syracuse University Research Corporation, November 1968.

5. N. N. Bojarski, "The Improvement of Electromagnetic Inverse Scattering Approximations," Contract No. AF30(602)-3961, Syracuse University Research Corporation, February 1967.

6. R. M. Lewis, "Physical Optics Inverse Diffraction," IEEE Trans. on Antennas and Propagation, Vol. AP-17, No. 3, pp. 308-314, May 1969.

7. R. D. Mager and Norman Bleistein, "An Examination of the Limited Aperture Problem of Physical Optics Inverse Scattering," Contract N0014-76-C-0039, Univeristy of Denver, Denver Research Institute, August 1976.

8. R. D. Mager and Norman Bleistein, "An Approach to the Limited Aperture Problem of Physical Optics Far Field Inverse Scatttering," Contract N0014-76-C-0039, University of Denver, Denver Research Institute, August 1976.

9. W. L. Perry, "On the Bojarski-Lewis Inverse Scattering Method," IEEE Trans. on Antennas and Propagation.

10. R. A. Day and W. M. Boerner, "On Radar Target Shape Estimation Using Algorithms for Reconstruction from Projections," IEEE Trans. on Antennas and Propagation, Vol. AP-26, #2, March 1978.

11. E. M. Kennaugh and R. L. Cosgriff, "The Use of Impulse Response in Electromagnetic Scatteirng Problems, IRE National Convention Record, Part 1, 1958.

12. E. M. Kennaugh and D. L. Moffatt, "Transient and Impulse Response Approximations," Proc. IEEE, Vol. 53, pp. 898-901, August 1965.

13. E. K. Walton and J. D. Young, "Radar Scattering Measarements of Cones and Computation of Transient Response," submitted to IEEE, APS, March 1979.

14. E. K. Walton, "Broadband Scattering Signature Measurements of Cone Targets," Report 784785-4, July 1979, The Ohio State University ElectroScience Laboratory, Department of Electrical Engineering; prepared under Contract DASG60-77-C-0133 for Ballistic Missile Defense Systems Command.

15. J. E. Keys and R. I. Primich, "The Radar Cross Sections of Right Circular Metal Cones - I," Defense Res. and Telecommun. Estab. Repts., p. 1010, 1959.

16. J. B. Keller, "Backscattering from a Finite Cone," IRE Trans. on Antennas and Propagation, 8, p. 175, 1960.

17. W. D. Burnside and L. Peters, Jr., "Axial-radar Cross Section of Finite Cones by the Equivalent-Current Concept with Higher-Order Diffraction," Radio Science, 7, No. 10, pp. 943-948, 1972.

18. F. R. Gross, "Application of Physical Optics Inverse Diffraction to the Identification of Cones from Limited Scattering Data," Report 784785-3, July 1979, The Ohio State University ElectroScience Laboratory, Department of Electrical Engineering; prepared under Contract DASG60-77-C-0133 for Ballistic Missile Defense Systems Command,

19. J. D. Young, "Radar Imaging from Ramp Response Signatures," IEEE Transactions on Antennas and Propagation, Vol. AP-24, #3, May 1976.

20. R. A. Day, "Automated Imaging of Cone-Like Targets from Transient Signature Data," Report 784785-2, March 1979, The Ohio State University ElectroScience Laboratory, Department of Electrical Engineering; prepared under Contract DASG60-77-C-0133 for Ballistic Missile Defense Systems Command.

Part 10
Special Topics

INTEGRAL REPRESENTATION TECHNIQUE FOR CALCULATING MODES OF COUPLED ELECTROMAGNETIC DIELECTRIC WAVEGUIDES

Leonard Eyges
Rome Air Development Center
Deputy for Electronic Technology
Hanscom AFB, Massachusetts 01731

ABSTRACT

An integral representation technique, closely related to the extended boundary condition method, has previously been used to calculate the modes of a dielectric electromagnetic waveguide which consists of a cylinder of arbitrary cross section and index of refraction n' embedded in a uniform medium of index n. The technique is here extended to calculate the modes of composite guides which consist of two or more single guides in proximity. Selected numerical results are given for two circular and two square guides.

INTRODUCTION

In a previous work[1] an integral representation technique was used to calculate the modal properties of electromagnetic dielectric (fiber optic) guides of various cross sectional shapes. The method is here extended to the calculation of modes of composite guides, i.e., guides which consist of two or more parallel single guides. From a slightly different viewpoint this calculation can be considered to show how modes of the single guides mix as the guides are brought together to form a composite guide.

We recall first the basic formulae for a single guide, assumed to be a homogeneous cylinder, parallel to the z-axis, with irregular cross section A, which cross section is bounded by a curve L. The guide has index of refraction n' and is embedded in a uniform medium of index n. The case of greatest practical importance is the weakly guiding one for which $n \approx n'$, and for simplicity we confine attention to it, although the method presented here can be extended to arbitrary n and n'. For weak guiding the vectorial character of the electromagnetic field ceases to be important[2], and one can work with an amplitude ψ which can be taken to be any of the transverse electromagnetic field components E_x, E_y, B_x, B_y. For time harmonic modes varying as $e^{-i\omega t}$ and with $k_0 = \omega/c$, the wave equation for Ψ is

$$\left(\frac{\partial^2}{\partial x^2} + \frac{\partial^2}{\partial y^2} + \frac{\partial^2}{\partial z^2} + n'^2 k_0^2\right)\Psi = 0$$

inside the guide, and a similar one with n' replaced by n outside the guide. With assumed propagation dependence as $e^{ik_g z}$, i.e., with

$$\Psi(x,y,z) = \Phi(x,y)e^{ik_g z}$$

the equation for Φ inside the guide is

$$(\nabla^2 + K'^2)\Phi = 0 \qquad \text{Inside Guide} \tag{1}$$

where $K'^2 = n'^2 k_0^2 - k_g^2$ and ∇^2 is the two dimensional Laplacian, and similarly

$$(\nabla^2 + K^2)\Phi = 0 \qquad \text{Outside Guide} \tag{2}$$

where $K^2 = n^2 k_0^2 - k_g^2$. Equations (1) and (2) are jointly equivalent to the integral representation

$$\Phi(\vec{\rho}) = (K^2 - K'^2) \int_A \Phi(\vec{\rho}\,')g_K(\vec{\rho},\vec{\rho}\,')dA' \tag{3}$$

where $\vec{\rho}$ is a position vector in the x-y plane, and the Green function

$$g_K(\vec{\rho},\vec{\rho}\,') = -(i/4)H_0(K|\vec{\rho}-\vec{\rho}\,'|),$$

satisfies

$$(\nabla^2 + K^2)g_K(\vec{\rho},\vec{\rho}\,') = \delta(\vec{\rho}-\vec{\rho}\,')$$

The equivalence of (1) and (2) to (3) is easily shown by operating on both sides of (3) with $\nabla^2 + K^2$. In addition to satisfying (1) and (2) both Φ and its normal derivative $\partial\Phi/\partial n$ must be continuous across the boundary of A. These conditions are satisfied by the Φ defined by (3).

The integral representation (3) is useful for perturbation calculations for guides which are almost circular[2]. Given the known solutions for the circular guide, the area integral in (3) can be written as one over a circle (inscribed, circumscribed, of the same area, etc.), and a remainder or perturbation integral, and useful formulae can thereby be derived for almost circular guides. For the present problem it is however convenient not to use (3) but rather use a line integral version of it. This is obtained by using (1) and (2) to write (3) as

$$\Phi(\vec{\rho}) = \int_A \{\Phi(\vec{\rho}\,')(-\nabla^2 g_K(\vec{\rho},\vec{\rho}\,') + \delta(\vec{\rho}-\vec{\rho}\,')) + g_K\nabla^2\Phi(\vec{\rho}\,')\}dA'$$

For $\vec{\rho}$ inside A this becomes, on using Green's Theorem to convert the area integral to a line integral,

$$0 = \int_L (\Phi_- \frac{\partial g_K}{\partial n'} - g_K \frac{\partial\Phi}{\partial n'}\Big|_-)dL' \tag{4}$$

where subscript - (+) indicates a point just inside (outside) the boundary. Equation (4) is also derivable from the bottom line of the standard Huyghen's principle representation

$$\left.\begin{matrix}\Phi(\vec{\rho}) \\ 0\end{matrix}\right\} = \int_L \left(\Phi_+ \frac{\partial g_K}{\partial n'} - g_K \frac{\partial \Phi}{\partial n'}\Big|_+\right) dL' \qquad \begin{matrix}\vec{\rho} \text{ outside A} \\ \vec{\rho} \text{ inside A}\end{matrix}$$

on using the fact that $\Phi_+ = \Phi_-$ and $\partial\Phi/\partial n|_+ = \partial\Phi/\partial n|_-$. Equation (4) is of course the homogeneous counterpart of the basic integral representation used in Waterman's formulation[3] of scattering problems.

We recall that the technique[1,4] for using (3) or (4) is to assume an expansion for Φ in the interior of the guide of the form

$$\Phi = \sum_{l=0}^{\infty} J_l(K'\rho)(A_l \cos l\varphi + B_l \sin l\varphi) \tag{5}$$

and combine this with an expansion of the Green function for small

$$g_K(\vec{\rho}, \vec{\rho}\,') = -(i/4) \sum_{l=0}^{\infty} \varepsilon_l J_l(K\rho) H_l(K\rho') \cos l(\varphi - \varphi') \tag{6}$$

whereupon Eq. (4) becomes a homogeneous set of equations for the A_l and B_l. By setting the determinant of this set to zero we get an equation for the propagation constants. For circular guides, for example, this procedure yields the well known dispersion relation

$$K' J_l'(K'a) H_l(ka) = K J_l(K'a) H_l'(Ka) \tag{7}$$

It is useful to recall here the connection of the present problem with an analogous quantum mechanical bound state problem, since it is sometimes convenient to call on quantum mechanical results or terminology. To see this connection we recall that the Schrodinger equation (units $\hbar = 2m = 1$) for the wave function Φ of a particle of energy E in a two dimensional potential $V(\vec{\rho})$ is

$$(-\nabla^2 + V(\vec{\rho}))\,\Phi = E\Phi \tag{8}$$

If the potential is an attractive one that is constant throughout some irregular cross section A, i.e., if

$$V(\vec{\rho}) = \begin{cases} -V_0 & , \quad \vec{\rho} \text{ in A} \\ 0 & , \quad \vec{\rho} \text{ outside A} \end{cases}$$

then Eq. (8) is equivalent to

$$(\nabla^2 + (V_0 + E))\,\Phi = 0 \quad , \quad \vec{\rho} \text{ in A}$$

$$(\nabla^2 + E)\Phi = 0 \quad , \quad \vec{\rho} \text{ outside A}$$

By comparing these equations with (1) and (2) we can make the identification $V_0 + E = K'^2$, $E = K^2$ so the equivalent strength of the potential is

$$V_0 = K^{2\prime} - K^2$$

In tabulating the modes of single guides, two parameters P^2 and B closely related to E and V_0 are used. For a circular guide of radius a they are

$$P^2 = |E|/V_0 = \frac{(k_g/k_0)^2 - n^2}{n^{2\prime} - n^2}$$

$$B^2 = V_0 a^2/\Pi^2 = (n^{2\prime} - n^2)k_0^2 a^2/\Pi^2 \tag{9}$$

OUTLINE OF METHOD FOR TWO COUPLED GUIDES

The above method is here extended to two guides; the extension to more than two is straightforward. The guides, both uniform and of index n' are embedded in a medium of index n as shown in Fig. 1.

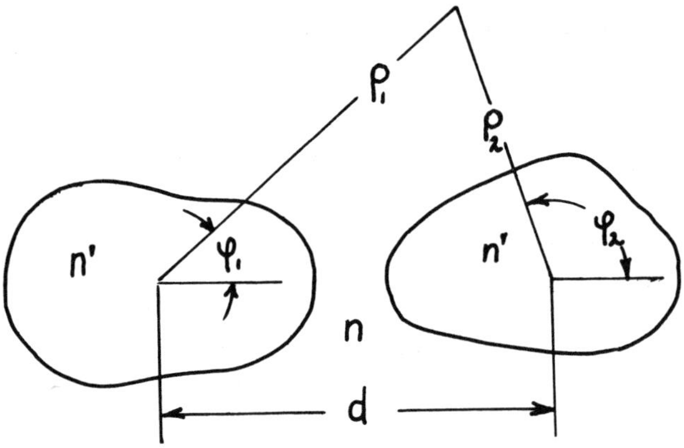

Fig. 1 Two irregular guides, a distance d apart

A coordinate system is associated with each guide and the distance between the origins of the systems is d. We want to apply Eq. (4) to the present case. Since (4) holds for arbitrary cross section it will hold for the geometry of Fig. 1. The fact that the two guides of Fig. 1 are disjoint is of no consequence. One can imagine them connected by a very thin dielectric fin which makes them into a single guide but which clearly does not affect the modal properties of the system. In the application of (4) to the present case, the integral goes over both boundaries and that equation becomes

$$0 = \int_{L_1} \left(\Phi_- \frac{\partial g_K}{\partial n_1'} - g_K \frac{\partial \Phi}{\partial n_1'}\Big|_-\right) dL_1' + \int_{L_2} \left(\Phi_- \frac{\partial g_K}{\partial n_2'} - g_K \frac{\partial \Phi}{\partial n_2'}\Big|_-\right) dL_2' \tag{10}$$

The technique for exploiting (10) is an extension of that for single boundaries. Namely, in terms of the two coordinate systems shown in Fig. 1, we write expansions like (5) for ϕ inside <u>each</u> guide

$$\Phi^{(1)} = \sum_{l=0}^{\infty} J_l(K'\rho_1)(A_l^{(1)}\cos l\varphi_1 + B_l^{(1)}\sin l\varphi_1) \tag{11}$$

$$\Phi^{(2)} = \sum_{l=0}^{\infty} J_l(K'\rho_2)(A_l^{(2)}\cos l\varphi_2 + B_l^{(2)}\sin l\varphi_2) \tag{12}$$

These are put into (10) and the Green function expansion (6), and Bessel function translation formulae, are used to evaluate Eq. (10) first at small ρ_1 then at small ρ_2, whereupon it becomes a set of homogeneous equations for the $A_l^{(1)}, A_l^{(2)}, B_l^{(1)}, B_l^{(2)}$. The vanishing determinant of this set yields the desired propagation constants.

The detailed treatment of two perfectly arbitrary guides can become somewhat lengthy, and hence less than transparent. The principles remain the same, but the details become much clearer if the individual guides have some symmetry and are symmetrically arrayed, as is often the case in practice. For the sake of clarity, we shall choose to analyze in detail composite guides with such symmetry properties. The generalization to arbitrary guides will then be straightforward.

<u>COMPOSITE GUIDES WITH TWO SYMMETRY AXES</u>

For the reasons above we now consider two identical guides, each with an axis of reflection symmetry. The guides are so placed (Fig. 2) that this axis is the x-axis and so that there is also reflection symmetry in the y-axis.

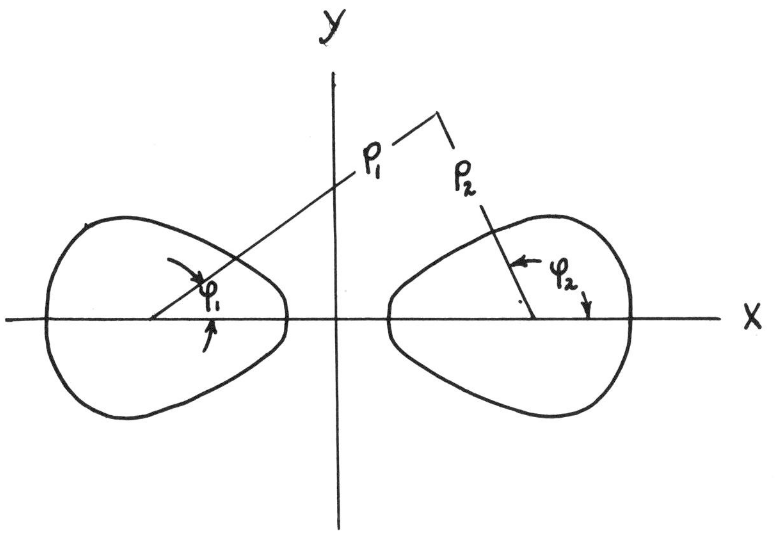

Fig. 2 Two guides that form a composite guide with reflection symmetry in both the x and y-axes.

For a composite guide of this symmetry, we know from the usual parity arguments of quantum mechanics that the amplitude Φ must be either even or odd with respect to reflection in the x-axis, and similarly with respect to the y-axis, i.e.

$$\Phi(x,y) = \pm\Phi(-x,y) \quad , \quad \Phi(x,y) = \pm\Phi(x,-y)$$

There are, in short, four kinds of states of the system which, with E for even and O for odd, can be designated EE, EO, OE, OO, where the first and second letters of the pair refer respectively to the left-right and up-down reflection properties. For any one of these states the general expansions of Eq. (11) and (12) simplify. For example, for the EE state, even with respect to reflection in the x-axis, there can be no sine terms in either expansion. Then possible forms are

$$\Phi^{(1)} = \sum_{l=0}^{\infty} A_l^{(1)} J_l(K'\rho_1)\cos l\psi_1 \tag{13}$$

$$\Phi^{(2)} = \sum_{l=0}^{\infty} A_l^{(2)} J_l(K'\rho_2)\cos l\psi_2 \tag{14}$$

Left-right evenness means of course that the amplitude Φ at a point must equal the amplitude at the reflected point. Consider a point inside cylinder #1 for which $\rho_1 = a, \psi_1 = \alpha$; from (13) the wave function there is $\Phi^{(1)} = \sum A_l^{(1)} J_l(K'a)\cos l\alpha$ The reflected point in guide #2 has coordinates $\rho_2 = a$, $\psi_2 = \pi - \alpha$ and with $\cos l(\pi-\alpha) = (-)^l \cos l\alpha$ the wave function, there is $\Phi^{(2)} = \sum (-)^l A_l^{(2)} J_l(K'a)\cos l\alpha$. We have then the requirement for the even state that

$$A_l^{(1)} = (-)^l A_l^{(2)} \equiv A_l \qquad \text{EE state}$$

We consider now the application of the above to <u>circular</u> guides. Figure 2 applies with the understanding that the two cross sections shown there are now to be taken as circles of radius a. We want to apply (4) to the EE state of this composite system for which we have by the above arguments

$$\Phi^{(1)} = \sum_{l=0}^{\infty} A_l J_l(K'\rho_1)\cos l\psi_1 \tag{15}$$

$$\Phi^{(2)} = \sum_{l=0}^{\infty} (-)^l A_l J_l(K'\rho_2)\sin l\psi_2 \tag{16}$$

These are to be put into (4) and the result evaluated at, say, small ρ_1. For doing the integral over circle #1, we use the expansion (6) of the Green function $g_K(\vec{\rho}_1, \vec{\rho}_1{}')$ for small ρ_1. For integrating over circle #2, we use the Green function in ρ_2 coordinates, $g_K(\vec{\rho}_2, \vec{\rho}_2{}')$, but with transformation formulae to express the ρ_2 dependence in ρ_1 coordinates. Now for ρ_1 small, ρ_2 is necessarily larger than d so that we can use the expansion

$$g_K(\vec{\rho}_2, \vec{\rho}_2{}') = -(i/4)\sum_{n=0}^{\infty} \varepsilon_n J_n(K'\rho_2) H_n(K\rho_2)(\sin n\psi_2 \sin n\psi_2' + \cos n\psi_2 \cos n\psi_2') \tag{17}$$

We combine this with the translation formula valid for $\rho_2 < d$

$$H_n(K\rho_2)\cos n\phi_2 = (-)^n \sum_{l=0}^{\infty} J_l(K\rho_1)\cos l\phi_1(H_{l+n}(Kd)-(-)^n H_{l-n}(Kd)) \tag{18}$$

Equations (15) through (18) are now inserted in (4) and the coefficients of $J_l(K\rho_l)\cos l\phi_1$ are equated to zero to yield the following set of equations for the A_l

$$0 = A_l\varepsilon_l X_l + \sum_{n=0}^{\infty} \varepsilon_n A_n \Gamma_n (H_{l+n}(Kd)-(-)^n H_{l-n}(Kd)) \ , \quad l=0,1,2... \tag{19}$$

where with $X = Ka$, $Y = K'a$,

$$X_l = XH_l'(X)J_l(Y) - YJ_l'(Y)H_l(X)$$

$$\Gamma_n = XJ_n'(X)J_n(Y) - YJ_n'(Y)J_n(X)$$

Since K is imaginary for bound modes ($K^2 = -|K|^2$) the function $H_s(Kd)$, which for large d behaves as e^{-Kd}, goes to zero for infinite d and Eq. (19) reduces to Eq. (7) for a single circular guide.

For finite d Eq. (10) can be solved by various truncation procedures. The clearest way of going about this is to start with some large d, where essentially only one A_l is different from zero, and then let d become smaller and follow the change of A_l and the growth of the other coefficients. We have done this for, among others, that EE mode of the pair of guides which becomes the dominant (ground state) mode of the single guide for infinite d. This dominant mode is the solution of Eq. (7) for $l = 0$ with the largest energy eigenvalue E, or in terms of P^2 of Eq. (9), with the largest value of P^2. The results are given in Table 1. Similar results have been obtained by Wijngaard[5] by a conventional matching method and our results, when the comparison is possible, are in excellent agreement with his.

TABLE 1 Propagation parameter P^2 of Eq. (9) and expansion coefficient A_l of Eqs. (15) and (16) for an EE mode of two coupled circular guides of radius a, with distance d between centers. A_0 is normalized to unity. B = 2.00

d/a	P^2	A_0	A_1	A_2	A_3	A_4
∞	.8916	1.00	0	0	0	0
3.00	.8916	1.00	.0007	.0008	.0017	.005
2.50	.8919	1.00	.007	.007	.013	.029
2.25	.8931	1.00	.034	.036	.066	.159
2.10	.8963	1.00	.101	.114	.217	.552
2.00	.9033	1.00	.254	.309	.625	1.698

As a further example we have applied the method above to an EE mode of two square guides, and again to that mode which becomes a dominant mode of the single guide for infinite separation. Expansions like (15) and (16) are substituted into (4). Much as for the circular case, one is led to a transcendental equation like (19) but somewhat more complicated since certain matrix elements that were diagonal for the circular case are no longer diagonal. The transcendental equation is, however, readily solved numerically, and again it is expedient to start with large d, since the solution for a single square guide is known from previous work[1]. Unlike the circular case, the expansions of Eqs. (15) and (16) do not reduce to a single term but they are dominated by a single term so that the convergence is satisfactory. What is of interest is the change of this convergence as d decreases from infinity and this is illustrated in Table 2.

TABLE 2 Propagation parameter P^2 of Eq. (9) and expansion coefficients A_l of Eqs. (15) and (16) for an EE mode of two coupled square guides, each of side 2b, and with distance d between centers. A_0 is normalized to unity, B = 2.00.

d/b	P^2	A_0	A_1	A_2	A_3	A_4
∞	.9072	1.00	0	0	0	-2.013
3.00	.9073	1.00	.0007	.0006	.0007	-2.014
2.50	.9079	1.00	.015	.014	.015	-2.018
2.25	.9104	1.00	.078	.074	.089	-1.984
2.10	.9182	1.00	.268	.279	.377	-1.768
2.00	.9418	1.00	.897	1.252	2.297	-1.165

An interesting check on these results can be made. As the distance between the guide decreases, there comes a point where they touch and in effect become a single rectangular guide of width 4b and height 2b. We have previously calculated[1] the dominant mode for this rectangular guide and hence can compare the result with that calculated in the present context, i.e., as the limit of two touching sqaure guides, d/b = 2. For the previous calculation, for B = 2.00, we found[1] P^2 = 0.9402. For the present one we have the reassuring result P^2 = 0.9418.

The programming and computational work that went into the above results was ably done by Dr. Peter Wintersteiner.

References

1. L. Eyges, P. Gianino and P. Wintersteiner, Modes of Dielectric Guides of Arbitrary Cross Sectional Shape, Jour. Opt. Soc. Am. (In Press)

2. L. Eyges, Fiberoptic Guides of Non-Circular Cross Section, Applied Optics 17, 1673 (1978)

3. P.C. Waterman, New Formulation of Acoustic Scattering, J. Acoust. Soc. Amer., 5, 1417-1429 (1969)

4. L. Eyges, Solution of Schrodinger and Related Equations for Irregular and Composite Regions, Ann. Phys. 81, 567 (1973)

5. W.W. Wijngaard, Guided Normal Modes of Two Parallel Circular Dielectric Rods, J. Opt. Soc. Am 63, 1518 (1973)

A SCATTERING MODEL FOR DETECTION OF TUNNELS USING VIDEO PULSE RADAR SYSTEMS

G. A. Burrell, J. H. Richmond, L. Peters, Jr., H.B. Tran
The Ohio State University ElectroScience Laboratory
Department of Electrical Engineering
Columbus, Ohio 43212

I. INTRODUCTION

The video pulse radar is now recognized as a viable tool for many applications. Some of the more successful applications include the Terrascan pipe detection system (1,2,3) observation of geological targets (4,5,6) and archaelogical exploration (7). Many of the targets to be detected have a two dimensional nature, i.e., they can be modelled without significant error by an infinite cylinder of some shape. An example of this is a mine shaft. The presence of abandoned mine shafts leads to serious subsidence problems, and if these can be detected from the surface, then substantial loss of property could be avoided. Also many geological structures such as fault lines, some oil and gas traps can be represented as non circular cylindrical geometries. The antennas used to launch and receive pulse to be used to detect such targets are finite in extent. The antenna system used in the pipe detector video pulse radar is a pair of loaded crossed linear antenna elements. The transmit-receive (T-R) isolation is achieved by the orthogonal properties of the antenna. This orthogonal antenna system will respond to a linear target such as a thin wire that is responsive to only one linear polarization. It will also respond to any target which is located off the axis of symmetry of the antenna system. When the target is represented as an infinite cylinder, the antenna cannot be located in the far zone of the target. Neither the echo width concept for the infinite cylinder for two dimensional scatterers nor the echo area for three dimensional scatterers are applicable.

A new model is introduced to obtain the voltage generated at the receiving antenna located at the same position in space as the transmitting antenna. Both receiving and transmitting antennas are finite length dipoles. This model treats separately the propagation phenomena between two dipoles separated twice the target distance in a homogeneous lossy media and the local scattering mechanisms, e.g., the infinite cylinder. The results are then combined to obtain the voltage generated at the terminals of the receiving antenna due to the scatterer.

Alternatively the Green's function using an infinitesimal dipole at a finite distance from the cylinder could be involved to obtain a solution for the scattered fields. A hybrid technique for the antenna currents which combines the moment method for the thin wire with the eigenfunction solution for the circular cylinder. Such a solution has been recently developed by Wang and Richmond (8). In general, the target is immersed in a lossy medium and the incident and reflected waves are attenuated as they propagate to and from the target. Thus, any of the conventional concepts for characterizing a scatterer would require a separate analysis for every target range, every antenna type and every set of electrical properties of the medium. Obviously, this leads to a prohibitively complex set of data to be obtained if a general underground radar system design is to be achieved.

In a previous paper (9) we have discussed the means of obtaining the propagation of pulses between the transmit and receive antennas in a lossy media. The basic model is represented as the transmission between a pair of antennas immersed in the medium as is shown in Fig. 1c. This transmission is represented in this paper as V_R'/V_T where $2V_T$ is the Thevenin voltage of the generator driving the transmitting antenna and V_R' is the voltage generated across the load impedance attached to the receiving antenna. We introduce the scattering from cylindrical targets into the model and calculate typical received signals. As discussed in a prior paper (9), we have modelled the pulse generator by its Thevenin equivalent circuit. The Thevenin generator voltage is twice the output voltage, because in antenna engineering it is common to refer to the output voltage of a generator as the voltage delivered to a load whose impedance equals the generator source impedance. This voltage is V_T. In general the voltage appearing across the antenna terminals will not be V_T because the antenna impedance will vary with frequency. To obtain both the transmission and reflection characteristics of the pair of antennas in Fig. 1(c) to the input pulse, the z-matrix of the two port network comprising the dipoles and the medium was computed for 512 equally spaced frequencies up to a maximum frequency which was about 2.5 times the highest frequency for which the input pulse had significant energy. The integral equation was solved for each frequency using a piecewise sinusoidal expansion for the current distribution (12). From the z matrix data computed as a function of frequency, the signal reflected from the transmitting dipole and the signal generated across the load impedance terminating the receiving dipole can be computed (9).

II. SCATTERING MODELS

The ratio V_R'/V_T represents the relative terminal voltage obtained at the receiving antenna (across the load) for a perfect planar reflector at a depth d. If the scatterer is a simple planar lithologic contrast, this result would be modified by the reflection coefficient at the interface and the result would be exact. If the scatterer is a planar layer, the reflection coefficient of the layer would be used. If, in general, we define the Scattering Attenuation Function (SAF) as the modification of the scattered fields of the perfect planar reflector, then these reflection coefficients represent the SAF for these geometries. A curved interface requires an additional correction factor.

A. Reflection from Convex Contrasts of Large Radii of Curvature

A geometrical optics approach (10,11) can be used to provide data for the scattered fields to a radar placed a distance d above a large curved surface to determine if such a discontinuity would be detectable from the surface. Geometrical optics is basically a high frequency concept. Energy propagates along a straight line path in a homogeneous medium; in our case, from the source to the target. At the target, these rays are scattered in the specular direction. The magnitude of the fields is obtained from the conservation of energy for a point source for this example and after reflection these fields may be represented by a virtual source a distance ρ from the reflecting surface as given by (10,11)

$$\rho = \frac{R\ d}{R + 2d} \tag{1}$$

This gives rise to a scattered field at the radar

$$|U| = \frac{R}{2}\left[\frac{1}{R + d}\right] U_o\ e^{-2\gamma d} \tag{2}$$

where $U_o e^{-\gamma d}$ is the field incident at the curved surface. The attenuation and phase delay introduced by the lossy medium are displayed separately for convenience. Equation (2) reduces to the case of the planar interface discussed earlier as the radius R goes to infinity. The net loss introduced by the curvature is given by

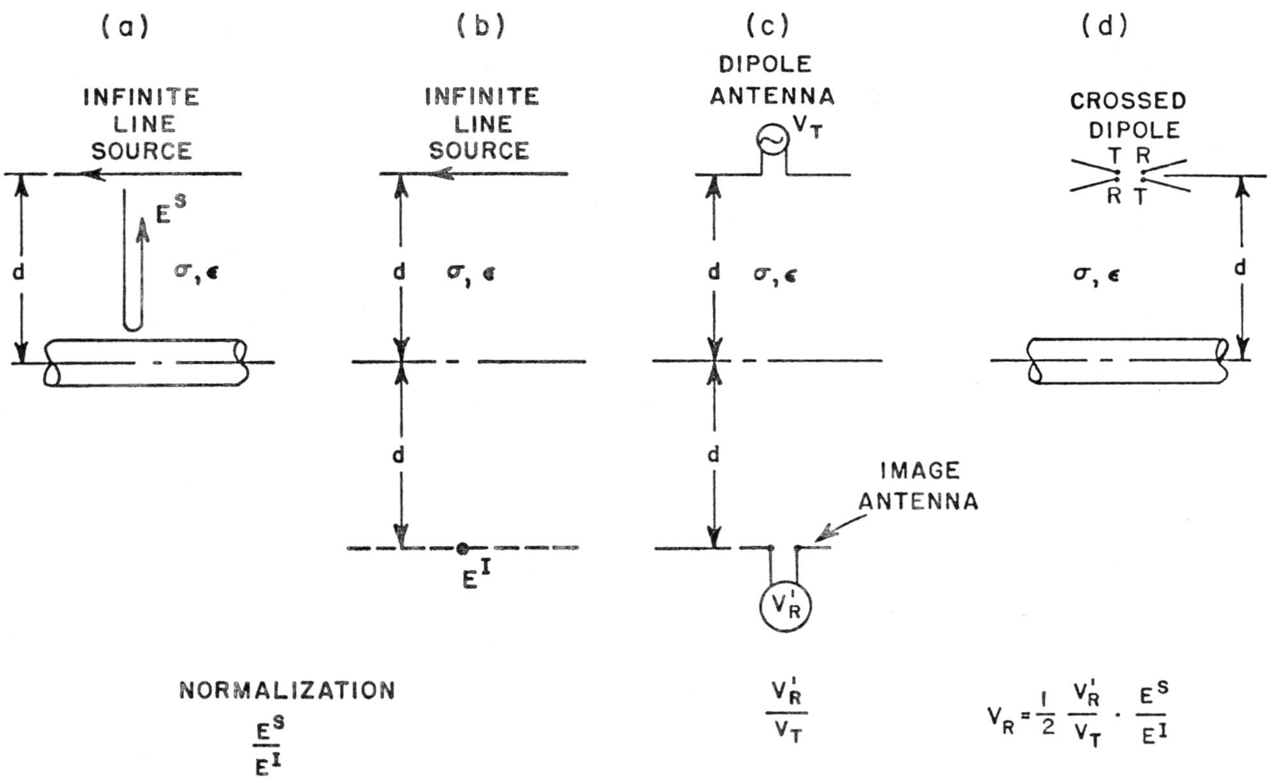

Fig. 1. Introducing the scattering attenuation function of a cylinder into the radar model.

$$\frac{U(R)}{U(R=\infty)} = \frac{R}{R+d} \quad . \tag{3}$$

When $R \approx d$, the scattered field is reduced by approximately a factor of two, and that when $R << d$, the scattered field reduction is of the order of R/d.

The reflecting surface need not be spherical and indeed can be represented by two principle radii of curvature R_1 and R_2. In this case, there are two virtual source positions, ρ_1, ρ_2 corresponding to the two principle radii of curvature. They are given by

$$\rho_i = \sqrt{\frac{R_i d}{R_i + 2d}} \tag{4}$$

These give rise to a reflected field of the form

$$|U| = \frac{1}{2}\sqrt{\frac{R_1}{R_1 + d} \cdot \frac{R_2}{R_2 + d}}\; U_0\, e^{-2\ d} \tag{5}$$

and a net loss of the form

$$\frac{U(R_1,R_2)}{U(R=\infty)} = \sqrt{\frac{R_1}{R_1 + d}\quad \frac{R_2}{R_2 + d}} \quad . \tag{6}$$

The spherical "like" model of Fig. 2 is not the only type of scatterer that may be of interest for geological exploration. Another possibility is that of a two dimensional or cylindrical (not necessarily circular) body. In this case, the apparent source position is different in the plane containing the axis of the cylinder and the plane orthogonal to it. This is represented by the general astigmatic ray tube of geometrical optics. The apparent source position in the plane containing the axis of the cylinder is the same as was the case for scattering from the planar surface, i.e., $R_2 \to \infty$ in Equation (5).

The ratio given by Equation (5) now reduces to

$$\frac{U(R)}{U(R=\infty)} = \sqrt{\frac{R}{R+d}} \quad . \tag{7}$$

B. Model for Cylindrical Contrasts

The computation required for the singly curved or infinitely long cylindrical contrast does not require the use of geometrical optics and in fact can be obtained using any other method for computing the value of $U(R)/U(\infty)$. When the cylinder radius is small, the SAF can be computed conveniently by a modal solution or by a moment method solution using polarization currents. Such a solution contains both the spatial divergence of the field reflected from the cylinder, the reflection coefficient introduced by a change in material at the interface, and the resonances which are characteristic of the cylinder. Figure 1 shows how we can obtain the radar return from an infinite cylinder using a scattering solution. First, we calculate the backscattered fields E^S a distance d meters from the cylinder. This field is then related to the field which would be scattered by an infinite ground plane replacing the cylinder. Image theory is invoked, and as in Fig. 1(b), we calculate the field E^I of an infinite line source 2d meters deep. E^S is now normalized by E^I to get the SAF(E^S/E^I), which is the representation of $U(R)/U(\infty)$.

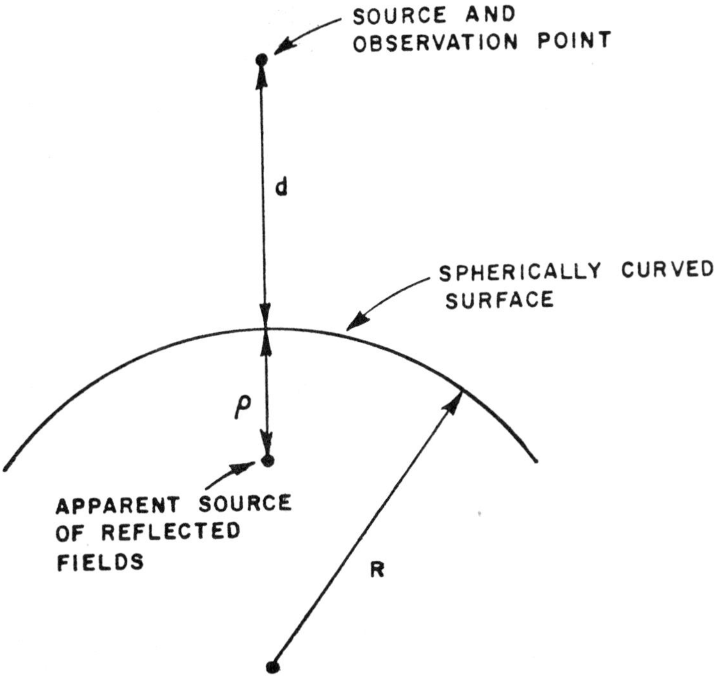

Fig. 2. Depicting the apparent source of reflection from a spherically curved surface.

From the model of the radar system with the perfectly reflecting interface (9), shown in Fig. 1(c), we have obtained the medium loss, the antenna frequency response, and the spherical wave attenuation of the wave radiated by the dipole antenna. The effect of the cylinder is now simply introduced by multiplying the system response of the model by the SAF (E^S/E^I), i.e.,

$$V_T' = \frac{V_R'}{V_T} \frac{E^S}{E^I} . \tag{8}$$

V_T' is now the received voltage on the terminals of the transmitting antenna due to the scatterer. This procedure is valid because the cylinder does not introduce additional wave divergence in the plane of the paper.

If the orthogonal antenna pair is used as shown in Fig. 1(d), the voltage on the receiving antenna is

$$V_R = \frac{1}{2} \frac{V_R'}{V_T} \frac{E^S}{E^I} . \tag{9}$$

The additional factor of 1/2 is introduced by the use of the crossed dipole geometry. This is introduced since the incident field component is oriented at 45^0 from the angle (0^0) required for optimum excitation of the thin cylinder resulting in a 3 dB loss, and the receiving antenna is oriented as 45^0 to the scattered field resulting in another 3 dB loss. It is observed that these 3 dB factors depend on the linear target scattering being polarization dependent. This has led to much confusion for three dimensional targets and this will be discussed later in more detail. The major objective here is to introduce a model for the scattered field that extends this previous analysis to include the target scattering.

It may appear that this model is valid only for thin cylindrical contrasts primarily because the position of a reflecting plane is chosen to coincide with the position of the scatterer. For a tunnel sufficiently large that different scattering mechanisms can occur, such as those from the top and the bottom of the tunnel, then different reference planes could be required for the different scattering mechanisms in the same model. This would drastically increase the amount of data required for design and require a much more complex analysis. Fortunately as is shown in Appendix I, the choice of the position of this reflecting reference plane does not significantly influence the results and thus the technique should be reasonably valid for a thick cylindrical contrast.

C. A Check on the Validity of the Model for Thin Cylindrical Contrasts

The received voltage V_R of Fig. 1d has been computed by a direct moment method solution to check the normalization procedure. For this comparison, an electrically small radius perfectly conducting circular cylinder was used because it could be analyzed by both methods. The moment method solution was by the sinusoidal reaction technique (12), so that the buried cylinder was required to be finite in length. The lossy medium attenuates the current on the cylinder as the distance from the excitation point becomes large so that the cylinder can be truncated without introducing significant error. We used a pair of 50 m long orthogonal dipoles 200 m from the center of a 1000 m long, 0.01 m radius wire. The antenna system was oriented for maximum signal (the plane of the antenna system was parallel to the plane containing the wire and the transmitting antenna was rotated 45^0 from the direction of the wire). The medium parameters were σ=0.01 mhos/m and ϵ_r= 4. Figure 3 shows V_R when V_T is a 50 microsecond (μs) 1 volt gaussian input pulse (9) for both the moment method solution and the modal solution modified in the way outlined above. The agreement in shape is almost exact and the amplitude difference of 0.8 dB (trivial considering the dynamic range of ~140 dB) is attributed to the moment method solution because the cylinder was truncated. Note also that for the example a high conductivity was used, and the depth was 200 m, and yet this is a detectable pulse for a properly designed system.

III. SCATTERED FIELD PROPERTIES

There are a number of computer solutions now available for obtaining the scattered fields from infinite cylinders in a conducting medium when illuminated by a line source. Figure 4 shows the scattered fields normalized in the form E^S/E^I or H^S/H^I for 1 meter radius circular cylinders. Included in Fig. 4 are curves for both perfectly conducting and hollow cylinders. The hollow or dielectric cylinders have been analyzed both by a model solution and by a moment method solution commonly designated as the polarization current technique. The two solutions produce results within 0.01 dB for the circular cylinder, and the polarization current technique is used to obtain E^S/E^I for non circular hollow cylinders. When the distance from the line source to the cylinder is large the scattered fields H^S/H^I can be taken to be the same as E^S/E^I since $E^S/E^I = ZH^S/ZH^I$ where Z is the intrinsic impedance of the media. H^S/H^I is designated $E_\perp$. E^S/E^I for polarization parallel to the cylinder is designated $E_\parallel$.

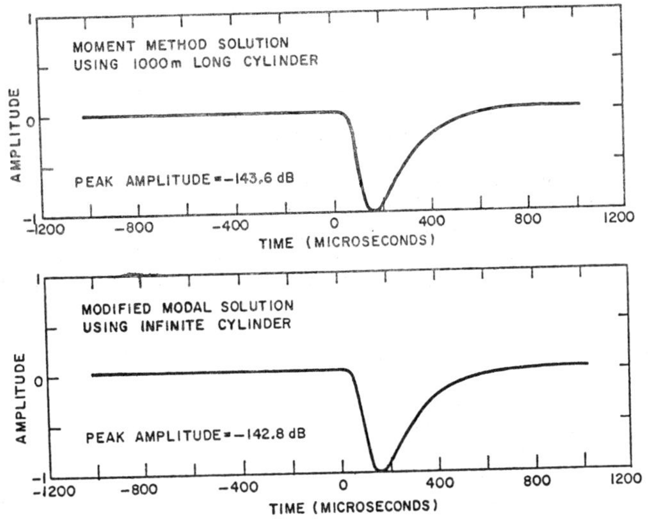

Fig. 3. Comparison of received pulse obtained from complete moment method and modified modal solution. The antenna system was a pair of 50 m orthogonal dipoles, and the target a 0.01 m radius wire 200 m from the antenna system. The constitutive parameters of the medium were σ=0.01 mhos/m, ε_r=4. The input pulse was a 50 μs gaussian.

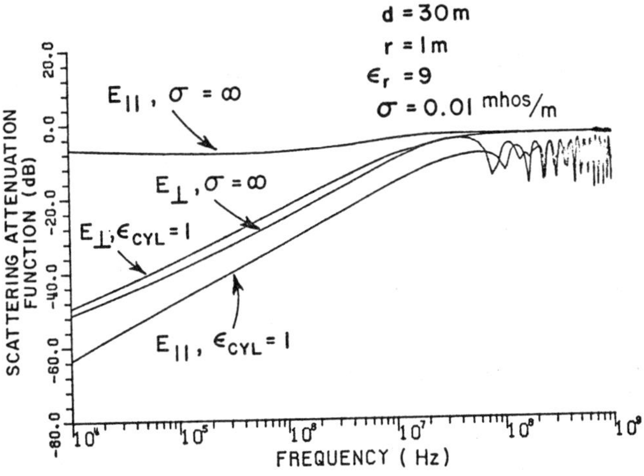

Fig. 4. Scattering attenuation functions for a cylinder immersed in a ground like medium. Results are presented both for an air filled cylinder (ε_{CYL}=1) and a perfectly conducting cylinder (σ=∞).

A. Conducting Cylinders

For the conducting cylinder (which may be representative of a flooded tunnel) the normalized scattered field is almost frequency independent over a wide frequency range, and the reflected field is only about 5 dB less than that which would be obtained from a perfect reflecting plane. Consequently such a target would be very visible to a parallel polarized electromagnetic radar. If the radar is polarized perpendicular to the cylinder then the cylinder would be less visible to a radar operating at low frequencies than would the same cylinder oriented parallel to the radar antenna. Note that in Fig. 4 no creeping wave resonances are observed in $E_\perp$ for the perfectly conducting cylinder. This is due to the lossy medium attenuating this wave. This resonance was observed when the conductivity is reduced to σ = 0.0001 mhos/m.

The results for the conducting cylinder are similar to those which would be obtained for a flooded tunnel. Figure 5 shows the normalized scattered fields for lossy cylinders of varying coductivity. It is observed that the lossy cylinder behaves similarly to a conducting cylinder until the frequency decreases so that the tunnel diameter equals the skin depth. The conductivity of the water in a flooded tunnel could vary within wide limits, but would be of the order of 1 mho/m if significant quantities of mineral salts were dissolved.

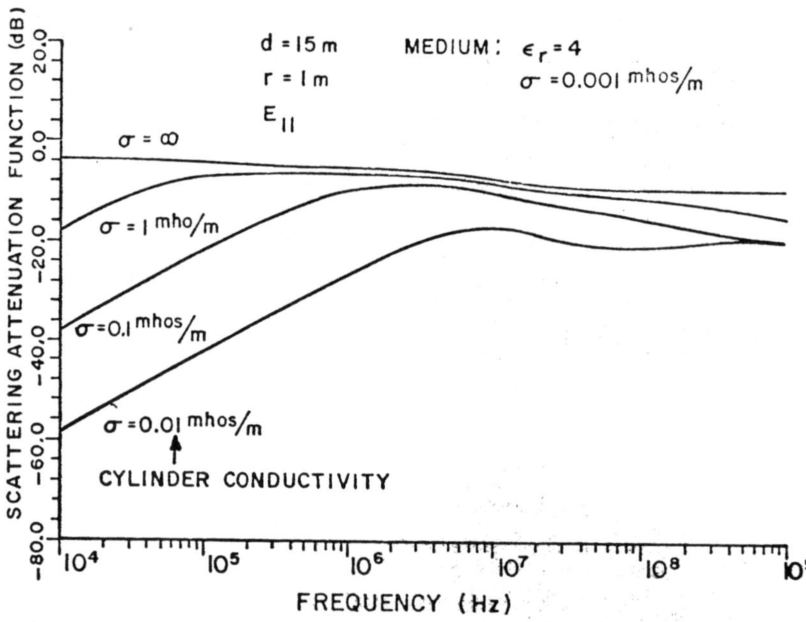

Fig. 5. Scattering attenuation function for lossy dielectric cylinders in a lossy medium for parallel electric polarization.

B. Air Filled Circular Cylinders (Tunnels)

The scattered fields for the air filled infinite cylinders (tunnels) shown in Fig. 4 are characterized by internal resonances. At frequencies higher than about the fourth resonance the scattered fields are the same for both polarizations: consequently at these frequencies the tunnel would not be detectable by electromagnetic means with an orthogonal antenna system. At frequencies lower than about the fourth resonance the fields scattered by the tunnel are different for the two polarizations. Thus, in this frequency range, the tunnel is potentially detectable by a radar with an orthogonal antenna system. At low frequencies, when the tunnel is electrically small, the tunnel is 12-16 dB more visible to a perpendicularly polarized radar than it is to a parallel polarized radar. This is the reverse of the case of the conducting cylinder, and occurs because a transverse cylinder obstructs the current flow in the medium more than a parallel cylinder does.

For $E_{\parallel}$ the first resonance occurs when the tunnel diameter is approximately a half wavelength (a 1 m radius tunnel resonates at about 50 MHz). The resonances can be predicted quite accurately using Geometrical Optics (10). For $E_{\parallel}$ only two rays need be considered for good engineering accuracy down to the first resonance (10): These are the direct reflection from the tunnel roof and the ray which travels into the tunnel to be reflected from the tunnel floor. This simple representation, which is polarization independent, gives good accuracy above the fourth resonance for $E_{\perp}$. For $E_{\perp}$ below the fourth resonance, additional rays (which are polarization dependent) need to be included in the G.O. solution.

An important characteristic of the scattered fields of the air filled tunnels is that the resonance behavior is independent of depth and the electrical parameters of the ground. This first resonant region for 1 m radius tunnels lies between 40 MHz and 60 MHz. This has been verified for a range of:

conductivities	$.0001 \text{ mhos/m} \leq \sigma \leq 0.01 \text{ mhos/m}$,
permittivities	$4 \leq \varepsilon_r \leq 16$ and
depths	$15 \text{ m} \leq d < 120 \text{ m}$.

Only the level of the curve changes significantly as these parameters are changed This remarkable behavior is a result of the fact that the resonances are caused by internal reflections. This is extremely important in terms of identification, i.e., a signal received by an electromagnetic radar operating in the resonance region will contain a signature of a hollow circular tunnel, and this signature will be relatively unaffected by the depth or the ground parameters. The only effect of the ground parameters on the shape of the curve in the resonant region is to alter the depth of the nulls: this occurs because a change in the Fresnel reflection coefficient alters the amplitudes of the direct reflection from the tunnel roof and the internal reflection from the tunnel floor.

The values of ε_r, σ and d affect the amplitude of the curve in the following ways.

1) Increasing ε_r increases the value of the Fresnel reflection coefficient, so that the direct reflection from the top of the tunnel is increased, and the scattered field is increased.

2) Increasing σ (or decreasing ε_r) increases the attenuation constant α of the medium (9). The scattered fields are normalized by the fields E^I scattered by a perfectly reflecting plane located at the axis of the cylinder, so that the E^I is lowered due to extra attenuation introduced because the propagation distance involved is increased by twice the tunnel radius. Thus the normalized

scattered fields E^S/E^I are increased. This effect can be very pronounced, as shown in Fig. 6 where the conductivity is changed by an order of magnitude. The level change of 14.1 dB is exactly accounted for by the change in path loss over a distance equal to the radius of the cylinder.

3) As the depth is changed, the change in level of the curve is described by Equation (7). This is ilustrated also in Fig. 6 where a depth change from 15 m to 120 m results in an 8.79 dB change in level.

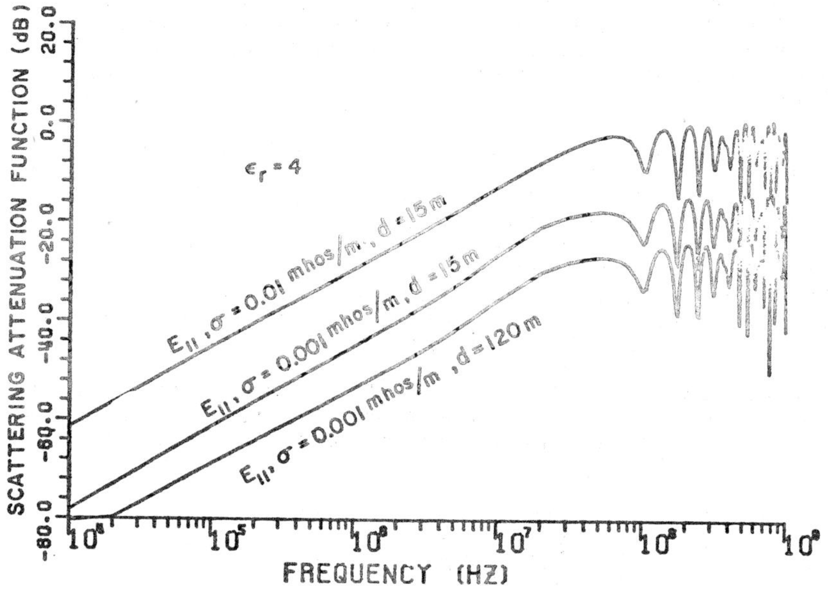

Fig. 6. Comparison of scattering attenuation function for 1 m radius tunnel for different depths (d) and conductivity (σ) shows that the primary effect of these parameters is to change the amplitude of the curve.

Figure 7 shows the SAF for a square tunnel as would be observed using a cross dipole system where both $E_{\perp}$ and $E_{\parallel}$ have been incorporated in the solution. The SAF's in this case are obtained using the polarization currents in a moment method solution.

IV. SYSTEM PROPERTIES

The goal of this paper is to include scattered fields from targets other than the planar interface in the analysis of the response of a Video Pulse radar system. In the previous section we obtained the SAF for several cylinderical targets. In the following section we use that data to obtain the video pulse response for a tunnel.

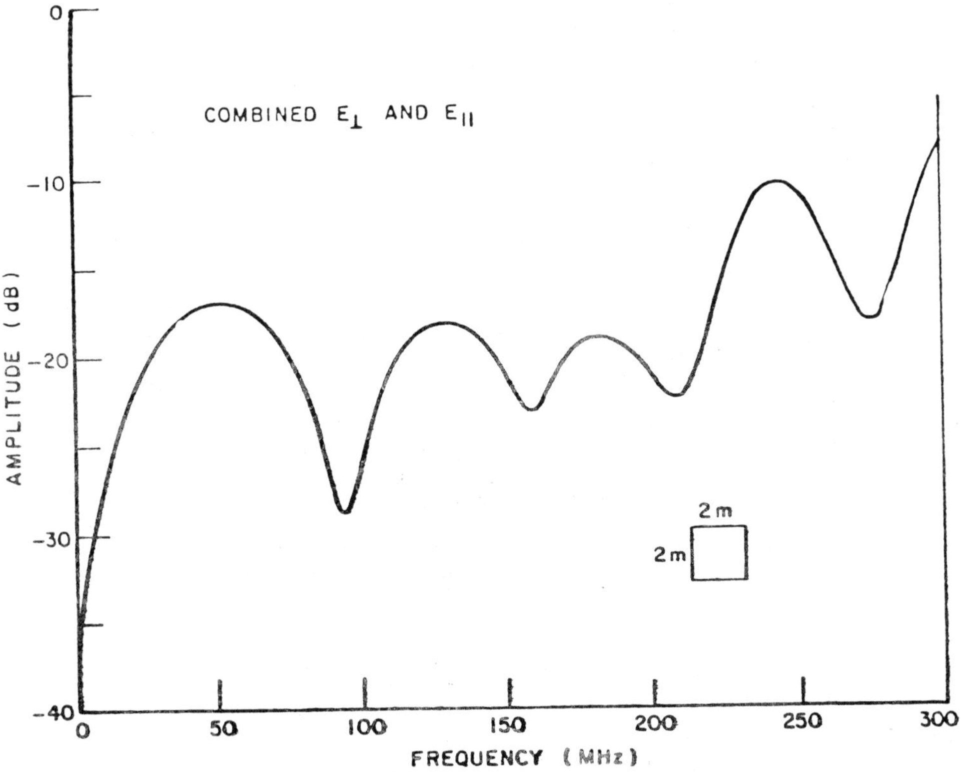

Fig. 7. Scattering attenuation function for a 2 m square tunnel immersed in a media with ϵ_r=4, σ=0.001 mhos/m at a 40 foot depth.

A. Signals Received from Tunnels

The effect of the various features of the scattered fields of the tunnels on typical received radar signals is illustrated with some examples. Fig. 8 shows the system response for a 1 m long dipole 30 m from and parallel to, a 1 mm radius infinite circular tunnel. The diple has a 0.0015 m radius coating of polythene (ϵ_r=2.3). The time domain received pulse after being reflected from the cylinder is shown in Fig. 8. The transmitted signal was an 8 ns doublet, and the parameters of this radar, were chosen for illumination of the tunnel in the resonant region (9). The tunnel resonances are clearly observed when Fig. 8 is compared to Fig. 9 which does not include the tunnel SAF. The tunnel resonances cause a ripple with an 8 ns period on the tail of the received signal, and this kind of signature is suitable as input to an identification scheme based on the complex natural resonances of the cylinder (13).

B. Direct Mode Operation

A crossed dipole antenna system, with one dipole used for transmit and the other for receive, can provide about 100 dB of isolation between the transmitted and received signals. However only targets which couple energy from one dipole to the other can be observed (e.g., a parallel interface cannot be observed). However if the same dipole is used both for transmitting and receiving (called a direct mode system) then targets in which the scattered signal is polarized in the same sense as the incident signal can be observed. The direct mode system is practical if there is a large time separation between the transmitted signal and the received signal (called a clear range window). From Fig. 8, it is observed that the received pulse is delayed by approximately 70 transmitted pulse widths. Clearly a pulser could be designed so that the transmitted pulse would decay well below the received pulse level when it arrives back at the antenna.

Second it may also be necessary to use the direct mode to detect planar interfaces and other targets that do not exhibit the desired polarization characteristics.

Figure 4 reveals that at frequencies such that the scattered field from the tunnel is above the fourth resonance, the scattered fields become independent of polarization. This means that the orthogonal mode geometry would not observe the tunnel when it is placed directly over the tunnel. In this case, it would be necessary to either use the direct mode system or to displace the antenna from the position directly above the tunnel. This latter concept is easily illustrated if the tunnel is placed in the plane of the crossed dipole pair as shown in Fig. 10. Here the electric field incident (E^i) on the cylinder is now parallel to the cylinder. The scattered electric field E^s has a component parallel to receive dipole as shown. This applies to any target, for example, a spherical scatterer would produce no significant received signal if it is placed on the axis of symmetry of the cross dipole antenna symmetry. It would be observed however if it is displaced from this axis in planes not containing either antenna.

C. <u>Influence of the Polarization Properties of the Target on the Received Signal</u>

In order to better visualize the various situations that can occur for this radar system using the crossed dipole concept for isolation, let us represent the radiated field as being that caused by a z oriented dipole at the origin of the conventional spherical coordinate system. The radiated field is

$$\overline{E}_d = \hat{e}_\theta \, K \, \frac{e^{-jkr}}{r} \, \sin\theta \tag{10}$$

The voltage obtained on the receive crossed dipole are represented by those of an x oriented dipole at the position $(2r,\theta,\phi)$. If the reflection process does not alter the polarization of the wave then the received voltage is proportional to $\overline{E}_e \cdot e_x$ where

$$\hat{e}_x = \hat{e}_\theta \cos\theta \cos\phi \tag{11}$$

or

$$V_r \propto \frac{e^{-jk2r}}{2} \, \frac{1}{2} \sin 2\theta \cos\phi \tag{12}$$

In this case, the target would be any geometry that can be represented by a nondepolarizing scattering center or flare spot located at the position (r,θ,ϕ). If on the other hand, the target is polarization sensitive as is the case for a thin wire or thin cylinder then the wave incident on the target must be decomposed into its appropriate components.

For example, assume the scatterer to be a thin conducting cylinder. Such a target is an effective scatterer for waves polarized parallel to the cylinder.

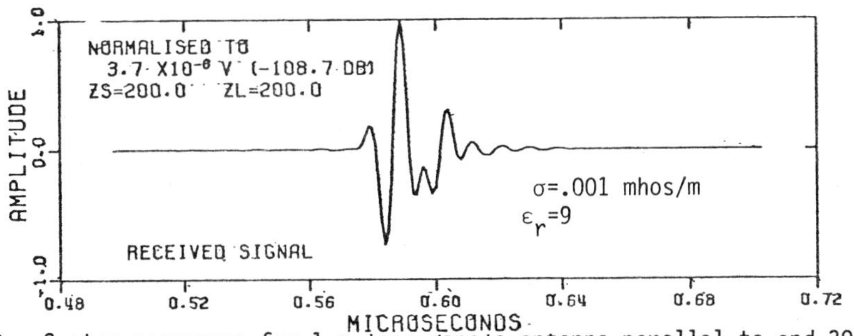

Fig. 8. System responses for 1 m long dipole antenna parallel to and 30 m from 1 m radius circular air-filled tunnel. The input pulse is an 8 ns gaussian doublet. The dipole is coated with a 0.0015 m radius layer of polythene insulation (ε_r=2.3).

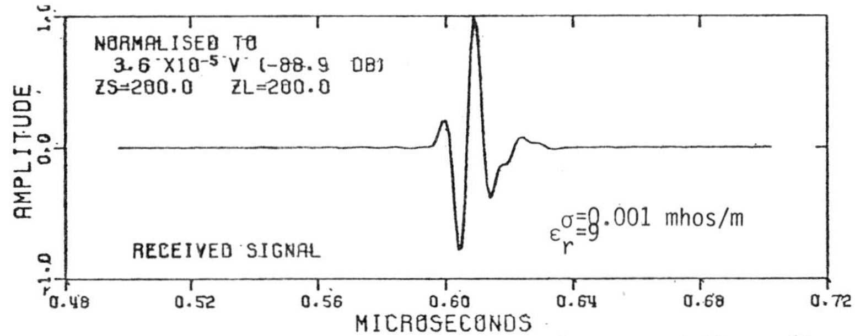

Fig. 9. System responses for 1 m parallel dipoles spaced 60 m. The system parameters are the same as for Fig. 8 except that the SAF for the tunnel has not been included.

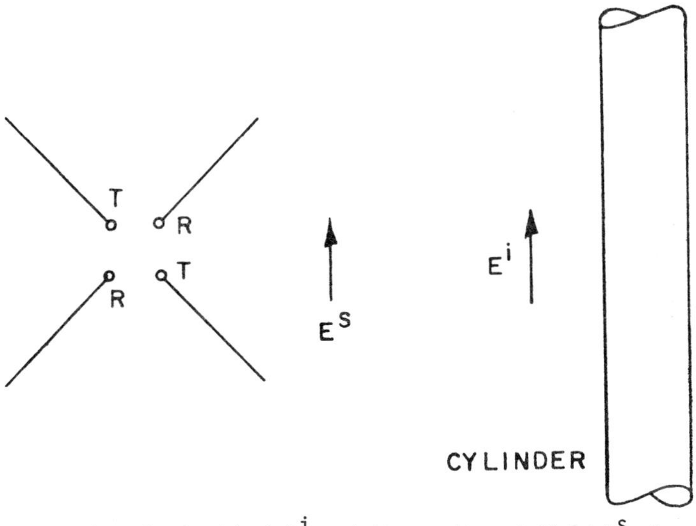

Fig. 10. Illustrating the incident E^i and the scattered field E^S when the cylinder is in the plane containing the orthogonal dipole antenna system.

Thus the incident wave should be decomposed into two waves whose electric field is given by

$$\bar{E}_{inc} = \hat{e}_{\parallel} E_{\parallel} + \hat{e}_{\perp} E_{\perp} \tag{13}$$

Now the thin conducting cylinder is placed so that

$$y = d, \text{ and } x = z \tan \theta' \tag{14}$$

where θ' is the angle between the cylinder and the x axis. The field scattered by the cylinder is now proportional to cos θ', and the received voltage at the x oriented dipole at the origin

$$V_r^{\parallel} = \frac{1}{2} \sin 2\theta'. \tag{15}$$

Similarly if the target scatters only the component of the wave polarized perpendicular to the cylinder,

$$V_r^{\perp} \propto - \frac{1}{2} \sin 2\theta' \tag{16}$$

where the negative sign indicates a reversal of polarity in comparison to that for parallel polarization.

There is also the possibility that the scattering caused by the two components of the incident wave satisfies neither of the above conditions i.e., neither scattering component vanishes. This was the case for the tunnels for frequencies up through the first few resonances. We define a complex proportionality constant

$$a_{\perp} = \frac{E_{\perp}^{S}}{E_{\parallel}^{S}} \tag{17}$$

where $E_{\parallel}^{S}$, $E_{\perp}^{S}$ are the scattered fields for the cylinder for incident polarizations parallel and perpendicular to the cylinder respectively.

Now Equation (23) may be written as

$$\bar{E}_{inc} = \hat{e}_{\parallel} E_D \cos \theta' + \hat{e}_{\perp} E_d \sin \theta' \tag{18}$$

The received voltage is now

$$V_R' \propto \frac{1}{2} E_d \sin 2\theta' - \frac{1}{2} E_d a_{\perp} \sin 2\theta' \tag{19}$$

$$\propto \frac{1}{2} E_d \sin 2\theta' (1 - a_{\perp})$$

For a = 1, $V_R' = 0$ corresponding to the case where the reflected field is orthogonal to the receiving dipole and is parallel to the transmitting dipole.

V. CONCLUDING REMARKS

Various cylindrical scattering models have been used to outline the expected performance of video pulse radar systems for the detection of cylindrical geometries. When incorporated with previous results for transmission between dipole antennas immersed in conducting media, these results provide the necessary design information for video pulse radars to detect tunnels.

APPENDIX I. DISCUSSION OF POTENTIAL ERRORS IN THE MODEL IF FIGURE 1

The reflecting plane of Fig. 1 is positioned at the axis of the cylinder and yet it has been noted that for large hollow cylinders, the dominant scattering occurs at the front interface (or top of the tunnel). This Appendix shows that this reflector position does not modify the results.

First, it is observed that it makes no difference where the reflecting plane is placed in so far as the model of Fig. 4a is concerned. However, it will influnce both the model of Fig. 4b (for the normalization E^I) and the separation of the dipole and its image shown in Fig. 4c. Recall that the model of Fig. 4c was introduced to treat the finite source and also to treat the influence of the ray divergence on the received voltage both in the plane of and perpendicular to the plane of the paper. If it is assumed that reflection occurs from the front interface as indeed appears reasonable, then our reflecting plane should be placed at that position. The major mechanism that could contribute to errors would be the loss introduced by the media. It could be argued that the energy enters the tunnel at the top and the tunnel (being free space) is lossless. We are concerned with the changes introduced in both the models of Figs. 4b and 4c when the reflecting plane is so moved. The ray divergence in the plane of the paper of Fig. 4c can be represented by a line source perpendicular to the paper. The fields of that line source are given by

$$U = U_o\, e^{-\gamma\rho_1} \sqrt{\frac{\rho_1}{\rho_1 + r}}\; e^{-\gamma r} \tag{A1}$$

where the factor $U_o e^{\gamma\rho_1}$ equals the fields incident of the ground plane at a distance $\rho_I = d$, the factor

$$\sqrt{\frac{\rho_1}{\rho_1 + r}} = \sqrt{\frac{d}{d + r}} \tag{A2}$$

is the spatial divergence factor and represents the decay in the fields caused by ray spreading as the observation distance r moves away from the source point, the factor $e^{-\gamma r}$ represents the usual phase delay factor plus the attenuation introduced by the medium. The fields scattered by a plane at the axis of the cylinder is represented by r = d whereas the fields scattered by a plane at the front interface is represented by r = d - 2a. Thus the factor

$$\frac{U(d-2a)}{U(d)} = e^{2\gamma a}\sqrt{\frac{d}{d - a}} \tag{A3}$$

represents the error introduced in the model of Fig. 4c.

The factor E^I would increase by $e^{2\gamma a}$ if the reference plane is shifted to the top of the tunnel. This cancels the factor $e^{-2\gamma a}$ in Equation (A3). The resulting error in Equations (10) and (11) is then only

$$\sqrt{\frac{d}{d-a}} \tag{A4}$$

This result implies that the reference plane can be taken at any convenient position as long as it is consistent with both the models of Figs. 4(b) and 4(c), and the amplitude of the result so obtained is modified by the $\sqrt{d/d-a}$ factor. This was confirmed by recomputing the result of Fig. 8 with the reference plane moved to a point one half of the distance from the target to the source. The result was exact in shape and differed by 3 dB in amplitude.

An important advantage of this result is that for an HFW radar the results for tunnels buried at different depths can be computed from one frequency response calculation for the model of Fig. 4(c).

REFERENCES

1. J. D. Young and R. Caldecott, "Underground Pipe Detector," U. S. Patent 3, 967,282, June 29, 1976.

2. J. D. Young, "A Transient Underground Radar for Buried Pipe Location," USNC/URSI Meeting, Boulder, Colorado, October 23, 1975.

3. A. C. Eberle and J. D. Young, "Development and Field Testing of a New Locator for Buried Plastic and Metallic Utility Lines," presented at 56th Annual meeting of the Transportation Research Board, Washington, D.C., January 24-28, 1977.

4. D. L. Moffatt and L. Peters, Jr., "An Electromagnetic Pulse Hazard Detection System," Proc. 1972 North American Rapid Excavation and Tunnelling Conference, Chapter 4, June 1972.

5. D. L. Moffatt, R. J. Puskar and L. Peters, Jr., "Electromagnetic Pulse Sounding for Geological Surveying with Application in Rock Mechanics and the Rapid Excavation Program," Report 3408-2, The Ohio State Univversity ElectroScience Laboratory, Department of Electrical Engineering; prepared under Contract H0230009 for Advanced Research Projects Agency, 1973.

6. D. L. Moffatt and R. J. Puskar, "A Subsurface Electromagnetic Pulse Radar," Geophysics, vol. 41, pp. 506-518, June 1976.

7. R. S. Vickers and L. T. Dolphin, "Subsurface Radar Sounding Experiments in Archaeology," USNC/URSI Meeting, Boulder, Colo., October 23, 1975.

8. N. N. Wang, L. Ersoy, and W. D. Burnside, "GTD Analysis and the Hybrid Techniques for the Investigation of the Surface Current and Charge Densities Induced on Aircraft," Report 4172-1, December 1976, The Ohio State University ElectroScience Laboratory, Department of Electrical Engineering; prepared under Contract F29601-75-C-0086 for Kirtland Air Force Base, New Mexico.

9. G. A. Burrell, L. Peters, Jr. and A. J. Terzuoli, Jr., "The Propagation of Electromagnetic Video Pulses with Application to Subsurface Radar for Tunnel Detection," Report 4460-2, December 1976, The Ohio State University ElectroScience Laboratory, Department of Electrical Engineering; prepared under Contract DAAG53-76-C-0179 for U. S. Army Mobility Equipment Research and Development Command.

10. R. G. Kouyoumjian, L. Peters, Jr., and D. T. Thomas, "A Modified Geometrical Optics Method for Scattering by Dielectric Bodies," IEEE Trans., Vol. AP-11, 1963, pp. 690-703.

11. L. Peters, Jr., T. Kawano and W. G. Swarner, "Applications for Dielectric or Plasma Scatterers," Proc. IEEE, Vol. 53, 1965, pp. 882-892.

12. J. H. Richmond, "Radiation and Scattering by Thin-Wire Structures in a Homogeneous Conducting Medium," IEEE Trans., Vol. AP-22, p. 365, March 1974.

13. C. W. Chuang and D. L. Moffatt, "Natural Resonances of Radar Targets via Prony's Method and Target Discrimination," IEEE Transactions on Aerospace and Electronic Systems, Vol. AES-12, p. 5, September 1976, pp. 583-589.

ACKNOWLEDGMENTS

The work reported in this paper was supported in part by The Ohio State University Research Foundation Contracts DAAG53-76-C-0179 with U. S. Army Mobility Equipment Research and Development Command, Ft. Belvoir, Virginia, and Joint Services Electronics Program N00014-78-C-0049 with Department of the Navy, Office of Naval Research, Arlington, Virginia.

SUMMARY OF PANEL DISCUSSIONS

(i) 25 June 1979 : Moderator - Professor Raj Mittra, University of Illinois, Urbana

Professor Mittra began the panel discussion with general comments on the day's talks and put forth specific questions to the speakers and participants.

A considerable amount of time was spent on how well the T-matrix method can handle scatterers with edges or corners. Professor Mittra commented that if the coordinate system chosen is not suitable for the particular scatterer then the matrices become highly ill conditioned and if special basis functions are used other than circular and spherical functions then there is considerable numerical difficulty involved in their computation. Professor Bates remarked that the use of special functions that could take care of the singularities at the edges would result in well conditioned matrices. Dr. Wall added that the use of regularization techniques would also result in matrices with good condition numbers. The same effect could also be achieved by the use of subsectional basis functions and he had tested this for the thin wire problem. Professor Wilton remarked that special functions were not needed to handle edges with the moment method but Professor Bates felt that the resulting matrices would then be ill conditioned.

Professor V.K. Varadan stated that good results had been obtained for the strip and the penny shaped crack although they are slender and long. Dr. Waterman said that these two cases are very special since the limit of zero aspect ratio exists for the ellipse and spheroid so that the computer programs developed for these geometries still provide good results even for crack like scatterers. He felt that these two examples should be excluded from a discussion on obstacles with edges.

Professor Mittra observed that the T-matrix formulation seemed naturally suited to handle multiple scattering problems. Professor Ström who originally solved such problems added that the framework of the theory is such that if one is familiar with its use for acoustic waves then one can easily extend it to electromagnetic and elastic wave problems. The inhomogeneous scatterer could also be handled by this method by modelling the scatterer as being comprised of several homogeneous layers. Good numerical results have already been obtained for such scatterers.

Professor Kouyoumjian made some general remarks on the relationship between the T-matrix and moment methods and was of the opinion that they are first cousins. Professor Baum also noted that in both cases one matricizes integral equations. Professor Kouyoumjian felt that one of the great advantages of the T-matrix approach was that only one integration had to be performed on the actual surface of the scatterer since the testing was done on a fictitious surface on which the basis functions were orthogonal. The key to the whole method was the explicit use of the extinction theorem to eliminate the unknown surface or interior fields.

Professor V.K. Varadan stated that atleast for elastic wave scattering problems, the T-matrix approach was the only one available at present for non-spherical obstacles. The moment method was applied only to two dimensional elastic wave problems and not well explored. Professor Baum felt that before making comments on which method had more success it was important to specify exactly what integral equation one was talking about. Professor Bates replied that when the T-matrix method was first introduced by Dr. Waterman for acoustic and electromagnetic waves, the moment method was already being used widely for scattering problems. But for elastic waves, the T-matrix approach was tried before the moment method and the response was enthusiastic since it quickly provided numerical results for a number of new problems, and it is natural to follow along a successful path.

Professor Mittra questioned Professor Achenbach on the apparent discrepancy between theoretical results using the geometrical theory of diffraction for cracks in elastic solids and experimental results. Professor Achenbach replied that this was due to the fact that in the laboratory one was dealing with real materials and real cracks. There was definite damping of the waves, some of the interfaces were not plane interfaces as was assumed in the theory and finally the discrepancy was only in the amplitudes of the scattered fields and not in the frequency dependence. He felt that it was the frequency dependence that mattered in inverse problems, since the amplitudes could easily be matched better by introducing damping in the theory. Professor Baum agreed and commented that the singularity expansion method (SEM) was based on the fact that the frequency dependence associated with a set of resonances characterizing the scatterer was source independent.

Finally Professor Mittra asked Professor Mei to comment on the unimoment method and the matching of boundary conditions off the actual surface of the scatterer and on a fictitious surface that was chosen to conform to the coordinate system being used. Unfortunately, Professor K.K. Mei's response was not recorded clearly.

(ii) 27 June 1979 : Moderator Professor R.H.T. Bates, University of Canterbury, New Zealand

Professor Bates posed three questions to the panel and the audience -

(1) Should the method introduced by Dr. Waterman which is the focus of this Symposium be called the T-matrix method, the null field method or the extended boundary condition method ?

(2) What are the conclusions to be drawn from the talks and discussions at the Symposium ?

(3) What are the new problems to be looked into where the T-matrix method can be conveniently applied ?

He then summarized his impressions from the Symposium -

(1) The T-matrix method seems to have made the maximum impact in elastic wave scattering problems and was being used by the people in the Engineering Mechanics Department at OSU and Professor Staffan Ström's group in Sweden.

(2) The method seems particularly well suited for multiple scattering problems especially in conjunction with ensemble averaging techniques of statistical mechanics and lastly,

(3) The application of the method to waveguide problems seems to establish the usefulness of the method unequivocally.

He then asked Professor V.V. Varadan if one could conclude that in elastic wave problems the T-matrix method has yielded answers superior to those by other methods. Professor V.V. Varadan stated that it would be unfair to come to that conclusion since the moment method that yields answers for the same range of wavelengths has been unexplored for 3-D problems and suggested that the group at OSU should look into this method. Dr. Thompson who had been involved with the Rockwell ARPA/AFML program for non-destructive testing for the past five years summarized his impressions of the different methods that have been used for elastic wave scattering.Simple but not so accurate approximations of scattering problems like the Born, the quazi static and the extended Born methods have provided good understanding and insight and more accurate numerical methods have provided results that agree more closely with experimental results. Ultimately one has to deal with extremely complex flaws in structural materials which perhaps can be approached with finite element calculations.

The panel was asked if transient problems like the scattering of seismic waves from buildings in which the building actually moves had been studied using the T-matrix method. Professor V.K. Varadan replied that it was possible if one could integrate the T-matrix solution over frequencies. At the end of this part of the discussion it was concluded that it was pointless to discuss the superiority of one method over another, A particular method may be very well suited for a certain problem and not so for another.

Next the use of the finite element method for elastic wave problems was discussed. Professor V.V. Varadan summarized the situation in this regard. Professor C.C. Mei had been using finite element methods for scalar diffraction problems i.e. the scattering was by rigid obstacles in water. The method has yet to be applied to the full elastodynamic problem. Usually the scatterers are modelled by approximate theories such as beam and shell models. But it would be very interesting to apply the uni-moment method which is related to the finite element approach for elastic wave problems.

The question of truncation sizes and convergence criteria in the use of the T-matrix method was raised by a member of the audience. Professor V.V. Varadan replied that at present there are none available and such questions were basically decided by checking the symmetry and unitary properties of the scattering matrix. Professor Ramm was of the opinion that although as Engineers and Physicists we are interested in the application of certain mathematical techniques for the solution of physical problems, it is important for us to be concerned about existence and convergence proofs. Professor Bates commented that the time was ripe for mathematicians to look into these questions in the context of the T-matrix.

There was an interesting discussion on whether the T-matrix method is a suitable name for the method introduced by Waterman. Professors Bates, Ström, V.K. Varadan, V.V. Varadan and Boerner participated in this discussion. Various names have been used in the papers published so far - Transition matrix, T-matrix, Transfer matrix, the extended boundary condition and the null field method. Since the term T-matrix is already widely used in quantum mechanics, this name applied to the Waterman method would cause some confusion. The name extended boundary condition method (EBC) is vague and misleading. It creates the impression that boundary conditions are not satisfied on the actual surface of the scatterer whereas the boundary conditions are satisfied in a perfectly natural way. It was the general opinion of the panel that the null field method was the most suitable name, since the key to the whole technique is the nulling of the incident fields at the points interior to the scatterer. Professor V.K. Varadan was concerned that researchers in mechanics who are now familiar with the name T-matrix approach should not be confused by the name null field method.

Some of the new problems that are ready to be studied by the null field approach are cluttering problems - scattering from moving and 'noisy' obstacles, multiple scattering from trace gases, aerosols, dust and other types of atmospheric contaminants, scattering from rough surfaces like the surface at the bottom of the ocean, scattering from flaws in structural materials in the presence of nearby boundaries to include surface wave effects. Professor V.K. Varadan mentioned that Dr. De Santo at Naval Research Laboratory had already started some nice work on rough surfaces and Professor Ström and his group have already started work on surface wave effects.

The last subject for discussion was the inverse problem. Professor Ksienski began this discussion by calling on theoreticians to look more directly at the end objective instead of trying to improve scattering theory and making it mathematically pleasing. In actual application, the identification of an unknown flaw at an unknown location is the important problem not if one has an elegant theory to describe the scattering from a model flaw. Professor Boerner agreed with him saying that even for inverse problems researchers are too concerned with theorems and proofs. He noted that Professor Young presented during his lecture a relationship that was derived for high frequency scattering but which seemed to apply very nicely in the low frequency resonant scattering region. Professor V.V. Varadan added that Dr. Rose using the inverse Born approximation, which really applies well only to weak contrasts between flaw and host material, seemed to have obtained excellent results for a cavity in a solid where the contrast is infinite. Professor Bates suggested that since human vision works on 90 % a priori information, in dealing with the inverse problem one should take advantage of all the a priori information available since there are too many unknowns to handle mathematically. He mentioned that the use of cross polarized channels in experiments had yielded tremendous information. Professor Seliga commented that this has had tremendous application in measuring rainfall rates. The actual variations in the magnitudes of the cross section are not important but it is the differentials relative to polarization, preferred orientation etc. that are important.

The panel discussion and Symposium concluded with a vote of thanks by Professor V.K. Varadan and Dr. N.L. Basdekas.

The two panel discussions were summarized from tape recordings that were sometimes totally unclear. Any misinterpretation of comments made by the panel and participants is totally unintentional.

Subject Index